Engineering Mathematics A programm[illegible]

C.W. Evans

This bookmark is intended to be used as a page mark. It should stop your eyes from glancing inadvertently down the page when you are trying a worked example on your own. On the reverse side of the bookmark is a table of standard integrals and derivatives together with several formulas. These are frequently needed in applications to engineering and science.

From time to time as you read through the text you will encounter a think line like this:

Normally this represents a breathing space; it means that you can stop at this point if you wish without interrupting the flow of the text too much.

All examples begin with the symbol □ and end with the symbol ■. These should help you to judge whether the length and the likely complexity of an example will fit into the time you have allocated for study.

QUICK REFERENCE GUIDE

$f'(x)$	$f(x) = F'(x)$	$F(x)$
nx^{n-1}	x^n	$x^{n+1}/(n+1)\ (n \neq -1)$
$-\sin x$	$\cos x$	$\sin x$
$\sec^2 x$	$\tan x$	$\ln \lvert\sec x\rvert$
$\sec x \tan x$	$\sec x$	$\ln \lvert\sec x + \tan x\rvert$
$-\operatorname{cosec}^2 x$	$\cot x$	$\ln \lvert\sin x\rvert$
$-\operatorname{cosec} x \cot x$	$\operatorname{cosec} x$	$\ln \lvert\operatorname{cosec} x - \cot x\rvert$
$1/x$	$\ln x$	$x \ln x - x$
$a^x \ln a$	$a^x\ (a > 0)$	$a^x / \ln a$
$\cosh x$	$\sinh x$	$\cosh x$
$\operatorname{sech}^2 x$	$\tanh x$	$\ln \cosh x$
$-\operatorname{sech} x \tanh x$	$\operatorname{sech} x$	$2\tan^{-1}(e^x)$
$-\operatorname{cosech}^2 x$	$\coth x$	$\ln \lvert\sinh x\rvert$
$-\operatorname{cosech} x \coth x$	$\operatorname{cosech} x$	$\ln \lvert\tanh (x/2)\rvert$
$2x/(1-x^2)^2$	$1/(1-x^2)$	$\frac{1}{2} \ln \lvert(1+x)/(1-x)\rvert$
$-2x/(1+x^2)^2$	$1/(1+x^2)$	$\tan^{-1} x$
$x/(1-x^2)^{3/2}$	$1/\sqrt{(1-x^2)}$	$\sin^{-1} x$
$-x/(1+x^2)^{3/2}$	$1/\sqrt{(1+x^2)}$	$\ln \lvert x + \sqrt{(1+x^2)}\rvert$
$-x/(x^2-1)^{3/2}$	$1/\sqrt{(x^2-1)}$	$\ln \lvert x + \sqrt{(x^2-1)}\rvert$
$-x/(1-x^2)^{1/2}$	$\sqrt{(1-x^2)}$	$\{x\sqrt{(1-x^2)} + \sin^{-1} x\}/2$
$x/(1+x^2)^{1/2}$	$\sqrt{(1+x^2)}$	$\{x\sqrt{(1+x^2)} + \ln[x + \sqrt{(x^2+1)}]\}/2$
$x/(x^2-1)^{1/2}$	$\sqrt{(x^2-1)}$	$\{x\sqrt{(x^2-1)} - \ln \lvert x + \sqrt{(x^2-1)}\rvert\}/2$
$e^{ax}(a\cos bx - b\sin bx)$	$e^{ax}\cos bx$	$e^{ax}(a\cos bx + b\sin bx)/(a^2+b^2)$
$e^{ax}(a\sin bx + b\cos bx)$	$e^{ax}\sin bx$	$e^{ax}(a\sin bx - b\cos bx)/(a^2+b^2)$

1 Binominal expansion $(a+b)^n = \sum_{r=0}^{n} \binom{n}{r} a^r b^{n-r}\ (n \in \mathbb{N})$

2 Taylor's expansion $f(a+h) = \sum_{r=0}^{n-1} \frac{h^r}{r!} f^{(r)}(a) + R_n$

where $R_n = \frac{h^n}{n!} f^{(n)}(a + \theta h) \quad \theta \in (0, 1)$

3 Newton's formula $x_{n+1} = x_n - \frac{f(x_n)}{f'(x_n)}$

4 Triple vector product $\mathbf{a} \times (\mathbf{b} \times \mathbf{c}) = (\mathbf{a} \cdot \mathbf{c})\mathbf{b} - (\mathbf{a} \cdot \mathbf{b})\mathbf{c}$

5 Leibniz's theorem $(uv)_n = \sum_{r=0}^{n} \binom{n}{r} u_r v_{n-r}\ (n \in \mathbb{N})$

6 Integration by parts $\int u\,dv = uv - \int v\,du$

Engineering Mathematics

Engineering Mathematics

A PROGRAMMED APPROACH

Third edition

C.W. Evans

School of Mathematical Studies
University of Portsmouth

CHAPMAN & HALL

London · Weinheim · New York · Tokyo · Melbourne · Madras

Published by Chapman & Hall, 2–6 Boundary Row, London SE1 8HN, UK

Chapman & Hall, 2–6 Boundary Row, London SE1 8HN, UK

Chapman & Hall GmbH, Pappelallee 3, 69469 Weinheim, Germany

Chapman & Hall USA, 115 Fifth Avenue, New York, NY 10003, USA

Chapman & Hall Japan, ITP-Japan, Kyowa Building, 3F, 2-2-1 Hirakawacho, Chiyoda-ku, Tokyo 102, Japan

Chapman & Hall Australia, 102 Dodds Street, South Melbourne, Victoria 3205, Australia

Chapman & Hall India, R. Seshadri, 32 Second Main Road, CIT East, Madras 600 035, India

---

First edition 1989

Reprinted 1991

Second edition 1992

Reprinted 1993, 1994

Third edition 1997

Printed in Great Britain by St Edmundsbury Press, Bury St Edmunds, Suffolk

ISBN 0 412 75260 3

A catalogue record for this book is available from the British Library

∞ Printed on permanent acid-free text paper, manufactured in accordance with ANSI/NISO Z39.48-1992 and ANSI/NISO Z39.48-1984 (Permanence of Paper).

To my family, friends and students;
past, present and future.

Contents

Preface

In this country today there are two conflicting forces acting on the mathematical curriculum and these are thrown into sharp contract when we consider Engineering Mathematics. Engineering Mathematics consists of a large body of material and techniques which is traditionally used by Engineers and Scientists in order to develop their theoretical work. As more of this work is developed the pressure is increased for students to acquire the necessary mathematical skills and techniques earlier. Set against this, there has been a general reduction in the numbers of young people who choose to study A-level mathematics at school. The examination boards have responded to this unpopularity by reducing the quantity of material which is included in the A-level syllabus and the level of skill required. Consequently, knowledge and facility, regarded as routine 10 years ago, is now not generally acquired until the student becomes an undergraduate.

The 'one still point in this turning world' is the unfortunate student. It can be argued that students are getting brighter but, due to the pressures which have already been mentioned, their mathematical experience on entering University does not reflect this. This new edition attempts, in some measure, to resolve these opposing forces by adopting on the one hand a very elementary starting point and including, on the other, some relatively advanced material. The first two chapters have been rewritten to make them more accessible. In this way an intelligent student, by sheer dint of determination and effort, should be able to raise the level of his or her individual expertise.

Thirty years ago Fourier series and Laplace transforms were regarded as part of the second year syllabus for most Engineering courses. Nowadays Engineers and Scientists require them early on and therefore these topics have been included and the text written to make them accessible to the well motivated student.

A number of extra exercises have been also provided and most of the chapters have the benefit of these additions. As usual, F. Smiley painstakingly worked through all of them. Needless to say any errors or omissions are entirely the responsibility of the author.

The supplementary material needed for this edition was produced in LaTeX and the author is greatly indebted to his friend and colleague Dr David Divall for using his considerable skill and effort to develop a LaTeX house style which resembles that which had been adopted by the publisher in the earlier editions. Any success which has been achieved in this regard is due entirely to his efforts.

Naturally all students wish to pass their mathematics examinations and this book aims to help them to achieve this. However this is not its sole purpose. Another aim is to give students the competence and the confidence to use mathematical ideas and techniques in their chosen field. We are trying not to 'cap' the flow of oil but to harness it and to make use of it. Consequently this book is not for the dull witted or the pig ignorant. It is not an end but a beginning for it is an invitation to acquire skill and power. We are not trying to switch out lights, we are trying to turn them on!

The author hopes that this new edition will prove popular with both staff and students.

C.W.E.

Acknowledgements

Inevitably with a book of this size there are many people who have assisted in one way or another. Although it is invidious to mention just a few, some have been particularly helpful. T. Mayhew of Mayhew Telonics supplied an updated version of the company's SCIWAYS ROM; a scientific microchip. The bulk of the manuscript was prepared initially on a wordprocessor using this ROM. F. Smiley checked through the examples and produced solutions to all the exercises. Many anonymous readers commented on the initial draft of the manuscript and without exception their comments proved useful. Dr Dominic Recaldin commissioned the book and has always been most supportive. Indeed many staff, both past and present, of Van Nostrand Reinhold and Chapman and Hall have assisted enthusiastically in the production of this book. Lastly, the author's family must not go unmentioned for supplying a unique blend of patience, exasperation and forbearance. To these, and all others who have assisted in the preparation of this book, the author extends his warmest thanks.

To the student

There are essentially two different ways in which you can use this book. Each of them depends on your past experience of the topic; whether it is a new topic or one with which you are familiar.

NEW TOPICS

- ☐ Work your way through the chapter with the aid of a note pad, making sure that you follow the worked examples in the text.
- ☐ When you come to a workshop be resolute and do not read the solutions until you have tried to work them out.
- ☐ Attempt the assignment at the end of the chapter. If there are any difficulties return to the workshop.
- ☐ Spend as much time as possible on the further exercises.

FAMILIAR TOPICS

- ☐ Start with the assignment, which follows the text, and see how it goes.
- ☐ If all is well continue with the further exercises.
- ☐ If difficulties arise with the assignment backtrack to the workshop.
- ☐ If difficulties arise in the workshop backtrack to the text.
- ☐ Read through the chapter to ensure you are thoroughly familiar with the material.

Basic ideas 1

This opening chapter is designed to lay the foundations of the work which we have to do later. We will therefore describe some notation and examine both arithmetic and algebraic processes.

After completing this chapter you should be able to

- ☐ Approximate calculations to a given number of decimal places and to a given number of significant figures;
- ☐ Apply the rules of elementary algebra correctly;
- ☐ Distinguish between identities and equations;
- ☐ Evaluate binomial coefficients and apply the binomial theorem.

At the end of this chapter we shall solve a practical problem involving the force on a magnetic pole.

1.1 ARITHMETIC

We are all familiar with the basic operations of arithmetic. These are addition and multiplication. The first numbers which we encounter are the 'whole' numbers, which we shall call the **natural** numbers

$$1,\ 2,\ 3,\ 4,\ \ldots$$

You will notice that we have not included zero as one of the natural numbers; some people do and some people don't!

We can add or multiply any two natural numbers together without obtaining results which go beyond this set of numbers.

$$\begin{aligned} 12 + 13 &= 25 \\ 12 \times 13 &= 156 \end{aligned}$$

Both 25 and 156 are natural numbers too.

When we introduce the operation of subtraction it is necessary to widen the concept of number to include the negative whole numbers and zero. These numbers are known as **integers**

$$\ldots, -2, -1,\ 0,\ 1,\ 2,\ 3,\ \ldots$$

We observe of course that every natural number is an integer.

Although the integers are sufficient for simple barter of discrete (individually distinct) objects, they are unable to cope with division. The operation of division forces us to extend the concept one stage further.

Any number which can be expressed in the form p/q, where p and q are integers is known as a **rational** number.

□ 3 is a rational number since $3 = 3/1$, 0.25 is a rational number since $0.25 = 1/4$, $1/3$ is a rational number and is $0.333\ldots$.

The process of division leads to decimal expansions. The digits in the decimal expansion fall either to the left or the right of the decimal point.

□ In the number 7.2386941, 2 is in the first decimal place, 3 is in the second decimal place and 9 is in the fifth decimal place.

Decimal expansions fall into two classes; finite decimal expansions and infinite decimal expansions. When finite decimal expansions are read from the left to the right, every digit in a decimal place beyond a certain point is zero. When infinite decimal expansions are read from the left to the right, however large the decimal place, there is always a digit with a larger decimal place which is non-zero.

Sometimes a set of digits in an infinite decimal expansion repeats without end. This set of digits is said to recur and in these circumstances the expansion corresponds to a rational number. However some infinite decimal expansions never recur and these correspond to **irrational** numbers. The collection of all the numbers we have been describing is usually referred to as the collection of **real** numbers.

1.2 REPRESENTATION

Although theoretically we can add, multiply, subtract and divide decimal fractions in whatever way they are expressed, in practice this becomes awkward.

□ Multiply 998954.32 by -0.0001334684

To facilitate calculations we introduce the *scientific* notation. In the scientific notation each number is written as a number greater than -10 but less than 10 and the decimal point is adjusted by multiplying or dividing by an appropriate power of 10.

Therefore, in the example, 998954.32 and -0.0001334684 become

$$9.9895432 \times 10^5 \text{ and } -1.334684 \times 10^{-4}$$

respectively.

It is easy to convert a number into scientific notation. Remember that each time we multiply a number by 10 we move the decimal point one step to the right. Consequently each time we divide by 10 (or equivalently multiply by 10^{-1}) we move the decimal point one step to the left.

For the first number we needed to move the decimal point 5 steps to the left. This is equivalent to multiplying by 10^{-5} and so to correct this, we multiply the number we have obtained by 10^5.

$$998954.32 = 9.9895432 \times 10^5$$

For the second number we needed to move the decimal point 4 steps to the right. This is equivalent to multiplying by 10^4 and so to correct this, we multiply the number we have obtained by 10^{-4}.

$$-0.0001334684 = -1.334684 \times 10^{-4}$$

These can now be multiplied together and the result expressed in scientific notation too

$998954.32 \times -0.0001334684$

$$\begin{aligned} &= (9.9895432 \times 10^5) \times (-1.334684 \times 10^{-4}) \\ &= -(9.9895432 \times 1.334684) \times 10 \\ &= -13.33288348 \times 10 \\ &= -1.333288348 \times 10^2 \end{aligned}$$

■

Hand calculators automatically make use of scientific notation. They normally record the power of 10 (but not the number 10 itself) on the far right of the display. Some calculators give the option of expressing every number in scientific form.

1.3 DECIMAL PLACES AND SIGNIFICANT FIGURES

It is not normally possible to perform all calculations exactly. In such circumstances it is usual to perform arithmetic using numbers accurate to a fixed number of decimal places or a given number of significant figures.

Significant figures are determined by locating the first non-zero digit from the left of the number. From this position we count to the right to determine where the number needs to be truncated. When this is done notice is taken of the part which is to be discarded and some adjustment to the last digit in the truncated number may be made. It may be necessary to include some trailing zeros when giving a number correct to a given number of significant figures.

Decimal places are counted to the right of the decimal point. The rule for rounding numbers is complicated to describe but fairly simple to apply.

- If the first digit in the discarded part is 6 or more then the number is 'rounded up' and the last digit is increased by 1.
- If the first digit in the discarded part is 4 or less then the number is truncated without change.
- If the first digit in the discarded part is 5 and it contains other non-zero digits then the number is rounded up so the last digit is increased by 1.

There is one case remaining. This is the vexed question of what to do when the first digit of the discarded part is 5 and all other digits in it are zero. Here there is no general consensus. Those who always round up introduce a bias automatically. To compensate for this, numerical analysts favour rounding-up the truncated decimal where necessary so that the last digit becomes even. This is the approach we shall adopt.

The following examples illustrate all the situations which arise.

☐ Express each of the following numbers correct to 5 significant figures

$$\begin{array}{lll} 0.0001234, & 1,548,796,854, & 1,548,746,854, \\ 0.100005, & 0.100015, & 11.11 \end{array}$$

When a number is to be expressed correct to 5 significant figures then the 5 figures are given, even if some of them are zero. Therefore we obtain

0.00012340	we need a trailing zero.
1,548,800,000	96854 discarded so number rounded up
1,548,700,000	46854 discarded so number unchanged
0.10000	5 is discarded but 0 is even
0.10002	5 is discarded, 1 is odd, round-up
11.110	a trailing zero is needed

■

☐ Give the following decimal fractions correct to 5 decimal places

$$\begin{array}{lll} 2.3657842, & 34.574836242, & 3.769845, \\ 3.769835, & 3.769845000001, & 0.000005 \end{array}$$

2.36578	discarded digits 42, so no round-up
34.57484	discarded digits 6242, so round-up
3.76984	discarded digit 5, but 4 is even
3.76984	discarded digit 5, round-up 3 to 4
3.76985	discarded digits 5000001 so round-up
0.00000	discarded digit 5 and 0 is even.

■

1.4 PRECEDENCE

When performing numerical calculations it is necessary to establish an order of precedence. Without this, an expression such as $2 + 4 \times 5$ would be ambiguous, for we could first add 2 to 4 and then multiply the result by 5 to obtain 30 or alternatively we could add 2 to the result of multiplying 4 and 5 to obtain 22. One way round this problem is to use brackets to distinguish the two situations. We can do this provided we understand that things in brackets must always be worked out first.

We then have the two situations

$$\begin{aligned}(2+4)\times 5 &= 6\times 5\\ &= 30\\ 2+(4\times 5) &= 2+20\\ &= 22\end{aligned}$$

In fact an order of precedence for these elementary mathematical operations is well established. It is simply that multiplication and division take precedence over addition and subtraction. However, as we have said, anything in brackets must be calculated first.

Brackets are very important and must never be discarded lightly. They should always be introduced whenever any ambiguity could arise.

□ The expression $144 \div 16 \div 3$ is meaningless because it is ambiguous. To see this we merely need to consider its two possible meanings

$$\begin{aligned}(144\div 16)\div 3 &= 9\div 3\\ &= 3\\ 144\div(16\div 3) &= 144\div(\frac{16}{3})\\ &= 27\end{aligned}$$

■

Many years ago somebody coined an acronym BODMAS to help students to learn the order of precedence. The letters stand for Brackets, Of, Division, Multiplication, Addition, Subtraction.

The word 'of' occurs in calculations such as '3% of £35 is £1.05p' but is really only there to make the acronym memorable.

BODMAS has its uses but implies that division takes precedence over multiplication whereas in fact division and multiplication have equal status. Similarly addition and subtraction have equal status. Fortunately this can cause no error. To illustrate this point note that the BODMAS rule gives

$$\begin{aligned} 3 \times 12 \div 9 &= 3 \times (12 \div 9) \\ &= 3 \times (\frac{4}{3}) \\ &= 4 \end{aligned}$$

Whereas the alternative is

$$\begin{aligned} 3 \times 12 \div 9 &= (3 \times 12) \div 9 \\ &= 36 \div 9 \\ &= 4 \end{aligned}$$

1.5 SET NOTATION

We sometimes represent the set of all the natural numbers

$$1,\ 2,\ 3,\ 4,\ \ldots$$

by $\mathbb{N}$ and then use the notation $x \in \mathbb{N}$ to indicate that x 'is a member of the set' $\mathbb{N}$. Likewise we write $y \notin \mathbb{N}$ to indicate that y 'is not a member of the set' $\mathbb{N}$

$$\text{So } 3 \in \mathbb{N} \text{ but } 2.5 \notin \mathbb{N}.$$

We denote the set which contains all the natural numbers and also 0 by $\mathbb{N}_0$. We shall occasionally find this notation quite useful.
The set of all the integers

$$\ldots, -2, -1,\ 0,\ 1,\ 2,\ 3,\ \ldots$$

is denoted by $\mathbb{Z}$. We know that every natural number is an integer and this is expressed by saying that $\mathbb{N}$ is a *subset* of $\mathbb{Z}$. In symbols this is

$$\mathbb{N} \subset \mathbb{Z}$$

In general a set A is a subset of a set B if whenever x is a member of the set A then x is also a member of the set B. In symbols:

$$A \subset B \text{ if and only if whenever } x \in A \text{ then } x \in B$$

We can also write $B \supset A$ to denote the same property. It follows that if both $A \subset B$ and $B \subset A$ then the two sets A and B have precisely the same elements and therefore we can write $A = B$.

We have seen that a rational number is any number which can be expressed in the form p/q where p and q are integers. The set of rational numbers is represented by $\mathbb{Q}$ and the set of all the real numbers is represented by $\mathbb{R}$.

We therefore have

$$\mathbb{N} \subset \mathbb{Z} \subset \mathbb{Q} \subset \mathbb{R}$$

We shall extend this notation a little further in Chapter 2.

1.6 DEDUCTIONS

In some ways mathematics is rather like a very large but incomplete jigsaw puzzle. Over the years mathematicians have been able to put some of the pieces together but there are an infinity of pieces and an infinity of gaps. The key concept which distinguishes mathematics from other subjects is the notion of **proof**. We shall have more to say about proof in Chapter 2 but we shall shortly begin to experience the idea at first hand. We shall consider, in the first instance, two types of proof - direct proof and indirect proof.

Let us consider for the moment what is meant by a direct proof. Suppose it is known that a particular machine, which uses water pressure as its power source, is in perfect working order. Suppose it is also known that water at the correct pressure is being supplied. We can then deduce that the machine is functioning correctly. This deduction is an example of a direct proof.

For an indirect proof suppose that we know that water at the correct pressure is being supplied but that the machine is not functioning correctly. We can deduce that the machine is **not** in perfect working order. How? Well, if it were, then by the direct proof we have already given, we should be able to deduce that the machine is functioning correctly. However we know that it is not functioning correctly and consequently it is not in perfect working order.

In general if we wish to deduce something using an indirect proof we take the opposite statement and see if we can deduce a contradiction of some kind. This is a rather more belligerent approach to the deductive process. In practical terms, if we wished to show that a machine was essential to the manufacturing process we could see what happened if we shut it down completely. We should either be able to continue manufacturing, however inefficiently, or we should not. If the manufacturing process comes to a halt then we can deduce that the machine was essential to the manufacturing process.

It is important also to realize that we must not jump to false conclusions. For example, in the case of the water powered machine, if we know that water at the correct pressure is being supplied and that the machine is functioning correctly we are **not** able to deduce that the machine is in perfect working order. In fact, we cannot deduce very much in these circumstances; the machine might be in perfect working order or it might not be. To see this you only have to imagine a machine which was required to rotate and to irrigate crops. It may be doing this but some of its outlets could be obstructed so that it is not in fact in perfect working order.

In the next section we shall begin to prove things directly and indirectly but it is not until Chapter 2 that we put things into a more formal setting.

1.7 ALGEBRA

We use the same conventions as we used in arithmetic when we use algebra. In algebra we use symbols to represent various things. To start with, we use these symbols to represent numbers, and this is called elementary algebra. We can add two numbers a and b and the result will be represented by $a+b$. If we multiply two numbers a and b we can represent the result, known as the product of the two numbers, in several ways; $a \times b$, $a \cdot b$ or even ab. So if we see the symbol uv and if we know that u and v are numbers then we know that

$$uv = u \times v$$

We write $a \times a$ as a^2 and $a \times a \times a$ as a^3. In general the positive exponent tells us how many times a occurs in the product.

We now list the algebraic rules which we shall need. Some of these will be listed under rules of addition and rules of multiplication since these are the two principal operations which we perform on numbers. Then we shall list the rule which interlinks these two operations.

1.8 RULES OF ELEMENTARY ALGEBRA

These rules are easily verified for a few numbers but what we are saying is that they are true for *all* numbers.

ADDITION RULES

1 Given any two numbers a and b their sum $a+b$ is also a number.

2 Given any three numbers a, b and c then

$$a + (b + c) = (a + b) + c.$$

3 There is a number 0 such that for every number a,

$$a + 0 = 0 + a = a.$$

4 To each number a there corresponds another number, designated by $-a$, such that

$$a + (-a) = (-a) + a = 0.$$

5 Given any two numbers a and b then $a + b = b + a$.

We can use these rules to carry out our first logical deduction.

□ The number $-a$ is known as the additive inverse of a. Show that no number can have two additive inverses.

We do this by means of an indirect proof. That is, we shall suppose that there is some number a which has two additive inverses $-a_1$, and $-a_2$ and then deduce that $-a_1 = -a_2$.

Now
$$\begin{aligned} -a_1 &= (-a_1) + 0 && \text{using rule 3} \\ &= (-a_1) + (a + (-a_2)) && \text{using rule 4} \\ &= ((-a_1) + a) + (-a_2) && \text{using rule 2} \\ &= 0 + (-a_2) && \text{using rule 4} \\ &= -a_2 && \text{using rule 3} \end{aligned}$$

■

The fact that there is only one additive inverse corresponding to each number justifies our use of the symbol $-a$ for the additive inverse of a.

□ Show that if $a + x = a + y$ then $x = y$.

It is clear how we must do this. We must 'take away' a from each side of the equation. In other words we must add the additive inverse of a to each side.

Here are the steps. See if you can see which rule is applied in each case.

$$\begin{aligned} (-a) + (a + x) &= (-a) + (a + y) \\ ((-a) + a) + x &= ((-a) + a) + y \\ 0 + x &= 0 + y \\ x &= y \end{aligned}$$

First we used rule 2 (the associative rule), then rule 3 (the definition of an additive inverse) and finally rule 4 (the property of 0) led to the required conclusion. ■

We now know that if we subtract the same number from each side of an equation then the equation remains true.

□ You may care to prove that $-(-a) = a$; the technique employed is very similar.

Since $-a$ is a number, it has an additive inverse $-(-a)$. Moreover $(-a) + (-(-a)) = 0$. But $(-a) + a = 0$ and therefore

$$(-a) + (-(-a)) = (-a) + a$$

By the previous example we have $-(-a) = a$, as required. ■

We write $a - b$ instead of $a + (-b)$ and thereby extend our algebraic operations to include subtraction.

We now turn our attention to the multiplication rules and you will observe that they follow a similar pattern to those of addition.

MULTIPLICATION RULES

1 Given any two numbers a and b, their product ab is also a number.

2 Given any three numbers a, b and c then

$$a \cdot (b \cdot c) = (a \cdot b) \cdot c$$

3 There is a number 1 such that for every number a,

$$a \times 1 = 1 \times a = a$$

4 To each number a ($\neq 0$) there corresponds another number designated by a^{-1} such that

$$a \cdot a^{-1} = a^{-1} \cdot a = 1$$

5 Given any two numbers a and b then $ab = ba$.

The number a^{-1} is known as the multiplicative inverse of a. A similar argument that we used before can be used to show that each number has a unique multiplicative inverse and that if $a \cdot y = a \cdot z$ and if $a \neq 0$ then $y = z$.

This is known as the cancellation law.

If, when $b \neq 0$, we write $a \div b$ (or a/b) for ab^{-1} we can extend our algebraic operations to include division. We remark that when $a \neq 0$ we write $a^0 = 1$ and a^{-2} instead of $1/a^2$ and so forth.

Lastly we need to state the rule which enables the operations of multiplication and division to interact with one another. This rule is known as the distributive rule.

Given any three numbers a, b and c then $a(b + c) = ab + ac$

□ Show that $a \cdot 0 = 0$ for every number a.

For any number a we have

$$a \cdot 0 = a \cdot (0 + 0) = a \cdot 0 + a \cdot 0$$

and so $a \cdot 0 = 0$. ■

□ Show that $a(-b) = -(ab)$.

We need to show that $a(-b)$ when added to ab gives 0, for then $a(-b)$ will be the additive inverse of ab and this is $-(ab)$. Now

$$ab + a(-b) = a[b + (-b)] = a \cdot 0 = 0$$

as required. ■

□ Show that the distributive rule works with a negative by showing $a(b - c) = ab - ac$.

$$\begin{aligned} a(b-c) &= a(b + [-c]) && \text{definition of '}-\text{'} \\ &= ab + a[-c] && \text{distributive rule} \\ &= ab + [-(ac)] && \text{previous example} \\ &= ab - ac && \text{definition of '}-\text{'} \end{aligned}$$

■

We need to become very familiar with these algebraic rules so that we can expand out brackets quickly and accurately without batting an eyelid.

□ The English poet W. H. Auden reported that he had learnt in mathematics the extraordinary rhyme

> 'Minus times minus is equal to plus
> the reason for this we need not discuss!'

Discuss the reason for this!

It is not clear exactly what is meant by these words; there are two possibilities. However we have already shown that $-(-a) = a$ and so we need only to justify the equation

$$(-1) \times (-1) = 1$$

Now $1 + (-1) = 0$ and so multiplying by a

$$a \times [1 + (-1)] = a \times 0 = 0$$

Therefore

$$a \times 1 + a \times (-1) = 0$$

so

$$a + a \times (-1) = 0$$

and consequently

$$a \times (-1) = -a$$

Replacing a by -1 we now have

$$(-1) \times (-1) = -(-1) = 1$$

consequently

$$(-1) \times (-1) = 1$$

as required. ■

This is a useful rule because it enables us to simplify complicated expressions involving minus signs.

Now it is time for you to try a few exercises. We shall tackle these step by step. Only move on to the next step when you have completed each one to the best of your ability.

1.9 Workshop

Here are several problems to try. We shall solve them one after another. However if you find you can do the first one, why not try them all before looking ahead to see if you have them right?

Alternatively if you feel you need to take things rather slower to build up your confidence then just go ahead one step at a time.

▷ **Exercise** Multiply out each of the following

1 $(a+b)(b+c)$
2 $(a+2b)(b-2a)$
3 $a(b-c)+b(c-a)+c(a-b)$
4 $(a-b)^2+2ab$

1 We can treat $a+b$ as a single number to obtain

$$\begin{aligned}(a+b)(b+c) &= (a+b)\cdot b+(a+b)\cdot c\\ &= ab+b^2+ac+bc\end{aligned}$$

2 Again $a+2b$ can be treated as a single number initially. So that

$$\begin{aligned}(a+2b)(b-2a) &= (a+2b)\cdot b-(a+2b)\cdot(2a)\\ &= ab+2b^2-2a^2-4ab\\ &= 2b^2-2a^2-3ab\end{aligned}$$

3 We multiply out each term in turn to obtain

$a(b-c)+b(c-a)+c(a-b)$

$$\begin{aligned}&= ab-ac+bc-ba+ca-cb\\ &= ab-ca+bc-ab+ca-bc\\ &= 0\end{aligned}$$

It helps to write the square as two brackets 4

$$\begin{aligned}(a-b)^2+2ab &= (a-b)(a-b)+2ab\\ &= (a-b)\cdot a-(a-b)\cdot b+2ab\\ &= a\cdot a-b\cdot a-a\cdot b+b\cdot b+2ab\\ &= a^2+b^2\end{aligned}$$

■

Check the following relationships carefully and then commit them to memory; forwards and backwards. They are very useful and often arise in algebraic work.

$$\begin{aligned}(a+b)^2 &= a^2+b^2+2ab\\ (a-b)^2 &= a^2+b^2-2ab\\ a^2-b^2 &= (a+b)(a-b)\\ a^3-b^3 &= (a-b)(a^2+ab+b^2)\\ a^3+b^3 &= (a+b)(a^2-ab+b^2)\end{aligned}$$

Now check that you have understood and learnt them by attempting the following example. 5

▷ **Exercise** Express each of the following as a product of algebraic factors.

1 x^2-1
2 x^3-1
3 x^4-1
4 x^3+8
5 $(x+1)^2-(x-1)^2$

How many factors did you get for each one? There should have been 2, 2, 3, 2, 1 respectively. 6
Here are the results.

---

1 $x^2-1=x^2-1^2=(x+1)(x-1)$ 7

2 $x^3-1=(x-1)(x^2+x+1)$
3 $x^4-1=(x^2)^2-(1)^2=(x^2-1)(x^2+1)$
$=(x-1)(x+1)(x^2+1)$
4 $x^3+8=x^3+2^3=(x+2)(x^2-2x+4)$
5 $(x+1)^2-(x-1)^2=[(x+1)-(x-1)][(x+1)+(x-1)]$
$=2[2x]=4x$

You could also obtain this by multiplying out each expression.
Now for a tricky one.

▷ **Exercise** Express $x^4 + 1$ as a product of algebraic factors. Here is a hint

$$(x^2 + 1)^2$$

Does that help?

8

$$\begin{aligned} x^4 + 1 &= (x^2 + 1)^2 - 2x^2 \\ &= (x^2 + 1)^2 - (\sqrt{2}x)^2 \\ &= [(x^2 + 1) - \sqrt{2}x][(x^2 + 1) + \sqrt{2}x] \\ &= [x^2 - \sqrt{2}x + 1][x^2 + \sqrt{2}x + 1] \end{aligned}$$

■

We have already seen that it is possible to use the rules of algebra to expand certain expressions. For example we saw that

$$(a + b)^2 = a^2 + 2ab + b^2$$

Sometimes it is necessary to reverse this process and collect algebraic terms together. There is only one way to acquire this skill and that is by repeated practice.

▷ **Exercise** Simplify each of the following expressions.

1 $1 + 4x + 4x^2$

2 $9 - 6x + x^2$

3 $x + 10x^2 + 25x^3$

9

The keys to the problem are pattern and factorization.

1 $1 + 4x + 4x^2 = (1 + 2x)(1 + 2x) = (1 + 2x)^2 = (2x + 1)^2$

2 $9 - 6x + x^2 = 3^2 - 2 \cdot 3 \cdot x + x^2$
$= (3 - x)(3 - x) = (3 - x)^2 = (x - 3)^2$

3 $x + 10x^2 + 25x^3 = x(1 + 10x + 25x^2)$
$= x(1 + 2 \cdot 1 \cdot [5x] + [5x]^2)$
$= x \cdot (1 + 5x)^2 = (5x + 1)^2 x$

We notice that in this example there are several different but equivalent answers. ■

1.10 IDENTITIES AND EQUATIONS

Whenever we have a expression which involves the sign '=' we say we have an equation.

For example

$$x+7=0, \quad x(x+2)=3, \quad x^3+8=0$$

are all examples of equations.

Very often the equation contains an unknown quantity which we are required to determine. The process of determining the unknown is called *solving* the equation and the unknown itself is called a solution or root of the equation. Some equations have many roots.

We have already seen how to solve some equations. For example the equation

$$x+a=0$$

where a is a constant (a number which is known) and x is the unknown number has the solution $x=-a$.

We can generalize this very slightly to solve any linear equation. A linear equation is an equation of the form

$$ax+b=0$$

where a and b are constants ($a \neq 0$) and x is the unknown. We have $ax=-b$ and so $x=-b/a$.

Occasionally an equation is true for all the numbers for which it is defined. We have already seen examples of this such as

$$a^2-b^2=(a-b)(a+b)$$

Such an equation is called an *identity* and sometimes the equals sign is replaced by '≡' to emphasize this. So we could write

$$a^2-b^2 \equiv (a-b)(a+b)$$

☐ In each of the following equations decide which are equations and which are identities. Solve the equations.

1 $4x^4-x^2=x^2\cdot(2x-1)(2x+1)$
2 $x(x+1)(x+2)=(x+1)(x+2)(x+3)$
3 $(x+1)(x-2)=x^2-x-2$
4 $(x+3)(x-2)=x^2-5x-6$

1 This is an identity

$$4x^4-x^2=x^2\cdot(4x^2-1)=x^2\cdot(2x-1)(2x+1)$$

2 This is an equation. We can only cancel out

$$(x+1)(x+2)$$

if it is non-zero. We should then obtain $x = x + 3$, which has no solution. Consequently

$$(x+1)(x+2) = 0$$

and so either $x + 1 = 0$ or $x + 2 = 0$.
Consequently we have the two roots $x = -1$ and $x = -2$.

3 This is an identity. Multiplying out we obtain,

$$(x+1)(x-2) = x^2 - 2x + x - 2 = x^2 - x - 2$$

4 Be careful! This is an equation.

$$(x+3)(x-2) = x^2 - 5x - 6$$

so

$$\begin{aligned} x^2 + 3x - 2x - 6 &= x^2 - 5x - 6 \\ x^2 + x - 6 &= x^2 - 5x - 6 \\ \text{so } x &= -5x \\ \text{and } 6x &= 0 \\ \text{therefore } x &= 0 \end{aligned}$$

■

1.11 SIMULTANEOUS EQUATIONS

Some of the language which is used to describe algebraic expressions can be a little confusing at first. Two words in particular can cause confusion, these are 'term' and 'coefficient'. We shall illustrate how these words are used by referring to a specific expression. The expression we shall take is

$$x^4 - 3x^3 + 11x^2 - 25x + 16$$

The 'term in x^2' refers to all that part of the expression which contributes to x raised to the power of 2. In the example, the term in x^2 is $11x^2$. Likewise the term in x is $-25x$. Notice that the sign must also be included. The constant term is 16, the x^3 term is $-3x^3$ and the term of the fourth degree in x is x^4.

The 'coefficient of x^2' is the number which must be multiplied by x^2 to give the term in x^2. In other words, the coefficient of x^2 is the term in x^2 divided by x^2.

In the example the coefficient of x^4 is 1, the coefficient of x^3 is -3, the coefficient of x^2 is 11 and the coefficient of x is -25.

Right! Now that we have got that clear we shall proceed to consider the solution of simultaneous equations. Naturally we shall use the rules of elementary algebra which we have described to solve them.

For instance, we may be given a pair of linear equations which are known to hold simultaneously. Specifically, suppose that

$$\left.\begin{array}{rcl} ax+by &=& h \\ cx+dy &=& k \end{array}\right\}$$

where a, b, c and d are constants. We are required to obtain x and y.

We shall have quite a lot to say about such systems of simultaneous equations when we discuss matrices in Chapter 13 but for the moment we shall simply discuss how to solve these equations.

The technique which is employed is known as the elimination method. We multiply through each equation by suitable numbers so that the coefficients of one of the unknowns become the same. We then subtract the equations we have obtained to produce a single linear equation which is easily solved.

□ Solve each of the following pairs of simultaneous equations.

1 $x+2y=11, x-y=2$

2 $3x-4y=10, 5x+2y=34$

1

$$\left.\begin{array}{rcl} x+2y &=& 11 \\ x-y &=& 2 \end{array}\right\}$$

Subtract the second equation from the first to eliminate x

$$2y-(-y)=11-2$$

So that $3y=9$ and so $y=3$. Substituting back into the second of the equations now gives $x=y+2$ and so $x=5$.

2

$$\left.\begin{array}{rcl} 3x-4y &=& 10 \\ 5x+2y &=& 34 \end{array}\right\}$$

Here we can multiply the second equation by 2 so that apart from a change of sign the coefficients of y will be the same. We can then add the equations together to eliminate y.

$$\left.\begin{array}{rcl} 3x-4y &=& 10 \\ 10x+4y &=& 68 \end{array}\right\}$$

Adding these equations we now obtain $13x=78$ from which $x=6$. The first equation then gives $4y=3x-10=18-10=8$ from which $y=2$. ■

It is always worth checking that the values of x and y which have been obtained do in fact satisfy the equation. We have only considered a pair of simultaneous algebraic equations in two unknowns but the technique clearly extends to a system of n equations in n unknowns. Naturally the more equations the more tedious the elimination is likely to become. We shall consider systematic methods of solving such sets of simultaneous equations in Chapter 13.

1.12 RATIONAL EXPRESSIONS

One of the most useful algebraic processes involves collecting terms together in a single expression. Here some of the processes of elementary arithmetic can be mimicked. For example, to simplify $1/2 + 1/4 + 1/6$, we take the terms over a 'common denominator'. That is, we look for a natural number which is exactly divisible by $2, 3$ and 6. Preferably we look for the lowest common multiple of these numbers but if we can't spot it we can always multiply all these denominators together to obtain one. Here we see that 12 is the lowest common multiple. This means that, if we ensure that each of these numbers is expressed with 12 as a denominator, we can collect them all together over a common denominator.

$$\frac{1}{2} + \frac{1}{4} + \frac{1}{6} = \frac{6}{12} + \frac{3}{12} + \frac{2}{12} = \frac{6+3+2}{12} = \frac{11}{12}$$

It is usual to leave out the second step of this process because, once the common denominator is known, the corresponding numerators can be found by dividing the individual denominators into it.

We can apply the same principle to algebraic quotients but before we do so we shall look at one more numerical example to establish the pattern.

☐ Without using a calculator (or using mental arithmetic) simplify $2/3 + 1/2 + 5/6$.

Here the lowest common multiple of all these denominators is 6 and so we proceed as follows

$$\frac{2}{3} + \frac{1}{2} + \frac{5}{6} = \frac{4+3+5}{6} = \frac{12}{6} = 2$$

■

Now let's apply the same process to algebraic symbols.

☐ Simplify $1/(x+1) + 1/(x-1)$

We have $(x+1)(x-1)$ as a common denominator so

$$\begin{aligned}\frac{1}{x+1}+\frac{1}{x-1} &= \frac{x-1}{(x+1)(x-1)}+\frac{x+1}{(x+1)(x-1)}\\ &= \frac{(x-1)+(x+1)}{(x+1)(x-1)}\\ &= \frac{2x}{x^2-1}\end{aligned}$$

■

It is necessary to be able to simplify expressions of this form confidently. The basic principle underlying what we are doing can be expressed by

$$\frac{a}{c}+\frac{b}{c}=\frac{a+b}{c}$$

In other words expressions can be collected over a common denominator.

This is a direct consequence of the distributive rule, from which

$$a\cdot c^{-1}+b\cdot c^{-1}=(a+b)c^{-1}$$

You need to be alert to the fact that expressions cannot be collected under a common numerator. In other words

$$\frac{a}{b}+\frac{a}{c}\neq\frac{a}{b+c}$$

This is because

$$a\cdot b^{-1}+a\cdot c^{-1}\neq a(b+c)^{-1}$$

This is an error which occurs surprisingly frequently. Do avoid making this mistake!

We shall also need to be able to reverse the process we have described. This will be one of our studies in Chapter 2 when we describe how to put a rational expression into *partial* fractions.

□ Obtain a and b if

$$\frac{x}{x^2-4}\equiv\frac{a}{x-2}+\frac{b}{x+2}$$

The expression on the right is

$$\frac{a(x+2)+b(x-2)}{(x-2)(x+2)}=\frac{a(x+2)+b(x-2)}{x^2-4}$$

We shall have an identity only if the corresponding numerators are identically equal

$$x\equiv a(x+2)+b(x-2)$$

In order to obtain a and b we are now entitled to put in any values of x, including $x=2$ and $x=-2$. If we do this we deduce $2=4a$ and $-2=-4b$. So $a=1/2$ and $b=1/2$. ■

1.13 REARRANGING EQUATIONS

Sometimes the quantity which we wish to calculate is concealed in an equation and we need to rearrange the equation to obtain it. Under such circumstances we say that the unknown is given *implicitly* by the equation and we wish to obtain it *explicitly*. The equations we have been solving have expressed the unknown implicitly and the process of solving the equation has been to make these unknowns explicit.

□ The equation $PV = RT$ relates pressure P, volume V and temperature T. R is a known constant. Express P, V and T explicitly in terms of the other variables.

From $PV = RT$ we obtain, by division,

$$P = \frac{RT}{V}$$
$$V = \frac{RT}{P}$$
$$T = \frac{PV}{R}$$

■

□ The period of a simple pendulum is given by $T = a \cdot \sqrt{(l/g)}$ where a and g are constant and l is the length of the pendulum. Obtain l explicitly.

From $T = a \cdot \sqrt{(l/g)}$ we have on squaring

$$T^2 = a^2 \cdot (l/g) = (a^2 \cdot l)/g$$

so $T^2 \cdot g = a^2 \cdot l$ therefore $l = (T^2 \cdot g)/a^2 = g \cdot (T/a)^2$. ■

1.14 QUADRATIC EQUATIONS

We have already seen how to solve any linear equation in a single unknown. This is an equation of the form

$$ax + b = 0$$

where a and b are constants $(a \neq 0)$.

We now turn our attention to the quadratic equation

$$ax^2 + bx + c = 0$$

where a, b and c are constants $(a \neq 0)$.

Before we deal with the general case, we shall consider one or two simple cases which can be solved easily.

□ Solve $x^2 - 10x + 25 = 0$.

We may factorize the left side of the equation to obtain

$$(x-5)(x-5) = (x-5)^2 = 0$$

The cancellation law now shows that if $x - 5 \neq 0$ then $x - 5 = 0$.

We conclude that in any event $x - 5 = 0$ and so $x = 5$.

This is known as a *repeated* root since, in a sense, it satisfies the equation twice. ■

□ Solve $x^2 - 9x + 14 = 0$

We shall solve this in two ways. The second method will lead into the general method.

1 Factorizing we obtain

$$(x-2)(x-7) = 0$$

By the cancellation law one of these factors must be zero and so we deduce that either $x - 2 = 0$ or $x - 7 = 0$. Consequently the solutions are $x = 2$ and $x = 7$.

2 We can rearrange the equation in the form

$$(x+a)^2 = k$$

where a is half the coefficient of x.

In this case we obtain

$$\left(x - \frac{9}{2}\right)^2 - \frac{81}{4} + 14 = 0$$

$$\left(x - \frac{9}{2}\right)^2 = \frac{81}{4} - 14 = \frac{25}{4} = \left(\frac{5}{2}\right)^2$$

Now we have an equation of the form $X^2 = A^2$ from which we deduce

$$X^2 - A^2 = (X - A)(X + A) = 0$$

and so $X = A$ or $X = -A$.

In this case

$$x - \frac{9}{2} = \frac{5}{2} \text{ or } x - \frac{9}{2} = -\frac{5}{2}$$

Consequently

$$x = \frac{14}{2} = 7 \text{ or } x = \frac{4}{2} = 2.$$

■

Of course, in this particular example, it was much easier to factorise the quadratic at the outset. However this method provides the springboard for a general idea known as *completing the square.*

We now turn our attention to the more general quadratic equation

$$ax^2 + bx + c = 0$$

where a, b and c are constants ($a \neq 0$).
First divide through by a so that the coefficient of x^2 is 1, so that

$$x^2 + \frac{b}{a}x + \frac{c}{a} = 0$$

Remember that to 'complete' the square we need to add to x half the coefficient of x and then adjust the algebra to maintain the equation. Here we obtain

$$\left(x + \frac{b}{2a}\right)^2 - \left(\frac{b}{2a}\right)^2 + \frac{c}{a} = 0$$

So

$$\begin{aligned}\left(x + \frac{b}{2a}\right)^2 &= \frac{b^2}{(2a)^2} - \frac{c}{a} \\ &= \frac{b^2}{(2a)^2} - \frac{4ac}{(2a)^2} \\ &= \frac{b^2 - 4ac}{(2a)^2}\end{aligned}$$

Taking the square root we now have

$$x + \frac{b}{2a} = \frac{\pm\sqrt{(b^2 - 4ac)}}{2a}$$

From which

$$x = \frac{-b \pm \sqrt{(b^2 - 4ac)}}{2a}$$

This is known as the formula for solving a quadratic equation.

It is important to note that the equation has equal roots when the 'discriminant', $b^2 - 4ac$, is zero and that, when the discriminant is negative, there are no real roots. We shall return to this point in Chapter 10 when we consider complex numbers.

1.15 POLYNOMIALS

In general a polynomial P is defined by

$$P(x) = a_n x^n + a_{n-1}x^{n-1} + \cdots + a_1 x + a_0$$

where the 'a's, which are known as the coefficients, are constants and n is a natural number. The expression $P(x)$ is known as a polynomial in x.

☐ If P is defined by

$$P(x) = x^3 + 4x^2 + x + 1$$

then P is a polynomial and $x^3 + 4x^2 + x + 1$ is the corresponding polynomial in x.

If the leading coefficient a_n is non-zero then the polynomial is said to have *degree* n. ■

☐ $x^2 + 1$ and $2x^5 - x^3 + 1$ are polynomials in x of degree 2 and 5 respectively.

■

If the degree of one polynomial is less than the degree of another it is possible to divide one into the other to obtain a quotient and a remainder.

☐ Divide $x^2 + 1$ into $2x^5 - x^3 + 1$ to obtain a quotient $Q(x)$ and a remainder $R(x)$.

We wish to obtain polynomials in x, $Q(x)$ and $R(x)$, where the degree of $R(x)$ is less than that of $x^2 + 1$, such that

$$2x^5 - x^3 + 1 = (x^2 + 1) \cdot Q(x) + R(x)$$

We begin by observing that to obtain the leading term $2x^5$ we must multiply $x^2 + 1$ by $2x^3$. We do this and then add and subtract terms as necessary to maintain the equality. This procedure is repeated as many times as are necessary in order to achieve the required result. So

$$2x^5 - x^3 + 1 = 2x^3(x^2 + 1) - 3x^3 + 1$$

Notice that $-3x^3$ has degree greater than $x^2 + 1$ and so we can repeat the process.

$$\begin{aligned} 2x^5 - x^3 + 1 &= 2x^3(x^2 + 1) - 3x^3 + 1 \\ &= 2x^3(x^2 + 1) - 3x(x^2 + 1) + 3x + 1 \\ &= (x^2 + 1) \cdot (2x^3 - 3x) + 3x + 1 \end{aligned}$$

Therefore the quotient $Q(x) = 2x^3 - 3x$ and the remainder $R(x) = 3x + 1$. It follows that

$$\frac{2x^5 - x^3 + 1}{x^2 + 1} = 2x^3 - 3x + \frac{3x + 1}{x^2 + 1}$$

This process is sometimes known as the method of 'short' division. ■

An expression which is the quotient of two polynomials in x is known as a *rational* expression in x. If the degree of the numerator is n and the degree of the denominator is m then the degree of the rational expression is defined to be $n - m$.

THE REMAINDER THEOREM

We have considered what happens in general when one polynomial is divided by another of smaller degree. However we obtain a useful result by considering what happens when a polynomial $P(x)$ of degree at least 1 is divided by $x - a$. In general we obtain

$$P(x) = (x - a)Q(x) + R$$

where $Q(x)$ is the quotient and the constant R is the remainder.

We observe that putting $x = a$ we obtain

$$P(a) = R$$

so that the remainder, when the polynomial $P(x)$ is divided by $x - a$, can be obtained by evaluating the polynomial at $x = a$.

It follows therefore that if $P(a) = 0$ then the remainder is zero and so the polynomial has $x-a$ as a factor. This result can be extremely useful if we are required to factorise a polynomial or to solve a polynomial equation.

1.16 THE BINOMIAL THEOREM

It is easy to verify, by direct multiplication, that

$$\begin{aligned}
(1+x)^0 &= 1 \\
(1+x)^1 &= 1+x \\
(1+x)^2 &= 1+2x+x^2 \\
(1+x)^3 &= 1+3x+3x^2+x^3 \\
(1+x)^4 &= 1+4x+6x^2+4x^3+x^4
\end{aligned}$$

A pattern is emerging for these coefficients which is often referred to as **Pascal's triangle**. Let's look at it.

1				
1	1			
(1)	(2)	1		
1	(3)	3	1	
1	4	6	4	1

Each entry consists of the sum of the entry above, and the entry above and to the left. The circled numbers illustrate this feature; 3 is the sum of 2 and 1.

Using this idea, see if you can write down the next line.

Here it is:

$$1 \quad 5 \quad 10 \quad 10 \quad 5 \quad 1$$

So we may conjecture that

$$(1+x)^5 = 1 + 5x + 10x^2 + 10x^3 + 5x^4 + x^5$$

and indeed direct multiplication will verify this fact.

Before we take this story any further we shall introduce some notation which you may not have come across before. Suppose n is a natural number then the symbol $n!$ is known as **factorial** n and is defined by

$$n! = n \times (n-1) \times \cdots \times 3 \times 2 \times 1$$

So that

$$5! = 5 \times 4 \times 3 \times 2 \times 1 = 120$$

One way to think of $n!$ is that it is the number of ways in which we can arrange n books on a shelf. For example, if $n = 3$ and we have three books A, B and C, the six possible arrangements are

$$ABC, \quad BCA, \quad CAB, \quad ACB, \quad BAC, \quad CBA$$

This interpretation is even consistent with 0!, which we define to be 1, for if there are no books, there is only one way to arrange them and that is to leave the shelf empty!

Using the factorial symbol we can now define another symbol which is known as the symbol for the binomial coefficients. Suppose that n and r are both positive integers and that r is less than or equal to n. We define

$$\binom{n}{r} = \frac{n!}{(n-r)!\,r!}$$

Using this notation we see that

$$\begin{aligned}\binom{3}{2} &= \frac{3!}{(3-2)!\,2!} = \frac{3!}{1!\,2!} = \frac{3 \times 2 \times 1}{1 \times 2 \times 1} = 3 \\ \binom{5}{2} &= \frac{5!}{(5-2)!\,2!} = \frac{5!}{3!\,2!} = \frac{5 \times 4 \times 3 \times 2 \times 1}{3 \times 2 \times 1 \times 2 \times 1} = 10\end{aligned}$$

If we calculate $\binom{n}{r}$ for n and r less than 5, we see that Pascal's triangle can be written

$$\binom{0}{0}$$

$$\binom{1}{0} \quad \binom{1}{1}$$

$$\binom{2}{0} \quad \binom{2}{1} \quad \binom{2}{2}$$

$$\binom{3}{0} \quad \binom{3}{1} \quad \binom{3}{2} \quad \binom{3}{3}$$

$$\binom{4}{0} \quad \binom{4}{1} \quad \binom{4}{2} \quad \binom{4}{3} \quad \binom{4}{4}$$

This enables us to conjecture the general formula for $(1+x)^n$. We have

$$(1+x)^n = \binom{n}{0} + \binom{n}{1} x + \binom{n}{2} x^2 + \cdots + \binom{n}{n} x^n$$

This is known as the **binomial theorem**. We shall extend it later to other values of n.

□ Obtain the coefficient of x^7 in the binomial expansion of $(1+x)^{12}$.

The coefficient of x^7 in the expansion of $(1+x)^n$ is $\binom{n}{7}$. So the coefficient of x^7 in the expansion of $(1+x)^{12}$ is

$$\binom{12}{7} = \frac{12!}{5!\,7!} = \frac{12 \times 11 \times 10 \times 9 \times 8}{5 \times 4 \times 3 \times 2 \times 1} = 792$$

■

If we glance back at Pascal's triangle we observe that each row appears to be symmetrical about its midpoint. In other words reading from left to right or reading from right to left seems to give an identical sequence of numbers. We have previously observed that each entry in the table appears to be the sum of two entries in the row above. These are two significant features of the binomial coefficients which we can now prove.

□ Deduce the following identities which are satisfied by the binomial coefficients.

a $\binom{n}{r} = \binom{n}{n-r}$

b $\binom{n}{r} = \binom{n-1}{r} + \binom{n-1}{r-1}$

a $\begin{pmatrix} n \\ n-r \end{pmatrix} = \dfrac{n!}{[n-(n-r)]!\,(n-r)!} = \dfrac{n!}{r!\,(n-r)!} = \begin{pmatrix} n \\ r \end{pmatrix}$

b $\begin{pmatrix} n-1 \\ r \end{pmatrix} + \begin{pmatrix} n-1 \\ r-1 \end{pmatrix}$

$$= \frac{(n-1)!}{(n-1-r)!\,r!} + \frac{(n-1)!}{[n-1-(r-1)]!\,(r-1)!}$$

$$= \frac{(n-1)!}{(n-r-1)!\,r!} + \frac{(n-1)!}{(n-r)!\,(r-1)!}$$

$$= \frac{(n-1)!}{(n-r-1)!\,(r-1)!}\left[\frac{1}{r} + \frac{1}{n-r}\right]$$

$$= \frac{(n-1)!}{(n-r-1)!\,(r-1)!}\left[\frac{n}{r(n-r)}\right]$$

$$= \frac{n!}{(n-r)!\,r!} = \begin{pmatrix} n \\ r \end{pmatrix}$$

■

THE $\sum$ AND $\prod$ NOTATION

Sooner or later we shall have to introduce the 'sigma' notation and there is no time like the present! This symbol, which is also known as the summation symbol $\sum$, provides a useful method for writing a large sum of terms in a compact form. There is also a corresponding symbol, $\prod$, which can be used to represent a product of terms and so while we are about it we shall introduce this too.

Suppose we write

$$\sum_{r=0}^{n} \begin{pmatrix} n \\ r \end{pmatrix}$$

This means that we allow r to take on all integer values from 0 to n (including 0 and n) in the symbol $\begin{pmatrix} n \\ r \end{pmatrix}$ and add the results. Consequently

$$\sum_{r=0}^{n} \begin{pmatrix} n \\ r \end{pmatrix} = \begin{pmatrix} n \\ 0 \end{pmatrix} + \begin{pmatrix} n \\ 1 \end{pmatrix} + \begin{pmatrix} n \\ 2 \end{pmatrix} + \cdots + \begin{pmatrix} n \\ n \end{pmatrix}$$

The binomial expansion can therefore be written compactly as

$$(1+x)^n = \sum_{r=0}^{n} \begin{pmatrix} n \\ r \end{pmatrix} x^r$$

In a similar way we define $\prod$ to be a product, rather than a sum. Using this symbol the product of the same set of terms could be written as

$$\prod_{r=0}^{n} \binom{n}{r}$$

So that

$$\prod_{r=0}^{n} \binom{n}{r} = \binom{n}{0} \times \binom{n}{1} \times \binom{n}{2} \times \cdots \times \binom{n}{n}$$

So, for example,

$$\prod_{r=1}^{n} r = 1 \times 2 \times 3 \cdots \times (n-1) \times n = n!$$

☐ Show that if n is a positive integer then

$$(a+b)^n = \sum_{r=0}^{n} \binom{n}{r} a^{n-r} b^r = \sum_{r=0}^{n} \binom{n}{r} a^r b^{n-r}$$

We have

$$(a+b)^n = \left(a\left[1+\frac{b}{a}\right]\right)^n = a^n \left(1+\frac{b}{a}\right)^n$$

This is now in a form in which we can apply the binomial theorem with $x = b/a$. So

$$\begin{aligned}(a+b)^n &= a^n \sum_{r=0}^{n} \binom{n}{r} \left(\frac{b}{a}\right)^r \\ &= \sum_{r=0}^{n} \binom{n}{r} a^{n-r} b^r\end{aligned}$$

Clearly if we interchange a and b the result will be unchanged and so

$$(a+b)^n = \sum_{r=0}^{n} \binom{n}{r} a^{n-r} b^r = \sum_{r=0}^{n} \binom{n}{r} a^r b^{n-r}$$

Now it's time for you to take a few steps on your own just to make sure you can handle problems involving the binomial theorem and the binomial coefficients.

1.17 Workshop

▷ **Exercise** Show that if n is any natural number 1

$$\binom{n}{0}+\binom{n}{1}+\binom{n}{2}+\cdots+\binom{n}{n}=2^n$$

that is

$$\sum_{r=0}^{n}\binom{n}{r}=2^n$$

Have a go at this and then see if you did the right thing.

We merely need to use the binomial theorem 2

$$(1+x)^n=\sum_{r=0}^{n}\binom{n}{r}x^r$$

and put $x=1$ to obtain the required identity.

If you were right then move ahead to step 4. If you didn't quite manage it then try this.

▷ **Exercise** If we choose any row in Pascal's triangle and alternate the signs, the sum is 0. For example, if $n=5$ then

$$1-5+10-10+5-1=0$$

Prove that this property holds for all natural numbers n.

As soon as you have done this, take the next step.

Again we must use the binomial theorem 3

$$(1+x)^n=\sum_{r=0}^{n}\binom{n}{r}x^r$$

However, this time we put $x=-1$ to deduce

$$0=\sum_{r=0}^{n}\binom{n}{r}(-1)^r$$

which is precisely what we wished to show.

▷ **Exercise** Obtain a relationship between h and k if the constant terms 4 in the binomial expansions of

$$\left(hx^2-\frac{k}{x^2}\right)^8 \quad \text{and} \quad \left(hx+\frac{k}{x}\right)^4$$

are equal.

Give this a little thought and as soon as you are ready, step forward!

---

5 We notice that in each case the constant term will be the middle term in the binomial expansion of $(a + b)^n$. It is only in these that all the powers of x will cancel out.

The constant term in the first occurs when $r = 4$ and the constant term in the second occurs when $r = 2$.

In the first we have $a = hx^2$ and $b = -k/x^2$ so the constant term is

$$\binom{8}{4}(hx^2)^4\left(-\frac{k}{x^2}\right)^4 = \frac{8 \times 7 \times 6 \times 5}{1 \times 2 \times 3 \times 4}h^4k^4 = 70h^4k^4$$

Whereas in the second we obtain

$$\binom{4}{2}(hx)^2\left(\frac{k}{x}\right)^2 = \frac{4 \times 3}{1 \times 2}h^2k^2 = 6h^2k^2$$

Therefore

$$\begin{aligned} 70h^4k^4 &= 6h^2k^2 \\ 35h^2k^2 &= 3 \end{aligned}$$

This is the relationship which we were required to obtain.

If you made a mistake or feel you would like some more practise then do the next exercise. Otherwise you may move through to the top of the steps.

▷ **Exercise** Obtain the approximate value of

$$y = \left(1 + x + \frac{x^2}{2}\right)^5$$

if x is so small that terms in x of degree higher than 3 may be neglected.

We use the notation $\approx$ instead of the equality sign $=$ if two things are 'approximately' equal to one another.

Do your best with this one and then step ahead.

---

6 We have

$$y = \left[1 + \left(x + \frac{x^2}{2}\right)\right]^5$$

Since terms of degree higher than 3 in x may be neglected, we obtain

$$\begin{aligned} y &\approx 1+\binom{5}{1}\left(x+\frac{x^2}{2}\right)+\binom{5}{2}\left(x+\frac{x^2}{2}\right)^2 \\ &\qquad +\binom{5}{3}\left(x+\frac{x^2}{2}\right)^3 \\ &\approx 1+5\left(x+\frac{x^2}{2}\right)+\frac{5\times 4}{1\times 2}\left(x^2+\frac{2x\,x^2}{2}\right)+\frac{5\times 4\times 3}{1\times 2\times 3}x^3 \\ &= 1+5x+\frac{5x^2}{2}+10x^2+10x^3+10x^3 \\ &= 1+5x+\frac{25x^2}{2}+20x^3 \end{aligned}$$

1.18 THE GENERAL BINOMIAL THEOREM

We introduced a special notation for the binomial coefficients and we defined

$$\binom{n}{r}=\frac{n!}{(n-r)!\,r!}=\frac{n\times(n-1)\times\cdots\times(n-r+1)}{1\times 2\times\cdots\times r}$$

This second form can be used to define $\binom{n}{r}$ even when n is not a natural number.

This is particularly important because although we have inferred the binomial expansion from a pattern which we observed when n is a natural number, it is in fact valid for all real numbers n, provided $-1 < x < 1$. The expansion then becomes

$$(1+x)^n=\sum_{r=0}^{\infty}\binom{n}{r}x^r$$

The symbol ∞ which appears above the summation sign indicates that we no longer have a finite expansion but instead have an infinite series. We shall discuss infinite series in some detail in Chapters 8 and 9 but for the moment it will be sufficient to think of them as adding terms indefinitely.

The sum of the first N terms will either approach some fixed number, as N increases, in which case the series is said to **converge**, or it will not, in which case the series is said to **diverge**.

The binomial theorem states that provided $-1 < x < 1$ then the series

$$\sum_{r=0}^{\infty} \binom{n}{r} x^r$$

converges and moreover that it converges to

$$(1+x)^n$$

The binomial theorem in its general form can prove quite useful if we wish to determine an approximate value for an expression involving powers of $1+x$ where $-1 < x < 1$. When x is numerically small we may be able to neglect all but very small powers of x. It is easy to convince yourself, by using a calculator, that when $-1 < x < 1$, large powers of x are extremely small.

It must be stressed that these properties are not self-evident and in fact require quite advanced mathematics to put them on a rigorous footing. Nevertheless we shall be content, for the moment, to apply them and we conclude this chapter by giving a practical application.

1.19 Practical

ELECTRICAL FORCE

A magnetic pole, distance x from the plane of a coil of radius a, and on the axis of the coil, is subject to a force

$$F = \frac{kx}{(a^2+x^2)^{5/2}} \quad (k \text{ constant})$$

when a current flows in the coil. Show that:
a if x is small compared with a then

$$F \approx \frac{kx}{a^5} - \frac{5kx^3}{2a^7}$$

b if x is large compared with a then

$$F \approx \frac{k}{x^4} - \frac{5ka^2}{2x^6}$$

Try **a**, then move ahead for the solution.

---

a We have x/a is small, and so we rearrange the expression for F so that we can expand it by the binomial theorem

$$\begin{aligned} F &= \frac{kx}{(a^2+x^2)^{5/2}} \\ &= \frac{kx}{a^5[1+(x/a)^2]^{5/2}} \\ &= \frac{kx}{a^5}\left[1+\left(\frac{x}{a}\right)^2\right]^{-5/2} \\ &\approx \frac{kx}{a^5}\left[1-\frac{5}{2}\left(\frac{x}{a}\right)^2\right] \end{aligned}$$

neglecting terms in x/a of degree higher than 2. So

$$F \approx \frac{kx}{a^5} - \frac{5kx^3}{2a^7}$$

Now see if you can do the second part. Remember that here x is large compared with a.

---

b We rearrange the expression for F in terms of a/x, which is small, with a view to using the binomial expansion in a very similar way to that of **a**.

$$\begin{aligned} F &= \frac{kx}{x^5[1+(a/x)^2]^{5/2}} \\ &= \frac{k}{x^4}\left[1+\left(\frac{a}{x}\right)^2\right]^{-5/2} \\ &\approx \frac{k}{x^4}\left[1-\frac{5}{2}\left(\frac{a}{x}\right)^2\right] \end{aligned}$$

neglecting terms in a/x of degree higher than 2. So

$$F \approx \frac{k}{x^4} - \frac{5ka^2}{2x^6}$$

SUMMARY

- □ We have seen how to classify real numbers into natural numbers $\mathbb{N}$, integers $\mathbb{Z}$, rational numbers $\mathbb{Q}$ and real numbers $\mathbb{R}$.

$$\mathbb{N} \subset \mathbb{Z} \subset \mathbb{Q} \subset \mathbb{R}$$

- □ We have seen how to approximate numbers to a given number of decimal places or a given number of significant figures.

- □ We have examined the rules of elementary algebra, distinguished between identities and equations, and seen how to solve the quadratic equation

$$ax^2 + bx + c = 0$$

to obtain

$$x = \frac{-b \pm \sqrt{(b^2 - 4ac)}}{2a}$$

- □ We investigated the general expansion of $(1 + x)^n$, where n is a natural number.

$$(1 + x)^n = \sum_{r=0}^{n} \binom{n}{r} x^r$$

EXERCISES

1 For each of the following pairs of numbers obtain the sum and the product in scientific notation giving the answers correct to (1) 3 decimal places (2) 4 significant figures

a 6.23509, 11.4731

b 16.2536, 0.0001124

c 0.00045792, 0.000059634

d 1.0000523, 154.000002

2 Give each of the following numbers (1) correct to 5 significant figures (2) correct to 5 decimal places

a 217.385, **b** 0.0002843, **c** 11.1 **d** 432.495,

e 1.0000472, **f** 1.00005 **g** 1.000050001

3 Multiply out each of the following algebraic expressions

a $(a + 3b)(a - 2b)$

b $(u - v)(v - w)(w - u)$

c $(x + 2y)(y + 2z)(z + 2x)$
d $(a^2 + b^2)(a - b)(a + b)$

4 Factorize each of the following expressions
a $a^3 + 3a^2b + 2ab^2$
b $x^3 + x^2y - 4xy^2 - 4y^3$
c $u^3 - 7u^2v + 7uv^2 - v^3$
d $x^4y^2 - x^2y^4$

5 Rearrange the following equations to give x explicitly in terms of y
a $1/x + 1/y = 1$
b $y = (x + 1)/(x - 1)$
c $y = 1/\sqrt{(1 + y/x)}$
d $1/(1 - x) - 1/(1 - y) = x - y$

6 Solve the following sets of simultaneous equations
a $x + 2y = 7, 3x - 4y = 1$
b $2u + 3v = 21, 3u + 2v = 19$
c $3p - 2q = 11, 2p + 3q = 29$
d $5h - 3k = 7, 3h + 5k = 11$

7 Prove the following identities
a $(a + b)(b + c)(c + a)$
$$= a(b^2 + c^2) + b(c^2 + a^2) + c(a^2 + b^2) + 2abc$$
b $a^2(b - c) + b^2(c - a) + c^2(a - b) + (a - b)(b - c)(c - a) = 0$

8 Classify the real roots of the following equations into natural numbers, integers, rational numbers, real numbers:
a $u^2 + u - 2 = 0$
b $u^2 + u + 2 = 0$
c $u^2 + 2u - 2 = 0$
d $u^2 - 3u + 2 = 0$
e $u^2 + 3u + 2 = 0$
f $2u^2 - 3u + 1 = 0$

9 Write down the first three terms in the binomial expansion, in ascending powers of x, of
a $(1 - 2x)^5$
b $(x + 3)^7$
c $(2x - 3)^8$
d $(4 - 5x)^{1/2}$
e $(3 + 5x)^{2/3}$
f $(x - 3)^{-2}$

ASSIGNMENT

1 Find all real solutions of the following equations and classify them into real numbers, rational numbers, integers, natural numbers:

a $x^2 - x = 6$
b $3y^2 - 7y + 2 = 0$
c $u^2 + 1 = 2/u^2$
d $v^2 + 15/v^2 = 8$
e $4x^2 - 4x + 1 = 0$

2 Simplify

$$\frac{1 - x^4}{(1 - x)(1 + x^2)}$$

3 A resistance r is given by the formula $1/r = 1/r_1 + 1/r_2$, where r_1 and r_2 are other resistances. Obtain an explicit rational expression for r in terms of r_1 and r_2.

4 By first putting $x = 1/u$ and $y = 1/v$, or otherwise, solve the following equations

a $2/u - 1/v = 4, 1/u + 3/v = 9$
b $4/u + 3/v = 22, 3/u - 2/v = 8$
c $5v - 3u = 7uv, 3v + 4u = 10uv$
d $u = v - uv, 6uv = 4v - 1$

5 Decide, in each case,which of the following are identities and which are equations. If they are equations then solve them.

a $(2x + 3)^2 - 2(x + 3)(2x + 3) + (x + 3)^2 = x^2$
b $(x + 3)/(x^2 + 3x + 2) - 1/(x + 1) + 1/(x + 2) = x$
c $(x - 1)^2 + (x - 4)^2 = (x - 2)^2 + (x - 3)^2 + 2^2$

6 Determine the value of k if the coefficient of x^{12} in the binomial expansion of $(1 + kx^3)^{15}$ is known to be 455/27.

7 Obtain the constant term in the binomial expansion of $[x-(1/x)]^8$.

8 The binomial expansions of $(1 + ax^3)^4$ and $(1 - bx^2)^6$ both have the same coefficient of x^6. Show that

$$3a^2 + 10b^3 = 0.$$

FURTHER EXERCISES

1 For each of the following equations, classify the real solutions into natural numbers, integers, rational numbers, real numbers.

a $(2x - 1)(2x - 4) + 2 = 0$
b $x - 1/x = 1 - 5/x$
c $(x - 3)^2 + 6(x - 3) + 6 = 0$
d $(\sqrt{x} - 1/\sqrt{x})^2 = 1 - 1/x$

2 Factorize each of the following algebraic expressions completely
a $a^4 - 5a^2b^2 + 4b^4$
b $4u^4 - 17u^2v^2 + 4v^4$
c $[(u+v)^2 - (u-v)^2][(v+w)^2 - (v-w)^2]$
d $x^5 - 10x^4 + 35x^3 - 50x^2 + 24x$

3 Show that twice the coefficients of x^3 in $(3x+2)^5$ is equal to three times the coefficient of x^3 in $(2x+3)^5$.

4 Write down the first four terms in the binomial expansion of
a $(1+x)^{10}$
b $(1+x)^{-1}$
c $(1-x)^{-1/2}$
d $(1-3x)^{-5/3}$
e $(4-7x)^{3/2}$

5 A trapezium has height h and parallel sides of length a and b. If the distance d of the centre of mass from the side of length a is given by

$$d = \frac{h}{3}\,\frac{(2b+a)}{(a+b)}$$

express b explicitly in terms of a, d and h.

6 The surface area of a rubber tyre is given by $S = 4\pi^2 ab$ where a is the radius of a circular cross-section of the tyre and b is the distance of the centre of this section from the centre of the ring. Express the area S in terms of the internal radius r and the external radius R of the ring.

7 When two resistances r_1 and r_2 are arranged in parallel their combined resistance r is given by the formula $1/r = 1/r_1 + 1/r_2$ but when they are arranged in series their combined resistance r is given by the formula $r = r_1 + r_2$. If an electromotive force (EMF) E is applied to a resistance r the current i is given by $E = ir$.

Two resistances r_1 and r_2 are first arranged in series and then in parallel. In each case a constant potential difference E is applied and the current to the combined resistance is measured. If these measurements are i_1 and i_2 respectively obtain formulae for r_1 and r_2 explicitly in terms of E, i_1 and i_2.

8 Obtain the displacement x in terms of the velocity v if

$$v^2 = 2k\left(\frac{1}{x} - \frac{1}{a}\right)$$

where k and a are known constants.

9 The torque T exerted by an induction motor is given by

$$T = \frac{ARs}{R^2 + X^2s^2}$$

Obtain the ratio s/R explicitly in terms of A, X and T only.

10 A hemispherical shell has inner and outer radii a and b respectively. It is found that the distance d of the centre of mass from the centre of the bounding sphere is given by

$$d = \frac{3(a+b)(a^2+b^2)}{8(a^2+ab+b^2)}$$

If $a+b=h$, express a/b explicitly in terms of d and h.

11 By putting $y = x - h$, or otherwise, show that, if powers of $(x-y)/x$ higher than degree 2 may be neglected, then

$$\frac{x^2-y^2}{2xy} \approx \left(\frac{x-y}{x}\right) + \frac{1}{2}\left(\frac{x-y}{x}\right)^2$$

12 The flow of water through a pipe is given by $G = \sqrt{[(3d)^5\ H/L]}$. if d decreases by 1% and H by 2%, use the binomial theorem to estimate the decrease in G.

13 The resonant frequency of a circuit of inductance L and capacitance C with negligible resistance is given by $f = 1/[2\pi\sqrt{(LC)}]$. If L and C increase respectively by 1% and 2%, estimate the percentage error in f.

14 The safe load W that can be carried by a beam of breadth b, depth d and length l is proportional to bd^3/l. Use the binomial theorem to estimate the percentage change in W if for a given beam the breadth is increased by 1%, the depth is decreased by 3% and the length is decreased by 3%.

15 The field strength of a magnet at a point on the x-axis at distance x from the centre is given by

$$H = \frac{M}{2a}\left[\frac{1}{(x-a)^2} - \frac{1}{(x+a)^2}\right]$$

where M is the moment and $2a$ is the length of the magnet. Show that if x is large compared with a then $H \approx 2M/x^3$.

16 A string is stretched between two points A and B distance l apart. A point P on the string distance d from A is pulled transversely through a small distance x. Show that the increase in the length of the string is approximately $lx^2/2d(1-d)$.

Further concepts 2

In Chapter 1 we examined the algebraic rules which numbers obey. In this chapter we shall discuss some useful mathematical concepts including inequalities and the laws of logarithms.

After completing this chapter you should be able to

- ☐ Apply the laws of indices and logarithms;
- ☐ Solve simple inequalities;
- ☐ Resolve a rational expression into partial fractions;
- ☐ Construct examples of direct proofs and indirect proofs;
- ☐ Use the method of proof known as 'mathematical induction'.

At the end of this chapter we shall solve a practical problem concerning a gas cylinder.

2.1 INDICES AND LOGARITHMS

Years ago all calculations of any difficulty had to be performed using tables of logarithms. It was therefore essential to become skilled in the use of these tables. With modern calculators this is no longer necessary, but it remains important to have a clear understanding of the rules which underpin them.

INDICES

We first explain what we mean by a^r, where r is any real number and a is a strictly positive real number. We shall build up to the general idea in stages and so we start with $r = n$, a natural number.

In fact a^n where n is a natural number will be defined when a is *any* real number. This is a luxury which we shall not be able to afford for general r.

$$a^1 = a, \quad a^2 = aa$$

and in general

$$a^n = aa^{n-1} \quad \text{where } n > 1$$

It follows that a^n is simply the product of a with itself n times.

If $a \neq 0$, we define $a^0 = 1$ and, if n is a natural number,

$$a^{-n} = \frac{1}{a^n}$$

This extends the definition to a^r, where r is any integer.

When we consider the definition carefully, we obtain when $a \neq 0$:

1 $a^p\, a^q = a^{p+q}$
2 $(a^p)^q = a^{pq}$
where p and q are any integers.

These two rules are often referred to as the **laws of indices**, and as we extend the definition of a^n we shall require these rules to remain true. One of the prices we have to pay for this is that we will have to restrict the definition of a^r to $a > 0$. Henceforward we shall suppose therefore that $a > 0$.

Suppose next that r is a *rational* number. So $r = p/q$, where p and q are integers. We define $a^{p/q}$ to be the positive real number x which satisfies the equation $x^q = a^p$.

There is in fact one and only one such real number but the proof of this, which relies on a theorem known as the intermediate value theorem, is outside the scope of our work.

□ $2^{1/2}$ is therefore the positive real number x which satisfies the equation $x^2 = 2$, so that $2^{1/2} = \sqrt{2} \approx 1.414$.

Lastly, for those who are interested, we extend the definition of a^r to the case where r is *any* real number. If r is a rational number then we have already defined a^r, so we may suppose r is an irrational number. Consider the non-recurring non-terminating infinite decimal expansion which corresponds to r and let

$$r_1, r_2, r_3, \cdots, r_k, \cdots$$

denote successive rational approximations to r, so that r_k is obtained by truncating the decimal expansion for r after k decimal places.

We now consider the numbers

$$a^{r_1}, a^{r_2}, a^{r_3}, \cdots, a^{r_k}, \cdots$$

Each one of these has been defined because r_k is a rational number. The number to which these are successive approximations is the number we call a^r.

It will be appreciated that this idea is quite sophisticated and it leaves a number of questions. For instance, how do we know that the numbers

$$a^{r_1}, a^{r_2}, a^{r_3}, \cdots, a^{r_k}, \cdots$$

are successive approximations to a number at all? Such questions are rather subtle and will not be discussed here. It will be sufficient for our purposes to know that a^r has been defined when $a > 0$ and that the laws of indices hold.

1 $a^r \; a^s = a^{r+s}$
2 $(a^r)^s = a^{rs}$
where r and s are any real numbers.

☐ Using a calculator and employing six successive approximations to π, obtain six successive approximations to 2^π.

We have $\pi = 3.141\,59\ldots$, from which we obtain the successive approximations

$$2^3, 2^{3.1}, 2^{3.14}, 2^{3.141}, 2^{3.1415}, 2^{3.14159}$$

That is,

$$8, 8.574, 8.815, 8.821, 8.824, 8.825$$

■

LOGARITHMS

From the laws of indices we obtain the laws of logarithms. Logarithms are important because they provide a transformation which enables the arithmetical processes of multiplication and division to be replaced by those of addition and subtraction.

Suppose $a = b^c$. Then c is said to be the power to which b has been raised to produce a. We then write

$$c = \log_b a$$

which is called the **logarithm** of a to the base b.

So the equations

$$a = b^c \qquad \text{and} \qquad c = \log_b a$$

are equivalent to one another.

In words, the logarithm of a number is the power to which the base must be raised to obtain the number.

Any positive number, except 1, is suitable as a base. In practice two bases are used:

1 Base 10: this produces the **common** logarithms
2 Base e: this produces the **natural** logarithms (also known as Naperian logarithms).

The number e (≈ 2.71828) is an irrational number. The reason why it is chosen and called the natural base will become clearer when we deal with differentiation (Chapter 4).

It is usual to write $y = \log x$ instead of $y = \log_{10} x$, and $y = \ln x$ instead of $y = \log_{\mathrm{e}} x$.

When the laws of indices are transformed into logarithmic notation, they result in the **laws of logarithms**:

1 $\log_c(ab) = \log_c a + \log_c b$
2 $\log_c(a/b) = \log_c a - \log_c b$
3 $\log_c(a^r) = r \log_c a$
4 $\log_a b = \log_c b / \log_c a \quad (a \neq 1)$

The last rule is usually called the **formula for a change of base**.

□ Use the laws of indices to deduce

$$\log_c(ab) = \log_c a + \log_c b$$

Let $\log_c a = x$ and $\log_c b = y$. Then $a = c^x$ and $b = c^y$, and so

$$ab = c^x c^y = c^{x+y}$$

Consequently $x + y = \log_c(ab)$. ■

Now one for you to try.

□ Deduce, using the laws of indices,

$$\log_c(a/b) = \log_c a - \log_c b$$

Have a go at this; it's very similar to the previous example.

This is what you should write. Let $\log_c a = x$ and $\log_c b = y$. Then $a = c^x$ and $b = c^y$, and so

$$\frac{a}{b} = \frac{c^x}{c^y} = c^x \ (c^y)^{-1} = c^x \ c^{-y} = c^{x-y}$$

Therefore

$$\log_c(a/b) = x - y = \log_c a - \log_c b$$

■

Was all well? If you would like some more practice then try this.

□ Deduce, using the laws of indices,

a $\log_c(a^r) = r \log_c a$
b $\log_a b = \log_c b / \log_c a$

Here is the working:

a Suppose $\log_c a = b$. Then $a = c^b$, and therefore $a^r = (c^b)^r = c^{rb}$. Consequently, $\log_c(a^r) = rb = r \log_c a$.

b Suppose $\log_c b = y$ and $\log_c a = x$. Then $b = c^y$ and $a = c^x$, and so

$$b^x = (c^y)^x = c^{xy} = (c^x)^y = a^y$$

It follows that

$$b = (b^x)^{1/x} = (a^y)^{1/x} = a^{y/x}$$

from which

$$\log_a b = \frac{y}{x} = \frac{\log_c b}{\log_c a}$$

Note that $\log_c a \neq 0$, since if $\log_c a = 0$ then $a = 1$. ■

Logarithms are very important for solving algebraic equations in which indices are present. However, it is easy to make mistakes. One of the commonest errors is to assume that the logarithm of a sum is the sum of the logarithms. This kind of rule is known as a **linearity rule**; unfortunately logarithms do not comply with it. Let us be specific and examine how the error is usually made.

□ Solve the equation

$$4^x + 2^x - 2 = 0$$

You can try this first and then examine the correct working afterwards.

The following working is the correct working. First,

$$4^x + 2^x - 2 = 0$$

So

$$\begin{aligned} 2^{2x} + 2^x - 2 &= 0 \\ (2^x)^2 + 2^x - 2 &= 0 \end{aligned}$$

Consequently

$$(2^x + 2)(2^x - 1) = 0$$

So either $2^x + 2 = 0$ or $2^x = 1$. Now $2^x > 0$ for all real numbers x. So the only possibility is $2^x = 1$, from which $x = 0$. ■

Now let's examine some *incorrect* working of the type which is frequently seen by examiners. In order not to mislead the unwary we shall avoid the equality sign - it cuts against the grain to use it - and instead use the symbol $||$. See how many errors you can spot in the following incorrect working of the previous example.

Taking logarithms

$$x \ln 4 + x \ln 2 - \ln 2 || 0$$

So

$$\begin{aligned} 2x \ln 2 + x \ln 2 - \ln 2 &\quad || \quad 0 \\ (3x - 1) \ln 2 &\quad || \quad 0 \end{aligned}$$

So $x || 1/3$.

The trouble is that on rare occasions nonsense like this can even lead to the correct answer! There are two glaring errors:

1 You cannot 'take logarithms' of both sides in the way that has been shown. The temptation to saunter through the equation from left to right dispensing logarithmic transformations on every term encountered seems to be so strong that many people find it irresistible. However, as we have remarked, logarithmic transformations are not linear and so the procedure is not valid.

2 Without even bothering to mention it, the assumption has been made that $\ln 0 = 0$. However, since $e^0 = 1$, $e^0 \neq 0$. In fact $\ln 0$ has no meaning.

Here is an example for you to try. Do be careful!

☐ Obtain all real solutions of the equation

$$2^{2x} - 2^{x+5} + 256 = 0$$

When you have solved the problem, look to see if you are right.

The easiest way to solve this is to put $u = 2^x$ and obtain a quadratic equation in u. For if $u = 2^x$,

$$2^{2x} = (2^x)^2 = u^2$$

and

$$2^{x+5} = 2^x\ 2^5 = 32u$$

So the equation becomes

$$\begin{aligned} u^2 - 32u + 256 &= 0 \\ u^2 - 2^5\ u + 2^8 &= 0 \\ (u - 2^4)(u - 2^4) &= 0 \end{aligned}$$

Therefore $u = 2^x = 2^4$, and so $x = 4$. ■

2.2 INEQUALITIES

In Chapter 1 we considered identities and equations and used the algebraic properties of the real numbers to solve some of them. When we come to examine inequalities we need some further algebraic rules. First, though, we describe the notation.

We write $a > b$ if and only if the number a is greater than the number b. This is known as a strict inequality. The symbol '$>$' is called 'greater than'.

We write $a \geqslant b$ if and only if the number a is greater than, or possibly equal to, the number b. The symbol '$\geqslant$' is called 'greater than or equal to'.

□ For the numbers -4, 0, x^2 write down the 8 correct statements which can be written using the symbols $\geqslant$ and $>$ and just two of the numbers.

We need to choose two numbers (not necessarily distinct) and one of the symbols. We have

$$\begin{array}{ccc} -4 \geqslant -4 & 0 \geqslant 0 & x^2 \geqslant x^2 \\ 0 \geqslant -4 & 0 > -4 & x^2 \geqslant 0 \\ x^2 \geqslant -4 & x^2 > -4 & \end{array}$$

■

You may be a little puzzled by the inclusion of $0 \geqslant -4$ and $x^2 \geqslant -4$ since we know that $0 > -4$ and $x^2 > -4$. However if you consider the meaning of the symbol $\geqslant$ carefully you should appreciate that 0 is indeed 'greater than or equal to' -4. You need to be very strict about the precise meanings of words and expressions and in that way avoid misunderstandings and errors.

In like manner we define the symbols $<$ and $\leqslant$; called 'less than' and 'less than or equal to' respectively.

We write $a < b$ if and only $b > a$. Similarly we write $a \leqslant b$ if and only if $b \geqslant a$.

In general inequalities are difficult to solve. However there are some which are quite amenable and these are the ones which we shall consider shortly.

One of the ideas which we shall employ is that if we take the product or quotient of two numbers of the same sign then the result is positive, whereas if we take the product or quotient of two numbers of opposite sign then the result is negative. This relatively simple idea, that 'two negatives cancel one another out', can be used to solve a number of inequalities.

When it comes to the algebraic processes, which we shall explore, it is necessary to state three simple rules. These are known as the order axioms for the real number system.

2.3 RULES FOR INEQUALITIES

1 For any real number a, just one of the following is true:

$$a > 0 \quad a = 0 \quad a < 0$$

2 If $a > 0$ and $b > 0$ then $a + b > 0$ and $ab > 0$

3 $a > b$ if and only if $a - b > 0$

You may feel that the first of these rules is self-evident. However, we shall meet many other mathematical objects for which inequalities are meaningless. Relatively simple examples of these are complex numbers (Chapter 10), matrices (Chapter 11) and vectors (Chapter 14). There is often a human hankering to compare things. People ask who was the greatest composer or the greatest film star, for instance, but these questions are without answer. Anything with more than one attribute is unlikely to satisfy the first rule (the law of trichotomy).

The second rule gives us the method for dealing algebraically with inequalities. However we need to be rather careful because although some of the rules which we apply to equations will also work with inequalities they do not all work. Unfortunately this can be a source of much error.

We deduce two properties. The first one is familiar to us because we know that it applies to equations.

Property 1 If $a > b$ and c is any real number then

$$a + c > b + c$$

Proof By rule 3 we deduce that $a + c > b + c$ if and only if

$$(a + c) - (b + c) > 0$$

But this is true if and only if $a - b > 0$. Again by rule 3 this holds if and only if $a > b$. Now we were given that $a > b$ and, since every step in the argument is an 'if and only if' condition, we deduce that

$$a + c > b + c$$

This means that we can add or subtract any number from each side of an inequality and still preserve the inequality. This is intuitively clear. If one body is hotter than another and their temperatures are both increased or decreased by the same amount then the hotter body remains hotter.

Property 2 If $a > b$ and c is any real number then

1 $ac > bc$ if $c > 0$

2 $ac < bc$ if $c < 0$

It is the second part of this property which is often overlooked and which results in errors.

Proof

1 We have $a - b > 0$ and $c > 0$ and so by rule 2, $(a - b)c > 0$ from which $ac - bc > 0$. Therefore by rule 3, $ac > bc$.

2 We have $a - b > 0$ and $-c > 0$ and so by rule 2, $(a - b)(-c) > 0$ from which $bc - ac > 0$. Therefore by rule 3, $bc > ac$ and so $ac < bc$.

In plain language this means that, when we are multiplying an inequality by a positive number, the direction of the inequality is preserved. However, when we are multiplying an inequality by a negative number, the direction of the inequality is reversed.

There are a number of methods which can be applied to solve inequalities and we shall illustrate them by solving the same problem in a variety of ways. It is important to study each of these methods carefully in order to understand fully how each is applied.

□ Determine those real numbers x which satisfy

$$x > 1/x$$

Method 1 (Algebraic)
We would like to multiply through by x but we need to be aware of the possibility that x could be negative. There are therefore two cases to consider.
Case 1 ($x > 0$) We obtain $x^2 > 1$ and so $x^2 - 1 > 0$. Therefore

$$(x - 1)(x + 1) > 0$$

We deduce that $x - 1$ and $x + 1$ must both have the same sign. So either both $x > 1$ and $x > -1$ or both $x < 1$ and $x < -1$.

The first condition is satisfied if and only if $x > 1$, and the second if and only if $x < -1$. However $x > 0$ and so the second can be discounted. Therefore we deduce that $x > 1$ is a solution.
Case 2 ($x < 0$) Here we obtain, on multiplying through by x,

$$x^2 < 1$$

and so

$$x^2 - 1 < 0$$

Consequently,

$$(x - 1)(x + 1) < 0$$

Therefore $x - 1$ and $x + 1$ must both have the opposite sign. So either $x > 1$ and $x < -1$ (impossible) or $x < 1$ and $x > -1$. We can express this by the compound symbol $-1 < x < 1$.
Note We can only sandwich inequalities together in this way when they all have the same direction.

However $x < 0$ and so we deduce $-1 < x < 0$. Therefore we have solved the inequality by obtaining the set of numbers which satisfy it.

$$x > 1/x \text{ if and only if } x > 1 \text{ or } -1 < x < 0.$$

This algebraic method is of great generality and can be used effectively to solve many inequalities. ■
There is a second approach we can use and we illustrate this method by solving the same inequality again.

☐ Determine those real numbers x which satisfy

$$x > 1/x$$

Method 2 (Analytic)
We employ rule 3 for inequalities and show how effective it can be.

Therefore we write $E = x - 1/x$ and determine when $E > 0$.

Now

$$E = x - \frac{1}{x} = \frac{x^2 - 1}{x} = \frac{(x-1)(x+1)}{x}$$

Each of the terms in this quotient, $x-1, x+1$ and x, changes sign just once as x increases through negative numbers to positive numbers.

The critical numbers at which the sign changes occur are $x = 1, x = -1$ and $x = 0$. We arrange these critical numbers in ascending order and investigate the sign of the quotient E by constructing a table in which the signs of the constituent terms are shown.

Term	$x < -1$	$-1 < x < 0$	$0 < x < 1$	$x > 1$
$x+1$	$-$	$+$	$+$	$+$
x	$-$	$-$	$+$	$+$
$x-1$	$-$	$-$	$-$	$+$
E	$-$	$+$	$-$	$+$

In the table, we merely need to count up to see if there is an odd number of negative signs or an even number of negative signs. An odd number of signs means that E is negative whereas an even number of signs means that E is positive. We conclude therefore that the solution of the inequality is either $-1 < x < 0$ or $x > 1$. ■

There is a third approach we can use and so we illustrate this method by solving the same inequality yet again.

☐ Determine those real numbers x which satisfy

$$x > 1/x$$

Method 3 (Graphical)
If you are unfamiliar with drawing graphs you will have to delay considering this method until you have read Chapter 3.

We draw graphs of $y = x$ and $y = 1/x$ on the same diagram. The basic idea is to compare the two graphs (Fig. 2.1) and to find those values of x for which $y = x$ is 'above' $y = 1/x$.

The graph of $y = x$ is a straight line and the graph of $y = 1/x$ is a rectangular hyperbola, but we do not need to know the names; we only need to be able to plot them in order to solve the problem. Graphics calculators provide a method of doing this automatically. We shall consider ideas for obtaining rough sketches of graphs in Chapter 5.

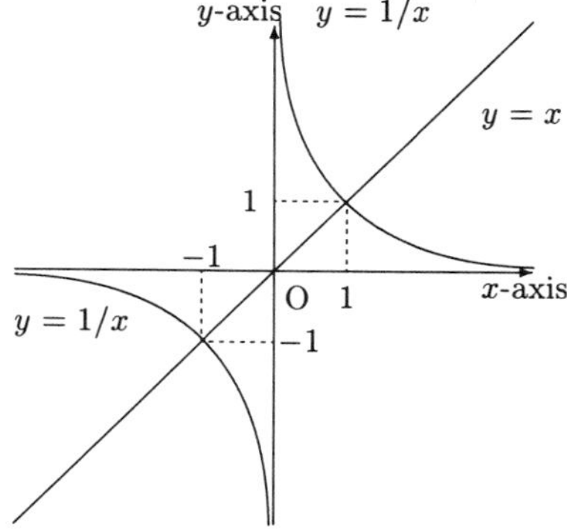

Fig. 2.1 Graphical solution of $x > 1/x$.

We observe that the graphs intersect when $x = 1$ and when $x = -1$ and from them deduce immediately that the inequality holds when $x > 1$ and $-1 < x < 0$. ■

In situations where the possibility of equality is included it is necessary to examine the sign of the inequality at the critical numbers too. For instance the inequality $x \geqslant 1/x$ has solutions $-1 \leqslant x < 0$ and $x \geqslant 1$. Note in particular that $1/x$ is meaningless when $x = 0$.

□ Show that if $a > 0$ and $b > 0$ and $a^2 > b^2$ then $a > b$. In words, the inequality is preserved when we take the positive square root.

We prove this by an indirect method: we show that only $a > b$ is possible because if any of the alternatives were to hold then a contradiction would result. There are just three possibilities: **a** $a = b$ **b** $a < b$ **c** $a > b$.

a We can reject $a = b$ immediately, since then

$$a^2 - b^2 = (a + b)(a - b) = 0$$

which contradicts $a^2 - b^2 > 0$.

b If $a < b$ then $b - a > 0$ and we know that $b + a > 0$. Therefore $(b + a)(b - a) > 0$ and so $b^2 - a^2 > 0$ which is a contradiction.

Only case **c** remains, and we deduce that $a > b$. ■

A similar method can be used to show that if $a \geqslant 0$ and $b \geqslant 0$ and $a^2 \geqslant b^2$ then $a \geqslant b$.

It is essential to note that in order to apply this property both a and b must be positive. For example $(-3)^2 > 2^2$ but $-3 < 2$. Trouble soon occurs if you overlook considerations of this kind.

THE TRIANGLE INEQUALITY

One useful notation which we shall employ from time to time is the **modulus** symbol. We write $|a|$ for the absolute value of a. Thus if a is any real number,

$$\begin{aligned} |a| &= a \quad \text{when } a \geqslant 0 \\ |a| &= -a \quad \text{when } a < 0 \end{aligned}$$

For example, $|-3| = 3$ and $|5| = 5$.

An inequality involving the modulus sign which we shall encounter occasionally is the **triangle inequality**. It can be interpreted physically as saying that the sum of the lengths of two sides of a triangle is always greater than or equal to the length of the third side. Geometrically this is obvious, but algebraically it is not quite so clear. Here it is:

The triangle inequality

$$|a+b| \leqslant |a| + |b|$$

whenever a and b are real numbers.

If you substitute a few numbers, you will soon convince yourself of the truth of this assertion.

To prove it we proceed as follows:

$$\begin{aligned} (|a| + |b|)^2 &= |a|^2 + 2|a||b| + |b|^2 \\ &= a^2 + 2|a||b| + b^2 \\ &\geqslant a^2 + 2ab + b^2 \\ &= (a+b)^2 \\ &= |a+b|^2 \end{aligned}$$

So that taking the positive square root,

$$|a| + |b| \geqslant |a+b|$$

or

$$|a+b| \leqslant |a| + |b|$$

■

2.4 PARTIAL FRACTIONS

In Chapter 1 we saw how we could collect a sum of rational expressions together over a common denominator and thereby simplify it. If we reverse this process we say we have put the rational expression into partial fractions. We shall now see how this is done.

To resolve a rational expression into partial fractions we first ensure that the numerator is of degree less than that of the denominator. If this is not already the case we must *divide* the denominator into the numerator. For example, take the expression

$$R = \frac{2x^3 - 1}{x^3 + x^2 - x - 1}$$

Here the denominator needs to be divided into the numerator. We can use 'short' division by observing that

$$2x^3 - 1 = 2(x^3 + x^2 - x - 1) - 2x^2 + 2x + 1$$

so that

$$\begin{aligned} R &= \frac{2x^3 - 1}{x^3 + x^2 - x - 1} \\ &= 2 - \frac{2x^2 - 2x - 1}{x^3 + x^2 - x - 1} \end{aligned}$$

Next we *factorize* the denominator as far as possible:

$$x^3 + x^2 - x - 1 = (x - 1)(x^2 + 2x + 1) = (x - 1)(x + 1)^2$$

To each factor of the denominator there corresponds a partial fraction. There are two cases:

1 If the factor is not repeated then the numerator of the partial fraction has degree less than its denominator.

For example, if the factor is $x - 1$ (that is, a polynomial of degree 1) then the corresponding numerator will be a constant, A say (that is, a polynomial of degree 0). Again, if the factor is $x^2 + 1$ (that is, a polynomial of degree 2) then the corresponding numerator will have the form $Ax + B$ (that is, a polynomial of degree 1).

2 If the factor is repeated r times then there correspond r partial fractions, one to each power of the factor.

For example, if the denominator was $(2x^2 + 1)^3$, we should obtain three corresponding partial fractions with denominators $2x^2 + 1$, $(2x^2 + 1)^2$, and $(2x^2 + 1)^3$ respectively. The form of each of the numerators is then identical and is determined by the factor itself; each numerator has degree less than that of the factor. So in this example we should obtain

$$\frac{Ax + B}{2x^2 + 1} + \frac{Cx + D}{(2x^2 + 1)^2} + \frac{Ex + F}{(2x^2 + 1)^3}$$

Finally, the unknown constants are obtained by using the fact that we require an identity.

□ Resolve into partial fractions

$$R = \frac{2x^2 - 1}{x^3 + x^2 - x - 1}$$

We have already shown that

$$\begin{aligned} R &= 2 - \frac{2x^2 - 2x - 1}{x^3 + x^2 - x - 1} \\ &= 2 - \frac{2x^2 - 2x - 1}{(x-1)(x+1)^2} \\ &= 2 - \left[\frac{A}{x-1} + \frac{B}{x+1} + \frac{C}{(x+1)^2}\right] \end{aligned}$$

We therefore require

$$\begin{aligned} \frac{2x^2 - 2x - 1}{(x-1)(x+1)^2} &\equiv \frac{A}{x-1} + \frac{B}{x+1} + \frac{C}{(x+1)^2} \\ &= \frac{A(x+1)^2 + B(x-1)(x+1) + C(x-1)}{(x-1)(x+1)^2} \end{aligned}$$

So we require

$$2x^2 - 2x - 1 \equiv A(x+1)^2 + B(x-1)(x+1) + C(x-1)$$

We may either equate coefficients or put in values of x. In any case our aim is to determine A, B and C as easily as possible. Putting $x = 1$ gives $2 - 2 - 1 = 4A$, so $A = -1/4$. Putting $x = -1$ gives $2 + 2 - 1 = -2C$, so $C = -3/2$. Finally, putting $x = 0$ gives $-1 = A - B - C$, so $B = A - C + 1 = -1/4 + 3/2 + 1 = 9/4$.
Consequently

$$\begin{aligned} R &= 2 - \left[\frac{(-1/4)}{x-1} + \frac{(9/4)}{x+1} + \frac{(-3/2)}{(x+1)^2}\right] \\ &= 2 + \frac{1}{4(x-1)} - \frac{9}{4(x+1)} + \frac{3}{2(x+1)^2} \end{aligned}$$

■

It is always possible to check your working by recombining the partial fractions. In the previous example we obtain:

$$\begin{aligned} \text{RHS} &= 2 + \frac{(x+1)^2 - 9(x-1)(x+1) + 6(x-1)}{4(x-1)(x+1)^2} \\ &= 2 + \frac{(x^2 + 2x + 1) - 9(x^2 - 1) + 6(x-1)}{4(x-1)(x+1)^2} \end{aligned}$$

$$= \frac{8(x-1)(x+1)^2 - 8x^2 + 8x + 4}{4(x-1)(x+1)^2}$$

$$= \frac{8(x^2-1)(x+1) - 8x^2 + 8x + 4}{4(x-1)(x+1)^2}$$

$$= \frac{8(x^3 - x + x^2 - 1) - 8x^2 + 8x + 4}{4(x-1)(x+1)^2}$$

$$= \frac{8x^3 - 4}{4(x-1)(x+1)^2}$$

$$= \frac{2x^3 - 1}{(x-1)(x+1)^2} = \text{LHS}$$

THE COVER-UP RULE

Although the technique always works, as you can see it can be rather long. A short cut is available for obtaining the partial fractions corresponding to factors which are *linear* (that is, polynomials of degree 1). This method is known as the **cover-up rule** and is simple to apply.

First we ensure, by dividing out if necessary, that the numerator of the rational expression has degree less than that of the denominator. Secondly we factorize the denominator and select the required factor. We cover up this factor and imagine that it has been put equal to zero. This will give a value for (say) x. Then we substitute this value for x in that part of the rational expression which remains uncovered. This procedure produces the required constant.

□ Take the rational expression

$$R = \frac{2x^2 - 2x - 1}{(x-1)(x+1)^2}$$

The denominator is already factorized and is of greater degree than the numerator. Suppose we require the constant numerator corresponding to the factor $x - 1$. We cover up $x - 1$ and imagine it has been put equal to 0, and thus obtain $x = 1$. This is the value of x which we must substitute into the remnant to give the required constant:

$$\frac{2x^2 - 2x - 1}{>\!\!<\ (x+1)^2} \rightarrow \frac{2 \times 1 - 2 - 1}{>\!\!<\ (1+1)^2} = -\frac{1}{4}$$

■

It is interesting to notice that, in the case of a repeated linear factor, the cover-up technique produces the constant numerator corresponding to the denominator of highest degree. In the previous example,

$$\frac{2x^2 - 2x - 1}{(x-1)\ >\!\!<} \rightarrow \frac{2(-1)^2 - 2(-1) - 1}{(-1-1)\ >\!\!<} = -\frac{3}{2}$$

The cover-up rule can be quite useful for cutting down the amount of algebra that would otherwise be necessary. Let's summarize the rule.

To obtain the constant numerator corresponding to a distinct linear factor $ax + b$, where a and b are constant:

1 Cover up the factor $ax + b$ in the denominator and imagine that it has been put equal to zero;

2 Substitute the value of x obtained in this way into the rest of the rational expression, and the result is the required constant.

Most students use a finger to cover up the linear factor but any other convenient part of the anatomy will do.

Now for some more steps.

2.5 Workshop

1 ▷ **Exercise** Use the cover-up rule to resolve into partial fractions

$$\frac{3x + 1}{x(x - 2)}$$

First find the numerator corresponding to the fraction with denominator x and then take step 2.

---

2 For the numerator corresponding to x we put $x = 0$ into $(3x + 1)/(x - 2)$, which gives $-1/2$.

Did you manage that all right? If you did then complete the resolution into partial fractions. If you made a mistake then take great care when obtaining the numerator corresponding to $x-2$, and check algebraically by recombining your answer that it is correct.

As soon as you are ready, take another step.

---

3 For the numerator corresponding to $x - 2$ we put $x = 2$ into $(3x+1)/x$ which gives $(6+1)/2 = 7/2$. Therefore

$$\frac{3x + 1}{x(x - 2)} = -\frac{1}{2x} + \frac{7}{2(x - 2)}$$

If you are still making mistakes, you should read the section on the cover-up method again to make sure you understand how to apply it correctly.

Now go on to this exercise.

▷ **Exercise** Resolve into partial fractions

$$\frac{x^4 + 2x - 3}{x^3 + 2x^2 + x}$$

First use 'short' division to divide the denominator into the numerator and then move to step 4 to see if you have the right answer.

4

$$
\begin{aligned}
x^4 + 2x - 3 &= x(x^3 + 2x^2 + x) - 2x^3 - x^2 + 2x - 3 \\
&= x(x^3 + 2x^2 + x) - 2(x^3 + 2x^2 + x) + 3x^2 + 4x - 3 \\
&= (x - 2)(x^3 + 2x^2 + x) + 3x^2 + 4x - 3
\end{aligned}
$$

So we have

$$\frac{x^4 + 2x - 3}{x^3 + 2x^2 + x} = x - 2 + \frac{3x^2 + 4x - 3}{x^3 + 2x^2 + x}$$

$$\text{and} \qquad Q = \frac{3x^2 + 4x - 3}{x^3 + 2x^2 + x}$$

remains to be resolved.

Now factorize the denominator D of Q as far as possible and write down the form of the resolution. Then step ahead.

5

$$x^3 + 2x^2 + x = x(x^2 + 2x + 1) = x(x + 1)^2$$

Here the factors of the denominator are x and $x+1$ (repeated). Therefore we shall obtain partial fractions with denominators x, $x + 1$ and $(x + 1)^2$, and the numerators will all be constant. So

$$Q = \frac{3x^2 + 4x - 3}{x^3 + 2x^2 + x} = \frac{A}{x} + \frac{B}{x + 1} + \frac{C}{(x + 1)^2}$$

Without using the cover-up method, obtain the constants A, B and C. Then go on to step 6.

6

If the partial fractions are recombined, the two numerators must be identically equal. Now

$$\frac{A}{x} + \frac{B}{x + 1} + \frac{C}{(x + 1)^2} = \frac{A(x + 1)^2 + Bx(x + 1) + Cx}{x(x + 1)^2}$$

Therefore we require

$$3x^2 + 4x - 3 \equiv A(x + 1)^2 + Bx(x + 1) + Cx$$

The constants A, B and C can be obtained either by substituting values of x into the identity, or by comparing the coefficients of powers of x on each side of it. In practice a mixture of the methods is usually the quickest.

Here if we put $x = 0$ we obtain $A = -3$. If we put $x = -1$ we obtain $3 - 4 - 3 = -C$, so $C = 4$. If we examine the coefficient of x^2 on each side of the identity we obtain $3 = A + B$, and so $B = 6$. Therefore

$$\frac{x^4 + 2x - 3}{x^3 + 2x^2 + x} = x - 2 - \frac{3}{x} + \frac{6}{x+1} + \frac{4}{(x+1)^2}$$

2.6 SET NOTATION

We have already described some standard sets of numbers $\mathbb{N}, \mathbb{Z}, \mathbb{Q}$ and $\mathbb{R}$. In fact sets often arise in one form or another, so before we proceed any further we shall outline the set notation that is commonly employed.

A set can be described best by using the notation

$$\{x|P(x)\}$$

where $P(x)$ is some statement about x, for example $x \in \mathbb{Q}$. The notation means 'the set of all things x which satisfy the condition $P(x)$'

□ $\{x|x \in \mathbb{Z}, x > 0\}$

This is the set of all elements x satisfying the two conditions

1 x is an integer

2 x is strictly positive

We already have a name for this set - the natural numbers. So

$$\mathbb{N} = \{x|x \in \mathbb{Z}, x > 0\}$$

■

If modulus signs are in use, then to avoid confusion the vertical line which appears in the notation $\{x|P(x)\}$ is usually replaced by a colon, so that we write

$$\{x : P(x)\}$$

Of course small finite sets can usually best be described by displaying their elements. For example $\{1, 2, 3, 4, 5\}$ is the set consisting of the first five natural numbers.

There are one or two other pieces of general set notation which we shall use occasionally.

The **difference** $A \setminus B$ of two sets is defined by

$$A \setminus B = \{x | x \in A \text{ and } x \notin B\}$$

The difference $A \setminus B$ is the set of all elements which are elements of A but not elements of B,

□ $\mathbb{I} = \mathbb{R} \setminus \mathbb{Q}$ is the set of all real numbers which are not rational numbers; that is, it is the set of irrational numbers.

■

If a and b are real numbers, $a < b$, we define

$$[a, b] = \{x | x \in \mathbb{R}, a \leqslant x \leqslant b\}$$

$$(a, b) = \{x | x \in \mathbb{R}, a < x < b\}$$

These are called **real** intervals. The first one is called the **closed** interval between a and b, while the second is called the **open** interval between a and b.

The important theoretical distinction between a closed interval and an open interval is that a closed interval includes the two end points a and b whereas an open interval does not include either a or b.

Note that in Chapter 3 we shall introduce the symbol (a, b) as an ordered pair of numbers to represent a *point*, but here we are using it to represent an open interval. Surely this is unsatisfactory; what are we going to do about it?

Well, we shall adopt the view that the context should make clear whether (a, b) is an ordered pair of real numbers or an open interval. Some books have introduced the symbol $]a, b[$ to represent an open interval, but in mathematical work it is quite common to use the same notation in different contexts for different things, and we should be sufficiently broad-minded to be flexible.

2.7 FUNCTIONS

Mathematics is concerned with relationships between things, and it is through the generality of these relationships that it is possible to apply mathematics to a variety of situations. For instance there is a relationship between the force applied to the centre of a beam which is freely supported at each end and the deflection at that point. The force of course could arise from many different sources - a heavy weight suspended from it, or a person standing on it.

Equations often relate two or more variables. For example:

$$y = x^2 + 2$$

$$y = x + \frac{1}{x}$$

$$x^2 + y^2 = 1$$

When the relationship between two variables x and y is such that given any x there corresponds at most one y, we say we have a **function** and write $y = f(x)$. The set of numbers x for which $f(x)$ is defined is called the **domain** of the function, and each element in the domain is called an argument of the function.

A function therefore has two essential ingredients:

1 The domain, the set of arguments of the function, the possible values for x;

2 The rule f which assigns to each element in the domain a unique value $f(x)$.

Strictly there is another essential ingredient: the set consisting of all possible values of the function.

Although when we specify the rule and domain, we may not be able to say precisely what the values of the function will be, we are normally able to state some set which *includes* all the possible values. For example we may know that all the values are real numbers. Therefore when we give a formal definition of a function we shall also specify a set which includes all the values of f. This set we shall call the **codomain**.

We write $f : A \to B$ to indicate that f is a function with domain A and codomain B. Then:

1 If $f : A \to B$ and $B \subset \mathbb{R}$ then f is said to be a **real-valued function**.

2 If $f : A \to B$ and both $A \subset \mathbb{R}$ and $B \subset \mathbb{R}$ then f is said to be a **real function**.

In this chapter we shall confine our attention to real functions.

□ $f : \mathbb{R} \to \mathbb{R}$ defined by $f(x) = x^2 + 2 \quad (x \in \mathbb{R})$

Here the rule is 'square the number and add 2'. This rule can be applied to all real numbers and there is no ambiguity about the result, so we have a function. ■

□ $f : \mathbb{N} \to \mathbb{R}$ defined by $f(x) = x + x^{-1} \quad (x \in \mathbb{N})$

Here the rule 'add the number to its reciprocal' can certainly be applied to every natural number, since 0 is not a natural number. The result in each case is a unique real number and consequently we have a function.

On the other hand if we attempted to extend this rule to $\mathbb{Z}$ we should no longer have a function because the number 0 in the domain has not been assigned a value. Indeed the rule could be applied to other sets of real numbers (in fact to any number except 0), but we specified the domain as $\mathbb{N}$ and if we change the domain we change the function. ■

THE MAXIMAL DOMAIN

In some cases an equation which defines a function is given but no indication is provided of either the domain or the codomain. Strictly speaking the definition is then deficient. One way round the difficulty is to take the codomain as $\mathbb{R}$ and the domain to be all those real numbers for which the rule is valid. That is, $f : A \to \mathbb{R}$ and A is the maximal subset of $\mathbb{R}$ which satisfies the condition

$$\text{if } x \in A \text{ then } f(x) \in \mathbb{R}$$

This convention is sometimes called the **convention of the maximal domain**.

□ Using the convention of the maximal domain, the equation $f(x) = x + x^{-1}$ defines a function $f : A \to \mathbb{R}$ where $A = \mathbb{R} \setminus \{0\}$. ■

In the formal notation for a function

$$f : A \to B$$

A is the domain, f is the rule, $f(A)$ is the image set and B is the codomain.

In practice it is often useful to write $y = f(x)$ and so to specify a variable y in terms of a variable x. The same function would be determined by using x and t, say, instead of y and x respectively. For this reason y and x are sometimes called **dummy variables**.

When the notation $y = f(x)$ is used it is customary to call x the **independent variable** and y the **dependent variable**. The reasoning behind this is that y is determined once x is known. Sometimes the fact that y is given in terms of x is indicated by writing $y = y(x)$ and saying that 'y is a function of x'. This notation has its uses and consequently its adherents.

2.8 METHODS OF PROOF

Mathematics is founded on the idea of proof. One method of proof is known as the **axiomatic** method and requires three essential ingredients - axioms, logic and theorems. **Axioms** are statements which we accept as being true and so do not require proof. **Logic** is the set of rules which enables us to deduce further statements from the axioms and **theorems** are the statements which have been deduced.

We shall represent simple statements by letters in lower case such as p, q, r, s and t. A *simple* statement is a statement which is either true or false but not both. You are probably aware that there are some statements which are neither true nor false and other statements which are both true and false. The statements: 'this computer is fast' and 'this sentence has too many letters' are examples of statements which we should exclude from our system. We should include statements such as '3 is greater than 2' which we know to be true and 'the product of two negative integers is a negative integer' which we know to be false.

We write $p \Rightarrow q$ if and only if by using the logic the statement q can be deduced from the statement p. p is sometimes called a **premise** and q a **conclusion**. When $p \Rightarrow q$ is expressed in words we say 'p implies q'.

We shall begin by considering two standard methods of proof - **direct** proof and **indirect** proof.

DIRECT PROOF

A direct proof involves what is known as a 'chain of argument'. In complicated proofs it is often necessary to apply several chains of arguments before the required conclusion can be drawn. A chain of argument relies on the following logical principle:

$$\text{If } p \Rightarrow q \text{ and } q \Rightarrow r \text{ then } p \Rightarrow r$$

You can argue this principle in the following way. We know that if p is true then q is true. We also know that if q is true then r is true. Consequently if p is true, q is true and r is true. This idea is basic to human activity and you will be able to think of many situations where this principle applies in life. For example, an employer pays the worker and the worker does the job. It is important to realize that no conclusion can be drawn if r is true. In the example, if the job is done it may not have been done by the worker and even if it was, we cannot deduce that the employer has paid the worker.

This is an important point because to some extent this is the place where mathematics departs from other subjects. Many subjects use 'evidence' to support a particular hypothesis, and given enough evidence the theory will be accepted. Even statistics which is closely related to mathematics does not claim to prove anything. A statistician will say that some hypothesis or another is true at the 95% confidence level but never with certainty. You are probably aware that 'circumstantial evidence' is used in legal circles too to gain convictions and many appeals result from challenging this kind of evidence.

Mathematics itself uses circumstantial evidence in order to make conjectures but distinguishes clearly between a conjecture and a theorem. Conjectures are 'guesses' which may be true but which, if they are to be accepted, require proof. There have been some notorious conjectures; two which were around for many years and were only proved in modern times after considerable effort were Fermat's last theorem (1637-1993) and the four-colour theorem (1852-1976). There are many other conjectures still to be settled!

Science and Technology are founded on experimentation and insight. Successful participants do not make wild generalizations but instead assemble evidence and then design experiments to test their theories. Such activity is good science but it is not mathematics.

So that we get a clear idea of a direct proof we shall look at two examples. These involve the ideas of odd and even numbers. You will probably know that an even number is an integer which can be divided exactly by 2. In other words if p is an even number then $p = 2r$ where r is an integer. Any integer which is not even is called odd. Any odd number q can therefore be expressed as $q = 2r+1$ where r is an integer. For the record we note that 0 is an even number.

☐ Prove that the square of every even number is an even number.

Proof Suppose that p is an even number then $p = 2r$ where r is an integer. We now have

$$\begin{aligned} p^2 &= (2r)^2 \\ &= 4r^2 \\ &= 2(2r^2) \end{aligned}$$

Now $2r^2$ is also an integer and so, by definition, we can deduce that p^2 is an even number. ■

Now see if you can prove a similar property for odd numbers.

▷ **Exercise** Prove that the square of every odd number is an odd number.

Proof Suppose p is an odd number then $p = 2r + 1$ where r is an integer. We now have

$$\begin{aligned} p^2 &= (2r+1)^2 \\ &= 4r^2 + 4r + 1 \\ &= 2(2r^2 + 2r) + 1 \end{aligned}$$

Now $2r^2 + 2r$ is also an integer and so, by definition, we can deduce that p^2 is an odd number. ■

INDIRECT PROOF

The **negation** of a statement is the logical opposite of the statement. So that whenever the statement p is true, the negation of p is false and whenever the statement p is false, the negation of p is true. We shall represent the negation of p by $\sim p$.

Suppose we wish to prove the statement q, which we believe is a consequence of the statement p; which is known to be true. We need to prove $p \Rightarrow q$. An indirect proof is obtained by taking the two statements p and $\sim q$ and by means of a chain of argument obtaining a contradiction. This contradiction can take many forms but in all circumstances there is a statement r, which is the result of the logical argument, and which is both true and false. Given that the logic is correct we conclude that $\sim q$ is false, and so q is true.

In England in the seventeenth century there was a series of trials for witchcraft. The method of testing a suspect for witchcraft has many of the ingredients of an indirect proof. The unfortunate individual was thrown into deep water. If the person floated then witchcraft was 'proved' and the individual was then hanged. If the person sank (and thereby drowned) then this was a 'proof' of innocence!

As a simple algebraic example of an indirect proof we shall prove the converse of the property which we have just proved concerning odd and even numbers.

☐ Prove that p is an even number if and only if p^2 is an even number.

Proof We have already shown that if p is even then p^2 is even. It remains to show that if p^2 is even then p is even.

Now an integer is either even or odd and so let us suppose that p is odd. We seek a contradiction. By the previous exercise we can deduce that p^2 is also odd. This means that p^2 is both odd and even which is impossible. Consequently p must be even. ■

A rather more interesting example of an indirect proof is provided by the next example.

☐ Prove that $\sqrt{2}$ is an irrational number.

Proof We suppose that, on the contrary, $\sqrt{2}$ is a rational number and seek a contradiction. Suppose then

$$\sqrt{2} = \frac{p}{q}$$

where p and q are integers. We can suppose further that there is no integer greater than 1 which divides both p and q, for if there were we could cancel it out and thereby reduce p and q to smaller integers.

Squaring the equation gives

$$2 = \frac{p^2}{q^2}$$

so that

$$p^2 = 2q^2$$

Now this implies that p^2 is even, and so p must be even too. Therefore if we put $p = 2r$ then r is an integer. We then obtain, substituting for p,

$$\begin{aligned}(2r)^2 &= 2q^2 \\ 4r^2 &= 2q^2 \\ 2r^2 &= q^2\end{aligned}$$

Now this implies that q^2 is even, and so q must be even too. However this is the crunch! We have deduced that both p and q are even, and yet we know that p and q have no common factor. This contradiction shows that our initial assumption that $\sqrt{2}$ was a rational number must be false. ■

It is interesting to remark that when the irrationality of $\sqrt{2}$ was first discovered, by the ancient Greeks (circa 420 BC), it caused a major philosophical upset.

MATHEMATICAL INDUCTION

We must not disguise the fact that the rules we have given for dealing with real numbers do not tell the whole story. For a complete description we would need one further axiom known as the **axiom of completeness**. We shall not describe this axiom because it is requires somewhat sophisticated mathematics but instead we remark that one of the consequences is a method of proof which is known as **induction**.

Suppose we have some statement $S(n)$ which we wish to prove is true for all natural numbers n. The principle of mathematical induction states that it is only necessary to prove two things:

1 $S(1)$ is true;
2 If $S(k)$ is true then $S(k+1)$ is true.

We often experience inductive processes in practice. One example is that of a petrol engine. In order for the engine to fire it must first be 'turned over'. This corresponds to condition 1. Each time the engine turns over it generates just enough electricity to fire the engine again. This corresponds to condition 2. Provided condition 2 continues to hold the engine will run even if the battery is flat. If you have ever had the experience of driving a car with a flat battery you will appreciate the need for both these conditions!

Once you get the idea of mathematical induction you will find it quite straightforward. We shall look at one or two examples to see how it works.

☐ Show that for all natural numbers n

$$1+2+3+\cdots+n=\frac{1}{2}n(n+1)$$

Suppose $S(n)$ is the statement

$$1 + 2 + 3 + \cdots + n = \frac{1}{2}n(n+1)$$

We wish to prove that $S(n)$ is true for every natural number n.

By the principle of mathematical induction we are required to prove just two things:

1 $S(1)$ is true;

2 If $S(k)$ is true then $S(k+1)$ is true.

We proceed as follows:

1 To show $S(1)$ is true we must show that

$$1 = \frac{1}{2}1(1+1)$$

This is done by simply evaluating the right-hand side.

2 Suppose $S(k)$ is true for some natural number k. Then

$$1 + 2 + 3 + \cdots + k = \frac{1}{2}k(k+1)$$

(This statement, $S(k)$, is often known as the induction **hypothesis**.)

We are required to show that $S(k+1)$ is true. In other words

$$1 + 2 + 3 + \cdots + (k+1) = \frac{1}{2}(k+1)[(k+1)+1]$$

Notice that we write down $S(k+1)$ simply by replacing n by $k+1$ in $S(n)$.

Remember that we are entitled to use the induction hypothesis to show that $S(k+1)$ is true.

To accomplish this we shall take the left-hand side and demonstrate that using the induction hypothesis we can deduce the right-hand side.

$$\begin{aligned}
\text{LHS} &= 1 + 2 + 3 + \cdots k + (k+1) \\
&= [1 + 2 + \cdots + k] + (k+1) \\
&= \frac{1}{2}k(k+1) + (k+1) \\
&\qquad \text{(using the induction hypothesis)} \\
&= \frac{1}{2}(k+1)(k+2) \\
&= \frac{1}{2}(k+1)[(k+1)+1] \\
&= \text{RHS}
\end{aligned}$$

■

That's all there is to it! Of course a proof by induction differs from a deductive proof because we need to know the formula that we wish to prove. However, it is a very useful technique because it is often possible to spot a pattern in mathematical work, infer a formula and then use induction to settle the matter.

There will be various occasions when we shall point out where a proof by induction would be appropriate. Now here is an example for you to try.

□ Show that if n is any natural number then

$$1^2 + 3^2 + 5^2 + \cdots + (2n-1)^2 = \frac{n(4n^2-1)}{3}$$

In other words, we have a formula for the sum of the squares of the first n odd numbers.

We have the statement $S(n)$, so you can write down the induction hypothesis $S(k)$ and also the statement, $S(k+1)$, which we must deduce. Don't forget that we must check that $S(1)$ is true as well.

Try it yourself and see how it goes.

$S(n)$ is the statement

$$1^2 + 3^2 + 5^2 + \cdots + (2n-1)^2 = \frac{n(4n^2-1)}{3}$$

and so $S(k)$ is the statement

$$1^2 + 3^2 + 5^2 + \cdots + (2k-1)^2 = \frac{k(4k^2-1)}{3}$$

1 To show that $S(1)$ is true we merely need to check that

$$1^2 = \frac{(1)(4(1)^2-1)}{3}$$

Each side of this equation has the value 1 and so $S(1)$ is true.

2 We must prove that

$$1^2 + 3^2 + 5^2 + \cdots + [2(k+1)-1]^2 = \frac{(k+1)[4(k+1)^2-1]}{3}$$

and of course we must expect to use the induction hypothesis to do this.

We work on the left-hand side

$$\begin{aligned}
\text{LHS} &= 1^2 + 3^2 + 5^2 + \cdots + [2(k+1)-1]^2 \\
&= 1^2 + 3^2 + 5^2 + \cdots + (2k+1)^2 \\
&= [1^2 + 3^2 + 5^2 + \cdots + (2k-1)^2] + (2k+1)^2 \\
&\quad \text{(here we have written down the last two terms)} \\
&= \frac{k(4k^2-1)}{3} + (2k+1)^2 \\
&\quad \text{(using the induction hypothesis)}
\end{aligned}$$

To complete this we must use some algebra to reduce this expression to the right-hand side of $S(k+1)$. Continuing we obtain

$$\begin{aligned}
&= \frac{k(2k-1)(2k+1)}{3} + (2k+1)^2 \\
&= (2k+1)\left[\frac{k(2k-1)}{3} + (2k+1)\right] \\
&= (2k+1)\,\frac{k(2k-1)+3(2k+1)}{3} \\
&= (2k+1)\,\frac{2k^2-k+6k+3}{3} \\
&= (2k+1)\,\frac{2k^2+5k+3}{3}
\end{aligned}$$

A glance at the expression we wish to obtain gives us the clue to factorizing:

$$\begin{aligned}
&= (2k+1)\,\frac{(k+1)(2k+3)}{3} \\
&= (k+1)\,\frac{(2k+1)(2k+3)}{3} \\
&= (k+1)\,\frac{4k^2+8k+3}{3} \\
&= (k+1)\,\frac{4(k+1)^2-1}{3} \\
&= \frac{(k+1)[4(k+1)^2-1]}{3} \\
&= \text{RHS}
\end{aligned}$$

We have shown the two parts
1 $S(1)$ is true;
2 If $S(k)$ is true then $S(k+1)$ is true.
Therefore, by induction, we have shown that $S(n)$ holds for every natural number n. ■
Finally we remark that although we used only $S(k)$, our induction hypothesis, we would have been entitled to use $S(r)$, for all $r \leqslant k$. When this is done it is usually known as using **strong** induction.

Lastly in this chapter we consider a problem which shows how the theory of partial fractions can be combined with the binomial theorem to produce an approximate formula.

2.9 Practical

LEAKING FUEL
The fuel reserve contained in a leaking gas cylinder is known to be given by the following formula:

$$R = 2P\left[\frac{(t+1)^2+t^2}{(2+t)(1+t^2)}\right]$$

where t represents time and P is the initial reserve. Express R in partial fractions, and show that it can be approximated by

$$R \approx \frac{P}{4}\left[(t+3)^2-5\right]$$

provided t is small.

Try this on your own first. If you are successful then look to see if you have everything correctly. If you are unsuccessful then read just enough to get going again and try once more. The full solution follows.

We have

$$\begin{aligned} R &= 2P\,\frac{2t^2+2t+1}{(2+t)(1+t^2)} \\ &= 2P\left[\frac{A}{2+t}+\frac{Bt+C}{1+t^2}\right] \end{aligned}$$

where A, B and C are constants. So

$$R = 2P\left[\frac{A(1+t^2)+(2+t)(Bt+C)}{(2+t)(1+t^2)}\right]$$

We obtain the identity

$$2t^2+2t+1 \equiv A(1+t^2)+(2+t)(Bt+C)$$

It follows that

$$1 = A + 2C \quad 2 = A + B \quad 2 = 2B + C$$

from which $A = 1$, $B = 1$ and $C = 0$. Consequently,

$$\begin{aligned}
R &= 2P\left[\frac{1}{2+t} + \frac{t}{1+t^2}\right] \\
&= 2P\left[\frac{1}{2}(1+\frac{t}{2})^{-1} + t(1+t^2)^{-1}\right] \\
&\approx 2P\left[\frac{1}{2}(1-\frac{t}{2}+\frac{t^2}{4}) + t\right]
\end{aligned}$$

Here we have neglected terms in t of degree higher than 2. So

$$\begin{aligned}
R &\approx 2P\left[\frac{1}{2} - \frac{t}{4} + \frac{t^2}{8} + t\right] \\
&= 2P\left[\frac{1}{2} + \frac{3t}{4} + \frac{t^2}{8}\right] \\
&= \frac{P}{4}\left[4 + 6t + t^2\right] \\
&= \frac{P}{4}\left[(t+3)^2 - 5\right]
\end{aligned}$$

SUMMARY

These are the things you should be able to do after completing this chapter:

- ☐ Apply the laws of indices and logarithms correctly
- ☐ Solve inequalities using algebraic, analytic and graphical techniques
- ☐ Resolve a rational expression into partial fractions
- ☐ Distinguish between direct and indirect proofs and supply examples of them.
- ☐ Prove statements for all natural numbers n using the principle of mathematical induction.

EXERCISES

1 Simplify, using the laws of indices

a

$$8^{5/3} \times 32^{1/5} \div 16^{3/4}$$

b

$$12^{5/2} \times 27^{1/4} \div 6^{1/2}$$

c

$$\frac{(1+x)^4(1-x^4)}{(1-x^2)^3}$$

d

$$\frac{(a^2-b^2)^5(a^4-b^4)}{(a^2+b^2)(a+b)^2}$$

2 Solve the equations, where x is real,

a $e^{2x} = 4e^x + 5$

b $2^{2x} - 5 \times 2^x + 4 = 0$

c $6^x - 9 \times 2^x - 8 \times 3^x + 72 = 0$

d $15^x + 15 = 3^{x+1} + 5^{x+1}$

3 Decide, in each case, which of the following are identities and which are equations. If they are equations then solve them.

a $e^{2x} + 2 = 2(2e^x - 1)$

b $\ln(x-1) + \ln(x+1) = \ln(2x)$

c $\ln(1 - 1/x) + \ln(1 + 1/x)^{-1} = \ln[1 - 2/(x+1)]$

4 Express x in terms of a, as simply as possible, in each of the following:

a $\ln x = 2\ln(a+1) - \ln(a^2-1)$

b $e^x = \{e^{a+1} \cdot e^{a-1}\}^2$

c $\ln(x^2-1) - \ln(x-1) = \ln(a^2-1) - \ln(a+1)$

d $e^x \cdot (e^{-x})^2 = (e^{-a})^3\,(e^a)^4$

5 Obtain those real numbers for which the following inequalities hold:

a $(x-2)(x+2) > 0$

b $(x-3)(2x+3) < 0$

c $1/x + 1/(x-1) > 0$

6 Resolve the following into partial fractions:

a

$$\frac{1}{(x-1)(x-3)}$$

b

$$\frac{x}{x^2+7x+12}$$

c

$$\frac{2x+1}{x^3+5x^2+6x}$$

d

$$\frac{x-7}{x^3-3x^2-9x-5}$$

7 Prove, using a direct proof, that

a the sum of two even numbers is always even

b the sum of two odd numbers is always even

c the sum of an odd number and an even number is always an odd number.

8 Use mathematical induction to prove that for all natural numbers n,

a $1+3+5+\cdots+(2n-1) = n^2$

b $1^3+3^3+5^3+\cdots+(2n-1)^3 = n^2(2n^2-1)$

ASSIGNMENT

1 Use the laws of indices to show that

$$\frac{(16)^{3/4}\,(25)^{1/2}}{(81)^{1/4}\,(125)^{1/3}} = \frac{8}{3}$$

2 Simplify

$$\frac{(1-x^2)^3(1-3x+2x^2)^4}{(1-x-2x^2)^6}\;\frac{(1+x)^3}{(1-x)^7}$$

3 Solve the equation

$$2^{3x} - 3 \times 2^{2x+1} + 2^{x+3} = 0$$

4 If $x = \log_a b$, $y = \log_b c$ and $z = \log_c a$, show that $xyz = 1$.

5 If $x = \log_a(bc)$, $y = \log_b(ca)$ and $z = \log_c(ab)$, show that

$$xyz = x + y + z + 2$$

6 If $x = \log_a(b/c)$, $y = \log_b(c/a)$ and $z = \log_c(a/b)$, show that

$$xyz + x + y + z = 0$$

7 Solve the following inequalities for real x

a $x(x^2 - 4) > 0$

b $x^2 - 5x - 6 \leq 0$

c $1/(x-1) + 1/(x^2-1) < 1/(x+1)$

8 Resolve into partial fractions:

a

$$\frac{x^2 + 2x - 1}{x^3 - x}$$

b

$$\frac{3x^3 + 5x^2 - x - 1}{x^3 + x^2}$$

c

$$\frac{4x^4 + x^3 - 11x^2 + x - 20}{x^4 - 3x^2 - 4}$$

9 Show by means of a direct proof that the sum of two rational numbers is always a rational number.

10 Show by means of an indirect proof that the sum of a rational number and an irrational number is always an irrational number. By considering $3 - \sqrt{2}$ and $\sqrt{2}$ (or any other suitable example) show that the sum of two irrational numbers is not necessarily an irrational number.

11 Use mathematical induction to prove that for every natural number n,

a $1^2 + 2^2 + 3^2 + \cdots + n^2 = n(n+1)(2n+1)/6$

b $1^3 + 2^3 + 3^3 + \cdots n^3 = [n(n+1)/2]^2$

FURTHER EXERCISES

1 Use the laws of indices to show that

$$\frac{(25)^{1/2}(8)^{1/3}}{(27)^{1/3}(16)^{1/4}} = \frac{5}{3}$$

2 Solve the equation

$$e^{2x} + e^{x} + e^{-2x} + e^{-x} = 3(e^{-2x} + e^{x})$$

3 Decide which of the following are identities and which equations. If they are equations then solve them.
a $\ln(x^4 - 1) = \ln(x - 1) + \ln(x + 1) + \ln(x^2 + 1)$
b $\ln(x + x^2) = \ln x + \ln x^2$
c $\ln(x^2 - 1)^3 = 3[\ln(x - 1) + \ln(x + 1)]$
d $e^{x^2}e^{x} = (e^{x})^3$
e $e^{x}e^{2x}e^{3x} = (e^{x})^6$

4 Solve the inequality

$$\frac{2x}{(x+2)(x-1)} > 1$$

5 Express in partial fractions
a

$$\frac{x^3 - 5}{(x-1)^2(x^2 - 3x + 2)}$$

b

$$\frac{(x^2+1)^2}{(x^2-1)^3}$$

6 Use the convention of the maximal domain to write down the domain of the real function f defined by
a $f(x) = (x^2 - 1)^{-1}$
b $f(x) = (x^2 - 1)^{-1/2}$
c $f(x) = (x^2 - 1)^{-1/2} + (1 - x^2)^{-1/2}$

7 Prove that the product of two rational numbers is always a rational number. By means of an example show that the product of two irrational numbers is not necessarily an irrational number.

8 Prove that the product of an irrational number and a non-zero rational number is always an irrational number.

9 Prove that for all natural numbers n
a $1 \cdot 2 + 2 \cdot 3 + \cdots + n(n+1) = n(n+1)(n+2)/3$
b $1 \cdot 2^2 + 2 \cdot 3^2 + \cdots n(n+1)^2 = n(n+1)(n+2)(3n+5)/12$
c

$$\frac{1}{1 \times 2} + \frac{1}{2 \times 3} + \cdots + \frac{1}{n(n+1)} = \frac{n}{n+1}$$

10 By first expressing

$$\frac{1}{(n+1)(n+2)}$$

in partial fractions show that

$$\sum_{n=1}^{N} \frac{1}{(n+1)(n+2)} = \frac{N}{2(N+2)}$$

11 The Heaviside unit function H is defined by

$$\begin{aligned} H(t) &= 1 \quad \text{when } t > 0 \\ &= 0 \quad \text{when } t < 0 \end{aligned}$$

a Write down the domain and image set of H.

b Show that if a constant voltage E is applied to a circuit, between time $t = 0$ and $t = 1$ only, then the voltage at time t is given by $E(t) = E[H(t) - H(t-1)]$.

c Express by means of a single equation, using the Heaviside unit function, the current $i(t)$ in a circuit satisfying

$$\begin{aligned} i(t) &= t \qquad \text{when } 0 < t < 1 \\ &= 2 - t \qquad \text{when } 1 < t < 2 \\ &= 0 \qquad \text{otherwise} \end{aligned}$$

12 The number n of terminals on a circuit board is known to satisfy the inequality $n^3 - 7n^2 + 5n - 35 < 0$. What is the maximum number?

13 The proportion p of purified oil which can be produced by an oil filter is known to satisfy $2(p^3 + 1) \le (p-2)^2$. Show that it can purify at most 50% of the oil.

14 Verify that

$$(x^4 + x\sqrt{2} + 1)(x^2 - x\sqrt{2} + 1) \equiv x^4 + 1$$

and thereby resolve $1/(x^4 + 1)$ into partial fractions.

15 Use the fact that $(a - b)^2 \ge 0$ for all real numbers a and b to deduce

$$\frac{a^2 + b^2}{2} \ge ab$$

so that if x and y are positive,

$$\frac{x + y}{2} \ge \sqrt{(xy)}$$

(The arithmetic mean of two numbers is greater than or equal to the geometric mean.)

Show that if two resistors are combined in series the total resistance is always greater than if they are combined in parallel.

16 The volume of a spherical raindrop of liquid decreases due to evaporation by one-half in 1 hour. The radius of the drop is given by $r = -kt + r_0$, where r_0 is the initial radius. If k is a constant and t denotes time, show that the time taken for the drop to evaporate completely is $(1 - 2^{-1/3})^{-1}$ hours.

17 The law governing radioactive decay is $p = p_0 \mathrm{e}^{-kt}$, where p is the intensity at time t and p_0 is the initial intensity. Show that if $p = p_0/2$ when $t = h$ then the time taken for the initial radioactivity to decay 99% is $2h \log_2 10$.

18 The depth x to which a drill applied under constant pressure will sink into rock over time t is given by

$$x = \frac{1}{w} \ln \left[\frac{pvt}{w} + 1 \right]$$

where w, p and v are constants. Show that the time T taken to drill from a depth x to a depth $x + h$ is

$$T = \frac{w}{vp} \mathrm{e}^{wx}(\mathrm{e}^{wh} - 1)$$

19 The charge on a leaking capacitor is given by

$$Q = \frac{2Q_0}{(1+t)(2+t)}$$

where t is time (seconds) and Q_0 is the initial charge (farads). Express Q in partial fractions, and show that it is approximately $(1 - 3t/2 + 7t^2/4)Q_0$ provided t is small.

20 The output of a system at time t is given by

$$A = 1 - [1 + t^4/(s+1)]^{-1}$$

where s is the imposed signal and t is time in seconds. If $s = t(1+t)^2$ at time t, resolve A into partial fractions and show that if terms in t of degree greater than 4 may be neglected then $A \approx t^4$.

Trigonometry and geometry 3

In the last two chapters we described some of the basic terminology which we need. We also picked up a few techniques which should prove useful later on. Soon we shall begin to develop the differential calculus, but before we do that we must make sure that we can handle any geometrical or trigonometrical problem that arises.

After working through this chapter you should be able to
- ☐ Use circular functions, recognize their graphs and be able to determine their domains;
- ☐ Solve equations involving circular functions;
- ☐ Recognize the equations of standard geometrical curves;
- ☐ Transform equations involving polar coordinates into those involving cartesian coordinates.

At the end of this chapter we shall solve practical problems in surveying and in circuits.

This chapter contains background work, and so it is possible that much of it will be familiar to you. If this is the case, then it is best to regard it as revision material. We shall be reviewing work on elementary trigonometry and coordinate geometry. If any section is very well known to you then simply read it through and devote your attention to that which is less familiar.

3.1 COORDINATE SYSTEMS

You are probably quite familiar with the **cartesian coordinate system**. In this system every point in the plane is determined uniquely by an ordered

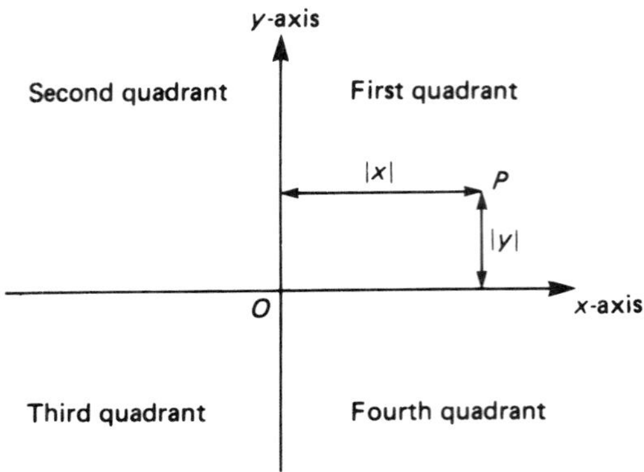

Fig. 3.1 The cartesian system.

pair of numbers (x, y). To do this, two fixed straight lines are laid at right angles to one another; these are called the x-axis and the y-axis. Their point of intersection is represented by O and is called the origin (Fig. 3.1). The quadrants so formed are labelled anticlockwise as the first quadrant, second quadrant, third quadrant and fourth quadrant respectively.

Given any point P, the absolute values of x and y are then obtained from the shortest distance of P to the y-axis and the x-axis respectively. The following conventions then hold:

First quadrant $x \geqslant 0, y \geqslant 0$
Second quadrant $x \leqslant 0, y \geqslant 0$
Third quadrant $x \leqslant 0, y \leqslant 0$
Fourth quadrant $x \geqslant 0, y \leqslant 0$

In this way, given any point in the plane we obtain a unique ordered pair (x, y) of real numbers. Conversely, given any ordered pair (x, y) of real numbers we obtain a unique point in the plane. We therefore identify the point P with the ordered pair (x, y) and refer to the point (x, y).

If P is the point (x, y), x and y are known as the cartesian coordinates of the point P; x is called the abscissa and y is called the ordinate.

This simple idea was initially due to the famous French philosopher Descartes and enabled algebra and geometry, two hitherto separate branches of mathematics, to be united. It is difficult to overestimate the benefits of this unification for science and technology, but Descartes threw it out almost as an afterthought to his philosophical treatise. The name 'cartesian system' comes from the latinized form of Descartes.

The cartesian system is not the only system which can be used to represent points in the plane. Another is the **polar coordinate system**.

In the polar coordinate system there is a fixed point O, called the origin,

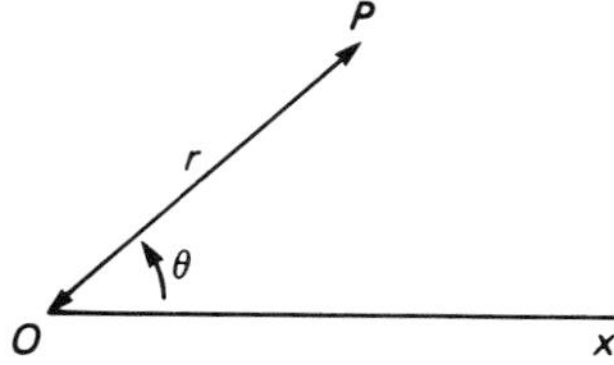

Fig. 3.2 The polar system.

and a fixed line emanating from O called the initial line OX (Fig. 3.2). It is convenient to identify the initial line with the positive x-axis, although this identification is by no means essential. A point P is then determined by r, its distance from O, and by θ, the angle XOP measured anticlockwise. In this way, given any point in the plane we obtain an ordered pair of real numbers (r, θ) where $r \geqslant 0$. Of course if we increase θ by 2π, a whole revolution, then we shall obtain the same point as before. In order to establish a unique representation we restrict θ so that $0 \leqslant \theta < 2\pi$.

There is a minor problem when $r = 0$, since we then lose our one-to-one correspondence between points in the plane and ordered pairs of real numbers of the form (r, θ). For example $(0, \pi)$ and $(0, \pi/2)$ both correspond to the origin. One way of avoiding this problem is to insist that if $r = 0$ then the origin will be the unique point $(0, 0)$: $r = 0$, $\theta = 0$. However, we shall not do this as the procedure creates more difficulties than it resolves. Instead we shall avoid representing the origin and insist that $r > 0$.

In fact the convention $0 \leqslant \theta < 2\pi$ is only used occasionally in coordinate geometry. Unfortunately we shall adopt a different convention, namely $-\pi < \theta \leqslant \pi$, when we deal with complex numbers (Chapter 10). The causes for this are historical and not mathematical, and this goes some way towards explaining why they are illogical.

3.2 CIRCULAR FUNCTIONS

It is possible to define the circular functions $\cos \theta$ and $\sin \theta$ for any angle θ by using cartesian coordinate geometry and a circle centred at the origin with radius r. In Fig. 3.3, let X be the point where the circle crosses the positive x-axis, and let the point P on the circle be such that $\angle XOP = \theta$. If P is the point (x, y) then $OP = r > 0$ and

$$\cos \theta = \frac{x}{r} \qquad \sin \theta = \frac{y}{r}$$

It follows immediately that

1 If $0 < \theta < \pi/2$ then $\cos \theta > 0$ and $\sin \theta > 0$;

2 If $\pi/2 < \theta < \pi$ then $\cos \theta < 0$ and $\sin \theta > 0$;

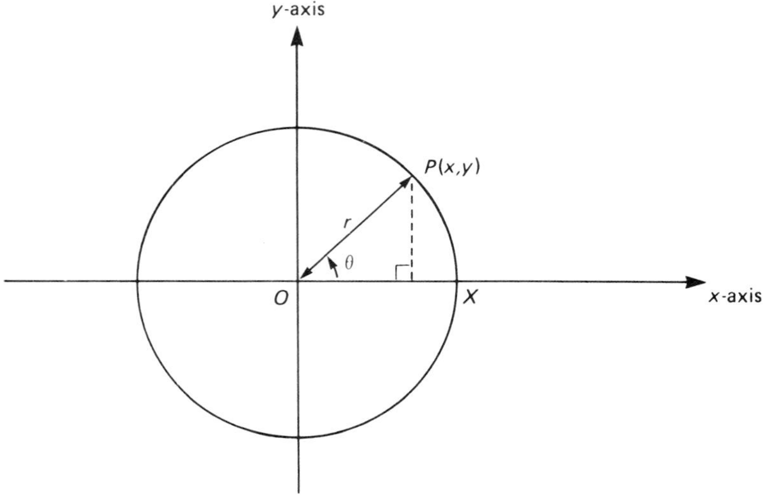

Fig. 3.3 The generating circle.

3 If $\pi < \theta < 3\pi/2$ then $\cos \theta < 0$ and $\sin \theta < 0$;
4 If $3\pi/2 < \theta < 2\pi$ then $\cos \theta > 0$ and $\sin \theta < 0$.

□ Use the definition to evaluate $\cos \theta$ and $\sin \theta$ when $\theta \in \{0, \pi/2, \pi, 2\pi\}$.
When
a $\theta = 0$ then $x = r$ and $y = 0$, so that

$$\cos \theta = r/r = 1 \text{ and } \sin \theta = 0/r = 0$$

b $\theta = \pi/2$ then $x = 0$ and $y = r$, so that

$$\cos \theta = 0/r = 0 \text{ and } \sin \theta = r/r = 1$$

c $\theta = \pi$ then $x = -r$ and $y = 0$, so that

$$\cos \theta = -r/r = -1 \text{ and } \sin \theta = 0/r = 0$$

d $\theta = 2\pi$ then P is in the same position as when $\theta = 0$, so that

$$\cos 2\pi = \cos 0 = 1 \text{ and } \sin 2\pi = \sin 0 = 0$$ ■

Now $\cos(\theta + 2\pi) = \cos \theta$ and $\sin(\theta + 2\pi) = \sin \theta$, so the circular functions are said to be **periodic functions**. In fact $T = 2\pi$ is the smallest positive number such that both $\cos(\theta + T) \equiv \cos \theta$ and $\sin(\theta + T) \equiv \sin \theta$. Consequently $T = 2\pi$ is called the **period** of the circular functions. In other words, if we increase the argument by 2π then the same value is obtained.

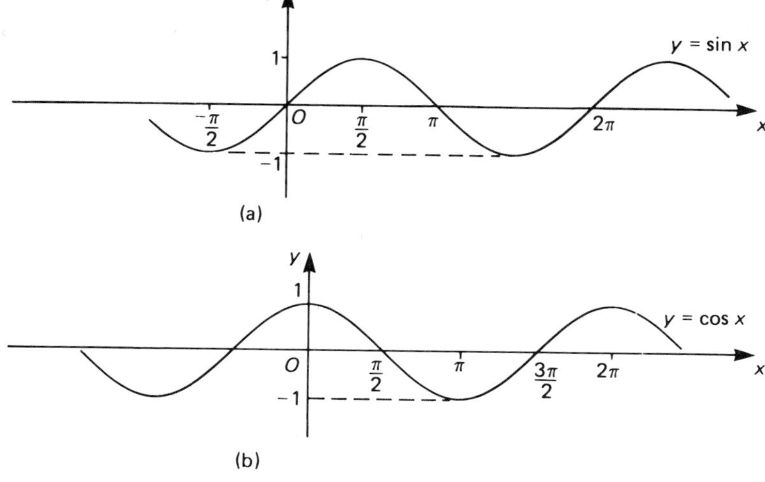

Fig. 3.4 (a) The sine function (b) The cosine function.

We can use these general definitions to draw the graphs of the circular functions. In fact once their values are known for arguments in the interval $[0, \pi/2]$ the rest can be deduced by symmetry. You have probably seen the graphs of the sine and cosine functions before (Fig. 3.4).

The other circular functions, known as tangent, cotangent, secant and cosecant, can be defined in terms of cosine and sine. In fact

$$\tan\theta = \frac{\sin\theta}{\cos\theta} \qquad \cot\theta = \frac{\cos\theta}{\sin\theta}$$

$$\sec\theta = \frac{1}{\cos\theta} \qquad \operatorname{cosec}\theta = \frac{1}{\sin\theta}$$

However, whereas cosine and sine have the real numbers $\mathbb{R}$ as their domain, these subsidiary functions are not defined for all real numbers.

□ Obtain the domain of each of these subsidiary circular functions by using the convention of the maximal domain.

a $\tan\theta$ is defined whenever $\cos\theta \neq 0$. From Fig. 3.4 we see that this is when θ is not an odd multiple of $\pi/2$. Any odd number can be written in the form $2n+1$ where $n \in \mathbb{Z}$. Therefore the domain of the tangent function is

$$A = \{x \mid x \in \mathbb{R},\ x \neq (2n+1)\,\pi/2,\ n \in \mathbb{Z}\}$$

b $\cot\theta$ is defined whenever $\sin\theta \neq 0$. So θ must not be a multiple of π. Therefore the domain of the cotangent function is

$$A = \{x \mid x \in \mathbb{R},\ x \neq n\pi\ ,\ n \in \mathbb{Z}\}$$

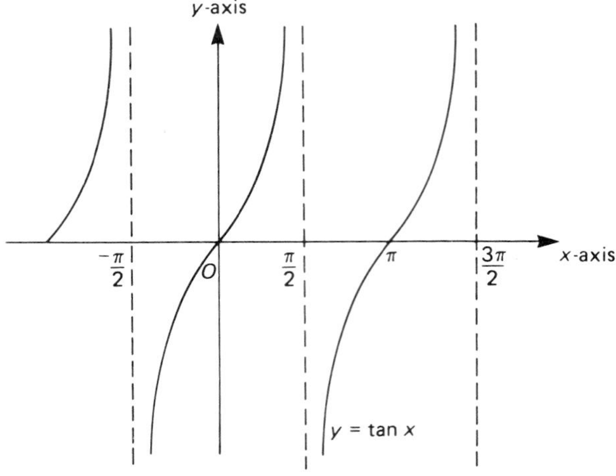

Fig. 3.5 The tangent function.

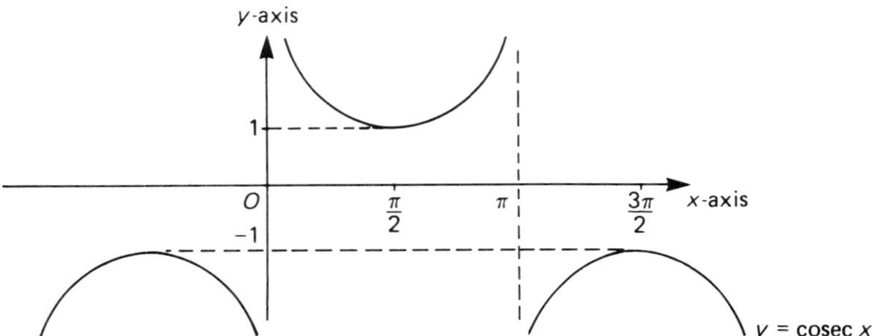

Fig. 3.6 The cosecant function.

The domains of the secant and cosecant are the same as those of the tangent and cotangent respectively. ■

The graph of $y = \tan x$ shows that the tangent function has period π (Fig. 3.5).

The graph of the sine, cosine and tangent functions can be used to draw the graphs of the cosecant (Fig. 3.6), secant (Fig. 3.7) and cotangent functions. The graph of $y = \sec x$ has the same shape as the graph of $y = \operatorname{cosec} x$. To obtain the graph of $y = \operatorname{cosec} x$ from the graph of $y = \sec x$ we merely need to relocate the y-axis through $x = \pi/2$ and relabel.

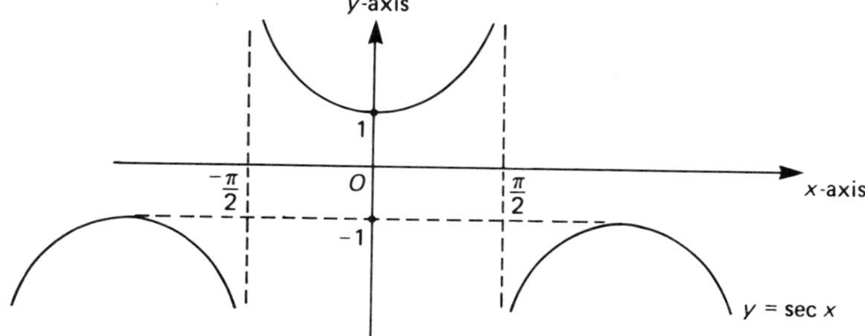

Fig. 3.7 The secant function.

You will remember that we write $\cos^n \theta$ instead of $(\cos \theta)^n$ when n is a natural number. This must not be confused with $\cos (n\theta)$, and you should be alert to the fact that this notation does not hold good when n is a negative integer. In particular,

$$\cos^{-1} \theta \neq (\cos \theta)^{-1}$$

We know that $(\cos \theta)^{-1}$ is $\sec \theta$, and in fact $\cos^{-1} \theta$ has a totally different meaning. Do watch out for this; it is a common mistake!

You may well have spent a long time in the past establishing identities between circular functions. We can deduce one well-known identity straight away:

$$\cos^2 \theta + \sin^2 \theta = 1$$

To show this we evaluate the expression on the left:

$$\cos^2 \theta + \sin^2 \theta = \frac{x^2}{r^2} + \frac{y^2}{r^2} = \frac{x^2 + y^2}{r^2} = 1$$

This is an identity; it holds for all θ.

All the remaining identities involving circular functions can be deduced from the expansion formula

$$\sin (A + B) = \sin A \cos B + \cos A \sin B$$

□ Deduce from the expansion formula for $\sin (A + B)$ the expansion formulas for **a** $\sin (A - B)$ **b** $\cos (A + B)$.

a We have

$$\begin{aligned} \sin (A - B) &= \sin (A + [-B]) \\ &= \sin A \cos [-B] + \cos A \sin [-B] \end{aligned}$$

Now from the definitions (or from the graphs) we have

$$\cos[-B] = \cos B \quad \text{and} \quad \sin[-B] = -\sin B$$

from which we have

$$\sin(A - B) = \sin A \cos B - \cos A \sin B$$

b Putting $A = \pi/2$ enables us to deduce first

$$\sin(\pi/2 - B) = \sin \pi/2 \cos B - \cos \pi/2 \sin B = \cos B$$

Therefore

$$\begin{aligned}\cos(A + B) &= \sin(\pi/2 - [A + B]) \\ &= \sin([\pi/2 - A] - B) \\ &= \sin(\pi/2 - A)\cos B - \cos(\pi/2 - A)\sin B \\ &= \cos A \cos B - \sin[\pi/2 - (\pi/2 - A)]\sin B \\ &= \cos A \cos B - \sin A \sin B\end{aligned}$$

Of course all this is rather algebraic and in some ways rather contrived, but the point is that starting with very little we can build up a host of identities. ■

3.3 TRIGONOMETRICAL IDENTITIES

Here is a list of most of the trigonometrical identities that you will have met:

1 $\cos(A + B) = \cos A \cos B - \sin A \sin B$
2 $\cos(A - B) = \cos A \cos B + \sin A \sin B$
3 $\sin(A + B) = \sin A \cos B + \cos A \sin B$
4 $\sin(A - B) = \sin A \cos B - \cos A \sin B$
5 $\tan(A + B) = \dfrac{\tan A + \tan B}{1 - \tan A \tan B}$
6 $\tan(A - B) = \dfrac{\tan A - \tan B}{1 + \tan A \tan B}$
7 $\cos 2\theta = \cos^2\theta - \sin^2\theta = 1 - 2\sin^2\theta = 2\cos^2\theta - 1$
8 $\sin 2\theta = 2 \sin\theta \cos\theta$
9 $\tan 2\theta = \dfrac{2\tan\theta}{1 - \tan^2\theta}$
10 $\cos C + \cos D = 2\cos\dfrac{C + D}{2}\cos\dfrac{C - D}{2}$
11 $\cos C - \cos D = -2\sin\dfrac{C + D}{2}\sin\dfrac{C - D}{2}$
12 $\sin C + \sin D = 2\sin\dfrac{C + D}{2}\cos\dfrac{C - D}{2}$
13 $\sin C - \sin D = 2\cos\dfrac{C + D}{2}\sin\dfrac{C - D}{2}$

14 $1 + \tan^2 \theta = \sec^2 \theta$
15 $1 + \cot^2 \theta = \operatorname{cosec}^2 \theta$

You might like to have a go at deducing these from the identities we already have. If you need any hints then observe that identity 5 can be deduced by dividing 3 by 1. Similarly, 6 can be deduced by dividing 4 by 2. The identities 7, 8 and 9 are obtained from 1, 3 and 5 respectively by putting $A = B = \theta$. It is possible to deduce 10 by the addition of 1 and 2, whereas subtracting these identities results in 11. Similarly 12 and 13 can be deduced from 3 and 4. Lastly the identities 14 and 15 can be obtained by dividing $\cos^2 \theta + \sin^2 \theta = 1$ by $\cos^2 \theta$ and $\sin^2 \theta$ respectively.

3.4 THE FORM $a \cos \theta + b \sin \theta$

You probably already know that it is possible to express $a \cos \theta + b \sin \theta$ in the form $R \cos (\theta - \alpha)$. This is used quite frequently, and so we shall describe briefly how it is done (see Fig. 3.8).

1 We put the point $P\ (a, b)$ in the plane using cartesian coordinate geometry.
2 The angle α which can be read directly from the diagram is $\angle XOP$.
3 R is the distance OP.

It is easy to see why this works because we have

$$\begin{aligned} a \cos \theta + b \sin \theta &= R[(a/R) \cos \theta + (b/R) \sin \theta] \\ &= R(\cos \alpha \cos \theta + \sin \alpha \sin \theta] \\ &= R \cos (\theta - \alpha) \end{aligned}$$

□ Express $\sin \theta - \cos \theta$ in the form $R \cos (\theta - \alpha)$.

We begin by expressing $\sin \theta - \cos \theta$ in the form $a \cos \theta + b \sin \theta$. $\sin \theta - \cos \theta = -\cos \theta + \sin \theta$, and so $a = -1$ and $b = 1$. Putting the

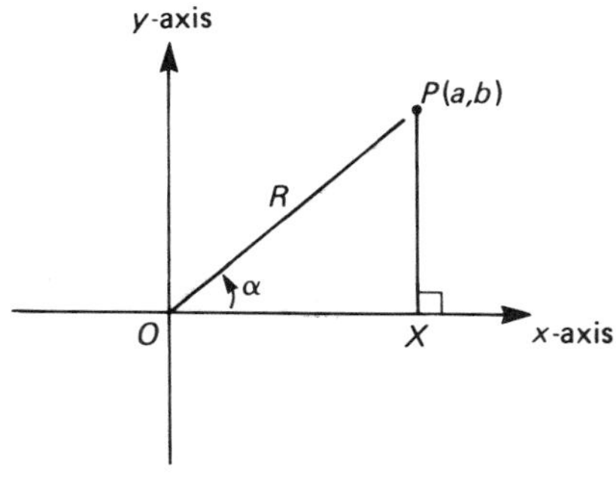

Fig. 3.8 Triangle relating a, b, R and α.

point $(-1, 1)$ on Fig. 3.8 shows that $R = \sqrt{2}$ and $\alpha = 3\pi/4$. Consequently

$$\sin\theta - \cos\theta = \sqrt{2}\cos(\theta - 3\pi/4)$$ ■

3.5 SOLUTIONS OF EQUATIONS

To solve the equation $\sin\theta = \sin\alpha$, where α is constant, we need a formula which will express θ in terms of α. Of course $\theta = \alpha$ is one solution but in fact there are many others. The graph of $y = \sin x$ enables us to determine this formula (Fig. 3.9).

As we observed, $\theta = \alpha$ is one solution of the equation, and since the sine function has period 2π we can deduce that $\theta = 2\pi + \alpha$ is also a solution. Generalizing, we deduce that $\theta = 2k\pi + \alpha$ is a solution, where k is any integer. This provides a whole set of solutions.

However, we have not finished because the symmetry of the sine function gives another solution, $\theta = \pi - \alpha$. Moreover we can add any integer multiple of 2π to this and always obtain another solution. So $\theta = 3\pi - \alpha$ is a solution, and in general $\theta = (2k + 1)\pi - \alpha$ is a solution, where k is any integer. This provides a second set of solutions. If we glance at the graph we can see how all these solutions arise and also that there are no more.

We can write the general solution in the form

$$\theta = n\pi + (-1)^n\alpha$$

where n is any integer. We see that when n is even we obtain the first set of solutions, whereas if n is odd we obtain the second set.

Similar arguments can be used to show that:

1 If $\cos\theta = \cos\alpha$, where α is a constant, then $\theta = 2n\pi \pm \alpha$, where n is any integer.
2 If $\tan\theta = \tan\alpha$, where α is a constant, then $\theta = n\pi + \alpha$, where n is any integer.

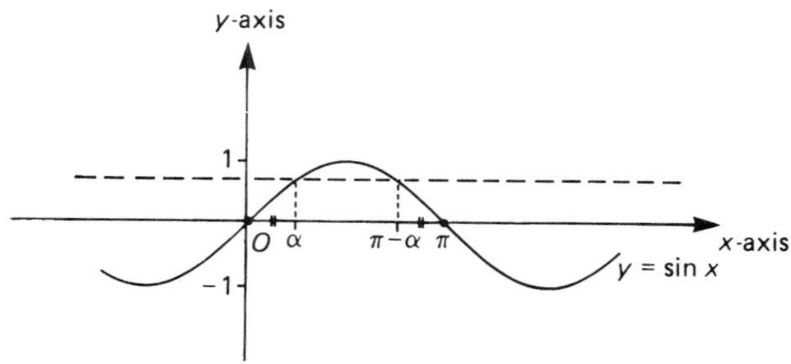

Fig. 3.9 Solutions of $\sin\theta = \sin\alpha$.

□ Obtain all the solutions of the equation $\sin 2x = \cos x$ in the interval $[0, 2\pi]$.

We have $2 \sin x \cos x = \cos x$, so $\cos x = 0$ or $2 \sin x = 1$. Remember to allow for the case $\cos x = 0$. If you don't you will lose some solutions and this may be very important.

Now $\cos \pi/2 = 0$ and $\sin \pi/6 = 1/2$, so we have reduced the equation to two cases: $\cos x = \cos \pi/2$, and $\sin x = \sin \pi/6$.

1 If $\cos x = \cos \pi/2$ then $x = 2n\pi \pm \pi/2$. Now we must pick out those solutions in the required interval:

$$\begin{aligned} n &= 0 \Rightarrow x = \pm\pi/2,\ \pi/2 \text{ is in range} \\ n &= 1 \Rightarrow x = 2\pi \pm \pi/2,\ 3\pi/2 \text{ is in range} \\ n &= -1 \Rightarrow x = -2\pi \pm \pi/2, \text{ out of range} \end{aligned}$$

Clearly other integer values for n will be out of range.

2 If $\sin x = \sin \pi/6$ then $x = n\pi + (-1)^n (\pi/6)$. Again we pick out those solutions which are in range:

$$\begin{aligned} n &= 0 \Rightarrow x = \pi/6, \text{ which is in range} \\ n &= 1 \Rightarrow x = \pi - \pi/6 = 5\pi/6, \text{ which is in range} \\ n &= 2 \Rightarrow x = 2\pi + \pi/6, \text{ out of range} \\ n &= -1 \Rightarrow x = -\pi + \pi/6, \text{ out of range} \end{aligned}$$

All other integer values for n will be out of range.

Finally we state the set of solutions in the interval $[0, 2\pi]$:

$$\{\pi/2,\ 3\pi/2,\ \pi/6,\ 5\pi/6\}$$ ■

Have you met the symbol ⇒ before? It is the one-way implication symbol; it means 'implies'. It is quite useful; you sometimes see it on traffic signs!

Now it's time for you to solve some problems. If you are unsure of the material, this is a good time to revise it. When you are ready, step ahead.

3.6 Workshop

▷**Exercise** Solve the equation 1

$$\cos 2x = 3 \cos x - 2$$

to obtain all solutions in the interval $[-\pi, \pi]$.

You need to remember your trigonometrical identities. There is a lot to be said for knowing them inside out.

If we use $\cos 2x = 2 \cos^2 x - 1$ we reduce the equations to a quadratic equation in $\cos x$: 2

$$2\cos^2 x - 1 = 3\cos x - 2$$

so that

$$2\cos^2 x - 3\cos x + 1 = 0$$

Now this factorizes to give

$$(2\cos x - 1)(\cos x - 1) = 0$$

from which either $\cos x = 1/2$ or $\cos x = 1$. However, $\cos \pi/3 = 1/2$ and $\cos 0 = 1$. So we can use the general solution of the equation $\cos\theta = \cos\alpha$, that is $\theta = 2n\pi \pm \alpha$, to obtain the general solution of this equation in the two cases:

1 $\cos x = \cos \pi/3 \Rightarrow x = 2n\pi \pm \pi/3$ where $n \in \mathbb{Z}$. We have to select those values of n which give solutions in the interval $[-\pi, \pi]$. We shall consider the positive and negative signs separately. If $x = 2n\pi + \pi/3$ then

$$\begin{aligned} n = -1 &\Rightarrow x = -5\pi/3, \text{ out of range} \\ n = 0 &\Rightarrow x = \pi/3, \text{ in range} \\ n = 1 &\Rightarrow x = 7\pi/3, \text{ out of range} \end{aligned}$$

If $x = 2n\pi - \pi/3$ then

$$\begin{aligned} n = -1 &\Rightarrow x = -7\pi/3, \text{ out of range} \\ n = 0 &\Rightarrow x = -\pi/3, \text{ in range} \\ n = 1 &\Rightarrow x = 5\pi/3, \text{ out of range} \end{aligned}$$

2 $\cos x = \cos 0 \Rightarrow x = 2n\pi$, and so $x = 0$ is the only solution in range.

Therefore the solution set is $\{-\pi/3, 0, \pi/3\}$.

If you managed that, then go on to the next exercise.

▷**Exercise** Obtain the general solution of the equation

$$\sin 2\theta = 2\cos\theta + \sin\theta - 1$$

Try this one carefully. Don't forget those identities. Then step ahead.

We use $\sin 2\theta = 2\sin\theta\cos\theta$ to obtain

$$\begin{aligned} 2\sin\theta\cos\theta &= 2\cos\theta + \sin\theta - 1 \\ 2\sin\theta\cos\theta - 2\cos\theta - \sin\theta + 1 &= 0 \end{aligned}$$

and this factorizes to give

$$(\sin\theta - 1)(2\cos\theta - 1) = 0$$

from which $\sin\theta = 1$ or $\cos\theta = 1/2$. There are therefore two sets of solutions:

1 $\sin\theta = \sin\pi/2 \Rightarrow \theta = n\pi + (-1)^n\pi/2$, where $n \in \mathbb{Z}$;

2 $\cos\theta = \cos\pi/3 \Rightarrow \theta = 2n\pi \pm \pi/3$, where $n \in \mathbb{Z}$.

3.7 COORDINATE GEOMETRY

Coordinate geometry is an algebraic description of geometry. It is essential for us to be able to recognize certain geometrical objects when they are expressed in algebraic form. Straight lines, circles and other curves can be represented by equations and we shall study the simplest of these.

We begin by obtaining the coordinates of a point midway between two others.

□ If P_1 and P_2 are the points (x_1, y_1) and (x_2, y_2) respectively, obtain the cartesian coordinates of the point M, the midpoint of P_1P_2 (Fig. 3.10).

Let M be the point (x, y). Then, using parallels,

$$x - x_1 = x_2 - x$$
$$y - y_1 = y_2 - y$$

So

$$x = \frac{x_1 + x_2}{2}, \qquad y = \frac{y_1 + y_2}{2}$$ ■

For example, the midpoints of the sides of the triangle with vertices (2, 6), (4, 0) and (−6, 6) are given by

$$(\tfrac{1}{2}[2 + 4], \tfrac{1}{2}[6 + 0]) = (3, 3)$$
$$(\tfrac{1}{2}[2 - 6], \tfrac{1}{2}[6 + 6]) = (-2, 6)$$
$$(\tfrac{1}{2}[4 - 6], \tfrac{1}{2}[0 + 6]) = (-1, 3)$$

LOCUS PROBLEMS

We shall use the methods of coordinate geometry to obtain the equations of several curves. To do this we consider a general point $P(x, y)$ on the

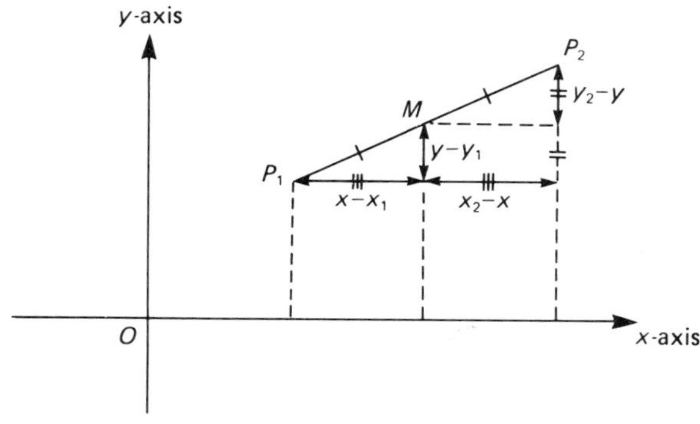

Fig. 3.10 The midpoint M of the line P_1P_2.

curve and obtain an equation relating x and y such that

1 If P is on the curve the equation holds;

2 If the equation holds then P is on the curve.

An equation which satisfies this condition is often called the **locus** of the point P; the Latin word *locus* means 'place'.

CHANGE OF AXES

When we identify a point P with an ordered pair of real numbers we must appreciate that this is relative to the cartesian coordinate system we have chosen. The same curve can have a very different equation if the axes are transformed in some way.

A **translation** is a change of axes in such a way that the new x-axis and the new y-axis are respectively parallel to the old ones. A **rotation** is a change of axes in which the origin remains fixed and axes rotate anti-clockwise through some angle θ.

Any movement of axes in the plane can be regarded as a translation followed by a rotation. We shall therefore consider the effects of these two transformations.

TRANSLATION

Suppose new axes X and Y are chosen which are parallel to the x and y axes. Suppose also the new origin O' is the point (h, k) relative to the system Oxy (see Fig. 3.11).

Then if P is the general point we may suppose that P is the point (x, y) relative to Oxy and (X, Y) relative to $O'XY$. We obtain

$$x = X + h, \quad y = Y + k$$

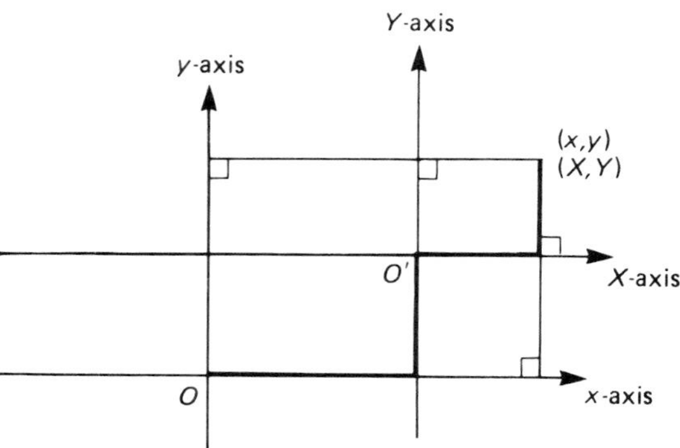

Fig. 3.11 A translation.

This is the change of coordinates corresponding to a translation of the origin to the point (h, k).

□ If the origin is translated to the point (1, 3), obtain the corresponding equation for the curve $x^2 + xy = y^2$.

Denoting the new axes by X and Y we have

$$x = X + 1, \qquad y = Y + 3$$

so that

$$(X + 1)^2 + (X + 1)(Y + 3) = (Y + 3)^2$$
$$X^2 + 2X + 1 + XY + Y + 3X + 3 = Y^2 + 6Y + 9$$
$$X^2 - Y^2 + XY + 5X - 5Y = 5$$

We may now drop the X and Y in favour of the usual x and y since we have done with the old coordinate system for good. Therefore the new equation is

$$x^2 - y^2 + xy + 5x - 5y = 5$$ ■

ROTATION

Suppose the axes Oxy are rotated anticlockwise through θ to produce OXY and that P is a general point. Let $OP = r$ and suppose that $\angle XOP = \alpha$ (Fig. 3.12).

Relative to OXY, P is the point $(r \cos \alpha, r \sin \alpha)$, whereas relative to Oxy, P is the point $(r \cos [\theta + \alpha], r \sin [\theta + \alpha])$. Therefore

$$X = r \cos \alpha, \qquad Y = r \sin \alpha$$
$$x = r \cos (\theta + \alpha), \qquad y = r \sin (\theta + \alpha)$$

So

$$\begin{aligned} x &= r(\cos \theta \cos \alpha - \sin \theta \sin \alpha) \\ &= X \cos \theta - Y \sin \theta \end{aligned}$$

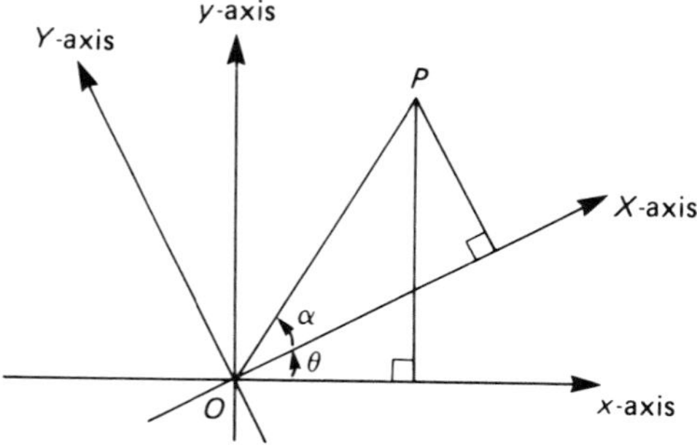

Fig. 3.12 A rotation.

$$\begin{aligned} y &= r(\sin\theta\cos\alpha + \cos\theta\sin\alpha) \\ &= X\sin\theta + Y\cos\theta \end{aligned}$$

We therefore have the change of coordinates

$$\begin{aligned} x &= X\cos\theta - Y\sin\theta \\ y &= X\sin\theta + Y\cos\theta \end{aligned}$$

for an anticlockwise rotation of the axes through an angle θ.

□ Obtain the equation of the curve $4x^2 + 6y^2 = 25$ if the axes are rotated anticlockwise through $\pi/4$.

Here $\theta = \pi/4$ and so the change of coordinates is

$$\begin{aligned} x &= X(1/\sqrt{2}) - Y(1/\sqrt{2}) = (X - Y)/\sqrt{2} \\ y &= X(1/\sqrt{2}) + Y(1/\sqrt{2}) = (X + Y)/\sqrt{2} \end{aligned}$$

Substituting into the equation gives

$$\begin{aligned} &2(X - Y)^2 + 3(X + Y)^2 = 25 \\ &2(X^2 - 2XY + Y^2) + 3(X^2 + 2XY + Y^2) = 25 \\ &5X^2 + 2XY + 5Y^2 = 25 \end{aligned}$$

So that reverting to x and y we have finally

$$5x^2 + 2xy + 5y^2 = 25$$ ■

3.8 THE STRAIGHT LINE

A straight line can be fixed in the plane in several ways. First, we can specify a point P_1 (x_1, y_1) on the straight line and also the slope m of the line. If P (x, y) is a general point on the line, we have

$$m = \tan\theta = \text{slope } PP_1 = \frac{PM}{P_1M} = \frac{y - y_1}{x - x_1}$$

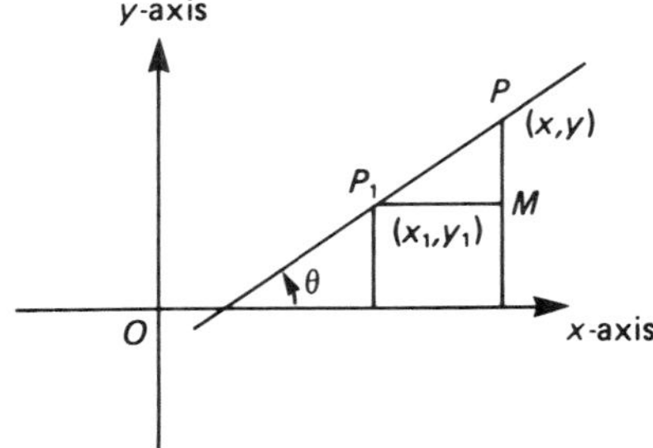

Fig. 3.13 Straight line; fixed slope through fixed point.

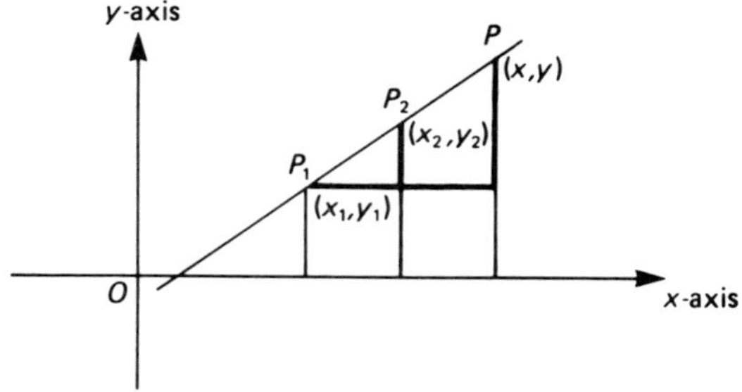

Fig. 3.14 Straight line; two fixed points.

Therefore

$$y - y_1 = m(x - x_1)$$

A second way of fixing a straight line is to specify two points $P_1\ (x_1, y_1)$ and $P_2\ (x_2, y_2)$ on the line (Fig. 3.14). Then if $P\ (x, y)$ is a general point on the line we have

$$\text{slope } PP_1 = \text{slope } P_2P_1$$

$$\frac{y - y_1}{x - x_1} = \frac{y_2 - y_1}{x_2 - x_1}$$

or

$$\frac{y - y_1}{y_2 - y_1} = \frac{x - x_1}{x_2 - x_1}$$

It is interesting to note that there are several other equivalent forms for this equation, and these can be obtained by equating the slopes of any two distinct pairs of points chosen from P_1, P_2 and P.

□ Putting slope PP_1 = slope PP_2 yields the equation

$$\frac{y - y_1}{y - y_2} = \frac{x - x_1}{x - x_2}$$

Show that this equation can be rewritten in the form

$$\frac{y - y_1}{y_2 - y_1} = \frac{x - x_1}{x_2 - x_1}$$

We have

$$(x - x_2)\,(y - y_1) = (x - x_1)\,(y - y_2)$$

So

$$xy - x_2y - xy_1 + x_2y_1 = xy - xy_2 - x_1y + x_1y_2$$

$$-x_2y - xy_1 + x_2y_1 = -xy_2 - x_1y + x_1y_2$$
$$x(y_2 - y_1) - x_1y_2 + x_1y_1 = y(x_2 - x_1) - y_1x_2 + y_1x_1$$
$$(x - x_1)(y_2 - y_1) = (y - y_1)(x_2 - x_1)$$

Therefore

$$\frac{y - y_1}{y_2 - y_1} = \frac{x - x_1}{x_2 - x_1}$$

as required. ■

□ Obtain the equation of the straight line joining the points $(-3, 7)$ to $(5, 1)$.

We may use the formula

$$\frac{y - y_1}{y_2 - y_1} = \frac{x - x_1}{x_2 - x_1}$$

where $(x_1, y_1) = (-3, 7)$ and $(x_2, y_2) = (5, 1)$. So

$$\frac{y - 7}{1 - 7} = \frac{x - (-3)}{5 - (-3)}$$
$$\frac{y - 7}{-6} = \frac{x + 3}{8}$$
$$8(y - 7) = -6(x + 3)$$
$$8y + 6x = 56 - 18$$
$$4y + 3x = 19$$

This is the required equation. ■

EQUATION OF A STRAIGHT LINE

Any equation of the form $ax + by = c$, where a, b and c are real constants, represents the equation of a straight line. Conversely, any straight line has an equation of the form $ax + by = c$ for some real constants a, b and c.

There are two other forms of the equation of a straight line which are often useful. One is for the straight line with slope m which has an intercept c on the y-axis. In other words, this is the line through $(0, c)$ with slope m:

$$y - c = m(x - 0)$$

Therefore

$$y = mx + c$$

This is the most commonly used equation of a straight line.

Another form is for the straight line which has intercepts a and b on the x-axis and y-axis respectively. In other words, we are looking for the straight line which passes through the points $(a, 0)$ and $(0, b)$ (Fig. 3.15). Therefore

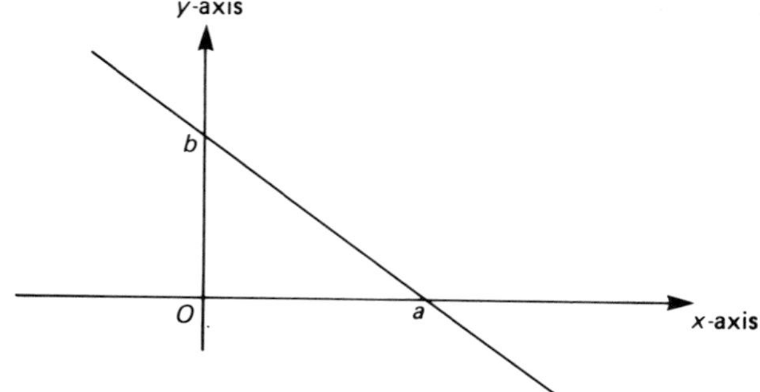

Fig. 3.15 Straight line; fixed intercepts.

$$\frac{y-b}{0-b} = \frac{x-0}{a-0}$$

$$-\frac{y}{b} + 1 = \frac{x}{a}$$

So

$$\frac{x}{a} + \frac{y}{b} = 1$$

ANGLE BETWEEN TWO STRAIGHT LINES

Suppose we have two straight lines with slopes m_1 and m_2 respectively (Fig. 3.16). Then if $m_1 = \tan\theta_1$ and $m_2 = \tan\theta_2$ the angle θ between the lines is given by $\theta = \theta_1 - \theta_2$. So

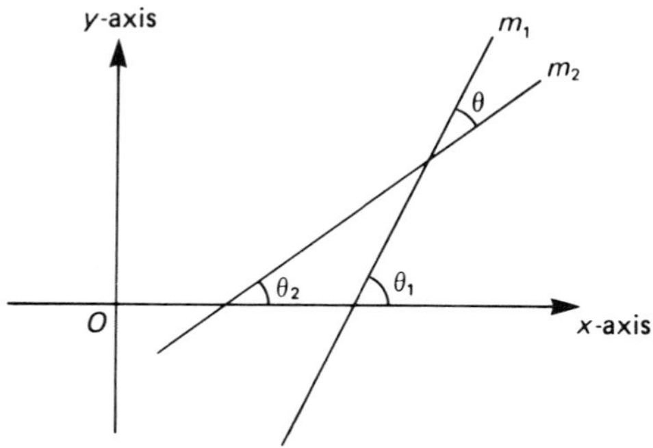

Fig. 3.16 Angle between two straight lines.

$$\tan\theta = \tan(\theta_1 - \theta_2) = \frac{\tan\theta_1 - \tan\theta_2}{1 + \tan\theta_1\tan\theta_2} = \frac{m_1 - m_2}{1 + m_1m_2}$$

This is valid provided $1 + m_1m_2 \neq 0$. If $m_1 = m_2$ then $\tan\theta = 0$ as the straight lines are parallel.

Also

$$\cot\theta = \cot(\theta_1 - \theta_2) = \frac{\cot\theta_1\cot\theta_2 + 1}{\cot\theta_1 - \cot\theta_2} = \frac{(1/m_1)(1/m_2) + 1}{(1/m_1) - (1/m_2)} = \frac{1 + m_1m_2}{m_2 - m_1}$$

If the lines are mutually perpendicular then $\theta = \pi/2$, so $\cot\theta = 0$ and therefore $m_1m_2 = -1$.

One small point needs to be made. We have been considering straight lines with slope m. What happens if the line is parallel to the y-axis? We know that $\tan\theta$ is not defined when $\theta = \pi/2$, so what do we do? We divide through by m and note that, as θ approaches $\pi/2$, $1/m = \cot\theta$ approaches 0.

□ Obtain the equation of the straight line parallel to the y-axis through the point (3, 7).

The equation of a straight line with slope m through the point (3, 7) is

$$y - 7 = m(x - 3)$$

So

$$\frac{y - 7}{m} = x - 3$$

Now as m gets larger and larger, $x - 3$ gets closer and closer to 0. So the required line is $x = 3$. ■

3.9 THE CIRCLE

A circle has the property that every point on it is at the same distance from a fixed point C, its centre. We begin by obtaining a formula for the distance between two points P and Q (Fig. 3.17).

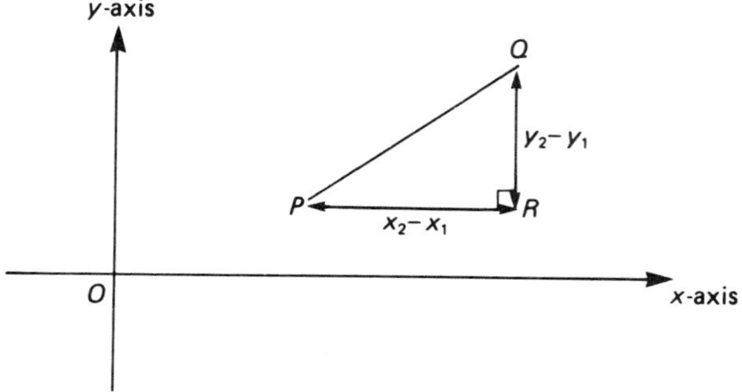

Fig. 3.17 Two points P and Q.

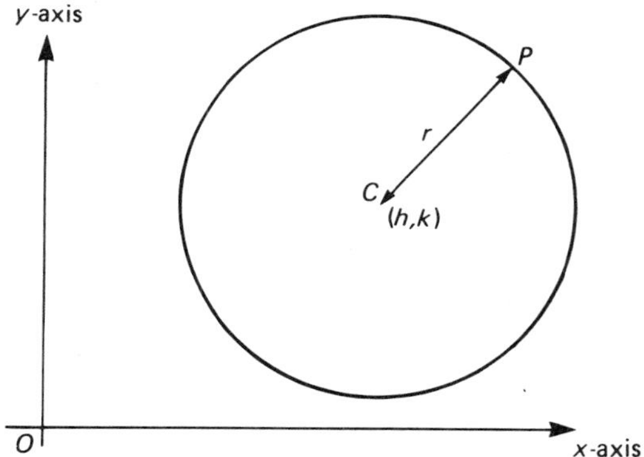

Fig. 3.18 Circle; centre (h, k), radius r.

Suppose P is (x_1, y_1) and Q is (x_2, y_2). Then

$$\begin{aligned} PQ^2 &= PR^2 + RQ^2 \\ &= (x_2 - x_1)^2 + (y_2 - y_1)^2 \end{aligned}$$

So

$$PQ = \sqrt{[(x_2 - x_1)^2 + (y_2 - y_1)^2]}$$

Although we have shown this only in the case where P and Q are as on the diagram, the formula is valid wherever P and Q are positioned.

Suppose now we have a circle radius r with its centre at the point (h, k) (Fig. 3.18). Then if P (x, y) is a general point on the circle, we have $PC^2 = r^2$. So

$$(x - h)^2 + (y - k)^2 = r^2$$

Conversely, any point (x, y) which satisfies this equation lies on the circle.

If we expand this equation we obtain

$$x^2 - 2hx + h^2 + y^2 - 2ky + k^2 = r^2$$

So

$$x^2 + y^2 - 2hx - 2ky + h^2 + k^2 - r^2 = 0$$

We can use this to obtain criteria for an equation to be the equation of a circle:

1 The equation must be of degree 2 in the two variables x and y;
2 The coefficient of x^2 must equal the coefficient of y^2;
3 There must be no xy term.

Such an equation is traditionally written in the form

$$x^2 + y^2 + 2gx + 2fy + c = 0$$

so that completing the square we obtain

$$(x + g)^2 + (y + f)^2 = g^2 + f^2 - c$$

Therefore provided $g^2 + f^2 - c > 0$ we have a circle. The circle has centre $(-g, -f)$ and radius $\sqrt{(g^2 + f^2 - c)}$.

□ **a** Obtain the centre and radius of the circle

$$x^2 + y^2 - 4x + 6y + 8 = 0$$

b Obtain the equation of the circle with centre $(-1, 5)$ and radius 7 in the standard form

$$x^2 + y^2 + 2gx + 2fy + c = 0$$

The procedures are as follows:

a If we compare the given equation with the standard equation of the circle

$$x^2 + y^2 + 2gx + 2fy + c = 0$$

We have $g = -2$, $f = 3$ and $c = 8$. So the centre is the point $(2, -3)$ and the radius is

$$\sqrt{(g^2 + f^2 - c)} = \sqrt{(4 + 9 - 8)} = \sqrt{5}$$

b The equation of the circle is

$$(x - h)^2 + (y - k)^2 = r^2$$

where (h, k) is the centre and r is the radius. Here $h = -1$, $k = 5$ and $r = 7$, so that

$$\begin{aligned}(x + 1)^2 + (y - 5)^2 &= 49\\ x^2 + 2x + 1 + y^2 - 10y + 25 &= 49\\ x^2 + y^2 + 2x - 10y - 23 &= 0\end{aligned}$$

■

PARAMETRIC FORM

Another way of representing a curve is to express each of the two variables x and y in terms of some third variable θ which is known as a **parameter**. This is done in such a way that

1 Every value of θ corresponds to a unique point on the curve; and
2 Every point on the curve corresponds to a unique value of θ.

We can therefore talk about the point θ.

If we were to eliminate θ we should obtain the cartesian equation of the curve.

□ For the circle $x^2 + y^2 = a^2$ we have a parametric form $x = a \cos \theta$, $y = a \sin \theta$. ■

3.10 THE CONIC SECTIONS

If we take a right circular cone and cut it through in various positions we obtain standard curves known as the parabola, the ellipse and the hyperbola. Each of these curves can be defined as a locus in much the same way as we defined the circle as a locus.

We consider a fixed straight line called the **directrix** and a fixed point S called the **focus** (Fig. 3.19). If P is a general point, suppose L is a point on the directrix such that the line PL and the directrix are perpendicular to one another. If the ratio PS/PL is a constant then the locus of P is one of the conic sections. The ratio $e = PS/PL$ is known as the **eccentricity**. We consider the three cases $e = 1$, $e < 1$ and $e > 1$.

THE PARABOLA ($e = 1$)

Suppose we choose the x-axis to be the line through the focus S perpendicular to the directrix, and the origin to be the point on the x-axis midway between the focus and the directrix (Fig. 3.20). If S is the point $(a, 0)$ then

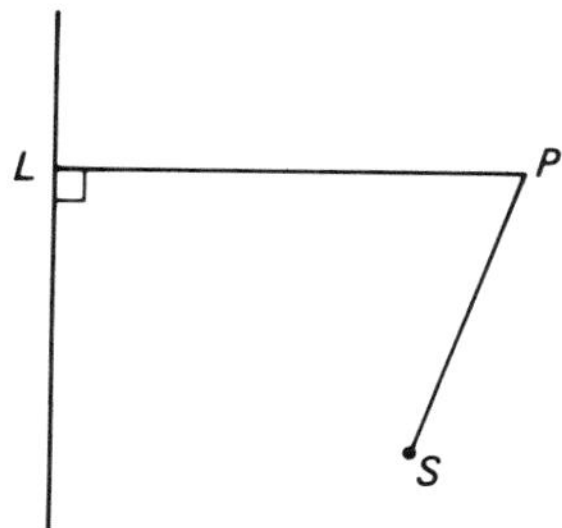

Fig. 3.19 Directrix and focus.

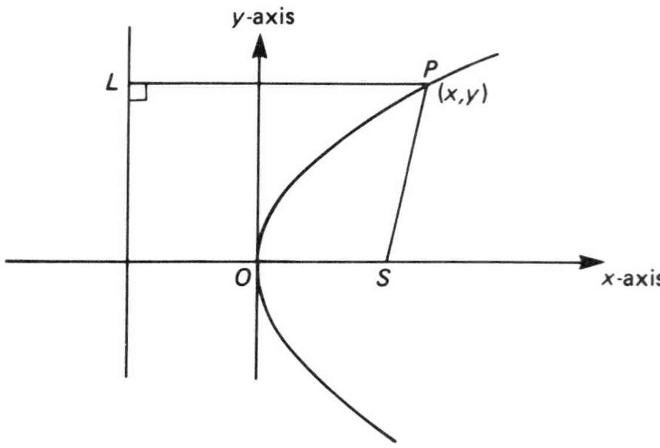

Fig. 3.20 The parabola.

the directrix has the equation $x = -a$. Now if P is a general point on the parabola we have

$$PS = \sqrt{[(x - a)^2 + y^2]}$$
$$PL = x + a$$

However, $PS = PL$ and so we have

$$(x - a)^2 + y^2 = (x + a)^2$$

Therefore we obtain the standard cartesian form for the parabola as

$$y^2 = (x + a)^2 - (x - a)^2 = 4ax$$

☐ Obtain the equation of the directrix and the position of the focus for the parabola $y^2 = 16x$.

We compare with the standard equation $y^2 = 4ax$ and obtain $a = 4$. Consequently the focus S is the point $(4, 0)$ and the equation of the directrix is $x + 4 = 0$. ■

The usual parametric form of the parabola $y^2 = 4ax$ is $x = at^2$ and $y = 2at$. Clearly if we eliminate t we obtain

$$y^2 = 4a^2t^2 = 4a(at^2) = 4ax$$

So a general point t on the parabola is $(at^2, 2at)$.

THE ELLIPSE ($e < 1$)

It is convenient to choose our focus to be $(-ae, 0)$ and the directrix as the line $x = -a/e$ (Fig. 3.21). Then

$$PS^2 = (x + ae)^2 + y^2$$
$$PL^2 = (x + a/e)^2$$

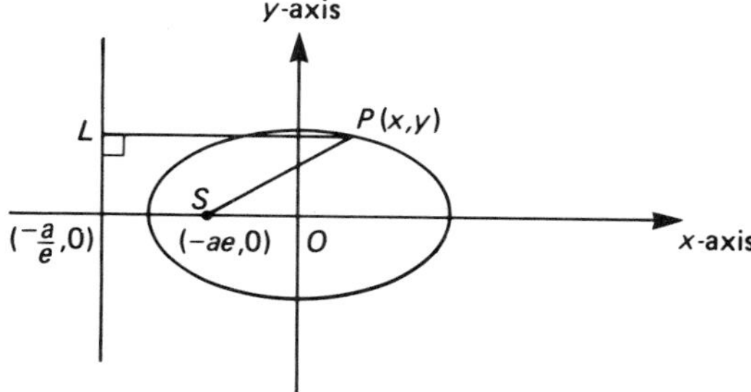

Fig. 3.21 The ellipse.

Now $PS/PL = e$, so that

$$(x + ae)^2 + y^2 = e^2(x + a/e)^2 = (ex + a)^2$$

Therefore

$$(x + ae)^2 - (ex + a)^2 + y^2 = 0$$
$$x^2(1 - e^2) - a^2(1 - e^2) + y^2 = 0$$
$$\frac{x^2}{a^2} + \frac{y^2}{a^2(1 - e^2)} = 1$$

Now $e < 1$, and so we may put $b^2 = a^2(1 - e^2)$ to obtain the standard cartesian form for the ellipse as

$$\frac{x^2}{a^2} + \frac{y^2}{b^2} = 1$$

The axes of symmetry are known as the major axis and the minor axis. The major axis has length $2a$ and the minor axis has length $2b$.

The symmetry of this curve suggests that it must be possible to define it in terms of another focus and another directrix. In fact the point $(ae, 0)$ provides a second focus, and the line $x = a/e$ the corresponding directrix. From Fig. 3.22

$$\frac{PS_1}{PL_1} = e = \frac{PS_2}{PL_2}$$

so that

$$\begin{aligned} PS_1 + PS_2 &= ePL_1 + ePL_2 \\ &= e(PL_1 + PL_2) = eL_1L_2 \\ &= e(2a/e) = 2a \end{aligned}$$

This says that at any point on the ellipse the sum of the distances to the foci is constant and equal to the length of the major axis. This property has practical uses. For example, gardeners sometimes use it to mark out the

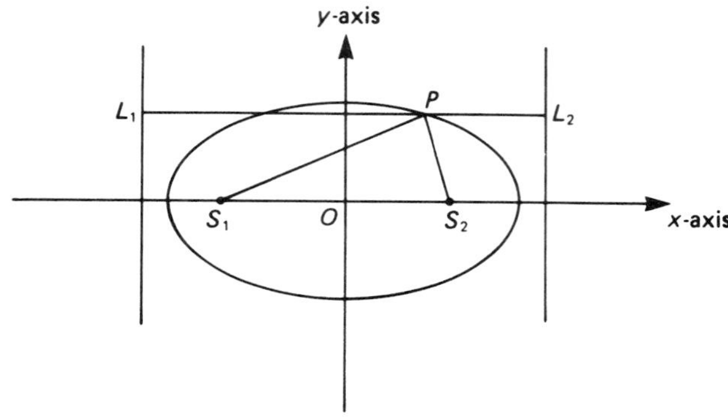

Fig. 3.22 The two foci of an ellipse.

boundary of an elliptical flower-bed. To do this two pegs are secured at the foci and a piece of rope equal in length to the major axis joins the two pegs. When the rope is held taut along the ground an ellipse can be traced out.

There are many possible parametric forms for the ellipse. The one which is usually employed is $x = a \cos \theta$ and $y = b \sin \theta$. Eliminating θ using $\cos^2 \theta + \sin^2 \theta = 1$ gives the ellipse in cartesian form.

THE HYPERBOLA ($e > 1$)

We choose the focus to be $(-ae, 0)$ and the directrix to be the line $x = -a/e$. However, since $e > 1$ the position of the focus is to the left of the directrix (Fig. 3.23). Then

$$PS^2 = (x + ae)^2 + y^2$$
$$PL^2 = (x + a/e)^2$$

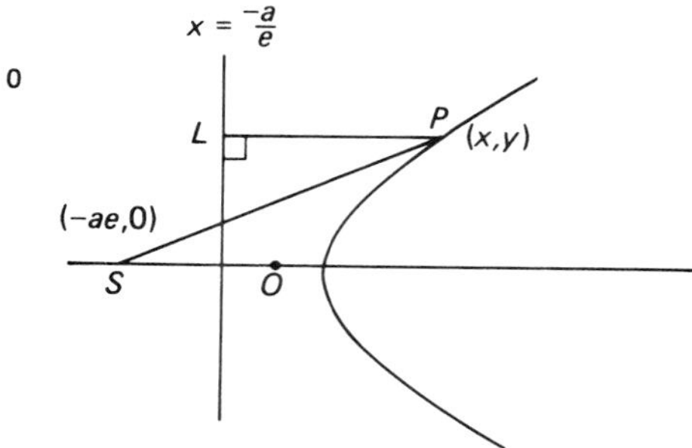

Fig. 3.23 The hyperbola.

Now $PS/PL = e$, so that

$$(x + ae)^2 + y^2 = e^2(x + a/e)^2 = (ex + a)^2$$

Therefore

$$\begin{aligned}(x + ae)^2 - (ex + a)^2 + y^2 &= 0\\ x^2(1 - e^2) + y^2 &= a^2(1 - e^2)\\ x^2(e^2 - 1) - y^2 &= a^2(e^2 - 1)\end{aligned}$$

Now $e > 1$, and so we may put $b^2 = a^2(e^2 - 1)$ to obtain the standard cartesian form for the hyperbola as

$$\frac{x^2}{a^2} - \frac{y^2}{b^2} = 1$$

There are several parametric forms for the hyperbola. The one which is usually chosen is $x = a \cosh u$, $y = a \sinh u$, which involves hyperbolic functions. Until we study these functions (Chapter 5) we shall have to be content with another parametric form, such as $x = a \sec \theta$, $y = b \tan \theta$. Note that

$$\frac{x^2}{a^2} - \frac{y^2}{b^2} = \sec^2 \theta - \tan^2 \theta = 1$$

The straight lines $y = \pm(b/a)x$ are the **asymptotes** of the hyperbola. The tangents approach these straight lines as $|x|$ increases in magnitude. Some books refer to them as 'tangents at infinity', but this does not really mean very much. If $b = a$ then the asymptotes are the straight lines $y = x$ and $y = -x$, which are mutually perpendicular, and the hyperbola is called a **rectangular hyperbola**. Moreover, if we rotate the curve anticlockwise through $\pi/4$ we can use the asymptotes as axes. This implies that the axes have been rotated clockwise through $\pi/4$.

Suppose $b = a$. Then we have $x^2 - y^2 = a^2$. For a rotation of $-\pi/4$ we have

$$\begin{aligned}x &= X \cos(-\pi/4) - Y \sin(-\pi/4) = (X + Y)/\sqrt{2}\\ y &= X \sin(-\pi/4) + Y \cos(-\pi/4) = (-X + Y)/\sqrt{2}\end{aligned}$$

So

$$\begin{aligned}x^2 - y^2 &= [(X + Y)^2 - (X - Y)^2]/2\\ &= 2XY = a^2\end{aligned}$$

So writing $c = a/\sqrt{2}$ and changing the notation X and Y to the more usual x and y, we have $xy = c^2$.

The usual parametric representation for the rectangular hyperbola $xy = c^2$ is $x = ct$ and $y = c/t$.

Now it's time to take a few more steps.

3.11 Workshop

1 **Exercise** Identify the polar equation

$$r^2 = 8 \operatorname{cosec} 2\theta$$

by transforming it into cartesian coordinates, or otherwise.

The phrase 'or otherwise' is used quite often in examination questions. Theoretically it means that if you can think of a different method you are at liberty to use it. In practice it often means 'or otherwise try another question'!

---

2 We use $x = r \cos \theta$ and $y = r \sin \theta$, from which $r = \sqrt{(x^2 + y^2)}$ and $\tan \theta = y/x$. Given the equation $r^2 = 8 \operatorname{cosec} 2\theta$, if we multiply through by $\sin 2\theta$ we obtain

$$r^2 \sin 2\theta = 8$$

and since $\sin 2\theta = 2 \sin \theta \cos \theta$ we have

$$2r^2 \sin \theta \cos \theta = 8$$
$$r^2 \sin \theta \cos \theta = 4$$

That is, $xy = 4$.

We should now recognize this as the equation of a rectangular hyperbola in which the axes coincide with the asymptotes.

If that went well, then move ahead to step 4. Otherwise, try the next exercise. Remember that to transform from polar coordinates to cartesian coordinates we must use $x = r \cos \theta$ and $y = r \sin \theta$. Once r and θ have been eliminated it is then just a question of identifying the curve.

▷ **Exercise** Identify the curve which has the equation in polar coordinates

$$r^2(\cos 2\theta - 3) + 10 = 0$$

When you have done this move forward.

---

3 If we remember the identity $\cos 2\theta = 2 \cos^2 \theta - 1$ it will help. In fact any of the three identities expressing $\cos 2\theta$ in terms of $\sin \theta$ and/or $\cos \theta$ will do. We obtain

$$\begin{aligned} r^2(2\cos^2\theta - 1 - 3) + 10 &= 0 \\ r^2(\cos^2\theta - 2) + 5 &= 0 \\ r^2\cos^2\theta - 2r^2 + 5 &= 0 \\ x^2 - 2(x^2 + y^2) + 5 &= 0 \\ x^2 + 2y^2 &= 5 \end{aligned}$$

We recognize this as the equation of an ellipse:

$$\frac{x^2}{5} + \frac{y^2}{(5/2)} = 1$$

The major axis has length $2a = 2\sqrt{5}$ and the minor axis has length $2b = 2\sqrt{(5/2)} = \sqrt{10}$.

If there are any problems remaining, then make sure you follow all the stages. You will have another chance to tackle one of these when you work through the problems at the end of the chapter. Now step ahead.

4

▷**Exercise** Identify the curve which has the equation

$$(x + y)^2 = 4(xy + 1)$$

Be just a little careful here. Try it, then step forward.

5

This is one of those problems where you can be too clever! You may think that the equation has the form $Y^2 = 4X$, where $Y = x + y$ and $X = xy + 1$, and be led by this to conclude that the equation was that of a parabola. However, the change of coordinates does not correspond to a movement of axes and so does not preserve geometrical shapes. Instead we do something much more mundane. We multiply out and rearrange the equation:

$$\begin{aligned} (x + y)^2 &= 4(xy + 1) \\ x^2 + 2xy + y^2 &= 4xy + 4 \\ x^2 - 2xy + y^2 &= 4 \\ (x - y)^2 - 4 &= 0 \\ [(x - y) - 2][(x - y) + 2] &= 0 \end{aligned}$$

So $x - y - 2 = 0$ or $x - y + 2 = 0$, and we therefore have the equation of a pair of parallel straight lines $y = x - 2$ and $y = x + 2$.

If you were right, then on you stride to step 7. Otherwise try this exercise.

▷**Exercise** Describe the geometrical curve which has the equation

$$(x + y)^2 = (x + 8)(x + 2y) + 8(x - 2y)$$

This should cause you no trouble.

We do the obvious thing and multiply out the equation with a view to simplifying it:

$$x^2 + 2xy + y^2 = x^2 + 8x + 2xy + 16y + 8x - 16y$$

Almost everything cancels out, and we are able to reduce the equation to

$$y^2 = 16x$$

We recognize this as the equation of a parabola in the standard form $y^2 = 4ax$, where $a = 4$. So the focus is the point $(4, 0)$ and the directrix is the line $x = -4$.

Exercise Obtain the condition that the line $y = mx + c$ intersects the parabola $y^2 = 4ax$ in two coincident points. Thereby obtain the equation of the tangent to the parabola $y^2 = 4ax$ with slope m.

We shall see in Chapter 6 how we can obtain the equation of the tangents to each of the conics at a general point by using calculus, but for the moment we shall restrict ourselves to algebraic methods.

The two 'curves' $y^2 = 4ax$ and $y = mx + c$ intersect when

$$\begin{aligned} (mx + c)^2 &= 4ax \\ m^2x^2 + 2mcx + c^2 &= 4ax \\ m^2x^2 + 2(mc - 2a)x + c^2 &= 0 \end{aligned}$$

In general, if $(mc - 2a)^2 > m^2c^2$, there will be two real solutions for x and so two points where the straight line intersects the parabola. However, in the special case $(mc - 2a)^2 = m^2c^2$ the roots coincide and we have a tangent. This gives

$$m^2c^2 - 4amc + 4a^2 = m^2c^2 \text{ and so } mc = a$$

that is $c = a/m$. So the equation of the tangent is

$$y = mx + c = mx + \frac{a}{m}$$

If that went well, then finish with this exercise.

▷**Exercise** Obtain the equation of the tangent with slope m to the circle $x^2 + y^2 = r^2$.

This is just like the last one, and so there should be no problems. Try it, then take the final step.

We use the equation of the straight line in the form $y = mx + c$ and we wish to obtain c, given that this straight line is a tangent. Substituting into $x^2 + y^2 = r^2$ we have

$$\begin{aligned} x^2 + (mx + c)^2 &= r^2 \\ x^2 + m^2x^2 + 2mcx + c^2 &= r^2 \\ (m^2 + 1)x^2 + 2mcx + c^2 - r^2 &= 0 \end{aligned}$$

If this quadratic equation is to have equal roots then

$$\begin{aligned} 4m^2c^2 &= 4(m^2 + 1)(c^2 - r^2) \\ m^2c^2 &= m^2c^2 - m^2r^2 + c^2 - r^2 \\ c^2 &= (m^2 + 1)r^2 \end{aligned}$$

Therefore $c = \pm r\sqrt{(m^2 + 1)}$, and so there are two tangents:

$$y = mx \pm r\sqrt{(m^2 + 1)}$$

You didn't overlook the minus sign, did you?

Here now are a couple of problems which arise in applications.

3.12 Practical

TOWER HEIGHT

A surveyor finds that from the foot of a tower the elevation of a mast is 9θ but that from the top of the tower the elevation is only 8θ. The tower and the mast are both built on ground at the same horizontal level, and the height of the tower is h.

Show that the horizontal distance from the tower to the mast is

$$d = h \operatorname{cosec} \theta \cos 8\theta \cos 9\theta$$

Obtain an expression for l, the height of the mast, in terms of h and θ.

You should be able to try this on your own. When you have made an attempt, read on and examine the solution.

The arrangement is shown in Fig. 3.24. We have

$$l/d = \tan 9\theta$$

and

$$\begin{aligned} (l - h)/d &= \tan 8\theta \\ l/d - h/d &= \tan 8\theta \end{aligned}$$

From these equations,

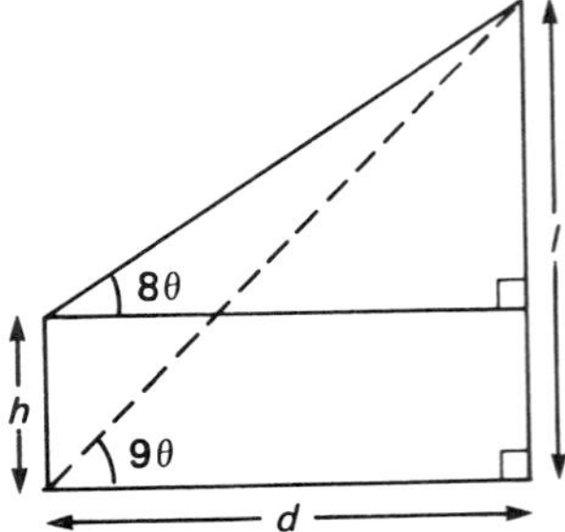

Fig. 3.24 Representation of tower and mast.

$$\begin{aligned} h/d &= \tan 9\theta - \tan 8\theta \\ &= \frac{\sin 9\theta}{\cos 9\theta} - \frac{\sin 8\theta}{\cos 8\theta} \\ &= \frac{\sin 9\theta \cos 8\theta - \cos 9\theta \sin 8\theta}{\cos 9\theta \cos 8\theta} \\ &= \frac{\sin (9\theta - 8\theta)}{\cos 9\theta \cos 8\theta} \\ &= \frac{\sin \theta}{\cos 9\theta \cos 8\theta} \end{aligned}$$

Therefore

$$d = h \operatorname{cosec} \theta \cos 8\theta \cos 9\theta$$

Also

$$l = d \tan 9\theta = h \operatorname{cosec} \theta \cos 8\theta \sin 9\theta$$

CIRCUIT ADMITTANCE

This problem uses some of the geometry we have developed.

The admittance of an RC series circuit may be represented by the point $P(x, y)$, where

$$x = R \bigg/ \left(R^2 + \frac{1}{\omega^2 C^2}\right)$$

$$y = \left(\frac{1}{\omega C}\right) \bigg/ \left(R^2 + \frac{1}{\omega^2 C^2}\right)$$

Eliminate ω to determine the admittance locus – the equation relating x and y. Show how P moves on this curve as ω increases from 0 without bound.

It is worthwhile seeing if you can sort this out for yourself before you move on.

We have

$$\frac{R}{x} = R^2 + \frac{1}{\omega^2 C^2} \tag{1}$$

$$\frac{1}{\omega C y} = R^2 + \frac{1}{\omega^2 C^2} \tag{2}$$

So

$$\frac{R}{x} = \frac{1}{\omega C y} \quad \text{or} \quad \frac{1}{\omega C} = \frac{Ry}{x}$$

Substituting back into (1) we have

$$\frac{R}{x} = R^2 + \frac{R^2 y^2}{x^2}$$

$$Rx = R^2 x^2 + R^2 y^2$$

Therefore

$$x^2 + y^2 - \frac{x}{R} = 0$$

$$\left(x - \frac{1}{2R}\right)^2 + y^2 = \frac{1}{4R^2}$$

So P is on a circle of centre $(1/2R, 0)$ and radius $1/2R$ (Fig. 3.25).

When $\omega \neq 0$ we have

$$x = \frac{R\omega^2 C^2}{R^2\omega^2 C^2 + 1} \qquad y = \frac{\omega C}{R^2\omega^2 C^2 + 1}$$

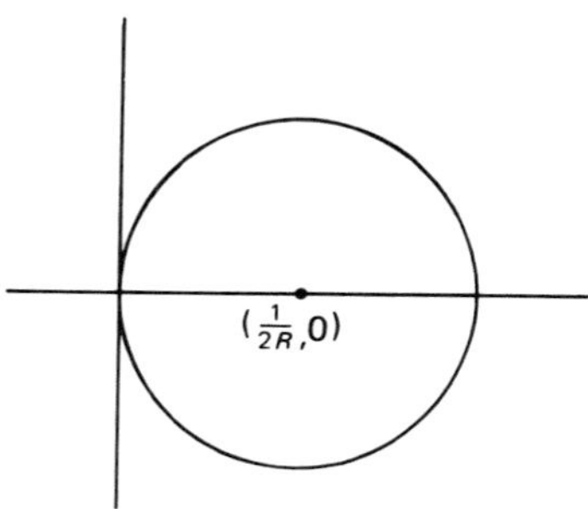

Fig. 3.25 The admittance locus.

As $\omega \to 0$ we have $(x, y) \to (0, 0)$, the origin. For $\omega > 0$ we have $y > 0$ and, as ω increases without bound, $(x, y) \to (1/R, 0)$. The movement of P is therefore confined to the upper semicircle (Fig. 3.26).

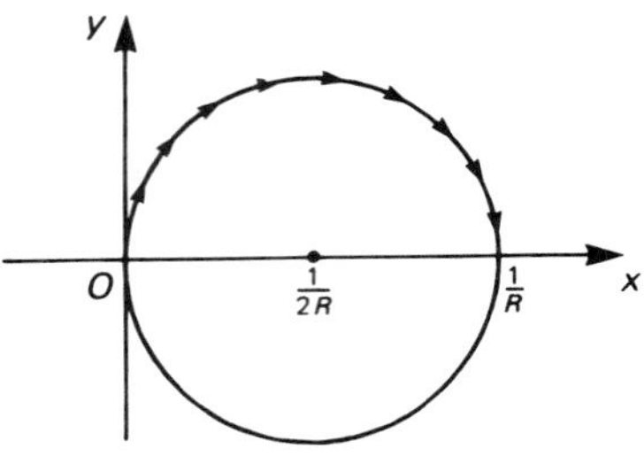

Fig. 3.26 The path of P.

SUMMARY

☐ We have defined the circular functions, drawn their graphs and deduced some of their properties. Two key identities are

$$\cos^2 \theta + \sin^2 \theta = 1$$
$$\sin (A + B) = \sin A \cos B + \cos A \sin B$$

☐ We have obtained the general solution of equations involving circular functions

$$\sin \theta = \sin \alpha \Rightarrow \theta = n\pi + (-1)^n\alpha$$
$$\cos \theta = \cos \alpha \Rightarrow \theta = 2n\pi \pm \alpha$$
$$\tan \theta = \tan \alpha \Rightarrow \theta = n\pi + \alpha$$

where n is any integer.

☐ We have obtained the standard equations of the straight line

a $y - y_1 = m(x - x_1)$ (slope m, through (x_1, y_1))

b $\dfrac{y - y_1}{y_2 - y_1} = \dfrac{x - x_1}{x_2 - x_1}$ (through (x_1, y_1) and (x_2, y_2))

c $y = mx + c$ (slope m, y-intercept c)

d $\dfrac{x}{a} + \dfrac{y}{b} = 1$ (x-intercept a, y-intercept b)

☐ We have shown that the angle between two straight lines with slopes m_1 and m_2 respectively is given by

$$\tan \theta = \frac{m_1 - m_2}{1 + m_1 m_2}$$

The lines are parallel if and only if $m_1 = m_2$.
The lines are mutually perpendicular if and only if $m_1 m_2 = -1$.

☐ We have obtained the equations in standard form of the conic sections

$x^2 + y^2 + 2gx + 2fy + c = 0$ (the circle)

$y^2 = 4ax$ (the parabola)

$\dfrac{x^2}{a^2} + \dfrac{y^2}{b^2} = 1$ (the ellipse)

$\dfrac{x^2}{a^2} - \dfrac{y^2}{b^2} = 1$ (the hyperbola)

☐ We have transformed polar equations into cartesian equations using the relationships

$$x = r \cos \theta \qquad y = r \sin \theta$$

EXERCISES

1 Establish the following identities:

a $\cos 3\theta = \cos\theta\,(\cos 2\theta - 2\sin^2\theta)$

b $\sin 4\theta = 4(\sin\theta\cos^3\theta - \cos\theta\sin^3\theta)$

c $\tan\theta + \cot\theta = \sec\theta\,\mathrm{cosec}\,\theta$

d $\cos 2\theta + \sin 2\theta = (\cos\theta + \sin\theta)^2 - 2\sin^2\theta$

e $\tan 3\theta = \dfrac{\tan\theta\,(3 - \tan^2\theta)}{(1 - \tan^2\theta)(1 - \tan\theta\tan 2\theta)}$

2 Solve the following equations in the interval $[0, 2\pi)$:

a $\sin 3\theta = 1$

b $\tan 4\theta = -1$

c $\cos\theta + \sin\theta = 1$

d $\cot^2\theta = 1$

e $\cos\theta = 2\cos^2\theta - 1$

f $\tan 3\theta = \dfrac{2\tan\theta}{1 - \tan^2\theta}$

3 Express in the form $R\cos(\theta - \alpha)$

a $\cos\theta + 2\sqrt{2}\sin\theta$

b $3\sin\theta - \sqrt{7}\cos\theta$

c $4\cos\theta - 3\sin\theta$

4 Identify each of the following curves:

a $(x + 4)^2 + (y + 3)^2 = 8x + 6y + 50$

b $\dfrac{1}{x} + \dfrac{5}{y} + \dfrac{10}{xy} = 1$

c $(y - 1)^2 - (x - 1)^2 = 2x - 1$

d $x^2 + 4y = y^2 + 2x + 19$

5 Identify these polar equations by transforming them to cartesian form:

a $\dfrac{\left(r + \dfrac{1}{r}\right)}{2} = \cos\theta - \sin\theta$

b $r^2\cos 2\theta + 2r(\sin\theta - 2\cos\theta) = 1$

c $r^2(1 + \cos^2\theta) = 4$

d $r\sin 2\theta + 2\sin\theta - 2\cos\theta = \dfrac{10}{r}$

6 Determine the equation of each of the following:

a the straight line through $(-1, 2)$ with slope 3

b the straight line through $(1, -4)$ with slope -5

c the straight line through $(2, 5)$ and $(-4, 6)$

d the straight line through $(-1, 4)$ and $(3, 2)$

e the straight line with x intercept -3 and y intercept 5

f the straight line with x intercept 2 and y intercept -3

g the straight line with slope 3 and y intercept -5

h the straight line with slope -2 and y intercept 4

i the circle centre $(1, 2)$ with radius 4
j the circle centre $(2, -3)$ with radius 5

7 For each of the following straight lines determine the slope, the x intercept and the y intercept:
a $x + 4y = 12$
b $2x + 3y + 6 = 0$
c $2(x + 3) + 5(y - 2) = 7$
d $4(x - 2) + 3(y + 1) = 9$

8 For each of the following circles determine the centre and the radius:
a $x^2 + y^2 + 4x + 6y + 9 = 0$
b $x^2 + y^2 + 6x + 8y + 21 = 0$
c $x^2 + y^2 - 2x + 4y - 4 = 0$
d $(x - y)^2 + (x + y)^2 = 12x + 4y + 30$

ASSIGNMENT

1 Show that

$$\frac{\cos \theta}{1 + \sin \theta} + \frac{1 + \sin \theta}{\cos \theta} = \frac{2}{\cos \theta}$$

2 Use the expansion formula for $\sin (A + B)$ to express $\sin 3\theta$ entirely in terms of $\sin \theta$. Hence, or otherwise, solve the equation

$$6 - 8 \sin^2 \theta = \operatorname{cosec} \theta$$

3 Obtain all solutions in the interval $[0, 2\pi]$ of the equation

$$2 \cos^3 \theta + \cos 2\theta = \cos \theta$$

4 Obtain the general solution of the equation

$$\sin 4\theta + 2 \sin 2\theta + 2 \sin^2 \theta = 2$$

5 Simplify and thereby identify each of the following equations as curves in the cartesian coordinate system:
a $(2x + y)^2 + (x - 2y)^2 = 16$
b $(x + y)(x + 5y) - 6x(y - 5) = 0$
c $(y + x)^2 = x(2y + 1) + 16$
d $(y + x)^2 = x(2y + x) + 16$
e $(3x - y)(x + y - 1)(x - y + 2) = 0$
In each case give a rough sketch.

6 Express in the form $R \cos (\theta - \alpha)$
a $\cos \theta + \sin \theta$
b $\sin \theta + \sqrt{3} \cos \theta$

7 By expressing the polar equation

$$r^2(1 - 7\cos 2\theta) = 10$$

in cartesian form, or otherwise, identify the curve.

FURTHER EXERCISES

1 Establish the following identities:
a $\sin\theta/(1 + \cos\theta) = \tan(\theta/2)$
b $(1 + \cos 2\theta)/(1 - \cos 2\theta) = \cot^2\theta$
c $\cot\theta - \tan\theta = 2\cot 2\theta$
d $\operatorname{cosec} 2\theta - \cot 2\theta = \tan\theta$

2 Solve each of the following equations to obtain $\theta \in \mathbb{R}$:
a $\sec^2\theta = 1 + \tan\theta$
b $\tan^4\theta = 9$
c $1 + \sin\theta + \sin^2\theta = 0$
d $2 - \cos\theta + 2\cos^2\theta - \cos^3\theta = 0$
e $1 + \sin\theta + \cos\theta = 0$

3 Show that

$$\frac{\cos 2n\theta + \sin 2(n+1)\theta}{\cos(2n+1)\theta + \sin(2n+1)\theta} = \frac{\cos 2(n-1)\theta + \sin 2n\theta}{\cos(2n-1)\theta + \sin(2n-1)\theta}$$

Hence or otherwise show that

$$\frac{\cos 12\theta + \sin 14\theta}{\cos 13\theta + \sin 13\theta} = \frac{1 + \sin 2\theta}{\cos\theta + \sin\theta}$$

4 Show that the equation of the chord joining two points (a_1, b_1) and (a_2, b_2) on the rectangular hyperbola $xy = c^2$ is

$$\frac{x}{a_1 + a_2} + \frac{y}{b_1 + b_2} = 1$$

5 Show that

$$\frac{\sin\theta + \sin 3\theta + \sin 5\theta + \sin 7\theta}{\cos\theta + \cos 3\theta + \cos 5\theta + \cos 7\theta} = \tan 4\theta$$

6 A surveyor stands on the same horizontal level as a television mast at a distance d from its base. The angle of elevation of a point P on the mast is θ and the angle of elevation of the top of the mast is ϕ. Show that the distance from the point P to the top of the mast is $d\sin(\phi - \theta)/\cos\theta\cos\phi$.

7 A simple pendulum of length L swings so that it subtends an angle θ with the vertical. Show that the height of the pendulum bob above its lowest position is $2L\sin^2(\theta/2)$.

8 Obtain the axis of symmetry and the position of the focus of the conic $2y^2 = x + 4y$.

9 A symmetrical parabolic arch has a span of 24 metres and a height of 20 metres. Determine the height of the arch at a distance 3 metres from the axis of symmetry.

10 A symmetrical road bridge has the shape of half an ellipse. Its span is 30 metres and its height is 20 metres. Determine the height at a distance of 12 metres from the axis of symmetry.

11 A symmetrical parabolic bridge has a height of 4 metres and a span of 8 metres. A vehicle is 4 metres broad and has a height just over 3 metres. Can the vehicle pass under the bridge? Determine the maximum head height which a vehicle 3 metres wide can have to pass under the bridge without contact.

12 A beam of length l lies in a vertical plane and rests against a cylindrical drum of radius a which is lying on its side. The foot of the beam is a distance x from the point of contact of the cylinder and the ground. Calculate the height of the top of the beam above the ground.

13 In a plane representing an electric field, O and A denote the cross-sections of charged wires. The distance OA is 8 units. When the point P moves in this plane in such a way that $OP = 3AP$ then P moves on an equipotential surface. Show that the equipotential surface is a circle of radius 3 units.

14 Two rods AB and AC of length p and q $(p > q)$ respectively are jointed together at one end A. The other ends, B and C, are secured to a wall with B vertically above C so that the distance BC is h. If C is moved a distance x down the wall away from B, show that A drops or rises by an amount $x/2 - (p^2 - q^2)x/2h(h + x)$.

15 When a surveyor is at a radial distance r from a church spire, the angle of elevation of the base is θ and the angle of elevation of the top is $\theta + \alpha$. Obtain $\tan \beta$ where β is the difference in the angles of elevation between the top and the bottom if the surveyor is at a distance $r + t$.

16 A cliff of height h above sea level is being eroded by wind and sea. A surveyor stands on the edge and finds that the angle of declination of a rock on the horizon is θ. One year later the height of the cliff is unchanged but the angle of declination of the rock has been decreased by α. Ignoring the curvature of the earth determine the distance d that the cliff has been eroded.

17 A symmetrical arch of height h is in the shape of a parabola and its span at ground level is $2r$. A ladder rests tangentially against the arch, in its plane, in such a way that the top of the ladder just touches the arch. If the foot of the ladder is at a horizontal distance d from the centre of the arch, determine the length l of the ladder in terms of h, r if $d = 5r/4$.

Limits, continuity and differentiation 4

Now that we have acquired the basic algebraic and geometrical tools that we need, we can begin to develop the calculus.

After completing this chapter you should be able to

- ☐ Evaluate simple limits using the laws of limits;
- ☐ Decide, in simple cases, whether a function is continuous or not;
- ☐ Perform the processes of elementary differentiation;
- ☐ Obtain higher-order derivatives of a product using Leibniz's theorem;
- ☐ Apply differentiation to calculate rates of change.

At the end of this chapter we shall solve practical problems concerning cylinder pressure and the seepage of water into soil.

4.1 LIMITS

One of the most important concepts in mathematics and therefore in its applications is that of a **limit**. We are often concerned with the long-term effects of things, or with what is likely to happen at a point of crisis – profitability, state of health, buoyancy, or stability of a structure. Such considerations often involve a limiting process.

We shall meet this idea in several ways. In the first instance we consider the limit of a function at a *point*. Suppose $y = f(x)$ has the property that $f(x)$ can be arbitrarily close to l just by choosing x ($\neq a$) sufficiently close to a. If so, we say that $f(x)$ tends to a limit l as x tends to a, and write

$$f(x) \to l \quad \text{as} \quad x \to a$$

Alternatively we say that f has a limit l at a, and write

$$l = \lim_{x \to a} f(x)$$

We do not insist that $f(a)$ is defined, or, if it is defined, that its value shall be equal to l. In other words, the point a need not be in the domain of the function and, even if it is, the value at the point a need not be l. Indeed we are not interested at all in what happens at $x = a$; we are interested only in what happens when x is *near* a.

□ Suppose

$$y = f(x) = \frac{x^2 - 3x + 2}{x - 1}$$

then the domain of this function consists of all real numbers other than $x = 1$; so $f(x)$ is not defined when $x = 1$. On the other hand, when $x \neq 1$ we may simplify the expression for y to

$$y = \frac{(x - 2)(x - 1)}{x - 1} = x - 2$$

The function is shown in Fig. 4.1: we use a hollow circle to represent a missing point. Now $f(x)$ can be made arbitrarily close to -1 just by choosing x sufficiently near to 1. Therefore

$$\lim_{x \to 1} f(x) = \lim_{x \to 1} (x - 2) = -1$$

However, we cannot make $f(x)$ equal to -1 because $f(x)$ is not defined when x is equal to 1. ■

□ Obtain $\lim_{x \to a} f(x)$ in each of the following cases:

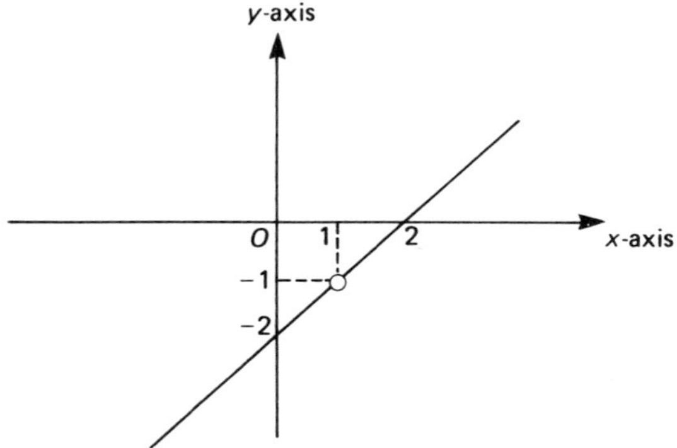

Fig. 4.1 The graph of f.

a $f(x) = \dfrac{x^2 - 4x + 3}{x^2 - 2x - 3} \qquad a = 3$

b $f(x) = \dfrac{e^{2x} - 1}{e^x - e^{-x}} \qquad a = 0$

a We have

$$f(x) = \frac{(x - 3)(x - 1)}{(x - 3)(x + 1)}$$

so that when $x \neq 3$

$$f(x) = \frac{x - 1}{x + 1}$$

Therefore as $x \to 3$,

$$f(x) \to \frac{3 - 1}{3 + 1} = \frac{2}{4} = \frac{1}{2}$$

b We have

$$f(x) = \frac{e^{2x} - 1}{e^x - e^{-x}}$$

If we try to put $x = 0$ straight away we get

$$f(x) \to \frac{0}{0}$$

which is undefined. Therefore we must be more subtle. If we multiply numerator and denominator by e^x (which is always non-zero) we obtain

$$f(x) = \frac{e^x(e^{2x} - 1)}{e^{2x} - 1} = e^x$$

provided $x \neq 0$. So that as $x \to 0$, $f(x) \to e^0 = 1$. ■

4.2 THE LAWS OF LIMITS

The following rules are often known as **the laws of limits**:

1 $\lim\limits_{x \to a} [f(x) + g(x)] = \lim\limits_{x \to a} f(x) + \lim\limits_{x \to a} g(x)$

2 $\lim\limits_{x \to a} [kf(x)] = k \lim\limits_{x \to a} f(x) \qquad k \in \mathbb{R}$

3 $\lim\limits_{x \to a} [f(x)\, g(x)] = \lim\limits_{x \to a} f(x) \lim\limits_{x \to a} g(x)$

4 If $\lim\limits_{x \to a} g(x) \neq 0$, then $\lim\limits_{x \to a} [f(x)/g(x)] = \left[\lim\limits_{x \to a} f(x)\right] \Big/ \left[\lim\limits_{x \to a} g(x)\right]$

These rules are to be interpreted carefully in the following way. If the right-hand side exists then the left-hand side exists and the two are equal. If the right-hand side does not exist, then the rule cannot be applied.

□ From the graphs of the circular functions (Figs 3.4, 3.5) it is clear that

$$\lim_{x \to 0} \sin x = 0$$

$$\lim_{x \to 0} \cos x = 1$$

$$\lim_{x \to 0} \tan x = 0$$

Obtain

a $\displaystyle\lim_{x \to 0} \frac{\sin x - 1}{2 \cos x}$

b $\displaystyle\lim_{x \to 0} \frac{\cos x - 1}{2 \sin x}$

a Using the laws of limits,

$$\lim_{x \to 0} \frac{\sin x - 1}{2 \cos x} = \frac{\lim_{x \to 0} (\sin x - 1)}{\lim_{x \to 0} (2 \cos x)}$$

$$= \frac{\lim_{x \to 0} (\sin x - 1)}{2 \lim_{x \to 0} (\cos x)} = \frac{0 - 1}{2 \times 1} = -\frac{1}{2}$$

Here the procedure is justified by the result; if you like, the end justifies the means! However, if the application of the rules produces at any stage an expression which is meaningless, we shall need to think again!

b If we go straight into the laws of limits we shall meet a problem:

$$\lim_{x \to 0} \frac{\cos x - 1}{2 \sin x} = \frac{\lim_{x \to 0} (\cos x - 1)}{\lim_{x \to 0} (2 \sin x)}$$

$$= \frac{\lim_{x \to 0} (\cos x) - 1}{2 \lim_{x \to 0} (\sin x)} = \frac{1 - 1}{2 \times 0} = \frac{0}{0}$$

At each stage we were able to carry out the simplification only on the understanding that the expression which resulted would be meaningful. However, 0/0 is indeterminate and so the procedure fails. The problem must be tackled differently. One way to sort things out is to try to express $f(x)$ in an alternative form when $x \neq 0$:

$$f(x) = \frac{\cos x - 1}{2 \sin x} = \frac{(\cos x - 1)(\cos x + 1)}{2 \sin x(\cos x + 1)}$$

$$= \frac{\cos^2 x - 1}{2\sin x(\cos x + 1)} = \frac{-\sin^2 x}{2\sin x(\cos x + 1)}$$

$$= \frac{-\sin x}{2(\cos x + 1)} \qquad \text{provided } x \neq 0$$

Therefore

$$\lim_{x\to 0} f(x) = \lim_{x\to 0} \frac{-\sin x}{2(\cos x + 1)}$$

$$= \frac{-\lim_{x\to 0} \sin x}{2\lim_{x\to 0} \cos x + 2} = \frac{0}{2 + 2} = 0$$

■

4.3 Workshop

1 **Exercise** Here are two limits for you to try; they are very similar to the ones we have just done.

a $\lim_{x\to 0} \frac{(\sin x - 1)\tan x}{2\cos x}$

b $\lim_{x\to 0} \frac{(\cos x - 1)\cot x}{2\sin x}$

When you have completed **a**, take the next step and see if you are right.

---

2 For **a** we have

$$\lim_{x\to 0} \frac{(\sin x - 1)\tan x}{2\cos x} = \frac{\lim_{x\to 0}(\sin x - 1)\tan x}{\lim_{x\to 0}(2\cos x)}$$

$$= \frac{\lim_{x\to 0}(\sin x - 1)\lim_{x\to 0}\tan x}{2\lim_{x\to 0}\cos x}$$

$$= \frac{-1 \times 0}{2 \times 1} = 0$$

How did you get on with that? If you made a mistake, look carefully at the laws of limits and see how they are applied. Problem **b**, like the example we have just done, requires some work before we take limits. Try it and see how it goes; then step ahead.

---

3 In **b** we cannot apply the laws of limits directly because the result is meaningless. However,

$$\frac{(\cos x - 1)\cot x}{2\sin x} = \frac{(\cos x - 1)\cos x}{2\sin^2 x}$$
$$= \frac{(\cos x - 1)\cos x}{2(1 - \cos^2 x)}$$
$$= \frac{(\cos x - 1)\cos x}{2(1 - \cos x)(1 + \cos x)}$$
$$= \frac{-\cos x}{2(1 + \cos x)}$$

Therefore

$$\lim_{x\to 0}\frac{(\cos x - 1)\cot x}{2\sin x} = \lim_{x\to 0}\frac{-\cos x}{2(1 + \cos x)}$$
$$= \frac{-1}{2(1 + 1)} = -\frac{1}{4}$$

Did you manage that?

The important fact that we need to remember about the limit of a function is that we are not at all concerned with the values of the function at the point. We are only interested in the values of the function *near* the point. In Fig. 4.2

$$\lim_{x\to a} f(x) = l \neq f(a)$$

Although the 'limit l of $f(x)$ as x tends to a' is only meaningful if l is a real number, we shall allow a slight extension of the notation. It is convenient but slightly absurd to write

$$\lim_{x\to a} f(x) = \infty$$

provided $f(x)$ can be made arbitrarily large just by choosing x sufficiently

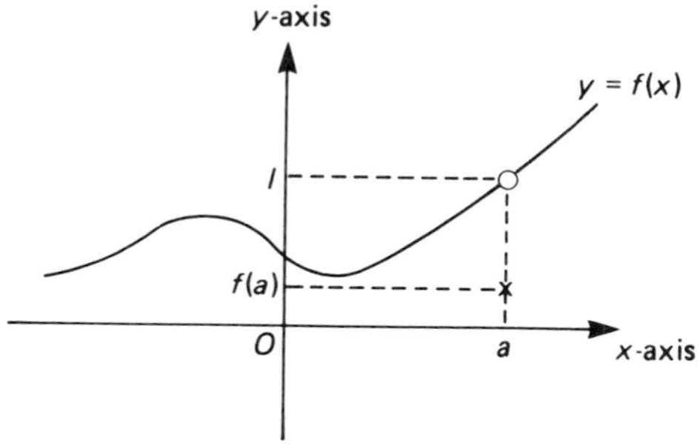

Fig. 4.2 The limit of a function.

close to a ($x \neq a$). That is, $f(x)$ can be made larger than any pre-assigned real number merely by choosing x close enough to the point a. Similarly,

$$\lim_{x \to a} f(x) = -\infty$$

means that $f(x)$ can be made less than any pre-assigned number merely by choosing x sufficiently close to a ($x \neq a$).

□ We have

$$\lim_{x \to 0} \operatorname{cosec}^2 x = \infty$$

The notation is slightly misleading because of course there is no limit! ■

Likewise we write

$$\lim_{x \to \infty} f(x) = l$$

if $f(x)$ can be made arbitrarily close to l merely by choosing x sufficiently large, and

$$\lim_{x \to -\infty} f(x) = l$$

if $f(x)$ can be made arbitrarily close to l merely by choosing the magnitude of x sufficiently large, where $x < 0$.

4.4 RIGHT AND LEFT LIMITS

We can extend the idea of a limit in a number of ways. One way, which is quite useful for applications, arises when $f(x)$ can be made arbitrarily close to r by choosing x ($>a$) sufficiently close to a. Here we are considering values of x greater than a, and so the limit is obtained as we approach the point a from the right-hand side. We call it a **right-hand limit** (Fig. 4.3). We write

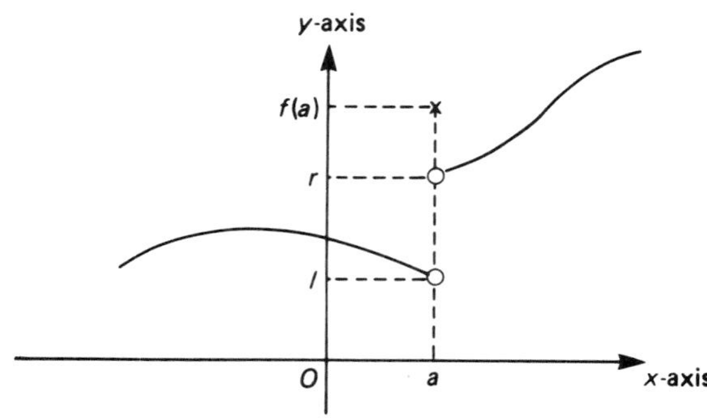

Fig. 4.3 Right-hand and left-hand limits.

$$\lim_{x \to a+} f(x) = r$$

Similarly, if $f(x)$ can be made arbitrarily close to l by choosing x ($<a$) sufficiently close to a, we have a **left-hand limit** (Fig. 4.3). We write

$$\lim_{x \to a-} f(x) = l$$

If

$$\lim_{x \to a+} f(x) = \lim_{x \to a-} f(x) = k$$

then

$$\lim_{x \to a} f(x) = k$$

INEQUALITIES

There is one further property of limits which we shall find particularly useful later and which we now describe briefly. It enables us to compare limits by comparing the functions which give rise to them.

Suppose

1 For all x in some open interval containing the point a, $0 \leqslant f(x) \leqslant g(x)$;
2 Both $\lim_{x \to a} f(x)$ and $\lim_{x \to a} g(x)$ exist.

Then

$$0 \leqslant \lim_{x \to a} f(x) \leqslant \lim_{x \to a} g(x)$$

It must be stressed that both requirements must be met before we assert confidently that one limit is bounded above by another. Here are the conditions in words:

1 Each function must be positive, and one must be greater than the other;
2 Both limits must exist.

An analogous property holds for right-hand and left-hand limits.

4.5 CONTINUITY

Most of the functions which we have met in our mathematical work have the property

$$\lim_{x \to a} f(x) = f(a)$$

This in effect says that the: e are no breaks in the graph of the function. Specifically, if for some point a we have

$$\lim_{x \to a} f(x) = f(a)$$

then the function f is said to be **continuous** at a. Moreover, if the function f is continuous at all points of its domain we say it is a **continuous function**.

Intuitively, then, a continuous function has its graph all in one piece. However, this statement can be a little misleading.

□ $y = \tan x$ is defined whenever x is not an odd multiple of $\pi/2$. It is continuous at all points where it is defined, and so is a continuous function. However, the graph is certainly not in one piece (Fig. 4.4). ■

The function in Fig. 4.5, although satisfying the requirements of a

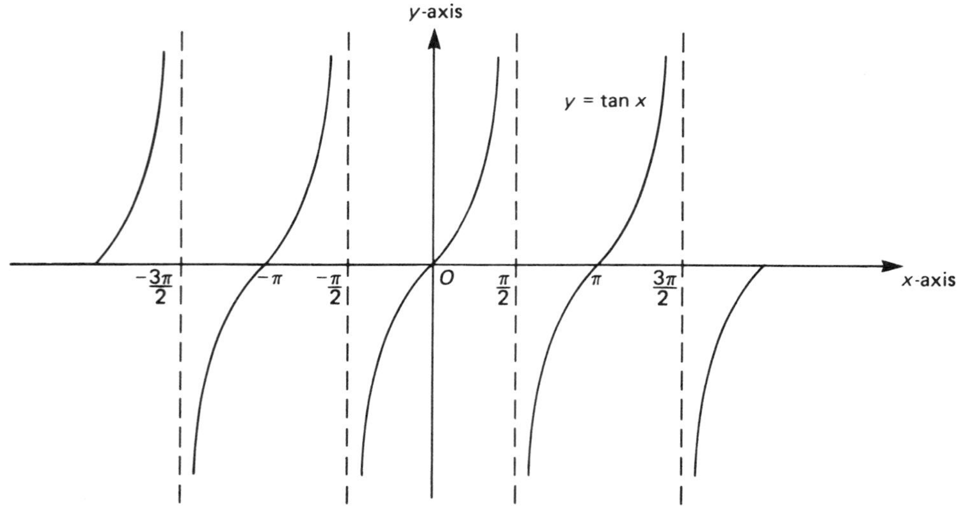

Fig. 4.4 The tangent function.

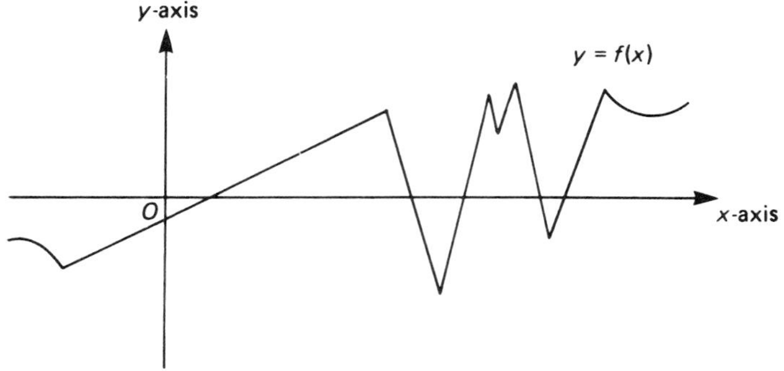

Fig. 4.5 A continuous function.

continuous function, is not the kind of function with which we are familiar. It isn't smooth, and there are several points at which it is impossible to draw a tangent.

4.6 DIFFERENTIABILITY

Suppose $y = f(x)$ is a smooth curve and that x determines a general point P on it (Fig. 4.6). Suppose also that h is small and Q corresponds to $x = a + h$. Using the notation shown in the diagram, the slope of the chord PQ is given by

$$\text{slope } PQ = \frac{QR}{PR} = \frac{f(a+h) - f(a)}{(a+h) - a} = \frac{f(a+h) - f(a)}{h}$$

Suppose now we consider what happens as h is made small ($h \neq 0$). Q moves closer to P, and intuitively the slope of the chord PQ becomes arbitrarily close to the slope of the tangent at P. So

$$\text{slope of tangent at } P = \lim_{h \to 0} \frac{f(x+h) - f(x)}{h}$$

If this limit exists then the function f is said to be differentiable at the point a. If f is differentiable at all its points then f is said to be a **differentiable function**. We write

$$\frac{dy}{dx} = \lim_{h \to 0} \frac{f(x+h) - f(x)}{h}$$

We call this the **derivative** of f at x, and represent it by $f'(x)$. The process by which $f'(x)$ is calculated from $f(x)$ is called **differentiation** with respect to x.

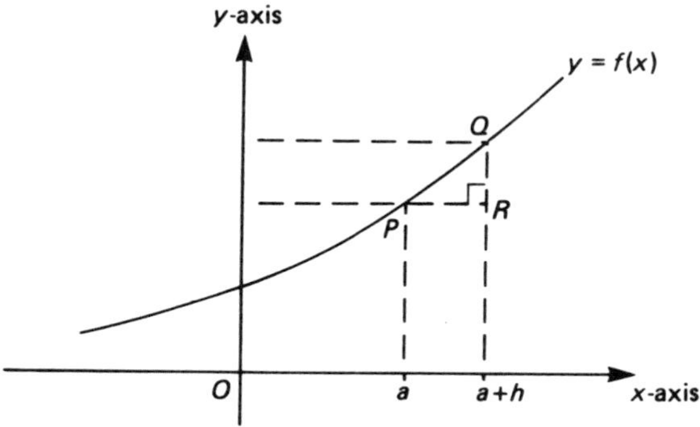

Fig. 4.6 Two neighbouring points on a smooth curve.

DIFFERENTIALS

Although we have used the notation dy/dx for the derivative of f at x, we have given no meaning to dy and dx which would enable them to be used separately. It is important to appreciate that the derivative is a *limit*; we may write $\delta x = h$ and $\delta y = f(x+h) - f(x)$ so that

$$\frac{dy}{dx} = \lim_{\delta x \to 0} \frac{\delta y}{\delta x}$$

but once the limit has been taken dy and dx become welded together and cannot be separated.

Nevertheless it is convenient to have an interpretation for dy and dx so that they can be used separately and are consistent with this definition of a derivative. Accordingly we define dx to be any change in x (not necessarily small) and define dy by the formula $dy = f'(x)\,dx$. This is consistent with the definition of a derivative because when $dx \neq 0$ we have $dy/dx = f'(x)$ as before. When dx and dy are used in this way they are called **differentials**. Note that if dx is a change in x, dy is not the corresponding change in y. However, if dx is numerically small then dy does *approximate* to the corresponding change in y.

RULES

The laws of limits enable us to deduce **rules for differentiation**. In what follows it will be supposed that $u = u(x)$ and $v = v(x)$ can be differentiated with respect to x and that k is a real constant.

1 $\dfrac{d}{dx}(u + v) = \dfrac{du}{dx} + \dfrac{dv}{dx}$ (the sum rule)

2 $\dfrac{d}{dx}(ku) = k\dfrac{du}{dx}$ (the factor rule)

3 $\dfrac{d}{dx}(uv) = u\dfrac{dv}{dx} + v\dfrac{du}{dx}$ (the product rule)

4 Suppose $y = y(u)$ and $u = u(x)$ are both differentiable. Then

$\dfrac{dy}{dx} = \dfrac{dy}{du}\dfrac{du}{dx}$ (the chain rule)

These are the basic rules for differentiation and, together with the derivatives of a few functions, they can be used to obtain the derivative of any function you are likely to need. Here is a short list of **derivatives**; if you were cast away on a desert island these would be sufficient for you to deduce all the standard forms:

1 $\dfrac{d}{dx}(x^n) = nx^{n-1}$ $(n \in \mathbb{R})$

2 $\dfrac{d}{dx}(e^x) = e^x$

3 $\dfrac{d}{dx}(\sin x) = \cos x$

Imagine for the moment that we are marooned on a desert island. There is a ship ready but the captain has gone mad and will not rescue us unless we can supply some simple derivatives. Let's try to build up some of the standard forms. See also Table 15.1.

□ Deduce the quotient rule

$$\frac{d}{dx}\left(\frac{u}{v}\right) = \left(v\frac{du}{dx} - u\frac{dv}{dx}\right)\Big/ v^2$$

We use the product rule and the chain rule to achieve the required form:

$$\begin{aligned}
\text{LHS} = \frac{d}{dx}\left(\frac{u}{v}\right) &= \frac{d}{dx}(uv^{-1}) \\
&= u\frac{d}{dx}(v^{-1}) + v^{-1}\frac{d}{dx}(u) && \text{(product rule)} \\
&= u\frac{d}{dv}(v^{-1})\frac{dv}{dx} + v^{-1}\frac{du}{dx} && \text{(chain rule)} \\
&= u(-v^{-2})\frac{dv}{dx} + v^{-1}\frac{du}{dx} && \text{(derivative 1)} \\
&= \left(v\frac{du}{dx} - u\frac{dv}{dx}\right)\Big/ v^2 = \text{RHS} && \text{(elementary algebra)}
\end{aligned}$$

■

□ Use the rules and the standard forms to obtain

a $\dfrac{d}{dx}(\cos x)$

b $\dfrac{d}{dx}(\tan x)$

c $\dfrac{d}{dx}(\sec x)$

d $\dfrac{d}{dx}(\ln x)$

e $\dfrac{d}{dx}(a^x) \qquad (a > 0)$

a We have $\cos x = \sin(\pi/2 - x)$. So

$$\frac{\mathrm{d}}{\mathrm{d}x}(\cos x) = \frac{\mathrm{d}}{\mathrm{d}x}[\sin(\pi/2 - x)]$$

Put $u = \pi/2 - x$; then

$$\begin{aligned}
\frac{\mathrm{d}}{\mathrm{d}x}(\cos x) &= \frac{\mathrm{d}}{\mathrm{d}x}(\sin u) \\
&= \frac{\mathrm{d}}{\mathrm{d}u}(\sin u)\,\frac{\mathrm{d}u}{\mathrm{d}x} && \text{(chain rule)} \\
&= \cos u \frac{\mathrm{d}}{\mathrm{d}x}(\pi/2 - x) && \text{(derivative 3)} \\
&= \cos u \left[\frac{\pi}{2}\frac{\mathrm{d}}{\mathrm{d}x}(x^0) - \frac{\mathrm{d}}{\mathrm{d}x}(x)\right] \\
&= \cos u\,(0 - 1) = -\cos(\pi/2 - x) = -\sin x
\end{aligned}$$

b We have

$$\begin{aligned}
\frac{\mathrm{d}}{\mathrm{d}x}(\tan x) &= \frac{\mathrm{d}}{\mathrm{d}x}\left(\frac{\sin x}{\cos x}\right) \\
&= \left[\cos x \frac{\mathrm{d}}{\mathrm{d}x}(\sin x) - \sin x \frac{\mathrm{d}}{\mathrm{d}x}(\cos x)\right] \Big/ \cos^2 x \\
&= \frac{\cos^2 x + \sin^2 x}{\cos^2 x} = \frac{1}{\cos^2 x} = \sec^2 x
\end{aligned}$$

c We have

$$\frac{\mathrm{d}}{\mathrm{d}x}(\sec x) = \frac{\mathrm{d}}{\mathrm{d}x}[(\cos x)^{-1}]$$

Put $u = \cos x$. Then

$$\begin{aligned}
\frac{\mathrm{d}}{\mathrm{d}x}(u^{-1}) &= \frac{\mathrm{d}}{\mathrm{d}u}(u^{-1})\,\frac{\mathrm{d}u}{\mathrm{d}x} \\
&= -u^{-2}\frac{\mathrm{d}}{\mathrm{d}x}(\cos x) \\
&= \frac{-1}{\cos^2 x}(-\sin x) = \frac{\sin x}{\cos^2 x} = \sec x \tan x
\end{aligned}$$

d Let $y = \ln x$. Then $x = \mathrm{e}^y$. So

$$\frac{\mathrm{d}}{\mathrm{d}x}(x) = \frac{\mathrm{d}}{\mathrm{d}x}(\mathrm{e}^y) = \frac{\mathrm{d}}{\mathrm{d}y}(\mathrm{e}^y)\,\frac{\mathrm{d}y}{\mathrm{d}x}$$

Therefore

$$1 = \mathrm{e}^y \frac{\mathrm{d}y}{\mathrm{d}x} = x \frac{\mathrm{d}y}{\mathrm{d}x}$$

$$\frac{dy}{dx} = \frac{1}{x}$$

e Let $y = a^x$. Then $\ln y = \ln a^x = x \ln a$. So

$$\frac{d}{dx}(\ln y) = \frac{d}{dx}(x \ln a)$$

Therefore

$$\frac{d}{dy}(\ln y)\frac{dy}{dx} = \ln a$$

So

$$\frac{1}{y}\frac{dy}{dx} = \ln a$$

$$\frac{dy}{dx} = (\ln a)\, a^x$$ ■

If $y = f(x)$ then $dy/dx = f'(x)$, and we say that $f(x)$ has been differentiated with respect to x. We may consider differentiating again with respect to x, and so we define the **higher-order derivatives**:

$$\frac{d^2y}{dx^2} = \frac{d}{dx}\left(\frac{dy}{dx}\right) = f''(x) = f^{(2)}(x)$$

$$\frac{d^3y}{dx^3} = \frac{d}{dx}\left(\frac{d^2y}{dx^2}\right) = f^{(3)}(x)$$

In general,

$$\frac{d^ny}{dx^n} = f^{(n)}(x)$$

is the result of differentiating n times with respect to x, and is known as the nth-order derivative of $f(x)$ with respect to x.

4.7 LEIBNIZ'S THEOREM

One rule which generalizes to higher-order derivatives is the product rule. The first few derivatives will establish the pattern. For the purposes of this section only we shall use a special subscript notation y_n to represent d^ny/dx^n.

The product rule is

$$\frac{d}{dx}(uv) = u\frac{dv}{dx} + \frac{du}{dx}v$$

so that

$$(uv)_1 = uv_1 + u_1 v$$

Now

$$\frac{d^2}{dx^2}(uv) = \frac{d}{dx}\left(u\frac{dv}{dx} + \frac{du}{dx}v\right)$$
$$= \left(u\frac{d^2v}{dx^2} + \frac{du}{dx}\frac{dv}{dx}\right) + \left(\frac{du}{dx}\frac{dv}{dx} + \frac{d^2u}{dx^2}v\right)$$

(using the product rule again)

$$= u\frac{d^2v}{dx^2} + 2\frac{du}{dx}\frac{dv}{dx} + \frac{d^2u}{dx^2}v$$

So

$$(uv)_2 = uv_2 + 2u_1v_1 + u_2v$$

□ Show by applying the product rule yet again that

$$(uv)_3 = uv_3 + 3u_1v_2 + 3u_2v_1 + u_3v$$

When you have managed this, read on. ■

Here is the pattern which is emerging as we apply the product rule repeatedly. Look at it and see if it reminds you of anything:

$$\begin{aligned}(uv)_1 &= uv_1 + u_1v \\ (uv)_2 &= uv_2 + 2u_1v_1 + u_2v \\ (uv)_3 &= uv_3 + 3u_1v_2 + 3u_2v_1 + u_3v\end{aligned}$$

Look at the coefficients. Yes! They are the binomial coefficients. Remember

$$\binom{n}{r} = \frac{n!}{(n-r)!\, r!} \qquad (r \in \mathbb{N}_0,\ n \in \mathbb{N}_0;\ r \leqslant n)$$

So if we put them in we obtain

$$(uv)_1 = \binom{1}{0} uv_1 + \binom{1}{1} u_1v$$

$$(uv)_2 = \binom{2}{0} uv_2 + \binom{2}{1} u_1v_1 + \binom{2}{2} u_2v$$

$$(uv)_3 = \binom{3}{0} uv_3 + \binom{3}{1} u_1v_2 + \binom{3}{2} u_2v_1 + \binom{3}{3} u_3v$$

In general

$$(uv)_n = \binom{n}{0} u_0v_n + \binom{n}{1} u_1v_{n-1} + \binom{n}{2} u_2v_{n-2}$$

$$+ \ldots + \binom{n}{r} u_rv_{n-r} + \ldots + \binom{n}{n} u_nv_0 \qquad (n \in \mathbb{N})$$

$$= \sum_{r=0}^{n} \binom{n}{r} u_rv_{n-r} \qquad \text{where } u_0 = u,\ v_0 = v$$

Remember that the summation sign simply tells us to let r take on all integer values between 0 and n and then add up the results.

This formula, which enables us to differentiate a product n times, is known as **Leibniz's theorem**. It can be proved by mathematical induction.

□ If $y = f(x) = x^3e^{2x}$, obtain $f^{(n)}(x)$.

Here we have a product, and so we use Leibniz's theorem. We must decide which factor to designate as u and which as v. The expansion will terminate after a few terms if we put $u = x^3$ because after differentiating u three times with respect to x the result is zero:

$$u = x^3, \qquad u_1 = 3x^2, \qquad u_2 = 6x, \qquad u_3 = 6$$

Now $v = e^{2x}$, so $v_1 = 2e^{2x}$, $v_2 = 2^2e^{2x}$, and in general $v_n = 2^ne^{2x}$. Therefore

$$(uv)_n = u_0v_n + \binom{n}{1} u_1v_{n-1} + \binom{n}{2} u_2v_{n-2} + \ldots$$

$$= x^3\,2^n\,e^{2x} + \binom{n}{1} (3x^2)\,(2^{n-1}\,e^{2x})$$

$$+ \binom{n}{2} (6x)\,(2^{n-2}\,e^{2x}) + \binom{n}{3} (6)\,(2^{n-3}\,e^{2x})$$

$$= x^3\,2^n\,e^{2x} + n(3x^2)\,(2^{n-1}\,e^{2x})$$

$$+ \frac{n(n-1)}{2} 6x\,2^{n-2}\,e^{2x} + \frac{n(n-1)\,(n-2)}{1 \times 2 \times 3} 6\,2^{n-3}\,e^{2x}$$

$$= [x^3\,2^n + 3nx^2\,2^{n-1} + 3n(n-1)x\,2^{n-2} + n(n-1)\,(n-2)2^{n-3}]\,e^{2x}$$

■

□ Obtain the nth derivative of x^2y with respect to x, where $y = y(x)$.

Here we take $u = x^2$, so $u_1 = 2x$, $u_2 = 2$ and $v = y$. So

$$(x^2y)_n = x^2y_n + \binom{n}{1} 2xy_{n-1} + \binom{n}{2} 2y_{n-2}$$

$$= x^2y_n + 2nxy_{n-1} + n(n-1)\,y_{n-2}$$

■

□ If

$$x\frac{dy}{dx} + y = x^2$$

show that

$$x\frac{d^{n+1}y}{dx^{n+1}} + (n+1)\frac{d^n y}{dx^n} = 0 \qquad \text{for } n \geqslant 3$$

We differentiate each side of the equation with respect to x using Leibniz's theorem:

$$\left(x\frac{dy}{dx}\right)_n = (xy_1)_n$$

$$= xy_{n+1} + \binom{n}{1} 1y_n = xy_{n+1} + ny_n$$

Now

$$(x^2)_1 = 2x,\ (x^2)_2 = 2,\ (x^2)_n = 0 \qquad (n \geqslant 3)$$

Consequently for $n \geqslant 3$ we have

$$xy_{n+1} + ny_n + y_n = 0$$

So

$$x\frac{d^{n+1}y}{dx^{n+1}} + (n+1)\frac{d^n y}{dx^n} = 0 \qquad \text{for } n \geqslant 3$$ ■

4.8 TECHNIQUES OF DIFFERENTIATION

Do not forget that very often if you pause and think for a few moments you can save yourself a lot of needless work. This is particularly true when it comes to differentiation, where a little algebraic simplification at the outset can make things very much easier.

□ Differentiate with respect to x

$$\left[\frac{1 - \sin x}{1 + \sin x}\right]^{1/2}$$

where $x \in (-\pi/2, \pi/2)$.

We could of course hit this head on and give it the full works, differentiating using the chain rule and the quotient rule. Instead we shall tame it first by multiplying numerator and denominator by $1 - \sin x$ inside the root. Algebraically this leaves everything the same, but from our point of view it will help greatly. So

$$\left[\frac{1 - \sin x}{1 + \sin x}\frac{(1 - \sin x)}{(1 - \sin x)}\right]^{1/2} = \left[\frac{(1 - \sin x)\,(1 - \sin x)}{(1 + \sin x)\,(1 - \sin x)}\right]^{1/2}$$

$$= \left[\frac{(1 - \sin x)^2}{(1 - \sin^2 x)}\right]^{1/2}$$

$$= \left[\frac{(1 - \sin x)^2}{\cos^2 x}\right]^{1/2}$$

$$= \left[\frac{(1 - \sin x)}{\cos x}\right]$$

$$= \sec x - \tan x$$

After all that excitement we mustn't forget to differentiate:

$$\frac{dy}{dx} = \sec x \tan x - \sec^2 x$$

Whenever we have a complicated expression to differentiate, it is worth looking to see if it can be simplified algebraically first. ■

Before we consider any further techniques, here are a few steps to get you used to using the chain rule without making a formal substitution.

4.9 Workshop

1

▷**Exercise** Differentiate with respect to t:

a $\ln(2t^2 + 1)$
b $\sin^3 t$
c $\sin 3t$
d $\sin t^3$

Try each one of these. Remember that the idea is to avoid having to write out all the details of a substitution. We differentiate with respect to 'the thing in brackets', then multiply by the derivative of 'the thing in brackets' with respect to t.

2

For **a** we have

$$\frac{d}{dt}[\ln(2t^2 + 1)] = \frac{d}{d(\ )}\ln(2t^2 + 1)\frac{d(\ )}{dt}$$

$$= \frac{1}{2t^2 + 1}\frac{d}{dt}(2t^2 + 1)$$

$$= \frac{4t}{2t^2 + 1}$$

If you made a mistake, check your working for **b**, **c** and **d** before you step ahead for the solutions.

3

For **b** we obtain

$$\begin{aligned}\frac{d}{dt}[\sin^3 t] &= \frac{d}{dt}(\sin t)^3\\ &= \frac{d}{d(\ )}(\sin t)^3\frac{d(\ )}{dt}\\ &= 3\,(\sin t)^2\frac{d}{dt}(\sin t)\\ &= 3\sin^2 t\cos t\end{aligned}$$

If you made a mistake here, possibly you confused $\sin^3 t$ with either $\sin 3t$ or $\sin t^3$. It is important to realize that these are three different expressions.

4 Next, for **c** we have

$$\begin{aligned}\frac{d}{dt}[\sin 3t] &= \frac{d}{d(\ )}\sin(3t)\frac{d}{dt}(3t)\\ &= (\cos 3t)\,3 = 3\cos 3t\end{aligned}$$

5 Finally, for **d** we obtain

$$\begin{aligned}\frac{d}{dt}[\sin t^3] &= \frac{d}{d(\ )}\sin(t^3)\frac{d(\ )}{dt}\\ &= \cos(t^3)\frac{d}{dt}(t^3) = 3t^2\cos t^3\end{aligned}$$

If you managed all those without difficulty you should be able to skip up the steps with ease. For further practice here is another problem.

▷ **Exercise** Differentiate with respect to u:

a $\exp(\sin 2u)$
b $\ln[1 + \sin^2(2u + 1)]$
c $\exp 2u \cos(3u + 1)$

Try them all and check your answers step by step.

6 For **a** we have

$$\begin{aligned}\frac{d}{du}[e^{\sin 2u}] &= \frac{d}{d(\ )}[e^{(\sin 2u)}]\frac{d}{du}(\sin 2u)\\ &= e^{\sin 2u}\frac{d}{du}(\sin 2u)\\ &= e^{\sin 2u}\frac{d}{d(\ )}[\sin(2u)]\frac{d}{du}(2u)\end{aligned}$$

$$= e^{\sin 2u} \cos 2u \; 2$$
$$= 2e^{\sin 2u} \cos 2u$$

Next, for **b** we obtain 7

$$\frac{d}{du} \ln [1 + \sin^2 (2u + 1)]$$

$$= \frac{d}{d[\ ]} \ln [1 + \sin^2 (2u + 1)] \frac{d}{du}[\ ]$$
$$= \frac{1}{1 + \sin^2 (2u + 1)} \frac{d}{du}[1 + \sin^2 (2u + 1)]$$
$$= \frac{1}{1 + \sin^2 (2u + 1)} \frac{d}{du}[\sin (2u + 1)]^2$$
$$= \frac{1}{1 + \sin^2 (2u + 1)} \frac{d}{d[\ ]}[\sin (2u + 1)]^2 \frac{d}{du}[\ ]$$
$$= \frac{1}{1 + \sin^2 (2u + 1)} 2 [\sin (2u + 1)] \frac{d}{du} \sin (2u + 1)$$
$$= \frac{1}{1 + \sin^2 (2u + 1)} 2 \sin (2u + 1) \cos (2u + 1) \; 2$$
$$= \frac{4 \sin (2u + 1) \cos (2u + 1)}{1 + \sin^2 (2u + 1)}$$

Finally, for **c** we have 8

$$\frac{d}{du}[e^{2u} \cos (3u + 1)]$$

$$= e^{2u} \frac{d}{du}[\cos (3u + 1)] + \cos (3u + 1) \frac{d}{du}[e^{2u}]$$
$$= e^{2u} [-\sin (3u + 1)] \; 3 + \cos (3u + 1) \; e^{2u} \; 2$$
$$= e^{2u} [-3 \sin (3u + 1) + 2 \cos (3u + 1)]$$

Now we are ready to continue.

4.10 LOGARITHMIC DIFFERENTIATION

It is not always possible to simplify an expression, but if it is a product or a quotient it may help to 'take logarithms' before differentiating. By this means we avoid the use of the product rule, but more importantly we avoid the very awkward quotient rule. Two examples will illustrate this technique.

□ Differentiate with respect to x

$$y = \frac{(1 + x)\sin^2 x}{(1 + 4x)(1 - x)^3}$$

Before taking logarithms we should be assured that both sides are positive. This is certainly true if $x \in (-1, 1)$, and so we shall suppose that we are within this interval. Using the laws of logarithms (see Chapter 1)

$$\begin{aligned}\ln y &= \ln(1 + x) + \ln(\sin^2 x) - \ln(1 + 4x) - \ln(1 - x)^3 \\ &= \ln(1 + x) + 2\ln(\sin x) - \ln(1 + 4x) - 3\ln(1 - x)\end{aligned}$$

Now differentiating throughout with respect to x gives

$$\begin{aligned}\frac{1}{y}\frac{dy}{dx} &= \frac{1}{1 + x} + \frac{2\cos x}{\sin x} - \frac{4}{1 + 4x} - \frac{3(-1)}{1 - x} \\ \frac{dy}{dx} &= \left(\frac{1}{1 + x} + \frac{2\cos x}{\sin x} - \frac{4}{1 + 4x} + \frac{3}{1 - x}\right)\frac{(1 + x)\sin^2 x}{(1 + 4x)(1 - x)^3}\end{aligned}$$ ■

□ Differentiate x^x with respect to x (>0).
Let $y = x^x$. Then

$$\ln y = \ln(x^x) = x \ln x$$

So

$$\frac{d}{dx}(\ln y) = \frac{d}{dx}(x \ln x) = x\frac{d}{dx}(\ln x) + (\ln x)\,1$$

Therefore

$$\frac{1}{y}\frac{dy}{dx} = x\frac{1}{x} + \ln x$$

So

$$\frac{dy}{dx} = y(1 + \ln x) = x^x(1 + \ln x)$$ ■

4.11 IMPLICIT DIFFERENTIATION

Occasionally when y is given in terms of x this is not expressed explicitly; instead, y and x are related by an equation. We sometimes say that y is given **implicitly** in terms of x. For example if x and y are related by the equation

$$x^2 + 3xy + y^3 = 5$$

then y is given implicitly in terms of x.

To differentiate y with respect to x it is not necessary first to express y explicitly in terms of x. Instead we can differentiate both sides of the equation with respect to x and use the chain rule. Here

$$\frac{d}{dx} f(y) = f'(y) \frac{dy}{dx}$$

□ Obtain the first derivative of y with respect to x at the point $(1, 1)$ if

$$x^2 + 3xy + y^3 = 5$$

We should check that the point $(1, 1)$ lies on the curve. It does because $x = 1$ and $y = 1$ satisfy the equation. Now we go through the equation, differentiating with respect to x and using the chain rule:

$$2x + 3x \frac{dy}{dx} + 3y + 3y^2 \frac{dy}{dx} = 0$$

$$3(x + y^2) \frac{dy}{dx} = -2x - 3y$$

$$\frac{dy}{dx} = \frac{-2x - 3y}{3(x + y^2)}$$

When $x = 1$ and $y = 1$ we obtain

$$\frac{dy}{dx} = \frac{-5}{6}$$ ■

4.12 PARAMETRIC DIFFERENTIATION

If a function is defined parametrically then it is better to use the chain rule to obtain its derivative in terms of the parameter. Of course theoretically we could eliminate the parameter and differentiate in the ordinary way. However, in practice this may not be possible.

Therefore if $y = y(t)$ and $x = x(t)$ we have

$$\frac{dy}{dx} = \frac{dy}{dt} \frac{dt}{dx}$$

and because

$$1 = \frac{dx}{dx} = \frac{dx}{dt} \frac{dt}{dx}$$

we obtain

$$\frac{dy}{dx} = \frac{dy}{dt}\bigg/\frac{dx}{dt}$$

There is one very important point to watch out for, and it is a frequent cause of error. Although

$$\frac{dy}{dx} = \frac{dy}{dt}\frac{dt}{dx}$$

a similar result does *not* hold for second-order derivatives. In symbols,

$$\frac{d^2y}{dx^2} \neq \frac{d^2y}{dt^2}\frac{d^2t}{dx^2}$$

In fact if you look carefully you will see that not even the notation leads you to believe this will work. Nevertheless many examination scripts contain attempts at solutions to differentiation problems which try to use this. It is a very popular mistake!

In order to obtain the second-order derivative it is necessary to use the chain rule again because the first derivative will be in terms of t:

$$\begin{aligned}\frac{d^2y}{dx^2} = \frac{d}{dx}\left(\frac{dy}{dx}\right) &= \frac{d}{dt}\left(\frac{dy}{dx}\right)\frac{dt}{dx}\\ &= \frac{d}{dt}\left(\frac{dy}{dt}\frac{dt}{dx}\right)\frac{dt}{dx}\end{aligned}$$

□ If $y = \cos t + \sin t$ and $x = \tan t$, obtain the first-order and second-order derivatives of y with respect to x.

As you can see, it would not be easy to eliminate t. So we use the chain rule

$$\frac{dy}{dx} = \frac{dy}{dt}\frac{dt}{dx}$$

Now

$$\begin{aligned}\frac{dy}{dt} &= \frac{d}{dt}(\cos t + \sin t)\\ &= -\sin t + \cos t\end{aligned}$$

Also

$$\frac{dx}{dt} = \frac{d}{dt}(\tan t) = \sec^2 t$$

Therefore the first-order derivative is

$$\begin{aligned}\frac{dy}{dx} = \frac{dy}{dt}\frac{dt}{dx} &= \frac{-\sin t + \cos t}{\sec^2 t}\\ &= -\sin t\cos^2 t + \cos^3 t\end{aligned}$$

We use the chain rule again to obtain the second-order derivative:

$$\frac{d^2y}{dx^2} = \frac{d}{dt}(\cos^3 t - \sin t \cos^2 t)\frac{dt}{dx}$$
$$= \{3 \cos^2 t (-\sin t) - [\sin t\, 2 \cos t (-\sin t) + \cos^2 t \cos t]\} \cos^2 t$$

See how we use the chain rule here without the formal substitution. To differentiate $\cos^3 t$ with respect to t, we first differentiate $\cos^3 t$ with respect to $\cos t$ to obtain $3 \cos^2 t$ and then multiply this by $-\sin t$, the derivative of $\cos t$ with respect to t.

Now we simplify:

$$\frac{d^2y}{dx^2} = (-3 \cos^2 t \sin t + 2 \sin^2 t \cos t - \cos^3 t) \cos^2 t$$
$$= (-3 \cos t \sin t + 2 \sin^2 t - \cos^2 t) \cos^3 t$$
$$= [-3 \cos t \sin t + 2(1 - \cos^2 t) - \cos^2 t] \cos^3 t$$
$$= (-3 \cos t \sin t + 2 - 3 \cos^2 t) \cos^3 t$$
$$= 2 \cos^3 t - 3 \cos^4 t(\cos t + \sin t)$$

If you use the chain rule correctly and don't invent your own version, nothing should go wrong. ■

4.13 RATES OF CHANGE

We can apply differentiation to obtain the rates at which variables change.

□ A spherical balloon is pumped up at a constant rate of 1 m^3/s. Obtain the rate of change of the radius at the instant when it is 0.5 m.

We may denote the volume of the balloon by V and the radius by r. As time increases, r and V change. We have the following relationship between V and r:

$$V = \frac{4}{3}\pi r^3$$

We know that dV/dt is constant at 1 m^3/s, and we wish to determine dr/dt. This is a simple application of the chain rule:

$$\frac{dV}{dt} = \frac{4}{3} 3\pi r^2 \frac{dr}{dt}$$
$$= 4\pi r^2 \frac{dr}{dt}$$

Therefore substituting into this equation we obtain

$$\frac{dr}{dt} = \frac{1}{4\pi(0.5)^2} = \frac{1}{\pi} \text{ m/s}$$ ■

Are you ready for some steps? If not then read through the chapter again to familiarize yourself with it.

4.14 Workshop

1

Exercise Evaluate the following limit:

$$\lim_{x \to \pi/2} \left(\frac{2 \tan x}{1 - \tan^2 x} \right)$$

As soon as you have had a crack at this, move on and see if you are right.

2

We cannot put $x = \pi/2$ since $\tan \pi/2$ is not defined. Therefore some other approach must be used. Here is one:

$$\tan 2x = \frac{2 \tan x}{1 - \tan^2 x}$$

So

$$\lim_{x \to \pi/2} \left(\frac{2 \tan x}{1 - \tan^2 x} \right) = \lim_{x \to \pi/2} (\tan 2x) = 0$$

If this has worked out well you may proceed directly to step 4. If you didn't get the limit correct then you will need to make sure you follow what has been done before you proceed. If you feel you would like some more practice then here is another problem.

▷**Exercise** Determine the following limit:

$$\lim_{x \to \pi/2} \left(\frac{1 - \sin x}{2 \cos^2 x} \right)$$

When you are ready, take the next step.

3

If we attempt to put $x = \pi/2$ we obtain the undefined expression 0/0. Here is one way of proceeding:

$$\frac{1 - \sin x}{2 \cos^2 x} = \frac{1 - \sin x}{2(1 - \sin^2 x)} = \frac{1}{2(1 + \sin x)}$$

$$\to \frac{1}{2(1 + 1)} = \frac{1}{4} \text{ as } x \to \pi/2$$

Now step ahead.

4

▷**Exercise** A function f is defined by

$$f(x) = \frac{2 \sin x}{\sec x \tan x} \qquad (x \neq 0)$$
$$f(x) = 2 \qquad (x = 0)$$

Use the convention of the maximal domain to describe the domain. Then decide whether or not the function is continuous at 0.

Make a good attempt at this one, then move to step 5 for the answer.

5

By the convention of the maximal domain (Chapter 2) the function is defined at all points where $\sec x \tan x$ is not zero. Now $\sec x$ is never zero, and $\tan x$ is only zero when x is a multiple of π. So the domain consists of all the real numbers except for a non-zero multiple of π. ($f(0)$ is defined separately.) Notationally the domain is

$$A = \{r \mid r \in \mathbb{R}, r \neq n\pi, \text{ where } n \in \mathbb{Z}, n \neq 0\}$$

If

$$\lim_{x \to 0} f(x) = f(0)$$

then the function is continuous at 0. We have

$$\begin{aligned}\lim_{x \to 0} f(x) &= \lim_{x \to 0} \left(\frac{2 \sin x}{\sec x \tan x} \right) \\ &= \lim_{x \to 0} \left(\frac{2 \sin x \cos^2 x}{\sin x} \right) \\ &= \lim_{x \to 0} (2 \cos^2 x) = 2\end{aligned}$$

Also by definition $f(0) = 2$, and so the function f is continuous at 0.

If you had trouble with the description of the domain it may repay you to concentrate some attention on Chapter 2.

Before we leave limits and continuity here is one more problem. It should cause no difficulty now.

▷**Exercise** The real function f is defined by

$$f(x) = \frac{e^{2x} - 1}{e^x - 1} \qquad (x \neq 0)$$
$$f(x) = k \qquad (x = 0)$$

Calculate the value of k if it is known that f is continuous.

Best foot forward!

6 For continuity we require $f(x) \to f(0)$ as $x \to 0$, so that

$$\begin{aligned} k &= \lim_{x\to 0} \frac{e^{2x} - 1}{e^x - 1} \\ &= \lim_{x\to 0} \frac{(e^x - 1)(e^x + 1)}{e^x - 1} \\ &= \lim_{x\to 0} (e^x + 1) = 1 + 1 = 2 \end{aligned}$$

Now move on to the next exercise.

▷**Exercise** Obtain the nth derivative with respect to x of

$$x^3 \frac{d^3y}{dx^3}$$

where $y = y(x)$ and n is any natural number.

Leibniz was one of the great philosophers, but you shouldn't need to ponder too deeply about this.

7 We must calculate $(x^3y_3)_n$, where we are using the subscript notation to denote differentiation with respect to x. Now

$$(uv)_n = uv_n + n\,u_1v_{n-1} + \ldots + u_nv$$

Put $u = x^3$: then $u_1 = 3x^2$, $u_2 = 6x$, $u_3 = 6$. It follows that $u_r = 0$ for $r > 3$. Put $v = y_3$: then $v_1 = y_4$, $v_2 = y_5$, $\ldots$, $v_r = y_{r+3}$. Substituting into Leibniz's formula, we obtain

$$\begin{aligned} (x^3y_3)_n &= x^3y_{n+3} + n\,3x^2y_{n+2} + \frac{n(n-1)}{1\times 2}6xy_{n+1} + \frac{n(n-1)(n-2)}{1\times 2\times 3}6y_n \\ &= x^3y_{n+3} + n\,3x^2y_{n+2} + 3n(n-1)xy_{n+1} + n(n-1)(n-2)y_n \\ &= x^3\frac{d^{n+3}y}{dx^{n+3}} + 3nx^2\frac{d^{n+2}y}{dx^{n+2}} + 3n(n-1)x\frac{d^{n+1}y}{dx^{n+1}} \\ &\qquad + n(n-1)(n-2)\frac{d^ny}{dx^n} \end{aligned}$$

Did that go well? If it didn't, here is another problem. If it did, you can miss this one out and step ahead to step 9.

▷**Exercise** Show that if

$$x\frac{dy}{dx} + y = e^{2x}$$

then

$$x \frac{d^{n+1}y}{dx^{n+1}} + (n+1) \frac{d^n y}{dx^n} = 2^n e^{2x} \qquad (n \in \mathbb{N}_0)$$

Step ahead for the answer.

8

Using the subscript notation, we have

$$xy_1 + y = e^{2x}$$

If throughout we differentiate n times with respect to x we obtain

$$(xy_1)_n + y_n = (e^{2x})_n$$

Now $(e^{2x})_1 = 2e^{2x}$, so $(e^{2x})_2 = 2^2 e^{2x}$ and in general $(e^{2x})_n = 2^n e^{2x}$. Using Leibniz's formula on the first term in the equation, we have

$$[xy_{n+1} + n(1)y_n] + y_n = 2^n e^{2x}$$

So

$$x \frac{d^{n+1}y}{dx^{n+1}} + (n + 1) \frac{d^n y}{dx^n} = 2^n e^{2x} \qquad (n \in \mathbb{N}_0)$$

One last exercise will reinforce much that we have covered.

9

>**Exercise** Suppose that $y = e^{2x} + e^{3x}$ and $z = x^3y$. Calculate $d^{12}z/dx^{12}$ when $x = 0$.

Think about this and then try it out for yourself before stepping on.

10

We could evaluate x^3y and differentiate the result twelve times, but that would be tedious. Instead we apply Leibniz's formula to differentiate z n times:

$$\begin{aligned} z_n &= (x^3y)_n \\ &= x^3y_n + n\,3x^2y_{n-1} + \frac{n(n-1)}{1 \times 2} 6xy_{n-2} + \frac{n(n-1)(n-2)}{6} 6y_{n-3} \end{aligned}$$

Now $y = e^{2x} + e^{3x}$, so $y_1 = 2e^{2x} + 3e^{3x}$, and in general $y_r = 2^r e^{2x} + 3^r e^{3x}$. So putting $x = 0$ in the expression for z_n produces

$$\begin{aligned} z_n &= 0 + 0 + 0 + n(n-1)(n-2)(2^{n-3}1 + 3^{n-3}1) \\ &= n(n-1)(n-2)(2^{n-3} + 3^{n-3}) \end{aligned}$$

Finally when $n = 12$ we obtain

$$\frac{d^{12}z}{dx^{12}} = 12 \times 11 \times 10\,(2^9 + 3^9) = 26\,657\,400$$

Now for two applications.

4.15 Practical

CYLINDER PRESSURE

The pressure inside a cylinder is given by

$$P = \frac{k}{\pi a^2 x}$$

where a and k are constants and x is allowed to change; initially $x = a$. Obtain the pressure gradient dP/dt, in terms of the initial pressure P_0, at the instant when x has doubled its initial value if x is moving at a constant rate of 1 m/s.

Try this; it is not at all difficult.

We require dP/dt, and so we differentiate through the equation using the chain rule. We obtain

$$\frac{dP}{dt} = \frac{-k}{\pi a^2 x^2}\,\frac{dx}{dt}$$

Now $dx/dt = 1$, and so when $x = 2a$

$$\frac{dP}{dt} = \frac{-k}{4\pi a^4}$$

The initial pressure is given by $P_0 = k/\pi a^3$, so $k = P_0 \pi a^3$. Therefore the pressure gradient when $x = 2a$ is given by

$$\frac{dP}{dt} = -\frac{P_0 \pi a^3}{4\pi a^4} = -\frac{P_0}{4a}$$

WATER SEEPAGE

Let's apply differentiation to solve another problem.

A crater, in the shape of part of a sphere of radius r, has been dug in porous soil by construction workers. The work has been interrupted by heavy rain. Water is falling at a constant rate w m^3/s per unit horizontal surface area and is seeping into the surrounding soil at a constant rate of p m^3/s per unit area of soil–water contact. When the water pool has depth h and surface diameter $2a$ it can be shown that the area of soil–water contact is $\pi(h^2 + a^2)$ and that the volume of water then present is $\pi h(h^2 + 3a^2)/6$.

Obtain the rate at which the depth h is increasing. Deduce that if $p = w/2$ then

$$\frac{\mathrm{d}h}{\mathrm{d}t} = \frac{w(a^2 - h^2)}{2a^2}$$

Deduce also that, in the steady state (when $\mathrm{d}h/\mathrm{d}t = 0$),

$$p = wa^2/(a^2 + h^2)$$

Try this problem, and then follow the solution through stage by stage when you are ready.

The water pool is shown in Fig. 4.7. We have

$$\begin{aligned} r^2 &= (r - h)^2 + a^2 \\ &= r^2 - 2hr + h^2 + a^2 \end{aligned}$$

So $2hr - h^2 = a^2$.

Now the rate at which the volume is increasing can be obtained by considering the water which comes in and subtracting the water which seeps out. The amount which comes in each second is proportional to the air–surface area: the amount which seeps out is proportional to the soil–water area. Therefore

$$\frac{\mathrm{d}V}{\mathrm{d}t} = w\pi a^2 - p\pi(a^2 + h^2)$$

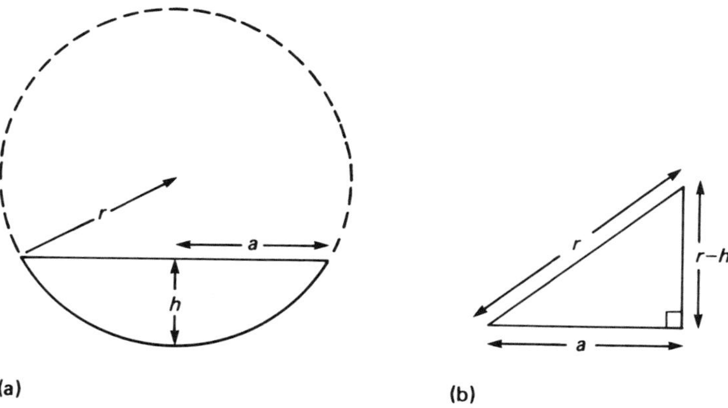

Fig. 4.7 (a) Representation of the crater (b) Triangle relating r, h and a.

If you didn't get to this stage, see if you can make the next move on your own before you check ahead.

We have

$$\begin{aligned} V &= \frac{\pi h^3}{6} + \frac{\pi a^2 h}{2} \\ &= \frac{\pi h^3}{6} + \frac{\pi h}{2}(2hr - h^2) \\ &= \pi r h^2 - \frac{\pi h^3}{3} \end{aligned}$$

So

$$\begin{aligned} \frac{dV}{dt} &= 2\pi r h \frac{dh}{dt} - \pi h^2 \frac{dh}{dt} \\ &= \pi(2rh - h^2)\frac{dh}{dt} \\ &= \pi h(2r - h)\frac{dh}{dt} \end{aligned}$$

If you have been unsuccessful to this stage, see if you can take over the problem now. It's simply a question of substitution to find the rate of increase of water depth.

Therefore

$$\begin{aligned} \pi h(2r - h)\frac{dh}{dt} &= \pi(wa^2 - pa^2 - ph^2) \\ h(2r - h)\frac{dh}{dt} &= (w - p)a^2 - ph^2 \\ a^2\frac{dh}{dt} &= (w - p)a^2 - ph^2 \\ \frac{dh}{dt} &= w - p - p\left(\frac{h}{a}\right)^2 \end{aligned}$$

If $p = w/2$ then we deduce that

$$\frac{dh}{dt} = \frac{w}{2} - \frac{w}{2}\left(\frac{h}{a}\right)^2 = \frac{w(a^2 - h^2)}{2a^2}$$

Finally we require the formula for the steady-state seepage rate. We have

$dh/dt = 0$, and so

$$0 = w - p - p\left(\frac{h}{a}\right)^2$$

$$w = p + p\left(\frac{h}{a}\right)^2$$

$$p = \frac{wa^2}{a^2 + h^2}$$

SUMMARY

- ☐ To evaluate $\lim_{x \to a} f(x)$ we examine the behaviour of $f(x)$ near the point a but not at the point a.
- ☐ We can use the laws of limits freely provided the result is meaningful: $0/0$, ∞/∞, and $0\,\infty$ are not meaningful.
- ☐ A function is continuous if $\lim_{x \to a} f(x) = f(a)$ for all points a in its domain.
- ☐ We define

$$\frac{dy}{dx} = \lim_{h \to 0} \frac{f(x + h) - f(x)}{h}$$

and from this definition the rules of elementary differentiation follow. We considered some of the techniques of differentiation too.
- ☐ Leibniz's theorem

$$(uv)_n = uv_n + \binom{n}{1} u_1 v_{n-1} + \ldots + \binom{n}{n} u_n v$$

can be used to differentiate a product n times.

EXERCISES

1 Differentiate each of the following with respect to x:

a $3x^2 + 5x + 1$
b $x^3 - 2x^2$
c $x^{1/2} + x^{-1/2}$
d $(x + 2)^6$
e $\sin(3x + 4)$
f $\tan^2 3x$
g $\ln(2x^2 + 1)$
h $x^2 \sin x^2$

2 Differentiate each of the following with respect to t:

a $\dfrac{(t^2 + 1)(t + 2)}{(t^2 + 2)(t + 1)}$

b $\dfrac{(t + 1)^3(t + 2)^3}{(t + 3)^2}$

c $\dfrac{\sin t \cos 2t}{\sec t}$

d $\dfrac{e^t + 1}{e^t - 1}$

3 If x varies with t, obtain an expression for dy/dt in terms of x and the variable $\alpha = dx/dt$ in each of the following:

a $y = x^3$

b $y = \sin x^2$

c $y = \ln(\sin x)$

d $y^2 = e^x$

4 Obtain

a $\lim_{x \to \infty} \left\{\dfrac{e^x + 1}{e^x - 1}\right\}$

b $\lim_{x \to 0} \left\{\dfrac{e^x - 1}{\ln(x + 1)^2}\right\}$

c $\lim_{x \to 0} \left\{\dfrac{\tan 2x}{\sin 3x}\right\}$

d $\lim_{x \to \infty} \left\{\dfrac{3 \ln x + 1}{2 \ln x - 5}\right\}$

ASSIGNMENT

1 Obtain

a $\lim_{x \to \pi/4} \dfrac{\sin x - \cos x}{\cos 2x}$

b $\lim_{x \to \pi/3} \dfrac{2 \cos x - 1}{\sin 3x}$

c $\lim_{x \to \pi/2} \dfrac{\tan x}{\cos x - 1}$

d $\lim_{x \to \infty} \dfrac{2e^x + 1}{3e^x - 1}$

2 Obtain the value of $f(0)$ if the following functions are known to be continuous at 0:

a $f(x) = \dfrac{\cos x - 1}{2 \sin x}$

b $f(x) = \dfrac{\sin^2 x}{\cos^3 x - 1}$

3 Show that if $y = \sin^2 (2x)$ then

$$\frac{d^2y}{dx^2} = 8 - 16y$$

4 Differentiate each of the following with respect to x:

a $x^{\ln x}$

b $e^{3x} \tan 2x$

5 If $x^2 + 2xy - y^2 = 16$, show that

$$\frac{dy}{dx} = \frac{y + x}{y - x}$$

6 If $x = t + \sin t$ and $y = t + \cos t$, obtain dy/dx and show that

$$(1 + \cos t)^3 \frac{d^2y}{dx^2} = \sin t - \cos t - 1$$

FURTHER EXERCISES

1 Differentiate each of the following with respect to x:

a $(2x + 1)(4x - 7)$

b $(x - 1)(x - 2)(x - 3)$

c $(x^2 - 1)^{1/2}$

d $(x^2 + 3)/(x^2 - 3)$

e $(a + bx^m)^n$, a, b, m and n constant.

f $\tan (ax + b)$, a and b constant.

g $\sqrt{(\operatorname{cosec} x^2)}$

2 Obtain the first four derivatives with respect to x of

a x^9

b $\sqrt{(x + 1)}$

c $\cos^2 x$

d $x^2 e^x$

3 If $y^2 = \sec 2x$, show that $d^2y/dx^2 = 3y^5 - y$

4 If $y = \sin 2t$ and $x = \cos t$, obtain d^2y/dx^2

5 If $x = \cos^3 \theta$ and $y = \sin^3 \theta$, obtain dy/dx and show that

$$\frac{d^2y}{dx^2} = \frac{1}{3 \cos^4 \theta \sin \theta}$$

6 Use the chain rule to show that

$$\frac{d^2y}{dx^2}\left(\frac{dx}{dy}\right)^3 + \frac{d^2x}{dy^2} = 0$$

7 If y is real and satisfies the equation $y^3 + y = x$, show that $\mathrm{d}y/\mathrm{d}x = y/(3x - 2y)$.

8 Show that

a $\lim_{x\to 3} [(x^2 - x - 6)/(x^2 + x - 12)] = 5/7$

b $\lim_{x\to 0} \{[(1 + x)^2 - (1 - x)^2]/2x\} = 2$

c $\lim_{x\to 0} \{[(a + x)^3 - (a - x)^3]/2x\} = 3a^2$

9 Evaluate

a $\lim_{x\to 0} \{[1/(x - 1) - 1/x]/[1/(x - 2) + 1/x]\}$

b $\lim_{x\to 1} \{[1/(x - 1) - 1/x]\,[1/(x - 2) + 1/x]\}$

10 Show that, for $x > 0$,

$$\sqrt{(x^2 + x + 1)} < x + 1/2 + 3/8x$$
$$\sqrt{(x^2 - x + 1)} < x - 1/2 + 1/2x$$

Deduce that

$$\lim_{x\to\infty} [\sqrt{(x^2 + x + 1)} + \sqrt{(x^2 - x + 1)} - 2x] = 0$$

11 The rate at which the surface area of a bubble is increasing is kA, where A is its surface area and k is a constant. If the bubble is spherical, obtain the corresponding rate at which the volume is increasing.

12 Show that if a probe moves in a straight line in such a way that its speed is proportional to the square root of its distance from a fixed point on the line, then its acceleration is constant.

13 The retaining strut on a step ladder breaks, and as the ladder collapses the vertical angle increases at a constant rate. Show that the rate of increase of the distance between the feet is proportional to the height.

14 A uniform beam is clamped horizontally at one end and carries a variable load $w = w(x)$, where x is the distance from the fixed end. If the transverse deflection of the beam is $y(x) = -x^2\mathrm{e}^{-x}$, obtain an expression for w, given that $w = EI\,\mathrm{d}^4y/\mathrm{d}x^4$ and EI is the flexural rigidity of the beam.

15 A mooring buoy in the shape of a right circular cone, with the diameter of its base equal to its slant height, is submerged in the sea. Marine mud is deposited on it uniformly across the surface at a constant rate ϱ. Calculate the rate at which the surface area is increasing in terms of the height of the cone.

16 The content V cm^3 and the depth p cm of water in a vessel are connected by the relationship $V = 3p^2 - p^3$ ($p > 2$). Show that if water is poured in at a constant rate of Q cm^3/s then, at the moment when the depth is p, it is increasing at a rate $\sigma = Q[3p(2 - p)]^{-1}$.

17 A beam of length l m has one end resting on horizontal ground and the other leaning against a vertical wall at right angles to it. It begins to slip downwards. Show that when the foot of the beam is x m from the wall and moving at h m/s away from it, the top is descending at a rate $xh/(l^2 - x^2)^{1/2}$ m/s.

18 A rope l m long is attached to a heavy weight and passed over a pulley h m above the ground ($2h < l$). The other end of the rope is tied to a vehicle which moves at a constant rate u in a radial direction away from the vertical line of the weight and the pulley. Calculate (a) the rate at which the weight is rising when the vehicle is x m from the vertical line of the weight and the pulley, and (b) the rate at which the angle between the rope and the ground is changing.

19 Sand falls from a chute and forms a conical pile in such a way that the vertical angle remains constant. Suppose r is the base radius and h is the height at time t.

a Show that if r is increasing at a rate α cm/s then the volume is increasing at a rate $\pi rh\alpha$ cm^3/s.

b Show that if the height h is increasing at a rate β cm/s then the exposed surface area is increasing at a rate $2\pi r\beta\sqrt{(h^2 + r^2)}/h$ cm^2/s.

(The volume of the cone is $\pi r^2h/3$ and the surface area is πrl, where l is the slant height.)

20 A body moves in a straight line in accordance with the equation $s = t^2/(1 + t^2)$, where t is time in seconds and s is the distance travelled in metres. Show that $0 \leqslant s < 1$. Show also that the speed u and acceleration f are given at time t by

$$u = \sin\theta\,(1 + \cos\theta)/2$$
$$f = (2\cos\theta - 1)\,(1 + \cos\theta)^2/2$$

where $t = \tan(\theta/2)$.

21 A landmark on a distant hill is x metres from a water tower. The angle of elevation from the top of the tower is observed to be θ degrees, whereas the angle of elevation from the foot of the tower is observed to be $\theta + h$ degrees.

a How high is the water tower?

b Show that if h is small then the height of the water tower is approximately $\pi xh/180\cos^2\theta$.

22 Obtain dy/dx in each of the following

a $y = x^3 \sin 2x$

b $x = t\sin t,\ y = \cos t$

c $x^2 + y^2 = y\,e^y$

23 The volume of a rubber tyre is given by $V = 2\pi^2a^2b$ and its surface area is given by $S = 4\pi^2ab$ where a and b are related to the internal radius r and the external radius R by the equations $R = b + a$, $r = b - a$. The tyre is inflated in such a way that the internal radius r remains constant. Show that the rate of increase in volume of the tyre, at the instant at which the rate of increase of the surface area is $(a + b)^2$, is $a(a + b)(a + 2b)/2$.

5 Hyperbolic functions

Although we have now explored some of the basic terminology of mathematics and developed the techniques of the differential calculus, we need to pause to extend our algebraic knowledge. In this chapter we shall describe a class of functions known as the hyperbolic functions which are very similar in some ways to the circular functions. We shall use the opportunity to consider in detail what is meant by an inverse function.

After studying this chapter you should be able to

- ☐ Use the hyperbolic functions and their identities;
- ☐ Solve algebraic equations which involve hyperbolic functions;
- ☐ Differentiate hyperbolic functions;
- ☐ Decide when a function has an inverse function;
- ☐ Express inverse hyperbolic functions in logarithmic form.

We shall also consider a practical problem concerning the sag of a chain.

5.1 DEFINITIONS AND IDENTITIES

The hyperbolic functions are in some ways very similar to the circular functions. Indeed when we deal with complex numbers (Chapter 10) we shall see that there is an algebraic relationship between the two. Initially we shall discuss the hyperbolic functions algebraically, but later we shall see that one of them arises in a physical context.

The functions cosine and sine are called circular functions because $x = \cos\theta$ and $y = \sin\theta$ satisfy the equation $x^2 + y^2 = 1$, which is the equation of a circle. The functions known as the hyperbolic cosine (cosh) and the hyperbolic sine (sinh) are called **hyperbolic functions** because $x = \cosh u$ and $y = \sinh u$ satisfy the equation $x^2 - y^2 = 1$, which is the equation of a rectangular hyperbola.

We shall define the hyperbolic functions and use these definitions to sketch their graphs. Here then are the definitions:

$$\cosh u = \frac{e^u + e^{-u}}{2} \qquad \sinh u = \frac{e^u - e^{-u}}{2}$$

Now the exponential function has domain $\mathbb{R}$ and consequently both $\cosh u$ and $\sinh u$ are defined for all real numbers u.

To obtain a sketch of the graphs of $y = \cosh x$ and $y = \sinh x$ we can use the graphs of $y = e^x$.

$y = \cosh x$

If the graphs of $y = e^x$ and $y = e^{-x}$ are both drawn on the same diagram (Fig. 5.1) then chords can be drawn parallel to the y-axis between these two curves. The midpoints of these chords then lie on the curve $y = \cosh x$.

The hyperbolic cosine curve is one which arises in practice. It is often called the **catenary**. If a heavy rope or chain is freely suspended between two fixed points, the shape it assumes is that of the catenary. This has to be taken into account when, for example, suspension bridges are designed. At one time surveyors had to make a 'catenary correction' when using steel tape measures, but with modern electronic measuring devices this is not necessary.

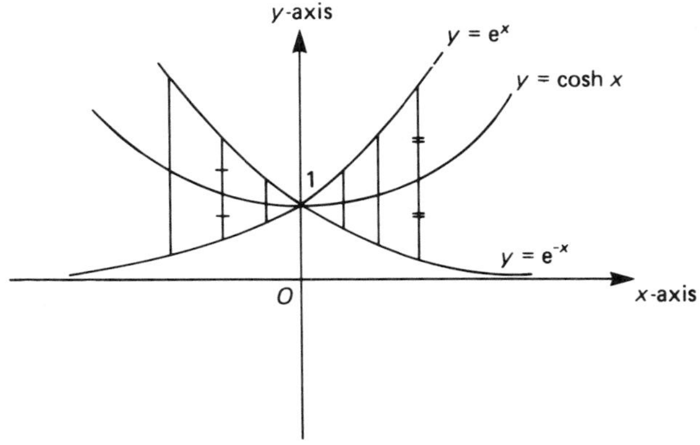

Fig. 5.1 The graph of $y = \cosh x$, by construction.

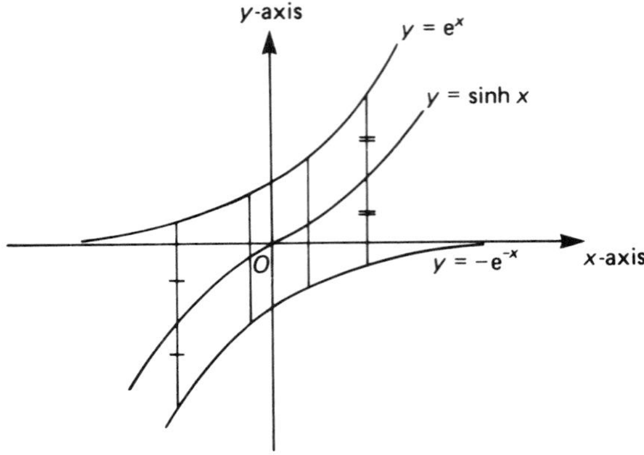

Fig. 5.2 The graph of $y = \sinh x$, by construction.

$y = \sinh x$

We draw the graphs of $y = e^x$ and $y = -e^{-x}$ on the same diagram (Fig. 5.2) and draw chords parallel to the y-axis between the two curves. The mid-points of the chords then lie on the curve $y = \sinh x$.

We define the hyperbolic tangent, cotangent, secant and cosecant by imitating the definitions for circular functions:

$$\tanh x = \frac{\sinh x}{\cosh x} \qquad \coth x = \frac{\cosh x}{\sinh x}$$

$$\operatorname{sech} x = \frac{1}{\cosh x} \qquad \operatorname{cosech} x = \frac{1}{\sinh x}$$

Strictly it is not necessary to know how to pronounce these new functions, but for completeness we shall indicate the standard practice. The hyperbolic functions cosh and coth are pronounced as they are written; sinh is pronounced 'shine'; tanh is pronounced 'than' but with a soft 'th' as in 'thank'; sech is pronounced 'sheck' and cosech as 'cosheck'.

From the definitions we obtain

$$\cosh x + \sinh x = \frac{e^x + e^{-x}}{2} + \frac{e^x - e^{-x}}{2} = e^x$$

$$\cosh x - \sinh x = \frac{e^x + e^{-x}}{2} - \frac{e^x - e^{-x}}{2} = e^{-x}$$

Therefore

$$\begin{aligned}\cosh^2 x - \sinh^2 x &= (\cosh x + \sinh x)(\cosh x - \sinh x) \\ &= e^x e^{-x} = 1\end{aligned}$$

So we have the identity

$$\cosh^2 x - \sinh^2 x = 1$$

(this corresponds to the circular identity $\cos^2 x + \sin^2 x = 1$).

We can either appeal to the symmetry of the graphs or use the definitions to deduce that

$$\begin{aligned}\cosh(-x) &= \cosh x \\ \sinh(-x) &= -\sinh x\end{aligned}$$

and therefore

$$\tanh(-x) = -\tanh x$$

We have already come across one identity involving hyperbolic functions:

$$\cosh^2 x - \sinh^2 x = 1$$

In fact corresponding to every identity involving circular functions there is an identity involving hyperbolic functions. There is a rule for converting identities involving circular functions into those involving hyperbolic functions. The rule is if there is a product of two sines or an *implied* product of two sines, the term changes sign. Although this can be applied in reverse it is easy to make mistakes, and so we shall avoid it altogether.

Here is a list of the main identities:

$$\begin{aligned}\cosh(x + y) &= \cosh x \cosh y + \sinh x \sinh y \\ \cosh(x - y) &= \cosh x \cosh y - \sinh x \sinh y \\ \sinh(x + y) &= \sinh x \cosh y + \cosh x \sinh y \\ \sinh(x - y) &= \sinh x \cosh y - \cosh x \sinh y\end{aligned}$$

□ Use the basic definitions to establish the identity

$$\cosh(x + y) = \cosh x \cosh y + \sinh x \sinh y$$

We have

$$\begin{aligned}\text{LHS} &= \frac{e^x + e^{-x}}{2}\frac{e^y + e^{-y}}{2} + \frac{e^x - e^{-x}}{2}\frac{e^y - e^{-y}}{2} \\ &= \frac{1}{4}(e^x e^y + e^x e^{-y} + e^{-x} e^y + e^{-x} e^{-y} \\ &\quad + e^x e^y - e^x e^{-y} - e^{-x} e^y + e^{-x} e^{-y})\end{aligned}$$

$$= \frac{1}{4}(2e^{x+y} + 2e^{-(x+y)})$$

$$= \frac{e^{x+y} + e^{-(x+y)}}{2} = \text{RHS}$$ ■

Now one for you to try!

□ Use the basic definitions to establish the identity

$$\sinh(x - y) = \sinh x \cosh y - \cosh x \sinh y$$

Make a good effort. Check carefully before you begin that you follow all the stages in the one we have just done – and then best foot forward!

We have

$$\begin{aligned}
&\sinh x \cosh y - \cosh x \sinh y \\
&= \frac{e^x - e^{-x}}{2}\frac{e^y + e^{-y}}{2} - \frac{e^x + e^{-x}}{2}\frac{e^y - e^{-y}}{2} \\
&= \frac{1}{4}[e^x e^y + e^x e^{-y} - e^{-x}e^y - e^{-x}e^{-y} \\
&\quad - (e^x e^y - e^x e^{-y} + e^{-x}e^y - e^{-x}e^{-y})] \\
&= \frac{1}{4}(e^x e^y + e^x e^{-y} - e^{-x}e^y - e^{-x}e^{-y} \\
&\quad - e^x e^y + e^x e^{-y} - e^{-x}e^y + e^{-x}e^{-y}) \\
&= \frac{1}{4}(2e^{x-y} - 2e^{-x+y}) \\
&= \frac{e^{x-y} - e^{-(x-y)}}{2} = \sinh(x - y)
\end{aligned}$$ ■

Good! If you need any more practice you can always try the other two identities.

We defined the hyperbolic functions in terms of exponential functions, and so it is perhaps not surprising that we need our work on logarithms (Chapter 1) when solving equations involving hyperbolic functions.

□ Obtain all the real solutions of the equation

$$\cosh x + \sinh x = 1$$

There are two approaches, each valid and so we shall solve the equation in two different ways.

1 We know $\cosh^2 x - \sinh^2 x = 1$, and so

$$(\cosh x - \sinh x)(\cosh x + \sinh x) = 1$$

Here $\cosh x + \sinh x = 1$, and therefore $\cosh x - \sinh x = 1$. So

$$(\cosh x + \sinh x) + (\cosh x - \sinh x) = 2$$

Therefore $2 \cosh x = 2$, so $\cosh x = 1$ and $x = 0$ is the only solution. Here we have used a hyperbolic identity to sort out the problem.

2 From the definitions,

$$\frac{e^x + e^{-x}}{2} + \frac{e^x - e^{-x}}{2} = 1$$

So $e^x = 1$ and therefore $x = 0$. ■

In this example it was much easier and more direct to use the definitions at the outset. However, this is not always the case.

□ Solve the equation

$$8 \sinh x = 3 \operatorname{sech} x$$

We begin by writing the equation in terms of $\sinh x$ and $\cosh x$:

$$8 \sinh x = \frac{3}{\cosh x}$$

So $\sinh x \cosh x = 3/8$. Therefore $2 \sinh x \cosh x = 3/4$, and $\sinh 2x = 3/4$.

Now $(e^{2x} - e^{-2x})/2 = 3/4$. Therefore $2e^{2x} - 2e^{-2x} = 3$. Multiplying by e^{2x} gives

$$\begin{aligned} 2(e^{2x})^2 - 2 &= 3e^{2x} \\ 2(e^{2x})^2 - 3e^{2x} - 2 &= 0 \\ (2e^{2x} + 1)(e^{2x} - 2) &= 0 \end{aligned}$$

So either $2e^{2x} + 1 = 0$ or $e^{2x} - 2 = 0$. Since $e^{2x} > 0$, only $e^{2x} - 2 = 0$ is a possibility. So $2x = \ln 2$ and therefore $x = (\ln 2)/2 = \ln \sqrt{2}$. ■

Why not try one yourself?

□ Solve the equation

$$3 \sinh 3x = 13 \sinh x$$

There are many approaches to this problem. If you obtain the correct answer it is probable that your working is basically correct.

Here is one solution:

$$3 \sinh 3x = 13 \sinh x$$
$$3(\sinh 3x - \sinh x) = 10 \sinh x$$

By identity,

$$3(2 \cosh 2x \sinh x) = 10 \sinh x$$

So either $\sinh x = 0$, from which $x = 0$; or $6 \cosh 2x = 10$, from which $3(e^{2x} + e^{-2x}) = 10$. For the latter case, multiply by e^{2x} to obtain

$$3(e^{2x})^2 + 3 = 10e^{2x}$$
$$(3e^{2x} - 1)(e^{2x} - 3) = 0$$

Therefore $e^{2x} = 1/3$ or $e^{2x} = 3$. So $2x = \ln(1/3) = -\ln 3$ or $2x = \ln 3$.

The three solutions are therefore $x = 0$ and $x = \pm\ln\sqrt{3}$. ■

5.2 DIFFERENTIATION OF HYPERBOLIC FUNCTIONS

We may use the basic rules of differentiation to obtain the derivatives of the hyperbolic functions. All we need to do is use the basic definitions, remembering that

$$\cosh x = \frac{e^x + e^{-x}}{2} \quad \text{and} \quad \sinh x = \frac{e^x - e^{-x}}{2}$$

You might like to try these on your own. Afterwards you can look to see if you were correct.

The derivative of $\cosh x$ is found as follows:

$$\begin{aligned}\frac{d}{dx}(\cosh x) &= \frac{d}{dx}\frac{e^x + e^{-x}}{2} \\ &= \frac{1}{2}\frac{d}{dx}e^x + \frac{1}{2}\frac{d}{dx}e^{-x} \\ &= \frac{1}{2}e^x + \frac{1}{2}(-1)e^{-x} \\ &= \frac{e^x - e^{-x}}{2} = \sinh x\end{aligned}$$

The derivative of $\sinh x$ is found as follows:

$$\begin{aligned}\frac{d}{dx}(\sinh x) &= \frac{d}{dx}\frac{e^x - e^{-x}}{2} \\ &= \frac{1}{2}\frac{d}{dx}e^x - \frac{1}{2}\frac{d}{dx}e^{-x}\end{aligned}$$

$$= \frac{1}{2} e^x - \frac{1}{2}(-1) e^{-x}$$

$$= \frac{e^x + e^{-x}}{2} = \cosh x$$

Therefore

1 When we differentiate the hyperbolic cosine we obtain the hyperbolic sine;

2 When we differentiate the hyperbolic sine we obtain the hyperbolic cosine.

Quite remarkable, isn't it?

The derivatives of the other hyperbolic functions can now be obtained from these by applying the rules for differentiation. Try some of these yourself. They are good exercise in differentiation and therefore well worth attempting.

Here is the working for each one.

$$\frac{d}{dx}(\tanh x) = \frac{d}{dx}\left(\frac{\sinh x}{\cosh x}\right)$$

$$= \frac{\cosh x \cosh x - \sinh x \sinh x}{\cosh^2 x}$$

$$= \frac{\cosh^2 x - \sinh^2 x}{\cosh^2 x}$$

$$= \frac{1}{\cosh^2 x} = \text{sech}^2 x$$

$$\frac{d}{dx}(\coth x) = \frac{d}{dx}\left(\frac{\cosh x}{\sinh x}\right)$$

$$= \frac{\sinh x \sinh x - \cosh x \cosh x}{\sinh^2 x}$$

$$= -\frac{\cosh^2 x - \sinh^2 x}{\sinh^2 x}$$

$$= \frac{-1}{\sinh^2 x} = -\text{cosech}^2 x$$

$$\frac{d}{dx}(\text{sech } x) = \frac{d}{dx}\left(\frac{1}{\cosh x}\right)$$

$$= \frac{d}{dx}(\cosh x)^{-1}$$

$$= -(\cosh x)^{-2} \sinh x$$

$$= -\frac{\sinh x}{\cosh^2 x} = -\text{sech } x \tanh x$$

$$\begin{aligned}\frac{\mathrm{d}}{\mathrm{d}x}(\operatorname{cosech} x) &= \frac{\mathrm{d}}{\mathrm{d}x}\left(\frac{1}{\sinh x}\right)\\ &= \frac{\mathrm{d}}{\mathrm{d}x}(\sinh x)^{-1}\\ &= -(\sinh x)^{-2}\cosh x\\ &= -\frac{\cosh x}{\sinh^2 x} = -\operatorname{cosech} x \coth x\end{aligned}$$

Did you try those with success?

5.3 CURVE SKETCHING

We are about to draw the graph of $y = \tanh x$, and so this is a good opportunity to refresh our memories about how to sketch curves which have equations expressed in cartesian form (Chapter 3). We have already used one method when we sketched the graphs of $y = \sinh x$ and $y = \cosh x$ (Figs 5.1, 5.2). There we were able to use a known graph $y = \mathrm{e}^x$.

There are several things we can do to gain pieces of information which help us to sketch curves:

1 Obtain the points where the curve crosses the axes. This will certainly help to locate the curve.

2 Look to see if there are any values of x or y where the curve is not defined. For example, if there are any values of y which make $x^2 < 0$, the curve doesn't appear for these values of y.

3 Look to see if there are any values of x which make y large or any values of y which make x large.

4 Look to see if the graph is symmetrical about either or both of the axes.

If when we replace x in the equation by $-x$ the same equation results, then the curve is symmetrical about the y-axis. Similarly if we replace y by $-y$ in the equation and the equation remains the same, then the curve is symmetrical about the x-axis.

□ $y^2 = 16x$ is symmetrical about the x-axis but not about the y-axis. This follows because $(-y)^2 = 16x$ but $y^2 \neq 16(-x)$. In the same way the curve $x^4 + 3x^2y = 4$ is symmetrical about the y-axis. ■

5 Look to see if the graph is **skew symmetrical**, that is symmetrical with respect to the origin.

In other words, if we join a point on the curve to the origin and produce an equal length, do we always obtain another point on the curve? There is a simple test for this. If we replace x and y *simultaneously* in the equation of the curve by $-x$ and $-y$ respectively, the equation will remain the same if and only if the graph is symmetrical with respect to the origin.

□ $x^2 + xy + y^4 = 16$ is symmetrical with respect to the origin because $(-x)^2 + (-x)(-y) + (-y)^4 = 16$. Similarly $y = \sin x$ is symmetrical with respect to the origin because $(-y) = \sin(-x)$. ■

Another way of thinking about symmetry with respect to the origin is that if we rotate the curve through π the graph will be unchanged.

6 Examine the behaviour of dy/dx, particularly near the origin and as $|x| \to \infty$.

7 See if there are any points at which the curve attains a local maximum, a local minimum or a point of inflexion. We shall see in Chapter 8 how to obtain and classify these points.

THE GRAPH OF $y = \tanh x$

We now turn our attention to the problem of drawing the graph of $y = \tanh x$. We begin by finding out more about $\tanh x$:

1

$$\tanh x = \frac{\sinh x}{\cosh x} = \frac{e^x - e^{-x}}{e^x + e^{-x}}$$

$$= \frac{1 - e^{-2x}}{1 + e^{-2x}}$$

(dividing top and bottom by e^x). Now as $x \to \infty$ we have $e^{-x} \to 0$ and so $e^{-2x} \to 0$. Consequently

$$\tanh x \to \frac{1 - 0}{1 + 0} = 1 \text{ as } x \to \infty$$

2

$$\tanh x = \frac{e^x - e^{-x}}{e^x + e^{-x}}$$

$$= \frac{e^{2x} - 1}{e^{2x} + 1}$$

(multiplying top and bottom by e^x)

$$= \frac{(e^{2x} + 1) - 2}{e^{2x} + 1} = 1 - \frac{2}{e^{2x} + 1}$$

Now e^{2x} is always positive, and so $\tanh x < 1$ for all x.

3

$$\tanh 0 = \frac{\sinh 0}{\cosh 0} = 0$$

4

$$\frac{d}{dx}(\tanh x) = \operatorname{sech}^2 x \leqslant 1$$

The maximum value of the slope is attained at the origin; it then decreases as x increases. In fact $dy/dx \to 0$ as $x \to \infty$.

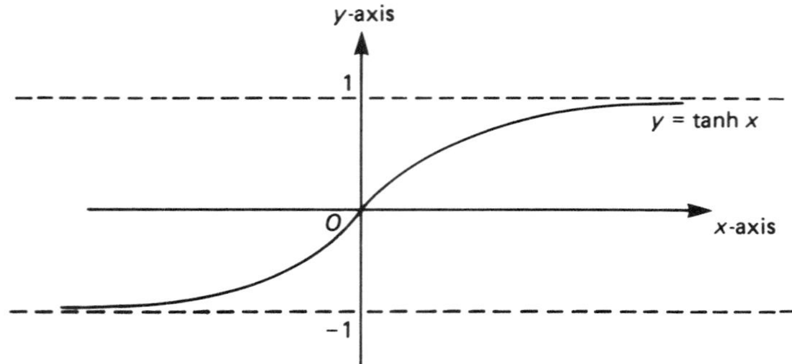

Fig. 5.3 The hyperbolic tangent function.

5 $$\tanh(-x) = \frac{\sinh(-x)}{\cosh(-x)} = \frac{e^{-x} - e^{x}}{e^{-x} + e^{x}} = -\tanh x$$

Consequently the curve is symmetrical with respect to the origin. If we put all this information together we obtain a good idea of the shape of the curve. This is shown in Fig. 5.3.

Now it's time for you to solve some problems. If you are unsure of the material this is a good time to look back once more. If you are ready, then here we go.

5.4 Workshop

1 **Exercise** Using the definitions of the hyperbolic functions, show that

$$\cosh 2x = 1 + 2\sinh^2 x$$

Don't move on until you have attempted this!

---

2 Notice that we have been asked to use the definitions, so we must do so. We must not assume the expansion formulas, for example. Now

$$\begin{aligned} \text{RHS} &= 1 + 2\left(\frac{e^{x} - e^{-x}}{2}\right)^2 \\ &= 1 + 2\left(\frac{e^{2x} - 2 + e^{-2x}}{4}\right) \\ &= \frac{e^{2x} + e^{-2x}}{2} = \text{LHS} \end{aligned}$$

Did you get that right? If so, then move on to step 4. If not, possibly the trouble arose because you did not apply the definitions correctly. Let's see if we can sort things out in the next exercise.

▷**Exercise** Use the definitions of the hyperbolic functions to show that

$$\tanh 2x = \frac{2 \tanh x}{1 + \tanh^2 x}$$

Remember: we must go back to the definitions.
Try it, then step ahead.

3

We have

$$\tanh x = \frac{\sinh x}{\cosh x} = \frac{(e^x - e^{-x})/2}{(e^x + e^{-x})/2}$$
$$= \frac{e^x - e^{-x}}{e^x + e^{-x}} = \frac{e^{2x} - 1}{e^{2x} + 1}$$

So

$$1 + \tanh^2 x = 1 + \left(\frac{e^{2x} - 1}{e^{2x} + 1}\right)^2$$
$$= \frac{(e^{2x} + 1)^2 + (e^{2x} - 1)^2}{(e^{2x} + 1)^2}$$
$$= \frac{e^{4x} + 2e^{2x} + 1 + e^{4x} - 2e^{2x} + 1}{(e^{2x} + 1)^2}$$
$$= \frac{2(e^{4x} + 1)}{(e^{2x} + 1)^2}$$

Therefore

$$\frac{2 \tanh x}{1 + \tanh^2 x} = 2\,\frac{e^{2x} - 1}{e^{2x} + 1}\,\frac{(e^{2x} + 1)^2}{2(e^{4x} + 1)}$$
$$= \frac{(e^{2x} - 1)\,(e^{2x} + 1)}{e^{4x} + 1}$$
$$= \frac{e^{4x} - 1}{e^{4x} + 1} = \tanh 2x$$

If you were unable to do this then look carefully at the working and go back to step 1. Otherwise step forward.

4

▷**Exercise** Use the expansion formula for $\sinh (x + y)$ and $\cosh (x + y)$ to obtain the expansion formula

$$\tanh(x+y) = \frac{\tanh x + \tanh y}{1 + \tanh x \tanh y}$$

You can take this in your stride.

---

5 The expansion formulas we need are

$$\sinh(x+y) = \sinh x \cosh y + \cosh x \sinh y$$
$$\cosh(x+y) = \cosh x \cosh y + \sinh x \sinh y$$

Therefore

$$\tanh(x+y) = \frac{\sinh(x+y)}{\cosh(x+y)} = \frac{\sinh x \cosh y + \cosh x \sinh y}{\cosh x \cosh y + \sinh x \sinh y}$$

So that dividing numerator and denominator by $\cosh x \cosh y$ we obtain

$$\tanh(x+y) = \frac{\tanh x + \tanh y}{1 + \tanh x \tanh y}$$

If you succeeded in getting this right, then move on to step 7. Otherwise, check carefully so that you see what has been done and then tackle the next problem.

▷**Exercise** Using the expansion formulas for $\sinh(x+y)$ and $\cosh(x+y)$, obtain the formula

$$\coth(x+y) = \frac{\coth x \coth y + 1}{\coth x + \coth y}$$

Try it, then move on.

---

6 As before we obtain

$$\coth(x+y) = \frac{\cosh x \cosh y + \sinh x \sinh y}{\sinh x \cosh y + \cosh x \sinh y}$$

So dividing numerator and denominator by $\sinh x \sinh y$ produces

$$\coth(x+y) = \frac{\coth x \coth y + 1}{\coth x + \coth y}$$

Now for another step!

---

7 **Exercise** Obtain all real solutions of the equation

$$13 \tanh 3x = 12$$

Try this and move on only when you have made a good attempt.

8

Here is the working:

$$\tanh 3x = \frac{e^{6x} - 1}{e^{6x} + 1} = \frac{12}{13}$$

So

$$13(e^{6x} - 1) = 12(e^{6x} + 1)$$

Consequently $e^{6x} = 25$ and therefore $6x = \ln 25 = 2 \ln 5$. So $x = (1/3) \ln 5$.

If you didn't get that right then you should check through each stage to make sure there are no misunderstandings. As soon as you are ready, try the next problem and take the final step.

▷**Exercise** Obtain all the real solutions of the equation

$$4 \sinh 4x - 17 \sinh 3x + 4 \sinh 2x = 0$$

You may need to think about this a little.

At first sight this might seem rather tricky – until you realize that it is possible to combine two of these hyperbolic sines together, using an identity as follows:

$$4 \sinh 4x - 17 \sinh 3x + 4 \sinh 2x = 0$$
$$4(2 \sinh 3x \cosh x) - 17 \sinh 3x = 0$$

Therefore either $\sinh 3x = 0$ or $8 \cosh x = 17$. If $\sinh 3x = 0$ then $x = 0$. If $8 \cosh x = 17$ then

$$4(e^x + e^{-x}) = 17$$
$$4(e^x)^2 + 4 - 17e^x = 0$$
$$(4e^x - 1)(e^x - 4) = 0$$

Therefore either $e^x = 1/4$ or $e^x = 4$. From this we obtain $x = \ln(1/4) = -2 \ln 2$ or $x = 2 \ln 2$.

So the three solutions are $x = 0$ and $x = \pm 2 \ln 2$.

5.5 INJECTIVE FUNCTIONS

You will remember from Chapter 2 how we defined a function $f: A \to B$ to be a rule which assigned to each element x in the domain A a unique element y in the codomain B. We wrote $y = f(x)$.

Now there is nothing in the definition to suggest that two different elements of A cannot be assigned to the same element of B. Indeed there are many functions which have this property.

□ Consider $y = x^2$. By the convention of the maximal domain we have domain $\mathbb{R}$: in other words, the domain consists of all the real numbers. Each real number x determines a unique value of y, but the same value of y is determined by two distinct arguments x. For instance

$$(-2)^2 = 4 = 2^2$$

so that when $x = 2$ or $x = -2$ we obtain the same value for y. ■

On the other hand there are some functions which do have the property that if $x_1 \neq x_2$ then $f(x_1) \neq f(x_2)$.

□ Consider $y = 1/x$. By the convention of the maximal domain this function has domain $\mathbb{R}\setminus\{0\}$: that is, the domain consists of all the real numbers except 0. In this instance if $x_1 \neq x_2$ then $f(x_1) \neq f(x_2)$.

To show this we simply show that if $f(x_1) = f(x_2)$ then it follows that $x_1 = x_2$. If $f(x_1) = f(x_2)$ then $1/x_1 = 1/x_2$, and so multiplying by x_1x_2 we obtain $x_2 = x_1$. ■

A function $f: A \to B$ which has the property that, for all $x_1, x_2 \in A$, if $x_1 \neq x_2$ then $f(x_1) \neq f(x_2)$ is called an **injection** (or a **one-one function**).

In practice injections are easy to recognize from their graphs since any line parallel to the x-axis must cut the curve at most once. Algebraically we can deduce a function is an injection by considering the implications of the equation $f(x_1) = f(x_2)$. If we can deduce that $x_1 = x_2$ then we have an injection, whereas if we can find x_1 and x_2 which are unequal and have $f(x_1) = f(x_2)$ then we do not have an injection.

□ Decide which, if either, of the following functions is an injection: $y = \sinh x$; $y = \cosh x$.

Notice how the language has been misused here. The equation identifies an equation with a function, which is rather like identifying a person with his occupation. However, provided we know what is meant there is no difficulty. Mathematics is a language, and we must get used to various dialects – and even on occasion tolerate bad grammar!

First, suppose $\sinh x_1 = \sinh x_2$. Then

$$\frac{e^{x_1} - e^{-x_1}}{2} = \frac{e^{x_2} - e^{-x_2}}{2}$$

So

$$e^{x_1} - e^{x_2} = e^{-x_1} - e^{-x_2}$$

$$= \frac{1}{e^{x_1}} - \frac{1}{e^{x_2}}$$

$$= \frac{e^{x_2} - e^{x_1}}{e^{x_1}e^{x_2}}$$

$$e^{x_1} - e^{x_2} = (e^{x_2} - e^{x_1})\, e^{x_1+x_2}$$

Now if $e^{x_1} - e^{x_2} \neq 0$ we have $e^{x_1+x_2} = -1$, which is impossible. Therefore $e^{x_1} = e^{x_2}$ and consequently $x_1 = x_2$. So $y = \sinh x$ defines an injection.

Secondly, suppose $\cosh x_1 = \cosh x_2$. Then

$$\frac{e^{x_1} + e^{-x_1}}{2} = \frac{e^{x_2} + e^{-x_2}}{2}$$

So

$$e^{x_1} - e^{x_2} = e^{-x_2} - e^{-x_1}$$

$$= \frac{1}{e^{x_2}} - \frac{1}{e^{x_1}}$$

$$= \frac{e^{x_1} - e^{x_2}}{e^{x_1}e^{x_2}}$$

$$e^{x_1} - e^{x_2} = (e^{x_1} - e^{x_2})\, e^{x_1+x_2}$$

Now if $e^{x_1} - e^{x_2} \neq 0$ we have $e^{x_1+x_2} = 1$ and so $x_1 + x_2 = 0$, that is $x_1 = -x_2$. In other words, $y = \cosh x$ does not define an injection because $\cosh(-u) = \cosh u$ for all real numbers u.

We could if we wished deduce the same results by looking at the graphs.

■

□ Determine which, if any, of the following equations define functions which are injections: (a) $y = x^3$ (b) $y = 1/x^2$ (c) $y = \tanh x$.

When you have had a try at these, move on to check if you have them correct.

---

a Suppose $x_1^3 = x_2^3$. Then $x_1^3 - x_2^3 = 0$, and so

$$(x_1 - x_2)(x_1^2 + x_1x_2 + x_2^2) = 0$$

If $x_1 \neq x_2$ then

$$x_1^2 + x_2^2 + x_1x_2 = 0$$

But

$$x_1^2 + x_2^2 + x_1x_2 = \tfrac{1}{2}(x_1 + x_2)^2 + \tfrac{1}{2}(x_1^2 + x_2^2)$$

is a sum of squares and is therefore only zero when both x_1 and x_2 are zero. Therefore we have an injection.

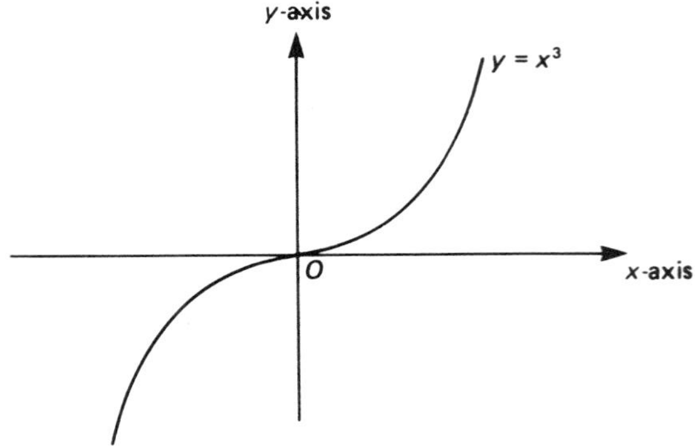

Fig. 5.4 The graph of $y = x^3$.

Alternatively a simple sketch of $y = x^3$ will establish the same result (Fig. 5.4).

b The domain of this function (by the convention of the maximal domain) is $\mathbb{R} \setminus \{0\}$. Moreover

$$\frac{1}{(-2)^2} = \frac{1}{4} = \frac{1}{(2)^2}$$

and so there are two points in the domain at which the value of the function is the same. Therefore the function is not an injection.

Again the graph $y = 1/x^2$ shows immediately that the function is not injective (Fig. 5.5).

c Suppose $\tanh x_1 = \tanh x_2$. Then

$$\frac{e^{2x_1} - 1}{e^{2x_1} + 1} = \frac{e^{2x_2} + 1}{e^{2x_2} + 1}$$

So

$$\begin{aligned}(e^{2x_1} - 1)(e^{2x_2} + 1) &= (e^{2x_2} - 1)(e^{2x_1} + 1)\\ e^{2x_1} - e^{2x_2} &= e^{2x_2} - e^{2x_1}\\ e^{2x_1} &= e^{2x_2}\end{aligned}$$

Therefore $e^{2(x_1 - x_2)} = 1$, so $x_1 - x_2 = 0$ and $x_1 = x_2$. So we have an injection. This property may be inferred directly from the graph of $y = \tanh x$ (Fig. 5.3). ■

Although a graph enables us to see whether or not a function is an injection, the algebraic approach is necessary to establish the fact.

The special feature possessed by an injection can be represented diagrammatically as in Fig. 5.6.

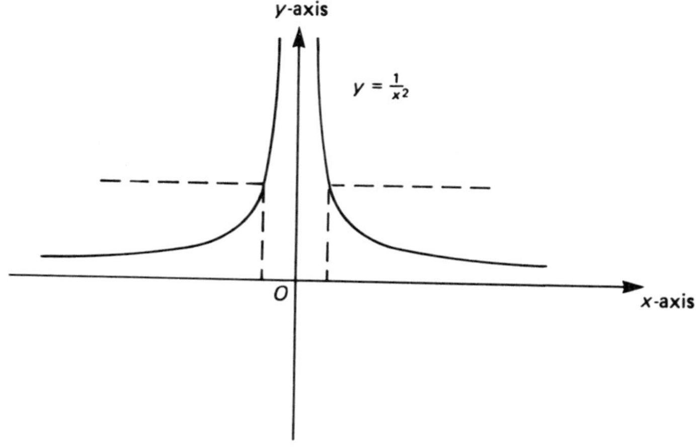

Fig. 5.5 The graph of $y = 1/x^2$.

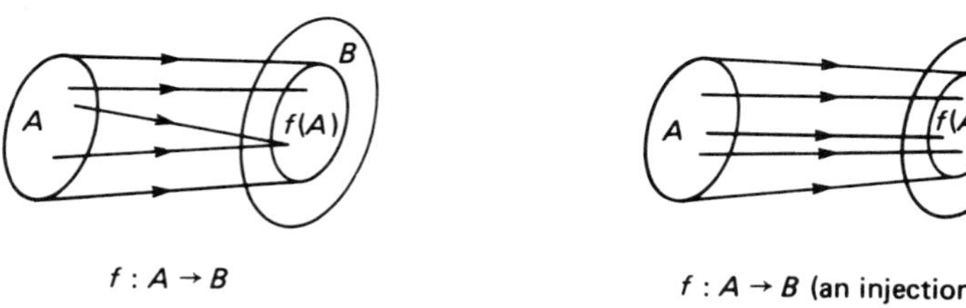

Fig. 5.6

5.6 SURJECTIVE FUNCTIONS

When we introduced the notion of a function (Chapter 2) we observed that if $f: A \to B$ then it is possible to have members of the codomain B which are not in fact values of the function at all.

□ $f: \mathbb{R} \to \mathbb{R}$ defined by $f(x) = \tanh x$ whenever $x \in \mathbb{R}$. Here we know that $-1 < \tanh x < 1$ and so there is no $x \in \mathbb{R}$ such that $\tanh x = 2$. ■

In fact we gave a special name to the set of values of a function. Do you remember what it is called? It is the image set (or range) of the function and is denoted by $f(A)$.

However, for some functions the image set is indeed the codomain. Such functions are somewhat unusual and are given a special name: they are called **surjections** (or **onto functions**). The test of whether or not a function

$f: A \to B$ is a surjection is whether or not $f(A) = B$. That is, whether or not for each $y \in B$ there exists some $x \in A$ such that $f(x) = y$.

□ Consider the functions with codomain $\mathbb{R}$ defined by each of the following equations: (a) $y = \tan x$ (b) $y = \cosh x$.

A graph can often be useful in helping to decide whether or not a function is a surjection.

a From the graph of $y = \tan x$ (Fig. 5.7) it is clear that every real number is a value of the function. In fact given any $y \in \mathbb{R}$ there exists some $x \in (-\pi/2, \pi/2)$ such that $y = \tan x$. We conclude that the tangent function is a surjection.

b From the graph of $y = \cosh x$ (Fig. 5.8) it is clear that there are some

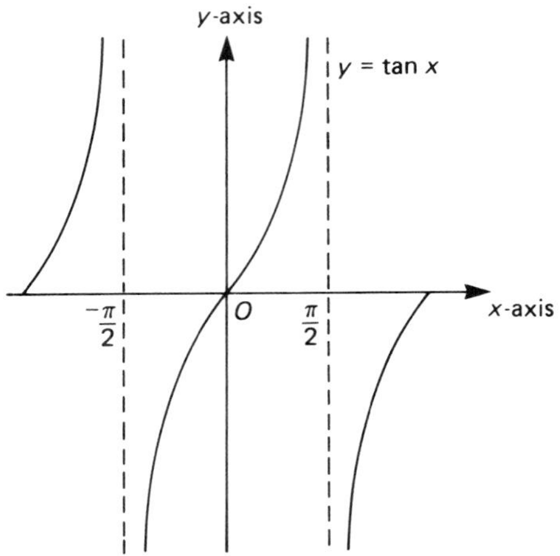

Fig. 5.7 The tangent function.

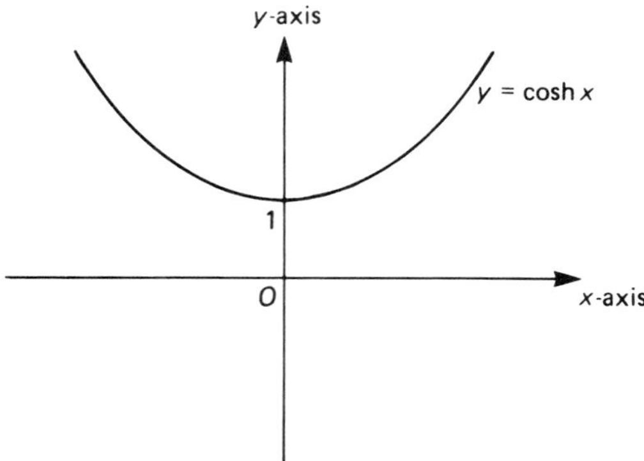

Fig. 5.8 The hyperbolic consine function.

real numbers y which are not values of the function. This is because $\cosh x \geqslant 1$ for all $x \in \mathbb{R}$, and therefore if $y < 1$ there is no real number x such that $y = \cosh x$. Consequently the hyperbolic cosine function is not a surjection. ■

□ For each of the following functions the convention of the maximal domain is to be used to obtain the domain and codomain. Decide in each case whether or not the function is a surjection.

a $y = \sinh x$
b $y = x^2$
c $y = x^3$.

Have a go at these. Don't be afraid to use the graphs to make your decisions.

---

a If we are given any real number y, it is possible to obtain a real number x such that $y = \sinh x$. This is clear from the graph (Fig. 5.9) and so we have a surjection.

b If x is any real number then $x^2 \geqslant 0$. Consequently if y is negative there is no real number x such that $y = x^2$ (Fig. 5.10). Therefore we do not have a surjection.

c From the graph (Fig. 5.11) it is clear that if y is any real number then there exists some real number x such that $y = x^3$. Therefore the function is indeed a surjection. ■

Once more we can use a diagram to represent the special property a function has when it is a surjection: see Fig. 5.12.

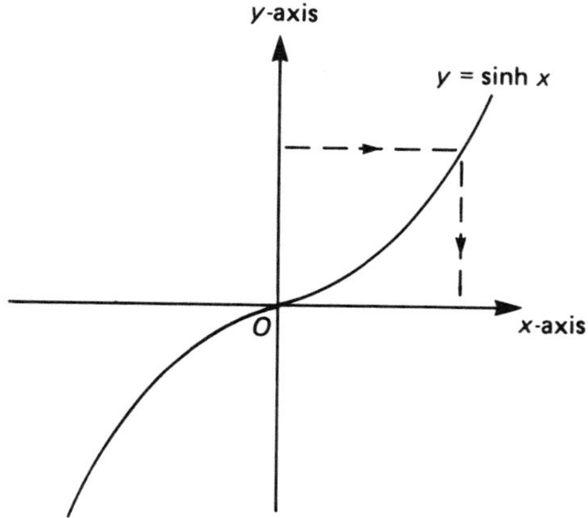

Fig. 5.9 The hyperbolic sine function.

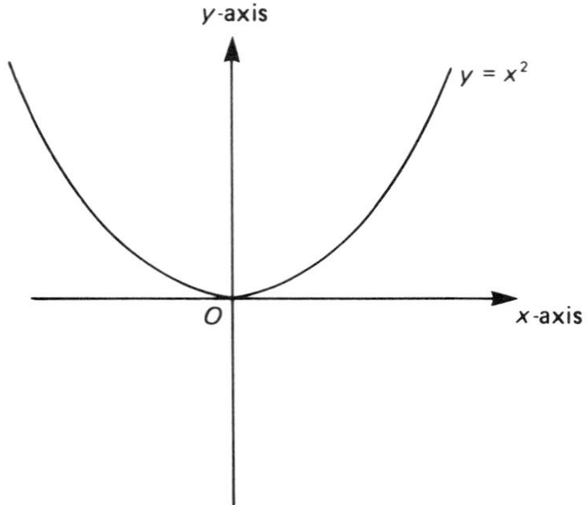

Fig. 5.10 The graph of $y = x^2$.

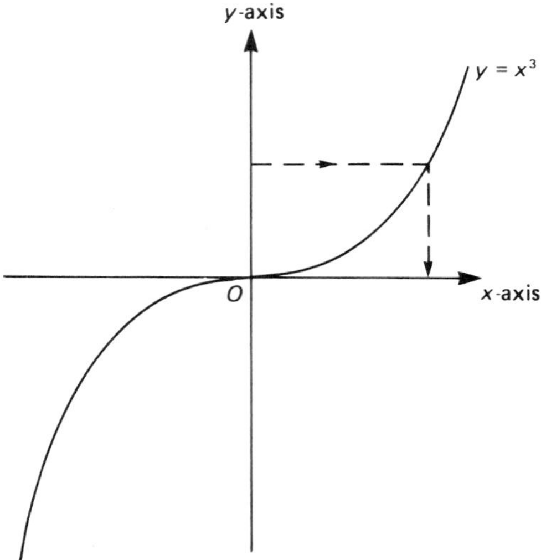

Fig. 5.11 The graph of $y = x^3$.

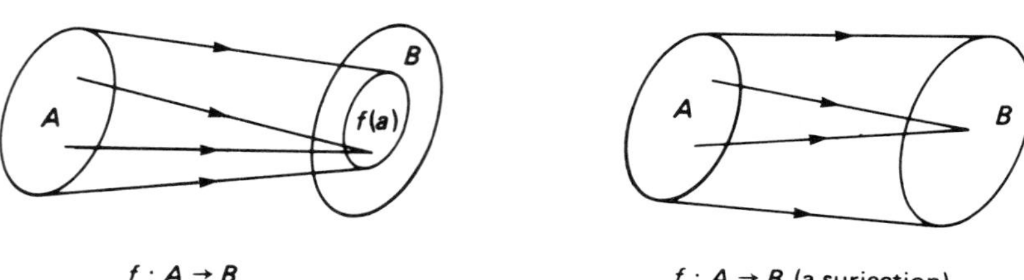

Fig. 5.12

5.7 BIJECTIVE FUNCTIONS

A function which is both an injection and a surjection is called a **bijection**. Such a function can be represented by Fig. 5.13.

Now if $f: A \to B$ is a bijection then to each $y \in B$ there corresponds a unique $x \in A$ such that $y = f(x)$. This means that the action of the function f can be reversed. Therefore there is a function $g: B \to A$ such that if $x \in A$ and $y = f(x)$ then $g(y) = x$.

The function g is called the **inverse function** of f and is usually represented by f^{-1}. Although there are good theoretical reasons for this notation, which we explore further in the context of linear operators (Chapter 22), it can cause problems to the unwary. You must remember that $f^{-1}(x)$ is not the same as $[f(x)]^{-1}$ and be vigilant about this, or nasty errors will be the result. You have been warned!

So if $f: A \to B$ is a bijection there exists an inverse function $f^{-1}: B \to A$ such that

1 If $x \in A$ then $f^{-1}[f(x)] = x$;
2 If $y \in B$ then $f[f^{-1}(y)] = y$.

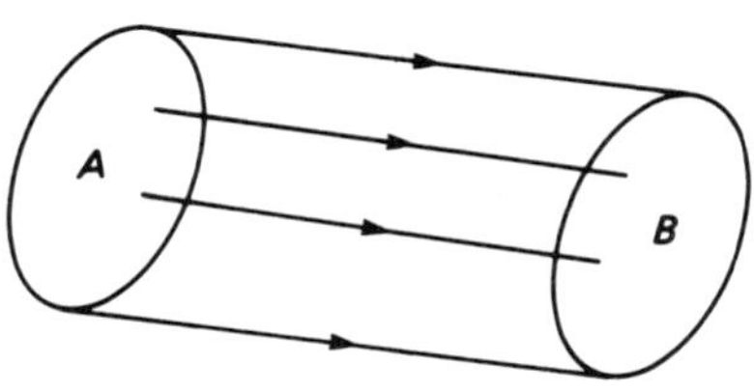

$f: A \to B$ (a bijection)

Fig. 5.13

□ Show that the function defined by $y = \sinh x$ is a bijection and give an explicit expression for its inverse function using logarithms.

By the convention of the maximal domain, the domain and codomain are both ℝ and we have already shown that this function is both an injection and a surjection. Consequently it is a bijection and so has an inverse function.

Suppose $y = \sinh x$. We must reverse this formula to express x in terms of y:

$$y = \sinh x = \frac{e^x - e^{-x}}{2}$$

So $e^x - e^{-x} = 2y$. Therefore

$$(e^x)^2 - 2y(e^x) - 1 = 0$$

This is a quadratic equation in e^x and so we can solve it:

$$e^x = y \pm \sqrt{(y^2 + 1)}$$

At first sight this might appear to give two solutions. However, $\sqrt{(y^2 + 1)} > y$ for all real numbers y, and so, since e^x is always positive, the negative sign must be rejected. Consequently

$$e^x = y + \sqrt{(y^2 + 1)}$$

and so $x = \ln [y + \sqrt{(y^2 + 1)}]$.

Interchanging the symbols x and y (since it is usual to use x for points in the domain and y for points in the codomain) we deduce the inverse function is defined by

$$y = \ln [x + \sqrt{(x^2 + 1)}]$$

or

$$\sinh^{-1} x = \ln [x + \sqrt{(x^2 + 1)}]$$ ■

5.8 PSEUDO-INVERSE FUNCTIONS

Bijections are comparatively rare, and so usually it is necessary to modify either the domain, the codomain or both in order to obtain a function which has an inverse. When this is done the inverse functions are not of course the inverses of the original functions, because the original functions are not bijections and so have no inverses. This fact is often obscured, but most people avoid the difficulty by giving these pseudo-inverse functions the name **principal inverse functions**.

An example will illustrate how this is done.

□ Obtain the principal inverse hyperbolic cosine function and express it in logarithmic form.

We already know that the function defined by $y = \cosh x$ is neither an injection nor a surjection (Fig. 5.14). We can obtain an injection by restricting the domain to $\mathbb{R}_0^+$, the positive real numbers including 0. The codomain must also be modified because, as we have observed, $\cosh x \geqslant 1$ for all real x.

Suppose now that $A = \mathbb{R}_0^+$, that $B = \{r \mid r \in \mathbb{R}, r \geqslant 1\}$ and that $f: A \to B$ is defined by $f(x) = \cosh x$ $(x \in A)$. Then f is a bijection and so has an inverse function $f^{-1}: B \to A$. To obtain $f^{-1}(y)$ explicitly we need to reverse the formula for $y = \cosh x$.

Suppose $y = \cosh x$. Then

$$y = \cosh x = \frac{e^x + e^{-x}}{2}$$

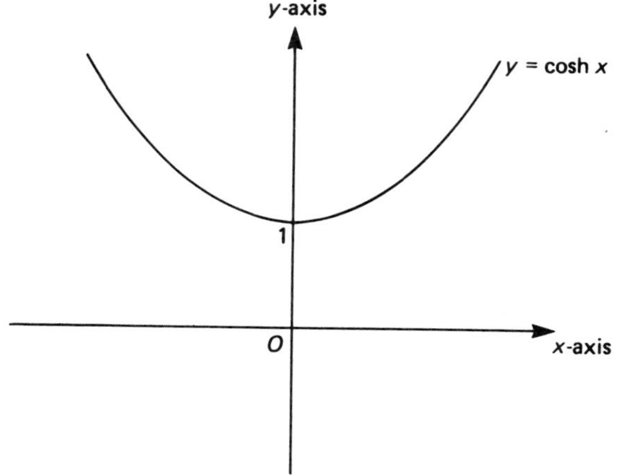

Fig. 5.14 The hyperbolic cosine function.

So $e^x + e^{-x} = 2y$. Therefore

$$(e^x)^2 - 2y(e^x) + 1 = 0$$

and so

$$e^x = y \pm \sqrt{(y^2 - 1)}$$

Now

$$\begin{aligned} y - \sqrt{(y^2 - 1)} &= [y - \sqrt{(y^2 - 1)}]\,\frac{y + \sqrt{(y^2 - 1)}}{y + \sqrt{(y^2 - 1)}} \\ &= \frac{[y - \sqrt{(y^2 - 1)}]\,[y + \sqrt{(y^2 - 1)}]}{y + \sqrt{(y^2 - 1)}} \\ &= \frac{y^2 - (y^2 - 1)}{y + \sqrt{(y^2 - 1)}} = \frac{1}{y + \sqrt{(y^2 - 1)}} \end{aligned}$$

So either

$$e^x = y + \sqrt{(y^2 - 1)}$$

or

$$\begin{aligned} e^x &= [y + \sqrt{(y^2 - 1)}]^{-1} \\ e^{-x} &= y + \sqrt{(y^2 - 1)} \end{aligned}$$

Therefore

$$\begin{aligned} \pm x &= \ln\,[y + \sqrt{(y^2 - 1)}] \\ x &= \pm \ln\,[y + \sqrt{(y^2 - 1)}] \end{aligned}$$

But $x \in A$ and so $x \geq 0$; therefore we must reject the negative value. So

$$x = \ln [y + \surd(y^2 - 1)]$$

The function defined in this way is called the principal inverse hyperbolic cosine function. Interchanging x and y we have

$$\cosh^{-1} x = \ln [x + \surd(x^2 - 1)]$$

Of course we could have chosen a different restriction such as $\mathbb{R}_0^-$ for the domain of the hyperbolic cosine. We have restricted the function so that continuity is not lost and selected positive numbers in preference to negative numbers. Until such a time as there is a campaign for equal rights for negative numbers, nobody is likely to object overmuch. ■

□ Show that $y = \ln x$ defines a bijection and obtain the inverse explicitly.
Try this. There is no need to modify the domain or codomain, but naturally you will need to use the convention of the maximal domain to obtain the domain and codomain.

The convention of the maximal domain gives the domain as $\mathbb{R}^+ = \{r \mid r \in \mathbb{R}, r > 0\}$ and the codomain as $\mathbb{R}$. The graph of $y = \ln x$ (Fig. 5.15) shows that we have a bijection. Now if $y = \ln x$ then $x = e^y$. Therefore the inverse function is the function $g : \mathbb{R} \to \mathbb{R}^+$ defined by $g(x) = e^x$ $(x \in \mathbb{R})$. ■

Observe how we can obtain the graph of an inverse function from the graph of the function itself. Imagine that the graph $y = f(x)$ is drawn on a sheet of glass. Lift the sheet of glass away from the paper, turn it over and put it

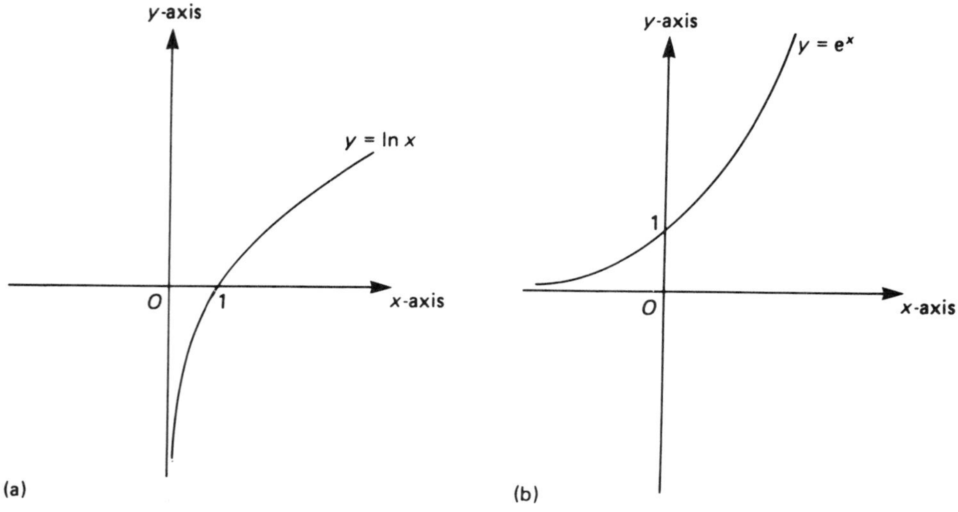

Fig. 5.15 (a) The logarithmic function (b) The exponential function.

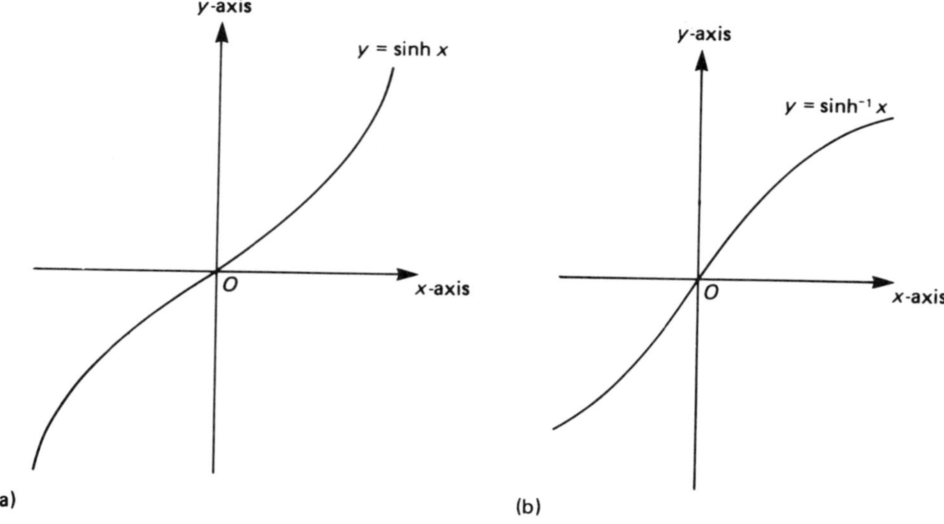

Fig. 5.16 (a) The hyperbolic sine function (b) The inverse hyperbolic sine function.

down with the x-axis where the y-axis was and the y-axis where the x-axis was. All that remains to be done is to relabel the x-axis and y-axis in the usual way.

□ Fig. 5.16 shows the function $y = \sinh x$ and its inverse. ■

5.9 DIFFERENTIATION OF INVERSE FUNCTIONS

In the case of the inverse hyperbolic functions we can differentiate them if we wish by using the logarithmic equivalent. However, this luxury is not generally available and when it isn't we must resort to the definition.

Suppose $y = f^{-1}(x)$. Then $f(y) = x$ and so, differentiating throughout with respect to x,

$$f'(y)\frac{dy}{dx} = 1$$

It is now simply a matter of eliminating y to obtain the derivative of the inverse function f^{-1}.

□ The inverse sine function is defined as the inverse of the bijection obtained by restricting the domain of the sine function to the interval $[-\pi/2, \pi/2]$. Show that

$$\frac{dy}{dx} = \frac{1}{\sqrt{(1 - x^2)}}$$

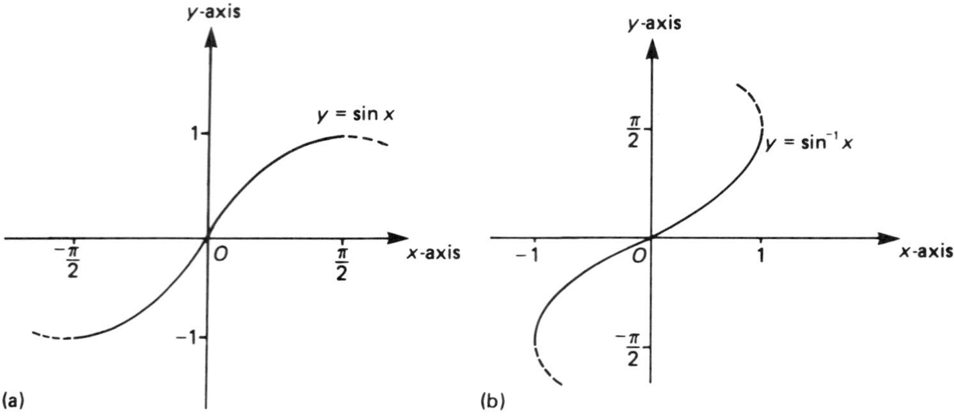

Fig. 5.17 (a) The graph of $y = \sin x$; $x \in [-\pi/2, \pi/2]$ (b) The graph of $y = \sin^{-1} x$.

and justify the choice of sign.

If $y = \sin^{-1} x$ then we know that $x = \sin y$. So differentiating throughout with respect to x we get

$$1 = \cos y \, \frac{dy}{dx}$$

Now $\cos^2 y + \sin^2 y = 1$, and so

$$\cos y = \pm\sqrt{(1 - \sin^2 y)} = \pm\sqrt{(1 - x^2)}$$

So we have

$$\frac{dy}{dx} = \frac{\pm 1}{\sqrt{(1 - x^2)}}$$

Now comes the crunch. If we had been sloppy about taking the square root and had ignored the negative sign, then we should be unaware that there was a crunch at all! A glance at the graph of $y = \sin^{-1} x$ (Fig. 5.17) tells us that the slope is always positive and so the negative can now be rejected with confidence. Naturally if we had taken a different restriction of the sine function to obtain our bijection, such as $[\pi/2, 3\pi/2]$, we could have obtained a negative slope instead! ■

5.10 THE INVERSE CIRCULAR FUNCTIONS

Here is a complete list of the principal inverse circular functions. As you can see, the domains and codomains of the circular functions have had to be modified to produce bijections.

1 If $A = [-\pi/2, \pi/2]$ and $B = [-1, 1]$,

$$f: A \to B \text{ defined by } f(x) = \sin x \ (x \in A)$$

is a bijection and its inverse is the principal inverse sine function. Both the domain and the codomain needed modification.

2 If $A = [0, \pi]$ and $B = [-1, 1]$,

$$f: A \to B \text{ defined by } f(x) = \cos x \ (x \in A)$$

is a bijection and its inverse is the principal inverse cosine function. Both the domain and the codomain needed modification.

3 If $A = (-\pi/2, \pi/2)$ and $B = \mathbb{R}$,

$$f: A \to B \text{ defined by } f(x) = \tan x \ (x \in A)$$

is a bijection and its inverse is the principal inverse tangent function. The domain needed modification.

4 If $A = [0, \pi] \setminus \{\pi/2\}$ and $B = \{r : r \in \mathbb{R}, |r| \geqslant 1\}$,

$$f: A \to B \text{ defined by } f(x) = \sec x \ (x \in A)$$

is a bijection and its inverse is the principal inverse secant function. Both the domain and the codomain needed modification.

5 If $A = [-\pi/2, \pi/2] \setminus \{0\}$ and $B = \{r : r \in \mathbb{R}, |r| \geqslant 1\}$,

$$f: A \to B \text{ defined by } f(x) = \operatorname{cosec} x \ (x \in A)$$

is a bijection and its inverse is the principal inverse cosecant function. Both the domain and the codomain needed modification.

6 If $A = [0, \pi]$ and $B = \mathbb{R}$,

$$f: A \to B \text{ defined by } f(x) = \cot x \ (x \in A)$$

is a bijection and its inverse is the principal inverse cotangent function. The domain needed modification.

Now it's time to take a few steps. As soon as you are ready, press ahead.

5.11 Workshop

▷**Exercise** Show that the function defined by $y = \tanh x$ is not a bijection, but that by restricting the codomain to $(-1, 1)$ a bijection is obtained. The principal inverse hyperbolic tangent function $\tanh^{-1}$ is the inverse of this modified function. Deduce that 1

$$\tanh^{-1} x = \frac{1}{2} \ln \left(\frac{1 + x}{1 - x} \right)$$

and give a rough sketch of the graph.

Try this carefully before you proceed.

---

2 We saw when we drew the graph of $y = \tanh x$ (Fig. 5.3) that, for all real x, $-1 < \tanh x < 1$. Therefore it is necessary to restrict the codomain to $(-1, 1)$ to obtain a surjection. The function is an injection, and therefore if $A = \mathbb{R}$ and $B = (-1, 1)$ the function

$$f: A \to B \text{ defined by } f(x) = \tanh x \ (x \in A)$$

is a bijection and has an inverse function $\tanh^{-1}$.

We obtain the graph of $y = \tanh^{-1} x$ by interchanging the positions of the x-axis and the y-axis. We need a three-dimensional transformation to achieve this. Another way of looking at this transformation is as a two-stage operation. First we twist the graph of $y = \tanh x$ anticlockwise by $\pi/2$. Then we flip it over, that is we reflect it in the x-axis which is now vertical. Finally we relabel the axes (Fig. 5.18).

If $y = \tanh x$ then

$$\begin{aligned} y &= \frac{e^{2x} - 1}{e^{2x} + 1} \\ y(e^{2x} + 1) &= e^{2x} - 1 \\ 1 + y &= e^{2x}(1 - y) \\ e^{2x} &= \frac{1 + y}{1 - y} \end{aligned}$$

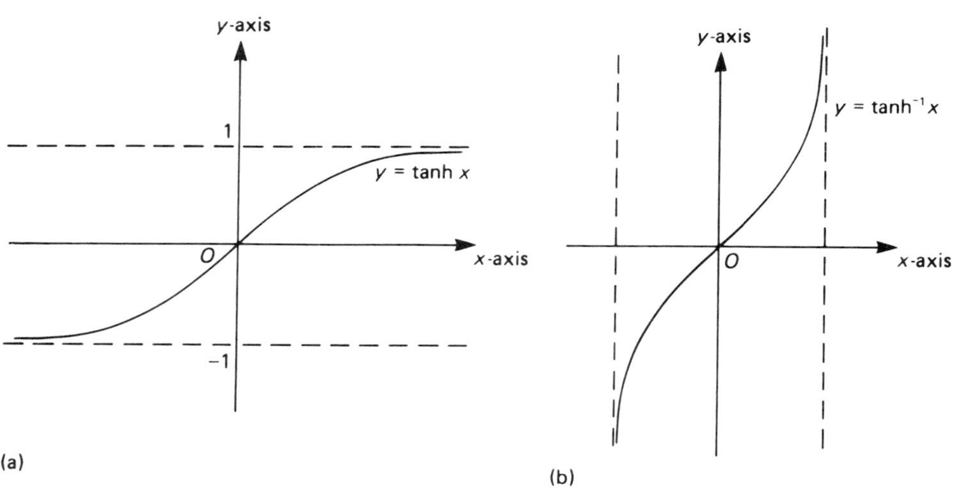

Fig. 5.18 (a) The graph of $y = \tanh x$ (b) The graph of $y = \tanh^{-1} x$.

Therefore

$$x = \frac{1}{2} \ln \left(\frac{1 + y}{1 - y} \right)$$

So

$$\tanh^{-1} x = \frac{1}{2} \ln \left(\frac{1 + x}{1 - x} \right)$$

If you managed to do all that correctly then you may move ahead to step 4. If there were unresolved problems then at this stage you should go through the theory of inverse functions once more. When you have smoothed out any difficulties, try the next exercise.

▷ **Exercise** Explain why the function defined by $y = \operatorname{cosech} x$ is not a bijection. Show that by removing a single point from both the domain and the codomain a bijection can be obtained. The principal inverse hyperbolic cosecant is the inverse of this modified function. Draw its graph and show that

$$\operatorname{cosech}^{-1} x = \ln \left[\frac{1}{x} + \sqrt{\left(\frac{1}{x^2} + 1 \right)} \right]$$

3

The graph of $y = \operatorname{cosech} x$ (Fig. 5.19) can be deduced easily from the graph of $y = \sinh x$ (Fig. 5.2).

There is no value of x for which $\operatorname{cosech} x = 0$, and so the function is not a surjection. Therefore there is no bijection, and consequently no inverse function. However, if we take

$$A = B = \{r \mid r \in \mathbb{R}, r \neq 0\}$$

then

$$f : A \to B \text{ defined by } f(x) = \operatorname{cosech} x \ (x \in A)$$

is a bijection. Its inverse can be drawn in the usual way (Fig. 5.20).

Now if $y = \operatorname{cosech} x$ then $y = 1/\sinh x$. So $\sinh x = 1/y$, from which

$$\begin{aligned} x &= \sinh^{-1} \left(\frac{1}{y} \right) \\ &= \ln \left[\frac{1}{y} + \sqrt{\left(\frac{1}{y^2} + 1 \right)} \right] \end{aligned}$$

so that

$$\operatorname{cosech}^{-1} y = \ln \left[\frac{1}{y} + \sqrt{\left(\frac{1}{y^2} + 1 \right)} \right] \ (y \neq 0)$$

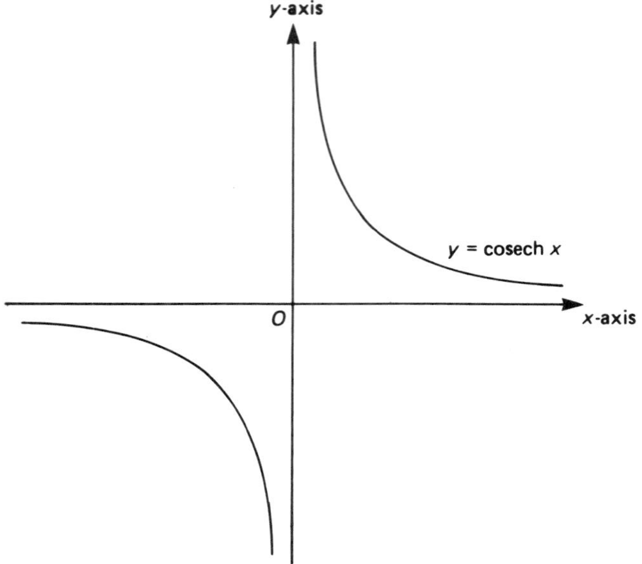

Fig. 5.19 The graph of $y = \text{cosech } x$.

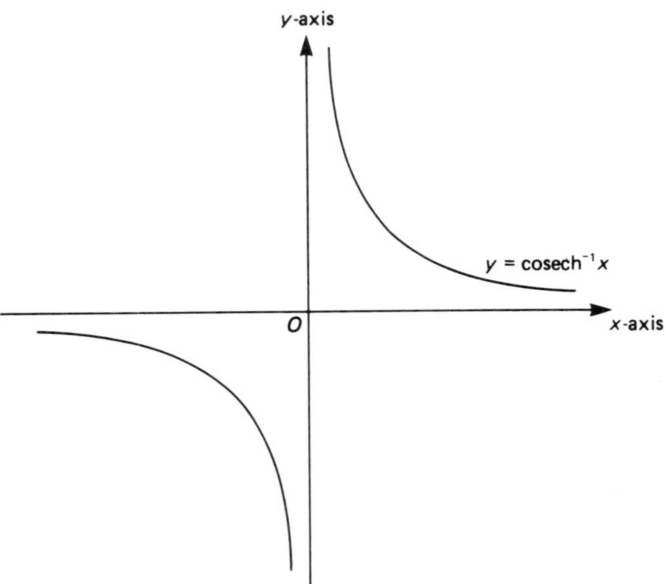

Fig. 5.20 The graph of $y = \text{cosech}^{-1} x$.

or in terms of x

$$\operatorname{cosech}^{-1} x = \ln\left[\frac{1}{x} + \sqrt{\left(\frac{1}{x^2} + 1\right)}\right]$$

Now we come to the problem of differentiation of inverse functions.

▷**Exercise** Differentiate $\operatorname{cosec}^{-1} x$ with respect to x.
Try this, then step forward. **4**

---

5 If $y = \operatorname{cosec}^{-1} x$ then $x = \operatorname{cosec} y$, so that differentiating with respect to x

$$1 = -\operatorname{cosec} y \cot y \frac{dy}{dx}$$

Now $1 + \cot^2 y = \operatorname{cosec}^2 y$, so

$$\cot y = \pm\sqrt{(\operatorname{cosec}^2 y - 1)} = \pm\sqrt{(x^2 - 1)}$$

Therefore

$$1 = \mp x\sqrt{(x^2 - 1)}\frac{dy}{dx}$$

$$\frac{dy}{dx} = \frac{\mp 1}{x\sqrt{(x^2 - 1)}}$$

It remains to decide which sign is the correct one. If we sketch the graph of $y = \operatorname{cosec}^{-1} x$ (Fig. 5.21, overleaf) we see that the slope is negative, and so the negative sign must be chosen:

$$\frac{dy}{dx} = \frac{-1}{x\sqrt{(x^2 - 1)}}$$

If you discussed the choice of sign and succeeded in obtaining the correct derivative, then try one last problem. If you omitted to consider the sign or if you made an error, take care with this one.

▷**Exercise** Differentiate $\operatorname{sech} x$ and $\operatorname{sech}^{-1} x$ with respect to x.
Have a go at both of these, then step ahead.

---

6 First, if $y = \operatorname{sech} x$ then $y = (\cosh x)^{-1}$. Therefore, using the chain rule (Chapter 4),

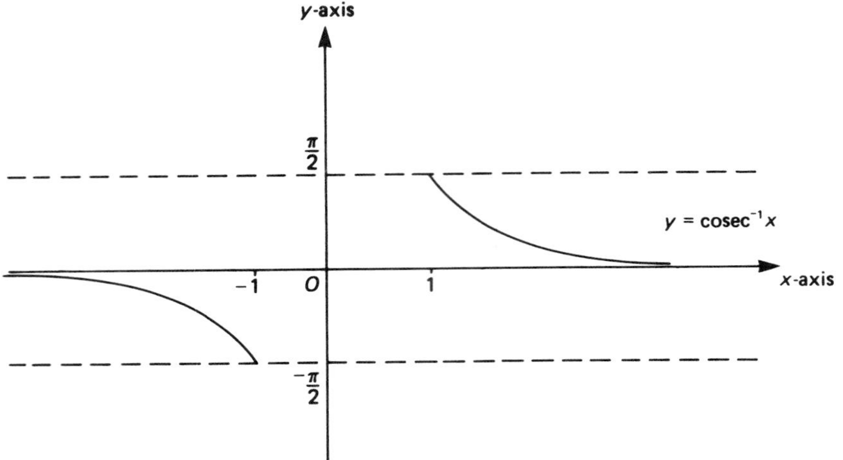

Fig. 5.21 The graph of $y = \operatorname{cosec}^{-1} x$.

$$\frac{dy}{dx} = -(\cosh x)^{-2} \sinh x$$

$$= -\frac{\sinh x}{\cosh^2 x} = -\operatorname{sech} x \tanh x$$

Secondly, if $y = \operatorname{sech}^{-1} x$ then $x = \operatorname{sech} y$. So

$$1 = -\operatorname{sech} y \tanh y \frac{dy}{dx}$$

Now $1 - \tanh^2 y = \operatorname{sech}^2 y$, so

$$\tanh y = \pm\sqrt{(1 - \operatorname{sech}^2 y)} = \pm\sqrt{(1 - x^2)}$$

Therefore

$$1 = \mp x\sqrt{(1 - x^2)} \frac{dy}{dx}$$

$$\frac{dy}{dx} = \frac{\mp 1}{x\sqrt{(1 - x^2)}}$$

It remains to decide which sign is the correct one. If we sketch the graph of $y = \operatorname{sech}^{-1} x$ (Fig. 5.22) we see that the slope is negative, and so the negative sign must be chosen:

$$\frac{dy}{dx} = \frac{-1}{x\sqrt{(1 - x^2)}}$$

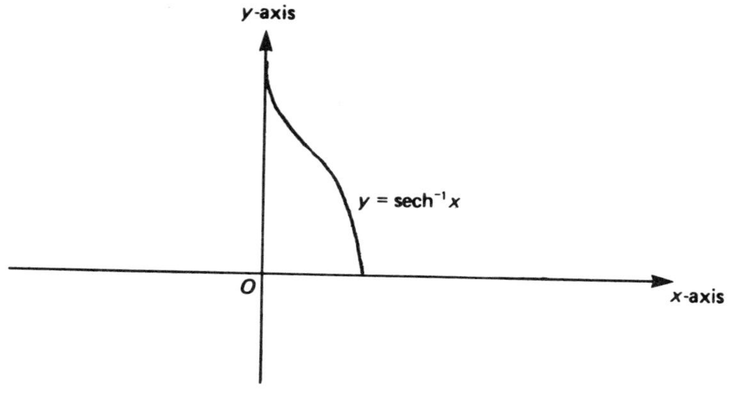

Fig. 5.22 The graph of $y = \operatorname{sech}^{-1} x$.

5.12 THE INVERSE HYPERBOLIC FUNCTIONS

For the sake of completeness we state the bijections which have inverses known as the principal inverse hyperbolic functions:

1 If $A = B = \mathbb{R}$,

$$f: A \to B \text{ defined by } f(x) = \sinh x \ (x \in A)$$

is a bijection and its inverse is the inverse hyperbolic sine function. No modification to the domain or the codomain was needed.

2 If $A = \mathbb{R}_0^+$ and $B = \{r \mid r \in \mathbb{R}, r \geqslant 1\}$,

$$f: A \to B \text{ defined by } f(x) = \cosh x \ (x \in A)$$

is a bijection and its inverse is the principal inverse hyperbolic cosine function. Both the domain and the codomain needed to be modified.

3 If $A = \mathbb{R}$ and $B = (-1, 1)$,

$$f: A \to B \text{ defined by } f(x) = \tanh x \ (x \in A)$$

is a bijection and its inverse is the principal inverse hyperbolic tangent function. The codomain needed modification.

4 If $A = \mathbb{R}_0^+$ and $B = \{r \mid r \in \mathbb{R}, 0 < r \leqslant 1\} = (0, 1]$,

$$f: A \to B \text{ defined by } f(x) = \operatorname{sech} x \ (x \in A)$$

is a bijection and its inverse is the principal inverse hyperbolic secant function. Both the domain and the codomain needed modification.

5 If $A = B = \mathbb{R} \setminus \{0\}$,

$$f: A \to B \text{ defined by } f(x) = \operatorname{cosech} x \ (x \in A)$$

is a bijection and its inverse is the principal inverse hyperbolic cosecant function. Both the domain and the codomain needed modification.

6 If $A = \mathbb{R}\setminus\{0\}$ and $B = \{r : r \in \mathbb{R}, |r| > 1\}$,

$$f : A \to B \text{ defined by } f(x) = \coth x \ (x \in A)$$

is a bijection and its inverse is the principal inverse hyperbolic cotangent function. Both the domain and the codomain needed modification.

Now it remains only to work through an application.

5.13 Practical

SAGGING CHAIN

A chain hangs in the shape of the curve

$$y = c \cosh (x/c)$$

It is suspended from two points at the same horizontal level and at distance $2d$ apart. Obtain an expression for the sag at the midpoint, if the angle of slope at the ends is $\theta°$ to the horizontal.

It is worthwhile seeing if you can make progress on your own. We shall solve the problem stage by stage, so try it first and then see how it goes.

The sagging chain is shown in Fig. 5.23. Using the diagram, we have $dy/dx = \tan \theta$ when $x = d$. So $\tan \theta = \sinh (d/c)$. Therefore, using the result in section 5.7,

$$\begin{aligned} d/c = \sinh^{-1} (\tan \theta) &= \ln [\tan \theta + \sqrt{(1 + \tan^2 \theta)}] \\ &= \ln (\tan \theta + \sec \theta) \end{aligned}$$

Consequently

$$c = d/\ln (\tan \theta + \sec \theta)$$

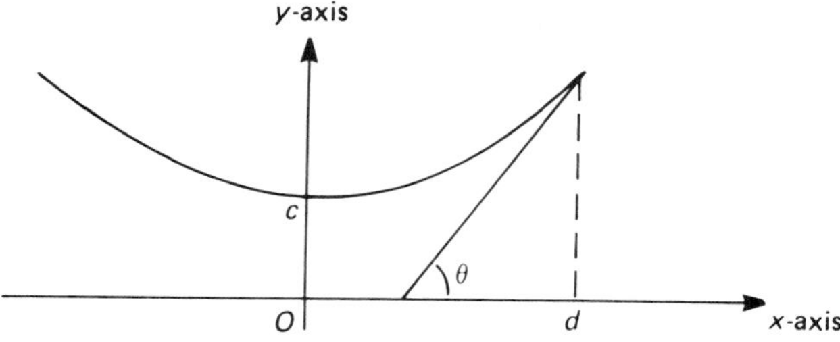

Fig. 5.23 The graph of $y = c \cosh (x/c)$.

Now that you know c, even if you weren't able to obtain it, you may be able to continue. Before doing so, make sure you follow all the stages.

The sag is the difference between the y value at $x = d$ and the y value at $x = 0$, namely c. Therefore

$$\text{sag} = c \cosh (d/c) - c = c\,[\cosh (d/c) - 1]$$

Now $\sinh (d/c) = \tan \theta$, and so

$$\cosh^2 (d/c) = 1 + \sinh^2 (d/c) = 1 + \tan^2 \theta = \sec^2 \theta$$

Therefore

$$\text{sag} = c\,(\sec \theta - 1) = d\,(\sec \theta - 1)/\ln (\tan \theta + \sec \theta)$$

Although in some ways this problem has been rather straightforward, it is not without practical significance. For instance, it would enable us to calculate the amount of clearance which a vehicle would have.

SUMMARY

□ We defined the hyperbolic functions

$$\cosh x = \frac{e^x + e^{-x}}{2} \qquad \sinh x = \frac{e^x - e^{-x}}{2}$$

and drew their graphs.

□ We obtained identities and solved equations involving hyperbolic functions.

□ We differentiated the hyperbolic functions

$$\frac{d}{dx}(\cosh x) = \sinh x \qquad \frac{d}{dx}(\sinh x) = \cosh x$$

□ We examined functions to see if they had inverses.

□ We defined the principal inverse hyperbolic functions and obtained logarithmic equivalents

$$\cosh^{-1} x = \ln [x + \sqrt{(x^2 - 1)}]$$
$$\sinh^{-1} x = \ln [x + \sqrt{(x^2 + 1)}]$$

EXERCISES

1 Establish each of the following identities:

a $2 \sinh^2 x = \cosh 2x - 1$

b $\coth(x + y) = \dfrac{\coth x \coth y + 1}{\coth y + \coth x}$

c $\tanh(x - y) = \dfrac{\tanh x - \tanh y}{1 - \tanh x \tanh y}$

d $8\sinh^4 u + 4\cosh 2u = 3 + \cosh 4u$

e $\cosh 4u = 8\cosh^4 u - 4\cosh 2u - 3$

2 Solve the following equations, where x is a real number:

a $1 + \sinh 2x = 10\cosh x - \cosh 2x$

b $x\cosh x - \sinh x + 1 = \cosh x - x\sinh x + x$

c $\cosh 2x - 7(\cosh x - 1) - 1 = 7(\sinh x - 1) - \sinh 2x + 1$

d $12(\operatorname{sech} x - 1) = 1 - 13\tanh^2 x$

3 Differentiate, with respect to t,

a $\operatorname{sech} 3t$

b $t^2 \sinh 2t$

c $\dfrac{\sinh t}{\cosh 2t}$

d $\operatorname{sech} t \operatorname{cosech} 2t$

e $\sqrt{(\operatorname{sech} t^2)}$

f $\operatorname{sech}^2 \sqrt{t}$

4 Differentiate, with respect to t,

a $\cosh^{-1}(2t^2 + 1)$

b $\tanh^{-1}(t^2 + 1)$

c $\sinh t \cosh^{-1} t$

d $\ln[\cosh^{-1} t]$

e $\cosh^{-1}[\ln t]$

f $\dfrac{1}{\sinh^{-1} t}$

ASSIGNMENT

1 Solve for real x the equations

a $2 \sinh 2x = 1 + \cosh 2x$

b $2 \sinh 6x = 5 \sinh 3x$

2 Prove that if $a = \cosh x + \sinh x$ and $b = \cosh x - \sinh x$ then

a $ab = 1$

b $a^2 + b^2 = 2 \cosh 2x$

c $a^2 - b^2 = 2 \sinh 2x$

3 Solve the equation

$$1 + \sinh 2x \sinh 3x = (4/3) \sinh 3x + (3/4) \sinh 2x$$

4 Obtain all the real numbers x which satisfy the equation

$$2 \sinh 2x - 4 \sinh x - 3 \cosh x + 3 = 0$$

5 If $u = \cosh x + \sinh x$ and $v = \cosh x \sinh x$, show that $u^4 = 4u^2v + 1$.
6 If $y = \ln x$, show that

$$\cosh y = \frac{x^2 + 1}{2x} \quad \text{and} \quad \sinh y = \frac{x^2 - 1}{2x}$$

Hence, or otherwise, show that
a If $\cosh y = a$ then $x = a \pm \sqrt{(a^2 - 1)}$; whereas
b If $\sinh y = a$ then $x = a \pm \sqrt{(a^2 + 1)}$.
7 If

$$y = \frac{\sin^{-1} x}{\sqrt{(1 + x^2)}}$$

show that

$$(1 + x^2)\frac{dy}{dx} + xy = \left(\frac{1 + x^2}{1 - x^2}\right)^{1/2}$$

FURTHER EXERCISES

1 By first simplifying each expression, or otherwise, differentiate with respect to x
a $\exp [\ln (x^{-1}) + 2 \ln x]$
b $\tan^{-1} [(1 - \cos x)/\sin x]$
Simplify your answer as far as possible.
2 If $y = x \tan^{-1} x - \ln (1 + x^2)^{1/2}$, show that

$$(1 + x^2)\frac{d^2y}{dx^2} = 1$$

3 Differentiate with respect to x
a $\cos^{-1} (3 \cos x)$
b $\tan^{-1} [(x^2 - 1)/2x]$
c $\tan^{-1} [(1 + \sin x)/(1 - \sin x)]$
4 If $a = \sinh 2x$ and $b = \tanh x$, show that $2b + ab^2 = a$.
5 Show that
a $\sinh (\sinh^{-1} a - \sinh^{-1} b) = a\sqrt{(b^2 + 1)} - b\sqrt{(a^2 + 1)}$
b $\cosh (\sinh^{-1} a - \sinh^{-1} b) = \sqrt{[(a^2 + 1)(b^2 + 1)]} - ab$
6 Establish each of the following from the definitions:
a $\text{cosech}^2 u = \coth^2 u - 1$
b $\cosh^2 u - \sinh^2 u = 1$
c $\cosh 2u = \cosh^2 u + \sinh^2 u$
d $\cosh (u + v) = \cosh u \cosh v + \sinh u \sinh v$
e $\sinh (u + v) = \sinh u \cosh v + \cosh u \sinh v$
7 Solve
a $\cosh 2x - 5 \cosh x + 3 = 0$

b $2\cosh x + \sinh x + \sinh 2x + 1 = 0$
c $\cosh x + \cosh 2x = 2$

8 A laser beam cuts a groove in a plate. The distance of the point of contact from a pivot is given at time t by $r = \alpha(t^2 - 2t + 2)$, where $0 \leqslant t \leqslant 10$ and α is positive.
a What is the shortest distance from the groove to the pivot?
b If the groove is in the shape of a straight line, determine the interval over which the beam etches the groove more than once.
c Show that the cutting process consists of three phases: clean plate is cut; plate is cut a second time; clean plate is cut. Determine the lengths of the time intervals for each phase.

9 An automatic paint spraying machine sprays paint at a height h (metres) at time t (seconds) given by $h = \sin 2t + \cos 2t + 2$.
a Determine the maximum and minimum heights at which the machine operates.
b How long should the machine be applied if each point is to be painted twice?
c At what time will the paint head be at its lowest height?

10 The input I and the output E of an experiment are related by $E = \cos 2I + \cos I + 2$. The experimenter wishes to be able to read the output and thereby determine the input uniquely. Practical considerations restrict possible inputs to $0 \leqslant I \leqslant 8$. What further restrictions should be imposed on the input given that the input must be an interval, and that small inputs are difficult to produce?

11 In a given volume of fluid an unknown number n of negatively charged particles of type A are present. It is proposed to count the particles by bombarding the fluid with positively charged particles of type B and type C. It is known that:
a Each particle of type A bonds with 11 particles of other types.
b Each particle of type B bonds with 7 particles of type A.
c Each particle of type C bonds with 5 particles of type A.
A mixture is made with 3 particles of type B to every 2 of type C. The mixture is introduced to the fluid until the overall mixture becomes stable and electrically neutral. This occurs when 605 particles have been introduced. Determine n.

Suppose particles of type A can be further classified into either β particles (those which bond with particles of type B only) or γ particles (those which bond with particles of type C only). How many β particles and how many γ particles are present?

12 Solve the equation

$$4\cosh(\ln x) - 2\sinh\left(\ln \frac{x}{2}\right) = 5$$

and hence show that the difference between the roots is 2/3.

13 Solve the equation

$$\frac{1}{\sinh^2 x} - \frac{1}{\cosh^2 x} = \frac{1}{2}$$

Further differentiation 6

In Chapter 4 we described how to differentiate simple functions. In this chapter we shall combine this knowledge with some of the geometrical ideas which we developed in Chapter 3 to obtain tangents and normals to plane curves.

After completing this chapter you should be able to

- ☐ Determine the equations of tangents and normals to plane curves;
- ☐ Use intrinsic coordinates and relate them to cartesian coordinates;
- ☐ Calculate the radius of curvature at a point on a curve and the position of the corresponding centre of curvature.

Finally in this chapter we shall solve a practical problem involving a moored dirigible.

6.1 TANGENTS AND NORMALS

We can apply differentiation directly to obtain the equations of the tangent and the normal at a general point (a, b) on a curve $f(x, y) = 0$. The normal is the straight line perpendicular to the tangent through the point of contact. Therefore if the slope of the tangent is m, the slope of the normal m' satisfies $mm' = -1$.

We know from our previous work (Chapter 4) that $\mathrm{d}y/\mathrm{d}x$ is the slope of the curve at a general point. Therefore we have a general method for obtaining the equations of tangents and normals to plane curves:

1 Differentiate, with respect to x, throughout the equation $f(x, y) = 0$ to obtain the slope $\mathrm{d}y/\mathrm{d}x$ at a general point (x, y).
2 Substitute $x = a$ and $y = b$ to obtain m, the slope of the curve at (a, b).
3 The equation of the **tangent** at (a, b) is then

$$y - b = m(x - a)$$

4 The equation of the **normal** at (a, b) is

$$y - b = m'(x - a)$$

where $mm' = -1$.

The only thing you have to be a little careful about is to make sure that the point (a, b) really is on the curve! You should check therefore that $x = a$, $y = b$ satisfy the equation $f(x, y) = 0$.

□ Determine the equations of the tangent and the normal at the point $(a, 2a)$ on the curve

$$xy^2 - x^3 = a^2y + ax^2$$

We follow the four stages of the general method:

1 Differentiating through the equation with respect to x gives

$$y^2 + x\,2y\,\mathrm{d}y/\mathrm{d}x - 3x^2 = a^2\,\mathrm{d}y/\mathrm{d}x + 2ax$$

So that

$$(2xy - a^2)\,\mathrm{d}y/\mathrm{d}x = 2ax - y^2 + 3x^2$$

Consequently

$$\mathrm{d}y/\mathrm{d}x = (2ax - y^2 + 3x^2)/(2xy - a^2)$$

2 At the point $(a, 2a)$ we therefore have

$$\begin{aligned}\mathrm{d}y/\mathrm{d}x &= [2a^2 - (2a)^2 + 3a^2]/[2a(2a) - a^2] \\ &= a^2/3a^2 = 1/3\end{aligned}$$

3 We have $m = 1/3$, and so the equation of the tangent is

$$\begin{aligned}y - 2a &= \tfrac{1}{3}(x - a) \\ 3(y - 2a) &= x - a \\ 3y - 6a &= x - a \\ 3y &= x + 5a\end{aligned}$$

4 For the normal we have the slope $m' = -3$, since $mm' = -1$. Therefore the equation of the normal is

$$\begin{aligned}y - 2a &= -3(x - a) \\ y - 2a &= -3x + 3a \\ y + 3x &= 5a\end{aligned}$$

■

If the curve is defined parametrically then the same principles apply. Naturally we shall obtain $\mathrm{d}y/\mathrm{d}x$ by using $\mathrm{d}y/\mathrm{d}x = (\mathrm{d}y/\mathrm{d}t)(\mathrm{d}t/\mathrm{d}x)$.

It is convenient to use a simplified notation, known as the **dot notation**. In this notation a derivative with respect to the parameter is indicated by the use of a dot over the variable: so $\dot{x} = \mathrm{d}x/\mathrm{d}t$. A second dot indicates a second-order derivative: so $\ddot{x} = \mathrm{d}^2x/\mathrm{d}t^2$. So we have shown $\mathrm{d}y/\mathrm{d}x = \dot{y}/\dot{x}$.

It is interesting to note that the dot is one of the few symbols to have survived from Newton's original work on the calculus. Much of the notation which we use today was introduced by the co-discoverer of the calculus, Leibniz.

□ Obtain the equations of the tangent and the normal at the general point p on the curve

$$x = p^2 + \sin 2p$$
$$y = 2p + 2\cos 2p$$

We have

$$\dot{x} = 2p + 2\cos 2p$$
$$\dot{y} = 2 - 4\sin 2p$$

So

$$\begin{aligned} m = \mathrm{d}y/\mathrm{d}x &= (2 - 4\sin 2p)/(2p + 2\cos 2p) \\ &= (1 - 2\sin 2p)/(p + \cos 2p) \end{aligned}$$

For the tangent,

$$(y - 2p - 2\cos 2p) = \frac{1 - 2\sin 2p}{p + \cos 2p}(x - p^2 - \sin 2p)$$

from which

$$(y - 2p - 2\cos 2p)(p + \cos 2p) = (1 - 2\sin 2p)(x - p^2 - \sin 2p)$$

So

$$\begin{aligned} &(p + \cos 2p)y - (1 - 2\sin 2p)x \\ &\quad = 2(p + \cos 2p)^2 - (1 - 2\sin 2p)(p^2 + \sin 2p) \\ &\quad = 2p^2 + 2\cos^2 2p + 4p\cos 2p - p^2 + 2p^2\sin 2p - \sin 2p + 2\sin^2 2p \\ &\quad = p^2 + 2 + 4p\cos 2p + 2p^2\sin 2p - \sin 2p \end{aligned}$$

For the normal,

$$(y - 2p - 2\cos 2p) = -\frac{p + \cos 2p}{1 - 2\sin 2p}(x - p^2 - \sin 2p)$$

from which

$$(y - 2p - 2\cos 2p)(1 - 2\sin 2p) = -(p + \cos 2p)(x - p^2 - \sin 2p)$$

So

$$\begin{aligned} &(1 - 2\sin 2p)y + (p + \cos 2p)x \\ &\quad = 2(p + \cos 2p)(1 - 2\sin 2p) + (p + \cos 2p)(p^2 + \sin 2p) \\ &\quad = (p + \cos 2p)(2 - 4\sin 2p + p^2 + \sin 2p) \\ &\quad = (p + \cos 2p)(2 - 3\sin 2p + p^2) \end{aligned}$$ ■

Here now are a few steps to make sure we have the ideas straight.

6.2 Workshop

1 **Exercise** For the curve $y^2 = x^3 + x + 1$, obtain the equations of the tangent and the normal at the point $(0, 1)$.

As soon as you have done this, take a step and see if you are right.

---

2 We check that the point $(0, 1)$ does in fact lie on the curve, and then proceed to differentiate to obtain the slope at a general point.

$$2y \, \mathrm{d}y/\mathrm{d}x = 3x^2 + 1$$

so that

$$\mathrm{d}y/\mathrm{d}x = (3x^2 + 1)/2y$$

For the tangent at the point $(0, 1)$ we have

$$m = \mathrm{d}y/\mathrm{d}x = (0 + 1)/2 = 1/2$$

The equation is therefore

$$(y - 1) = \tfrac{1}{2}(x - 0) = x/2$$
$$y = x/2 + 1$$

For the normal at the point $(0, 1)$ the slope m' satisfies $mm' = -1$, and so $m' = -2$. The equation is therefore

$$(y - 1) = -2(x - 0)$$
$$y = -2x + 1$$

If there are any difficulties here it may be necessary for you to revise your work on the equations of the straight line in Chapter 3.

Another exercise follows. Are you ready?

▷**Exercise** The parametric equations of a curve are given as

$$x = t + 1/t, \qquad y = t - 1/t$$

Obtain the equations of the tangent and the normal at a general point t, and at the point where $t = 1$.

When you have done it, step forward.

---

3 We must obtain $\mathrm{d}y/\mathrm{d}x$ at the point t. For this purpose we use the chain rule

$$\mathrm{d}y/\mathrm{d}x = (\mathrm{d}y/\mathrm{d}t)(\mathrm{d}t/\mathrm{d}x) = \dot{y}/\dot{x}$$

Now

$$\dot{x} = 1 - 1/t^2 = (t^2 - 1)/t^2$$
$$\dot{y} = 1 + 1/t^2 = (t^2 + 1)/t^2$$

So

$$m = \mathrm{d}y/\mathrm{d}x = (t^2 + 1)/(t^2 - 1)$$

at a general point t.

The equation of the tangent is therefore

$$[y - (t - 1/t)] = \frac{t^2 + 1}{t^2 - 1}[x - (t + 1/t)]$$

So

$$\begin{aligned}(t^2 - 1)[y - (t^2 - 1)/t] &= (t^2 + 1)[x - (t^2 + 1)/t] \\ (t^2 - 1)y - (t^2 + 1)x &= [(t^2 - 1)^2 - (t^2 + 1)^2]/t \\ &= (-2)(2t^2)/t = -4t\end{aligned}$$

using the algebraic identity $a^2 - b^2 \equiv (a - b)(a + b)$ for the difference of two squares. So the equation of the tangent at t is

$$(t^2 - 1)y - (t^2 + 1)x + 4t = 0$$

For the normal we use $mm' = -1$ and therefore

$$m' = -(t^2 - 1)/(t^2 + 1)$$

at a general point t. The equation is therefore

$$[y - (t - 1/t)] = -\frac{t^2 - 1}{t^2 + 1}[x - (t + 1/t)]$$

So

$$\begin{aligned}(t^2 + 1)[y - (t - 1/t)] + (t^2 - 1)[x - (t + 1/t)] &= 0 \\ (t^2 + 1)yt + (t^2 - 1)xt &= [(t^4 - 1) + (t^4 - 1)] \\ &= 2(t^4 - 1)\end{aligned}$$

The equation of the normal at t is therefore

$$(t^2 + 1)yt + (t^2 - 1)xt = 2(t^4 - 1)$$

Now when $t = 1$ we hit a slight snag: m is not defined. However, we can argue by continuity that these equations will hold for all t. Therefore we take the limit as $t \to 1$ throughout the equation

$$(t^2 - 1)y - (t^2 + 1)x + 4t = 0$$

which we have shown to be the equation of the tangent at a general point ($t^2 \neq 1$). We obtain straight away

$$0 - 2x + 4 = 0$$

and so the equation of the tangent is $x = 2$.

Similarly for the normal, from

$$(t^2 + 1)yt + (t^2 - 1)xt = 2(t^4 - 1)$$

by letting $t \to 1$ we obtain

$$2y + 0 = 0$$

and so the equation of the normal is $y = 0$ – which is, of course, the x-axis.

It is possible to give alternative arguments, but the conclusions should be the same.

Now try this final problem.

▷**Exercise** Show that if a light source is positioned at the focus of a parabolic mirror it casts a beam parallel to the axis.

Before solving this we remark that the design of a car headlamp utilizes this property. Further, the reverse action will concentrate light at the focus. Therefore if the sun's rays strike a parabolic mirror, parallel to the axis, they are reflected to the focus. The first engineer to make use of this fact is reputed to have been Archimedes, when he set fire to the sails of the Roman fleet.

---

4 We can use the equation of the parabola in standard form $y^2 = 4ax$ (Chapter 3), which we can regard as a cross-section through the mirror (Fig. 6.1). In parametric form this can be expressed by $x = at^2$, $y = 2at$. Therefore the slope of the tangent at a general point t is given by

$$m = \mathrm{d}y/\mathrm{d}x = \dot{y}/\dot{x} = 2a/2at = 1/t$$

So the slope of the normal at t is $-t$ (recall $mm' = -1$).

Now the basic property of light when it strikes a mirror is expressed by the equation

angle of incidence = angle of reflection

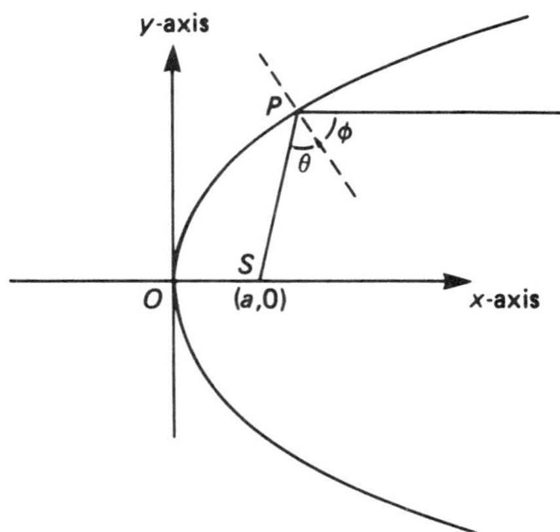

Fig. 6.1 The graph of $y^2 = 4ax$.

Let S be the focus $(a, 0)$ and let P be a general point on the parabola. We must show that the angle ϕ, between the normal at P and the x-axis, is equal to the angle θ, between PS and the normal at P. Since both angles are acute, it suffices to show that $\tan\theta = \tan\phi$. Now

$$\tan\phi = [(-t) - 0]/[1 + (-t)0] = -t$$

The slope of PS is given by

$$(2at - 0)/(at^2 - a) = 2t/(t^2 - 1)$$

(recall that $m = (y_1 - y_2)/(x_1 - x_2)$). So

$$\begin{aligned}\tan\theta &= \left[\frac{2t}{t^2 - 1} - (-t)\right]\Big/\left[1 + (-t)\frac{2t}{t^2 - 1}\right]\\ &= \frac{t^3 + t}{-t^2 - 1} = -t\end{aligned}$$

6.3 INTRINSIC COORDINATES

We are familiar with the two coordinate systems which are used to describe plane curves and regions. These are the cartesian coordinate system and the polar coordinate system (see Chapter 3). In each of these systems we may represent a point in the plane by an ordered pair of numbers. For the cartesian system this is (x, y) and for the polar coordinate system (r, θ) (Fig. 6.2).

In these systems a curve is represented by an equation. For example $x^2 + y^2 = 1$ and $r = 1$ are, in these two systems respectively, the equations of a circle of unit radius centred at the origin.

In the cartesian system, points are described relative to two fixed mutually perpendicular straight lines known as the axes. In the polar coordinate system, points are described relative to a point called the origin and a straight line emanating from the origin called the initial line.

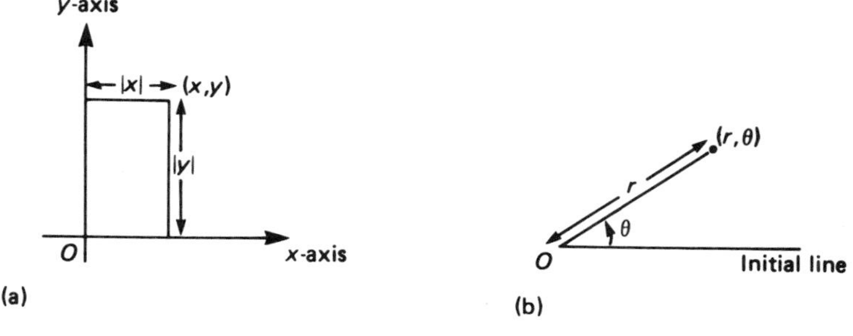

Fig. 6.2 (a) Cartesian coordinates (b) Polar coordinates.

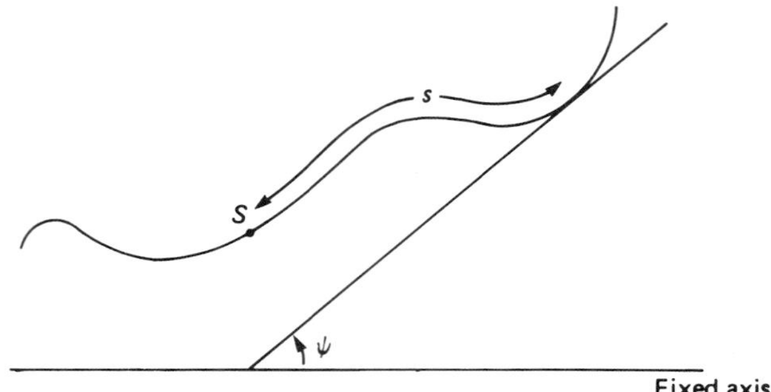

Fig. 6.3 Intrinsic coordinates.

We now describe another coordinate system, known as the **intrinsic coordinate system** (Fig. 6.3). Suppose we have a smooth curve, a fixed point S on the curve, and a fixed straight line. It will be convenient to think of the straight line as the x-axis. There are two possible ways in which we can move along the curve from S; we shall regard one as the positive direction, and the other as the negative direction. Given any real number s we therefore obtain a unique point on the curve by measuring a distance s (positive or negative) from S along the curve. The curve is smooth and so it has a tangent at all its points, and we shall suppose that there is an angle ψ at the point where the tangent meets the fixed axis.

A point on a curve in this system is then represented by an ordered pair (s, ψ), where s is the distance along the curve measured from S and ψ is the angle made by the tangent with the fixed axis.

This system, although useful, is not as versatile as the cartesian and polar coordinate systems, for it is not possible to represent a general point in the plane in terms of intrinsic coordinates. It is only possible to represent points on the curve.

6.4 THE CATENARY

Suppose a uniform chain or a heavy rope is freely suspended between two points; then the shape of the curve it assumes is known as the catenary (see section 5.1). Intrinsic coordinates enable us to determine the equation of the catenary quite easily. To do this we take the fixed line as the x-axis and S as the lowest point, and measure s positive to the right and negative to the left (Fig. 6.4). Suppose the mass per unit length is m. If P is a general point on the curve then P has coordinates (s, ψ).

We consider the forces on the portion of the rope SP (Fig. 6.5). There is a horizontal tension T_0 at S and a tension T in the direction of the tangent at P, and the rope is kept in equilibrium by its weight mgs which acts vertically downwards. We now resolve these forces vertically and horizontally to obtain

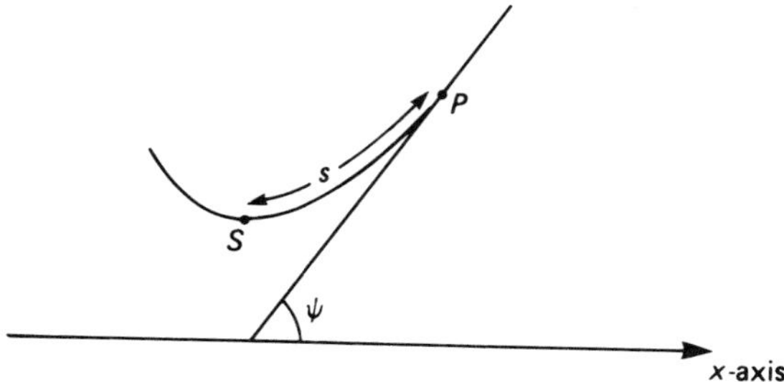

Fig. 6.4 The catenary.

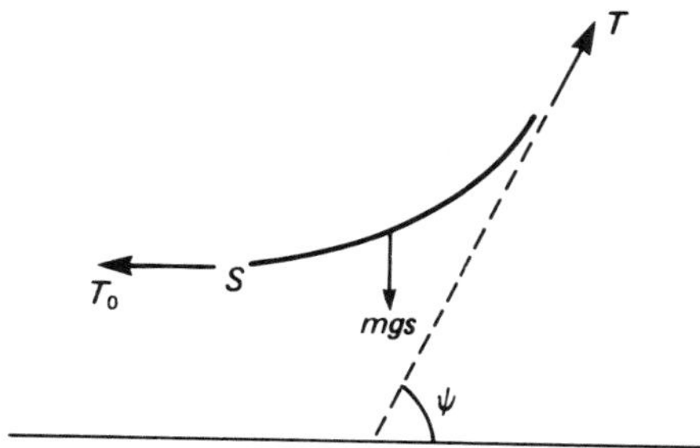

Fig. 6.5 Forces on the piece of rope.

$$mgs = T \sin \psi$$
$$T_0 = T \cos \psi$$

Eliminating T by dividing gives

$$mgs/T_0 = \tan \psi$$

which, on putting a constant $c = T_0/mg$, reduces to

$$s = c \tan \psi$$

This is the intrinsic equation of the catenary.

The catenary has many uses and needs to be considered whenever cables are strung between buildings. Although a light cable may not under normal circumstances be in the shape of a catenary, a severe winter's night with snow and ice on the cable can change the picture. When later we convert the equation of the catenary into cartesian coordinates, we shall find we are dealing with an old friend.

In order to link together the intrinsic coordinate system and the cartesian coordinate system, we shall need to locate the x and y axes. As we have said already, it is convenient to choose the x-axis as the fixed axis of the intrinsic coordinate system, and we shall choose the y-axis in such a way that S lies on it (Fig. 6.6).

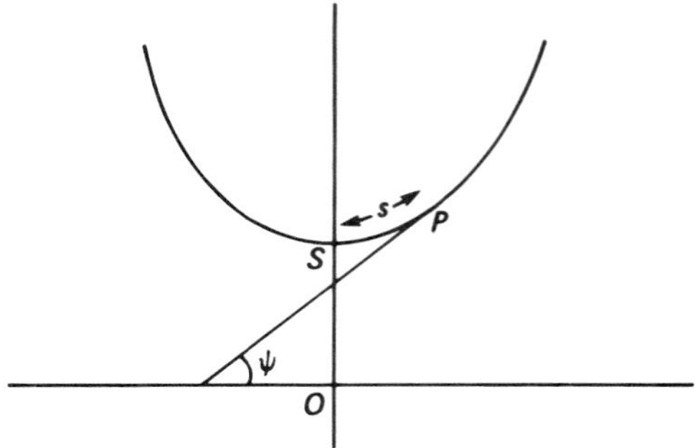

Fig. 6.6 Relating cartesian and intrinsic coordinates.

Then tan ψ is the slope of the curve at P, and so this is $\mathrm{d}y/\mathrm{d}x$. Therefore the *first linking equation* is

$$\tan \psi = \mathrm{d}y/\mathrm{d}x$$

Moreover, s is the length of the curve. So if δx and δy are small increases in x and y respectively, the corresponding increase in s is given by δs (Fig. 6.7). Therefore

$$(\delta x)^2 + (\delta y)^2 \simeq (\delta s)^2$$

It is reasonable to assume that as $\delta x \to 0$ the approximation will become good. Therefore dividing through by $(\delta x)^2$ and taking the limit as $\delta x \to 0$ we obtain

$$1 + (\delta y/\delta x)^2 \simeq (\delta s/\delta x)^2$$

So

$$1 + (\mathrm{d}y/\mathrm{d}x)^2 = (\mathrm{d}s/\mathrm{d}x)^2$$

If we choose s increasing with x we can take the positive square root to obtain the *second linking equation* as

$$\mathrm{d}s/\mathrm{d}x = [1 + (\mathrm{d}y/\mathrm{d}x)^2]^{1/2}$$

□ Transform the equation of the catenary $s = c \tan \psi$, in intrinsic coordinates, to an equation in cartesian coordinates.

It will be necessary to fix the catenary relative to the cartesian coordinate

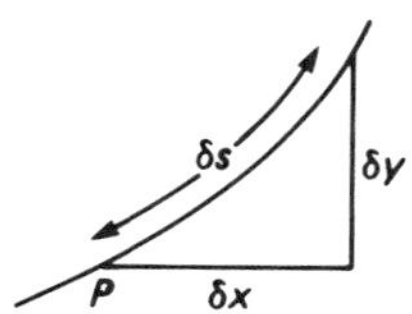

Fig. 6.7 Relating s, x and y.

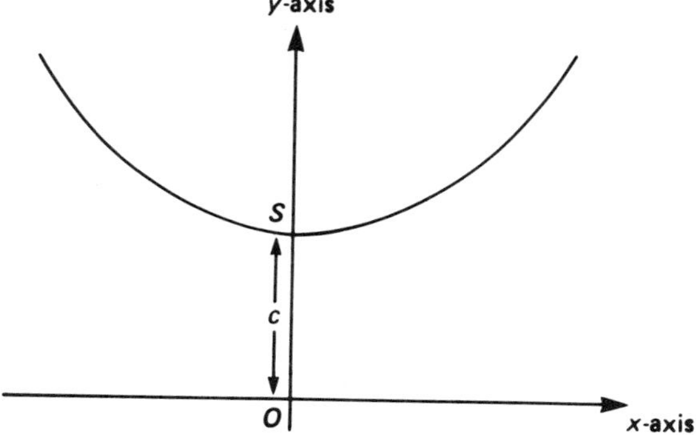

Fig. 6.8 Catenary relative to cartesian coordinates.

system, and so we shall choose S so that $OS = c$ (Fig. 6.8). The equation is $s = c \tan \psi$, and we have the linking equations

$$\tan \psi = \mathrm{d}y/\mathrm{d}x$$
$$\mathrm{d}s/\mathrm{d}x = [1 + (\mathrm{d}y/\mathrm{d}x)^2]^{1/2}$$

Now

$$s = c \tan \psi = c\, \mathrm{d}y/\mathrm{d}x = cu$$

where $u = \mathrm{d}y/\mathrm{d}x$. Therefore differentiating this equation with respect to x gives

$$\mathrm{d}s/\mathrm{d}x = c\, \mathrm{d}u/\mathrm{d}x$$

from which

$$\begin{aligned} c^2(\mathrm{d}u/\mathrm{d}x)^2 &= (\mathrm{d}s/\mathrm{d}x)^2 = 1 + (\mathrm{d}y/\mathrm{d}x)^2 \\ &= 1 + u^2 \\ \mathrm{d}x/\mathrm{d}u &= c/\sqrt{(1 + u^2)} \end{aligned}$$

Now we already know that if $x = \sinh^{-1} u$ then $\mathrm{d}x/\mathrm{d}u = 1/\sqrt{(1 + u^2)}$. So $x = c \sinh^{-1} u + A$, where A is a constant. When $x = 0$ we have $\mathrm{d}y/\mathrm{d}x = 0$, since this is the lowest point of the curve; consequently $A = 0$ and $x = c \sinh^{-1} u$.

However, $u = \mathrm{d}y/\mathrm{d}x$, and since we now have $u = \sinh x/c$ it follows that $\mathrm{d}y/\mathrm{d}x = \sinh x/c$. Consequently $y = c \cosh x/c + B$, where B is a constant. Finally when $x = 0$ we have $y = c$, and so $B = 0$. Therefore

$$y = c \cosh x/c$$

is the equation of the catenary in the cartesian coordinate system. ■

6.5 CURVATURE

The amount by which a curve bends determines the curvature of the curve. If the curve bends sharply then the curvature is large, whereas if the curve bends gently the curvature is small. In the intrinsic coordinate system we define the **curvature** $\varkappa$ by $\varkappa = d\psi/ds$. This is consistent with our intuitive idea because, for a small change in s, if ψ increases greatly then the curvature is high, whereas if ψ increases only gradually then the curvature is small.

The reciprocal of the curvature has the unit of length and is called the **radius of curvature** ϱ. So we have $\varrho = ds/d\psi$.

We now give a physical interpretation for the radius of curvature. Later we shall express it in terms of cartesian coordinates, and also determine a form suitable for calculating the radius of curvature if the curve is given parametrically.

Suppose that P is the point (s, ψ) and that Q is the point $(s + \delta s, \psi + \delta\psi)$ (Fig. 6.9). Suppose also that the normals to the curve at P and Q meet at C. Then since the angle between the tangents at P and Q is $\delta\psi$, the angle between the normals is also $\delta\psi$. Consequently $\angle PCQ = \delta\psi$, and because the length of the element of curve PQ is δs we conclude that

$$CP\,\delta\psi \simeq \delta s$$

The smaller that $\delta\psi$ becomes, the closer Q moves to P and the more CP comes to equalling CQ. Therefore as $\delta\psi \to 0$ the approximation $CP \simeq \delta s/\delta\psi$ becomes

$$CP = ds/d\psi = \varrho$$

that is, the radius of curvature.

The circle centred at C with radius ϱ is called the **circle of curvature**

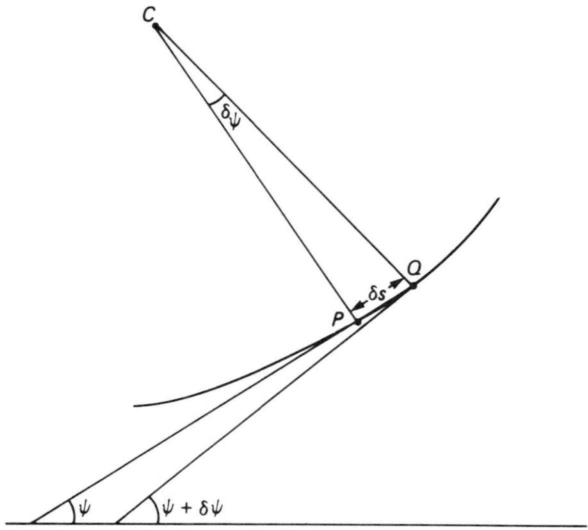

Fig. 6.9 Intersecting normals.

and the point C is called the **centre of curvature**. It is worth observing that if ϱ is negative this means that the curve is bending towards the x-axis (concave to the x-axis) and so in the opposite direction to the way shown in the diagram (convex to the x-axis).

THE CENTRE OF CURVATURE

To determine the coordinates of the centre of curvature it is best to use both cartesian coordinates and intrinsic coordinates at one and the same time.

Suppose that P is the point (x, y) in cartesian coordinates and also the point (s, ψ) relative to the curve in intrinsic coordinates. In Fig. 6.10 $\varrho > 0$, C has coordinates (X, Y) and T is the point (X, y). By similar triangles we have $\angle PCT = \psi$, and so

$$X = x - \varrho \sin \psi$$
$$Y = y + \varrho \cos \psi$$

Remarkably these formulas also work in the case $\varrho < 0$.

☐ Show that if $\varrho < 0$ then the centre of curvature (X, Y) is given by

$$X = x - \varrho \sin \psi$$
$$Y = y + \varrho \cos \psi$$

You will need a different diagram, but the working is very easy. It may help to put $p = |\varrho|$. Try it and see how you get on.

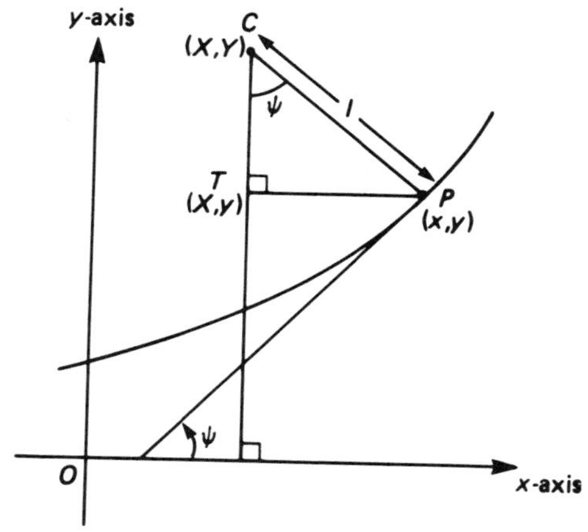

Fig. 6.10 The centre of curvature ($\varrho > 0$).

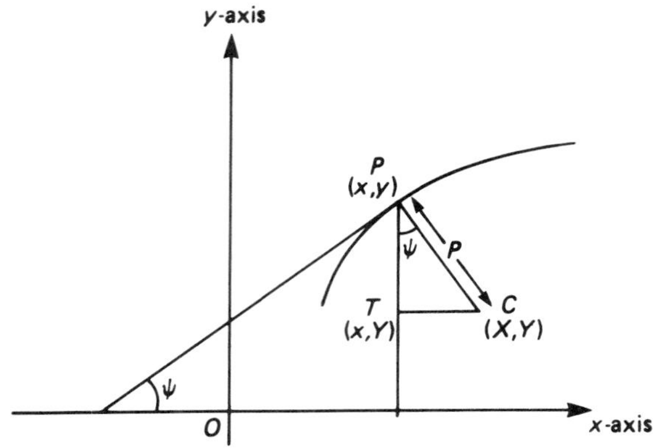

Fig. 6.11 The centre of curvature ($\varrho < 0$).

Put $p = |\varrho|$. Then $p > 0$ and we have a rather different figure, where now the curve bends towards the x-axis (Fig. 6.11). Now suppose T is the point (x, Y); then $\angle CPT = \psi$. So

$$X = x + p \sin \psi$$
$$Y = y - p \cos \psi$$

and therefore since $p = -\varrho$ we obtain

$$X = x - \varrho \sin \psi$$
$$Y = y + \varrho \cos \psi$$

as before. ■

We have seen how to determine the centre of curvature once the radius of curvature is known, and we have also seen how the sign of the radius of curvature can help us to decide which way the curve is bending. We now need a method of determining the radius of curvature without having to reduce a cartesian equation into one involving intrinsic coordinates.

THE RADIUS OF CURVATURE

Essentially there are two ways in which, using the cartesian coordinate system, a curve can be defined. It can be described directly by means of an equation involving x and y, or it can be described parametrically. In the parametric form x and y are each defined in terms of a third variable, for example t. Theoretically it could be argued that it is possible to eliminate t and thereby reduce the second case to the first one. However, in practice this may be very difficult to achieve. Therefore we shall deal with the two situations separately.

THE CARTESIAN FORM

We have $\varrho = \mathrm{d}s/\mathrm{d}\psi = (\mathrm{d}s/\mathrm{d}x)(\mathrm{d}x/\mathrm{d}\psi)$. Now

$$\mathrm{d}s/\mathrm{d}x = [1 + (\mathrm{d}y/\mathrm{d}x)^2]^{1/2}$$

and so it remains to obtain $\mathrm{d}x/\mathrm{d}\psi$ in cartesian form. Now $\mathrm{d}y/\mathrm{d}x = \tan\psi$, and so differentiating with respect to x

$$\begin{aligned}\mathrm{d}^2y/\mathrm{d}x^2 &= \sec^2\psi\,(\mathrm{d}\psi/\mathrm{d}x)\\ &= (1 + \tan^2\psi)(\mathrm{d}\psi/\mathrm{d}x)\\ &= [1 + (\mathrm{d}y/\mathrm{d}x)^2](\mathrm{d}\psi/\mathrm{d}x)\end{aligned}$$

Therefore the radius of curvature in cartesian form is

$$\varrho = \frac{[1 + (\mathrm{d}y/\mathrm{d}x)^2]^{1/2}\,[1 + (\mathrm{d}y/\mathrm{d}x)^2]}{\mathrm{d}^2y/\mathrm{d}x^2}$$

$$\varrho = \frac{[1 + (\mathrm{d}y/\mathrm{d}x)^2]^{3/2}}{\mathrm{d}^2y/\mathrm{d}x^2}$$

□ For the curve $y = c\cosh(x/c)$ obtain (a) the radius of curvature at a general point (x, y) and (b) the position of the centre of curvature at the point where $x = c\ln 2$.

a We have $y = c\cosh(x/c)$, and so

$$\mathrm{d}y/\mathrm{d}x = c\,(1/c)\sinh(x/c) = \sinh(x/c)$$

Therefore

$$\mathrm{d}^2y/\mathrm{d}x^2 = (1/c)\cosh(x/c)$$

It follows that

$$1 + (\mathrm{d}y/\mathrm{d}x)^2 = 1 + \sinh^2(x/c) = \cosh^2(x/c)$$

and so

$$[1 + (\mathrm{d}y/\mathrm{d}x)^2]^{3/2} = \cosh^3(x/c)$$

We now have

$$\begin{aligned}\varrho &= \frac{[1 + (\mathrm{d}y/\mathrm{d}x)^2]^{3/2}}{\mathrm{d}^2y/\mathrm{d}x^2}\\ &= \frac{\cosh^3(x/c)}{(1/c)\cosh(x/c)} = c\cosh^2(x/c)\end{aligned}$$

This is the radius of curvature at a general point. Observe that it is always positive; this is not surprising since the curve is always convex to the x-axis.

b When $x = c\ln 2$ we have $x/c = \ln 2$, and so

$$\mathrm{e}^{x/c} = \mathrm{e}^{\ln 2} = 2$$

So that, at $x = c \ln 2$,

$$\cosh (x/c) = \tfrac{1}{2}(e^{x/c} + e^{-x/c}) = \tfrac{1}{2}(2 + \tfrac{1}{2}) = 5/4$$
$$\sinh (x/c) = \tfrac{1}{2}(e^{x/c} - e^{-x/c}) = \tfrac{1}{2}(2 - \tfrac{1}{2}) = 3/4$$

The radius of curvature at $x = c \ln 2$ is therefore $25c/16$. Also, when $x = c \ln 2$ we have

$$y = c \cosh (x/c) = 5c/4$$

Furthermore

$$\tan \psi = dy/dx = \sinh (x/c) = 3/4$$

This gives $\sin \psi = 3/5$ and $\cos \psi = 4/5$.

Now C, the centre of curvature, is the point (X, Y), where

$$\begin{aligned} X = x - \varrho \sin \psi &= c \ln 2 - (25c/16)\,(3/5) \\ &= c \ln 2 - 15c/16 \\ Y = y + \varrho \cos \psi &= 5c/4 + (25c/16)\,(4/5) \\ &= 5c/4 + 5c/4 = 5c/2 \end{aligned}$$

So when $x = c \ln 2$ the centre of curvature is

$$(c \ln 2 - 15c/16,\ 5c/2)$$ ■

THE PARAMETRIC FORM

If x and y are each given in terms of a parameter t, then a small change δt in t will result in small changes δx and δy in x and y respectively. We have

$$(\delta s)^2 \simeq (\delta x)^2 + (\delta y)^2$$

As $\delta t \to 0$ both δx and δy tend to zero and this approximate formula becomes exact. Now

$$(\delta s/\delta t)^2 \simeq (\delta x/\delta t)^2 + (\delta y/\delta t)^2$$

so that as $\delta t \to 0$ we obtain

$$(ds/dt)^2 = (dx/dt)^2 + (dy/dt)^2$$

Therefore

$$\dot{s}^2 = \dot{x}^2 + \dot{y}^2$$

Now

$$\varrho = ds/d\psi = (ds/dt)(dt/d\psi)$$

We have seen how to find ds/dt; we need therefore to obtain $dt/d\psi$. The equation linking intrinsic coordinates with cartesian coordinates and which involves ψ is

$$\tan \psi = dy/dx = (dy/dt)(dt/dx) = \dot{y}/\dot{x}$$

We now differentiate throughout with respect to t and obtain

$$\sec^2 \psi \frac{d\psi}{dt} = \frac{\dot{x}\ddot{y} - \dot{y}\ddot{x}}{\dot{x}^2}$$

Now

$$\sec^2 \psi = 1 + \tan^2 \psi = 1 + (dy/dx)^2 = 1 + (\dot{y}/\dot{x})^2$$

so that

$$\dot{x}^2 \sec^2 \psi = \dot{x}^2 + \dot{y}^2 = \dot{s}^2$$

Substituting into the expression for ϱ we now have

$$\varrho = \frac{ds/dt}{d\psi/dt} = \frac{\dot{s}}{\dot{\psi}} = \frac{\dot{s}\dot{x}^2 \sec^2 \psi}{\dot{x}\ddot{y} - \dot{y}\ddot{x}}$$

$$= \frac{\dot{s}^3}{\dot{x}\ddot{y} - \dot{y}\ddot{x}} = \frac{(\dot{x}^2 + \dot{y}^2)^{3/2}}{\dot{x}\ddot{y} - \dot{y}\ddot{x}}$$

Therefore the radius of curvature in parametric form is

$$\varrho = \frac{(\dot{x}^2 + \dot{y}^2)^{3/2}}{\dot{x}\ddot{y} - \dot{y}\ddot{x}}$$

□ Obtain the radius of curvature at a general point, determined by the parameter θ, on the curve

$$x = \sin\theta + 2\cos\theta$$
$$y = \cos\theta - 2\sin\theta$$

We substitute into the formula

$$\varrho = \frac{(\dot{x}^2 + \dot{y}^2)^{3/2}}{\dot{x}\ddot{y} - \dot{y}\ddot{x}}$$

where the dot here indicates differentiation with respect to θ. We have

$$\dot{x} = \cos\theta - 2\sin\theta$$
$$\dot{y} = -\sin\theta - 2\cos\theta$$

So that squaring and adding,

$$\begin{aligned}\dot{x}^2 + \dot{y}^2 &= (\cos\theta - 2\sin\theta)^2 + (\sin\theta + 2\cos\theta)^2 \\ &= \cos^2\theta + 4\sin^2\theta - 4\sin\theta\cos\theta + \sin^2\theta + 4\cos^2\theta + 4\sin\theta\cos\theta \\ &= 5(\cos^2\theta + \sin^2\theta) = 5\end{aligned}$$

Further,

$$\ddot{x} = -\sin\theta - 2\cos\theta$$
$$\ddot{y} = -\cos\theta + 2\sin\theta$$

so that

$$\begin{aligned}\dot{x}\ddot{y} - \dot{y}\ddot{x} &= (\cos\theta - 2\sin\theta)(-\cos\theta + 2\sin\theta)\\ &\quad - (-\sin\theta - 2\cos\theta)(-\sin\theta - 2\cos\theta)\\ &= -\cos^2\theta - 4\sin^2\theta + 4\sin\theta\cos\theta\\ &\quad - (\sin^2\theta + 4\cos^2\theta + 4\sin\theta\cos\theta)\\ &= -5\cos^2\theta - 5\sin^2\theta = -5\end{aligned}$$

Substituting into the formula for ϱ gives

$$\varrho = 5^{3/2}/(-5) = -\sqrt{5}$$

Now this means that there is a constant radius of curvature, and the only curve which has a constant radius of curvature is a circle. Therefore these parametric equations must define a circle. It is easy to confirm this by eliminating θ. We have

$$\begin{aligned}x &= \sin\theta + 2\cos\theta\\ y &= \cos\theta - 2\sin\theta\end{aligned}$$

So

$$\begin{aligned}2x + y &= 5\cos\theta\\ x - 2y &= 5\sin\theta\end{aligned}$$

If we square and add we obtain

$$(2x + y)^2 + (x - 2y)^2 = 25$$

which is $x^2 + y^2 = 5$.

The fact that ϱ is negative suggests that the circle is concave to the x-axis, and indeed since the circle in question is centred at the origin we can confirm this. ■

Right! Are you ready for some steps?

6.6 Workshop

1 **Exercise** Express the equation of curve

$$s = \sec\psi\tan\psi + \ln(\sec\psi + \tan\psi)$$

in terms of cartesian coordinates, where the axes are to be chosen so that
1 when $x = 0$, $y = 0$;
2 when $x = 0$, $s = 0$.
As usual we suppose that the x-axis is parallel to the fixed axis of the intrinsic coordinate system.

Remember the linking equations, and see how you get on.

2 The linking equations are

$$\tan \psi = dy/dx$$
$$ds/dx = [1 + (dy/dx)^2]^{1/2}$$

So in general we have

$$ds/dx = [1 + (dy/dx)^2]^{1/2}$$
$$= (1 + \tan^2 \psi)^{1/2} = \sec \psi$$

since $1 + \tan^2 \psi = \sec^2 \psi$. Here

$$s = \sec \psi \tan \psi + \ln (\sec \psi + \tan \psi)$$

Therefore

$$\begin{aligned} ds/d\psi &= \sec \psi \sec^2 \psi + \tan \psi \sec \psi \tan \psi \\ &\quad + (\sec \psi + \tan \psi)^{-1} (\sec \psi \tan \psi + \sec^2 \psi) \\ &= \sec \psi (\sec^2 \psi + \tan^2 \psi) + \sec \psi \\ &= \sec \psi (\sec^2 \psi + \tan^2 \psi + 1) \\ &= 2 \sec^3 \psi \end{aligned}$$

since $1 + \tan^2 \psi = \sec^2 \psi$.

If you were stuck then try to get going from this point. Otherwise read on and see if you got everything right.

3

This means that

$$(ds/dx)(dx/d\psi) = 2 \sec^3 \psi$$

and we have already shown that

$$ds/dx = \sec \psi$$

Therefore

$$dx/d\psi = 2 \sec^2 \psi$$

It follows at once that

$$x = 2 \tan \psi + A$$

where A is a constant which we need to determine. We now have

$$dy/dx = \tan \psi = \tfrac{1}{2}(x - A)$$

and so

$$y = \tfrac{1}{4}x^2 - \tfrac{1}{2}Ax + B$$

where B is another constant which we need to determine.

See if you can determine these constants, and then take another step.

4

We have the initial conditions, and these will help us to fix A and B:

1 When $x = 0$, $y = 0$ and so $B = 0$;
2 When $x = 0$, $s = 0$.
Now we have shown that when $x = 0$, $\tan \psi = -A/2 = C$ (say), and so $\sec \psi = \pm(1 + C^2)^{1/2}$. From the equation for s,

$$0 = C[\pm(1 + C^2)^{1/2}] + \ln [C \pm (1 + C^2)^{1/2}]$$

This has no meaning if the negative sign is chosen because the argument of the logarithm is then negative. Consequently the positive root for $\sec \psi$ must apply.

We have to solve the equation

$$\ln [C + (1 + C^2)^{1/2}] = -C(1 + C^2)^{1/2}$$

Now this is a tricky business. You should be able to spot that $C = 0$ is one solution, but it is quite a different matter to show that $C = 0$ is the only solution. You would not normally be expected to do this, but you might care to try out your algebraic skills!

There are several possible approaches. You could use the work we have not yet covered on maxima and minima (Chapter 8) and examine

$$y = x\sqrt{(1 + x^2)} + \ln [x + \sqrt{(1 + x^2)}]$$

You could then argue that if there are two values of x for which y is zero, then by continuity there must be a local maximum or a local minimum. It would then follow that dy/dx would be zero at some point. However, it is possible to show that $dy/dx = 2\sqrt{(x^2 + 1)}$ and so is never zero.

Nevertheless there is a purely algebraic approach. See if you can finish the problem off. If it's too much, just read through the solution and try to appreciate what is involved. Whatever your decision, take another step when you are ready.

We shall show that $C = 0$ is the only possible solution of

$$\ln [C + (1 + C^2)^{1/2}] = -C(1 + C^2)^{1/2}$$

Suppose $C > 0$. Then the right-hand side of the equation is negative, whereas the left-hand side is positive:

$$C + (1 + C^2)^{1/2} > 1$$

Suppose $C < 0$. Then

$$\exp [-C(1 + C^2)^{1/2}] = C + (1 + C^2)^{1/2}$$

We can rearrange the left-hand side of this by multiplying throughout by $-C + (1 + C^2)^{1/2}$. We then have

$$[-C + (1 + C^2)^{1/2}] \exp [-C(1 + C^2)^{1/2}] = -C^2 + (1 + C^2) = 1$$

However, the left-hand side is greater than 1 because

$$-C + (1 + C^2)^{1/2} > 1$$

and the exponential value of any positive number is always greater than 1.

The only possibility which remains is $C = 0$.

Finally, then, the equation of the curve in cartesian coordinates is

$$y = \tfrac{1}{4}x^2$$

If you managed that all on your own, you have handled an awkward problem successfully.

Now for something rather different.

▷**Exercise** For the curve

$$y = \tfrac{1}{2}x^2 - \tfrac{1}{4}\ln(x + 1) + x$$

obtain, at the origin, the radius of curvature and the centre of curvature.

All you need to do is calculate the ingredients for the formulas and you are away! Work out the radius of curvature and take another step.

Here we go then. We need $\mathrm{d}y/\mathrm{d}x$ and later $\mathrm{d}^2y/\mathrm{d}x^2$. So 6

$$\begin{aligned} y &= \tfrac{1}{2}x^2 - \tfrac{1}{4}\ln(x + 1) + x \\ \mathrm{d}y/\mathrm{d}x &= x - \tfrac{1}{4}(x + 1)^{-1} + 1 \\ &= \frac{4(x + 1)^2 - 1}{4(x + 1)} \end{aligned}$$

Then

$$\begin{aligned} \left(\frac{\mathrm{d}s}{\mathrm{d}x}\right)^2 &= 1 + \left(\frac{\mathrm{d}y}{\mathrm{d}x}\right)^2 \\ &= 1 + \frac{[4(x + 1)^2 - 1]^2}{16(x + 1)^2} \\ &= \frac{16(x + 1)^2 + [4(x + 1)^2 - 1]^2}{16(x + 1)^2} \\ &= \frac{[4(x + 1)^2 + 1]^2}{16(x + 1)^2} \end{aligned}$$

Did you spot how to collect that together? If you multiply everything out you risk not being able to see the wood for the trees. It is always worth trying to stand back and see if there is a simple approach.

So we have

$$\frac{\mathrm{d}s}{\mathrm{d}x} = \frac{4(x + 1)^2 + 1}{4(x + 1)}$$

Now

$$\frac{d^2y}{dx^2} = 1 + \frac{1}{4(x+1)^2}$$
$$= \frac{4(x+1)^2 + 1}{4(x+1)^2}$$

Then

$$\varrho = (ds/dx)^3/(d^2y/dx^2)$$
$$= \frac{[4(x+1)^2+1]^3}{[4(x+1)]^3}\,\frac{4(x+1)^2}{4(x+1)^2+1}$$
$$= \frac{[4(x+1)^2+1]^2}{16(x+1)}$$
$$= \frac{25}{16} \qquad \text{when } x = 0$$

If you made a slip, check to find where you went wrong. Then see how you get on with the centre of curvature, and take another step.

---

7 We must obtain cos ψ and sin ψ. Now $\tan\psi = dy/dx = 3/4$ when $x = 0$ so $\cos\psi = 4/5$ and $\sin\psi = 3/5$. We now have all the information we need to obtain the position of the centre of curvature:

$$X = x - \varrho\sin\psi = 0 - (25/16)(3/5) = -15/16$$
$$Y = y + \varrho\cos\psi = 0 + (25/16)(4/5) = 5/4$$

So the centre of curvature is $(-15/16, 5/4)$.

It is a good idea to practise the parametric formula, and so here is another exercise for you to try.

▷ **Exercise** Determine the radius of curvature at the point where $t = 1$ on the curve defined parametrically by

$$x = t + t^2, \qquad y = 1 + t^4$$

When you have given this all you can, take the next step.

---

8 Don't forget the formula

$$\varrho = \frac{(\dot{x}^2 + \dot{y}^2)^{3/2}}{\dot{x}\ddot{y} - \dot{y}\ddot{x}}$$

Now $x = t + t^2$ and $y = 1 + t^4$. So $\dot{x} = 1 + 2t$ and $\dot{y} = 4t^3$. Therefore when $t = 1$ we have $\dot{x} = 3$ and $\dot{y} = 4$. Hence

$$(\dot{x}^2 + \dot{y}^2)^{3/2} = (9 + 16)^{3/2} = 125$$

Further, $\ddot{x} = 2$ and $\ddot{y} = 12t^2$. So when $t = 1$ we have $\ddot{x} = 2$ and $\ddot{y} = 12$. Hence

$$\dot{x}\ddot{y} - \dot{y}\ddot{x} = 3 \times 12 - 4 \times 2 = 36 - 8 = 28$$

Therefore $\varrho = 125/28$ when $t = 1$.

If you managed that then read through the next exercise and step. If there are still a few problems, then try the exercise yourself first.

▷**Exercise** Obtain the position at the origin of the centre of curvature for the parametric curve

$$x = p + \sinh p, \qquad y = -1 + \cosh p$$

As soon as you have done this, move on to step 9.

9

When $p = 0$ we have $x = 0$ and $y = 0$, and so we begin by obtaining ϱ when $p = 0$. We have

$$\dot{x} = 1 + \cosh p, \qquad \dot{y} = \sinh p$$
$$\ddot{x} = \sinh p, \qquad \ddot{y} = \cosh p$$

so that when $p = 0$,

$$\dot{x} = 2, \qquad \dot{y} = 0$$
$$\ddot{x} = 0, \qquad \ddot{y} = 1$$

The radius of curvature can now be found:

$$(\dot{x}^2 + \dot{y}^2)^{3/2} = (4 + 0)^{3/2} = 8$$
$$\dot{x}\ddot{y} - \dot{y}\ddot{x} = 2 \times 1 - 0 \times 0 = 2$$

Consequently $\varrho = 8/2 = 4$.

Is all well so far? If there are any problems, look through the work at this stage and then see if you can complete the problem by finding the position of the centre of curvature. Remember, you will need $\sin \psi$ and $\cos \psi$.

10

Here goes then. First, $\tan \psi = dy/dx = \dot{y}/\dot{x} = 0$ at the origin. We deduce that $\psi = 0$, and so $\sin \psi = 0$ and $\cos \psi = 1$. The centre of curvature is now the point $(x - \varrho \sin \psi, y + \varrho \cos \psi)$, and this is $(0 - 0, 0 + 4 \times 1) = (0, 4)$.

If any problems remain at this stage it is best to go back through the chapter again.

Now for a practical problem.

6.7 Practical

MOORED DIRIGIBLE

A dirigible is moored to a 200 m warp which is secured to a post. The tension at the post is equal to the weight of 50 m of warp and is inclined at

$\tan^{-1}(4/3)$ to the horizontal. Determine the tension and the direction of the warp at its upper end, and show that the dirigible is about 192 m above the post.

Make a real effort to solve this problem entirely on your own. As usual we shall give the solution stage by stage so that you can join in at whatever stage you can.

The warp is shown in Fig. 6.12. Let the tension at the top be T_1 inclined at an angle ϕ to the horizontal, and let the tension at the lower end be T_0 inclined at an angle θ to the horizontal.

Resolving the forces horizontally gives

$$T_1 \cos \phi = T_0 \cos \theta = 50w\,(3/5) = 30w$$

where w is the weight per metre. Resolving the forces vertically gives

$$200w + T_0 \sin \theta = T_1 \sin \phi$$

So

$$T_1 \sin \phi = 200w + 50w\,(4/5) = 240w$$

If you have not studied statics you may not have been able to obtain these equations. However, all should be well now we have obtained all the information we need.

Next we require ϕ, and we have shown

$$T_1 \cos \phi = 30w$$

$$T_1 \sin \phi = 240w$$

Therefore $\tan \phi = 240w/30w = 8$, and consequently $\phi = \tan^{-1} 8$ is the angle of inclination of the warp to the horizontal at the upper end.

We also require T_1. See if you can obtain this.

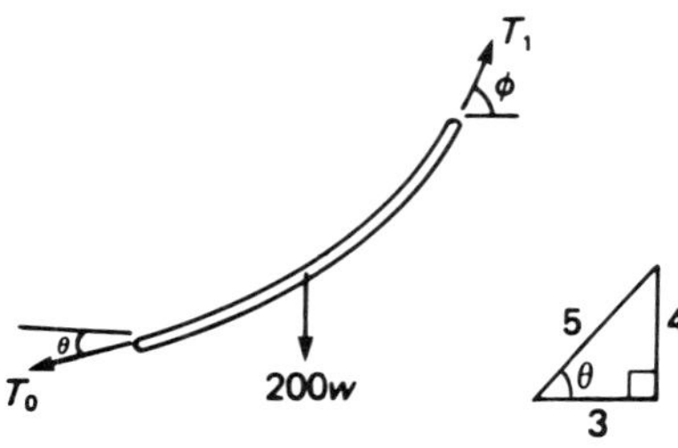

Fig. 6.12 Forces on the warp.

We have

$$T_1^2 (\sin^2 \phi + \cos^2 \phi) = (30w)^2 + (240w)^2$$

Therefore

$$T_1^2 = (30w)^2(1 + 64)$$
$$T_1 = 30w\sqrt{65}$$

Lastly we must obtain the height of the dirigible above the post. For this we need to use the equation $y = c \cosh (x/c)$.

From $y = c \cosh (x/c)$ we obtain

$$dy/dx = \tan \psi = \sinh (x/c) = s/c$$

Now

$$\cosh^2 (x/c) - \sinh^2 (x/c) = 1$$

Therefore

$$(y/c)^2 - (s/c)^2 = 1$$
$$y^2 = s^2 + c^2$$

This formula is not in any way dependent on the details of this problem, and so can always be used when we have a catenary.

The easiest way to proceed now is to consider the missing part of the catenary, length s_0, from the post to the lowest point (Fig. 6.13). Then

$$200 + s_0 = c \tan \phi$$
$$s_0 = c \tan \theta$$

See if you can complete things.

We have

$$200 = c(\tan \phi - \tan \theta) = c[8 - (4/3)] = 20c/3$$

Consequently $c = 30$ and $s_0 = c \tan \theta = 30(4/3) = 40$.

Now

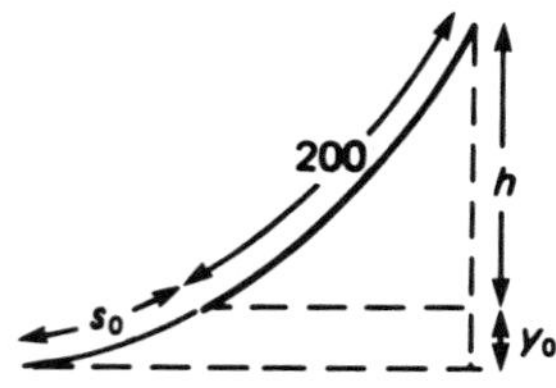

Fig. 6.13 Part of the catenary.

$$(h + y_0)^2 = (200 + s_0)^2 + c^2 = (240)^2 + (30)^2$$
$$h + y_0 = 30\sqrt{65}$$

Also

$$y_0^2 = s_0^2 + c^2 = (30)^2 + (40)^2 = (50)^2$$
$$y_0 = 50$$

Finally, $h = 30\sqrt{65} - 50 \simeq 192$ m.

Notice how we leave any approximation to the last possible stage. We should always avoid premature approximation because it usually leads to greater inaccuracy.

SUMMARY

- □ We have shown how to find the equations of tangents and normals to plane curves.
- □ We have introduced intrinsic coordinates (s, ψ) and seen how to link them to the cartesian coordinate system:

$$\tan \psi = \mathrm{d}y/\mathrm{d}x$$
$$\mathrm{d}s/\mathrm{d}x = [1 + (\mathrm{d}y/\mathrm{d}x)^2]^{1/2}$$

- □ We have derived the equation of the catenary in intrinsic coordinates in the standard form $s = c \tan \psi$, and shown that this can be written in cartesian form as $y = c \cosh (x/c)$.
- □ We have introduced the ideas of
  a curvature $\varkappa = \mathrm{d}\psi/\mathrm{d}s$
  b radius of curvature $\varrho = \mathrm{d}s/\mathrm{d}\psi$
  c centre of curvature $(x - \varrho \sin \psi, y + \varrho \cos \psi)$.
- □ We have given cartesian and parametric forms for the radius of curvature:

$$\varrho = \frac{[1 + (\mathrm{d}y/\mathrm{d}x)^2]^{3/2}}{\mathrm{d}^2y/\mathrm{d}x^2}$$
$$\varrho = \frac{(\dot{x}^2 + \dot{y}^2)^{3/2}}{\dot{x}\ddot{y} - \dot{y}\ddot{x}}$$

EXERCISES

1 Obtain the equation of the tangent to each of the following curves at the point where $x = 0$:
a $y = x^2 + \mathrm{e}^x$
b $y^2 = x^2 + \sqrt{(x^2 + 1)}$

c $y = \cos\left(\frac{\pi e^x}{2}\right)$
d $y = xy + x^2$

2 Obtain the equation of the normal to each of the following curves at the point where $y = 1$:
a $x^3 + y^3 = 2$
b $x^2y + y^2x = 6$
c $(x + y)^2 = x(x - y)$
d $x = \sin \pi y$

3 Obtain the radius of curvature of each of the following curves at the point $x = -1$:
a $(x + y)^2 = (x - y)^2 + 1$
b $y = xe^{1+x}$
c $x^2 + xy + y^2 = 1$

4 Obtain the radius of curvature at the point $t = 0$ on the curves
a $x = \sin t$, $y = \cosh t$
b $x = t + t^2$, $y = t - t^2$
c $x = t^3 + 1$, $y = t - 1$
d $x = \sin t$, $y = e^t$

ASSIGNMENT

1 Obtain the radius of curvature and the position of the centre of curvature of the curve $y = x^2 + 1$ at (0,1).

2 Show that for the curve described in intrinsic coordinates by $s = a\psi^2/2$ (where a is constant) the radius of curvature satisfies $\varrho^2 = 2as$.

3 Obtain the equations of the tangent and the normal at $x = 0$ for the curve $y = \exp x^2$.

4 Determine the radius of curvature of the curve $y = \exp x^2$ at (0,1).

5 Prove that at any point on the rectangular hyperbola $xy = c^2$ the radius of curvature $\varrho = r^3/2c^2$, where r is the distance of the point to the origin.

6 Determine the equation of the tangent and the normal at a general point where $y = t$ on the curve $y^2 + 3xy + y^3 = 5$.

7 Obtain the equations of the tangent and the normal for the parametric curve $x = \frac{1}{2}t^2 - t$, $y = \frac{1}{2}t^2 + t$ at a general point t.

FURTHER EXERCISES

1 The parabola $y^2 = 4ax$ and the ellipse

$$\frac{x^2}{a^2} + \frac{y^2}{b^2} = 1$$

intersect at right angles. Show that $2a^2 = b^2$.

2 Show that the equation of the tangent at the point (x_1, y_1) on the curve

$$ax^2 + bxy + cy^2 + dx + ey + f = 0$$

where a, b, c, d, e and f are constants, is given by

$$axx_1 + \tfrac{1}{2}b(xy_1 + x_1y) + cyy_1 + \tfrac{1}{2}d(x + x_1) + \tfrac{1}{2}e(y + y_1) + f = 0$$

(This equation is the general second-degree equation in x and y and includes the circle, parabola, ellipse and hyperbola. The transformations

$$uv \to \tfrac{1}{2}(uv_1 + u_1v)$$

$$u \to \tfrac{1}{2}(u + u_1)$$

where $u, v \in \{x, y\}$ enable the equation of the tangent at a point on any one of these curves to be written down straight away.)

3 Obtain the coordinates of the centre of curvature at the point $(1, 2)$ on the curve $(x - y)^2 = 2xy - x - y$.

4 Show that the perpendicular from the focus $(a, 0)$ on the parabola $y^2 = 4ax$ to any tangent intersects it on the y-axis.

5 P and Q are two points on the rectangular hyperbola $xy = 1$, constrained so that the line PQ is tangential to the parabola $y^2 = 8x$. Show that the locus of R, the midpoint of PQ, is also a parabola $y^2 + x = 0$.

6 Identify each of the following curves and give a rough sketch:

a $xy - 2y - x + 1 = 0$

b $xy + 12 = 3x + 4y$

c $x^2 + 2y^2 + 4x + 12y + 18 = 0$

d $2x^2 - y^2 - 4x + 6y = 15$

e $(x + y)^2 = 2(x + 3)(y + 4) - 33$

7 For the equation defined parametrically by

$$x = \sin^3\theta + 3\sin\theta$$
$$y = \cos^3\theta - 6\cos\theta$$

obtain the coordinates (X, Y) of the centre of curvature at a general point. Eliminate θ and thereby obtain an equation relating X and Y; this is the locus of the centre of curvature, known as the **evolute** of the curve.

8 Show that for the parabola $y^2 = 4ax$ the locus of the centre of curvature (the evolute) is the curve $4(x - 2a)^3 = 27ay^2$.

9 Show that if $y = ax^2 + bx^3$ then at the origin $\varrho = 1/2a$ and $d\varrho/dx = -3b/2a^2$.

10 Determine the radius of curvature of the parametric curve

$$x = \cos^2 p \sin p$$
$$y = \sin^2 p \cos p$$

at a general point with parameter p.

11 The bending moment M at a point on a uniform strut subjected to loading is given by $M = EI/\varrho$, where E and I are constants dependent on the material of the beam and ϱ is the radius of curvature at the point. When suitable axes are chosen the profile of such a strut is defined parametrically by $x = a(t - \sin t)$, $y = a(1 - \cos t)$. Show that $M = (EI/4a) \operatorname{cosec}(t/2)$.

12 A curve is defined by the equations $x = 3 \tan^2 \theta$, $y = 1 + 2 \tan^3 \theta$. Prove that the radius of curvature at a general point with parameter θ is $6 \tan \theta \sec^3 \theta$.

13 A curve is defined parametrically by the equations

$$x = 2 \cos \theta - 2 \cos^2 \theta + 1$$
$$y = 2 \sin \theta + 2 \sin \theta \cos \theta$$

Show that the normal at a general point is

$$x \sin(\theta/2) + y \cos(\theta/2) = 3 \sin(3\theta/2)$$

14 A curve is defined parametrically by $x = a \sin 2\theta$, $y = a \sin \theta$. Show that

$$\mathrm{d}^2y/\mathrm{d}x^2 = \sin \theta\,(1 + 2\cos^2 \theta)/4a \cos^3 2\theta$$

Obtain also the radius of curvature at the point $(0, a)$.

15 A uniform chain of length $2l$ and weight w per unit length is suspended between two points at the same level and has a maximum depth of sag d. Prove that the tension at the lowest point is $w(l^2 - d^2)/2d$, and that the distance between the points of suspension is

$$[(l^2 - d^2)/d] \ln [(l + d)/(l - d)]$$

16 When a body moves along a curve it experiences at any point an acceleration u^2/ϱ along the normal, where u is its speed and ϱ is the radius of curvature. Find this normal component of acceleration at the origin for the curve $y = x^2(x - 3)$ if the speed is a constant 12 m/s.

17 Determine $\mathrm{d}y/\mathrm{d}x$ for each of the following and list any real values of x for which $\mathrm{d}y/\mathrm{d}x$ is not defined.

a $$y = x^2 \ln(1 + \sinh x)$$

b $$y = \frac{1}{x} + \frac{2}{x^2} + \frac{3}{x^3}$$

c $$y = \frac{1 + x - x^2}{1 - x + x^2}$$

18 If, for $x > 0$,

$$y = \frac{1 - x^{1/2}}{1 + x^{1/2}}$$

obtain formulae for $\mathrm{d}y/\mathrm{d}x$, $\mathrm{d}^2y/\mathrm{d}x^2$ and $\mathrm{d}^3y/\mathrm{d}x^3$.

19 Suppose $y = 1/(1 + 1/x)$ and $z = 1/(1 + 1/y)$. Obtain $\mathrm{d}y/\mathrm{d}x$ and $\mathrm{d}z/\mathrm{d}x$.

Partial differentiation 7

Now that we can apply some of the techniques of differentiation to functions of a single variable we shall see to what extent we can generalize these ideas to functions of several variables.

After completing this chapter you should be able to

- ☐ Use the language and standard notation for functions of several variables;
- ☐ Obtain first-order and second-order partial derivatives;
- ☐ Use the formulas for a change of variables correctly;
- ☐ Calculate an estimate of accuracy in using a formula where the variables have known errors.

At the end of this chapter we tackle practical problems of tank volume and oil flow.

7.1 FUNCTIONS

We know that, given a real function f, we can draw a graph of it in the plane by writing $y = f(x)$. The set of arguments for which the function is defined is called the domain of f, and the set of values is called the range or image set of f (see Chapter 2).

What happens when we have a function of more than one variable? For example, suppose we consider the equation $z = (x + y) \sin x$. In this case, given any pair of real numbers x and y, we obtain a unique real number z. We have a function of *two* real variables, and we can write $f: \mathbb{R}^2 \to \mathbb{R}$ where

$$z = f(x, y) = (x + y) \sin x$$

This is just a generalization of the ideas we have already explored for a real function, which in this chapter we shall call a function of a single variable.

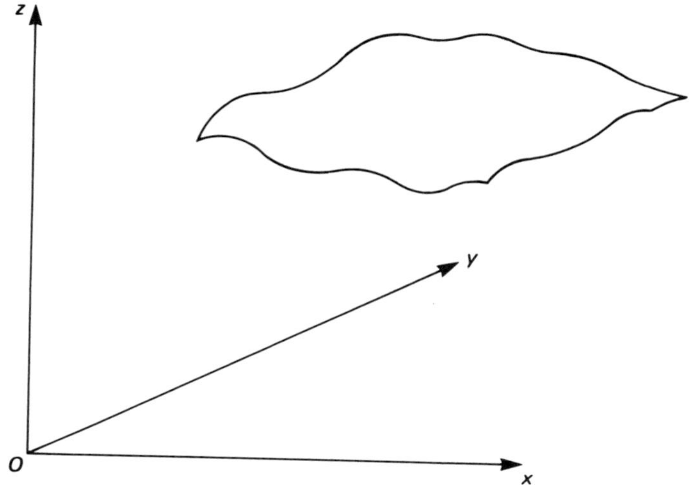

Fig. 7.1 A function of two variables.

It was a great asset when considering functions of a single variable to be able to draw the graph of the function. Here things are not quite so simple because in order to give a similar geometrical representation we shall need three axes Ox, Oy and Oz. Luckily we can represent situations like this by using a plane representation (Fig. 7.1). Instead of the 'curve' which we use to represent a function of a single variable, there corresponds a 'surface' for functions of two real variables.

However, once we extend the idea one stage further and consider functions of three real variables, we lose the picture altogether. Luckily we have the *algebraic* properties of the functions to enlighten us, and it is surprising how little we feel the loss of an adequate geometrical description. Nevertheless we can talk of 'hypersurfaces' for functions of more than two variables.

7.2 CONTINUITY

Intuitively a function of two variables is continuous at a point if the surface at the point has no 'cuts, holes or tears' in it. To put this a little more precisely, suppose (a, b) is a point in the domain of a function f; then f is continuous at the point (a, b) if $f(x, y)$ can be made arbitrarily close to $f(a, b)$ just by choosing (x, y) sufficiently close to (a, b) (Fig. 7.2).

Another way of thinking of this is that, whatever path of approach we use, as the point (x, y) approaches the point (a, b) the corresponding value $f(x, y)$ approaches the value $f(a, b)$.

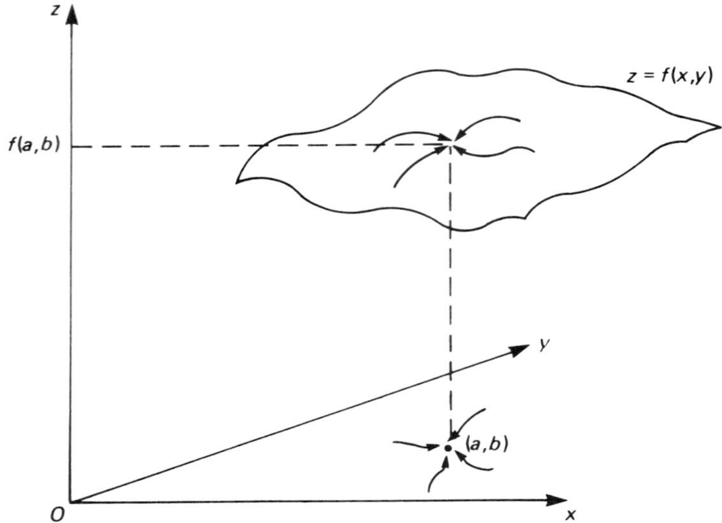

Fig. 7.2 Continuity at (a, b).

7.3 PARTIAL DERIVATIVES

When we considered functions of a single variable we saw that some functions which were continuous at a point were also differentiable there (Chapter 4). You probably recall the definition

$$\frac{\mathrm{d}f}{\mathrm{d}x} = f'(x) = \lim_{h \to 0} \frac{f(x + h) - f(x)}{h}$$

We make similar definitions for functions of several variables. For instance, suppose f is a function of two variables. Then we define

$$\frac{\partial f}{\partial x} = f_x(x, y) = \lim_{h \to 0} \frac{f(x + h, y) - f(x, y)}{h}$$

$$\frac{\partial f}{\partial y} = f_y(x, y) = \lim_{k \to 0} \frac{f(x, y + k) - f(x, y)}{k}$$

whenever these limits exist, and call these the **first-order partial derivatives** of f with respect to x and y respectively.

Notice the special symbol ∂ for partial differentiation. It must be carefully distinguished from the Greek delta δ and the d of ordinary differentiation.

At first sight these definitions seem rather formidable. However, when we examine them carefully we see that they tell us something very simple. If we look at the first one we notice that only the x part varies and that the

y part is unchanged. This gives us the clue. We treat y as if it is constant, and differentiate in the ordinary way with respect to x.

Similarly, inspection of the second expression reveals that to obtain the first-order partial derivative with respect to y we simply differentiate in the ordinary way, treating x as if it is constant.

Geometrically we can think of the first-order partial derivative of f with respect to x as the slope of the curve where the plane parallel to the Oyz plane through the point (a, b) cuts the surface defined by f (Fig. 7.3). Similarly the first-order partial derivative of f with respect to y is represented as the slope of the curve where the Oxz plane through (a, b) cuts the surface defined by f.

Although we have defined partial derivatives for a function of two variables only, the definition can be extended in a similar way to functions of several variables.

To see how very easy it is to perform partial differentiation we shall do some examples.

□ Suppose

$$z = f(x, y) = \sin 2x \cos y$$

Obtain $\partial z/\partial x$ and $\partial z/\partial y$.

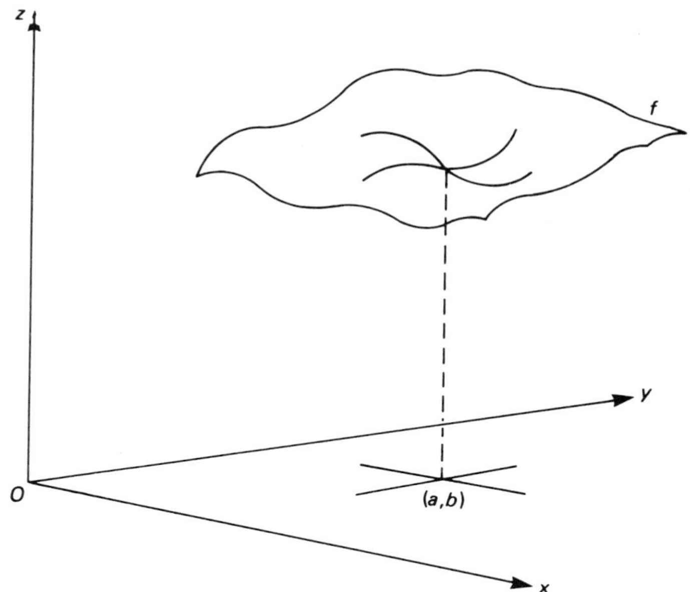

Fig. 7.3 Curves where planes cut f.

Remember that to differentiate partially with respect to a variable, all we need to do is to treat any other variables present as if they are constant. To begin with we wish to differentiate partially with respect to x, so we must treat y as if it is a constant. We obtain therefore

$$\frac{\partial z}{\partial x} = 2 \cos 2x \cos y$$

Likewise

$$\frac{\partial z}{\partial y} = -\sin y \sin 2x$$

It really is as simple as that! ■

Now you try one.

□ If

$$u = f(s, t) = s^2 + 4st^2 - t^3$$

obtain $\partial u/\partial s$ and $\partial u/\partial t$.

Notice here we are using different letters for the variables. This should cause no problems. When you have done this problem, read on and see if it is correct.

Using our simple procedure we obtain

$$\frac{\partial u}{\partial s} = 2s + 4t^2$$

Remember that we treat t as if it is constant. This does not mean that every term containing t automatically becomes zero when we differentiate. If you left out $4t^2$ you should think carefully about this point. Then check over your answer for $\partial u/\partial t$ before proceeding.

Differentiating with respect to t we obtain

$$\frac{\partial u}{\partial t} = 8st - 3t^2$$ ■

7.4 HIGHER-ORDER DERIVATIVES

Suppose we have a function $f = f(x, y)$ of two real variables x and y for which the first-order partial derivatives exist. We can consider differentiating these first-order partial derivatives again with respect to x and y. There are four possibilities, and we call these the **second-order partial derivatives** of f:

1 $\dfrac{\partial^2 f}{\partial x^2} = \dfrac{\partial}{\partial x}\left(\dfrac{\partial f}{\partial x}\right)$

2 $\dfrac{\partial^2 f}{\partial y\,\partial x} = \dfrac{\partial}{\partial y}\left(\dfrac{\partial f}{\partial x}\right)$

3 $\dfrac{\partial^2 f}{\partial x\,\partial y} = \dfrac{\partial}{\partial x}\left(\dfrac{\partial f}{\partial y}\right)$

4 $\dfrac{\partial^2 f}{\partial y^2} = \dfrac{\partial}{\partial y}\left(\dfrac{\partial f}{\partial y}\right)$

The definitions 1 and 4 are called the second-order partial derivatives of f with respect to x and y respectively. Definitions 2 and 3 are known as **mixed** second-order partial derivatives. It so happens that if these mixed derivatives are continuous at all points in a neighbourhood of the point (a, b), then they are equal at the point (a, b). It is very unusual to come across a function for which this condition does not hold, and so we shall assume that all the functions which we shall encounter have equal mixed second-order partial derivatives.

Example

□ Obtain all the first-order and second-order partial derivatives of the function f defined by

$$z = f(x, y) = x + \sin x^2y + \ln y$$

We have

$$\frac{\partial z}{\partial x} = 1 + 2xy \cos x^2y$$

$$\frac{\partial z}{\partial y} = x^2 \cos x^2y + y^{-1}$$

So therefore:

$$\begin{aligned}\frac{\partial^2 z}{\partial x^2} &= \frac{\partial}{\partial x}(1 + 2xy \cos x^2y)\\ &= 0 + 2xy(-2xy \sin x^2y) + 2y \cos x^2y\\ &= -4x^2y^2 \sin x^2y + 2y \cos x^2y\end{aligned}$$

$$\begin{aligned}\frac{\partial^2 z}{\partial y\,\partial x} &= \frac{\partial}{\partial y}(1 + 2xy \cos x^2y)\\ &= 2x \cos x^2y - 2xy \sin x^2y\,(x^2)\\ &= 2x \cos x^2y - 2x^3y \sin x^2y\end{aligned}$$

$$\frac{\partial^2 z}{\partial x\,\partial y} = \frac{\partial}{\partial x}(x^2 \cos x^2y + y^{-1})$$
$$= 2x \cos x^2y - x^2 \sin x^2y\,(2xy)$$
$$= 2x \cos x^2y - 2x^3y \sin x^2y$$
$$\frac{\partial^2 z}{\partial y^2} = \frac{\partial}{\partial y}(x^2 \cos x^2y + y^{-1})$$
$$= -x^2 \sin x^2y\,(x^2) - y^{-2}$$
$$= -x^4 \sin x^2y - y^{-2}$$

A partial check is provided by the fact that the two mixed derivatives are indeed equal. ■

So that you will acquire plenty of practice, we shall take a few steps before proceeding any further.

7.5 Workshop

1 **Exercise** Consider the following function of two real variables:

$$f(x, y) = \tan^{-1}\left(\frac{x}{y}\right)$$

Obtain the first-order partial derivatives f_x and f_y and thereby show that

$$x\frac{\partial f}{\partial x} + y\frac{\partial f}{\partial y} = 0$$

Try it first on your own and see how it goes. Then step ahead.

2 Differentiating partially with respect to x and with respect to y in turn gives

$$\frac{\partial f}{\partial x} = \frac{1}{[1 + (x/y)^2]}\frac{1}{y}$$
$$\frac{\partial f}{\partial y} = \frac{1}{[1 + (x/y)^2]}\frac{-x}{y^2}$$

Consequently

$$x\frac{\partial f}{\partial x} + y\frac{\partial f}{\partial y} = \frac{1}{[1 + (x/y)^2]}\frac{x - x}{y} = 0$$

If you managed that all right, you should move ahead to step 4. If something went wrong, then try this.

▷**Exercise** Suppose

$$f(x, y) = \cos^3 (x^2 - y^2)$$

Derive the first-order partial derivatives and show that

$$y\frac{\partial f}{\partial x} + x\frac{\partial f}{\partial y} = 0$$

Try it carefully and take the next step.

3

Given that

$$f(x, y) = \cos^3 (x^2 - y^2)$$

we have

$$\frac{\partial f}{\partial x} = -3 \cos^2 (x^2 - y^2) \sin (x^2 - y^2)(2x)$$

$$\frac{\partial f}{\partial y} = -3 \cos^2 (x^2 - y^2) \sin (x^2 - y^2)(-2y)$$

It follows at once that

$$y\frac{\partial f}{\partial x} + x\frac{\partial f}{\partial y} = 0$$

If things are still going wrong there are only two possibilities. Either you are having difficulty with the ordinary differentiation, or you are forgetting when partially differentiating to treat the other variable as if it is a constant. Go back and review what you have done to make certain you can manage this correctly.

Assuming that all is now well, we can move ahead. Now we are going to test the work on second-order derivatives.

4

▷**Exercise** Suppose $z = \ln \sqrt{(x^2 + y^2)}$. Show that **Laplace's equation** in two dimensions is satisfied, namely

$$\frac{\partial^2 z}{\partial x^2} + \frac{\partial^2 z}{\partial y^2} = 0$$

Laplace's equation has many applications and you will certainly come across it again from time to time. It is an example of a partial differential equation.

As soon as you have made a good attempt at this, move ahead to the next step.

5 It may help to simplify z before differentiating. Here we have

$$z = \tfrac{1}{2}\ln(x^2 + y^2)$$

using the laws of logarithms, and so

$$\frac{\partial z}{\partial x} = \frac{1}{2(x^2 + y^2)}(2x)$$
$$= \frac{x}{x^2 + y^2}$$

So

$$\frac{\partial^2 z}{\partial x^2} = \frac{(x^2 + y^2) - x(2x)}{(x^2 + y^2)^2}$$
$$= \frac{y^2 - x^2}{(x^2 + y^2)^2}$$

Now we don't need to do any more partial differentiation for this problem. Can you see why not? It is because of symmetry. Look back at the original expression for z; if we were to interchange x and y in it, it would not be altered. Therefore if we interchange x and y in this partial derivative we shall obtain a correct statement. So, by symmetry, we obtain

$$\frac{\partial^2 z}{\partial y^2} = \frac{x^2 - y^2}{(y^2 + x^2)^2}$$

Always keep an eye open for symmetry, it can save you a lot of time and effort!

Finally, adding these two second-order partial derivatives produces

$$\frac{\partial^2 z}{\partial x^2} + \frac{\partial^2 z}{\partial y^2} = \frac{y^2 - x^2 + x^2 - y^2}{(x^2 + y^2)^2} = 0$$

If all's well then you can move ahead to step 7. However, if it went wrong then try this problem first.

▷**Exercise** Given that

$$w = \sin(2x + y) - (x - 3y)^3$$

verify that the mixed partial derivatives are equal.

As soon as you have done it move on to the next step.

6 There is no symmetry here, so we have no alternative but to get our heads down and do the partial differentiation. Of course we only require the mixed second-order derivatives, and so we would be doing needless work if we obtained all four second-order derivatives. Did you? If you did, console yourself that you have at least practised some partial differentiation; your time was not completely wasted.

We have

$$\frac{\partial w}{\partial x} = 2 \cos (2x + y) - 3(x - 3y)^2$$

$$\frac{\partial w}{\partial y} = \cos (2x + y) - 3(x - 3y)^2 (-3)$$

$$= \cos (2x + y) + 9(x - 3y)^2$$

Differentiating again,

$$\frac{\partial^2 w}{\partial x\, \partial y} = -2 \sin (2x + y) + 18(x - 3y)$$

$$\frac{\partial^2 w}{\partial y\, \partial x} = -2 \sin (2x + y) - 6(x - 3y)(-3)$$

$$= -2 \sin (2x + y) + 18(x - 3y)$$

If trouble persists, make sure you can handle the chain rule for ordinary differentiation (Chapter 4).

▷ **Exercise** Show that if f is a differentiable real function and if $z = f(x/y)$ then **7**

$$x\frac{\partial z}{\partial x} + y\frac{\partial z}{\partial y} = 0$$

Try hard with this. Although it seems a little abstract we have already done all the necessary work before when we differentiated $\tan^{-1} (x/y)$ at the beginning of this workshop. As soon as you have made a good attempt read on.

We have straight away, using the chain rule for ordinary differentiation, **8**

$$\frac{\partial z}{\partial x} = f'\left(\frac{x}{y}\right)\left(\frac{1}{y}\right)$$

Remember: first differentiate with respect to the bracketed terms, then multiply the result by the derivative of the bracketed terms with respect to x. Similarly,

$$\frac{\partial z}{\partial y} = f'\left(\frac{x}{y}\right)\left(\frac{-x}{y^2}\right)$$

Consequently,

$$x\frac{\partial z}{\partial x} + y\frac{\partial z}{\partial y} = f'\left(\frac{x}{y}\right)\left(\frac{x - x}{y}\right) = 0$$

You may have not managed that, but even if you did it is a good idea to try another exercise like that. Before tackling it, look very carefully at the previous exercise to make sure you have a good start.

▷**Exercise** Suppose f is a twice differentiable real function. Deduce that if $z = f(r)$, where $r = \sqrt{(x^2 + y^2)}$, then z satisfies the equation

$$\frac{\partial^2 z}{\partial x^2} + \frac{\partial^2 z}{\partial y^2} = \frac{1}{r} f'(r) + f''(r)$$

When you have tried it, take the next step.

---

9 Using exactly the same idea as before, we obtain

$$\frac{\partial z}{\partial x} = f'\sqrt{(x^2 + y^2)}\,\frac{1}{2}\,(x^2 + y^2)^{-1/2}\,2x$$

$$= f'\sqrt{(x^2 + y^2)}\,\frac{x}{\sqrt{(x^2 + y^2)}}$$

Now differentiating again we obtain

$$\frac{\partial^2 z}{\partial x^2} = f'\sqrt{(x^2 + y^2)}\,\frac{\sqrt{(x^2 + y^2)} - x\,(1/2)\,(x^2 + y^2)^{-1/2}\,2x}{(x^2 + y^2)}$$

$$+ f''\sqrt{(x^2 + y^2)}\,\frac{x^2}{x^2 + y^2}$$

$$= f'\sqrt{(x^2 + y^2)}\,\frac{(x^2 + y^2) - x^2}{(x^2 + y^2)^{3/2}} + f''\sqrt{(x^2 + y^2)}\,\frac{x^2}{x^2 + y^2}$$

$$= f'\sqrt{(x^2 + y^2)}\,\frac{y^2}{(x^2 + y^2)^{3/2}} + f''\sqrt{(x^2 + y^2)}\,\frac{x^2}{x^2 + y^2}$$

Similarly

$$\frac{\partial^2 z}{\partial y^2} = f'\sqrt{(y^2 + x^2)}\,\frac{x^2}{(y^2 + x^2)^{3/2}} + f''\sqrt{(y^2 + x^2)}\,\frac{y^2}{y^2 + x^2}$$

Adding we obtain

$$\frac{\partial^2 z}{\partial x^2} + \frac{\partial^2 z}{\partial y^2} = f'\sqrt{(x^2 + y^2)}\,\frac{x^2 + y^2}{(x^2 + y^2)^{3/2}} + f''\sqrt{(x^2 + y^2)}\,\frac{x^2 + y^2}{x^2 + y^2}$$

$$= f'\sqrt{(x^2 + y^2)}\,\frac{1}{(x^2 + y^2)^{1/2}} + f''\sqrt{(x^2 + y^2)}$$

Therefore

$$\frac{\partial^2 z}{\partial x^2} + \frac{\partial^2 z}{\partial y^2} = \frac{1}{r}f'(r) + f''(r)$$

We shall soon see that there is in fact a much easier way to do this.

A word about the notation for partial differentiation is not out of place at this point. Much has been written about the problems inherent with the notation, particularly when (as in the previous workshop) more than two symbols for independent variables are involved. It is necessary to be clear which are the independent variables.

For example, suppose that $z = x^2 + y^3$ and that $x = r\cos\theta$ and $y = r\sin\theta$. Here z is expressed in terms of two independent variables x and y. It is also possible to express z in terms of two other independent variables r and θ. It would be a mistake to express z in terms of mixtures such as r, x and y, or x and θ, and to attempt to form partial derivatives, because the variables would not be independent and so the partial derivatives we attempted to find would be incorrect. Major errors have followed from misunderstandings of this nature; many attempts have been made to improve the notation, but the result is generally unattractive and difficult to follow.

7.6 THE FORMULA FOR A CHANGE OF VARIABLES: THE CHAIN RULE

In the final exercise in the previous workshop it would have been correct to use the abbreviated notation r instead of $\sqrt{(x^2 + y^2)}$, provided we obeyed the chain rule for ordinary differentation diligently. In fact we shall now state a version of the chain rule for functions of more than one variable.

If we write $f = f(x, y)$ where $x = x(u, v)$ and $y = y(u, v)$ we are using x and y in two ways: first as a dummy variable, and secondly as a function symbol. We have done this sort of thing before in the case of real functions, and shall find it particularly useful to avoid introducing many unwanted symbols. To express this in words: the function f is expressed in terms of the independent variables x and y, and the variables x and y are themselves expressed in terms of independent variables u and v.

Here now is the **chain rule**. Suppose that $f = f(x, y)$ and that $x = x(u, v)$ and $y = y(u, v)$. Then, if all the partial derivatives exist,

$$\frac{\partial f}{\partial u} = \frac{\partial f}{\partial x}\frac{\partial x}{\partial u} + \frac{\partial f}{\partial y}\frac{\partial y}{\partial u}$$

$$\frac{\partial f}{\partial v} = \frac{\partial f}{\partial x}\frac{\partial x}{\partial v} + \frac{\partial f}{\partial y}\frac{\partial y}{\partial v}$$

Notice how when we choose to differentiate with respect to one of the subsidiary variables we must be impartial. We differentiate with respect to each of the main variables in turn, multiplying the result in each case by the derivative of the main variable with respect to the subsidiary variable. This chain of products is then added together. The formula is reminiscent of the formula for differentiating a function of a single variable:

$$\frac{\mathrm{d}y}{\mathrm{d}x} = \frac{\mathrm{d}y}{\mathrm{d}u}\frac{\mathrm{d}u}{\mathrm{d}x}$$

Indeed, this is a special case of the more general rule.

However, there is one important point to watch. In the case of a single variable the formula looks right because we can imagine $\mathrm{d}u$ as cancelling out. Indeed, if we extend the definitions to allow $\mathrm{d}y$ and $\mathrm{d}x$ to be used separately as differentials the procedure becomes justifiable. However, we must *never* cancel out symbols such as ∂x or ∂y. You can see why if you look at the formulas for a change of variables: it would give $1 = 2$!

To avoid unnecessary complications we shall justify the formula for a change of variables in a special case only. Suppose $f = f(x, y)$ and that $x = x(t)$, $y = y(t)$. You may like to think of t as time or temperature; then as t changes so too do x and y and consequently the value of f.

If δt is a small non-zero change in t and δx and δy are the corresponding changes in x and y respectively, these in turn produce a change δf in f. Now

$$\begin{aligned}\delta f &= f(x+\delta x, y+\delta y) - f(x, y)\\ &= f(x+\delta x, y+\delta y) - f(x, y+\delta y) + f(x, y+\delta y) - f(x, y)\\ &= \frac{f(x+\delta x, y+\delta y) - f(x, y+\delta y)}{\delta x}\delta x + \frac{f(x, y+\delta y) - f(x, y)}{\delta y}\delta y\end{aligned}$$

Dividing through by δt,

$$\frac{\delta f}{\delta t} = \frac{f(x+\delta x, y+\delta y) - f(x, y+\delta y)}{\delta x}\frac{\delta x}{\delta t} + \frac{f(x, y+\delta y) - f(x, y)}{\delta y}\frac{\delta y}{\delta t}$$

Now we consider what happens as δt tends to zero. we have $\delta x \to 0$, $\delta y \to 0$ and also

$$\frac{\delta x}{\delta t} \to \frac{\mathrm{d}x}{\mathrm{d}t}, \qquad \frac{\delta y}{\delta t} \to \frac{\mathrm{d}y}{\mathrm{d}t}, \qquad \frac{\delta f}{\delta t} \to \frac{\mathrm{d}f}{\mathrm{d}t}$$

Moreover,

$$\frac{f(x, y + \delta y) - f(x, y)}{\delta y} \to \frac{\partial f}{\partial y}$$

$$\frac{f(x+\delta x, y+\delta y) - f(x, y+\delta y)}{\delta x} \to \frac{f(x+\delta x, y) - f(x, y)}{\delta x} \to \frac{\partial f}{\partial x}$$

Consequently,

$$\frac{\mathrm{d}f}{\mathrm{d}t} = \frac{\partial f}{\partial x}\frac{\mathrm{d}x}{\mathrm{d}t} + \frac{\partial f}{\partial y}\frac{\mathrm{d}y}{\mathrm{d}t}$$

Here we have made a number of assumptions which we have not justified. Principal among these is that the limits exist and that we can select the order in which to take these limits. Nevertheless we have given a justification for the formula in the case where x and y are both functions of a single variable t, and it is a simple matter to extend this to the more general case.

It is important that we learn how to apply the chain rule correctly. Unfortunately it is sometimes possible to misapply the chain rule and still obtain a correct result. Diligent examiners are always on the lookout for errors of this kind, and so marks will be lost if you are sloppy!

☐ If $z = (x + y)^2$ where $x = r \cos \theta$ and $y = r \sin \theta$, show that

$$2r^2 \frac{\partial^2 z}{\partial r^2} + \frac{\partial^2 z}{\partial \theta^2} = 4r^2$$

We begin by finding the first-order partial derivatives:

$$\begin{aligned}\frac{\partial z}{\partial r} &= \frac{\partial z}{\partial x}\frac{\partial x}{\partial r} + \frac{\partial z}{\partial y}\frac{\partial y}{\partial r} \\ &= 2(x + y) \cos \theta + 2(x + y) \sin \theta \\ &= 2(x + y)(\cos \theta + \sin \theta)\end{aligned}$$

Do you follow it so far?

See if you can obtain the other first-order partial derivative.

Here we are:

$$\begin{aligned}\frac{\partial z}{\partial \theta} &= \frac{\partial z}{\partial x}\frac{\partial x}{\partial \theta} + \frac{\partial z}{\partial y}\frac{\partial y}{\partial \theta} \\ &= 2(x + y)(-r \sin \theta) + 2(x + y)r \cos \theta \\ &= 2r(x + y)(\cos \theta - \sin \theta)\end{aligned}$$

There is not normally too much of a problem at this stage; it is when we differentiate again that some students, and unfortunately textbooks, overlook terms. It is important to remember that the chain rule must always be used when we differentiate with respect to a subsidiary variable.

So now

$$\frac{\partial^2 z}{\partial r^2} = \frac{\partial}{\partial r}\left(\frac{\partial z}{\partial r}\right)$$

$$= \frac{\partial}{\partial r}[2(x + y)](\cos\theta + \sin\theta) + 2(x + y)\frac{\partial}{\partial r}(\cos\theta + \sin\theta)$$

The second term is zero. Therefore, applying the chain rule to the first term,

$$\begin{aligned}\frac{\partial^2 z}{\partial r^2} &= 2(\cos\theta + \sin\theta)\left[\frac{\partial}{\partial x}(x + y)\frac{\partial x}{\partial r} + \frac{\partial}{\partial y}(x + y)\frac{\partial y}{\partial r}\right]\\ &= 2(\cos\theta + \sin\theta)(\cos\theta + \sin\theta)\\ &= 2(\cos^2\theta + 2\sin\theta\cos\theta + \sin^2\theta)\\ &= 2(1 + \sin 2\theta)\end{aligned}$$

Also

$$\begin{aligned}\frac{\partial^2 z}{\partial \theta^2} &= \frac{\partial}{\partial \theta}\left(\frac{\partial z}{\partial \theta}\right)\\ &= 2(x + y)\frac{\partial}{\partial \theta}[r(\cos\theta - \sin\theta)] + 2r(\cos\theta - \sin\theta)\frac{\partial}{\partial \theta}(x + y)\\ &= 2(x + y)[r(-\sin\theta - \cos\theta)] + 2r(\cos\theta - \sin\theta)(-r\sin\theta + r\cos\theta)\\ &= 2r^2[-(\sin\theta + \cos\theta)^2 + (\cos\theta - \sin\theta)^2]\\ &= 2r^2(-4\sin\theta\cos\theta)\\ &= -4r^2\sin 2\theta\end{aligned}$$

Therefore

$$2r^2\frac{\partial^2 z}{\partial r^2} + \frac{\partial^2 z}{\partial \theta^2} = 2r^2[2(1 + \sin 2\theta)] - 4r^2\sin 2\theta = 4r^2$$

As an exercise you may wish to check this by first eliminating x and y in the expression for z and then differentiating directly with respect to r and θ; it's much quicker!

7.7 THE TOTAL DIFFERENTIAL

In the case of a differentiable function f of a single variable we originally defined $\mathrm{d}y/\mathrm{d}x$ as a limit, so that the symbols $\mathrm{d}y$ and $\mathrm{d}x$ used on their own were meaningless. However, we found it useful to be able to use these symbols separately, and accordingly we extended the definition (Chapter 4). We call $\mathrm{d}x$ a differential, and we can think of it as a change in the value of x, not necessarily small. In fact it is any real number. The differential $\mathrm{d}y$

is then defined by $dy = f'(x)\, dx$. Of course if dx happens to be small then dy is an approximation to the corresponding change in y.

In the case of a function of several variables we adopt a similar approach. Specifically, suppose that $z = f(x, y)$. Then the differentials dx and dy are defined to be any real numbers. You may choose to think of them as changes in x and y respectively which are not necessarily small. The differential dz is then defined by

$$dz = \frac{\partial z}{\partial x}\, dx + \frac{\partial z}{\partial y}\, dy$$

dz is usually called the **total differential**.

When we derived the special case of the chain rule (section 7.6), we showed that when dx and dy are small the total differential is approximately the corresponding change in z. We can therefore use this to obtain a rough estimate of the error caused by inaccuracies in measurement.

☐ A surveyor estimates the area of a triangular plot of land using the formula

$$A = \tfrac{1}{2}\, ab \sin C$$

where a and b are the lengths of two sides and C is the included angle. If the sides are measured to an accuracy of 2% and the angle C, measured as 45°, is measured to within 1%, calculate approximately the percentage error in A.

You may have a go at this on your own if you wish.

Using the total differential,

$$dA = \frac{\partial A}{\partial a}\, da + \frac{\partial A}{\partial b}\, db + \frac{\partial A}{\partial C}\, dC$$

Now $da = \pm 0.02a$, $db = \pm 0.02b$ and $dC = \pm 0.01C$. Also

$$\frac{\partial A}{\partial a} = \frac{1}{2} b \sin C \qquad \frac{\partial A}{\partial b} = \frac{1}{2} a \sin C \qquad \frac{\partial A}{\partial C} = \frac{1}{2} ab \cos C$$

Substituting,

$$\begin{aligned} dA &= 1/2\, b \sin C\, (\pm 0.02a) + 1/2\, a \sin C\, (\pm 0.02b) \\ &\quad + 1/2\, ab \cos C\, (\pm 0.01C) \\ &= 1/2\, ab \sin C\, (\pm 0.02 \pm 0.02 \pm 0.01C \cot C) \end{aligned}$$

Now $C = 45° = \pi/4$, and so $\cot C = 1$. Therefore

$$dA = A\,[\pm 0.02 \pm 0.02 \pm 0.01\,(\pi/4)]$$

The greatest error occurs when all have the same sign, so that

$$|dA| \leqslant 0.01A\,[2 + 2 + (\pi/4)] \simeq 0.048A$$

So the error is not more than 5% approximately. ■

Finally, we mention a few of the notations which are sometimes used for partial differentiation. Suppose that $z = f(x, y)$. Then

$$\frac{\partial z}{\partial x} \qquad \frac{\partial f}{\partial x} \qquad f_x \qquad f_1 \qquad D_1 f$$

are all equivalent and

$$\frac{\partial^2 z}{\partial x\,\partial y} \qquad \frac{\partial^2 f}{\partial x\,\partial y} \qquad f_{xy} \qquad f_{12} \qquad D_{12} f$$

are also equivalent to one another. There are many variations of these, and it is necessary to determine which notation is being used at any time. Our notation is the one which is most widely used.

So now we are ready to take steps. We are going to tackle a problem which is sometimes incorrectly solved in textbooks.

7.8 Workshop

1 **Exercise** Transform the partial differential equation

$$\frac{\partial^2 z}{\partial x^2} + \frac{\partial^2 z}{\partial y^2} = 0$$

where z is expressed in terms of cartesian coordinates (x, y), into a partial differential equation in polar coordinates (r, θ), where $x = r\cos\theta$ and $y = r\sin\theta$.

Let us begin by writing down formulas which express the partial derivatives with respect to x and y in terms of the partial derivatives with respect to r and θ. When you have done this, take the next step.

We have

$$\frac{\partial z}{\partial x} = \frac{\partial z}{\partial r}\frac{\partial r}{\partial x} + \frac{\partial z}{\partial \theta}\frac{\partial \theta}{\partial x}$$

This is a straight application of the chain rule. Similarly,

$$\frac{\partial z}{\partial y} = \frac{\partial z}{\partial r}\frac{\partial r}{\partial y} + \frac{\partial z}{\partial \theta}\frac{\partial \theta}{\partial y}$$

If all is well, follow through the next step. If you made an error, write down formulas which express the partial derivatives with respect to r and θ in terms of those with respect to x and y. Don't move on until you have done this.

3

Here they are:

$$\frac{\partial z}{\partial r} = \frac{\partial z}{\partial x}\frac{\partial x}{\partial r} + \frac{\partial z}{\partial y}\frac{\partial y}{\partial r}$$

$$\frac{\partial z}{\partial \theta} = \frac{\partial z}{\partial x}\frac{\partial x}{\partial \theta} + \frac{\partial z}{\partial y}\frac{\partial y}{\partial \theta}$$

We now have a choice as to which pair to use. There are advantages and disadvantages either way. If we use the first pair we shall have to express r and θ explicitly in terms of x and y so that we can differentiate. If we use the second pair we can find the partial derivatives easily enough, but we shall then have to eliminate to obtain the differential equation and it might be difficult to find our way.

Let us be definite: we shall use the first pair. So we must obtain r and θ explicitly in terms of x and y, and then the partial derivatives. Do this and then take the next step.

4

We have $r^2 = x^2 + y^2$ and so $r = \sqrt{(x^2 + y^2)}$. Consequently,

$$\frac{\partial r}{\partial x} = \frac{x}{\sqrt{(x^2 + y^2)}}$$

$$\frac{\partial r}{\partial y} = \frac{y}{\sqrt{(x^2 + y^2)}}$$

Also $\tan\theta = y/x$ so that $\theta = \tan^{-1}(y/x)$. Therefore

$$\frac{\partial \theta}{\partial x} = \frac{-y}{(x^2 + y^2)}$$

$$\frac{\partial \theta}{\partial y} = \frac{x}{(x^2 + y^2)}$$

and so

$$\frac{\partial r}{\partial x} = \frac{x}{r} \qquad \frac{\partial r}{\partial y} = \frac{y}{r} \qquad \frac{\partial \theta}{\partial x} = \frac{-y}{r^2} \qquad \frac{\partial \theta}{\partial y} = \frac{x}{r^2}$$

Now substitute these expressions into the formulas for a change of variables, and take the next step.

5 We obtain

$$\frac{\partial z}{\partial x} = \frac{\partial z}{\partial r}\frac{\partial r}{\partial x} + \frac{\partial z}{\partial \theta}\frac{\partial \theta}{\partial x}$$
$$= \frac{\partial z}{\partial r}\frac{x}{r} + \frac{\partial z}{\partial \theta}\frac{(-y)}{r^2}$$

and

$$\frac{\partial z}{\partial y} = \frac{\partial z}{\partial r}\frac{\partial r}{\partial y} + \frac{\partial z}{\partial \theta}\frac{\partial \theta}{\partial y}$$
$$= \frac{\partial z}{\partial r}\frac{y}{r} + \frac{\partial z}{\partial \theta}\frac{x}{r^2}$$

Now we are ready for the second-order derivatives with respect to x and y respectively. You find the second-order partial derivative with respect to x – but be careful! This is the place at which we pass the bones of reputations bleached white by the sun.

6 Here we go! We must use the chain rule again and not overlook any terms:

$$\frac{\partial^2 z}{\partial x^2} = \frac{\partial}{\partial x}\left[\frac{\partial z}{\partial r}\frac{x}{r} + \frac{\partial z}{\partial \theta}\frac{(-y)}{r^2}\right]$$
$$= \frac{1}{r}\frac{\partial z}{\partial r}\frac{\partial x}{\partial x} + x\frac{\partial}{\partial x}\left(\frac{1}{r}\frac{\partial z}{\partial r}\right) - \frac{1}{r^2}\frac{\partial z}{\partial \theta}\frac{\partial y}{\partial x} - y\frac{\partial}{\partial x}\left(\frac{1}{r^2}\frac{\partial z}{\partial \theta}\right)$$

Here we have simply used the product rule but kept the two sets of variables apart. Since x and y are independent we can deduce that their partial derivatives with respect to each other are zero, and so the third term is zero. We must use the chain rule again to expand the remaining terms. So the second-order partial derivative of z with respect to x is

$$\frac{\partial^2 z}{\partial x^2} = \frac{1}{r}\frac{\partial z}{\partial r} + x\left[\frac{\partial}{\partial r}\left(\frac{1}{r}\frac{\partial z}{\partial r}\right)\frac{\partial r}{\partial x} + \frac{\partial}{\partial \theta}\left(\frac{1}{r}\frac{\partial z}{\partial r}\right)\frac{\partial \theta}{\partial x}\right]$$
$$- y\left[\frac{\partial}{\partial r}\left(\frac{1}{r^2}\frac{\partial z}{\partial \theta}\right)\frac{\partial r}{\partial x} + \frac{\partial}{\partial \theta}\left(\frac{1}{r^2}\frac{\partial z}{\partial \theta}\right)\frac{\partial \theta}{\partial x}\right]$$

This is where people make the error: they overlook the mixed derivative terms. As it happens they end up with the correct equation at the final stage, even though the second-order derivatives themselves are incorrect. To continue:

$$\frac{\partial^2 z}{\partial x^2} = \frac{1}{r}\frac{\partial z}{\partial r} + x\left[\left(\frac{1}{r}\frac{\partial^2 z}{\partial r^2} - \frac{1}{r^2}\frac{\partial z}{\partial r}\right)\frac{x}{r} + \left(\frac{1}{r}\frac{\partial^2 z}{\partial \theta\,\partial r}\right)\frac{(-y)}{r^2}\right]$$

$$- y\left[\left(\frac{-2}{r^3}\frac{\partial z}{\partial\theta}+\frac{1}{r^2}\frac{\partial^2 z}{\partial r\partial\theta}\right)\frac{x}{r}+\left(\frac{1}{r^2}\frac{\partial^2 z}{\partial\theta^2}\right)\frac{(-y)}{r^2}\right]$$

$$=\frac{1}{r}\frac{\partial z}{\partial r}+\frac{x^2}{r^2}\frac{\partial^2 z}{\partial r^2}-\frac{x^2}{r^3}\frac{\partial z}{\partial r}-\frac{xy}{r^3}\frac{\partial^2 z}{\partial\theta\,\partial r}+\frac{2xy}{r^4}\frac{\partial z}{\partial\theta}-\frac{xy}{r^3}\frac{\partial^2 z}{\partial r\partial\theta}+\frac{y^2}{r^4}\frac{\partial^2 z}{\partial\theta^2}$$

Well, there it is. Pretty tough going, isn't it? You have to keep a clear head and make sure you use the chain rule properly. If you didn't manage that then you are undoubtedly part of a huge majority, so you may take consolation in that. Also you may be relieved to know that the going seldom gets harder. Anyway, we still have the other second-order derivative to obtain: so off you go!

Hold on to your hats. **7**

$$\frac{\partial^2 z}{\partial y^2}=\frac{\partial}{\partial y}\left(\frac{\partial z}{\partial r}\frac{y}{r}+\frac{\partial z}{\partial\theta}\frac{x}{r^2}\right)$$

$$=\frac{1}{r}\frac{\partial z}{\partial r}\frac{\partial y}{\partial y}+y\frac{\partial}{\partial y}\left(\frac{1}{r}\frac{\partial z}{\partial r}\right)+\frac{1}{r^2}\frac{\partial z}{\partial\theta}\frac{\partial x}{\partial y}+x\frac{\partial}{\partial y}\left(\frac{1}{r^2}\frac{\partial z}{\partial\theta}\right)$$

$$=\frac{1}{r}\frac{\partial z}{\partial r}+y\left[\frac{\partial}{\partial r}\left(\frac{1}{r}\frac{\partial z}{\partial r}\right)\frac{\partial r}{\partial y}+\frac{\partial}{\partial\theta}\left(\frac{1}{r}\frac{\partial z}{\partial r}\right)\frac{\partial\theta}{\partial y}\right]$$

$$+x\left[\frac{\partial}{\partial r}\left(\frac{1}{r^2}\frac{\partial z}{\partial\theta}\right)\frac{\partial r}{\partial y}+\frac{\partial}{\partial\theta}\left(\frac{1}{r^2}\frac{\partial z}{\partial\theta}\right)\frac{\partial\theta}{\partial y}\right]$$

$$=\frac{1}{r}\frac{\partial z}{\partial r}+y\left[\left(\frac{1}{r}\frac{\partial^2 z}{\partial r^2}-\frac{1}{r^2}\frac{\partial z}{\partial r}\right)\frac{y}{r}+\left(\frac{1}{r}\frac{\partial^2 z}{\partial\theta\,\partial r}\right)\frac{x}{r^2}\right]$$

$$+x\left[\left(\frac{-2}{r^3}\frac{\partial z}{\partial\theta}+\frac{1}{r^2}\frac{\partial^2 z}{\partial r\partial\theta}\right)\frac{y}{r}+\left(\frac{1}{r^2}\frac{\partial^2 z}{\partial\theta^2}\right)\frac{x}{r^2}\right]$$

$$=\frac{1}{r}\frac{\partial z}{\partial r}+\frac{y^2}{r^2}\frac{\partial^2 z}{\partial r^2}-\frac{y^2}{r^3}\frac{\partial z}{\partial r}+\frac{xy}{r^3}\frac{\partial^2 z}{\partial\theta\,\partial r}-\frac{2xy}{r^4}\frac{\partial z}{\partial\theta}+\frac{xy}{r^3}\frac{\partial^2 z}{\partial r\partial\theta}+\frac{x^2}{r^4}\frac{\partial^2 z}{\partial\theta^2}$$

Lastly we must add the two expressions for the second-order derivatives and equate to zero.

This is what you should get if you remember that $x^2+y^2=r^2$: **8**

$$\frac{1}{r}\frac{\partial z}{\partial r}+\frac{\partial^2 z}{\partial r^2}+\frac{1}{r^2}\frac{\partial^2 z}{\partial\theta^2}=0$$

so that

$$\frac{\partial^2 z}{\partial r^2}+\frac{1}{r^2}\frac{\partial^2 z}{\partial\theta^2}+\frac{1}{r}\frac{\partial z}{\partial r}=0$$

This is **Laplace's equation** in two dimensions expressed in polar coordinates (see also section 7.5).

If you would like even more practice at this sort of work you can always try the other approach. The alternative method involves determining expressions for the second-order derivatives of z with respect to r and θ in terms of those with respect to x and y and then using Laplace's equation in cartesian coordinates to eliminate x and y. It helps to know the equation we are aiming to derive.

You probably feel that those steps were quite steep, but if you persisted and completed the exercise you will have gained some useful experience. Remember to pay particular attention to brackets. They are not there for decorative purposes: they play a vital part. Students who pay scant attention to brackets rarely succeed in mathematics.

7.9 Practical

VOLUME ERROR

The volume of a hydraulic tank, in the shape of a ring, is calculated using the formula

$$V = \pi r^2 h - \pi s^2 h$$

where h is the height and r and s are the external and internal radii respectively. If r and s were measured 3% too large, estimate the maximum error in V if h is correct to within 1%.

Try this. We need the formula for the total differential. Begin by writing this down for these symbols, obtain the partial derivatives and substitute into your equation.

This is correct:

$$\mathrm{d}V = \frac{\partial V}{\partial r}\,\mathrm{d}r + \frac{\partial V}{\partial s}\,\mathrm{d}s + \frac{\partial V}{\partial h}\,\mathrm{d}h$$

Now

$$\frac{\partial V}{\partial r} = 2\pi rh \qquad \frac{\partial V}{\partial s} = -2\pi sh \qquad \frac{\partial V}{\partial h} = \pi r^2 - \pi s^2$$

Also $\mathrm{d}r = 0.03r$, $\mathrm{d}s = 0.03s$, $\mathrm{d}h = \pm 0.01h$. This gives

$$\begin{aligned}\mathrm{d}V &= 2\pi r^2 h(0.03) - 2\pi s^2 h(0.03) \pm (\pi r^2 - \pi s^2)\,(0.01h)\\ &= V(0.06 \pm 0.01)\end{aligned}$$

So the calculated value of V is between 5% and 7% too large.

It's all quite simple really. Remember the chain rule, and remember to put in brackets whenever necessary, and everything should be fine.

OIL FLOW

Here is a problem about oil – just to make sure everything is running smoothly!

The motion of a light oil flowing with speed u past a cylindrical bearing of radius a may be roughly described by the equation

$$\phi = u \cos \alpha \left(r + \frac{a^2}{r}\right) \cos \theta + zu \sin \alpha$$

where α is a constant and r, θ and z are independent variables. Show that ϕ satisfies the equation

$$\frac{\partial^2 \phi}{\partial r^2} + \frac{1}{r}\frac{\partial \phi}{\partial r} + \frac{1}{r^2}\frac{\partial^2 \phi}{\partial \theta^2} + \frac{\partial^2 \phi}{\partial z^2} = 0$$

Try it yourself first, and then we will look at it stage by stage.

---

1 To solve this problem we merely need to show that ϕ satisfies the partial differential equation. Don't, whatever you do, take your starting-point as the partial differential equation and then try to deduce the expression for ϕ from it. We haven't been given enough information for that approach, even if we wished to attempt it that way.

We begin by finding the first-order partial derivatives:

$$\frac{\partial \phi}{\partial r} = u \cos \alpha \left(1 - \frac{a^2}{r^2}\right) \cos \theta$$

$$\frac{\partial \phi}{\partial \theta} = u \cos \alpha \left(r + \frac{a^2}{r^2}\right) (-\sin \theta)$$

$$\frac{\partial \phi}{\partial z} = u \sin \alpha$$

Check yours and see if they are right. The next step, of course, is to obtain the second-order derivatives we need.

---

2 Here we have

$$\frac{\partial^2 \phi}{\partial r^2} = u \cos \alpha \left(\frac{2a^2}{r^3}\right) \cos \theta$$

$$\frac{\partial^2 \phi}{\partial \theta^2} = -u \cos \alpha \left(r + \frac{a^2}{r}\right) \cos \theta$$

$$\frac{\partial^2 \phi}{\partial z^2} = 0$$

If you have these correct, it remains to substitute them into the left-hand side of the partial differential equation and confirm that the result is 0.

---

3 We obtain

$$\begin{aligned}\frac{\partial^2\phi}{\partial r^2} + \frac{1}{r}\frac{\partial\phi}{\partial r} + \frac{1}{r^2}\frac{\partial^2\phi}{\partial\theta^2} + \frac{\partial^2\phi}{\partial z^2} &= u\cos\alpha\left(\frac{2a^2}{r^3}\right)\cos\theta + \frac{u\cos\alpha}{r}\left(1 - \frac{a^2}{r^2}\right)\cos\theta \\ &\quad - \frac{u\cos\alpha}{r^2}\left(r + \frac{a^2}{r}\right)\cos\theta \\ &= 0\end{aligned}$$

SUMMARY

We have

☐ Introduced the notion of partial differentiation and seen how to obtain partial derivatives of the first and second order.

☐ Derived the chain rule

$$\frac{\partial f}{\partial u} = \frac{\partial f}{\partial x}\frac{\partial x}{\partial u} + \frac{\partial f}{\partial y}\frac{\partial y}{\partial u}$$

$$\frac{\partial f}{\partial v} = \frac{\partial f}{\partial x}\frac{\partial x}{\partial v} + \frac{\partial f}{\partial y}\frac{\partial y}{\partial v}$$

and practised its use.

☐ Used the total differential

$$\mathrm{d}z = \frac{\partial z}{\partial x}\,\mathrm{d}x + \frac{\partial z}{\partial y}\,\mathrm{d}y$$

to estimate errors in calculations which involve formulas with more than one independent variable.

EXERCISES

1 Obtain the first-order partial derivatives of the functions defined by the following formulas:

a $f(x, y) = x^3 + yx^2$

b $f(x, y) = \sin xy \cos (x + y)$

c $f(x, y) = (x + 2y)^3$

d $f(x, y) = \exp(x + y)\cos xy$
e $f(x, y) = \sqrt{\cosh(x + y)}$
f $f(x, y) = \cosh(x/y)$

2 Obtain the mixed second-order partial derivatives of each of the following functions defined by the formulas:
a $f(x, y) = \sqrt{(x^2 + y^2)} + \sin xy$
b $f(x, y) = \sqrt{(x + 2y)} + x^2y$
c $f(x, y) = x^2\sin(3x + 4y)$
d $f(x, y) = \sin(x^2/y^3)$

3 Obtain expressions in terms of u, v and partial derivatives with respect to u and v for $\partial z/\partial x$ and $\partial z/\partial y$ if
a $x = u^2v^2$, $y = u + v$
b $x = u + 2v$, $y = 2u + v$
c $x = u + uv$, $y = v - uv$
d $x = \sqrt{(u^2 + v^2)}$, $y = u/v$

ASSIGNMENT

1 Obtain the first-order partial derivatives of the functions defined by each of the following formulas:
a $f(x, y) = \sin x \cos y + xy^2$
b $f(x, y) = e^x \cos y + e^x \sin y$
c $z = \ln \sqrt{(x^2 + y^2)}$
d $z = (x + 2y)(x - 2y)^5$
e $z = \sin (u^2 - v^2)^2$

2 Verify the equality of the mixed second-order partial derivatives of the function f defined for all real x and y $(x^2 + y^2 \neq 0)$ by

$$f(x, y) = \ln(x^2 + y^2) + \cos(x + 2x^2y)$$

3 Obtain an expression in terms of u and v and partial derivatives with respect to u and v for

$$\frac{\partial z}{\partial x} + \frac{\partial z}{\partial y}$$

if $x = e^u \cos v$ and $y = e^u \sin v$.

4 The force on a body is calculated using the formula

$$F = k\frac{m_1m_2}{r^2}$$

where k is constant, m_1 and m_2 are masses and r is a distance. Calculate approximately the percentage error in F if the massses are measured to within 1% and the distance to within 5%.

FURTHER EXERCISES

1 If $x = r\cos\theta$ and $y = r\sin\theta$, and $z = f(x, y) = g(r, \theta)$, prove that
a $g_r = (xf_x + yf_y)/r$
b $g_\theta = xf_y - yf_x$
c $(g_r)^2 + \frac{1}{r^2}(g_\theta)^2 = (f_x)^2 + (f_y)^2$

2 Show that if $u = \ln x + \ln y$ and $v = xy$ then

$$\frac{\partial u}{\partial x}\frac{\partial v}{\partial y} = \frac{\partial v}{\partial x}\frac{\partial u}{\partial y}$$

3 If $z = f(x + y) + g(x - y)$, where f and g are both twice differentiable real functions, prove that

$$\frac{\partial^2 z}{\partial x^2} = \frac{\partial^2 z}{\partial y^2}$$

Hence or otherwise show that $z = \sin x \cos y$ and $z = e^x \sinh y$ each satisfy this partial differential equation.

4 If $z = x^m y^n$, where m and n are constants, show that

$$dz/z = m\,dx/x + n\,dy/y$$

5 If $z = \ln(x^2 + y^2)$, prove that $z_{xx} + z_{yy} = 0$. Show further that if $z = f(x^2 + y^2)$ then

$$z_{xx} + z_{yy} = 4tf''(t) + 4f'(t)$$

where $t = x^2 + y^2$.

6 If $z = f(y/x)$ show that $xz_x + yz_y = 0$. Hence or otherwise show that this partial differential equation is satisfied by each of the following:
a $z = \sin[(x^2 + y^2)/xy]$
b $z = \ln(y/x)$
c $z = \exp[(x - y)/(x + y)]$

7 A beam of very low weight, uniform cross-section and length l simply supported at both ends carries a concentrated load W at the centre. It is known that the deflection at the centre is given by $y = Wl^3/48EI$, where E is Young's modulus and I is a moment of inertia. E is constant but the following small percentage increases occur: 2ε in W, ε in l and 5ε in I. Show that the error in y is then negligible.

8 The second moment of area of a rectangle of breadth B and depth D about an axis through one horizontal edge is given by $I = BD^3/3$. If a small increase of δ% in D takes place, estimate the change in B required if the calculated value of I is to remain constant.

9 In telecommunications, the transmission line equations may be written as

$$Ri + L\frac{\partial i}{\partial t} = -\frac{\partial v}{\partial x}$$

$$Gv + C\frac{\partial v}{\partial t} = -\frac{\partial i}{\partial x}$$

where R, L, G and C are constant. Show that both i and v satisfy the telegraphists' equation

$$\frac{\partial^2 y}{\partial x^2} = LC\frac{\partial^2 y}{\partial t^2} + (GL + RC)\frac{\partial y}{\partial t} + RGy$$

In the case of a distortionless line, $RC = LG$. If $w^2 = 1/LC$ and $a^2 = RG$, show that telegraphists' equation becomes

$$\frac{\partial^2 y}{\partial x^2} = \frac{1}{w^2}\frac{\partial^2 y}{\partial t^2} + \frac{2a}{w}\frac{\partial y}{\partial t} + a^2 y$$

10 The ratio r of the magnetic moments of two magnets was evaluated using the formula $r = (t_2^2 + t_1^2)/(t_2^2 - t_1^2)$, where t_1 and t_2 are the times of oscillation of the magnets when like poles are in the same direction and when like poles are in the opposite direction respectively. If e_1 is the percentage error in t_1 and e_2 is the percentage error in t_2, show that the percentage error in r is approximately $4t_1^2t_2^2(e_1 - e_2)/(t_2^4 - t_1^4)$.

11 The natural frequency of oscillation of an LRC series circuit is given by

$$f = \frac{1}{2\pi}\sqrt{\left(\frac{1}{LC} - \frac{R^2}{4L^2}\right)}$$

If L is increased by 1% and C is decreased by 1%, show that the percentage increase in f is approximately $R^2C/(4L - R^2C)$.

12 The heat generated in a resistance weld is given by $H = Ki^2Rt$, where K is a constant, i is the current between the electrodes and t is the time for which current flows. H must not vary by more than 5% if the weld is to remain good. It is possible to control t to within 0.5% and R to within 2.5%. Estimate the maximum possible variation in current if the weld is to retain its quality.

13 Air is pumped into a rubber tyre which has a volume given by $V = 2\pi^2a^2b$. The internal radius r and the external radius R are related to a and b by $r = b - a$ and $R = b + a$. If the internal radius decreases by approximately 1% and the external radius increases by approximately 2% obtain approximately an expression for the percentage increase in volume in the tyre.

Series expansions and their uses 8

In previous chapters we have described the processes of elementary differentiation for functions of a single variable. In Chapter 7 we extended some of these ideas to functions of several variables. There is much more to calculus than this, and our next task is to consider some other applications. We shall see in particular that series expansions play a vital role.

After studying this chapter you should be able to

- ☐ Obtain Taylor's expansion of a function about a point;
- ☐ Expand $f(x)$ as a power series in x using Maclaurin's theorem;
- ☐ Apply l'Hospital's rule correctly;
- ☐ Classify stationary points;
- ☐ Determine points of inflexion.

At the end of this chapter we shall solve a practical problem concerning a valve.

8.1 THE MEAN VALUE PROPERTY

The graph of a smooth function is shown in Fig. 8.1. By smooth we mean that the function is differentiable everywhere. Geometrically this implies that the curve has a tangent at all its points.

A and B are two points where this smooth curve crosses the x-axis. It can be shown that there is at least one point P on the curve between A and B with the property that the tangent at P is parallel to the x-axis. In symbols we can express this by saying that if $f(a) = 0$ and $f(b) = 0$ then there exists some point $c \in (a, b)$ such that $f'(c) = 0$.

Although this property is intuitively obvious, its proof requires quite advanced mathematical ideas and so we shall omit it. Surprisingly perhaps

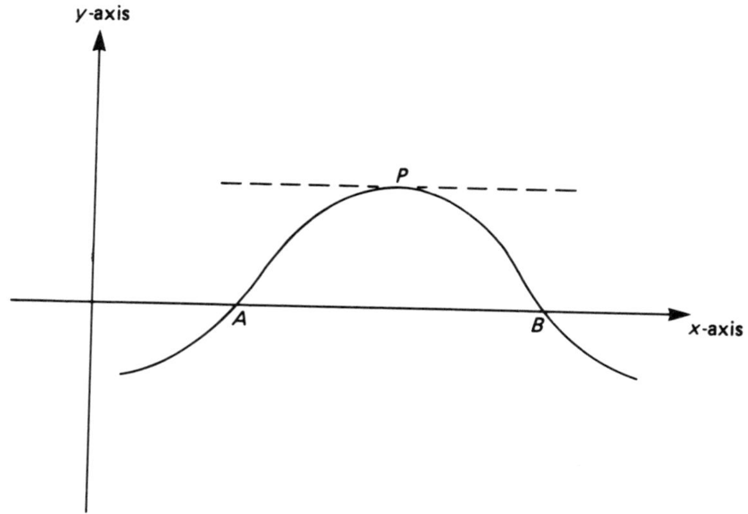

Fig. 8.1 A smooth function.

this simple theorem, known as **Rolle's theorem**, has quite profound consequences. In particular it leads to Taylor's expansion.

We shall deduce one simple generalization straight away; this is known as the mean value property.

Suppose we have a smooth curve and two points A and B on it (Fig. 8.2). The mean value property says that there is some point P on the curve

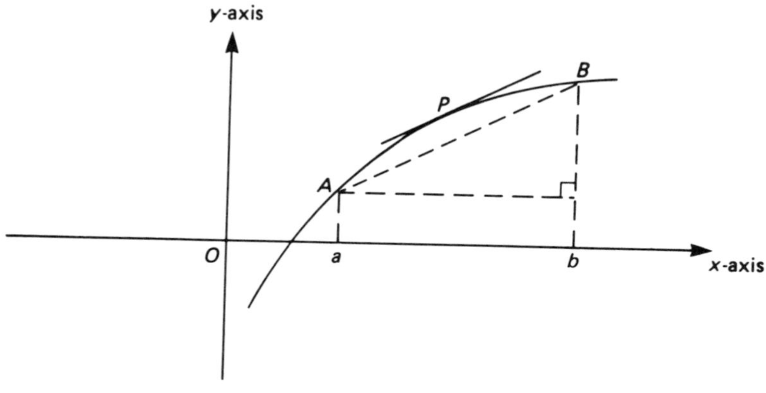

Fig. 8.2 The mean value property.

between A and B such that the tangent at P is parallel to the chord AB. Suppose A is the point $(a, f(a))$ and B is the point $(b, f(b))$. Then

$$\text{slope } AB = \frac{f(b) - f(a)}{b - a}$$

Consider $g = g(x)$ defined by

$$g(x) = f(x) - f(a) - \left[\frac{f(b) - f(a)}{b - a}\right](x - a)$$

Because g is a sum of differentiable functions, g is also differentiable. In fact

$$g'(x) = f'(x) - \frac{f(b) - f(a)}{b - a}$$

Now $g(a) = 0$ and $g(b) = 0$, and so by Rolle's theorem there exists $c \in (a, b)$ such that $g'(c) = 0$. That is,

$$f'(c) = \frac{f(b) - f(a)}{b - a}$$

for some c, $a < c < b$. This is precisely the **mean value property**.

8.2 TAYLOR'S THEOREM

If we rearrange the formula which describes the mean value property we obtain

$$f(b) - f(a) = (b - a)f'(c) \qquad c \in (a, b)$$

So if we write $b = a + h$ then $c = a + \theta h$ for some $\theta \in (0, 1)$. Therefore

$$f(a + h) = f(a) + hf'(a + \theta h) \qquad \theta \in (0, 1)$$

It is possible to generalize this result so that, if we have a function which can be differentiated twice everywhere in an open interval containing the point a, then

$$f(a + h) = f(a) + hf'(a) + \frac{h^2}{2} f''(a + \theta h) \quad \text{for some } \theta \in (0, 1)$$

Therefore

$$f(a + h) = f(a) + hf'(a) + R_2$$

where

$$R_2 = \frac{h^2}{2} f''(a + \theta h) \quad \text{for some } \theta \in (0, 1)$$

Generalizing still further, if f is a real function which can be differentiated n times at all points in an open interval containing the point a, then

$$f(a + h) = f(a) + hf'(a) + \frac{h^2}{2!} f''(a) + \ldots + \frac{h^{n-1}}{(n-1)!} f^{(n-1)}(a) + R_n$$

where

$$R_n = \frac{h^n}{n!} f^{(n)}(a + \theta h) \quad \text{for some } \theta \in (0, 1)$$

This expansion is known as **Taylor's expansion** of f about the point a, and R_n is called the **remainder** after n terms.

So we have

$$f(a + h) = \sum_{r=0}^{n-1} \frac{h^r}{r!} f^{(r)}(a) + R_n$$

where

$$R_n = \frac{h^n}{n!} f^{(n)}(a + \theta h) \quad \text{for some } \theta \in (0, 1)$$

TAYLOR'S SERIES

If f is *infinitely* smooth, which means it can be differentiated an arbitrary number of times, and if $R_n \to 0$ as $n \to \infty$, then we obtain **Taylor's series**. This is an infinite series representation for $f(a + h)$:

$$f(a + h) = \sum_{r=0}^{\infty} \frac{h^r}{r!} f^{(r)}(a)$$

If we use the variable x instead of h this becomes a power series in x:

$$f(a + x) = \sum_{r=0}^{\infty} \frac{x^r}{r!} f^{(r)}(a)$$

The special case where $a = 0$ is known as **Maclaurin's series**:

$$f(x) = \sum_{r=0}^{\infty} \frac{x^r}{r!} f^{(r)}(0)$$

Although you may not have a clear understanding of what is meant by an infinite series until you study them in Chapter 9, be content for the time being to derive Taylor and Maclaurin expansions for known functions. However, one word of warning is in order.

It is not true that if we obtain a Taylor or Maclaurin series from a function that the series expansion is always *valid*. Nor is it true that if the series converges then the expansion is valid. In fact we are only entitled to write equality in the case where $R_n \to 0$ as $n \to \infty$. We shall therefore write the equals sign on the understanding that we are restricting the domain of the function to those points for which $R_n \to 0$.

Here are two limits which can be quite useful in deciding whether or not $R_n \to 0$:

$$\lim_{n \to \infty} \frac{x^n}{n!} = 0 \qquad \text{for all } x \in \mathbb{R}$$

$$\lim_{n \to \infty} x^n = 0 \qquad \text{if } x \in (-1, 1)$$

□ Obtain the Maclaurin expansion for e^x.

Maclaurin's expansion is

$$f(x) = \sum_{n=0}^{\infty} \frac{x^n}{n!} f^{(r)}(0)$$

It is therefore necessary for us to calculate the values of successive derivatives of f at 0. If we put $f(x) = e^x$ we have $f(0) = e^0 = 1$. Moreover $f'(x) = e^x$, so that $f'(0) = e^0 = 1$. Clearly $f^{(n)}(x) = e^x$ for all $n \in \mathbb{N}$, and so $f^{(n)}(0) = 1$ for every n. Therefore

$$\begin{aligned} e^x &= \sum_{n=0}^{\infty} \frac{x^n}{n!} \\ &= 1 + x + \frac{x^2}{2!} + \frac{x^3}{3!} + \ldots + \frac{x^n}{n!} + \ldots \end{aligned}$$

Here

$$R_n = \frac{c^n}{n!}$$

which tends to zero as n tends to infinity for all $c \in \mathbb{R}$. This shows in fact that this expansion is valid for all $x \in \mathbb{R}$. ■

THE EXPONENTIAL FUNCTION

We introduced the exponential function by stating that when e^x is differentiated with respect to x the result is e^x (Chapter 4). We have now used this to derive a representation of e^x as a power series in x. There are several ways of introducing the exponential function. Another way is to define $\exp x$ by means of the power series in x, and then to use deep theory of power series (Chapter 9) to show that $\exp x$ obeys the laws of indices and is unchanged when differentiated with respect to x. Yet another way to define the exponential function is by means of a limit:

$$\lim_{n\to\infty}\left(1+\frac{x}{n}\right)^n$$

We shall give a very informal justification of this by showing how we can obtain the infinite series representation from it. First we expand $(1 + x/n)^n$ by means of the binomial theorem:

$$\begin{aligned}\left(1+\frac{x}{n}\right)^n &= 1+n\left(\frac{x}{n}\right)+\frac{n(n-1)}{1\times 2}\left(\frac{x}{n}\right)^2+\frac{n(n-1)(n-2)}{1\times 2\times 3}\left(\frac{x}{n}\right)^3+\ldots\\ &= 1+x+\frac{1(1-1/n)}{1\times 2}x^2+\frac{1(1-1/n)(1-2/n)}{1\times 2\times 3}x^3+\ldots\end{aligned}$$

Now as $n \to \infty$ we have for each fixed $r \in \mathbb{N}$

$$\left(1-\frac{1}{n}\right)\left(1-\frac{2}{n}\right)\left(1-\frac{3}{n}\right)\ldots\left(1-\frac{r}{n}\right)\to 1$$

In this way we see that

$$\begin{aligned}\lim_{n\to\infty}\left(1+\frac{x}{n}\right)^n &= 1+x+\frac{x^2}{2!}+\frac{x^3}{3!}+\ldots,+\frac{x^n}{n!}+\ldots\\ &= e^x\end{aligned}$$

as foretold.

We must not disguise the fact that once again we have used properties of infinite series which, although plausible, require proof. Unfortunately there are many properties which appear plausible in the context of infinite series but which are *false*.

□ Use Maclaurin's expansion to obtain the binomial series for $(1 + x)^n$, where n is any real number.

Notice how Maclaurin's expansion and series are different names for the same thing. Some books distinguish between the expansion, which is the formula with remainder, and the series, which is an infinite series. However, there is no consensus on this terminology.

First we should remark that if r is any real number we have defined a^r only for $a > 0$. Therefore we must presuppose that $1 + x > 0$, so that $x > -1$. Now

$$\begin{aligned}f(x) &= (1+x)^n && \text{so } f(0) = (1+0)^n = 1\\ f'(x) &= n(1+x)^{n-1} && \text{so } f'(0) = n\times 1 = n\\ f''(x) &= n(n-1)(1+x)^{n-2} && \text{so } f''(0) = n(n-1)\end{aligned}$$

In general,

$$f^{(r)}(x) = n(n-1)\ldots(n-r+1)(1+x)^{n-r}$$

So

$$f^{(r)}(0) = n(n - 1) \dots (n - r + 1)$$

Therefore in the Maclaurin expansion the coefficient of x^r is

$$\frac{n(n - 1) \dots (n - r + 1)}{1 \times 2 \times 3 \times \dots \times r}$$

But by definition this is the binomial coefficient $\binom{n}{r}$. Consequently we obtain

$$(1 + x)^n = \binom{n}{0} + \binom{n}{1}x + \binom{n}{2}x^2 + \dots + \binom{n}{r}x^r + \dots$$

Again we stress that in order to be justified in using this infinite series as a representation for $(1 + x)^n$ we should need to examine the remainder after r terms and show that it tends to 0 as $r \to \infty$. In fact this binomial expansion is only valid when $x \in (-1, 1)$, that is $-1 < x < 1$. ■

You may have come across some other power series. There are power series in x corresponding to the circular functions and the hyperbolic functions. Here are the main ones:

$$\cos x = \sum_{n=0}^{\infty} \frac{(-1)^n x^{2n}}{(2n)!} = 1 - \frac{x^2}{2!} + \frac{x^4}{4!} - \frac{x^6}{6!} + \dots$$

$$\sin x = \sum_{n=0}^{\infty} \frac{(-1)^n x^{2n+1}}{(2n + 1)!} = x - \frac{x^3}{3!} + \frac{x^5}{5!} - \frac{x^7}{7!} + \dots$$

$$\cosh x = \sum_{n=0}^{\infty} \frac{x^{2n}}{(2n)!} = 1 + \frac{x^2}{2!} + \frac{x^4}{4!} + \frac{x^6}{6!} + \dots$$

$$\sinh x = \sum_{n=0}^{\infty} \frac{x^{2n+1}}{(2n + 1)!} = x + \frac{x^3}{3!} + \frac{x^5}{5!} + \frac{x^7}{7!} + \dots$$

It can be shown that these are valid for all $x \in \mathbb{R}$. You will probably have observed the close similarity between the series expansions for the circular functions and their hyperbolic counterparts. We shall investigate this similarity when we consider complex numbers (Chapter 10).

□ Obtain the first four non-zero terms in the expansion of $\tan x$ as a power series in x.

Here we put

$$f(x) = \tan x$$

so $f(0) = \tan 0 = 0$.

$$f'(x) = \sec^2 x$$

so $f'(0) = \sec^2 0 = 1$.

$$f''(x) = 2 \sec x \sec x \tan x = 2 \sec^2 x \tan x$$

so $f''(0) = 0$.

$$\begin{aligned} f^{(3)}(x) &= 2 \sec^2 x \sec^2 x + 2 \tan x \, 2 \sec x \sec x \tan x \\ &= 2 \sec^4 x + 4 \sec^2 x \tan^2 x \\ &= 2 \sec^4 x + 4 \sec^2 x (\sec^2 x - 1) \\ &= 6 \sec^4 x - 4 \sec^2 x \end{aligned}$$

so $f^{(3)}(0) = 2$.

$$\begin{aligned} f^{(4)}(x) &= 24 \sec^3 x \sec x \tan x - 8 \sec x \sec x \tan x \\ &= 24 \sec^4 x \tan x - 8 \sec^2 x \tan x \end{aligned}$$

so $f^{(4)}(0) = 0$.

$$\begin{aligned} f^{(5)}(x) &= 24 \sec^4 x \sec^2 x + 24 \tan x \, 4 \sec^3 x \sec x \tan x \\ &\quad - 8 \sec^2 x \sec^2 x - 8 \tan x \, 2 \sec x \sec x \tan x \\ &= 24 \sec^6 x + 96 \sec^4 x (\sec^2 x - 1) - 8 \sec^4 x \\ &\quad - 16 \sec^2 x (\sec^2 x - 1) \\ &= 120 \sec^6 x - 120 \sec^4 x + 16 \sec^2 x \end{aligned}$$

so $f^{(5)}(0) = 16$.

$$\begin{aligned} f^{(6)}(x) &= 720 \sec^5 x \sec x \tan x - 480 \sec^3 x \sec x \tan x + 32 \sec x \sec x \tan x \\ &= 720 \sec^6 x \tan x - 480 \sec^4 x \tan x + 32 \sec^2 x \tan x \end{aligned}$$

so $f^{(6)}(0) = 0$.

$$\begin{aligned} f^{(7)}(x) &= 4320 \sec^5 x \sec x \tan x \tan x + 720 \sec^6 x \sec^2 x \\ &\quad - 1920 \sec^3 x \sec x \tan x \tan x - 480 \sec^4 x \sec^2 x \\ &\quad + 64 \sec x \sec x \tan x \tan x + 32 \sec^2 x \sec^2 x \\ &= 4320 \sec^6 x \tan^2 x + 720 \sec^8 x - 1920 \sec^4 x \tan^2 x \\ &\quad - 480 \sec^6 x + 64 \sec^2 x \tan^2 x + 32 \sec^4 x \end{aligned}$$

so $f^{(7)}(0) = 0 + 720 - 0 - 480 + 0 + 32 = 272$.

This is the fourth non-zero term, and so we can write down the expansion:

$$\begin{aligned} \tan x &= 0 + x + 0 + 2 \frac{x^3}{3!} + 0 + 16 \frac{x^5}{5!} + 0 + 272 \frac{x^7}{7!} + \ldots \\ &= x + \frac{x^3}{3} + \frac{2}{15} x^5 + \frac{17}{315} x^7 + \ldots \end{aligned}$$

This example shows that we should not always expect a discernible pattern to emerge. ■

Now for some steps – but make sure you have understood all the examples first.

8.3 Workshop

1 **Exercise** Differentiate $(\ln x)^x$ $(x > 1)$ with respect to x.

This is just a little differentiation to warm up on. Try it, then step forward.

2 Here we go then. It's easiest to put $y = (\ln x)^x$ and take logarithms:

$$\ln y = \ln [(\ln x)^x] = x \ln (\ln x)$$

so that

$$\frac{1}{y}\frac{dy}{dx} = \ln (\ln x) + x \frac{1}{\ln x}\frac{1}{x}$$

$$\frac{dy}{dx} = \ln (\ln x)(\ln x)^x + (\ln x)^{x-1}$$

If you were right then step ahead to step 4. If you were wrong, check carefully what you have done; then try the next problem.

▷ **Exercise** Obtain dy/dx if $y = (\cosh x)^x$.

You know the method so there should be no serious problems.

3 We obtain $\ln y = x \ln (\cosh x)$. So, differentiating both sides with respect to x and using the product rule and chain rule, we have

$$\frac{1}{y}\frac{dy}{dx} = \ln (\cosh x) + x \frac{1}{\cosh x}\frac{d}{dx}(\cosh x)$$

$$= \ln (\cosh x) + x \frac{\sinh x}{\cosh x}$$

$$\frac{dy}{dx} = [\ln (\cosh x) + x \tanh x] (\cosh x)^x$$

Now move on to the next step.

4 **Exercise** Obtain the first four terms of Taylor's expansion for $y = \sin x$ about the point $x = \pi/4$ as a power series in x.

The difficulty with a question like this is sorting out what is really wanted. The trouble is that x has been used in three different ways here: first, as a dummy variable in the definition of sine; secondly, as a specific point; and thirdly, as the dummy variable in an algebraic description of the power series. Let's separate all these things so that we can proceed.

In general given $y = f(x)$ we obtain, using Taylor's expansion, $f(a + h)$ as a power series in h. Here $y = f(x) = \sin x$, $a = \pi/4$, and we can therefore obtain a power series in h for $\sin(\pi/4 + h)$. To satisfy the question we shall at the final stage replace h by x.

Right! You know what you have to do, so see how it goes.

5

We have

$$f(a + h) = f(a) + hf'(a) + \dots$$

So we must evaluate successive derivatives at the point $\pi/4$:

$$\begin{aligned} f(x) &= \sin x &\Rightarrow\quad f(a) &= \sin \pi/4 = 1/\sqrt{2} \\ f'(x) &= \cos x &\Rightarrow\quad f'(a) &= \cos \pi/4 = 1/\sqrt{2} \\ f''(x) &= -\sin x &\Rightarrow\quad f''(a) &= -\sin \pi/4 = -1/\sqrt{2} \\ f^{(3)}(x) &= -\cos x &\Rightarrow\quad f^{(3)}(a) &= -\cos \pi/4 = -1/\sqrt{2} \end{aligned}$$

This will give the first four non-zero terms, and so we have

$$\begin{aligned} f(a + h) &= f(a) + hf'(a) + \tfrac{1}{2}h^2 f''(a) + \dots \\ \sin(\pi/4 + h) &= 1/\sqrt{2} + h/\sqrt{2} - h^2/2\sqrt{2} - h^3/6\sqrt{2} + \dots \end{aligned}$$

Finally, replacing h by x we have

$$\sin(\pi/4 + x) = 1/\sqrt{2} + x/\sqrt{2} - x^2/2\sqrt{2} - x^3/6\sqrt{2} + \dots$$

Now try another problem.

▷ **Exercise** Obtain Maclaurin's expansion for $f(x) = \sin(x + \pi/4)$ as a power series in x as far as the term in x^3.

This shows how closely Taylor's expansion and Maclaurin's expansion are to one another. Superficially Taylor's expansion seems more general. However, not only can we deduce Maclaurin's expansion from Taylor's, but it is also possible to deduce Taylor's expansion from Maclaurin's. Try it, then step ahead.

6

We have

$$f(x) = f(0) + xf'(0) + \tfrac{1}{2}x^2 f''(0) + \dots$$

We therefore need to obtain successive derivatives of f evaluated when $x = 0$. Now

$$\begin{aligned} f(x) &= \sin(x + \pi/4) &\Rightarrow\quad f(0) &= \sin \pi/4 = 1/\sqrt{2} \\ f'(x) &= \cos(x + \pi/4) &\Rightarrow\quad f'(0) &= \cos \pi/4 = 1/\sqrt{2} \\ f''(x) &= -\sin(x + \pi/4) &\Rightarrow\quad f''(0) &= -1/\sqrt{2} \end{aligned}$$

$$f^{(3)}(x) = -\cos(x + \pi/4) \Rightarrow f^{(3)}(0) = -\cos \pi/4 = -1/\sqrt{2}$$

Substituting into Maclaurin's expansion:

$$\sin(\pi/4 + x) = 1/\sqrt{2} + x/\sqrt{2} - x^2/2\sqrt{2} - x^3/6\sqrt{2} + \ldots$$

as before.

If you managed that successfully, then on you go to step 8. Otherwise, try a further exercise.

▷**Exercise** Obtain an expansion of $\ln(1 + x)$ as a power series in x.

---

7 We use Maclaurin's expansion and so we put $f(x) = \ln(1 + x)$. We shall need to obtain successive derivatives of f at 0 to substitute into the expansion formula

$$f(x) = f(0) + xf'(0) + \tfrac{1}{2}x^2f''(0) + \ldots$$

So:

$$f(x) = \ln(1 + x) \quad \Rightarrow \quad f(0) = \ln 1 = 0$$

$$f'(x) = \frac{1}{1 + x} \quad \Rightarrow \quad f'(0) = 1$$

$$f''(x) = \frac{-1}{(1 + x)^2} \quad \Rightarrow \quad f''(0) = -1$$

$$f^{(3)}(x) = \frac{(-1)(-2)}{(1 + x)^3} \quad \Rightarrow f^{(3)}(0) = 2$$

$$f^{(4)}(x) = \frac{(-1)(-2)(-3)}{(1 + x)^4} \quad \Rightarrow f^{(4)}(0) = -3!$$

$$f^{(5)}(x) = \frac{(-1)(-2)(-3)(-4)}{(1 + x)^5} \Rightarrow f^{(5)}(0) = 4!$$

We can see a pattern emerging:

$$f^{(n)}(0) = (-1)^{n+1}(n - 1)! \qquad \text{when } n \in \mathbb{N}$$

When we substitute into the Maclaurin expansion we obtain

$$f(0) + xf'(0) + \frac{x^2}{2!}f''(0) + \frac{x^3}{3!}f^{(3)}(0) + \ldots + \frac{x^n}{n!}f^{(n)}(0) + \ldots$$

$$= 0 + x(1) + \frac{x^2}{2!}(-1) + \frac{x^3}{3!}2! + \frac{x^4}{4!}(-3!) + \ldots$$

$$+ \frac{x^n}{n!}(-1)^{n+1}(n - 1)! \ldots$$

$$= x - \frac{x^2}{2} + \frac{x^3}{3} - \frac{x^4}{4} + \ldots + (-1)^{n+1}\frac{x^n}{n} + \ldots$$

We have

$$\ln (1 + x) = \sum_{n=1}^{\infty} (-1)^{n+1} \frac{x^n}{n}$$

By examining the remainder after n terms it is possible to show that this series represents $\ln (1 + x)$ whenever $-1 < x \leqslant 1$.

Now we go on to another problem.

▷**Exercise** Given $y = \sin^{-1} x$, show that 8

$$(1 - x^2) \frac{d^2y}{dx^2} - x \frac{dy}{dx} = 0$$

Differentiate n times using Leibniz's theorem to deduce

$$(1 - x^2) \frac{d^{n+2}y}{dx^{n+2}} - (2n + 1) x \frac{d^{n+1}y}{dx^{n+1}} - n^2 \frac{d^n y}{dx^n} = 0$$

Hence or otherwise obtain a power series expansion for $\sin^{-1} x$.

This problem will help you to revise your work on Leibniz's theorem. Let's do it in three steps. First, obtain the equation for the second derivative.

If $y = \sin^{-1} x$ then $\sin y = x$. So differentiating throughout with respect to x we have 9

$$\cos y \frac{dy}{dx} = 1$$

Squaring we have

$$\cos^2 y \left(\frac{dy}{dx}\right)^2 = 1$$

and since $\cos^2 y = 1 - \sin^2 y = 1 - x^2$ we have

$$(1 - x^2)\left(\frac{dy}{dx}\right)^2 = 1$$

Now differentiating again throughout with respect to x,

$$(1 - x^2)\, 2 \frac{dy}{dx} \frac{d^2y}{dx^2} + (-2x)\left(\frac{dy}{dx}\right)^2 = 0$$

Since dy/dx is not zero we obtain

$$(1 - x^2) \frac{d^2y}{dx^2} - x \frac{dy}{dx} = 0$$

If you put a foot wrong then locate your error and take the next step, which is to use Leibniz's theorem.

---

10 We must consider the two terms in the last equation separately, since each is a product and will need to be differentiated n times by Leibniz's theorem.

For the first term, put $u = 1 - x^2$ and $v = y_2$, where the subscript n denotes the nth-order derivative with respect to x. We have $u_1 = -2x$, $u_2 = -2$ and $u_3 = 0$, so that $u_n = 0$ for $n \geqslant 3$. We also have $v_1 = y_3$, $v_2 = y_4$ and in general $v_n = y_{n+2}$. Now Leibniz's theorem gives

$$(uv)_n = uv_n + n\,u_1v_{n-1} + \tfrac{1}{2}n(n-1)\,u_2v_{n-2} + \ldots$$

Now since all the other terms are zero we have

$$(1 - x^2)\frac{d^{n+2}y}{dx^{n+2}} + n(-2x)\frac{d^{n+1}y}{dx^{n+1}} + \frac{n(n-1)}{2}(-2)\frac{d^ny}{dx^n}$$
$$= (1 - x^2)\frac{d^{n+2}y}{dx^{n+2}} - 2nx\frac{d^{n+1}y}{dx^{n+1}} - n(n-1)\frac{d^ny}{dx^n}$$

For the second term, if we put $u = x$ and $v = y_1$ then $u_1 = 1$ and $u_n = 0$ if $n > 1$. Also $v_1 = y_2$ and in general $v_n = y_{n+1}$. Therefore

$$(uv)_n = uv_n + n\,u_1v_{n-1} + \ldots$$

and so we obtain

$$x\frac{d^{n+1}y}{dx^{n+1}} + n\frac{d^ny}{dx^n}$$

Finally we combine the two terms. So differentiating throughout the equation n times we obtain

$$(1 - x^2)\frac{d^{n+2}y}{dx^{n+2}} - 2nx\frac{d^{n+1}y}{dx^{n+1}} - n(n-1)\frac{d^ny}{dx^n} - x\frac{d^{n+1}y}{dx^{n+1}} - n\frac{d^ny}{dx^n} = 0$$

Therefore

$$(1 - x^2)\frac{d^{n+2}y}{dx^{n+2}} - (2n+1)\,x\frac{d^{n+1}y}{dx^{n+1}} - n^2\frac{d^ny}{dx^n} = 0$$

You did well if you managed to do that. Now you must think how you can use this to obtain Maclaurin's expansion.

---

11 If $y = f(x)$ then the equation is

$$(1 - x^2)\,f^{(n+2)}(x) - (2n+1)\,xf^{(n+1)}(x) - n^2f^{(n)}(x) = 0$$

but we require the values of the derivatives when $x = 0$. So the equation reduces to

$$f^{(n+2)}(0) - n^2 f^{(n)}(0) = 0$$
$$f^{(n+2)}(0) = n^2 f^{(n)}(0)$$

Now the equation was derived using Leibniz's theorem, and so is valid when $n \geqslant 1$. However, it also holds when $n = 0$ since it then reduces to the second-order equation. This means that the equation

$$f^{(n+2)}(0) = n^2 f^{(n)}(0)$$

can be used to generate all the coefficients in Maclaurin's expansion from $f(0)$ and $f'(0)$. Now $f(0) = \sin^{-1} 0 = 0$ and $f'(0) = 1$. So we deduce that $f^{(n)}(0) = 0$ if n is even, whereas

$$f^{(3)}(0) = 1^2 1$$
$$f^{(5)}(0) = 3^2 f^{(3)}(0) = 3^2 1^2$$
$$f^{(7)}(0) = 5^2 f^{(5)}(0) = 5^2 3^2 1^2$$

So we obtain the expansion

$$x + \frac{1^2}{3!}x^3 + \frac{3^2 1^2}{5!}x^5 + \ldots + \frac{(2r-1)^2 (2r-3)^2 \ldots 3^2 1^2}{(2r+1)!}x^{2r+1} + \ldots$$

Well, there it is. A bit of a monster, isn't it?

8.4 L'HOSPITAL'S RULE

L'Hospital's rule is extremely useful in the evaluation of a limit which might otherwise be difficult to obtain. Suppose we wish to evaluate

$$\lim_{x \to a} \frac{f(x)}{g(x)}$$

and either

$$\lim_{x \to a} f(x) = 0 \qquad \text{and} \qquad \lim_{x \to a} g(x) = 0$$

or

$$\lim_{x \to a} f(x) = \pm\infty \qquad \text{and} \qquad \lim_{x \to a} g(x) = \pm\infty$$

Then **l'Hospital's rule** says

$$\lim_{x \to a} \frac{f(x)}{g(x)} = \lim_{x \to a} \frac{f'(x)}{g'(x)}$$

provided the limit exists.

Although we shall not prove it to be true generally, we can give an informal justification for this remarkable rule in the case where $f(a) = g(a) = 0$ and f' and g' are continuous at the point a:

$$\lim_{x\to a} \frac{f(x)}{g(x)} = \lim_{x\to a} \frac{f(x) - f(a)}{g(x) - g(a)}$$

$$= \lim_{x\to a} \left[\frac{f(x) - f(a)}{x - a} \, \frac{x - a}{g(x) - g(a)} \right]$$

Now putting $x - a = h$ we obtain

$$\lim_{x\to a} \frac{f(x) - f(a)}{x - a} = \lim_{h\to 0} \frac{f(a + h) - f(a)}{h}$$

$$= f'(a) = \lim_{x\to a} f'(x)$$

Similarly

$$\lim_{x\to a} \frac{g(x) - g(a)}{x - a} = \lim_{x\to a} g'(x)$$

Therefore

$$\lim_{x\to a} \frac{f(x)}{g(x)} = \lim_{x\to a} \frac{f'(x)}{g'(x)}$$

□ Obtain

$$\lim_{x\to 0} \frac{\sin x - x}{x^3}$$

Here $f(x) = \sin x - x$ and $g(x) = x^3$. So

$$\lim_{x\to 0} f(x) = \lim_{x\to 0} [\sin x - x] = 0$$

$$\lim_{x\to 0} g(x) = \lim_{x\to 0} x^3 = 0$$

So we can use l'Hospital's rule:

$$f'(x) = \cos x - 1 \qquad g'(x) = 3x^2$$

but

$$\lim_{x\to 0} f'(x) = 1 - 1 = 0$$

$$\lim_{x\to 0} g'(x) = 3 \times 0 = 0$$

So we can use l'Hospital's rule again:

$$f''(x) = -\sin x \qquad g''(x) = 6x$$

As before,

$$\lim_{x\to 0} f''(x) = 0$$

$$\lim_{x\to 0} g''(x) = 0$$

So we use l'Hospital's rule once more:

$$f^{(3)}(x) = -\cos x \qquad g^{(3)}(x) = 6$$

So

$$\lim_{x\to 0}\frac{\sin x - x}{x^3} = \lim_{x\to 0}\frac{\cos x - 1}{3x^2}$$

$$= \lim_{x\to 0}\frac{-\sin x}{6x}$$

$$= \lim_{x\to 0}\frac{-\cos x}{6} = -\frac{1}{6}$$

The use of l'Hospital's rule at each stage is now justified because this final limit exists. ■

□ Obtain

$$\lim_{x\to \pi/2}\frac{1 - \sin x}{\cot x}$$

When you have done this, see if you are correct.

First, $1 - \sin \pi/2 = 1 - 1 = 0$ and $\cot \pi/2 = 0$. So we may use l'Hospital's rule:

$$\lim_{x\to \pi/2}\frac{1 - \sin x}{\cot x} = \lim_{x\to \pi/2}\frac{-\cos x}{-\operatorname{cosec}^2 x} = 0$$

Alternatively, if we wish we can avoid l'Hospital's rule and instead use algebraic simplification:

$$\lim_{x\to \pi/2}\frac{1 - \sin x}{\cot x} = \lim_{x\to \pi/2}\frac{(1 - \sin x)\sin x}{\cos x}$$

$$= \lim_{x\to \pi/2}\frac{(1 - \sin x)\sin x\cos x}{\cos^2 x}$$

$$= \lim_{x\to \pi/2}\frac{(1 - \sin x)\sin x\cos x}{1 - \sin^2 x}$$

$$= \lim_{x\to \pi/2}\frac{(1 - \sin x)\sin x\cos x}{(1 - \sin x)(1 + \sin x)}$$

$$= \lim_{x \to \pi/2} \frac{\sin x \cos x}{1 + \sin x}$$

$$= \frac{1.0}{1 + 1} = 0$$ ■

One very important thing to remember about l'Hospital's rule is that you must not use it unless you have an **indeterminate form**. In other words, one of the following must hold:

$$\lim_{x \to a} f(x) = 0 \quad \text{and} \quad \lim_{x \to a} g(x) = 0$$

or

$$\lim_{x \to a} f(x) = \pm\infty \quad \text{and} \quad \lim_{x \to a} g(x) = \pm\infty$$

Indeterminate forms can be represented by 0/0 or ∞/∞: these expressions are meaningless and so indeterminate.

Why not check that you understand this by taking a few steps?

8.5 Workshop

1 **Exercise** Evaluate the following limit:

$$\lim_{x \to 0} \frac{(e^x + 1)x}{e^{2x} - 1}$$

Move on only when you have done it – or when you think you can't do it.

---

2 If we put $x = 0$ straight away we obtain 0/0, which is indeterminate. However,

$$\frac{(e^x + 1)x}{e^{2x} - 1} = \frac{(e^x + 1)x}{(e^x + 1)(e^x - 1)}$$

$$= \frac{x}{e^x - 1}$$

Again this produces the indeterminate 0/0 if we try to substitute $x = 0$, but now we can easily use l'Hospital's rule:

$$\lim_{x \to 0} \frac{x}{e^x - 1} = \lim_{x \to 0} \frac{1}{e^x - 0}$$

$$= \lim_{x \to 0} \frac{1}{e^x} = 1$$

Did you get that right? If you did then you may go on to the next section. Otherwise, try this next exercise.

▷**Exercise** Evaluate the limit

$$\lim_{x\to 1} \frac{\sin \pi x}{\sin (\pi x + x - 1)}$$

Try very hard with this one. Then step forward.

If we attempt to put $x = 1$ we obtain the form 0/0, and so we shall use l'Hospital's rule: 3

$$\lim_{x\to 1} \frac{\sin \pi x}{\sin (\pi x + x - 1)} = \lim_{x\to 1} \frac{\sin \pi x}{\sin [(\pi + 1)x - 1]}$$

$$= \lim_{x\to 1} \frac{\pi \cos \pi x}{(\pi + 1) \cos [(\pi + 1)x - 1]}$$

$$= \frac{\pi(-1)}{(\pi + 1)(-1)} = \frac{\pi}{\pi + 1}$$

REPEATED USE OF L'HOSPITAL'S RULE

We can use Taylor's theorem to justify the repeated use of l'Hospital's rule. Suppose that both the real functions f and g have a Taylor expansion about the point a, and that both

$$\lim_{x\to a} f^{(r)}(x) \quad \text{and} \quad \lim_{x\to a} g^{(r)}(x)$$

are zero for all integers r such that $0 \leqslant r < n$, but that

$$\lim_{x\to a} g^{(n)}(x) \neq 0$$

By Taylor's theorem we have

$$f(a + h) = f(a) + hf'(a) + \ldots + \frac{h^{n-1}}{(n-1)!}f^{(n-1)}(a) + \frac{h^n}{n!}f^{(n)}(a + \theta h)$$

$$g(a + h) = g(a) + hg'(a) + \ldots + \frac{h^{n-1}}{(n-1)!}g^{(n-1)}(a) + \frac{h^n}{n!}g^{(n)}(a + \phi h)$$

where $\theta, \phi \in (0, 1)$.

The continuity of the derivatives at the point a gives

$$f^{(r)}(a) = 0 = g^{(r)}(a)$$

for $0 \leqslant r < n$. Therefore these Taylor series reduce to

$$f(a + h) = \frac{h^n}{n!}f^{(n)}(a + \theta h)$$

$$g(a + h) = \frac{h^n}{n!} g^{(n)}(a + \phi h)$$

So that

$$\frac{f(a + h)}{g(a + h)} = \frac{f^{(n)}(a + \theta h)}{g^{(n)}(a + \phi h)}$$

Writing $x = a + h$ we have

$$\begin{aligned}
\lim_{x \to a} \frac{f(x)}{g(x)} &= \lim_{h \to 0} \frac{f(a + h)}{g(a + h)} \\
&= \lim_{h \to 0} \frac{f^{(n)}(a + \theta h)}{g^{(n)}(a + \phi h)} \\
&= \frac{f^{(n)}(a)}{g^{(n)}(a)} \\
&= \lim_{x \to a} \frac{f^{(n)}(x)}{g^{(n)}(x)}
\end{aligned}$$

8.6 MAXIMA AND MINIMA

There are many situations in which we have an interest in those points where a function attains a maximum or a minimum value. For example:

1 A company may wish to maximize its profits, but increasing the price of its goods may decrease the demand. A reasonable question to ask is: 'What price will maximize profit?'

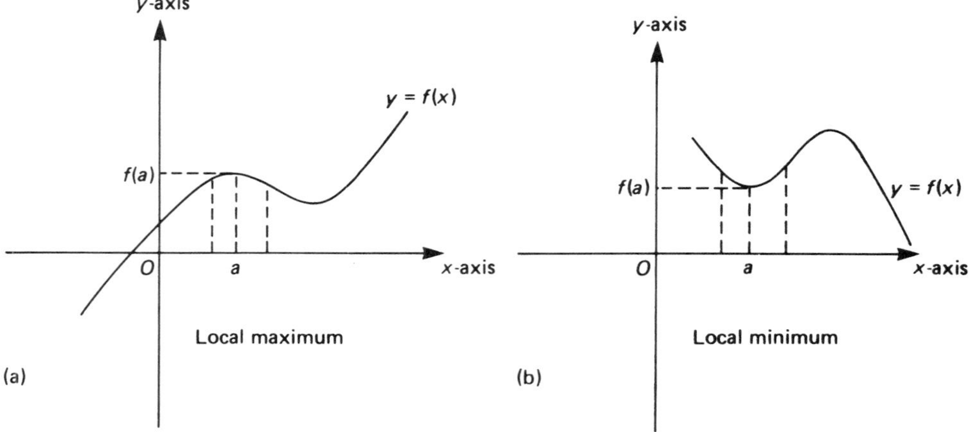

Fig. 8.3 (a) Local maximum (b) Local minimum.

2 An architect may be asked to design a library extension which, within a given budget, will maximize the available floor space.
3 An electrical engineer may wish to maximize the power in a circuit.

Suppose f is a real function which is defined at all points in some open interval containing the point a (Fig. 8.3). We say that f has a **local maximum** at the point a if and only if, for all h sufficiently small but non-zero,

$$f(a + h) - f(a) < 0$$

Similarly, f has a **local minimum** at the point a if and only if, for all h sufficiently small but non-zero,

$$f(a + h) - f(a) > 0$$

We shall confine our attention to functions which are infinitely smooth. As we have already remarked, this means that the function has derivatives of all orders.

It is intuitively obvious that if f is differentiable at either a local maximum or a local minimum then its derivative there is zero. Any point at which the derivative of f is zero is called a **stationary point** of f. The value of f at a stationary point is called a stationary value of f. (Stationary points are also sometimes known as turning points or critical points.)

A simple picture shows that not all stationary points are points at which f attains either a local maximum or a local minimum. Any point at which a curve crosses its tangent is called a **point of inflexion**. If we obtain the stationary points we shall obtain not only the points at which the function attains a local maximum or a local minimum but also some of the points of inflexion. In Fig. 8.4 A and I are local minima and E is a local maximum; A, C, E, G and I are stationary points, whereas B, C, D, F, G and H are points of inflexion.

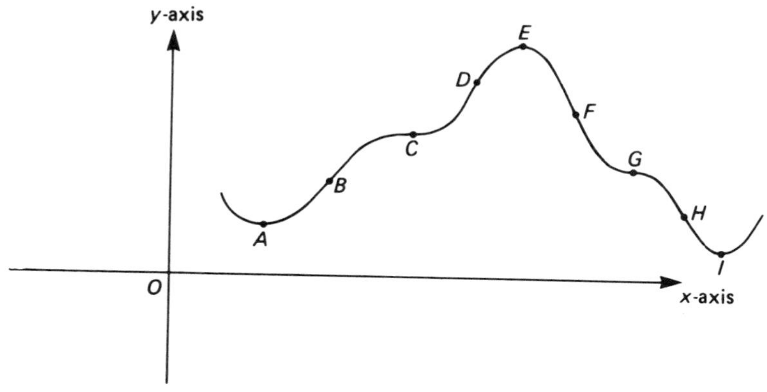

Fig. 8.4 Stationary points and points of inflexion.

As this makes clear, not every point of inflexion is a stationary point. We shall see later how to determine the points of inflexion of a function.

TESTING FOR MAXIMA AND MINIMA

There is an elementary method for determining local maxima and local minima which relies on the observation that at these points f' changes sign (Fig. 8.5):

1 As we pass through a local maximum, f' changes from positive to negative;

2 As we pass through a local minimum, f' changes from negative to positive.

It follows that if we examine the sign of f' on either side of the stationary point we should be able to classify it correctly. At a point of inflexion the sign is preserved.

However, it is possible to obtain a test for maxima and minima which does not involve examining the sign of f' on either side of the stationary point. Suppose that the function f has a stationary point at a, so that $f'(a) = 0$. By Taylor's theorem we know that

$$f(a + h) = f(a) + hf'(a) + \frac{h^2}{2!} f''(a) + \frac{h^3}{3!} f^{(3)}(a + \theta h)$$

where $0 < \theta < 1$. So in this case we have

$$f(a + h) - f(a) = \frac{h^2}{2!} f''(a) + \frac{h^3}{3!} f^{(3)}(a + \theta h)$$

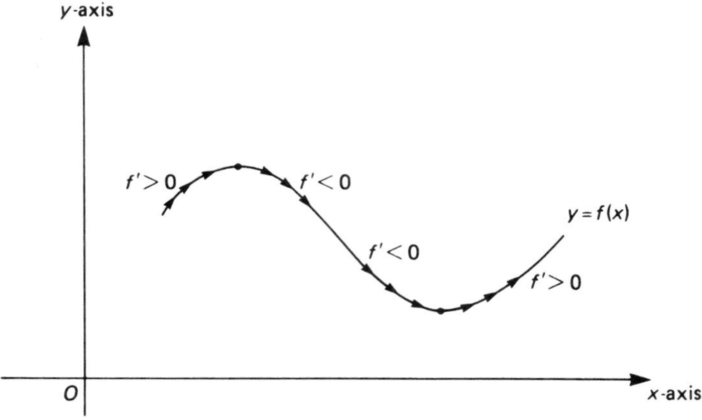

Fig. 8.5 The sign of f' near stationary points.

Now if $f''(a)$ is non-zero and $f^{(3)}(a + \theta h)$ is bounded, it is possible to choose h so small, $h \neq 0$, that the sign of the right-hand side of this equation is the same as the sign of the first term $(h^2/2!)f''(a)$, and the sign of this first term is of course the same as that of $f''(a)$. So if $f''(a) > 0$ we deduce that $f(a + h) - f(a) > 0$ for all h sufficiently small but non-zero, whereas if $f''(a) < 0$ then $f(a + h) - f(a) < 0$ for all h sufficiently small but non-zero. That is,

$$f''(a) > 0 \Rightarrow \text{local minimum}$$
$$f''(a) < 0 \Rightarrow \text{local maximum}$$

Therefore the following is the rule for obtaining the points at which $y = f(x)$ attains a local maximum or a local minimum:

1 Obtain all the stationary points, that is the points a at which $f'(a) = 0$.
2 For each stationary point a examine the sign of $f''(a)$:
  a if $f''(a) > 0$ then local minimum
  b if $f''(a) < 0$ then local maximum
  c if $f''(a) = 0$ then further testing is necessary.

□ Obtain and classify the stationary points of

$$y = x^2 e^x$$

We first obtain dy/dx and then equate it to 0 for the stationary points:

$$\frac{dy}{dx} = x^2e^x + 2xe^x = x(x + 2)e^x$$

Now $e^x \neq 0$: so $x(x + 2) = 0$, from which $x = 0$ and $x = -2$ are the stationary points.

Next we differentiate again and evaluate the second derivative at each stationary point to determine its sign there:

$$\frac{d^2y}{dx^2} = x^2e^x + 2xe^x + 2xe^x + 2e^x = (x^2 + 4x + 2)e^x$$

The sign of this is the same as the sign of

$$x^2 + 4x + 2 = (x + 2)^2 - 2$$

So

$$x = 0 \quad \Rightarrow \frac{d^2y}{dx^2} > 0 \Rightarrow \text{local minimum}$$

$$x = -2 \Rightarrow \frac{d^2y}{dx^2} < 0 \Rightarrow \text{local maximum}$$ ■

TESTING FOR INFLEXION

We now turn our attention to the problem of what to do when $f''(a) = 0$ at the stationary point. As before we can use Taylor's theorem to obtain

$$f(a + h) = f(a) + hf'(a) + \frac{h^2}{2!} f''(a) + \frac{h^3}{3!} f^{(3)}(a) + \frac{h^4}{4!} f^{(4)}(a + \theta h)$$

for some $\theta \in (0, 1)$.

Now $f'(a) = 0$ and $f''(a) = 0$, so that this reduces to

$$f(a + h) - f(a) = \frac{h^3}{3!} f^{(3)}(a) + \frac{h^4}{4!} f^{(4)}(a + \theta h)$$

If $f^{(3)}(a) \neq 0$ and $f^{(4)}(a + \theta h)$ is bounded, the same argument as before shows that the sign of the right-hand side is the same as the sign of $(h^3/3!)f^{(3)}(a)$, which changes sign with h. So the sign of $f(a + h) - f(a)$ changes sign as h changes sign so we have therefore a point of inflexion.

Consequently if at a stationary point we have $f''(a) = 0$ and $f^{(3)}(a) \neq 0$ we obtain a point of inflexion. On the other hand if $f^{(3)}(a)$ is zero too, then we need to continue with our analysis. This leads to a complete test for maxima and minima.

MAXIMA AND MINIMA: COMPLETE TEST

To obtain and classify those points at which $y = f(x)$ attains a local maximum or a local minimum:

1 Determine the stationary points of f. That is, obtain those points at which $\mathrm{d}y/\mathrm{d}x = 0$.

2 Obtain for each stationary point the smallest value of n for which $\mathrm{d}^n y/\mathrm{d}x^n \neq 0$.

3 If n is odd then the function has a point of inflexion at the stationary point.

4 If n is even then the function attains either a local maximum or a local minimum at a:

$\mathrm{d}^n y/\mathrm{d}x^n > 0$ implies a local minimum
$\mathrm{d}^n y/\mathrm{d}x^n < 0$ implies a local maximum

AT POINTS OF INFLEXION

If we examine the slope of the curve as we pass through a point of inflexion we see that two situations can occur (Fig. 8.6):

1 $\mathrm{d}y/\mathrm{d}x$ decreases as we approach the point of inflexion and increases thereafter;

2 $\mathrm{d}y/\mathrm{d}x$ increases as we approach the point of inflexion and decreases thereafter.

If we were driving a car along a road which went up a hill with the shape of Fig. 8.6 we should be conscious of this change in slope at the point of inflexion.

When we interpret mathematically what this means we see that f' itself has either a local maximum or a local minimum at a point of inflexion. It

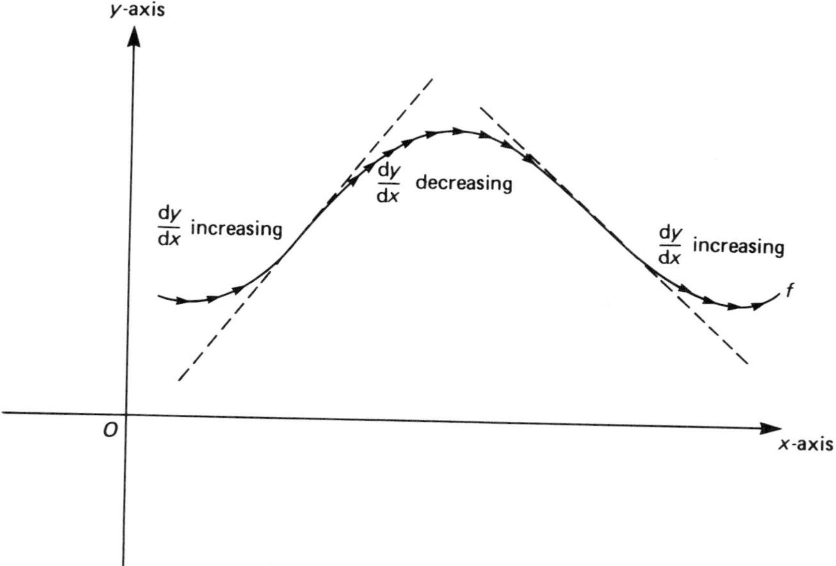

Fig. 8.6 The sign of dy/dx near points of inflexion.

follows that, at a point of inflexion, $d^2y/dx^2 = 0$. However, if the second order derivative is zero at a stationary point it does not necessarily follow that we have a point of inflexion. This is a very common misconception and it is possible that you too have fallen victim to it. In short:

$$\text{point of inflexion} \Rightarrow \frac{d^2y}{dx^2} = 0$$
$$\frac{d^2y}{dx^2} = 0 \nRightarrow \text{point of inflexion}$$

The next problem will help to reinforce this point.

□ Obtain and classify the stationary points of $y = x^4e^x$.

We begin by determining the stationary points:

$$\frac{dy}{dx} = x^4e^x + 4x^3e^x$$

Therefore we put $x^3\,(x + 4) = 0$ and deduce that the stationary points are $x = 0$ or $x = -4$. Now

$$\begin{aligned}\frac{d^2y}{dx^2} &= x^4e^x + 8x^3e^x + 12x^2e^x \\ &= x^2(x^2 + 8x + 12)e^x\end{aligned}$$

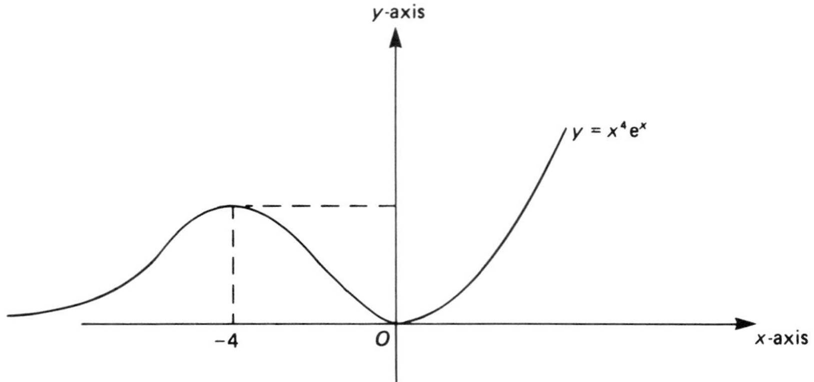

Fig. 8.7 The graph of $y = x^4e^x$.

When $x = 0$ we obtain $d^2y/dx^2 = 0$, and so further testing is necessary. When $x = -4$ we obtain

$$\frac{d^2y}{dx^2} = 16(16 - 32 + 12)e^{-4} < 0$$

so there is a local maximum at $x = -4$. Now

$$\frac{d^3y}{dx^3} = x^4e^x + 12x^3e^x + 36x^2e^x + 24xe^x$$

so that when $x = 0$, $d^3y/dx^3 = 0$ and still further testing is needed.

$$\frac{d^4y}{dx^4} = x^4e^x + 16x^3e^x + 72x^2e^x + 96xe^x + 24e^x$$

so that when $x = 0$ we obtain $d^4y/dx^4 = 24 > 0$, which corresponds to a local minimum. The stationary points are shown in Fig. 8.7.

Now you can be absolutely certain that if $d^2y/dx^2 = 0$ at a stationary point then there is no guarantee that it corresponds to a point of inflexion. In the words of the old song: 'It ain't necessarily so!' ■

Now for some more steps.

8.7 Workshop

1 **Exercise** Obtain and classify all the stationary points of

$$y = x^2(x + 3)^2e^x$$

This should cause very little difficulty provided you can determine the derivatives. Try it, then step ahead.

We have **2**

$$y = (x^2 + 3x)^2 e^x$$
$$\frac{dy}{dx} = (x^2 + 3x)^2 e^x + 2(x^2 + 3x)(2x + 3)e^x$$
$$= (x^2 + 3x)(x^2 + 3x + 4x + 6)e^x$$
$$= x(x + 3)(x + 6)(x + 1)e^x$$

Equating to zero we obtain the stationary points 0, −3, −6, −1.

Now we must differentiate again so that these can be classified as local maxima, local minima or points of inflexion. We could use the product rule, but there are five factors and algebraic simplification is tedious. We put

$$z = x(x + 3)(x + 6)(x + 1)e^x$$
$$\ln z = \ln x + \ln (x + 3) + \ln (x + 6) + \ln (x + 1) + x$$

So differentiating with respect to x,

$$\frac{1}{z}\frac{dz}{dx} = \frac{1}{x} + \frac{1}{x + 3} + \frac{1}{x + 6} + \frac{1}{x + 1} + 1$$

Multiplying through by z,

$$\frac{d^2y}{dx^2} = (x + 3)(x + 6)(x + 1)e^x + x(x + 6)(x + 1)e^x + x(x + 3)(x + 1)e^x$$
$$+ x(x + 3)(x + 6)e^x + x(x + 3)(x + 6)(x + 1)e^x$$

Now we examine each stationary point in turn:

1 $x = 0$: $d^2y/dx^2 = 18 > 0 \Rightarrow$ local minimum
2 $x = -3$: $d^2y/dx^2 = (-3)3(-2)e^{-3} > 0 \Rightarrow$ local minimum
3 $x = -6$: $d^2y/dx^2 = (-6)(-3)(-5)e^{-6} < 0 \Rightarrow$ local maximum
4 $x = -1$: $d^2y/dx^2 = (-1)2(5)e^{-1} < 0 \Rightarrow$ local maximum

If that was a personal success for you then skip through to step 4. Otherwise, try this exercise.

▷ **Exercise** Obtain and classify the stationary points of the function defined by

$$f(x) = x^3 e^{3x}$$

Have a go, then step forward.

We have **3**

$$f'(x) = x^3 3e^{3x} + 3x^2 e^{3x}$$

so that, equating to 0 for stationary points,

$$3x^2(x + 1)\,e^{3x} = 0$$

from which $x = 0$ (repeated) and $x = -1$ are the stationary points.

To classify these stationary points it is necessary to differentiate again:

$$\begin{aligned} f''(x) &= x^3 9e^{3x} + 9x^2 e^{3x} + 9x^2 e^{3x} + 6xe^{3x} \\ &= e^{3x}(9x^3 + 18x^2 + 6x) \\ &= 3xe^{3x}(3x^2 + 6x + 2) \end{aligned}$$

When $x = 0$ we obtain $f''(x) = 0$, and so further testing is necessary. When $x = -1$ we obtain

$$f''(x) = -3e^{-3}(3 - 6 + 2) > 0$$

so there is a local minimum. Differentiating again:

$$f^{(3)}(x) = e^{3x}(27x^2 + 36x + 6) + 3e^{3x}(9x^3 + 18x^2 + 6x)$$

When $x = 0$ we obtain $f^{(3)}(x) = 6 \neq 0$, and so when $x = 0$ there is a point of inflexion.

Now step ahead.

4

Exercise Obtain all the points of inflexion on the curve

$$y = x^6 - 5x^4 + 15x^2 + 1$$

Be careful: read the question. Try it, then step forward.

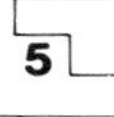

5

For a point of inflexion the slope has either a local maximum or a local minimum. Let s be the slope. Then

$$s = dy/dx = 6x^5 - 20x^3 + 30x$$

We must therefore examine this to see where s attains a local maximum or a local minimum.

We first obtain the stationary points of s:

$$ds/dx = 30x^4 - 60x^2 + 30$$

So equating to 0 we have

$$\begin{aligned} 30x^4 - 60x^2 + 30 &= 0 \\ x^4 - 2x^2 + 1 &= 0 \\ (x^2 - 1)^2 &= 0 \end{aligned}$$

So $x = 1$ or $x = -1$. These are the stationary points of s, and so are candidates for points of inflexion of y.

We must continue the test to be certain:

$$d^2s/dx^2 = 120x^3 - 120x$$

When $x = 1$ we obtain 0, and when $x = -1$ we obtain 0 too. Therefore we differentiate again:

$$d^3s/dx^3 = 360x^2 - 120$$

So s has a point of inflexion at $x = 1$ and $x = -1$. This means that s does not attain either a local maximum or a local minimum, and so y has no points of inflexion.

Now we shall solve a practical problem in which l'Hospital's rule is used.

8.8 Practical

VALVE RESPONSE

The response x of a valve when subject to a certain input is given for $t > 0$ by

$$dx/dt = \sqrt{3}\,[1 - (x^2/\tan^2 t)]^{1/2}$$

where $x(t) \to 0$ as $t \to 0+$. Show that

$$dx/dt \to \sqrt{3}/2 \quad \text{as} \quad t \to 0+$$

Try this yourself first, and then move through the solution stage by stage.

The physical situation enables us to assert that the limit exists. Suppose $dx/dt \to k$ as $t \to 0+$. We can see that the difficulty is centred on the term $x/\tan x$, for if we knew the limit of this as $t \to 0+$ we could calculate the limit of dx/dt using the laws of limits.

See if you can deal with the problem now.

We have, using l'Hospital's rule,

$$\lim_{t\to 0+} \frac{x}{\tan t} = \lim_{t\to 0+} \frac{dx/dt}{\sec^2 t}$$

Now $\sec^2 t \to 1$ as $t \to 0+$, and $dx/dt \to k$ as $t \to 0+$.

If you have been stuck, take over now.

Therefore we obtain

$$\lim_{t\to 0+} \frac{x}{\tan t} = \lim_{t\to 0+} \frac{dx/dt}{\sec^2 t} = \frac{\lim_{t\to 0+} (dx/dt)}{\lim_{t\to 0+} (\sec^2 t)} = \frac{k}{1} = k$$

So

$$\begin{aligned} k &= \sqrt{3}(1 - k^2)^{1/2} \\ k^2 &= 3(1 - k^2) \\ 4k^2 &= 3 \end{aligned}$$

It follows that $k = \pm\sqrt{3}/2$. However, $k \geqslant 0$ because $dx/dt > 0$. Consequently, as $t \to 0+$ we have shown that $dx/dt \to \sqrt{3}/2$.

We knew that the initial displacement was zero, and l'Hospital's rule has enabled us to determine the initial speed. Notice in particular that this is not an obvious result. Without l'Hospital's rule you might find it very difficult to confirm the limit.

SUMMARY

- ☐ We described the mean value theorem

$$\frac{f(b) - f(a)}{b - a} = f'(c)$$

for some $c \in (a, b)$.

- ☐ We generalized the mean value theorem to obtain Taylor's expansion about the point a

$$f(a + h) = \sum_{r=0}^{n-1} \frac{h^r}{r!} f^{(r)}(a) + R_n$$

where

$$R_n = \frac{h^n}{n!} f^{(n)}(a + \theta h) \quad \text{for some } \theta \in (0, 1)$$

- ☐ We obtained Maclaurin's expansion as the special case $a = 0$ of Taylor's expansion.
- ☐ We described l'Hospital's rule for determining a limit

$$\lim_{x\to a} \frac{f(x)}{g(x)} = \lim_{x\to a} \frac{f'(x)}{g'(x)}$$

provided $f(a)/g(a)$ is indeterminate.

- ☐ We used Taylor's expansion to deduce a complete test for maxima and minima.

EXERCISES

1 Obtain the first three non-zero terms of the Taylor expansion of each of the following:

a $\exp(\sin x)$

b $\operatorname{sech} x$

c $\dfrac{\ln(1+x)}{1+x}$

2 Obtain the stationary points of each of the following curves and classify them:

a $y = x^4 - 10x^2 + 1$

b $y = x(x-1)\exp x$

c $y = x^2 \exp(-x^2)$

d $y = x^2 \ln(1+x)$

3 Obtain the limit of x as t tends to 0, where

a $x = \dfrac{2\sinh t - \sin 2t}{\cosh 2t - \cos 2t}$

b $x = \dfrac{\exp(\sin t) - \cos t}{\cosh t - \exp(\sinh t)}$

c $x = (\sin^2 t)^t$

d $x = \dfrac{\ln t^3}{\ln(\sin t)}$

4 Obtain the limit of x as t tends to infinity, where

a $x = \dfrac{\ln(t+1)}{\ln(t^2+1)}$

b $x = (\exp t + 1)^{1/t}$

c $x = \dfrac{t}{\sqrt{(t^2+1)}}$

d $x = \dfrac{\sinh t + t^2}{\cosh t + t}$

ASSIGNMENT

1 Differentiate with respect to x:

a $\tanh^{-1}[\tan(x/2)]$

b $(\sin x)^x \qquad x \in (0, \pi)$

c $\tan^2 3x$

d $x \cos^{-1}(1 - 2x^2)$

2 Obtain dy/dx if

$$\exp(xy) + x \ln y = \sin 2x$$

3 Show that the first four terms in the Maclaurin expansion of $\tan(x + \pi/4)$ and $\ln(1 + \sin x)$ are respectively

a $1 + 2x + 2x^2 + 8x^3/3$

b $0 + x - x^2/2 + x^3/6$

4 Express $e^x \cos x$ as a power series in x as far as, and including, the term in x^3.

5 Obtain and classify the stationary points, and any points of inflexion, of

a $y = 3x^5 - 10x^3 + 15x + 1$

b $y = x^4 - 4x^3 + 4x^2 + 7$

6 A dangerous chemical has to be stored in a cylindrical container of a given volume. The mass of metal used in the construction of the cylinder and the thickness of the material are constant. The cylinder is to stand on one flat end in an open space. If the surface area exposed to the atmosphere is to be a minimum, calculate the relationship between its diameter and its height.

7 Show that the minimum value of $1 + x \ln x$ is $(e - 1)/e$.

8 If $y = \exp(\sinh^{-1} x)$, deduce

$$(1 + x^2)\frac{d^2y}{dx^2} + x\frac{dy}{dx} - y = 0$$

and

$$(1 + x^2)\frac{d^{n+2}y}{dx^{n+2}} + (2n + 1)x\frac{d^{n+1}y}{dx^{n+1}} + (n^2 - 1)\frac{d^ny}{dx^n} = 0$$

Hence verify Maclaurin's expansion:

$$x + \sqrt{(1 + x^2)} = 1 + x + \frac{x^2}{2!} - \frac{8x^4}{4!} + \ldots$$

9 Obtain each of the following limits as $x \to 0$:

a $\tan nx/\tan x$

b $(e^x - 1 - x)/x^2$

c $(1 - \cos \pi x)/x \tan \pi x$

d $[\tan^{-1}(x - 1) + \pi/4]/x$

10 Obtain each of the following limits as $x \to \infty$:

a $\sqrt{x} - \sqrt{(x - 1)}$

b $xa^{1/x} - x$ where $a > 0$

FURTHER EXERCISES

1 Show that

$$\lim_{x\to 0}\frac{\sin(a \tan^{-1}bx)}{\tan(c \sin^{-1}dx)} = \frac{ab}{cd}$$

2 Prove that

a $(x + 1)e^x = 1 + 2x + \dfrac{3x^2}{2!} + \ldots + \dfrac{nx^{n-1}}{(n-1)!} + \ldots$

b $(1 + 3x + x^2) \exp x = \sum_{n=0}^{\infty} \dfrac{(n+1)^2 x^n}{n!}$

3 Prove that

a $\lim_{x \to 0} \dfrac{\exp ax - 1}{\exp bx - 1} = \dfrac{a}{b}$

b $\lim_{x \to 0} \dfrac{\exp(\sin x) - 1}{x} = 1$

c $\lim_{x \to 0} (\cos x)^{\operatorname{cosec}^2 x} = \dfrac{1}{\sqrt{e}}$

4 Prove that

a $\lim_{x \to 0+} x^x = 1$

b $\lim_{x \to 0+} [\ln(1 + x)]^x = 1$

c $\lim_{x \to 0+} (e^x - 1)^x = 1$

5 When a unit cube of rubber is deformed into a cube of length λ by the action of temperature T and external forces, the energy E is given by

$$E = cT(\lambda^2 + 2\lambda^{-1})$$

where c is a constant. Prove that if T is constant the energy is a minimum when $\lambda = 1$.

6 A steel girder 7 m long is moved on rollers along a passage 3 m wide into another passage at right angles to the first. What is the minimum width of the second passage for which this manoeuvre is possible?

7 A prospector has a fixed length of fencing available and has to enclose a rectangular plot on three sides. One side is adjacent to a river. Determine the ratio of the length to the breadth if the enclosed area is to be a maximum.

8 A dish is made in the shape of a right circular cone. Calculate the ratio of the height to the diameter of the surface which will give a maximum volume if (a) the slant height is specified (b) the area of the curved surface is specified.

9 A greenhouse is to be made in the shape of a cylinder with a hemispherical roof. The material for the roof is twice as expensive per unit area as the material for the sides. Show that if it is to enclose a given air space and the total cost of the materials is to be a minimum, then the height of the cylindrical part must equal the diameter of the hemisphere.

10 A resistor is made up of two resistors in parallel. The first branch consists of wire of resistance 1/3 ohms/metre and the second consists of wire of resistance 1/4 ohms/metre. If 2 m of wire are available altogether, obtain the maximum possible total resistance.

11 When an EMF of Ee^{-t} is applied to an LR series circuit, the current i satisfies the equation

$$\frac{\mathrm{d}i}{\mathrm{d}t} + \frac{R}{L}i = \frac{E}{L}\mathrm{e}^{-t}$$

If $E = 1$ volt, $R = 1$ ohm, $L = 1$ henry and $i(0) = 0$, calculate the first three non-zero terms in the Taylor expansion of $i(t)$ about 0.

12 The force F exerted by a current moving on a circle of radius r on a unit magnetic pole on the polar axis of a circle is

$$F = kx/(r^2 + x^2)^{5/2}$$

where k is a constant and x is the distance of the magnetic pole from the centre of the circle.

a Show that the maximum force occurs when $x = r/2$.

b Show that the maximum force is $k(4/5)^{5/2}/2r^4$.

13 A compound pendulum of length $2l$ metres is pivoted x metres from the centre of mass and has a period

$$T = 2\pi\left(\frac{l^2}{3gx} + \frac{x}{g}\right)^{1/2}$$

Show that for minimum period $x = l/\sqrt{3}$.

14 A beam of length l and weight w per unit length is clamped horizontally at both ends. Its deflection y at a distance x from one end is given by

$$y = \frac{wx^2}{24EI}(l - x)^2$$

where E and I are constants. Show that the maximum deflection of the beam is $wl^4/384EI$.

The bending moment M at x is given by $M = EI/\varrho$, where ϱ is the radius of curvature of y at x. For small deflections ϱ is approximately $1/|y''|$. Show that under these circumstances the bending moment at the point of maximum deflection is approximately $wl^2/24$.

15 When an alternating EMF E sin nt is applied to a quiescent LC circuit, the current i at time t is given by

$$i = \frac{n\mathrm{E}}{\mathrm{L}(n^2 - w^2)}(\cos wt - \cos nt)$$

where $w^2 = 1/LC$ and w is not equal to n.

Show that when n is tuned to the natural frequency w of the circuit

$$i = \frac{Et \sin wt}{2L}$$

16 Show that the first two non-zero terms in the Maclaurin expansion of $\exp(\cos x - 1)$ are

$$1 - \frac{x^2}{2}$$

Hence or otherwise obtain

$$\lim_{x \to 0} \frac{\exp(\cos x - 1) - \exp(x^2)}{x^2}$$

17 Obtain all the stationary points of $y = (x - 1)^3 \, e^{2x}$ and calculate the minimum value. Determine also the equations of both the tangent and the normal at the point $(0, \ -1)$.

18 Evaluate each of the following limits

a $\displaystyle \lim_{x \to 0} \frac{\sin px}{\sin x}$ (p constant)

b $\displaystyle \lim_{x \to \infty} \frac{1 - \tanh x}{\frac{\pi}{2} - \tan^{-1} x}$

19 A wire is submerged in liquid and subjected to electro-chemical corrosion. The time t taken to corrode an amount of mass m is given by the equation

$$\frac{2t}{\lambda} = (m + \frac{M}{\theta - 1}) \ln[M + (\theta - 1)m] + (M - m) \ln(M - m)$$
$$- \frac{\theta M}{(\theta - 1)} \ln M$$

where M is the original mass of the wire and λ and θ are constant. Show that

$$\frac{dt}{dm} = \lambda \ln \sqrt{1 + \frac{\theta m}{M - m}}$$

Thereby deduce the rate of corrosion when only half of the wire remains.

20 The charge q on a capacitor is given by

$$q = Q_0 t e^{-t} + Q_1 (\sin 2t + \cos 2t)$$

where Q_0 and Q_1 are positive constants ($Q_0 > 8Q_1/3$). By first expressing $q = q_0 + q_1$, where q_0 and q_1 are to be chosen appropriately, or otherwise, show that the charge on the capacitor never exceeds $Q_0/e + Q_1\sqrt{2}$. The current in the circuit is obtained from the equation $i = dq/dt$. Show that, if t is so small that powers of t higher than those of degree 2 may be neglected, then the current drops to a minimum at time $t = 2(Q_0 + 2Q_1)/(3Q_0 - 8Q_1)$. Comment on this procedure.

9 Infinite series

In Chapter 8 we encountered several infinite series. In this chapter we shall clarify what we mean by infinite series and show that some of them behave rather unexpectedly.

After studying this chapter you should be able to
- ☐ Recognize an infinite series;
- ☐ Determine the sum to n terms of standard series;
- ☐ Examine for convergence directly by using the sum to n terms;
- ☐ Apply basic tests to examine a series for convergence or divergence;
- ☐ Determine the radius of convergence of a power series.

At the end of the chapter we shall solve practical problems concerning radioactive emission and a leaning tower.

9.1 SERIES

You have already come across infinite series. For example, the arithmetic series and the geometric series are quite well known. Here they are in standard notation.

The **arithmetic series** is of the form

$$a + (a + d) + (a + 2d) + (a + 3d) + \dots$$

where a and d are real numbers. We can represent this, using the summation notation, by

$$\sum_{m=0}^{\infty} (a + md)$$

Here a is the first term and d, the difference between any two consecutive terms, is known as the **common difference**.

The **geometric series** is of the form

$$a + ar + ar^2 + ar^3 + \ldots$$

where a and r are real numbers. Again this can be represented in a more compact form by

$$\sum_{m=0}^{\infty} ar^m$$

Here r is known as the **common ratio**, since it is the ratio of any two consecutive terms.

Now we know how to add together any finite collection of numbers, but we do not as yet have a clear idea as to what can be meant by an infinite sum. Of course we could imagine the situation in which we never stop calculating, but such a dream (or maybe a nightmare) is not really helpful. To begin to make sense of the idea we shall first obtain the sums of some finite series. In fact we shall obtain the **sum to *n* terms** of the arithmetic series and the geometric series.

For the arithmetic series, suppose

$$S = a + (a + d) + (a + 2d) + \ldots + (a + [n - 1]d)$$

Then if we reverse the order of the terms,

$$S = (a + [n - 1]d) + (a + [n - 2]d) + (a + [n - 3]d) \ldots + a$$

The reason for doing this now becomes clear, for if we add together the two expressions for S we obtain

$$2S = (2a + [n - 1]d) + (2a + [n - 1]d) + \ldots + (2a + [n - 1]d)$$

which is a sum of n equal terms. Consequently

$$2S = n(2a + [n - 1]d)$$

So the sum to n terms of the arithmetic series is

$$S = \tfrac{1}{2}n(2a + [n - 1]d)$$

□ Determine the sum of the first n natural numbers.

We require $1 + 2 + 3 + \ldots + n$, which is the sum of the first n terms of an arithmetic series where $a = 1$ and $d = 1$. Therefore

$$S = \tfrac{1}{2}n(2 + [n - 1]) = \tfrac{1}{2}n(n + 1)$$

You may be able to use this to impress younger members of your family by declaring, after Christmas lunch, that you can add up the first 100 (say) whole numbers in an instant. This formula gives 5050, and by the time they have checked it you should have had a few moments' peace. ■

For the geometric series, suppose

$$S = a + ar + ar^2 + \dots + ar^{n-1}$$

Then multiplying through by r we obtain

$$Sr = ar + ar^2 + \dots + ar^n$$

We have done this because, if we subtract Sr from S, terms cancel out in pairs and all that remains is the first term in S and the last term in Sr:

$$S - Sr = a - ar^n$$
$$S(1 - r) = a(1 - r^n)$$

So the sum to n terms of the geometric series is

$$S = a\,\frac{1 - r^n}{1 - r} \quad \text{provided } r \neq 1$$ ■

□ Determine the value of

$$1 + \tfrac{1}{2} + \tfrac{1}{4} + \dots + (\tfrac{1}{2})^n$$

Here we require the sum of the first $n + 1$ terms of a geometric series in which the first term is 1 and the common ratio is 1/2. Using the formula we obtain

$$S = \frac{1 - (1/2)^{n+1}}{1 - (1/2)} = 2(1 - [\tfrac{1}{2}]^{n+1}) = 2 - (\tfrac{1}{2})^n$$ ■

Each of these series is unusual in the sense that we have been able to determine formulas for S, the sum of the first n terms. Of course S depends upon n; therefore we shall in future denote the sum of the first n terms of an infinite series by s_n. In general we may write an **infinite series** in the form

$$\Sigma\, a_n = a_1 + a_2 + a_3 + \dots + a_r + \dots$$

Observe some of the general features. There are two sets of dots indicating missing terms. The first set of dots shows that terms occur between a_3 and the general term a_r. The second set of dots indicates that this is an infinite series and does not terminate.

We shall not normally adorn the summation sign Σ by writing $n = 1$ below and ∞ on top. However, if we wish to use a different dummy variable, or begin the sum at some other value of n, it is necessary to indicate this by an appropriate choice of labels.

9.2 CONVERGENCE AND DIVERGENCE

The two series which we have been considering display features which help us to describe the general situation.

We have seen that the sum of the first n natural numbers

$$1 + 2 + 3 + \ldots + n$$

is given by $s_n = \frac{1}{2}n(n + 1)$. We observe that in this instance $s_n \to \infty$ as $n \to \infty$. This means that the sum to n terms can be made arbitrarily large just by choosing n sufficiently large.

Again, the sum of the first n terms of the series

$$1 + \tfrac{1}{2} + \tfrac{1}{4} + \ldots + (\tfrac{1}{2})^n + \ldots$$

is given by $s_n = 2 - (1/2)^n$ and so $s_n \to 2$ as $n \to \infty$. This means that the sum to n terms can be made arbitrarily close to 2 just by choosing n sufficiently large.

In general, given an infinite series

$$\Sigma a_n = a_1 + a_2 + a_3 + \ldots + a_r + \ldots$$

suppose that s_n denotes the sum to n terms. Then:

1 If there exists a number s such that $s_n \to s$ as $n \to \infty$ then the series is said to **converge**. Moreover if s is known we can say that the series converges to s.

2 If there is no number s such that $s_n \to s$ as $n \to \infty$ then the series is said to **diverge**. If $s_n \to \infty$ as $n \to \infty$ then the series is said to diverge to ∞. If $s_n \to -\infty$ as $n \to \infty$ then the series is said to diverge to $-\infty$. A divergent series does not necessarily diverge either to ∞ or to $-\infty$; for example it might oscillate.

If it were always possible to obtain a straightforward formula for s_n it would be a relatively simple matter to examine a series to see if it converges or diverges. As it is, we can rarely obtain a simple formula for s_n and so tests have been devised to examine the behaviour of series which arise in practice.

In the examples the arithmetic series Σn diverges and the geometric series $\Sigma (1/2)^n$ converges (to 2). In fact *every* arithmetic series diverges; there really is a last straw which will break the camel's back!

For geometric series the situation is a little more subtle and we shall need to examine it closer. For the geometric series

$$\sum_{m=0}^{\infty} ar^m = a + ar + ar^2 + \ldots + ar^n + \ldots$$

we have

$$s_n = a\,\frac{1 - r^n}{1 - r} \quad \text{provided } r \neq 1$$

Then

1 If $|r| < 1$ we have $r^n \to 0$ as $n \to \infty$. So $s_n \to a/(1 - r)$ and the series converges.

2 If $r > 1$ then $r^n \to \infty$ as $n \to \infty$. So s_n does not tend to a limit as n tends to infinity and the series diverges.

3 If $r < -1$ then r^n increases in magnitude but alternates in sign as n tends to infinity. So once again s_n does not tend to a limit as n tends to infinity and the series consequently diverges.

4 It remains only to consider the cases $r = 1$ and $r = -1$. When $r = 1$ we obtain $s_n = na$; so the series diverges unless $a = 0$. When $r = -1$ we obtain $s_n = a$ if n is odd and $s_n = 0$ if n is even; again we conclude that the series diverges unless $a = 0$.

Consequently, if $a \neq 0$, the *geometric series*

$$\sum_{m=0}^{\infty} ar^m = a + ar + ar^2 + \ldots + ar^n + \ldots$$

converges when $|r| < 1$ and *diverges* when $|r| \geqslant 1$.

There are two types of series which arise in applications and which you are likely to encounter in theoretical work. These are power series and trigonometrical series.

A **power series** is a series of the form

$$a_0 + a_1x + a_2x^2 + a_3x^3 + \ldots + a_rx^r + \ldots$$

where the 'a's are constants.

A **trigonometrical series** is a series of the form

$$(1/2)a_0 + (a_1 \cos x + b_1 \sin x) + (a_2 \cos 2x + b_2 \sin 2x) + \ldots + (a_r \cos rx + b_r \sin rx) + \ldots$$

where the 'a's and 'b's are constants. (The (1/2) in the $(1/2)a_0$ term may seem strange, and strictly it is superfluous. However there are advantages in expressing the first term in this form and it is usual to do so.)

The discussion of trigonometrical series leads to Fourier series which we shall investigate in Chapter 21. Power series will be discussed later in this chapter.

We have remarked that, although it is sometimes possible to examine the limit of s_n directly, in general this is not possible. To cope with the general situation we need some tests for convergence and divergence, and these we now describe.

9.3 TESTS FOR CONVERGENCE AND DIVERGENCE

There are very many tests which have been devised to examine infinite series to determine whether or not they converge or diverge. It is reasonable to ask whether there is one test which will settle the matter once and for all. Unfortunately there is no supertest; whatever test we have there is always a series which can be produced on which the test will fail.

Before we take things any further we should point out that this is a subtle area of mathematics where it is easy to make mistakes. Mathematical operations which we carry out on finite sums do not necessarily work when we attempt them on infinite series. Infinite series should therefore be treated with respect and, if in theoretical work you should come across one, it may be advisable to consult a competent mathematician rather than try to handle it yourself.

Nevertheless we are going to describe some basic tests which will enable us to examine most of the series which we are likely to meet at the moment.

TEST 1: THE DIVERGENCE TEST

The infinite series

$$\Sigma a_n = a_1 + a_2 + a_3 + \ldots + a_r + \ldots$$

diverges if

$$\lim_{n \to \infty} a_n \neq 0$$

To show this we examine the situation when Σa_n converges and show that then $\lim_{n \to \infty} a_n = 0$. Suppose that Σa_n converges to s. Now

$$a_1 + a_2 + \ldots + a_n = s_n$$
$$a_1 + a_2 + \ldots + a_{n-1} = s_{n-1}$$

Subtracting,

$$a_n = s_n - s_{n-1}$$

Therefore

$$\lim_{n \to \infty} a_n = \lim_{n \to \infty} (s_n - s_{n-1}) = s - s = 0$$

Consequently if Σa_n converges the nth term tends to 0 as n tends to ∞. However, we are told that the nth term does not tend to 0 as n tends to ∞. Therefore Σa_n cannot converge and so must diverge.

□ Examine for convergence $\Sigma(1 + 1/n)$.

Here $a_n = 1 + 1/n$ and so as $n \to \infty$ we have $a_n \to 1$, which is non-zero. So by the divergence test the series diverges. ■

It is important to realize that this test is a divergence test; it can *never* be used to establish the convergence of a series. There are many divergent series which have their nth terms tending to zero.

TEST 2: THE COMPARISON TEST

Suppose Σa_n and Σb_n are real series such that $0 < a_n \leqslant b_n$. Then if Σb_n converges, so too does Σa_n.

We shall not justify this, but instead use it to examine the convergence of Σn^{-2}.

□ By considering s_n, the sum to n terms of the series $\Sigma[1/n(n+1)]$, examine the series for convergence. Hence or otherwise establish the convergence of Σn^{-2}.

Now

$$\frac{1}{n(n+1)} = \frac{1}{n} - \frac{1}{n+1}$$

So

$$\begin{aligned} s_n &= a_1 + a_2 + \ldots + a_n \\ &= \left(1 - \frac{1}{2}\right) + \left(\frac{1}{2} - \frac{1}{3}\right) + \ldots + \left(\frac{1}{n} - \frac{1}{n+1}\right) \end{aligned}$$

These cancel out in pairs, leaving

$$s_n = 1 - \frac{1}{n+1} \to 1 \quad \text{as } n \to \infty$$

Consequently $\Sigma[1/n(n+1)]$ is convergent.

Now if n is any natural number, $n < n+1$ and so $n(n+1) < (n+1)^2$. Therefore

$$0 < \frac{1}{(n+1)^2} < \frac{1}{n(n+1)}$$

Consequently by the comparison test $\Sigma(n+1)^{-2}$ is convergent. Now in what way does this series differ from Σn^{-2}? It has the first term missing, and it is surely inconceivable that this single omission can affect the convergence. Therefore we conclude Σn^{-2} is convergent.

Although this line of reasoning may seem convincing, the statement requires proof. Luckily we can tighten things up without much difficulty. Let s_n be the sum to n terms of the first series and t_n the sum to n terms of the second series. Then $t_n = 1 + s_n - (n+1)^{-2}$. Now s_n is known to converge to s (say), and consequently $t_n \to 1 + s - 0 = 1 + s$. ■

The comparison test is particularly useful once a collection of series have been produced which are *known* to converge or to diverge. Here is the test again:

Suppose Σa_n and Σb_n are real series such that $0 < a_n \leqslant b_n$. Then if Σb_n converges, so too does Σa_n.

It is worth remarking that if Σa_n diverges then so too does Σb_n. For if Σb_n were to converge then by the comparison test we could deduce the convergence of Σa_n.

Here is a series which at first sight looks very innocuous: $\Sigma\, 1/n$. Clearly the terms get smaller and smaller as n gets larger and larger, and it looks as if it is going to converge to a fairly small number. We might even be tempted to get a computer to estimate its value by, say, summing the first 1000 terms.

However, all is not as it seems. In fact the series *diverges* (very slowly), as we shall now show. We have

$$s_n = 1 + 1/2 + 1/3 + \ldots + 1/n$$

Now if $n = 2^m$ we have

$$s_n = 1 + \frac{1}{2} + \left(\frac{1}{3} + \frac{1}{4}\right) + \left(\frac{1}{5} + \frac{1}{6} + \frac{1}{7} + \frac{1}{8}\right) + \ldots + \left(\ldots + \frac{1}{2^m}\right)$$

Here we have grouped the terms together so that the last term in each bracket is a power of 2. Now in each bracket each term is greater than the last term in the bracket, and the number of terms in each bracket is a power of 2. Therefore

$$s_n > 1 + \frac{1}{2} + \left(\frac{1}{4} + \frac{1}{4}\right) + \left(\frac{1}{8} + \frac{1}{8} + \frac{1}{8} + \frac{1}{8}\right) + \ldots + \left(\ldots + \frac{1}{2^m}\right)$$

So

$$s_n > 1 + \tfrac{1}{2} + \tfrac{1}{2} + \tfrac{1}{2} + \ldots + \tfrac{1}{2} = 1 + \tfrac{1}{2}m$$

It follows that $s_n > 1 + m/2$ when $n = 2^m$. Now as m tends to ∞, n tends to ∞, and yet s_n is unbounded and so does not tend to a limit. Consequently $\Sigma\, 1/n$ is divergent.

This series is a member of the family $\Sigma\, 1/n^p$ where p is real. It can be shown that

1 When $p > 1$ the series converges;
2 When $p \leqslant 1$ the series diverges.

TEST 3: THE ALTERNATING TEST

Suppose $\Sigma\, a_n$ is an infinite series in which

1 The terms alternate in sign;
2 $|a_n| \geqslant |a_{n+1}|$ for all $n \in \mathbb{N}$;
3 $|a_n| \to 0$ as $n \to \infty$.

Then the series converges.

□ Show that $\Sigma\,(-1)^n/n$ is convergent.

We observe that each of the conditions of the alternating test is satisfied:

1 The terms alternate in sign;
2 $|a_n| = 1/n > 1/(n + 1) = |a_{n+1}|$;
3 $1/n \to 0$ as $n \to \infty$.

So the conclusion is that the series is indeed convergent. ■

TEST 4: THE RATIO TEST

Suppose $\Sigma\, a_n$ is an infinite series and that

$$l = \lim_{n\to\infty} \left|\frac{a_{n+1}}{a_n}\right|$$

exists. Then
1 If $l < 1$, $\Sigma\, a_n$ converges;
2 If $l > 1$, $\Sigma\, a_n$ diverges;
3 If $l = 1$, no conclusion can be reached.

□ Discuss the behaviour of the series $\Sigma\, 1/n!$.
Here $a_n = 1/n!$ and so $a_{n+1} = 1/(n+1)!$. Therefore

$$a_{n+1}/a_n = [1/(n+1)!]/[1/n!] = n!/(n+1)! = 1/(n+1)$$

so that $|a_{n+1}/a_n| = 1/(n+1) \to 0$ as $n \to \infty$. Of course $0 < 1$, and so we deduce that the series is convergent. ■

TEST 5: THE ABSOLUTE CONVERGENCE TEST

Suppose $\Sigma\, a_n$ is an infinite series such that $\Sigma\, |a_n|$ converges. Then $\Sigma\, a_n$ converges.

Any series $\Sigma\, a_n$, real or complex, which has the property that $\Sigma\, |a_n|$ converges is called an **absolutely convergent** series. This test tells us that if a series is absolutely convergent then it is convergent. There are many series which are convergent but which are not absolutely convergent. These series are called **conditionally convergent**.

□ We have seen that $\Sigma\, [(-1)^n/n]$ is convergent but that $\Sigma\, (1/n)$ is divergent. Since

$$\left|\frac{(-1)^n}{n}\right| = \frac{1}{n}$$

this implies that $\Sigma\, [(-1)^n/n]$ is conditionally convergent. ■

9.4 POWER SERIES

Consider the power series $\Sigma\, a_n x^n$. We shall show that if $|a_n/a_{n+1}| \to R \neq 0$ as $n \to \infty$ then the power series
1 Converges whenever $|x| < R$;
2 Diverges whenever $|x| > R$.
R is known as the **radius of convergence** of the power series. Every power series in x converges when $x = 0$, and if this is the only value of x for which

it converges we say it has zero radius of convergence and write $R = 0$. Some power series in x converge for all x, and we then say the power series has an infinite radius of convergence and write $R = \infty$.

If we were to extend these ideas to complex power series we should obtain a disc of convergence instead of an open interval (Chapter 10).

We apply the ratio test, but we have to be a little careful about the notation since a_n appears as the coefficient of x^n and not as the term itself. To avoid this confusion we shall call the nth term u_n. Now

$$u_{n+1}/u_n = a_{n+1}x^{n+1}/a_n x^n = x a_{n+1}/a_n$$

So

$$|u_{n+1}/u_n| = |x|\,|a_{n+1}/a_n|$$

Now $|a_n/a_{n+1}| \to R$ as $n \to \infty$, and since $R \neq 0$ we deduce that

$$|u_{n+1}/u_n| \to |x|/R \quad \text{as } n \to \infty$$

Consequently if $|x|/R < 1$ the series converges, whereas if $|x|/R > 1$ the series diverges. Finally, since $R > 0$ we have

$$\begin{aligned} |x| < R &\Rightarrow \text{convergence} \\ |x| > R &\Rightarrow \text{divergence} \end{aligned}$$

In fact the radius of convergence of a power series is very useful because if $x \in (-R, R)$ it is possible to differentiate and integrate with respect to x term by term and obtain correct results.

It is important to realize that in general any operation on an infinite series may disturb its convergence. Such operations include rearranging terms, inserting or removing brackets, differentiating and integrating. Convergence of the series is *not* enough to ensure that we can perform these operations and obtain the expected results. We need special forms of convergence to ensure that. For algebraic operations we need *absolute* convergence and for calculus operations we need *uniform* convergence. We shall not describe uniform convergence in this book.

Well, now it's time to take a few steps.

9.5 Workshop

▷ **Exercise** Discuss the behaviour of the series $\Sigma x^n/n!$ for all real x. 1
Try this and see how you get on.

This exercise is an application of the ratio test: 2

$$a_{n+1}/a_n = [x^{n+1}/(n + 1)!]/[x^n/n!] = x/(n + 1)$$

and so $|a_{n+1}/a_n| = |x|/(n + 1) \to 0$ as $n \to \infty$. Therefore the series is convergent for all real x. In fact you have seen this series before: it converges, if we start when $n = 0$, to e^x.

Did you manage that? Here is another exercise to try.

▷**Exercise** Discuss for all real x, $|x| \neq 1$, the convergence of the binomial series

$$\sum_{r=0}^{\infty} \binom{n}{r} x^r$$

where

$$\binom{n}{r} = \frac{n(n-1)(n-2)\ldots(n-r+1)}{1 \times 2 \times 3 \times 4 \times \ldots \times r}$$

Here of course the dummy variable is r; n is constant. Make an effort and then take the next step.

---

3 We obtain

$$\frac{a_{r+1}}{a_r} = \frac{n(n-1)(n-2)\ldots(n-[r+1]+1)x^{r+1}}{1 \times 2 \times 3 \times 4 \times \ldots \times [r+1]} \times \frac{1 \times 2 \times 3 \times 4 \times \ldots \times r}{n(n-1)(n-2)\ldots(n-r+1)x^r} = \frac{(n-r)x}{r+1}$$

Now

$$\begin{aligned}|a_{r+1}/a_r| &= |(n-r)x/(r+1)| \\ &= |([n/r]-1)x/(1+[1/r])| \\ &\to |-x/1| = |x| \quad \text{as } r \to \infty\end{aligned}$$

Consequently if $|x| < 1$ the series converges, whereas if $|x| > 1$ the series diverges.

You may have found that one rather too algebraic. If you did then the next one may be more to your taste.

▷**Exercise** Discuss, for all real x, the convergence of the series $\Sigma x^n/n$.

As soon as you have tested the series, take the next step.

---

4 Here $a_n = x^n/n$ and so

$$a_{n+1}/a_n = [x^{n+1}/(n + 1)][n/x^n] = xn/(n + 1)$$

so that

$$|a_{n+1}/a_n| = |x| \{1/(1 + [1/n])\} \to |x| \quad \text{as } n \to \infty$$

So the ratio test shows that if $|x| < 1$ the series converges, whereas if $|x| > 1$ the series diverges.

There only remain the cases $x = 1$ and $x = -1$. When $x = 1$ the series reduces to $\Sigma\, 1/n$, which we have already shown to be divergent. When $x = -1$ the series reduces to $\Sigma\,(-1)^n/n$, which we have already shown to be convergent.

We conclude therefore that $\Sigma\, x^n/n$ is convergent when $-1 \leqslant x < 1$ and divergent when $x \geqslant 1$ or $x < -1$. In fact this is the series expansion corresponding to $\ln(1 - x)$.

Now let us look at a few series. Although the ratio test is very useful, we should not forget the other tests.

▷**Exercise** Determine whether $\Sigma\, 1/(n^2 + 1)$ is convergent or divergent. 5
Attempt this carefully and then move on to see if all is well.

The ratio test is of no use to us here. However, we do know that $\Sigma\, 1/n^2$ is 6
convergent, and this series is only slightly different.

Can we use the comparison test? Well, $n^2 < n^2 + 1$ for any natural number n, and so we have

$$0 < 1/(n^2 + 1) < 1/n^2$$

The convergence of $\Sigma\, 1/(n^2 + 1)$ now follows.

How about this one?

▷**Exercise** Examine for convergence $\Sigma\,(1 + 1/n)^n$.

At first sight this appears to be a pretty fearsome series to test. However, a 7
bell should sound. It may be a rather distant, muffled bell but it should sound nevertheless. Haven't we seen $(1 + 1/n)^n$ somewhere before? We have, you know. We have found the limit of it as $n \to \infty$ (Chapter 8). The limit is e, the natural base of logarithms.

This observation is all that we need to dispose of the problem once and for all. $(1 + 1/n)^n \to \mathrm{e}$ as $n \to \infty$, and since $\mathrm{e} \neq 0$ the divergence test shows that the series diverges.

Finally let us look at a limit.

▷**Exercise** Obtain the limit as $n \to \infty$ of $(1 + 2 + 3 + \ldots + n)/n^2$.
Try it, but be careful.

8 Perhaps you proceeded in the following manner:

$$(1 + 2 + 3 + \ldots + n)/n^2 = 1/n^2 + 2/n^2 + 3/n^2 + \ldots + 1/n$$

Then possibly you argued that there are n terms each of which is tending to 0 as $n \to \infty$, and concluded that the limit itself is zero.

Unfortunately this argument is flawed. Although it is true that the terms are getting smaller and smaller, there are more and more of them! If you got it wrong then have another try and take another step.

9 We know that

$$1 + 2 + 3 + \ldots + n = n(n + 1)/2$$

So

$$\begin{aligned}(1 + 2 + 3 + \ldots + n)/n^2 &= n(n + 1)/2n^2 \\ &= (n + 1)/2n = (1 + 1/n)/2\end{aligned}$$

As $n \to \infty$ we obtain the limit 1/2.

This is as far as we are going to take the topic of infinite series. There is much more that can be said, but it is important to realize that this is a sensitive area where even otherwise competent mathematicians are prone to error.

It is sometimes quite alarming to see what the uninformed will do with infinite series. It is always possible that the results could be catastrophic: bridges could fall down, aircraft disintegrate, dance floors cave in, buildings collapse, and power plants get out of control. Every infinite series should carry a government health warning!

Now here are some practical problems for you to try.

9.6 Practical

RADIOACTIVE EMISSION

Radioactive material is stored in a thick concrete drum. It is believed to ingress, by the end of each year, into the uncontaminated surrounding material a depth $d = Q/n$, where n is the number of years and the quantity Q is a constant. At the end of the first year, $d = 0.5$ cm.

First, if the surrounding concrete is 4 m thick, will this contain the material for all time? Secondly, if the material remains hazardous for 1000 years, what would be a safe thickness of concrete?

See if you can handle this problem. We will go through it stage by stage.

---

1 We have $d = Q/n$. When $n = 1$, $d = 0.5$, so that $Q = 0.5$. Now the depth of penetration after n years is given by

$$d = Q + \frac{Q}{2} + \frac{Q}{3} + \frac{Q}{4} + \ldots + \frac{Q}{n} = Q \sum_{r=1}^{n} \frac{1}{r}$$

Now you are at this stage, take over the solution.

---

2 We know that $\Sigma\,(1/n)$ is a divergent series, and so as $n \to \infty$ we infer that $d \to \infty$. The conclusion we draw is that whatever the value of Q (>0), penetration will eventually occur; so 4 m is certainly not enough.

Luckily the second part of the problem accords more with reality. See if you can finish it off.

---

3 As an exceedingly crude estimate we have

$$Q + \frac{Q}{2} + \frac{Q}{3} + \frac{Q}{4} + \ldots + \frac{Q}{1000} < 1000Q$$

Therefore provided $1000Q$ is less than the thickness T of the surrounding material, everything will certainly remain safe. So $T > 1000Q = 500$ cm $= 5$ m will do.

We could get away with considerably less concrete. In fact if you add the first 1000 terms you obtain $7.5Q$, and so in fact 0.0375 m is good enough!

Here is another problem.

LEANING TOWER

A tower is built in such a way that shortly after its construction it begins to lean. It is believed that the angle of tilt is increased at the end of each year by an amount $K/(1 + n^2)$, where K is constant and n is the age in years of the tower. At the end of the first year the angle of tilt was 3°.

a Assuming that the formula is correct, show that the tower will not fall flat.

b Show that eventually the angle of tilt will satisfy $4° \leqslant \theta \leqslant 5°$.

c Use a calculator to determine how many years it will take for the angle to become 4°.

Solve part **a**. It is not unlike the previous problem.

---

1 For **a** we have, when $n = 1$, $\theta = 3° = K + K/2$ and so $K = 2$. After n years the angle of tilt will be

$$\theta = K + \frac{K}{2} + \frac{K}{5} + \frac{K}{10} + \ldots + \frac{K}{1 + n^2}$$

We therefore need to examine $\Sigma\, 1/(1 + n^2)$ for convergence. The comparison test can be applied, for we know that $\Sigma\, 1/n^2$ is convergent. We have $1 + n^2 > n^2$, and so

$$0 < \frac{1}{1 + n^2} < \frac{1}{n^2}$$

The convergence of $\Sigma\, 1/n^2$ now implies the convergence of $\Sigma\, 1/(1 + n^2)$. So, provided $K \Sigma\, 1/(1 + n^2)$ converges to a number less than 90°, the tower will not fall flat.

This matter will be settled provided we can sort out part **b**. Here is a hint:

$$n(n - 1) < n^2 + 1 < n(n + 1)$$

---

2 For **b**, using this inequality, we have for $n > 1$

$$\frac{1}{n(n + 1)} < \frac{1}{1 + n^2} < \frac{1}{n(n - 1)}$$

Add up the first N terms. Don't forget your work on partial fractions!

---

3 So

$$\sum_{n=2}^{N} \frac{1}{n(n + 1)} < \sum_{n=2}^{N} \frac{1}{1 + n^2} < \sum_{n=2}^{N} \frac{1}{n(n - 1)}$$

$$1 + \frac{1}{2} + \sum_{n=2}^{N} \frac{1}{n(n + 1)} < \sum_{n=0}^{N} \frac{1}{1 + n^2} < \frac{3}{2} + \sum_{n=2}^{N} \frac{1}{n(n - 1)}$$

$$\frac{3}{2} + \sum_{n=2}^{N} \left(\frac{1}{n} - \frac{1}{n + 1}\right) < \sum_{n=0}^{N} \frac{1}{1 + n^2} < \frac{3}{2} + \sum_{n=2}^{N} \left(\frac{1}{n - 1} - \frac{1}{n}\right)$$

$$2 - \frac{1}{N + 1} < \sum_{n=0}^{N} \frac{1}{1 + n^2} < \frac{5}{2} - \frac{1}{N}$$

Consequently

$$2K \leqslant \sum_{n=0}^{\infty} \frac{K}{1 + n^2} \leqslant \frac{5K}{2}$$

and so $4° \leqslant \theta \leqslant 5°$.

Lastly, start tapping the buttons on your calculator.

---

4 For **c** we must calculate

$$K\left(1 + \frac{1}{2} + \frac{1}{5} + \frac{1}{10} + \ldots + \frac{1}{1 + n^2} + \ldots\right)$$

until for some n the total exceeds 4°. In fact it takes 13 years for the tower to lean 4°.

SUMMARY

- ☐ We have seen how to represent an infinite series.
- ☐ We have explained what is meant by convergence and divergence.
- ☐ We have described some tests which can be applied to infinite series to see whether they converge or diverge. The tests we described were called
  - **a** the divergence test
  - **b** the comparison test
  - **c** the alternating test
  - **d** the ratio test
  - **e** the absolute convergence test.
- ☐ We have defined the radius of convergence of a power series.

EXERCISES

1 Obtain the limit of the nth term of each of the following series and so show that each is divergent:

a $\sum \frac{n}{\sqrt{(n^2 + 1)}}$

b $\sum \frac{(\cosh n + n)}{(\sinh n + n)}$

c $\sum \frac{\ln (n^2 + 1)}{\ln (n^3 + 1)}$

d $\sum (2^n + 1)^{1/n}$

2 Obtain s_n, the sum to n terms of each of the following series, and thereby test for convergence:

a $\sum \frac{1}{4n^2 - 1}$

b $\sum \frac{2n + 1}{n^2(n + 1)^2}$

c $\sum \dfrac{1}{n\sqrt{(n+1)} + (n+1)\sqrt{n}}$

d $\sum \dfrac{1}{\sinh n \sinh (n-1)}$

3 By using the comparison test show that each of the following series is divergent:

a $\sum \dfrac{1}{2n-1}$

b $\sum \dfrac{1}{n - \sqrt{n}}$

c $\sum \dfrac{1}{n \sin n}$

d $\sum \dfrac{1}{n \tanh n}$

4 By using the comparison test show that each of the following series is convergent:

a $\sum \dfrac{1}{n^2+3}$

b $\sum \dfrac{1}{\sqrt{(n^4+1)}}$

c $\sum \dfrac{n}{\sqrt{(n^6+1)}}$

d $\sum \dfrac{2n+1}{n^3+1}$

ASSIGNMENT

Examine each of the following series for convergence or divergence:

1 $\Sigma\,(n^2-1)/(n^2+1)$
2 $\Sigma\,1/(n^2+2n)$
3 $\Sigma\,(-1)^n n^2/2^n$
4 $\Sigma\,n^{5/2}/(n^2+1)$
5 $\Sigma\,\sin n/n^2$
6 $\Sigma\,n!/3^n$
7 $\Sigma\,\mathrm{e}^{nx}/n^2$
8 $\Sigma\,1/\sqrt{n}$
9 $\Sigma\,n!/(2n)!$
10 $\Sigma\,1/n(n^2+1)$

Determine the radius of convergence of each of the following power series:

11 $\Sigma\,x^n/2^n$
12 $\Sigma\,(n!)x^n/(2n)!$
13 $\Sigma\,x^{2n}/3^n$

14 $\Sigma x^n/n^3$
15 $\Sigma (x/n)^n$
16 $\Sigma (2x)^n/n$
17 $\Sigma (nx)^n$
18 $\Sigma (nx)^n/n!$
19 $\Sigma x^n/\sqrt{n}$
20 $\Sigma n^3 x^n$

FURTHER EXERCISES

1 Examine for convergence:

a $\displaystyle\sum_{n=0}^{\infty} \frac{n}{n+1}$

b $\displaystyle\sum_{n=1}^{\infty} \frac{2^{n-1}}{n^3}$

c $\displaystyle\sum_{n=0}^{\infty} \frac{n}{\sqrt{(n^2+1)}}$

d $\displaystyle\sum_{n=2}^{\infty} \frac{n}{\sqrt{(n^2-1)}}$

2 Classify each of the following series as absolutely convergent (AC) or conditionally convergent (CC):

a $\displaystyle\sum_{n=1}^{\infty} \frac{(-1)^n}{n^2}$

b $\displaystyle\sum_{n=0}^{\infty} \frac{(-1)^n n}{3^{n-1}}$

c $\displaystyle\sum_{n=0}^{\infty} \frac{(-1)^n n^2}{n^3+1}$

3 Show that if p and q are positive integers $(p < q)$ then

$$1 + (p/q) + (p/q)^2 + \ldots + (p/q)^N < q/(q - p)$$

Deduce that

$$1 + (3/4) + (3/4)^2 + \ldots + (3/4)^n + \ldots \leqslant 4$$

4 Test for convergence or divergence:

a $\displaystyle\sum_{n=0}^{\infty} \frac{1}{1+nx}$

b $\displaystyle\sum_{n=0}^{\infty} \frac{1}{n+x}$

c $\sum_{n=0}^{\infty}\left(x^n + \frac{1}{x^n}\right)$

5 By first showing that $\sqrt{(n^2 - 1)} < n < (n + 1)$ and $\sqrt{(n^2 + 1)} < n + 1$, show that

$$\sum_{n=1}^{\infty} [\sqrt{(n^2 + 1)} + \sqrt{(n^2 - 1)}]^{-1}$$

is divergent.

6 Show that the radius of convergence of each of the following power series is 1. Investigate the convergence of each when $|x| = 1$.

a $\sum_{n=1}^{\infty} \frac{(-1)^n x^{2n}}{n}$

b $\sum_{n=1}^{\infty} \frac{x^n}{n(n + 1)}$

c $\sum_{n=1}^{\infty} \frac{n + 2}{n(n + 1)} x^n$

d $\sum_{n=1}^{\infty} \frac{(n + 1)}{n^2} x^n$

7 The quantity of liquid p_n which is extracted from pulp in the nth cycle of a pressure pump is given approximately by $p_n = p/(n^2 + 1)$, where p is the initial quantity extracted. Prove that the total quantity Q of liquid extracted is bounded and that $2p \leqslant Q \leqslant 5p/2$. Show that after 13 cycles at least $2p$ (80% of the upper bound) has been extracted. (In fact $Q = (\pi \coth \pi + 1)/2$, so that after 13 cycles 96% is extracted and after only 5 cycles 80% is extracted.)

8 The electromotive force $e(t)$ of period $2\pi/3$ supplied by a half-wave rectifier is believed to be represented by the series

$$e(t) = \frac{E}{\pi} + \frac{E}{2} \sin \omega t - \frac{2E}{\pi} \sum_{n=1}^{\infty} \frac{1}{4n^2 - 1} \cos 2n\omega t$$

By considering the first N terms and using the triangle inequality $|x + y| \leqslant |x| + |y|$ repeatedly, or otherwise, show that for all t

$$|e(t)| \leqslant 2E(1 + \pi/4)/\pi$$

9 When the power is shut down from a vertical power pounder the piston continues for a time to strike. After the nth stroke the time taken until the next stroke is $p^n u/f$ and the height of the recoil is $(p^n u)^2/f$. The dimensionless quantity p, the speed u and the acceleration f are constant, and $0 < p < 1$. If distances and times are measured in metres and seconds respectively, and if after the first stroke the time taken for the pounder to come to rest is less than t seconds, show that $p < (1 + u/ft)^{-1}$. Show also that if $u = 1, f = 1$ and $t = 1$ and $p = 1/4$ then the total distance travelled by the piston after the first stroke is about 13.3 cm.

10 Sand is being eroded from a dune due to high winds. It is found that, due to conservation measures being enforced, each year the loss is only 50% of the loss in the previous year. The present volume of sand in the dune is V and the volume lost during the present year is v.

a Obtain an expression for the volume of sand in the dune after n further years.

b Obtain a relationship between v and V if, despite conservation measures, the dune will eventually become eroded completely.

c An expert has predicted that at the present rate of erosion the dune will reduce in size by 25% after 10 years. Obtain the relationship between v and V if this is so.

11 An extractor is designed to remove 40% of the humidity in a room every hour. The device is subjected to a test and is fitted to a sealed room which has air at 20% humidity level initially. Decide whether or not, on the basis of the manufacturer's claim, the humidity should be able to be reduced to an arbitrary low level.

The water which is extracted from the room is piped into a tank. If after the first hour a volume v cc of water is extracted determine the minimum volume of the tank if, on the basis of the design specification, it is to cope with all the water extracted from the room.

After 20 hours the humidity in the room is found to be 0.01% Estimate the constant K if in fact K% of the humidity has been extracted each hour and thereby decide whether or not the extractor meets the advertised specification.

Use the revised specification to recalculate the minimum volume of the tank which is to hold the water.

12 An under sea oil extraction process removes a fixed volume of liquid from the reservoir each year and replaces it with sea water to prevent structural collapse. It is assumed that the oil and water in the reservoir become uniformly mixed together by this process. The volume of the reservoir is V. Once the mixture has been obtained an extraction process can separate the oil from the water. In the first year a volume v of liquid was extracted and was 100% pure. Obtain an expression for the amount of oil extracted in the nth year and thereby the total amount of oil which has been extracted after n years.

Decide whether or not it would eventually be possible by this process to extract an arbitrarily large proportion of the oil from the reservoir.

10 Complex numbers

We have developed one-half of the calculus – differentiation. The other half is the reverse process, known as integration. However, before we consider that, we need to enlarge our algebraic knowledge. We have already mentioned, when we dealt with power series, that a familiarity with complex numbers would have enabled us to say more. Indeed this has not been the only occasion where the idea of a complex number has arisen. In this chapter we shall begin a short study of algebraic concepts that will lead via complex numbers, matrices and determinants to vectors. Only when we have done all this will we return to the calculus to gain the full stereophonic effect.

After working through this chapter you should be able to

- ☐ Solve equations involving complex numbers;
- ☐ Express a complex number in polar form;
- ☐ Represent sets of complex numbers as regions of the complex plane;
- ☐ Solve the equation $z^n = \alpha$ where $n \in \mathbb{N}$ and $\alpha \in \mathbb{C}$;
- ☐ Relate circular and hyperbolic functions using complex numbers.

At the end of this chapter we shall apply this work to the practical problem of an AC bridge.

10.1 GENESIS

If we consider the quadratic equation

$$ax^2 + bx + c = 0$$

where a, b and c are real numbers, $a \neq 0$, we obtain

$$x^2 + \frac{b}{a}x + \frac{c}{a} = 0$$

So, completing the square,

$$\left(x + \frac{b}{2a}\right)^2 - \frac{b^2}{4a^2} + \frac{c}{a} = 0$$

Notice how we add half the coefficient of x to complete the square. Then

$$\left(x + \frac{b}{2a}\right)^2 = \frac{b^2}{4a^2} - \frac{c}{a} = \frac{b^2 - 4ac}{4a^2}$$

If $b^2 - 4ac \geqslant 0$ then

$$x = \frac{-b \pm \sqrt{(b^2 - 4ac)}}{2a}$$

You will certainly have met this before (see Chapter 1). It is known as the formula for solving a quadratic equation, and we know that for real roots we require

$$b^2 - 4ac \geqslant 0$$

What are we to do if $b^2 - 4ac < 0$? Clearly $\sqrt{(b^2 - 4ac)}$ is not a real number, because whenever we square a real number the result is positive.

Suppose nevertheless that there is a number, which we shall represent by i, which behaves with respect to addition and multiplication exactly as if it were a real number but which has the special property that $\mathrm{i}^2 = -1$. If such a number exists then we can write

$$x = \frac{-b \pm \mathrm{i}\sqrt{(4ac - b^2)}}{2a}$$

and obtain two roots.

So if we start with the real numbers, and augment them with this new number i, the operations of addition and multiplication will generate such numbers as

$$(2 + \mathrm{i})(1 - 3\mathrm{i}) + (2 + 4\mathrm{i})^2 (\mathrm{i} - 1)(2\mathrm{i} + 1)$$

This is rather like adding an extra ingredient to a stew which is being cooked; the flavour permeates through.

Using the rules of elementary algebra, and the special property of i, namely $\mathrm{i}^2 = -1$, any number which we generate can be reduced to the form

$$a + \mathrm{i}b$$

where a and b are real numbers.

We define the set of **complex numbers** $\mathbb{C}$ to be those numbers which can be expressed in the form $a + ib$, where a and b are real numbers:

$$\mathbb{C} = \{a + ib \mid a \in \mathbb{R}, b \in \mathbb{R}\}$$

a is called the **real** part of $a + ib$, and b is called the **imaginary** part of $a + ib$. These are rather unsatisfactory names because each of them is in fact a real number!

If $b = 0$ then $a + ib$ is a real number, whereas if $a = 0$ then $a + ib = ib$. A number of the form ib where b is real is often called a **pure imaginary** number.

The complex numbers, with the usual operations of addition and multiplication, form a mathematical structure known as a **field**.

In the field of complex numbers, any quadratic equation

$$az^2 + bz + c = 0 \qquad a \neq 0$$

always has two roots. Of course if $b^2 = 4ac$ then the two roots are equal.

NOTATION

There are unwritten conventions about the use of letters to represent mathematical objects. These conventions are often broken, but here are some broad guidelines:

1 a, b, c and d are used for constants.
2 e is reserved for the natural base of logarithms.
3 f and g are used for functions.
4 i and j are reserved for complex numbers.
5 h, k, l, m, n, p, q, r, s and t are used for constants or variables.
6 u, v, w, x, y and z are used for functions or variables.

It is often convenient to use a single letter to represent a complex number, and so that no confusion can arise it is customary to reserve z and w for this purpose. Other letters such as α and β can be used provided it is clear that the number is not real but complex.

So if $z = a + ib$ then

1 a is the real part of z; we write $a = \text{Re}(z)$.
2 b is the imaginary part of z; we write $b = \text{Im}(z)$.

Some books use $\mathscr{R}(z)$ and $\mathscr{I}(z)$ instead of $\text{Re}(z)$ and $\text{Im}(z)$ respectively.

EQUATING REAL AND IMAGINARY PARTS

□ Show that if $a + ib = c + id$, where a, b, c and d are real, then $a = c$ and $b = d$.

Suppose $a + ib = c + id$. Then $a - c = -i(b - d)$, so that squaring

$$(a - c)^2 = i^2(b - d)^2 = -(b - d)^2$$

Therefore

$$(a - c)^2 + (b - d)^2 = 0$$

Now $(a - c)^2$ is a positive real number and so too is $(b - d)^2$, and the sum of these two positive numbers is zero. It therefore follows that each of these real numbers must be zero. Therefore $a - c = 0$ and $b - d = 0$, so $a = c$ and $b = d$.

Of course we know that the converse is always true. That is, if $a = c$ and $b = d$ then $a + ib = c + id$. ■

This example has important consequences. It means that, given an equation involving complex numbers, we can equate the real parts and the imaginary parts and thereby obtain two real equations from one complex equation.

☐ Obtain x and y in terms of a and b if

$$\frac{1}{x + iy} + \frac{1}{a + ib} = 1$$

We obtain

$$\frac{1}{x + iy} = 1 - \frac{1}{a + ib} = \frac{a + ib - 1}{a + ib}$$

$$x + iy = \frac{a + ib}{a + ib - 1}$$

$$= \frac{a + ib}{(a - 1) + ib}$$

Now if we multiply numerator and denominator by $(a - 1) - ib$ the denominator will become a real number:

$$x + iy = \frac{a + ib}{(a - 1) + ib}\,\frac{(a - 1) - ib}{(a - 1) - ib}$$

$$= \frac{(a + ib)(a - 1 - ib)}{(a - 1)^2 + b^2}$$

$$= \frac{a(a - 1) + b^2 + i[b(a - 1) - ab]}{(a - 1)^2 + b^2}$$

$$= \frac{a(a - 1) + b^2 - ib}{(a - 1)^2 + b^2}$$

So that, equating real and imaginary parts,

$$x = \frac{a(a - 1) + b^2}{(a - 1)^2 + b^2} \quad \text{and} \quad y = \frac{-b}{(a - 1)^2 + b^2}$$ ■

This example shows that an equation involving complex numbers produces two equations involving real numbers, and that to obtain these equations we can *equate real and imaginary parts*.

10.2 THE COMPLEX PLANE: ARGAND DIAGRAM

We can obtain a geometrical representation for complex numbers by using the conventions of coordinate geometry (Chapter 3).

To each complex number $a + ib$ there is a unique point (a, b) in the plane Oxy. Conversely, given any point (a, b) in the plane Oxy, there is a unique complex number $a + ib$. There is therefore a one-to-one correspondence between the points in the plane Oxy and the complex numbers (Fig. 10.1).

When the plane is used in this way it is often called an **Argand diagram** or the **complex plane**. The x-axis is then called the **real axis** and the y-axis is called the **imaginary axis**.

POLAR FORM

Of course we know that a point in the plane can be expressed in polar coordinates rather than in cartesian coordinates (Fig. 10.2). We obtain from elementary trigonometry

$$a = r \cos \theta, \qquad b = r \sin \theta$$

so that

$$\begin{aligned} a + ib &= r \cos \theta + ir \sin \theta \\ &= r(\cos \theta + i \sin \theta) \end{aligned}$$

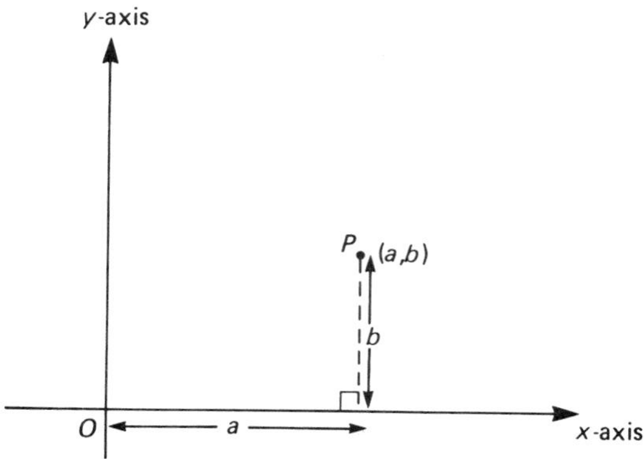

Fig. 10.1 Cartesian coordinates.

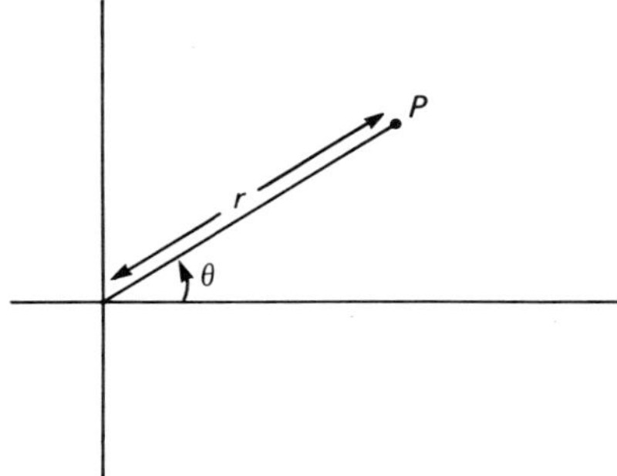

Fig. 10.2 Polar coordinates.

When a complex number is expressed in this way, we say it is expressed in **polar form**.

The easiest way to express a complex number in polar form is to put the complex number on the Argand diagram using the correspondence $a + ib \rightarrow (a, b)$, and then to read off the distance $r = \sqrt{(a^2 + b^2)}$ and the angle θ.

MODULUS AND ARGUMENT

The usual convention for polar coordinates is to take $r > 0$ and $0 \leqslant \theta < 2\pi$, so that a *unique* representation is obtained for every point other than the origin.

In the complex plane, the convention is slightly different. Here we take $r > 0$, as before, but $-\pi < \theta \leqslant \pi$. r is known as the **modulus** of the complex number: $r = \sqrt{(a^2 + b^2)}$. θ is known as the **argument** of the complex number: $\theta = \tan^{-1}(y/x)$ when $\theta \in (-\pi/2, \pi/2)$.

It follows that any non-zero complex number can be represented uniquely by the modulus r and the argument θ. The notation $r\angle\theta$ is often used to denote these essential ingredients. When a complex number is expressed in polar form where θ is the argument of the complex number, so that $-\pi < \theta \leqslant \pi$, it is said to be in **modulus-argument form**.

□ Express the complex number $1 - 2i$ in the form $r\angle\theta$.

We begin by representing the complex number by a point on the Argand diagram (Fig. 10.3): $1 - 2i \rightarrow (1, -2)$. We calculate the modulus straight away using Pythagoras's theorem:

$$r^2 = a^2 + b^2 = 1^2 + (-2)^2 = 5$$

and therefore the modulus is $\sqrt{5}$. The argument can be read from the diagram using a little trigonometry. The acute angle α is given by $\tan^{-1} 2 \simeq 63°26'$ or $63.44°$, so that $r = \sqrt{5}$ and $\theta = -\tan^{-1} 2$. ■

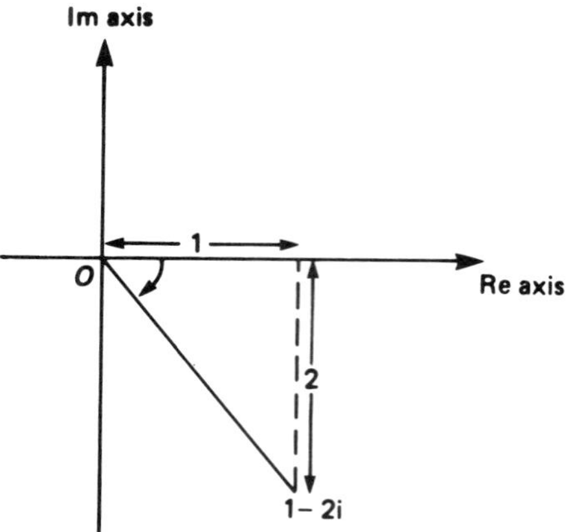

Fig. 10.3 Cartesian representation of 1 − 2i.

Note that we can express the argument in degrees if we wish, but we must indicate clearly that we have done so. In many ways it is best to get used to the natural measure of angle, the so-called **radian** (π radians = 180 degrees). For instance, in the series expansion for the circular function

$$\cos x = 1 - \tfrac{1}{2}x^2 + \ldots$$

x is the natural measure of angle, and it would be incorrect to attempt to use degrees.

COMPLEX CONJUGATE

If $z = a + ib$, we denote the modulus of z by $|z|$ and the argument of z by $\arg z$.

Another useful concept, which we have already used implicitly, is known as the **complex conjugate** of z and is denoted by $\bar{z}$. If $z = a + ib$ then we define $\bar{z} = a - ib$. It follows that

$$\begin{aligned} z\bar{z} &= (a + ib)(a - ib) = a^2 - (ib)^2 \\ &= a^2 + b^2 = |z|^2 \end{aligned}$$

which is a real number.

We can use this to reduce any rational expression involving complex numbers to the form $a + ib$, where a and b are real numbers. To achieve this we render the denominator real by multiplying numerator and denominator by the complex conjugate of the denominator. This is precisely what we did in a previous example. Here is another example to make the idea crystal clear.

☐ Express the following complex number in the standard cartesian form $a + ib$, where a and b are real numbers:

$$\frac{(2 + i)^3}{(3 + 4i)^3}$$

We can simplify the numerator and denominator separately

$$\frac{(2 + i)(4 + 4i + i^2)}{(3 + 4i)(9 + 24i + 16i^2)} = \frac{(2 + i)(3 + 4i)}{(3 + 4i)(9 + 24i - 16)}$$
$$= \frac{2 + i}{(-7 + 24i)}$$

We now multiply numerator and denominator by the conjugate of the denominator, since we know this will reduce the denominator to a real number:

$$\frac{2 + i}{-7 + 24i} = \frac{(2 + i)(-7 - 24i)}{(-7)^2 + (24)^2}$$
$$= \frac{-14 + 24 - 55i}{49 + 576} = \frac{10 - 55i}{625}$$
$$= \frac{2 - 11i}{125} = \frac{2}{125} + i\frac{-11}{125}$$

■

10.3 VECTORIAL REPRESENTATION

Another related geometrical method for representing complex numbers is to regard them as directed line segments emanating from the origin. More precisely, if $z = a + ib$ corresponds to P, the point (a, b) in the complex plane, then we represent z by the line segment OP (Fig. 10.4). It is easy to show that, with this representation, when two complex numbers are added together their sum is obtained by adding the corresponding line segments according to the parallelogram law.

To see this, suppose P and Q represent the complex numbers $a + ib$ and $c + id$ respectively. Then the sum of the complex numbers is

$$(a + ib) + (c + id) = (a + c) + i(b + d)$$

We show that this is represented by the point R, where R is obtained from P and Q by completing the parallelogram $POQR$.

If we complete the parallelogram $POQR$ as shown in Fig. 10.4 we have

$$OA = a, \qquad AP = b, \qquad OC = c, \qquad CQ = d$$

Moreover $OA = CD$, $AP = DB$ and $CQ = BR$ using parallels. So

$$OD = OC + CD = OC + OA = a + c$$
$$DR = DB + BR = AP + CQ = b + d$$

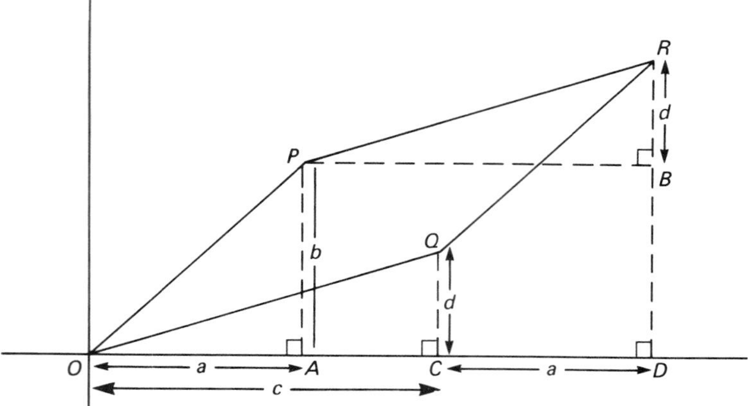

Fig. 10.4 Cartesian representation of $(a + ib) + (c + id)$.

This is the property we wished to show.

If the points P and Q represent the complex numbers z and α (Fig. 10.5) then since

$$z = (z - \alpha) + \alpha$$

the vector representing $z - \alpha$ is equal in length and parallel to PQ. This observation gives us a geometrical interpretation for $|z - \alpha|$ and $\arg(z - \alpha)$ which we shall find useful when describing sets of points.

We shall refer to the 'point z' or the 'vector z' rather than the more correct but awkward 'point representing the complex number z' and 'vector representing the complex number z' respectively.

We can carry out operations involving multiplication geometrically in the Argand diagram if we observe the following properties:

1 $|zw| = |z|\,|w|$

2 $\left|\dfrac{z}{w}\right| = \dfrac{|z|}{|w|}$ provided $w \neq 0$

3 $\arg(zw) = \arg z + \arg w$

4 $\arg\left(\dfrac{z}{w}\right) = \arg z - \arg w$ provided $w \neq 0$

To be strict, it may be necessary to add or subtract 2π to bring the argument in properties **3** and **4** within the range $-\pi < \theta \leqslant \pi$, that is the interval $(-\pi, \pi]$. However, if we add or subtract 2π from the polar angle of a complex number it has no geometrical effect in the complex plane. Therefore, if we are concerned solely with the geometrical effects of these operations, these algebraic details are irrelevant.

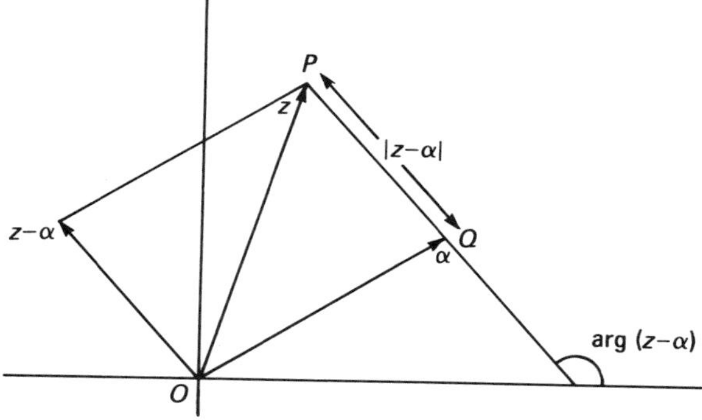

Fig. 10.5 Geometrical representation of $|z - \alpha|$.

To justify these properties, suppose $z = r(\cos\theta + \mathrm{i}\sin\theta)$ and $w = s(\cos\phi + \mathrm{i}\sin\phi)$. Then

$$\begin{aligned} zw &= rs(\cos\theta + \mathrm{i}\sin\theta)(\cos\phi + \mathrm{i}\sin\phi) \\ &= rs[\cos\theta\cos\phi - \sin\theta\sin\phi + \mathrm{i}(\cos\theta\sin\phi + \cos\phi\sin\theta)] \\ &= rs[\cos(\theta + \phi) + \mathrm{i}\sin(\theta + \phi)] \end{aligned}$$

so that properties **1** and **3** follow:

$$|zw| = rs = |z|\,|w|$$

$$\begin{aligned} \arg(zw) &= \theta + \phi \qquad (\text{mod } 2\pi) \\ &= \arg z + \arg w \qquad (\text{mod } 2\pi) \end{aligned}$$

The expression mod 2π indicates that it may be necessary to add or subtract multiples of 2π to bring the argument within range.

We can readily deduce properties **2** and **4**. First, using property **1**,

$$|z| = \left|\frac{z}{w}w\right| = \left|\frac{z}{w}\right|\,|w|$$

Then dividing through by $|w|$ gives property **2**. Next, using property **3**,

$$\begin{aligned} \arg z &= \arg\left(\frac{z}{w}w\right) \\ &= \arg\left(\frac{z}{w}\right) + \arg w \qquad (\text{mod } 2\pi) \end{aligned}$$

Then subtracting arg w from each side gives property **4**.

We have shown that

1 When two complex numbers are multiplied the moduli are multiplied and the arguments are added;

2 When two complex numbers are divided the moduli are divided and the arguments are subtracted.

□ What is the effect in the complex plane of multiplying a complex number by i?

When we put i in polar form we obtain

$$i = 1(\cos \pi/2 + i \sin \pi/2)$$

so that the modulus is 1 and the argument is $\pi/2$. Suppose z is any complex number with modulus r and argument θ. Then iz is a complex number with modulus r and argument $\theta + \pi/2$. Therefore geometrically the effect is to **rotate** the vector representing z anticlockwise through $\pi/2$. ■

□ A complex number satisfies the equation

$$|z - i| = |z + i|$$

Determine the locus of the point which represents z in the Argand diagram.

We shall discuss two ways of solving this problem.

Geometrical method In Fig. 10.6, $|z - i|$ is the distance between the point representing z and the point representing i. Likewise $|z + i|$ is the distance between the point representing z and the point representing $-i$. The equation tells us that these two distances are equal, and since this is the only constraint on z it follows that z lies on the perpendicular bisector of the line joining $-i$ to i. This is the real axis.

Algebraic method Put $z = x + iy$ and examine what can be deduced from the equation $|z - i| = |z + i|$. We obtain

$$|(x + iy) - i| = |(x + iy) + i|$$
$$|(x + iy) - i|^2 = |(x + iy) + i|^2$$

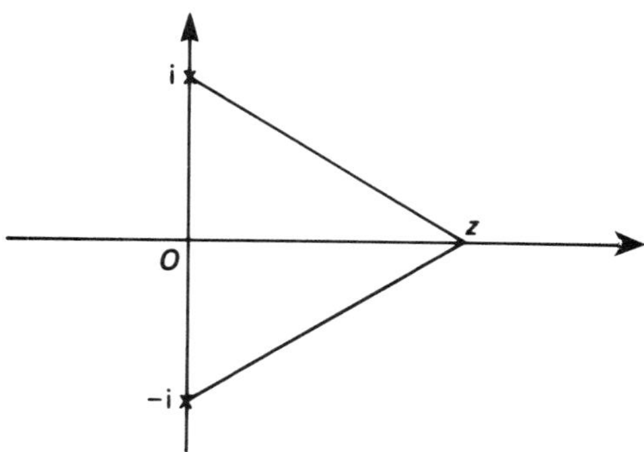

Fig. 10.6 Locus of z if $|z - i| = |z + i|$.

$$
\begin{aligned}
|x + i(y - 1)|^2 &= |x + i(y + 1)|^2 \\
x^2 + (y - 1)^2 &= x^2 + (y + 1)^2 \\
-2y &= 2y \\
y &= 0
\end{aligned}
$$

Therefore z lies on the real axis, as we deduced before. ■

In general there are two methods available for solving locus problems: the geometrical method and the algebraic method.

10.4 FURTHER PROPERTIES OF THE CONJUGATE

We have already seen that

$$z\bar{z} = |z|^2$$

and we have at once

$$
\begin{aligned}
z + \bar{z} &= (a + ib) + (a - ib) = 2a = 2\,\mathrm{Re}(z) \\
z - \bar{z} &= (a + ib) - (a - ib) = 2ib = 2i\,\mathrm{Im}(z)
\end{aligned}
$$

□ Show that the conjugate of a product is the product of the conjugates.

We have shown that if $z = r(\cos\theta + i\sin\theta)$ and $w = s(\cos\phi + i\sin\phi)$, then

$$zw = rs[\cos(\theta + \phi) + i\sin(\theta + \phi)]$$

Therefore

$$\overline{zw} = rs[\cos(\theta + \phi) - i\sin(\theta + \phi)]$$

Now

$$
\begin{aligned}
\bar{z} &= r(\cos\theta - i\sin\theta) \\
&= r[\cos(-\theta) + i\sin(-\theta)]
\end{aligned}
$$

and

$$
\begin{aligned}
\bar{w} &= s(\cos\phi - i\sin\phi) \\
&= s[\cos(-\phi) + i\sin(-\phi)]
\end{aligned}
$$

So

$$
\begin{aligned}
\bar{z}\bar{w} &= rs\,[\cos(-\theta - \phi) + i\sin(-\theta - \phi)] \\
&= rs[\cos(\theta + \phi) - i\sin(\theta + \phi)] = \overline{zw}
\end{aligned}
$$

as required. ■

You will recall that, with one exception, we have defined a^n where n is any integer and a is any real number (Chapter 1). The exception is $a = 0$, for we do not define 0^0. The reason for this omission is that whatever definition we were to choose we should violate the laws of indices, and we wish to preserve these at all costs.

With the exception that we do not define 0^0 we now define, in the obvious way, z^n where n is any integer and z is any complex number. We define

$$z^0 = 1 \quad \text{provided } z \neq 0$$
$$z^{n+1} = zz^n \qquad (n \in \mathbb{N})$$

Therefore z^n, when $n \in \mathbb{N}$, is a product of z with itself n times. Finally, we define

$$z^{-n} = 1/z^n \quad \text{when } n \in \mathbb{N},\ z \neq 0$$

10.5 DE MOIVRE'S THEOREM

We have already seen how to express a complex number z in polar form. One of the advantages of doing so is that it is then possible to calculate z^n very easily. This is a consequence of **De Moivre's theorem**, which says that if n is any integer

$$(\cos\theta + \mathrm{i}\sin\theta)^n = \cos n\theta + \mathrm{i}\sin n\theta$$

We shall accept this without proof. The usual method of proof is to prove it first for natural numbers and then to extend the proof to all integers. If you are familiar with the method of proof known as mathematical induction (Chapter 1) you should have no difficulty in supplying the details.

We should be very wary of trying to use De Moivre's theorem for other values of n. For example, we have defined (section 1.5) $1^{1/2}$ to be the positive real root of the equation $x^2 = 1$, and so $1^{1/2} = 1$. However, if

$$(\cos\theta + \mathrm{i}\sin\theta)^r = \cos r\theta + \mathrm{i}\sin r\theta$$

were to hold for *all* real numbers r then

$$1 = 1^{1/2} = (\cos 2\pi + \mathrm{i}\sin 2\pi)^{1/2} = \cos\pi + \mathrm{i}\sin\pi = -1$$

You may find books that claim that for all real numbers r

$$(\cos\theta + \mathrm{i}\sin\theta)^r = \cos r\theta + \mathrm{i}\sin r\theta$$

and some which purport to prove it! The best that can be said of this is that the expression on the right is *one* of the values of the expression on the left, where the expression on the left may have been somewhat loosely defined.

If $z \in \mathbb{C}$, we shall not need to define z^n except when n is an integer.

There are several uses for De Moivre's theorem.

□ Obtain $(1 + i)^{28}$.

We could expand by the binomial theorem, but this would be no easy task. Instead we begin by putting $z = 1 + i$ into polar form. If we put the point $(1, 1)$ on the Argand diagram we can read off the modulus and the argument straight away. We see that $r = \sqrt{2}$ and $\theta = \pi/4$. Therefore

$$z = \sqrt{2}(\cos \pi/4 + i \sin \pi/4)$$

So

$$\begin{aligned} z^{28} &= (\sqrt{2})^{28}(\cos \pi/4 + i \sin \pi/4)^{28} \\ &= 2^{14}(\cos 7\pi + i \sin 7\pi) \end{aligned}$$

using De Moivre's theorem. Therefore

$$z^{28} = 2^{14}(\cos \pi + i \sin \pi) = -2^{14}$$

■

Here we can see that De Moivre's theorem has helped us considerably. We can also use De Moivre's theorem to deduce trigonometrical identities.

□ Use De Moivre's theorem to deduce identities for $\sin 3\theta$ and $\cos 3\theta$ in terms of $\sin \theta$ and $\cos \theta$ respectively.

It helps to use a shorthand notation. We write $c = \cos \theta$ and $s = \sin \theta$, so that

$$c + is = \cos \theta + i \sin \theta$$

Now by De Moivre's theorem

$$\begin{aligned} \cos 3\theta + i \sin 3\theta &= (c + is)^3 \\ &= c^3 + 3c^2(is) + 3c(is)^2 + (is)^3 \\ &= c^3 + 3ic^2s - 3cs^2 - is^3 \end{aligned}$$

using $i^2 = -1$. So equating real and imaginary parts we have

$$\begin{aligned} \cos 3\theta &= c^3 - 3cs^2 \\ \sin 3\theta &= 3c^2s - s^3 \end{aligned}$$

Now $c^2 + s^2 = 1$, and so

$$\begin{aligned} \cos 3\theta &= c^3 - 3c(1 - c^2) \\ &= 4c^3 - 3c \\ &= 4 \cos^3 \theta - 3 \cos \theta \\ \sin 3\theta &= 3(1 - s^2)s - s^3 \\ &= 3s - 4s^3 \\ &= 3 \sin \theta - 4 \sin^3 \theta \end{aligned}$$

■

□ Simplify the expression

$$\frac{(\cos 3\theta + i \sin 3\theta)^6 (\cos 2\theta - i \sin 2\theta)^7}{(\sin 5\theta + i \cos 5\theta)^6 (\cos \theta - i \sin \theta)^8}$$

We begin by remarking that because

$$(\cos\theta + i\sin\theta)^n = \cos n\theta + i\sin n\theta$$

for any integer n, it follows that

$$(\cos\theta - i\sin\theta)^n = \cos n\theta - i\sin n\theta$$

There are many ways of seeing this. One way is to take the complex conjugate of each side of the equation using the property that the conjugate of a product is the product of the conjugates. Also

$$\begin{aligned}\sin\theta + i\cos\theta &= -i^2\sin\theta + i\cos\theta\\ &= i(\cos\theta - i\sin\theta)\end{aligned}$$

Using De Moivre's theorem we now have

$$\begin{aligned}&\frac{(\cos 3\theta + i\sin 3\theta)^6(\cos 2\theta - i\sin 2\theta)^7}{(\sin 5\theta + i\cos 5\theta)^6(\cos\theta - i\sin\theta)^8}\\ &\quad= \frac{(\cos\theta + i\sin\theta)^{18}(\cos\theta - i\sin\theta)^{14}}{i^6(\cos 5\theta - i\sin 5\theta)^6(\cos\theta - i\sin\theta)^8}\end{aligned}$$

Now $i^2 = -1$ and so $i^6 = (-1)^3 = -1$. Therefore the expression becomes

$$\begin{aligned}&= \frac{(\cos\theta + i\sin\theta)^{18}(\cos\theta - i\sin\theta)^{14}}{-(\cos\theta - i\sin\theta)^{30}(\cos\theta - i\sin\theta)^8}\\ &= \frac{-(\cos\theta + i\sin\theta)^{18}}{(\cos\theta - i\sin\theta)^{24}}\\ &= -(\cos\theta + i\sin\theta)^{42}\\ &= -(\cos 42\theta + i\sin 42\theta)\end{aligned}$$

■

It is now time for you to take some steps.

10.6 Workshop

1

Exercise Express the complex number

$$\frac{(\sqrt{3} + i)^2}{(i - \sqrt{3})^3}$$

in polar form.

Try this carefully before you take the next step.

2

There are essentially two ways of proceeding. One method is to multiply everything out, simplify it down to obtain the cartesian form, and then produce the polar form. This is routine but long.

The better alternative is to put the complex numbers which appear in the expression into polar form and use De Moivre's theorem to simplify it. Thus

$$\sqrt{3} + i = 2(\cos\frac{\pi}{6} + i\sin\frac{\pi}{6})$$

$$i - \sqrt{3} = -\sqrt{3} + i = 2\left(\cos\frac{5\pi}{6} + i\sin\frac{5\pi}{6}\right)$$

Then

$$\begin{aligned}
\frac{(\sqrt{3}+i)^2}{(i-\sqrt{3})^3} &= \frac{[2(\cos\pi/6 + i\sin\pi/6)]^2}{[2(\cos 5\pi/6 + i\sin 5\pi/6)]^3} \\
&= \frac{4(\cos\pi/3 + i\sin\pi/3)}{8(\cos 5\pi/2 + i\sin 5\pi/2)} \\
&= \frac{1}{2}\left(\cos\left[\frac{1}{3} - \frac{5}{2}\right]\pi + i\sin\left[\frac{1}{3} - \frac{5}{2}\right]\pi\right) \\
&= \frac{1}{2}\left(\cos\left[\frac{2-15}{6}\right]\pi + i\sin\left[\frac{2-15}{6}\right]\pi\right) \\
&= \frac{1}{2}\left(\cos\left[-\frac{13\pi}{6}\right] + i\sin\left[-\frac{13\pi}{6}\right]\right) \\
&= \frac{1}{2}\left(\cos\left[-\frac{\pi}{6}\right] + i\sin\left[-\frac{\pi}{6}\right]\right)
\end{aligned}$$

So $r = 1/2$ and $\theta = -\pi/6$ and the required polar form is

$$\frac{1}{2}\left(\cos\left[-\frac{\pi}{6}\right] + i\sin\left[-\frac{\pi}{6}\right]\right)$$

You can if you prefer express the complex number in the modulus-argument form as $(1/2)\angle(-\pi/6)$.

If you managed that, except for possibly a numerical slip, then proceed at full speed to step 4. Otherwise, try this exercise.

▷**Exercise** Express the complex number

$$\frac{(1 + 4i)(2 - 3i)}{(i + 6)(1 + 3i)}$$

in the cartesian form $a + ib$.

There is no need for polar form here.

All we need to do is to rationalize the expression by multiplying numerator and denominator by the conjugate of the denominator. You can do **3** this before you multiply out or afterwards; it's up to you to choose. So, multiplying out first, we have

$$\begin{aligned}\frac{(1+4\mathrm{i})(2-3\mathrm{i})}{(\mathrm{i}+6)(1+3\mathrm{i})} &= \frac{2+8\mathrm{i}-3\mathrm{i}-12\mathrm{i}^2}{\mathrm{i}+6+3\mathrm{i}^2+18\mathrm{i}} \\ &= \frac{2+5\mathrm{i}+12}{19\mathrm{i}+6-3} = \frac{14+5\mathrm{i}}{3+19\mathrm{i}}\,\frac{3-19\mathrm{i}}{3-19\mathrm{i}} \\ &= \frac{42+15\mathrm{i}-266\mathrm{i}+95}{9+361} = \frac{137-251\mathrm{i}}{370}\end{aligned}$$

So the number is

$$\frac{137}{370} - \mathrm{i}\frac{251}{370}$$

Now step ahead.

4

Exercise Describe the following set of points in the complex plane:

$$\{z : z \in \mathbb{C},\ \arg(z-1) < \pi/2\}$$

It is best to use a geometric method here because the algebraic method will involve you in work with inequalities which you may find too difficult.

5

Suppose we take a general point P in the set (Fig. 10.7). We know that if we join P to the point A, representing the complex number 1, and $\theta = \angle XAP$, then $\theta < \pi/2$. This follows because $\arg(z-1) = \angle XAP$.

Any point in the lower half of the complex plane satisfies this condition,

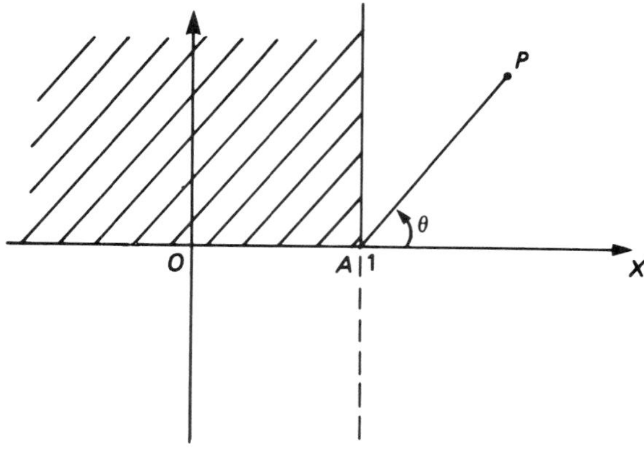

Fig. 10.7 $\{z : z \in \mathbb{C},\ \arg(z-1) < \pi/2\}$ (unshaded region).

and so too does any point in the upper plane to the right of the line defined by $\operatorname{Re}(z) = 1$. We see therefore that we must exclude all points corresponding to complex numbers which have their real part less than or equal to 1 and their imaginary part greater than or equal to 0. This region is shown in the diagram.

If you managed to get that right then move on to the next section. Otherwise, try one more exercise.

▷**Exercise** Describe the locus of the point z which moves in the complex plane in such a way that

$$|z - \mathrm{i}| = 2|z - 1|$$

Only when you have tried this should you move on.

6

We can use either the geometrical method or the algebraic method. For the geometrical method you need to know that the locus of a point which moves so that the ratio of its distances from two fixed points is a constant is a circle. If you are not aware of this fact you might like to prove it. The only exception is when the ratio is 1, in which case the locus is a straight line. Once you know the locus is a circle, the centre and radius can be deduced from a diagram.

The algebraic method is more straightforward in this instance. Let $z = x + \mathrm{i}y$. Then

$$|z - \mathrm{i}| = 2|z - 1|$$

So

$$\begin{aligned}
|(x + \mathrm{i}y) - \mathrm{i}| &= 2|(x + \mathrm{i}y) - 1| \\
|x + \mathrm{i}(y - 1)| &= 2|(x - 1) + \mathrm{i}y| \\
|x + \mathrm{i}(y - 1)|^2 &= 4|(x - 1) + \mathrm{i}y|^2 \\
x^2 + (y - 1)^2 &= 4[(x - 1)^2 + y^2] \\
x^2 + y^2 - 2y + 1 &= 4(x^2 - 2x + 1 + y^2) \\
3x^2 + 3y^2 - 8x + 2y + 3 &= 0 \\
x^2 + y^2 - 8x/3 + 2y/3 + 1 &= 0
\end{aligned}$$

This is the equation of a circle with centre $(4/3, -1/3)$ and radius $2\sqrt{2}/3$ (see Chapter 3).

10.7 THE nTH ROOTS OF A COMPLEX NUMBER

We are now in a position to solve the equation $z^n = \alpha$ where α is any complex number and $n \in \mathbb{N}$. The solutions of this equation are called the nth roots of α.

Suppose we have a polynomial in x

$$f(x) = c_n x^n + c_{n-1} x^{n-1} + \ldots + c_1 x + c_0 \qquad c_n \neq 0$$

If the coefficients

$$c_n, c_{n-1}, \ldots, c_1, c_0$$

are real numbers then we know there are at most n real roots of the equation $f(x) = 0$.

We began this chapter by looking at the quadratic equation and noticing that on occasion we did not have two real roots. This motivated the extension of the number system to complex numbers. We did this to ensure that every quadratic equation had two roots. It would not be surprising if when we turned our attention to polynomials of higher degree that further extensions of the number system would be required. However, it is a quite remarkable fact that when we allow complex numbers into the picture then the polynomial equation

$$f(z) = c_n z^n + c_{n-1} z^{n-1} + \ldots + c_1 z + c_0 = 0$$

always has n roots. Some of the roots may be repeated, but there are always n in total.

Unfortunately in general it is not possible to obtain formulas for solving polynomial equations of degree higher than 4. However, we can solve the equation $z^n = \alpha$ by using De Moivre's theorem, and this we now do.

We begin by expressing the number α in polar form:

$$\alpha = r(\cos\theta + \mathrm{i}\sin\theta)$$

We observe first that if

$$z_0 = r^{1/n}[\cos(\theta/n) + \mathrm{i}\sin(\theta/n)]$$

then

$$\begin{aligned} z_0^n &= (r^{1/n})^n[\cos(\theta/n) + \mathrm{i}\sin(\theta/n)]^n \\ &= r[\cos n(\theta/n) + \mathrm{i}\sin n(\theta/n)] \end{aligned}$$

by De Moivre's theorem. So

$$z_0^n = r[\cos\theta + \mathrm{i}\sin\theta] = \alpha$$

Therefore z_0 is one of the roots of the equation $z^n = \alpha$.

However, we can write α in the form

$$\alpha = r[\cos(\theta + 2k\pi) + \mathrm{i}\sin(\theta + 2k\pi)]$$

where k is any integer. Therefore by the same token, if we put

$$z_k = r^{1/n}[\cos(\theta + 2k\pi)/n + \mathrm{i}\sin(\theta + 2k\pi)/n]$$

it follows that z_k is one of the roots of the equation $z^n = \alpha$. Now this is true for *every* integer k, and so at first sight it might look as if we have an infinite number of solutions. However, if we allow k to take on $n + 1$ successive integer values the last one will be a repeat of the first. That is,

$$\begin{aligned} z_n &= r^{1/n}[\cos(\theta + 2n\pi)/n + \mathrm{i}\sin(\theta + 2n\pi)/n] \\ &= r^{1/n}[\cos(\theta/n + 2\pi) + \mathrm{i}\sin(\theta/n + 2\pi)] \\ &= r^{1/n}[\cos(\theta/n) + \mathrm{i}\sin(\theta/n)] = z_0 \end{aligned}$$

Therefore we obtain exactly n roots.

To sum up, the method for obtaining the nth roots of a complex number α is as follows:

1 Put the complex number α into polar form

$$\alpha = r(\cos\theta + \mathrm{i}\sin\theta)$$

2 Write α in the form

$$\alpha = r[\cos(\theta + 2k\pi) + \mathrm{i}\sin(\theta + 2k\pi)]$$

where k is an arbitrary integer.

3 By De Moivre's theorem one of the roots of the equation $z^n = \alpha$ is

$$z_k = r^{1/n}[\cos(\theta + 2k\pi)/n + \mathrm{i}\sin(\theta + 2k\pi)/n]$$

for every integer k.

4 Allow k to take on n successive integer values to determine the nth roots.

If we think in geometrical terms we see that each of the roots has the same modulus $r^{1/n}$ and the arguments increase by $2\pi/n$. This means that they are equally spaced around a circle centred at the origin (Fig. 10.8). This observation gives a geometrical method for obtaining the roots once the first one is known. Moreover, De Moivre's theorem, gives z_0 straight away:

$$z_0 = r^{1/n}[\cos(\theta/n) + \mathrm{i}\sin(\theta/n)]$$

□ Obtain the fifth roots of i.

We follow the method described. First, if we imagine i in the Argand diagram we see that $r = 1$ and $\theta = \pi/2$, so that

$$\mathrm{i} = 1(\cos\pi/2 + \mathrm{i}\sin\pi/2)$$

If k is any integer we can rewrite this as

$$\mathrm{i} = 1[\cos(\pi/2 + 2k\pi) + \mathrm{i}\sin(\pi/2 + 2k\pi)]$$

Using De Moivre's theorem we have that the fifth roots are

$$\begin{aligned} z_k &= 1^{1/5}[\cos(\pi/10 + 2k\pi/5) + \mathrm{i}\sin(\pi/10 + 2k\pi/5)] \\ &= \cos(\pi/10 + 2k\pi/5) + \mathrm{i}\sin(\pi/10 + 2k\pi/5) \end{aligned}$$

Finally, we let k take five consecutive integer values, for example -2, -1, 0, 1, and 2, to obtain the five roots:

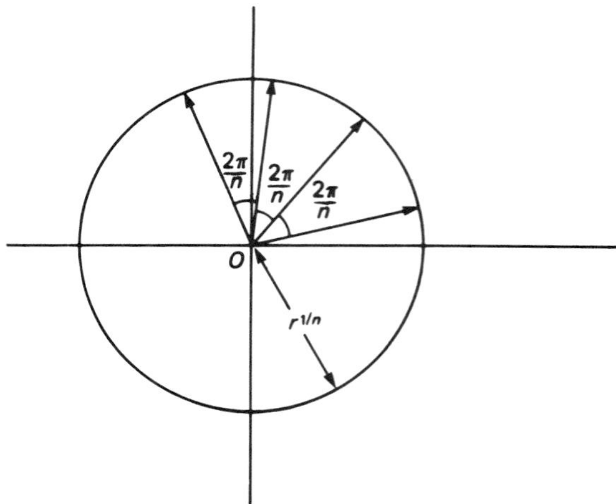

Fig. 10.8 The roots of $z^n = \alpha$.

$$\begin{aligned}
z_{-2} &= \cos(\pi/10 - 4\pi/5) + i\sin(\pi/10 - 4\pi/5) \\
&= \cos(7\pi/10) - i\sin(7\pi/10) \\
z_{-1} &= \cos(\pi/10 - 2\pi/5) + i\sin(\pi/10 - 2\pi/5) \\
&= \cos(3\pi/10) + i\sin(3\pi/10) \\
z_0 &= \cos(\pi/10) + i\sin(\pi/10) \\
z_1 &= \cos(\pi/10 + 2\pi/5) + i\sin(\pi/10 + 2\pi/5) \\
&= \cos(\pi/2) + i\sin(\pi/2) = i \\
z_2 &= \cos(\pi/10 + 4\pi/5) + i\sin(\pi/10 + 4\pi/5) \\
&= \cos(9\pi/10) + i\sin(9\pi/10)
\end{aligned}$$

We can easily check z_1:

$$i^5 = (i^2)^2 i = (-1)^2 i = i$$

You can if you prefer use the geometric method to write down z_0 and obtain the other roots using the fact that they are equally spaced around a circle centred at the origin. ■

10.8 POWER SERIES

You will have met the power series expansions for the exponential function and the circular functions (Chapter 8):

$$e^x = 1 + x + \frac{x^2}{2!} + \frac{x^3}{3!} + \ldots$$

$$\cos x = 1 - \frac{x^2}{2!} + \frac{x^4}{4!} - \frac{x^6}{6!} + \ldots$$

$$\sin x = x - \frac{x^3}{3!} + \frac{x^5}{5!} - \frac{x^7}{7!} + \ldots$$

In fact it is possible to take these series representations as the *definitions* of the functions themselves, since the series converge for all $x \in \mathbb{R}$. We should then of course have to derive all the usual properties of the functions.

CONVERGENCE

Suppose s_n is the sum to n terms of a complex series. Then we can extend the concept of convergence to infinite complex series. We say s_n converges to s if and only if $|s_n - s| \to 0$ as $n \to \infty$.

Since $|s_n - s|$ is the distance in the complex plane between the point s_n and the point s we see that, as in the real case, the series is convergent if and only if the distance between s_n and s can be made arbitrarily small merely by choosing n sufficiently large.

We shall use the real series for e^x, $\cos x$ and $\sin x$ to extend the definitions of these functions to complex arguments by defining

$$e^z = 1 + z + \frac{z^2}{2!} + \frac{z^3}{3!} + \ldots$$

$$\cos z = 1 - \frac{z^2}{2!} + \frac{z^4}{4!} - \frac{z^6}{6!} + \ldots$$

$$\sin z = z - \frac{z^3}{3!} + \frac{z^5}{5!} - \frac{z^7}{7!} + \ldots$$

It can be shown that each of these series converges for all complex numbers z and in such a way that the algebraic identities which we have stated for these functions remain valid.

The series for $\exp z = e^z$ can be used to extend the definition of a^r, where a is a positive real number and $r \in \mathbb{R}$, to a^z where $z \in \mathbb{C}$. We define

$$a^z = e^{z \ln a}$$

Clearly a^z is defined *uniquely* by this formula, and is consistent with our previous definition in the special case when $z = r$, a real number.

We shall avoid attempting a general definition of z^w where z and w are *both* complex numbers. The reason for this is that we must either make a rather arbitrary choice for the definition or extend the definition of a function to allow more than one value to each argument. Each course of action has its own problems, and since the concept of z^w is without practical applications we shall do well to avoid it altogether.

EULER'S FORMULA

We have then

$$e^z = \sum_{r=0}^{\infty} \frac{z^r}{r!}$$

$$\cos z = \sum_{r=0}^{\infty} \frac{(-1)^r z^{2r}}{(2r)!}$$

$$\sin z = \sum_{r=0}^{\infty} \frac{(-1)^r z^{2r+1}}{(2r+1)!}$$

If we replace z by iz in the series for e^z we obtain

$$e^{iz} = \sum_{k=0}^{\infty} \frac{(iz)^k}{k!} = \sum_{r=0}^{\infty} \frac{(iz)^{2r+1}}{(2r+1)!} + \sum_{r=0}^{\infty} \frac{(iz)^{2r}}{(2r)!}$$

Here we have assumed it is permissible, without affecting the convergence, to rearrange the terms in this series to sum the odd terms first and then the even terms. In fact the exponential series is a particularly tame one, and in this case the procedure can be justified. In general, however, (1) rearranging terms (2) removing or inserting brackets and (3) differentiating or integrating the terms can disturb the convergence of a series. The message as always is clear: infinite series can behave in unexpected ways and so they must be handled with care.

Luckily here we can throw caution to the wind and proceed!

$$e^{iz} = \sum_{k=0}^{\infty} \frac{(iz)^k}{k!} = \sum_{r=0}^{\infty} \frac{(iz)^{2r}}{(2r)!} + \sum_{r=0}^{\infty} \frac{(iz)^{2r+1}}{(2r+1)!}$$

Now

$$(iz)^{2r} = (i)^{2r}z^{2r} = (i^2)^r z^{2r} = (-1)^r z^{2r}$$
$$(iz)^{2r+1} = (i)^{2r+1}z^{2r+1} = (-1)^r i z^{2r+1}$$

So that

$$e^{iz} = \sum_{r=0}^{\infty} \frac{(-1)^r z^{2r}}{(2r)!} + i\sum_{r=0}^{\infty} \frac{(-1)^r z^{2r+1}}{(2r+1)!} = \cos z + i \sin z$$

That is

$$e^{iz} = \cos z + i \sin z$$

where z is any complex number. This is an important relationship known as **Euler's formula**, and it has many consequences.

For instance if θ is real we obtain

$$e^{i\theta} = \cos \theta + i \sin \theta$$

In particular, if $\theta = \pi$ we have

$$e^{i\pi} = \cos \pi + i \sin \pi = -1$$

This quite remarkable formula relates two transcendental numbers e and π. **Transcendental numbers** are numbers which do not satisfy any polynomial equation with integer coefficients.

CIRCULAR AND HYPERBOLIC FUNCTIONS

Replacing z by $-z$ in Euler's formula gives

$$\begin{aligned} e^{-iz} &= \cos(-z) + i \sin(-z) \\ &= \cos z - i \sin z \end{aligned}$$

So that

$$\begin{aligned} e^{iz} + e^{-iz} &= 2 \cos z \\ e^{iz} - e^{-iz} &= 2i \sin z \end{aligned}$$

Equivalently

$$\cos z = \frac{e^{iz} + e^{-iz}}{2}$$

$$\sin z = \frac{e^{iz} - e^{-iz}}{2i}$$

Now if you remember the definitions of the hyperbolic functions you will notice the striking similarity between these relationships and the definitions

$$\cosh z = \frac{e^{z} + e^{-z}}{2}$$

$$\sinh z = \frac{e^{z} - e^{-z}}{2}$$

Here of course we have extended the domain and codomain of the hyperbolic functions to include all the complex numbers.

In fact there are some simple algebraic relationships between the hyperbolic and the circular functions:

1 $\cosh(iz) = \cos z$
2 $\cos(iz) = \cosh z$

3 sinh $(iz) = i \sin z$
4 sin $(iz) = i \sinh z$

These relationships are easy to derive. For instance to establish **4**:

$$
\begin{aligned}
\sin(iz) &= \frac{e^{i(iz)} - e^{-i(iz)}}{2i} \\
&= \frac{e^{-z} - e^{z}}{2i} \\
&= \frac{i^2(e^{z} - e^{-z})}{2i} \qquad \text{(using } i^2 = -1\text{)} \\
&= \frac{i(e^{z} - e^{-z})}{2} = i \sinh z
\end{aligned}
$$

Why not have a go at the others? They are all very similar.

Here is the working. You can check and see if you have chosen the best way.

$$
\begin{aligned}
\cos(iz) &= \frac{e^{i(iz)} + e^{-i(iz)}}{2} \\
&= \frac{e^{-z} + e^{z}}{2} \\
&= \frac{(e^{z} + e^{-z})}{2} = \cosh z
\end{aligned}
$$

$$
\begin{aligned}
\cosh(iz) &= \cos(i[iz]) = \cos(-z) = \cos z \\
\sinh(iz) &= -i \sin(i[iz]) = -i \sin(-z) = i \sin z
\end{aligned}
$$

As a result of the relationships between hyperbolic and circular functions it is possible to translate identities between them. For example,

$$\cos 2\theta = 1 - 2\sin^2\theta$$

is a well-known identity involving circular functions. So

$$
\begin{aligned}
\cos(2iz) &= 1 - 2\sin^2(iz) \\
\cosh 2z &= 1 - 2[i \sinh z]^2 \\
&= 1 + 2\sinh^2 z
\end{aligned}
$$

That is, we have deduced the hyperbolic identity

$$\cosh 2u = 1 + 2\sinh^2 u$$

Now for some more steps.

10.9 Workshop

1

▷ **Exercise** Obtain the sixth roots of $32\sqrt{2}\,(1 - i)$.

Don't forget the easiest way to put a complex number into polar form is to draw a diagram and read off r and θ directly.

2

We put the complex number $32\sqrt{2}\,(1 - i)$ into polar form to obtain

$$32\sqrt{2}\,(1 - i) = 64[\cos(-\pi/4) + i \sin(-\pi/4)]$$

By De Moivre's theorem one of the sixth roots is

$$\begin{aligned} z_0 &= (64)^{1/6}[\cos(-\pi/24) + i \sin(-\pi/24)] \\ &= 2 \angle(-\pi/24) \end{aligned}$$

The equal spacing property now enables us to write down all the roots:

$$\begin{aligned} z_0 &= 2 \angle(-\pi/24) \\ z_1 &= 2 \angle(-\pi/24 + 2\pi/6) = 2 \angle(7\pi/24) \\ z_2 &= 2 \angle(7\pi/24 + 2\pi/6) = 2 \angle(15\pi/24) \\ z_3 &= 2 \angle(15\pi/24 + 2\pi/6) = 2 \angle(23\pi/24) \\ z_4 &= 2 \angle(23\pi/24 + 2\pi/6) = 2 \angle(31\pi/24) = 2 \angle(-17\pi/24) \\ z_5 &= 2 \angle(-17\pi/24 + 2\pi/6) = 2 \angle(-9\pi/24) \end{aligned}$$

We can check this by

$$\begin{aligned} z_6 &= 2 \angle(-9\pi/24 + 2\pi/6) \\ &= 2 \angle(-\pi/24) = z_0 \end{aligned}$$

If that went well then leap ahead to step 4. Otherwise, try this exercise.

▷ **Exercise** Obtain the seventh roots of -1.

There is nothing new about this problem. You can use the geometrical method or the algebraic method.

3

For a change we shall use the algebraic method. We begin by putting -1 in polar form:

$$-1 = 1(\cos \pi + i \sin \pi)$$

So that for any integer k

$$\begin{aligned} -1 &= \cos(\pi + 2k\pi) + i \sin(\pi + 2k\pi) \\ &= \cos(2k + 1)\pi + i \sin(2k + 1)\pi \end{aligned}$$

Using De Moivre's theorem we deduce that for every integer k

$$\begin{aligned} z_k &= \cos(2k + 1)\pi/7 + i \sin(2k + 1)\pi/7 \\ &= \exp(2k + 1)i\pi/7 \end{aligned}$$

is a solution of the equation $z^7 = -1$.

Finally, allowing k to take on seven successive integer values provides us with all the roots:

$$\begin{aligned}
z_{-3} &= \exp(-5i\pi/7)\\
z_{-2} &= \exp(-3i\pi/7)\\
z_{-1} &= \exp(-i\pi/7)\\
z_0 &= \exp(i\pi/7)\\
z_1 &= \exp(3i\pi/7)\\
z_2 &= \exp(5i\pi/7)\\
z_3 &= \exp(7i\pi/7) = \exp i\pi = -1
\end{aligned}$$

It's worth looking out for situations in which we can use this theory. For example, suppose we were required to solve the equation

$$(w - 1)^7 + (w + i)^7 = 0$$

We can rearrange this to give

$$\left(\frac{w - 1}{w + i}\right)^7 = -1$$

Then if we put

$$z = \frac{w - 1}{w + i}$$

we merely need to solve $z^7 = -1$ and then express w in terms of z:

$$\begin{aligned}
z(w + i) &= w - 1\\
w(z - 1) &= -iz - 1\\
w &= \frac{-iz - 1}{z - 1}
\end{aligned}$$

Now, if you can manage it, here is a further problem.

4

Exercise Obtain the general solution of the equation $\sinh z = -2$. We need to remember that we are in the field of *complex* numbers.

5

Putting $z = x + iy$ we have

$$\begin{aligned}
\sinh z &= \sinh(x + iy)\\
&= \sinh x \cosh iy + \cosh x \sinh iy\\
&= \sinh x \cos y + \cosh x\,(i \sin y)\\
&= \sinh x \cos y + i \cosh x \sin y
\end{aligned}$$

So that we have

$$-2 = \sinh x \cos y + \mathrm{i} \cosh x \sin y$$

and equating real and imaginary parts

$$-2 = \sinh x \cos y$$
$$0 = \cosh x \sin y$$

Now $\cosh x$ is never zero and so we conclude that $\sin y = 0$. So $y = n\pi$, where n is any integer. We then have $\cos y = \cos n\pi = (-1)^n$, and so

$$-2 = (-1)^n \sinh x$$

from which

$$\sinh x = 2(-1)^{n+1}$$

Now

$$x = \sinh^{-1}[2(-1)^{n+1}] = \ln [2(-1)^{n+1} + \sqrt{(4 + 1)}]$$

so that

$$z = \ln [2(-1)^{n+1} + \sqrt{5}] + \mathrm{i}n\pi \qquad \text{where } n \in \mathbb{N}$$

If you succeeded in solving that problem then you can read through the next one and gloat over your achievement. For those who fell short of the target there is one more hurdle.

▷**Exercise** Solve the equation $\tan z = \mathrm{i}$.

Don't attempt this until you feel confident that you understand the previous problem.

If we put $z = x + \mathrm{i}y$ then we have 6

$$\tan z = \tan (x + \mathrm{i}y)$$
$$= \frac{\tan x + \tan \mathrm{i}y}{1 - \tan x \tan \mathrm{i}y} = \frac{\tan x + \mathrm{i} \tanh y}{1 - \mathrm{i} \tan x \tanh y}$$

Therefore

$$(1 - \mathrm{i} \tan x \tanh y)\mathrm{i} = \tan x + \mathrm{i} \tanh y$$
$$\mathrm{i} + \tan x \tanh y = \tan x + \mathrm{i} \tanh y$$

Equating real and imaginary parts we obtain

$$\tan x \tanh y = \tan x$$
$$1 = \tanh y$$

We conclude that $\tanh y = 1$, which is impossible for $y \in \mathbb{R}$. Consequently there is no $z \in \mathbb{C}$ satisfying the equation $\tan z = \mathrm{i}$.

It is now time for us to apply some of this work. Check that you have understood all the material in this chapter. If there are any weak points then concentrate on them before you begin.

10.10 Practical

BALANCED BRIDGE

When the AC bridge shown in Fig. 10.9 is balanced, the complex impedances Z_1, Z_2, Z_3, Z_4 of the arms satisfy

$$Z_1Z_3 = Z_2Z_4$$

If the bridge is balanced, determine C and L in terms of R_1, R_2, R_3, C_1 and C_2.

Many electrical and electronic engineers prefer to reserve the symbol i to denote current, and consequently another symbol j is then used instead of the complex number i. We shall employ this notation in this example. We must be flexible about notation so that we can change it whenever the need arises.

If you are familiar with circuit theory you may be able to reach the first stage. If you are not then read it through to obtain the necessary equations.

---

1 We have the complex impedances

$$Z_1 = R_1 + \frac{1}{\mathrm{j}\omega C_1}$$

$$Z_2 = R_2 + \frac{1}{\mathrm{j}\omega C_2}$$

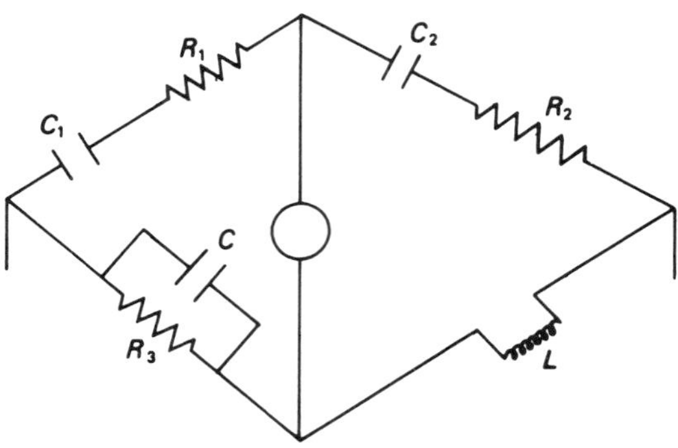

Fig. 10.9 An AC bridge.

$$Z_3 = \frac{1}{(1/R_3) + j\omega C}$$

$$Z_4 = j\omega L$$

Now that you have the necessary ingredients it should be possible for you to complete the solution.

---

2 We have

$$\left(R_1 + \frac{1}{j\omega C_1}\right) \frac{1}{(1/R_3) + j\omega C} = \left(R^2 + \frac{1}{j\omega C_2}\right) j\omega L$$

so that

$$R_1 + \frac{1}{j\omega C_1} = \left(R_2 + \frac{1}{j\omega C_2}\right)\left(\frac{1}{R_3} + j\omega C\right) j\omega L$$

Rationalizing terms:

$$R_1 - j\frac{1}{\omega C_1} = \left(R_2 - \frac{j}{\omega C_2}\right)\left(\frac{j\omega L}{R_3} - \omega^2 LC\right)$$

$$= -R_2\omega^2 LC + \frac{L}{R_3 C_2} + j\left(\frac{R_2\omega L}{R_3} + \frac{\omega LC}{C_2}\right)$$

Now, if you were stuck, take over the working at this stage.

---

3 Equating real and imaginary parts we obtain two equations:

$$R_1 = -R_2\omega^2 LC + \frac{L}{R_3 C_2} \qquad (1)$$

$$-\frac{1}{\omega C_1} = \frac{R_2\omega L}{R_3} + \frac{\omega LC}{C_2} \qquad (2)$$

From (1) we obtain

$$\frac{R_1}{C_2} = -R_2\omega\left(\frac{\omega LC}{C_2}\right) + \frac{L}{R_3 C_2^2}$$

So using (2) to eliminate C:

$$\frac{R_1}{C_2} = -R_2\omega\left(-\frac{1}{\omega C_1} - \frac{R_2\omega L}{R_3}\right) + \frac{L}{R_3 C_2^2}$$

$$= \frac{R_2}{C_1} + \frac{R_2^2\omega^2 L}{R_3} + \frac{L}{R_3 C_2^2}$$

Therefore to obtain L we have

$$\left(\frac{R_2^2\omega^2}{R_3} + \frac{1}{R_3C_2^2}\right)L = \frac{R_1}{C_2} - \frac{R_2}{C_1}$$

That is,

$$L = \frac{[(R_1/C_2) - (R_2/C_1)]R_3C_2^2}{1 + R_2^2C_2^2\omega^2}$$

Now see if you can determine C.

---

4 Using (1) we have

$$R_2\omega^2LC = \frac{L}{R_3C_2} - R_1$$

So

$$\begin{aligned}
C &= \frac{1}{\omega^2R_2R_3C_2} - \frac{R_1}{R_2\omega^2L} \\
&= \frac{1}{\omega^2R_2R_3C_2} - \frac{R_1(1 + R_2^2C_2^2\omega^2)}{R_2\omega^2[(R_1/C_2) - (R_2/C_1)]R_3C_2^2} \\
&= \frac{1}{\omega^2R_2R_3C_2}\left[1 - \frac{R_1(1 + R_2^2C_2^2\omega^2)}{R_1 - (R_2/C_1)C_2}\right] \\
&= \frac{1}{\omega^2R_2R_3C_2}\,\frac{R_1C_1 - R_2C_2 - R_1C_1(1 + R_2^2C_2^2\omega^2)}{R_1C_1 - R_2C_2} \\
&= \frac{1}{\omega^2R_2R_3C_2}\,\frac{R_2C_2 + R_1R_2^2C_1C_2^2\omega^2}{R_2C_2 - R_1C_1} \\
&= \frac{1}{\omega^2R_3}\,\frac{1 + R_1R_2C_1C_2\omega^2}{R_2C_2 - R_1C_1}
\end{aligned}$$

SUMMARY

- ☐ We defined the complex numbers and showed how to arrange them in the cartesian form $a + ib$.
- ☐ We expressed complex numbers in polar form $r(\cos\theta + i\sin\theta)$.
- ☐ We gave two geometrical interpretations of complex numbers: as points in the Argand diagram, and as vectors in the complex plane.
- ☐ We described and applied De Moivre's theorem

$$(\cos\theta + i\sin\theta)^n = \cos n\theta + i\sin n\theta \quad (n \in \mathbb{Z})$$

- ☐ We saw how to obtain the nth roots of a complex number. If $z^n = \alpha = r(\cos\theta + i\sin\theta)$ then

$$z = r^{1/n}[\cos(\theta + 2k\pi)/n + i\sin(\theta + 2k\pi)/n]$$

for some integer k.
- ☐ We obtained the relationships between the circular functions and the hyperbolic functions
  a $\cosh(iz) = \cos z$
  b $\cos(iz) = \cosh z$
  c $\sinh(iz) = i\sin z$
  d $\sin(iz) = i\sinh z$

EXERCISES

1 Express in cartesian form $a + ib$, where a and b are real:
a $(2 + 3i)(i - 4)^2$
b $\dfrac{3 + i}{(2 + i)^2}$
c $\dfrac{1}{2 + i} + \dfrac{i}{1 + 3i}$
d $\exp(2 + 4i) + i$

2 Express in polar form $r(\cos\theta + i\sin\theta)$ where r and θ are real numbers $(r > 0, -\pi < \theta \leqslant \pi)$:
a $2/(\cos\pi/4 + i\sin\pi/4)$
b $i(1 + i)^2$
c $(\cos\pi/4 + i\sin\pi/4)^2 + (\cos\pi/2 + i\sin\pi/2)^5$
d $\exp i$

3 Obtain all complex numbers z which satisfy the equation
a $z^2 + 4z + 5 = 0$
b $z^3 + i = (z - i)^3$
c $z^4 + 2z^2 + 9 = 0$
d $\dfrac{1}{z + i} + \dfrac{1}{z - i} = i$

4 Describe the set of points in the complex plane for which z satisfies

a $|z - 3i| = 5$

b $|z - i| + |z + i| = 4$

c $\left|\dfrac{z - 2i}{2z - i}\right| = 1$

ASSIGNMENT

1 Express in the form $a + ib$

$$\frac{(1 + i)(2 - 3i)(1 + 4i)}{(3 + 2i)(3 + 5i)}$$

2 Express in polar form

$$\frac{(1 + i)\, i\, (\sqrt{3} - i)}{(1 + i\sqrt{3})(2 - 2i)}$$

3 Solve the equation

$$(z + i)^3 = i(z - i)^3$$

4 Simplify the expression

$$\frac{(\cos\theta - i\sin\theta)^9 (\cos 2\theta + i\sin 2\theta)^4}{(\cos 3\theta + i\sin 3\theta)^6 (\cos 4\theta - i\sin 4\theta)^2}$$

5 Use De Moivre's theorem to express $\cos 4\theta$ and $\sin 4\theta$ in terms of $\cos\theta$ and $\sin\theta$ only.

6 Express the complex expression $\tan(x + iy)$, where x and y are real numbers, in the form $a + ib$.

7 Solve the equation $\sin z = i$.

8 Describe the set of points in the complex plane which satisfy

$$\operatorname{Re}\left(\frac{z + i}{z - i}\right) = 0$$

FURTHER EXERCISES

1 If P is a point in the Argand diagram representing the complex number z, interpret the following as loci:

a $\arg[(z - 4)/(z + 4)] = \pi/2$

b $|z - i\sqrt{7}| + |z + i\sqrt{7}| = 8$

Determine z if z satisfies both **a** and **b**.

2 **a** Show that if $|3z - 2| = |z - 6|$ then $|z| = 2$.

b Determine z if $\arg(z + 2) = \pi/3$ and $\arg(z - 2) = 5\pi/6$.

c The centre of a square in the Argand diagram is represented by the complex number $1 + 3i$. Suppose one of the vertices is represented by $3 + 6i$: determine the complex numbers which represent the remaining vertices.

3 a Determine all the solutions of the equation $z^4 + 16 = 0$.
b Obtain the roots of the equation $z^4 - 9z^2 + 400 = 0$.

4 a By expressing $1 + i\sqrt{3}$ in polar form, or otherwise, calculate $(1 + i\sqrt{3})^{12}$.
b Obtain all the solutions of the equation $z^4 = 1$ and hence, or otherwise, solve the equation $(w + 1)^4 = (w - 1)^4$.

5 a Solve the equation $z^4 - 11z^2 + 49 = 0$.
b Show that if $|2z - 5| = |z - 10|$ then $|z| = 5$.

6 Show that if $z = a + ib$ is any complex number,
a $\text{Re}(z) \leqslant |z|$
b $|z| = |\bar{z}|$
c $\text{Re}(z) = \frac{1}{2}\{z + \bar{z}\}$
Deduce that, whenever $z_1, z_2 \in \mathbb{C}$,

$$z_1\bar{z}_2 + z_2\bar{z}_1 = 2\,\text{Re}(z_1\bar{z}_2) \leqslant 2|z_1z_2|$$

By considering $(z_1 + z_2)(\bar{z}_1 + \bar{z}_2)$, or otherwise, deduce the triangle inequality

$$|z_1 + z_2| \leqslant |z_1| + |z_2|$$

7 The admittance Y of an RC series circuit is given by $1/Y = R + 1/j\omega C$. Show that as ω varies from 0 to ∞ the admittance locus is a circle of radius $1/2R$ which passes through the origin.

8 In a transmission line the voltage reflection equation is

$$(Z - Z_0)/(Z + Z_0) = K \exp j\theta$$

where K is real, $Z = R + jX$ and $Z_0 = R_0 + jX_0$. Show that if $X_0 = 0$ then

$$\tan\theta = 2XR_0/(X^2 + R^2 - R_0^2)$$

9 For a certain network the input impedance is w and the output impedance is z, where $w = (z + j)/(z + 1)$.
a Express z explicitly in terms of w.
b Put $z = x + jy$ and $w = u + jv$ and thereby express x and y in terms of u and v.
c Show that for z pure imaginary ($x = 0$) the input impedance w must lie on a circle.
d Show that for z real ($y = 0$) the input impedance must lie on a straight line.

10 The impedance Z of an RC parallel circuit is given by $1/Z = 1/R + j\omega C$. Show that as ω varies from 0 to ∞ the impedance locus is a semicircle below the real axis.

11 Matrices

In Chapter 10 we extended our algebraic knowledge by examining some of the properties of complex numbers. In this chapter we continue our algebraic studies by describing a widely used algebraic concept known as a matrix.

After studying this chapter you should be able to

- ☐ Use matrix notation;
- ☐ Perform the basic operations of matrix algebra;
- ☐ Apply the rules of matrix algebra correctly and use the zero matrices O and the identity matrices I;
- ☐ Write equations in matrix form and solve matrix equations.

At the end of this chapter we shall apply matrix methods to some practical problems in electrical theory.

11.1 NOTATION

Matrices are very useful, for example they are used extensively in the finite element method (structural engineering) and in network analysis (electrical engineering). The general availability of high-speed computers has generated considerable interest in numerical methods and many of these methods use matrices. Wherever large amounts of data need to be handled in a logical and easily accessible manner, matrices prove useful.

What are matrices and where do they come from? You will remember, in elementary algebra, solving sets of simultaneous linear equations. If we were to examine the underlying structure of systems of equations of this kind, we should discover matrix algebra.

Luckily for us we do not have to consider the origins of matrices. We need only learn what they are and how to handle them.

A matrix is a rectangular array of elements arranged in rows and columns. For example,

$$\begin{bmatrix} 2 & 3 & 1 \\ 4 & -2 & 6 \end{bmatrix}$$

is an example of a matrix with two **rows**, three **columns** and six **elements**.

Matrices will be denoted by capital letters, and in general a matrix can be written as

$$A = \begin{bmatrix} a_{11} & a_{12} & a_{13} & \cdots & a_{1n} \\ a_{21} & a_{22} & a_{23} & \cdots & a_{2n} \\ \vdots & \vdots & \vdots & & \vdots \\ a_{m1} & a_{m2} & a_{m3} & \cdots & a_{mn} \end{bmatrix}$$

where the dots indicate elements which have not been displayed.

You will observe that each element in the matrix has been given two subscripts. These subscripts indicate the **address** of the element: the first subscript gives the number of the row, and the second subscript gives the number of the column. We say that the matrix has **order** m by n, which we write as $m \times n$ (in much the same way as carpenters describe the sizes of pieces of wood). We do not evaluate the product $m \times n$, for this would merely give the total number of elements in the matrix and no indication of its shape. For example a carpenter might talk about pieces of 4 by 2, but he would not talk about pieces of 8; that's a different story altogether!

If $m = n$ we say we have a **square matrix** of order n.

The notation

$$A = [a_{i,j}]$$

is a useful shorthand notation when the order of the matrix is known. The element shown is a typical element, for as i and j take on all possible values, each element of the matrix is obtained.

The type of bracket used for matrices is largely a matter of personal choice; some books use parentheses.

A matrix which has just one row or column is called an **algebraic vector**; so a matrix which has just one row is called a **row vector**, and a matrix with just one column is called a **column vector**. Vector notation is sometimes employed, so that algebraic vectors are denoted by **x** or **y**.

□

$$\begin{bmatrix} 2 \\ 3 \\ 4 \end{bmatrix} \qquad [1, 5, 7, 1] \qquad [0, 0, 0]$$

These are respectively a column vector with three elements, a row vector with four elements and a row vector with three elements. ■

There is a convention which you may come across, known as the printer's convention (because it preserves text line spacing), in which a column vector is written horizontally as if it were a row vector! The fact that it is really a column vector is indicated by reserving curly brackets for the purpose. Thus $\{4, 2, 6, 1\}$ may be used to represent the column vector

$$\begin{bmatrix} 4 \\ 2 \\ 6 \\ 1 \end{bmatrix}$$

This convention should be used with caution, particularly if there is a possibility of confusion with a *set* of elements. We shall avoid its use altogether.

11.2 MATRIX ALGEBRA

EQUALITY

Two matrices A and B are **equal** and we write $A = B$ if and only if
1 A and B have the same order (i.e. the same size and shape);
2 Corresponding elements are equal.
We may write this more formally if we wish in the following way. Suppose that A and B are two matrices of order $m \times n$ and $r \times s$ respectively, and that $A = [a_{i,j}]$ and $B = [b_{i,j}]$. Then $A = B$ if and only if
1 $r = m$ and $s = n$;
2 $a_{i,j} = b_{i,j}$ for all i and j.
Don't be worried if you find this algebraic definition a little difficult at first. Compare it carefully with our informal definition and try to understand it.

The definition of equality of matrices enables us to write a set of several algebraic equations by means of a single matrix equation.

□ Deduce the values of x, y and z if

$$\begin{bmatrix} x + y \\ y + z \\ x + z \end{bmatrix} = \begin{bmatrix} 4 \\ 6 \\ 8 \end{bmatrix}$$

See if you can do this. The working is given below.

We obtain the three equations

$$\begin{aligned} x + y &= 4 \\ y + z &= 6 \\ x + z &= 8 \end{aligned}$$

If we add all three equations and divide by 2 we obtain

$$x + y + z = 9$$

so that $x = 3$, $y = 1$ and $z = 5$. ■

Did you manage that? Let's continue! Now that we have decided when two matrices are equal, we consider how to add two matrices.

ADDITION

Two matrices are compatible for addition if and only if they have the *same order*. Addition is then performed by adding corresponding elements.

In symbols, suppose $A = [a_{i,j}]$ and $B = [b_{i,j}]$ are of order $m \times n$ and $r \times s$ respectively. Then $A + B$ exists if and only if $r = m$ and $s = n$, and

$$A + B = [c_{i,j}]$$

where

$$c_{i,j} = a_{i,j} + b_{i,j}$$

Once more try to follow this algebraic definition by comparing it with the working definition which is given above it.

Clearly if $A + B$ exists then $B + A$ also exists and the two are equal. We emphasize this because later when we introduce the operation of multiplication of matrices we shall see that generally AB and BA are *not* equal.

□ Obtain x, y, z and w if

$$\begin{bmatrix} 2x & -y - 1 \\ y & x \end{bmatrix} + \begin{bmatrix} -4x & 2y \\ x & y - 1 \end{bmatrix} = \begin{bmatrix} y & -z \\ -w & 0 \end{bmatrix}$$

Give this example a try. It gives a test of equality and addition. The correct working is given below.

We obtain

$$\begin{bmatrix} -2x & -y - 1 + 2y \\ y + x & x + y - 1 \end{bmatrix} = \begin{bmatrix} y & -z \\ -w & 0 \end{bmatrix}$$

From which $-2x = y$, $y - 1 = -z$, $y + x = -w$ and $x + y - 1 = 0$. It follows that $x = -1$, $y = 2$, $z = -1$ and $w = -1$. ■

So far so good!

TRANSPOSITION

The transpose A^{T} (or A') of a matrix A is obtained from A by interchanging each of its rows with each of its corresponding columns. So if A is of order $m \times n$ then the **transpose** of A is of order $n \times m$.

□ If

$$A = \begin{bmatrix} 2 & 1 & 3 \\ 4 & 6 & 5 \end{bmatrix} \quad \text{then} \quad A^{\mathrm{T}} = \begin{bmatrix} 2 & 4 \\ 1 & 6 \\ 3 & 5 \end{bmatrix}$$ ■

In general we have $(A^{\mathrm{T}})^{\mathrm{T}} = A$. Note also that if $\mathbf{x}$ is a column vector then $\mathbf{x}^{\mathrm{T}}$ is a row vector. So the use of this concept obviates the need for the printer's convention.

More formally, if $A = [a_{i,j}]$ then $A^{\mathrm{T}} = [b_{i,j}]$, where $b_{i,j} = a_{j,i}$ for $i \in \{1, \ldots, n\}$ and $j \in \{1, \ldots, m\}$. If this algebraic form of the definition causes difficulties, try to understand it but don't be over-concerned.

SCALAR MULTIPLICATION

Any matrix A can be multiplied by any number. The multiplication is performed by multiplying every element in the matrix by the number. The numbers are often called scalars.

In symbols, if $A = [a_{i,j}]$ is a matrix of order $n \times m$ and k is a scalar then

$$kA = k[a_{i,j}] = [ka_{i,j}]$$

□ If

$$A = \begin{bmatrix} 2 & 1 \\ 3 & 4 \end{bmatrix} \qquad B = \begin{bmatrix} 1 & 5 \\ 2 & 3 \end{bmatrix}$$

then

$$3A = \begin{bmatrix} 6 & 3 \\ 9 & 12 \end{bmatrix} \qquad 4B = \begin{bmatrix} 4 & 20 \\ 8 & 12 \end{bmatrix}$$ ■

MATRIX MULTIPLICATION

The rule for matrix multiplication is more complicated, and so we introduce it in two stages.

Stage 1

Suppose $\mathbf{x}$ and $\mathbf{y}$ are a row vector and a column vector respectively, each with the same number of elements. Specifically, let us suppose

$$\mathbf{x} = [x_1, \ldots, x_n] \qquad \mathbf{y} = \begin{bmatrix} y_1 \\ y_2 \\ \vdots \\ y_n \end{bmatrix}$$

Then we define

$$\mathbf{xy} = [x_1, \ldots, x_n] \begin{bmatrix} y_1 \\ y_2 \\ \vdots \\ y_n \end{bmatrix} = x_1y_1 + x_2y_2 + \ldots + x_ny_n$$

□

$$[4, 3, 2] \begin{bmatrix} 6 \\ 7 \\ 5 \end{bmatrix} = 4 \times 6 + 3 \times 7 + 2 \times 5 = 24 + 21 + 10 = 55 \quad \blacksquare$$

The order in which we write down these vectors is important. We have defined the product **xy** but we have not yet defined the product **yx**. When we do so, we shall see that the two are *not* equal.

Stage 2

We are now in a position to consider the general rule for multiplying two matrices A and B together to form a product AB. As a precondition we require that the number of columns of A equals the number of rows of B. If this precondition is not satisfied then the product AB will not be defined.

Suppose then that A has order $r \times s$ and that B has order $s \times t$. We regard the matrix A as made up of row vectors and the matrix B as made up of column vectors. We shall initially use dashed lines to help us to visualize this.

□ Let

$$A = \begin{bmatrix} 1 & 4 & 7 \\ 2 & 3 & 5 \end{bmatrix} \qquad B = \begin{bmatrix} 1 & -1 & 2 \\ 2 & 1 & 0 \\ 3 & 0 & 4 \end{bmatrix}$$

We have

$$AB = \left[\begin{array}{ccc} 1 & 4 & 7 \\ \hdashline 2 & 3 & 5 \end{array}\right] \left[\begin{array}{c:c:c} 1 & -1 & 2 \\ 2 & 1 & 0 \\ 3 & 0 & 4 \end{array}\right]$$

Now the product AB has, in its ith row and jth column position, the product of the ith row of A with the jth column of B viewed as vectors.

For example, using the first row of A and the first column of B,

$$[1\ 4\ 7]\begin{bmatrix}1\\2\\3\end{bmatrix} = 1 \times 1 + 4 \times 2 + 7 \times 3 = 1 + 8 + 21 = 30$$

It follows that AB has 30 as the element in the first row and first column, that is the position (1, 1).

Consequently,

$$\begin{aligned} AB &= \begin{bmatrix}1 & 4 & 7\\ \hline 2 & 3 & 5\end{bmatrix}\left[\begin{array}{c|c|c}1 & -1 & 2\\2 & 1 & 0\\3 & 0 & 4\end{array}\right] \\ &= \begin{bmatrix}1+8+21 & -1+4+0 & 2+0+28\\2+6+15 & -2+3+0 & 4+0+20\end{bmatrix} \\ &= \begin{bmatrix}30 & 3 & 30\\23 & 1 & 24\end{bmatrix}\end{aligned}$$

■

So that in general, if A has order $r \times s$ and B has order $s \times t$, then AB has order $r \times t$.

We shall sometimes write A^2 for AA, and in general $A^{n+1} = A^nA$ when $n \in \mathbb{N}$, $n > 1$.

We can write the matrix multiplication rule in symbols if we wish. Suppose $A = [a_{ij}]$ and $B = [b_{ij}]$. Then

$$AB = [c_{ij}]$$

where

$$\begin{aligned} c_{ij} &= \sum_{k=1}^{s} a_{ik}b_{kj} \\ &= a_{i1}b_{1j} + a_{i2}b_{2j} + \ldots + a_{is}b_{sj}\end{aligned}$$

The summation sign Σ should cause no problems. Remember, it means we let k take on every possible integer value from 1 to s, and then add up all the terms.

As we can see, the rule is more involved and less intuitive than the other rules. However, it is quite simple to apply and you will be surprised how quickly you can get used to it.

One small point: you remember that when we multiplied a row vector by a column vector we obtained a number. However, if we use this definition we obtain a 1×1 matrix. One way round this slight contradiction is to say that we shall regard 1×1 matrices as numbers.

It can be shown that the **associative law**

$$A(BC) = (AB)C$$

holds whenever these products are defined. You should not assume that this rule is self-evident, however. It is the notation and the use of the word 'product' which may lead you to this erroneous conclusion. You have already met several examples of non-associative operations. For example, ordinary division for real numbers is non-associative: (3/2)/5 is not equal to 3/(2/5).

To reinforce the fact that we need to exercise caution when carrying out algebraic operations using objects with which we are unfamiliar, we remark that for matrices A and B the products AB and BA are not in general equal. In fact, we have a precondition that may not be satisfied in both cases, so that only one of the products may exist. If, however, A is of order $r \times s$ and B is of order $s \times r$ then AB is a square matrix of order r and BA is a square matrix of order s. Now matrices cannot be equal unless they have the same order, so that before we can even begin to consider equality we must have $r = s$. Even this is not enough to ensure equality! If A and B are both square matrices of order r then in general

$$AB \neq BA$$

Matrices for which $AB = BA$ are said to **commute**. This is a comparatively rare event!

Now we shall make sure we have our ideas straight.

11.3 Workshop

We shall use the following matrices to step through some exercises: 1

$$A = \begin{bmatrix} 1 & 3 \\ 6 & 1 \end{bmatrix} \quad B = \begin{bmatrix} 1 & 3 & 1 \\ 2 & 1 & 5 \end{bmatrix} \quad C = \begin{bmatrix} 1 & 3 \\ -1 & 1 \\ 2 & 4 \end{bmatrix}$$

$$D = \begin{bmatrix} 2 & 1 \\ 5 & 4 \end{bmatrix} \quad E = \begin{bmatrix} 1 & 1 \\ 2 & 1 \end{bmatrix}$$

▷ **Exercise** Without evaluating them, write down all possible products of pairs of distinct matrices from the list which include the matrix A. (For instance, if you think AC exists include AC in your list, but if you believe that CA does not exist exclude CA from your list.) In each case write the order of the product alongside.

When you have completed this, look at the next step to see if you have the right answers!

2 Here are the correct answers:

$$\begin{array}{c} AB\ (2 \times 3) \\ AD\ (2 \times 2) \\ AE\ (2 \times 2) \\ CA\ (3 \times 2) \\ DA\ (2 \times 2) \\ EA\ (2 \times 2) \end{array}$$

Did you manage to get them all right? If you did then move on to step 5. If you made some mistakes, check back carefully to see the precondition for matrix multiplication and the rule for calculating the order of a product. Then solve this next problem.

▷ **Exercise** Write down a list of all the products of two matrices which have B as one of them. As before, write alongside the order of each product.

As soon as you have finished, take the next step to see if you are right.

3 Here are the answers:

$$\begin{array}{c} BC\ (2 \times 2) \\ CB\ (3 \times 3) \\ DB\ (2 \times 3) \\ EB\ (2 \times 3) \end{array}$$

If they are all right then move on to step 5. If there are still a few difficulties you should go back carefully over what we have done and then try this problem.

▷ **Exercise** If we consider pairs of distinct matrices from our original list, there are still four which we have not listed. Say which these are and give the orders of these products.

When you have completed this, look at the list in the next step.

4 Here are the answers:

$$\begin{array}{c} CD\ (3 \times 2) \\ CE\ (3 \times 2) \end{array}$$

$$DE\ (2 \times 2)$$
$$ED\ (2 \times 2)$$

Now let's move on to matrix multiplication.

5

▷**Exercise** Obtain AB, AD, BC, AE, CD and DE.

Try these, then look at the next step to see if you have all the answers right.

6

Here are the correct results:

$$AB = \begin{bmatrix} 1 & 3 \\ 6 & 1 \end{bmatrix} \begin{bmatrix} 1 & 3 & 1 \\ 2 & 1 & 5 \end{bmatrix} = \begin{bmatrix} 7 & 6 & 16 \\ 8 & 19 & 11 \end{bmatrix}$$

$$AD = \begin{bmatrix} 1 & 3 \\ 6 & 1 \end{bmatrix} \begin{bmatrix} 2 & 1 \\ 5 & 4 \end{bmatrix} = \begin{bmatrix} 17 & 13 \\ 17 & 10 \end{bmatrix}$$

$$BC = \begin{bmatrix} 1 & 3 & 1 \\ 2 & 1 & 5 \end{bmatrix} \begin{bmatrix} 1 & 3 \\ -1 & 1 \\ 2 & 4 \end{bmatrix} = \begin{bmatrix} 0 & 10 \\ 11 & 27 \end{bmatrix}$$

$$AE = \begin{bmatrix} 1 & 3 \\ 6 & 1 \end{bmatrix} \begin{bmatrix} 1 & 1 \\ 2 & 1 \end{bmatrix} = \begin{bmatrix} 7 & 4 \\ 8 & 7 \end{bmatrix}$$

$$CD = \begin{bmatrix} 1 & 3 \\ -1 & 1 \\ 2 & 4 \end{bmatrix} \begin{bmatrix} 2 & 1 \\ 5 & 4 \end{bmatrix} = \begin{bmatrix} 17 & 13 \\ 3 & 3 \\ 24 & 18 \end{bmatrix}$$

$$DE = \begin{bmatrix} 2 & 1 \\ 5 & 4 \end{bmatrix} \begin{bmatrix} 1 & 1 \\ 2 & 1 \end{bmatrix} = \begin{bmatrix} 4 & 3 \\ 13 & 9 \end{bmatrix}$$

If you have these all correct then you can proceed to step 8. If some of the products didn't work out, try this next exercise.

▷**Exercise** Calculate the products CA, CB, DA, DB, ED.

When you have finished, check in step 7 to see if you have them right.

7

Here are the answers:

$$CA = \begin{bmatrix} 1 & 3 \\ -1 & 1 \\ 2 & 4 \end{bmatrix} \begin{bmatrix} 1 & 3 \\ 6 & 1 \end{bmatrix} = \begin{bmatrix} 1\times1+3\times6 & 1\times3+3\times1 \\ -1\times1+1\times6 & -1\times3+1\times1 \\ 2\times1+4\times6 & 2\times3+4\times1 \end{bmatrix} = \begin{bmatrix} 19 & 6 \\ 5 & -2 \\ 26 & 10 \end{bmatrix}$$

$$CB = \begin{bmatrix} 1 & 3 \\ -1 & 1 \\ 2 & 4 \end{bmatrix} \begin{bmatrix} 1 & 3 & 1 \\ 2 & 1 & 5 \end{bmatrix} = \begin{bmatrix} 7 & 6 & 16 \\ 1 & -2 & 4 \\ 10 & 10 & 22 \end{bmatrix}$$

$$DA = \begin{bmatrix} 2 & 1 \\ 5 & 4 \end{bmatrix} \begin{bmatrix} 1 & 3 \\ 6 & 1 \end{bmatrix} = \begin{bmatrix} 8 & 7 \\ 29 & 19 \end{bmatrix}$$

$$DB = \begin{bmatrix} 2 & 1 \\ 5 & 4 \end{bmatrix} \begin{bmatrix} 1 & 3 & 1 \\ 2 & 1 & 5 \end{bmatrix} = \begin{bmatrix} 4 & 7 & 7 \\ 13 & 19 & 25 \end{bmatrix}$$

$$ED = \begin{bmatrix} 1 & 1 \\ 2 & 1 \end{bmatrix} \begin{bmatrix} 2 & 1 \\ 5 & 4 \end{bmatrix} = \begin{bmatrix} 7 & 5 \\ 9 & 6 \end{bmatrix}$$

There we are! Now step ahead.

8

Exercise Among the matrices we have been considering, there are two that commute. Find out which they are!

When you have done this, read on and see if you are right. It shouldn't take too long!

Here is the solution. We have seen that for two matrices to commute, they must both be square and have the same order. Therefore the only ones we need to consider are A, D and E. The products AD, AE, DA, ED and DE have already been found, and so there remains only EA to find:

$$EA = \begin{bmatrix} 1 & 1 \\ 2 & 1 \end{bmatrix} \begin{bmatrix} 1 & 3 \\ 6 & 1 \end{bmatrix} = \begin{bmatrix} 7 & 4 \\ 8 & 7 \end{bmatrix} = AE$$

Consequently A and E commute.

11.4 MATRIX EQUATIONS

The simplest way to deal with this subject is by example.

Suppose we want to write the following equations in matrix form and thereby obtain u and v in terms of x, y and z:

$$\begin{aligned} u &= 3p + 4q \qquad & p &= 2x + 3y - z \\ v &= p - 3q \qquad & q &= x - y + 2z \end{aligned}$$

We have

$$\begin{bmatrix} u \\ v \end{bmatrix} = \begin{bmatrix} 3 & 4 \\ 1 & -3 \end{bmatrix} \begin{bmatrix} p \\ q \end{bmatrix}$$

and

$$\begin{bmatrix} p \\ q \end{bmatrix} = \begin{bmatrix} 2 & 3 & -1 \\ 1 & -1 & 2 \end{bmatrix} \begin{bmatrix} x \\ y \\ z \end{bmatrix}$$

These matrix equations may be verified easily by multiplying out and using the rule for equality. Observe that in each case the matrix consists of the coefficients and that it is multiplied by a column vector of the unknowns. So, writing

$$\mathbf{u} = \begin{bmatrix} u \\ v \end{bmatrix} \qquad \mathbf{p} = \begin{bmatrix} p \\ q \end{bmatrix} \qquad \mathbf{x} = \begin{bmatrix} x \\ y \\ z \end{bmatrix}$$

with

$$A = \begin{bmatrix} 3 & 4 \\ 1 & -3 \end{bmatrix} \qquad B = \begin{bmatrix} 2 & 3 & -1 \\ 1 & -1 & 2 \end{bmatrix}$$

we have

$$\mathbf{u} = A\mathbf{p} \quad \text{and} \quad \mathbf{p} = B\mathbf{x}$$

So substituting for **p** we have

$$\mathbf{u} = A(B\mathbf{x}) = (AB)\mathbf{x}$$

using the associative law. Now

$$AB = \begin{bmatrix} 3 & 4 \\ 1 & -3 \end{bmatrix} \begin{bmatrix} 2 & 3 & -1 \\ 1 & -1 & 2 \end{bmatrix} = \begin{bmatrix} 10 & 5 & 5 \\ -1 & 6 & -7 \end{bmatrix}$$

So that

$$\begin{bmatrix} u \\ v \end{bmatrix} = \begin{bmatrix} 10 & 5 & 5 \\ -1 & 6 & -7 \end{bmatrix} \begin{bmatrix} x \\ y \\ z \end{bmatrix}$$

Therefore

$$\begin{aligned} u &= 10x + 5y + 5z \\ v &= -x + 6y - 7z \end{aligned}$$

Of course, we know we could have solved this problem by using elementary algebra. The point is that we have now developed matrix algebra to such an extent that we have an alternative method. This is just the beginning of the story.

11.5 ZERO, IDENTITY AND INVERSE MATRICES

Any matrix that has all its elements zero is called a **zero matrix** or a **null matrix** and is denoted by O. Once we decide on the order of the zero matrix, for example 3×2, then the zero matrix is uniquely determined.

□ The 3×2 zero matrix is

$$\begin{bmatrix} 0 & 0 \\ 0 & 0 \\ 0 & 0 \end{bmatrix}$$

■

If A is any matrix and O is the zero matrix of the same order as A then $A + O = A$.

We do not usually need to emphasize the order of O since its context clarifies the position. However, when we do need to show the order we write it underneath, as in the next example.

□

$$\underset{3 \times 2}{A} \quad \underset{2 \times 5}{O} = \underset{3 \times 5}{O}$$

■

Observe how the rule for matrix multiplication determines the order of the zero matrix on the right once the order of the zero matrix on the left is given.

If we have a square matrix, the set of elements on the diagonal from the top left-hand corner to the bottom right-hand corner is known as the **leading diagonal**. Any square matrix which has its only non-zero elements on the leading diagonal is known as a **diagonal matrix**. Such a matrix is uniquely determined by the leading diagonal, so that

$$A = \text{diag}\,\{a, b, c, d\}$$

is a 4×4 matrix in which the elements on the leading diagonal are a, b, c and d respectively. That is,

$$A = \begin{bmatrix} a & 0 & 0 & 0 \\ 0 & b & 0 & 0 \\ 0 & 0 & c & 0 \\ 0 & 0 & 0 & d \end{bmatrix}$$

Any square matrix which has all the elements on the leading diagonal equal to 1 and all other elements 0 is known as an **identity matrix** and is represented by I.

□ The 3×3 identity matrix is

$$\begin{bmatrix} 1 & 0 & 0 \\ 0 & 1 & 0 \\ 0 & 0 & 1 \end{bmatrix} = \text{diag}\{1, 1, 1\}$$ ■

Suppose that A is an $n \times m$ matrix. Then we can easily show that

$$\underset{n \times m}{A} \quad \underset{m \times m}{I} = \underset{n \times n}{I} \quad \underset{n \times m}{A} = \underset{n \times m}{A}$$

□ Check this relation with the matrix

$$A = \begin{bmatrix} 2 & 5 & 6 \\ 4 & 3 & 1 \end{bmatrix}$$ ■

Another name for the identity matrix is the **unit matrix**. The equation $AI = A = IA$ shows that in matrix algebra I plays much the same role as the number 1 does in elementary algebra.

The consideration of algebraic equations leads to a further important type of matrix. Take, for example,

$$\begin{aligned} ax + by &= h \\ cx + dy &= k \end{aligned}$$

We have seen how to write these in matrix form:

$$\begin{bmatrix} a & b \\ c & d \end{bmatrix} \begin{bmatrix} x \\ y \end{bmatrix} = \begin{bmatrix} h \\ k \end{bmatrix}$$

or alternatively by the single matrix equation $A\mathbf{x} = \mathbf{h}$.

Suppose now that we can find another matrix B such that $BA = I$. Then if we pre-multiply both sides by B we obtain

$$\begin{aligned} B(A\mathbf{x}) &= B\mathbf{h} \\ (BA)\mathbf{x} &= B\mathbf{h} \end{aligned}$$

That is,

$$\begin{aligned} I\mathbf{x} &= B\mathbf{h} \\ \mathbf{x} &= B\mathbf{h} \end{aligned}$$

Consequently, this set of simultaneous equations can be solved, using matrices, if we can obtain the matrix B. The matrix B, for which $BA = I = AB$, is known as the **inverse** of the matrix A (see Chapter 13). Although we shall see later that not every square matrix has an inverse, the quest for inverses leads us to the study of determinants in the next chapter.

As you know, any point in the plane can be expressed uniquely using rectangular cartesian coordinates as an ordered pair of numbers (x, y). If we write this as a column vector and pre-multiply it by a 2×2 real matrix we obtain another point. So any 2×2 real matrix can be regarded as a transformation of the plane to itself. Similarly any 3×3 real matrix can be regarded as a transformation of three-dimensional space to itself.

11.6 ALGEBRAIC RULES

Finally we state without further ado the algebraic rules which matrices obey. In each case we shall suppose that the matrices which appear in the equations are of the correct order so that the equations are meaningful.

1 $(A + B) + C = A + (B + C)$: addition is associative.
2 $A + B = B + A$: addition is commutative.
3 $A + O = O + A = A$: zero matrices exist.
4 $A(BC) = (AB)C$: multiplication is associative.
5 $AI = IA = A$: unit matrices exist. (Unless A is square the two matrices denoted by I will have different orders.)
6 $A(B + C) = (AB) + (AC)$: multiplication is distributive over addition.
7 $k(AB) = (kA)B = A(kB)$ where k is a scalar.

Perhaps the most important rules to remember are the following:

8 In general $AB \neq BA$.
9 If $AB = O$ it does *not* necessarily follow that either $A = O$ or $B = O$.

□

$$\begin{bmatrix} 1 & 1 \\ 2 & 2 \end{bmatrix}\begin{bmatrix} -1 & 1 \\ 1 & -1 \end{bmatrix} = \begin{bmatrix} 0 & 0 \\ 0 & 0 \end{bmatrix}$$ ■

This means that we must be extra careful when multiplying by matrices to keep the order in which we are to perform the operations clear. We either **pre-multiply** or **post-multiply** by a matrix.

We cannot 'cancel out' matrix equations in the same way as we cancel out equations involving real or complex numbers.

□

$$\begin{bmatrix} -1 & 1 \\ 1 & -1 \end{bmatrix}\begin{bmatrix} 0 & 2 \\ 2 & 4 \end{bmatrix} = \begin{bmatrix} 2 & 2 \\ -2 & -2 \end{bmatrix}$$

Also

$$\begin{bmatrix} -1 & 1 \\ 1 & -1 \end{bmatrix}\begin{bmatrix} -2 & -1 \\ 0 & 1 \end{bmatrix} = \begin{bmatrix} 2 & 2 \\ -2 & -2 \end{bmatrix}$$

so that

$$\begin{bmatrix} -1 & 1 \\ 1 & -1 \end{bmatrix}\begin{bmatrix} 0 & 2 \\ 2 & 4 \end{bmatrix} = \begin{bmatrix} -1 & 1 \\ 1 & -1 \end{bmatrix}\begin{bmatrix} -2 & -1 \\ 0 & 1 \end{bmatrix}$$

but

$$\begin{bmatrix} 0 & 2 \\ 2 & 4 \end{bmatrix} \neq \begin{bmatrix} -2 & -1 \\ 0 & 1 \end{bmatrix}$$

■

Now let's apply some of the work we have done to a practical problem.

11.7 Practical

ELECTRICAL NETWORKS

Obtain the transmission matrices for the circuits shown in Fig. 11.1(a) and (b). Use these to obtain the transmission matrix for the network in Fig. 11.1(c).

We shall solve this problem stage by stage. If you are not an electrical engineering student you may not be concerned with the underlying theory. If this is the case you may skip over the derivations of the series and shunt transmission matrices.

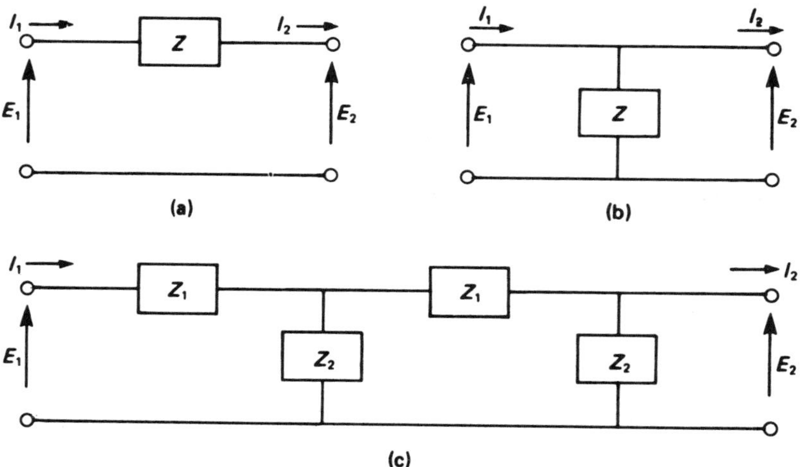

Fig. 11.1 (a) Series impedance (b) Shunt impedance (c) Cascade of impedances.

For the series impedance (Fig. 11.1(a)) we have

$$E_1 = E_2 + ZI_2$$
$$I_1 = I_2$$

When we write these equations in matrix form we obtain the transmission matrix:

$$\begin{bmatrix} E_1 \\ I_1 \end{bmatrix} = \begin{bmatrix} 1 & Z \\ 0 & 1 \end{bmatrix} \begin{bmatrix} E_2 \\ I_2 \end{bmatrix}$$

where

$$\begin{bmatrix} 1 & Z \\ 0 & 1 \end{bmatrix}$$

is the transmission matrix.

See if you can write down the transmission matrix for the shunt impedance before moving on.

For the shunt impedance (Fig. 11.1(b)) we have

$$E_1 = E_2$$
$$I_1 = (1/Z)E_2 + I_2$$

So in matrix form we have

$$\begin{bmatrix} E_1 \\ I_1 \end{bmatrix} = \begin{bmatrix} 1 & 0 \\ 1/Z & 1 \end{bmatrix} \begin{bmatrix} E_2 \\ I_2 \end{bmatrix}$$

where the transmission matrix is

$$\begin{bmatrix} 1 & 0 \\ 1/Z & 1 \end{bmatrix}$$

The network shown in Fig. 11.1(c) can be regarded as a cascade of series and shunt impedances as in Fig. 11.2. For each impedance we use the matrix equations

$$\begin{bmatrix} E_1 \\ I_1 \end{bmatrix} = \begin{bmatrix} a & b \\ c & d \end{bmatrix} \begin{bmatrix} E_2 \\ I_2 \end{bmatrix}$$

where

$$\begin{bmatrix} a & b \\ c & d \end{bmatrix}$$

is one of the two types of transmission matrix obtained.

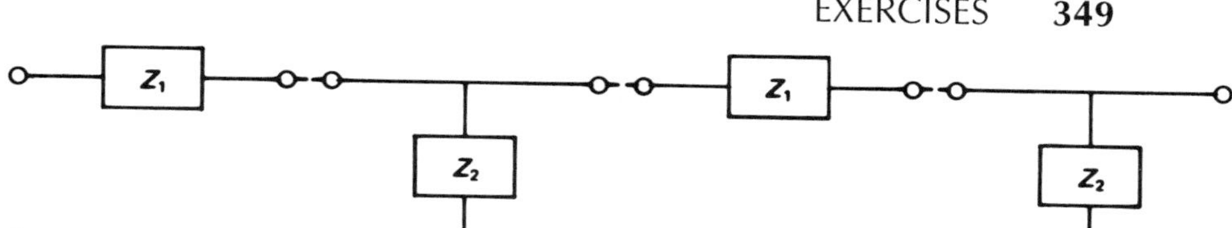

Fig. 11.2 Cascade of series and shunt impedances.

To obtain the transmission matrix of the whole network we put all the individual matrices together:

$$\begin{bmatrix} E_1 \\ I_1 \end{bmatrix} = \begin{bmatrix} 1 & Z_1 \\ 0 & 1 \end{bmatrix} \begin{bmatrix} 1 & 0 \\ 1/Z_2 & 1 \end{bmatrix} \begin{bmatrix} 1 & Z_1 \\ 0 & 1 \end{bmatrix} \begin{bmatrix} 1 & 0 \\ 1/Z_2 & 1 \end{bmatrix} \begin{bmatrix} E_2 \\ I_2 \end{bmatrix}$$

$$= \begin{bmatrix} 1 + Z_1/Z_2 & Z_1 \\ 1/Z_2 & 1 \end{bmatrix} \begin{bmatrix} 1 + Z_1/Z_2 & Z_1 \\ 1/Z_2 & 1 \end{bmatrix} \begin{bmatrix} E_2 \\ I_2 \end{bmatrix}$$

$$= \begin{bmatrix} (1 + Z_1/Z_2)^2 + Z_1/Z_2 & Z_1(1 + Z_1/Z_2) + Z_1 \\ (1/Z_2)(1 + Z_1/Z_2) + 1/Z_2 & (Z_1/Z_2) + 1 \end{bmatrix} \begin{bmatrix} E_2 \\ I_2 \end{bmatrix}$$

Therefore the transmission matrix is

$$\begin{bmatrix} (1 + Z_1/Z_2)^2 + Z_1/Z_2 & Z_1(2 + Z_1/Z_2) \\ (1/Z_2)(2 + Z_1/Z_2) & 1 + Z_1/Z_2 \end{bmatrix}$$

SUMMARY

- □ We have introduced matrices and explained the concepts of
  - **a** equality of matrices
  - **b** matrix addition
  - **c** transposition of matrices
  - **d** scalar multiplication
  - **e** matrix multiplication
  - **f** the matrices O and I.
- □ We have seen how to write a set of simultaneous linear equations in matrix form.
- □ We have listed the rules which matrices obey.
- □ We have seen that if a set of algebraic equations is expressed in the form $A\mathbf{x} = \mathbf{h}$ in which A is a square matrix, and if we can find a matrix B such that $BA = I$, then $\mathbf{x} = B\mathbf{h}$ and the equations are solved.

EXERCISES

The following exercises use the matrices

$$A = \begin{bmatrix} 1 & 2 \\ -1 & 3 \end{bmatrix} \qquad B = \begin{bmatrix} 1 & 3 \\ -2 & 4 \end{bmatrix}$$

$$C = \begin{bmatrix} -3 & 2 \\ 1 & -4 \end{bmatrix} \qquad D = \begin{bmatrix} 1 & -2 \\ -1 & 2 \end{bmatrix}$$

1 Calculate

a $A + B$ **b** $(A + B)C$ **c** $(A^{\mathrm{T}} + B^{\mathrm{T}})^{\mathrm{T}}$

d $AD + BD$ **e** $D^{\mathrm{T}}D$ **f** DD^{T}

2 If U, V and W are square matrices, obtain the expansion of $(U + V + W)^2$ and check your expansion in the case $U = A$, $V = B$, $W = C$.

3 Obtain the matrix X in each of the following:

a $A + X = B - X$

b $A + X = I$

c $(A + X)^{\mathrm{T}} = B$

d $(A + X)C = I$

e $(A - X)B = C^{\mathrm{T}}$

f $AX = BX + C$

4 Obtain the diagonal matrix X which satisfies

$$(AX)(BX)^{\mathrm{T}} = D$$

ASSIGNMENT

1 Obtain the values of x and y if the matrix

$$A = \begin{bmatrix} \cos w & x \\ \sin w & y \end{bmatrix}$$

satisfies the equation $AA^{\mathrm{T}} = I$, where A^{T} is the transpose of A.

2 By considering the matrices

$$A = \begin{bmatrix} 1 & 1 & 2 \\ 3 & 2 & -1 \\ 4 & 1 & 2 \end{bmatrix} \qquad B = \begin{bmatrix} 2 & -1 \\ 4 & 2 \\ 3 & -1 \end{bmatrix}$$

verify the identity $(AB)^{\mathrm{T}} = B^{\mathrm{T}}A^{\mathrm{T}}$.

3 a Identify the matrices diag $\{1, 1\}$ and diag $\{0, 0, 0\}$.

b Show that if $A = \text{diag}\,\{a, b, c, d\}$ and $B = \text{diag}\,\{h, k, l, m\}$ then

$$AB = BA = \text{diag}\,\{ah, bk, cl, dm\}$$

Obtain also A^n.

4 Expand (a) $(A + B)^2$ (b) $(A + 2B)^2$ and check the expansions obtained with

$$A = \begin{bmatrix} 1 & 3 \\ 2 & 6 \end{bmatrix} \qquad B = \begin{bmatrix} 1 & 4 \\ 3 & 2 \end{bmatrix}$$

5 A square matrix M is **symmetric** if and only if it is equal to its transpose. Obtain a, b and c if

$$A = \begin{bmatrix} 3 & a & -1 \\ 2 & 5 & c \\ b & 8 & 2 \end{bmatrix}$$

is symmetric.

6 A square matrix M is **skew symmetric** if and only if $M^T = -M$. Obtain a, b, c, d, e and f if

$$A = \begin{bmatrix} a & 3 & e \\ d & b & f \\ -2 & 6 & c \end{bmatrix}$$

is skew symmetric.

7 Given the matrices

$$A = \begin{bmatrix} 1 & x & 1 \\ x & 2 & y \\ 1 & y & 3 \end{bmatrix} \qquad B = \begin{bmatrix} 3 & -3 & z \\ -3 & 2 & -3 \\ z & -3 & 1 \end{bmatrix}$$

obtain x, y and z if AB is symmetric. Show that A and B commute.

FURTHER EXERCISES

1 If

$$A = \begin{bmatrix} 1 & 2 & 1 \\ 2 & 4 & 2 \\ 3 & 6 & 3 \end{bmatrix} \qquad B = \begin{bmatrix} 1 & -3 & -2 \\ -1 & 2 & 1 \\ 1 & -1 & 0 \end{bmatrix}$$

verify that $AB = O$.

2 If $AB = O$ and $BC = I$ prove that

$$(A + B)^2(A + C)^2 = I$$

3 Show that if

$$A = \begin{bmatrix} 1 & 2 & 3 \\ 3 & 7 & 9 \\ 4 & 8 & 13 \end{bmatrix} \qquad B = \begin{bmatrix} 19 & -2 & -3 \\ -3 & 1 & 0 \\ -4 & 0 & 1 \end{bmatrix}$$

then $AB = BA = I$. Hence, or otherwise, solve each of the following systems of equations:

a $x + 2y + 3z = 1$
$3x + 7y + 9z = 4$
$4x + 8y + 13z = 3$

b $x + 3y + 4z = 2$
$2x + 7y + 8z = 6$
$3x + 9y + 13z = 4$

4 In atomic physics the Pauli spin matrices are

$$S_1 = \begin{bmatrix} 0 & 1 \\ 1 & 0 \end{bmatrix} \qquad S_2 = \begin{bmatrix} 0 & -\mathrm{i} \\ \mathrm{i} & 0 \end{bmatrix} \qquad S_3 = \begin{bmatrix} 1 & 0 \\ 0 & -1 \end{bmatrix}$$

where $\mathrm{i}^2 = -1$. Verify

a $S_1S_2 = \mathrm{i}S_3$

b $S_2S_1 = -\mathrm{i}S_3$

c $S_1^2 = S_2^2 = S_3^2 = I$

5 Obtain the transmission matrices of the three cascade circuits shown in Fig. 11.3.

6 If there exists a positive integer n such that $A^n = 0$ then the square matrix A is said to be **nilpotent**. Verify that

$$A = \begin{bmatrix} 0 & 1 & a \\ 0 & 0 & 1 \\ 0 & 0 & 0 \end{bmatrix}$$

is nilpotent.

7 Show that if A is any square matrix then

a AA^{T} is symmetric;

b $A + A^{\mathrm{T}}$ is symmetric;

c $A - A^{\mathrm{T}}$ is skew symmetric.

Verify this general property by considering the matrix

$$A = \begin{bmatrix} 1 & 2 & 3 \\ 4 & 5 & 6 \\ 7 & 8 & 9 \end{bmatrix}$$

8 Suppose

$$A = \begin{bmatrix} 5 & 11 & 4 \\ 2 & 9 & 5 \\ 3 & 6 & 2 \end{bmatrix} \qquad B = \begin{bmatrix} -12 & 2 & 19 \\ 11 & -2 & -17 \\ -15 & 3 & 23 \end{bmatrix}$$

a Verify that $AB = I = BA$.

b Use **a** and the rules of matrices to write down $A^{\mathrm{T}}B^{\mathrm{T}}$.

c Use **a** and **b** to write each of the following systems in matrix form and thereby solve them:

i $-12x + 2y + 19z = 1$
$11x - 2y - 17z = 2$

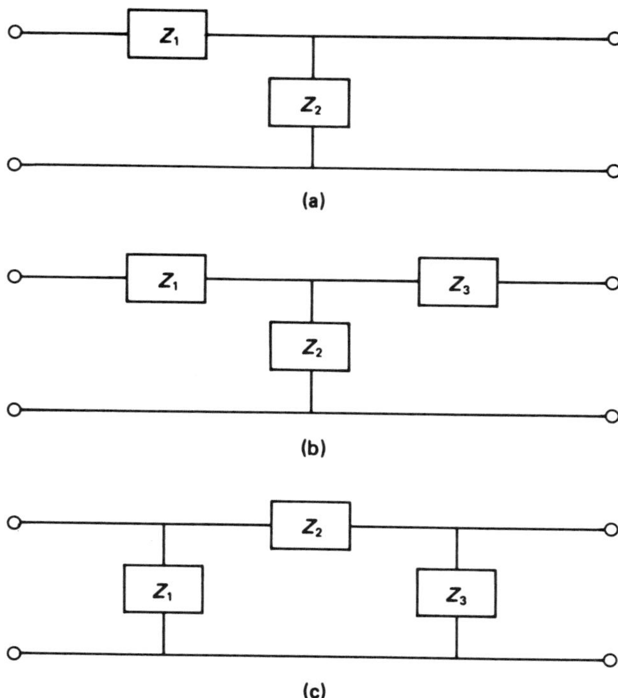

Fig. 11.3 Cascades of impedances.

$$-15x + 3y + 23z = 3$$

ii
$$\begin{aligned} -12u + 11v - 15w &= 1 \\ 2u - 2v + 3w &= 2 \\ 19u - 17v + 23w &= 3 \end{aligned}$$

9 If $1 + \alpha + \alpha^2 = 0$, show that $\alpha^3 = 1$. Hence or otherwise show that if

$$A = \begin{bmatrix} 1 & 1 & 1 \\ 1 & \alpha^2 & \alpha \\ 1 & \alpha & \alpha^2 \end{bmatrix} \qquad B = \frac{1}{3}\begin{bmatrix} 1 & 1 & 1 \\ 1 & \alpha & \alpha^2 \\ 1 & \alpha^2 & \alpha \end{bmatrix}$$

then $AB = I$.

Three inputs e_1, e_2 and e_3 are expressed in terms of three outputs E_1, E_2 and E_3 by

$$\begin{aligned} e_1 &= E_1 + E_2 + E_3 \\ e_2 &= E_1 + \alpha^2 E_2 + \alpha E_3 \\ e_3 &= E_1 + \alpha E_2 + \alpha^2 E_3 \end{aligned}$$

Express E_1, E_2 and E_3 explicitly in terms of e_1, e_2 and e_3.

12 Determinants

In Chapter 11 we explained what is meant by a matrix and examined the elementary properties of matrices. To take the story any further we shall need the concept of a determinant; the subject of this next chapter.

After you have completed this chapter you should be able to

- ☐ Distinguish between matrices and determinants;
- ☐ Evaluate determinants;
- ☐ Use Cramer's rule;
- ☐ Calculate minors and cofactors;
- ☐ Simplify determinants.

At the end of this chapter we shall apply determinants to the practical problem of a Wheatstone bridge.

12.1 NOTATION

A determinant is a number which is calculated from the elements in a square matrix. If we have a square matrix, we may represent its determinant using the row and column notation of matrices. To distinguish the two concepts we enclose the elements of a determinant between vertical parallel lines. We can write the determinant of the square matrix A as either $|A|$ or det A.

☐

$$A = \begin{bmatrix} 1 & 3 \\ 2 & 4 \end{bmatrix} \qquad \det A = \begin{vmatrix} 1 & 3 \\ 2 & 4 \end{vmatrix}$$ ■

The rule for evaluating a determinant is best introduced by considering the simplest cases. If A is a 1×1 matrix, then the determinant of A is merely the element itself.

□ If $A = [-3]$ then $\det A = -3$. ■

In this instance the notation $|A|$ is unfortunate because it could become confused with the modulus sign. Luckily 1×1 determinants are so trivial they seldom arise.

Things are slightly more straightforward when we consider determinants of order two. We define

$$\begin{vmatrix} a_{11} & a_{12} \\ a_{21} & a_{22} \end{vmatrix} = a_{11}a_{22} - a_{21}a_{12}$$

□

$$\begin{vmatrix} \cosh u & \sinh u \\ \sinh u & \cosh u \end{vmatrix} = \cosh^2 u - \sinh^2 u = 1$$ ■

The rule is: top left times bottom right minus bottom left times top right.

The rule itself may seem rather strange and arbitrary. To see why it is like this, consider the following pair of simultaneous equations:

$$\begin{aligned} ax + by &= h \\ cx + dy &= k \end{aligned}$$

If we perform elementary algebraic operations on these equations to express x and y explicitly in terms of h and k, we obtain

$$x = \frac{hd - kb}{ad - cb} \qquad y = \frac{ak - ch}{ad - cb}$$

We can write these equations in terms of second-order determinants as follows:

$$x = \frac{\begin{vmatrix} h & b \\ k & d \end{vmatrix}}{\begin{vmatrix} a & b \\ c & d \end{vmatrix}} \qquad y = \frac{\begin{vmatrix} a & h \\ c & k \end{vmatrix}}{\begin{vmatrix} a & b \\ c & d \end{vmatrix}}$$

12.2 CRAMER'S RULE

In fact these equations provide us with an algebraic method for solving equations, known as **Cramer's rule**:

1 Write down the equations with the constants on the right-hand side:

$$\begin{aligned} a_{11}x_1 + a_{12}x_2 &= h_1 \\ a_{21}x_1 + a_{22}x_2 &= h_2 \end{aligned}$$

2 Calculate Δ, the determinant of the coefficients of the unknowns:

$$\Delta = \begin{vmatrix} a_{11} & a_{12} \\ a_{21} & a_{22} \end{vmatrix} = a_{11}a_{22} - a_{21}a_{12}$$

3 To obtain one of the unknowns, cover up its coefficients in the equations and imagine them to have been replaced by the corresponding constants from the right-hand side. Evaluate the determinant of the fictitious coefficients and equate it to the product of the unknown with Δ.

If you look carefully at the rule you will be able to see how to apply it easily. Try this example; the working is given below.

☐ Obtain x and y explicitly in terms of u and v when

$$\begin{aligned} x \cos w + y \sin w &= u \\ -x \sin w + y \cos w &= v \end{aligned}$$

Using Cramer's rule, we obtain

$$\Delta = \begin{vmatrix} \cos w & \sin w \\ -\sin w & \cos w \end{vmatrix} = \cos^2 w + \sin^2 w = 1$$

Therefore

$$\Delta x = x = \begin{vmatrix} u & \sin w \\ v & \cos w \end{vmatrix} = u \cos w - v \sin w$$

$$\Delta y = y = \begin{vmatrix} \cos w & u \\ -\sin w & v \end{vmatrix} = v \cos w + u \sin w$$

So

$$\begin{aligned} x &= u \cos w - v \sin w \\ y &= v \cos w + u \sin w \end{aligned}$$

■

Cramer's rule extends to n equations in n unknowns and can be useful for dealing with purely algebraic systems. However, there are much more efficient ways of solving such systems of equations when the coefficients are numerical.

12.3 HIGHER-ORDER DETERMINANTS

We now turn our attention to determinants of order three:

$$\begin{vmatrix} a_{11} & a_{12} & a_{13} \\ a_{21} & a_{22} & a_{23} \\ a_{31} & a_{32} & a_{33} \end{vmatrix}$$

Before seeing how to evaluate this determinant, we shall consider an operation which can be performed either on a square matrix or on a determinant.

If the row and column in which an element is situated are deleted and the resulting determinant is evaluated, the result is known as the **minor** of the element. The minor of a_{ij} is represented by M_{ij}.

□

$$A = \begin{bmatrix} 3 & 2 & 1 \\ 4 & 6 & -1 \\ 2 & 1 & 3 \end{bmatrix}$$

Then

$$M_{22} = \begin{vmatrix} 3 & 1 \\ 2 & 3 \end{vmatrix} = 9 - 2 = 7$$

$$M_{31} = \begin{vmatrix} 2 & 1 \\ 6 & -1 \end{vmatrix} = -2 - 6 = -8$$

When every element is replaced by its minor we obtain the **matrix of minors** M. In this example,

$$M = \begin{bmatrix} 19 & 14 & -8 \\ 5 & 7 & -1 \\ -8 & -7 & 10 \end{bmatrix}$$

Check these calculations carefully and see if you agree! ■

An idea closely associated with that of a minor is that of a **cofactor**. This is sometimes known as the **signed minor** because it has the same absolute value as the minor. We define

$$A_{ij} = (-1)^{i+j} M_{ij}$$

This rule is by no means as complicated as it looks. We see that

1 If $i + j$ is odd then $(-1)^{i+j} = -1$;

2 If $i + j$ is even then $(-1)^{i+j} = 1$.

So that to obtain the cofactor we take the minor, and if $i + j$ is odd we change its sign, whereas if $i + j$ is even we leave it as it is. That is,

$$\begin{aligned} A_{ij} &= M_{ij} && \text{if } i + j \text{ is even} \\ A_{ij} &= -M_{ij} && \text{if } i + j \text{ is odd} \end{aligned}$$

An easy way to see whether or not to change the sign of a minor when calculating the cofactor is to note that the rule provides us with a sign convention:

$$\begin{bmatrix} + & - & + & - & + & \cdot & \cdot & \cdot \\ - & + & - & + & \cdot & \cdot & \cdot & \cdot \\ + & \cdot & \cdot & \cdot & \cdot & \cdot & \cdot & \cdot \\ \cdot & \cdot & & & & & & \\ \cdot & \cdot & & & & & & \end{bmatrix}$$

where + indicates that the sign is unchanged, and − indicates that the sign must be changed.

We now return to the evaluation of the third-order determinant:

$$\begin{vmatrix} a_{11} & a_{12} & a_{13} \\ a_{21} & a_{22} & a_{23} \\ a_{31} & a_{32} & a_{33} \end{vmatrix} = a_{11}M_{11} - a_{12}M_{12} + a_{13}M_{13} = a_{11}A_{11} + a_{12}A_{12} + a_{13}A_{13}$$

In fact we can evaluate this determinant in terms of the elements in any row or any column provided we multiply each element by its appropriate cofactor and add the results. So, for instance,

$$\begin{aligned} |A| &= a_{12}A_{12} + a_{22}A_{22} + a_{32}A_{32} \\ &= a_{31}A_{31} + a_{32}A_{32} + a_{33}A_{33} \end{aligned}$$

Check this carefully from the previous example.

It will help to fix the ideas of minor and cofactor in your mind so that you can be sure you know the difference. Here is an example for you to try.

□ Evaluate the determinant given below by using **a** the elements of the second row **b** the elements of the third column:

$$\begin{vmatrix} 8 & 7 & 6 \\ 3 & 9 & 1 \\ 2 & 2 & 4 \end{vmatrix}$$

Here is the working:

a $|A| = a_{21}A_{21} + a_{22}A_{22} + a_{23}A_{23}$

$$= -3 \times \begin{vmatrix} 7 & 6 \\ 2 & 4 \end{vmatrix} + 9 \times \begin{vmatrix} 8 & 6 \\ 2 & 4 \end{vmatrix} - \begin{vmatrix} 8 & 7 \\ 2 & 2 \end{vmatrix}$$

$$= -3(7 \times 4 - 6 \times 2) + 9(8 \times 4 - 2 \times 6) + (-1)\,(8 \times 2 - 2 \times 7)$$

$$= (-3) \times 16 + 9 \times 20 - 2 = -48 + 180 - 2 = 130$$

b $|A| = a_{13}A_{13} + a_{23}A_{23} + a_{33}A_{33}$

$$= 6 \times \begin{vmatrix} 3 & 9 \\ 2 & 2 \end{vmatrix} - \begin{vmatrix} 8 & 7 \\ 2 & 2 \end{vmatrix} + 4 \times \begin{vmatrix} 8 & 7 \\ 3 & 9 \end{vmatrix}$$

$$= 6(6 - 18) - (16 - 14) + 4(72 - 21)$$

$$= -72 - 2 + 204 = 130$$

■

Although we shall have little need to evaluate determinants of order higher than three, we remark that the same rule applies:

1 Select any row or any column;
2 Multiply each element in it by its cofactor;
3 Obtain the total.

As we can see, to evaluate a fourth-order determinant we shall have to evaluate four determinants of order three. This is equivalent to evaluating twelve determinants of order two. Similarly, to evaluate a fifth-order determinant requires the evaluation of sixty determinants of order two. The evaluation of determinants of high order is therefore a very inefficient way of using a computer, and should be avoided.

MIXED COFACTORS

One curious property of determinants is that if we multiply each element in a row by the corresponding cofactor in another row, the sum is always zero. We shall need this property a little later, and so it is worth illustrating it.

□ In the previous example the cofactors of the second row were -16, 20 and -2 respectively. Therefore using these with the elements in the first row gives

$$8 \times (-16) + 7 \times 20 + 6 \times (-2) = 0$$

Likewise with the elements in the third row we obtain

$$2 \times (-16) + 2 \times 20 + 4 \times (-2) = 0$$

as claimed.

■

□ Check this property for the columns by using the cofactors of the third column with the elements in columns 1 and 2 in turn. ■

12.4 RULES FOR DETERMINANTS

From time to time we shall need to evaluate third-order determinants. A number of rules have been devised which enable us to simplify determinants before we evaluate them. Of course, we know that if we change the elements in a matrix, we obtain a different matrix. However, we should remember that a determinant can be *evaluated*; it is a number, and it is possible for several ostensibly different determinants to have the same value. It is this property which is used to both theoretical and practical advantage when determinants are simplified before being evaluated.

Rule 1 The value of a determinant is the same as the value of its transpose.

In other words if we interchange each of the rows with each of the corresponding columns, the value of the determinant does not change.

□ Check that the determinants of these two matrices are the same:

$$\begin{bmatrix} 2 & 3 & 4 \\ 5 & 11 & 6 \\ 7 & -4 & 8 \end{bmatrix} \qquad \begin{bmatrix} 2 & 5 & 7 \\ 3 & 11 & -4 \\ 4 & 6 & 8 \end{bmatrix}$$ ■

Of course, this rule by itself is not going to simplify the numbers inside the determinant. However, it does tell us the mathematical equivalent of 'What's sauce for the goose is sauce for the gander', or in this case 'What's true for rows is true for columns'! So for any statement we make about the rows of a determinant there is a corresponding statement about the columns of a determinant.

Rule 2 If two rows of a determinant are interchanged then the determinant changes sign.

For example, if a determinant had the value 54 then, were we to interchange two of its rows, the new determinant would have the value -54.

For instance, if we evaluate the determinant

$$\begin{vmatrix} 1 & 3 & 7 \\ 2 & 9 & 6 \\ 5 & 4 & 9 \end{vmatrix}$$

we obtain the number -166. Again, if we interchange the second and third rows we obtain

$$\begin{vmatrix} 1 & 3 & 7 \\ 5 & 4 & 9 \\ 2 & 9 & 6 \end{vmatrix}$$

which you should be able easily to verify has the value 166.

One useful consequence of this rule is that if a determinant has two rows the same, then it must be zero. This is easy to see if we consider the effect of interchanging the equal rows. On the one hand the determinant has not been changed at all, and yet on the other hand, by rule 2, the determinant has changed signs. Consequently $D = -D$ and so $2D = 0$; therefore $D = 0$. This is a very useful rule, so let us repeat it:

If a determinant has two identical rows or two identical columns, it is zero.

Rule 3 A determinant may be multiplied by a number by selecting any single row (or column) and multiplying all its elements by the number. So

$$3 \times \begin{vmatrix} 2 & 1 & 2 \\ 1 & 2 & 3 \\ 4 & -1 & 6 \end{vmatrix} = \begin{vmatrix} 6 & 1 & 2 \\ 3 & 2 & 3 \\ 12 & -1 & 6 \end{vmatrix}$$

Here we have multiplied all the elements in the first column by 3.

There are two important points to note here:

1 The rule is in marked contrast to the rule for multiplying a matrix by a number (where every element in the matrix must be multiplied by the number).

2 The principal application of the rule is its reverse. That is, we simplify the arithmetic by taking out a common factor from a row or a column.

□

$$\begin{vmatrix} 18 & 9 & 4 \\ 6 & 15 & 12 \\ 12 & 3 & 16 \end{vmatrix} = 6 \times \begin{vmatrix} 3 & 9 & 4 \\ 1 & 15 & 12 \\ 2 & 3 & 16 \end{vmatrix}$$

$$= 6 \times 3 \times \begin{vmatrix} 3 & 3 & 4 \\ 1 & 5 & 12 \\ 2 & 1 & 16 \end{vmatrix}$$

$$= 6 \times 3 \times 4 \times \begin{vmatrix} 3 & 3 & 1 \\ 1 & 5 & 3 \\ 2 & 1 & 4 \end{vmatrix}$$

Here we have taken out factors from the first column, second column and third column in turn. ■

Rule 4 The value of a determinant is unchanged if a constant multiple of the elements of a chosen row is added to the corresponding elements of another row.

Of course, the constant multiple can be negative, so it is possible to subtract a multiple of one row from another row. This is illustrated by the following example: the new notation is explained after the example.

□

$$\begin{vmatrix} 1 & 2 & 3 \\ 2 & 3 & 4 \\ 4 & 5 & 6 \end{vmatrix} = \begin{vmatrix} 1 & 2 & 3 \\ 0 & -1 & -2 \\ 4 & 5 & 6 \end{vmatrix} \qquad (\mathrm{r2} = \mathrm{r2} - 2\mathrm{r1})$$

$$= \begin{vmatrix} 1 & 2 & 3 \\ 0 & -1 & -2 \\ 0 & -3 & -6 \end{vmatrix} \qquad (\mathrm{r3} = \mathrm{r3} - 4\mathrm{r1})$$

$$= 3 \times \begin{vmatrix} 1 & 2 & 3 \\ 0 & -1 & -2 \\ 0 & -1 & -2 \end{vmatrix} = 0 \qquad (\text{two equal rows})$$ ■

With practice it is often possible to perform several of these operations at the same time. In order to check back it is important to record, alongside the determinant, the operations which have been performed. In this way we can easily check for errors. Notice also the use of this strange notation. For example, r2 = r2 – 2r1 means the new row 2 is equal to the old row 2 minus twice the old row 1. The equality sign is used here in the same way as it is used in computer programming. Columns are referred to by the letter c.

One word of warning: when using this rule you must keep the row you are subtracting fixed. This can easily be overlooked if you attempt several operations of this kind together. The best way to check that your operations are valid is to see if they could all take place in a logical sequence, one at a time.

Here is an example for you to try. There are many ways of doing the problem; one way is shown below.

□ Simplify and thereby evaluate the determinant

$$\begin{vmatrix} 13 & 15 & 18 \\ 15 & 17 & 21 \\ 14 & 16 & 27 \end{vmatrix}$$

Here is a solution:

$$\begin{vmatrix} 13 & 15 & 18 \\ 15 & 17 & 21 \\ 14 & 16 & 27 \end{vmatrix} = \begin{vmatrix} 13 & 15 & 18 \\ 2 & 2 & 3 \\ 1 & 1 & 9 \end{vmatrix} \qquad \begin{matrix} \\ (\text{r2} = \text{r2} - \text{r1}) \\ (\text{r3} = \text{r3} - \text{r1}) \end{matrix}$$

$$= 3 \times \begin{vmatrix} 13 & 15 & 6 \\ 2 & 2 & 1 \\ 1 & 1 & 3 \end{vmatrix} \qquad (\text{c3} = \text{c3}/3)$$

$$= 3 \times \begin{vmatrix} 13 & 15 & 6 \\ 0 & 0 & -5 \\ 1 & 1 & 3 \end{vmatrix} \qquad (\text{r2} = \text{r2} - 2\text{r3})$$

$$= 3 \times [-(-5)] \times \begin{vmatrix} 13 & 15 \\ 1 & 1 \end{vmatrix} = -30$$ ■

□ Solve the equation

$$\begin{vmatrix} w^2 - 1 & w + 1 & w - 1 \\ w + 1 & w - 1 & w + 1 \\ w - 1 & 2w & 0 \end{vmatrix} = 0$$

We are looking for all values of w which satisfy this equation. We should be very unwise to multiply out the determinant straight away because this would result in an equation of the fourth degree and solutions might be difficult to spot. Instead we use the rules for simplifying determinants:

$$\begin{vmatrix} w^2 - 1 & 2w & w - 1 \\ w + 1 & 2w & w + 1 \\ w - 1 & 2w & 0 \end{vmatrix} = 0 \qquad (\text{c2} = \text{c2} + \text{c3})$$

$$2w \begin{vmatrix} w^2 - 1 & 1 & w - 1 \\ w + 1 & 1 & w + 1 \\ w - 1 & 1 & 0 \end{vmatrix} = 0 \qquad (\text{c2} = \text{c2}/2w)$$

$$2w \begin{vmatrix} w^2 - w & 0 & w - 1 \\ w + 1 & 1 & w + 1 \\ w - 1 & 1 & 0 \end{vmatrix} = 0 \qquad (\text{r1} = \text{r1} - \text{r3})$$

$$2w(w-1) \begin{vmatrix} w & 0 & 1 \\ w + 1 & 1 & w + 1 \\ w - 1 & 1 & 0 \end{vmatrix} = 0 \qquad (\text{c1} = \text{c1}/(w - 1))$$

$$2w(w-1) \begin{vmatrix} w - 1 & 0 & 1 \\ 0 & 1 & w + 1 \\ w - 1 & 1 & 0 \end{vmatrix} = 0 \qquad (\text{c1} = \text{c1} - \text{c3})$$

$$2w(w-1)^2\begin{vmatrix}1 & 0 & 1\\ 0 & 1 & w+1\\ 1 & 1 & 0\end{vmatrix} = 0 \qquad (\text{c}1 = \text{c}1/(w-1))$$

Expanding the determinant in terms of the first column gives

$$2w(w-1)^2[-(w+1)+(-1)] = 2w(w-1)^2(-w-2) = 0$$

So that $w = 0$, $w = 1$ (repeated root) or $w = -2$. ■

Before we leave this section we remark that if A and B are square matrices of order n then the product AB is also square of order n. The following rule holds:

$$\det AB = \det A \det B$$

Therefore the same rule that we used for multiplying matrices together can be used for multiplying determinants.

□ Check this property with the matrices

$$\begin{bmatrix}2 & 3 & -5\\ 1 & -1 & 6\\ 4 & 2 & 3\end{bmatrix} \qquad \begin{bmatrix}1 & 3 & 9\\ 4 & -2 & 6\\ -1 & 5 & 7\end{bmatrix}$$ ■

Here are some steps for you to take to check that everything is all right.

12.5 Workshop

1 **Exercise** Using Cramer's rule, express u and v explicitly in terms of $t(t \neq -1)$ from the following:

$$\begin{aligned} tu \quad + (2t+1)v &= (t+1)^2\\ (t-1)u + \quad 2tv \quad &= t(t+1)\end{aligned}$$

Try it, then move to step 2 for the solution.

2 We obtain

$$\begin{aligned}\Delta &= \begin{vmatrix} t & 2t+1\\ t-1 & 2t\end{vmatrix}\\ &= 2t^2 - (t-1)(2t+1)\\ &= 2t^2 - (2t^2 - t - 1)\\ &= t+1\end{aligned}$$

Now

$$\begin{aligned}\Delta u = (t+1)u &= \begin{vmatrix}(t+1)^2 & 2t+1\\ t(t+1) & 2t\end{vmatrix}\\ &= (t+1)^2 2t - t(t+1)(2t+1)\\ &= (t^2+2t+1)2t - t(2t^2+3t+1)\\ &= 2t^3+4t^2+2t-2t^3-3t^2-t\\ &= t^2+t = t(t+1)\end{aligned}$$

Consequently $u = t$. To obtain v, either substitute u into one of the equations or use Cramer's rule again. You should obtain $v = 1$.

If you were successful then move straight on to step 3. Otherwise, here is a similar problem to try.

▷ **Exercise** Solve, using Cramer's rule,

$$\begin{aligned}3x + 2y &= 1\\ 4x - y &= 5\end{aligned}$$

You should obtain $\Delta = -11$ and thereby the correct result $x = 1$ and $y = -1$.

If you came unstuck with the first exercise, go back and give it another go before going on to step 3.

▷ **Exercise** Evaluate the determinant 3

$$\begin{vmatrix}2 & 1 & -1\\ 3 & 0 & 1\\ 1 & 2 & 2\end{vmatrix}$$

in terms of the second row and the second column.

Try it, then step ahead.

First, expanding in terms of the second row the determinant is 4

$$\begin{aligned}-3\begin{vmatrix}1 & -1\\ 2 & 2\end{vmatrix} + 0\begin{vmatrix}2 & -1\\ 1 & 2\end{vmatrix} - 1\begin{vmatrix}2 & 1\\ 1 & 2\end{vmatrix}\\ = -3[2-(-2)] + 0 - (2\times 2 - 1\times 1)\\ = -3\times 4 - 4 + 1 = -12 - 4 + 1 = -15\end{aligned}$$

Next, expanding in terms of the second column the determinant is

$$\begin{aligned}(-1)\begin{vmatrix}3 & 1\\ 1 & 2\end{vmatrix} + 0\begin{vmatrix}2 & -1\\ 1 & 2\end{vmatrix} - 2\begin{vmatrix}2 & -1\\ 3 & 1\end{vmatrix}\\ = (-1)(3\times 2 - 1\times 1) + 0 - 2[2\times 1 - 3\times(-1)]\end{aligned}$$

$$= (-1)(6 - 1) - 2(2 + 3)$$
$$= (-1) \times 5 - 2 \times 5 = -5 - 10 = -15$$

Don't forget the sign convention. Of course, if there are zero elements in the row or column we do not normally bother to write down the minors which correspond to them, because their contribution to the value of the determinant is zero.

If you got this one right, move ahead to step 5. If not, try this similar problem.

▷**Exercise** Evaluate the following determinant in terms of the third row and the second column:

$$\begin{vmatrix} 3 & 5 & 1 \\ 2 & 1 & 0 \\ 1 & 4 & -1 \end{vmatrix}$$

The answer is 14.

Now step forward.

5 **Exercise** Simplify, and thereby evaluate, the determinant

$$\begin{vmatrix} 15 & 20 & 28 \\ 28 & 42 & 59 \\ 21 & 32 & 45 \end{vmatrix}$$

Have a go, then step ahead.

6 There are many ways of proceeding. Here is one of them.

Fix the first row and subtract multiples of it from the others:

$$\begin{vmatrix} 15 & 20 & 28 \\ -2 & 2 & 3 \\ 6 & 12 & 17 \end{vmatrix} \qquad \begin{matrix} \\ (r2 = r2 - 2r1) \\ (r3 = r3 - r1) \end{matrix}$$

Now fix the second column and subtract multiples from the others. We are trying to produce zeros and reduce the large numbers:

$$\begin{vmatrix} 35 & 20 & 8 \\ 0 & 2 & 1 \\ 18 & 12 & 5 \end{vmatrix} \qquad \begin{matrix} (c1 = c1 + c2) \\ (c3 = c3 - c2) \end{matrix}$$

We can produce a second zero in the second row:

$$\begin{vmatrix} 35 & 4 & 8 \\ 0 & 0 & 1 \\ 18 & 2 & 5 \end{vmatrix} \qquad (c2 = c2 - 2c3)$$

$$= (-1)\begin{vmatrix} 35 & 4 \\ 18 & 2 \end{vmatrix}$$

$$= (-1)(70 - 72) = 2$$

If all is well you may move on to step 7. If not, check back carefully. Are you remembering to keep a note of the operations you have been using?

Now try this exercise.

▷**Exercise** Simplify the following determinant and then evaluate it:

$$\begin{vmatrix} 18 & 23 & 32 \\ 32 & 41 & 59 \\ 25 & 32 & 47 \end{vmatrix}$$

The correct answer is 3.

Now let's check that we can use these operations algebraically.

7

▷**Exercise** Solve the equation

$$\begin{vmatrix} 2x - 3 & 3x - 5 & 4x - 8 \\ 3x - 5 & 5x - 9 & 6x - 12 \\ 4x - 6 & 6x - 10 & 9x - 19 \end{vmatrix} = 0$$

Try it, then step forward for the solution.

8

Here is one way of solving this problem. There are alternative approaches, but it would be unwise to evaluate the determinant straight away because this would result in a cubic equation which on occasions might be difficult to solve.

There are no obvious factors, and so we look to see whether or not subtracting rows or columns will produce any. Alternatively we should like to produce some zeros.

We have, using r2 = r2 − r1 and r3 = r3 − 2r1,

$$\begin{vmatrix} 2x - 3 & 3x - 5 & 4x - 8 \\ x - 2 & 2x - 4 & 2x - 4 \\ 0 & 0 & x - 3 \end{vmatrix} = 0$$

This enables us to take out the factor $x - 3$ from the third row and the factor $x - 2$ from the second row. We obtain

$$(x-2)(x-3)\begin{vmatrix} 2x-3 & 3x-5 & 4x-8 \\ 1 & 2 & 2 \\ 0 & 0 & 1 \end{vmatrix} = 0$$

Therefore

$$\begin{aligned}(x-2)(x-3)[2(2x-3)-(3x-5)] &= 0\\ (x-2)(x-3)(x-1) &= 0\end{aligned}$$

so that $x = 1$, $x = 2$ or $x = 3$.

If you got it right, move on to the next section. If not, have a go at another one of these exercises.

▷**Exercise** Solve the equation

$$\begin{vmatrix} 6u+1 & 3u+1 & 2u+1 \\ 9u+1 & 5u+1 & 3u+1 \\ 14u+4 & 7u+3 & 5u+3 \end{vmatrix} = 0$$

The correct answer is $u = 1$, $u = 0$ or $u = -1$.

Here now is a problem which can be solved by using determinants.

12.6 Practical

ELECTRICAL BRIDGE

A Wheatstone bridge (Fig. 12.1) has the following set of equations for the loop currents i_1, i_2 and i_3:

$$\begin{aligned} i_1r_1 + (i_1 - i_2)R_4 + (i_1 - i_3)R_3 &= E\\ (i_1 - i_3)R_3 - i_3R_1 - (i_3 - i_2)r_2 &= 0\\ (i_1 - i_2)R_4 + (i_3 - i_2)r_2 - i_2R_2 &= 0\end{aligned}$$

When the bridge is balanced, no current flows through r_2. Show that $R_1R_4 = R_2R_3$ when the bridge is balanced.

See if you can sort this out using Cramer's rule. We shall tackle the problem stage by stage.

The first thing to realize is that we are interested in i_1, i_2 and i_3, so that these are the 'unknowns' in our equations. Therefore we rewrite the equations in this form:

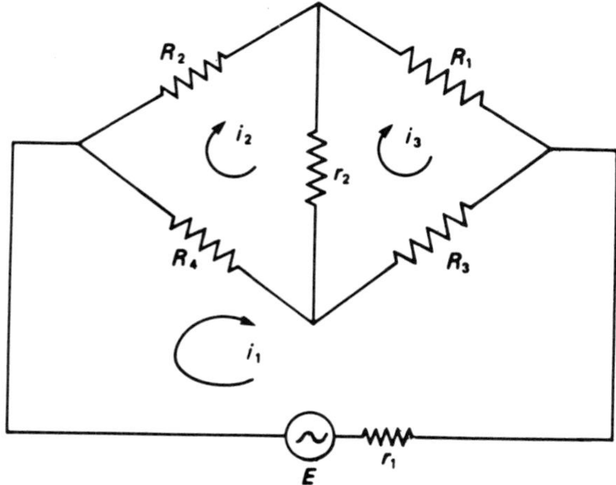

Fig. 12.1 A Wheatstone bridge.

$$\begin{array}{lcccl} (r_1 + R_4 + R_3)i_1 & - & R_4 i_2 & - & R_3 i_3 = E \\ R_3 i_1 & + & r_2 i_2 & - & (R_1 + R_3 + r_2)i_3 = 0 \\ R_4 i_1 & - & (R_2 + R_4 + r_2)i_2 & + & r_2 i_3 = 0 \end{array}$$

Now use Cramer's rule delicately. We don't need it with all its weight.

We can argue in the following way. When the bridge is balanced there is no current through r_2 and so $i_2 = i_3$. Therefore $\Delta i_2 = \Delta i_3$. We do not need to calculate Δ itself, only Δi_2 and Δi_3. Write down the determinants Δi_2 and Δi_3 and see if you can complete the solution.

By Cramer's rule

$$\Delta i_2 = \begin{vmatrix} r_1 + R_4 + R_3 & E & -R_3 \\ R_3 & 0 & -(R_1 + R_3 + r_2) \\ R_4 & 0 & r_2 \end{vmatrix}$$

$$\Delta i_3 = \begin{vmatrix} r_1 + R_4 + R_3 & -R_4 & E \\ R_3 & r_2 & 0 \\ R_4 & -(R_2 + R_4 + r_2) & 0 \end{vmatrix}$$

When the bridge is balanced we can equate these two determinants.

Do this and then check the final stage.

We expand the determinant Δi_2 in terms of the second column and the determinant Δi_3 in terms of the third column:

$$-E\begin{vmatrix} R_3 & -(R_1 + R_2 + r_2) \\ R_4 & r_2 \end{vmatrix} = E\begin{vmatrix} R_3 & r_2 \\ R_4 & -(R_2 + R_4 + r_2) \end{vmatrix}$$

That is,

$$-[R_3 r_2 + R_4(R_1 + R_3 + r_2)] = -R_3(R_2 + R_4 + r_2) - R_4 r_2$$

From which

$$R_1 R_4 = R_2 R_3$$

Determinants have their uses, independent of matrix work. Cramer's rule is one example of this, and we shall come across another occasion where a knowledge of determinants can be valuable. This will be when we derive a formula for the vector product of two vectors in terms of components (Chapter 14). We shall see then that this can be expressed using a determinant.

There are several other examples too. For instance, suppose we have a triangle in the plane with vertices (x_1, y_1), (x_2, y_2) and (x_3, y_3). The area of this triangle can be expressed neatly in terms of a determinant:

$$\pm\tfrac{1}{2}\begin{vmatrix} x_1 & y_1 & 1 \\ x_2 & y_2 & 1 \\ x_3 & y_3 & 1 \end{vmatrix}$$

Why not check this for yourself?

SUMMARY

In this chapter we have examined

- ☐ The definition of a minor and a cofactor and the relationship between them

$$A_{ij} = (-1)^{i+j} M_{ij}$$

- ☐ The procedure for evaluating determinants
  - **a** Select any row or any column;
  - **b** Multiply each element in it by its cofactor;
  - **c** Obtain the total.
- ☐ Cramer's rule for solving simultaneous equations.
- ☐ Rules for simplifying determinants
  - **a** The value of a determinant is the same as the value of its transpose.
  - **b** If two rows of a determinant are interchanged, then the determinant changes sign.
  - **c** A determinant may be multiplied by a number. Select any row (or column) and multiply all its elements by the number.
  - **d** The value of a determinant is unchanged if a constant multiple of the elements of a chosen row are added to the corresponding elements of another row.

EXERCISES

1 Solve the following equations:

a $\begin{vmatrix} x+1 & 2 \\ 6 & x-3 \end{vmatrix} = 0$ **b** $\begin{vmatrix} x-1 & 3 \\ 6 & x-4 \end{vmatrix} = 0$

c $\begin{vmatrix} x-2 & x \\ x & x-6 \end{vmatrix} = 0$ **d** $\begin{vmatrix} x+1 & x+2 \\ x+2 & x+4 \end{vmatrix} = 0$

2 Evaluate the following determinants:

a $\begin{vmatrix} 2 & 0 & 1 \\ -1 & 1 & 2 \\ 3 & 0 & 4 \end{vmatrix}$ **b** $\begin{vmatrix} 1 & 2 & 1 \\ -5 & 1 & -1 \\ 4 & 3 & 0 \end{vmatrix}$

c $\begin{vmatrix} 2 & 1 & 1 \\ 1 & 3 & 5 \\ 4 & 1 & 2 \end{vmatrix}$ **d** $\begin{vmatrix} x+1 & x+2 & x+3 \\ x+2 & x+3 & x+1 \\ x+3 & x+1 & x+2 \end{vmatrix}$

3 Evaluate, by simplifying first,

a $\begin{vmatrix} 21 & 15 & 14 \\ 18 & 45 & 7 \\ 24 & 40 & 21 \end{vmatrix}$ **b** $\begin{vmatrix} 75 & 48 & 90 \\ 125 & 64 & 75 \\ 50 & 32 & 45 \end{vmatrix}$

4 Use Cramer's rule to solve the simultaneous equations where a, u, v and w are known:

a $x\surd(1 + a^2) + ya = (1 + a^2)^{-1/2}$
$xa + y\surd(1 + a^2) = 1/a$

b $x - y\exp(u + v) + z\exp(u - w) = \exp u$
$x\exp(v - u) + y\exp 2v - z\exp(v - w) = \exp v$
$x\exp(w - u) - y\exp(w + v) - z = -\exp w$

ASSIGNMENT

1 Solve the equation

$$\begin{vmatrix} x - 2 & 3 \\ 4 & x + 2 \end{vmatrix} = 0$$

2 Simplify and thereby evaluate

$$\begin{vmatrix} 19 & 18 & 25 \\ 22 & 21 & 29 \\ 20 & 28 & 32 \end{vmatrix}$$

3 Solve the equation

$$\begin{vmatrix} x + 1 & x & x - 4 \\ 2 & 1 & -4 \\ 3 & 5 & 1 \end{vmatrix} = 0$$

4 Obtain x if

$$\begin{vmatrix} x - 4 & x - 2 & x \\ x - 3 & x - 1 & x + 2 \\ x - 2 & x + 4 & 3x \end{vmatrix} = 0$$

5 Obtain M, the matrix of minors of the matrix

$$A = \begin{bmatrix} 7 & 6 & 9 \\ 4 & 5 & 6 \\ 7 & 8 & 10 \end{bmatrix}$$

6 Obtain C, the matrix of cofactors of the matrix

$$A = \begin{bmatrix} 5 & 2 & 4 \\ 4 & 3 & 5 \\ 2 & 2 & 3 \end{bmatrix}$$

FURTHER EXERCISES

1 Obtain w if

$$\begin{vmatrix} w-6 & w-4 & w-2 \\ w-5 & w-2 & w+2 \\ w-4 & w & w+6 \end{vmatrix} = 0$$

2 Show that

$$\begin{vmatrix} x^2 & 2x & -2 \\ 2x & 2-x^2 & 2x \\ 2 & -2x & -x^2 \end{vmatrix} = (x^2+2)^3$$

Hence or otherwise obtain the possible values of x if the determinant is known to have the value 27.

3 Obtain an expression for k in terms of a, b, c and d if

$$\begin{vmatrix} 1+a & 1 & 1 & 1 \\ 1 & 1+b & 1 & 1 \\ 1 & 1 & 1+c & 1 \\ 1 & 1 & 1 & 1+d \end{vmatrix} = abcd(k+1)$$

4 It is given that the arithmetic mean of the three numbers a, b and c is zero and that the root mean square (RMS) value is the square root of 8. Solve the equation

$$\begin{vmatrix} x+a & c & b \\ c & x+b & a \\ b & a & x+c \end{vmatrix} = 0$$

Note: the RMS value is the square root of the arithmetic mean of the squares of the numbers.

5 Show that if A is the matrix

$$\begin{bmatrix} 0 & \cos u & \sin u \\ \cos v & -\sin u \sin v & \cos u \sin v \\ \sin v & \sin u \cos v & -\cos u \cos v \end{bmatrix}$$

then $\det A = 1$.

6 Solve the equation

$$\begin{vmatrix} x+1 & 2x+3 & 3x+5 \\ 3x+3 & 5x+7 & 7x+11 \\ 5x+6 & 8x+12 & 14x+24 \end{vmatrix} = 0$$

7 Simplify and then evaluate

$$\begin{vmatrix} 27 & 37 & 42 \\ 11 & 15 & 16 \\ 23 & 31 & 34 \end{vmatrix}$$

8 Without evaluating them, show that the following determinants are zero:

a $\begin{vmatrix} 1 & yz & yz(y+z) \\ 1 & zx & zx(z+x) \\ 1 & xy & xy(x+y) \end{vmatrix}$

b $\begin{vmatrix} a+x & x+y & a-y \\ x+w & x-a & a+w \\ y+w & w-x & x+y \end{vmatrix}$

9 Use Cramer's rule to express u and v explicitly in terms of w where

$$\begin{aligned} u + v \sin w &= \sec w \\ u \tan w + v \cos w &= 2 \sin w \end{aligned}$$

10 The loop currents in a circuit satisfy

$$\begin{aligned} R_1 i_1 + R_2(i_1 - i_3) &= 0 \\ R_3 i_2 + R_4(i_2 - i_3) &= 0 \\ R_1 i_1 + R_3 i_2 &= 0 \end{aligned}$$

where R_1, R_2, R_3 and R_4 are resistances. Show that if not all the currents are zero then

$$\frac{1}{R_1} + \frac{1}{R_2} + \frac{1}{R_3} + \frac{1}{R_4} = 0$$

11 Determine those values of λ such that

$$\begin{vmatrix} 1-\lambda & 2 & 1 \\ -3 & \lambda & -2 \\ 2 & -2 & 1+\lambda \end{vmatrix} = 0$$

12 In a Wheatstone Bridge, i denotes the current through the galvanometer. Given that

$$\begin{aligned} R_2 i_1 + R_4(i_1 - i) &= E \\ R_4(i_1 - i) - R_3(i_2 + i) - Ri &= 0 \\ R_2 i_1 - R_1 i_2 + Ri &= 0 \end{aligned}$$

use Cramer's rule to obtain the value of i. Thereby deduce the usual condition for the bridge to be balanced.

13 Inverse matrices

We have seen what determinants are and how to expand and simplify them. Our next task is to see what part they play in matrix algebra.

When you have completed this chapter you should be able to

- ☐ Decide when a matrix is non-singular;
- ☐ Calculate the inverse of a non-singular matrix by the formula;
- ☐ List the elementary row transformations;
- ☐ Calculate the inverse of a non-singular matrix using row transformations;
- ☐ Apply the method of systematic elimination to solve simultaneous equations.

At the end of this chapter we shall solve a practical problem involving a binary code.

13.1 THE INVERSE OF A SQUARE MATRIX

We now turn our attention to the problem of finding the inverse of a square matrix A. This is the nearest we get to an operation of division for matrices. We are looking for a matrix B such that

$$AB = BA = I$$

where I is the identity matrix.

The fact that there is at most one such matrix can be deduced as follows. Suppose there is a square matrix A for which there are two matrices B and C such that

$$\begin{aligned} AB &= BA = I \\ AC &= CA = I \end{aligned}$$

We have

$$C = CI = C(AB) = (CA)B = IB = B$$

So we have deduced $C = B$. That is, we have shown that if the square matrix A has an inverse then the inverse is *unique*. We denote the inverse of A, if it has one, by A^{-1}.

Suppose A is a square matrix. Then we have seen how to form from A a matrix C of cofactors (Chapter 12). That is, if

$$A = \begin{bmatrix} a_{11} & a_{12} & a_{13} \\ a_{21} & a_{22} & a_{23} \\ a_{31} & a_{32} & a_{33} \end{bmatrix}$$

then

$$C = \begin{bmatrix} A_{11} & A_{12} & A_{13} \\ A_{21} & A_{22} & A_{23} \\ A_{31} & A_{32} & A_{33} \end{bmatrix}$$

The transpose of C is known as the **adjoint** (or **adjugate**) matrix of A and is denoted by adj A. So

$$\text{adj } A = \begin{bmatrix} A_{11} & A_{21} & A_{31} \\ A_{12} & A_{22} & A_{32} \\ A_{13} & A_{23} & A_{33} \end{bmatrix}$$

From what we have done before, we deduce that

$$A(\text{adj } A) = (\text{adj } A)A = (\det A)I$$

We illustrate this property in the case of a 3×3 matrix:

$$\begin{bmatrix} a_{11} & a_{12} & a_{13} \\ a_{21} & a_{22} & a_{23} \\ a_{31} & a_{32} & a_{33} \end{bmatrix} \begin{bmatrix} A_{11} & A_{21} & A_{31} \\ A_{12} & A_{22} & A_{32} \\ A_{13} & A_{23} & A_{33} \end{bmatrix}$$

$$= \begin{bmatrix} a_{11}A_{11}+a_{12}A_{12}+a_{13}A_{13} & a_{11}A_{21}+a_{12}A_{22}+a_{13}A_{23} & a_{11}A_{31}+a_{12}A_{32}+a_{13}A_{33} \\ a_{21}A_{11}+a_{22}A_{12}+a_{23}A_{13} & a_{21}A_{21}+a_{22}A_{22}+a_{23}A_{23} & a_{21}A_{31}+a_{22}A_{32}+a_{23}A_{33} \\ a_{31}A_{11}+a_{32}A_{12}+a_{33}A_{13} & a_{31}A_{21}+a_{32}A_{22}+a_{33}A_{23} & a_{31}A_{31}+a_{32}A_{32}+a_{33}A_{33} \end{bmatrix}$$

$$= \begin{bmatrix} |A| & 0 & 0 \\ 0 & |A| & 0 \\ 0 & 0 & |A| \end{bmatrix} = |A|I$$

Consequently, if $|A|$ is non-zero then the inverse of A exists and is given by

$$A^{-1} = \frac{1}{|A|}\text{adj } A$$

Any square matrix with a zero determinant is called a **singular matrix**. (Singular is used here in the sense of 'unusual', as when Sherlock Holmes remarks to Dr Watson on a singular occurrence.) Consequently when a square matrix has a non-zero determinant, it is called a **non-singular matrix**. From what we have done we can now assert that every non-singular square matrix has a unique inverse given by the formula

$$A^{-1} = \frac{1}{|A|}\text{adj } A$$

It follows therefore that for such matrices if $AB = I$ then $BA = I$. However, this is a consequence of the algebraic structure satisfied by the elements of the matrices. Real and complex numbers are examples of what mathematicians call **fields**. If the elements in the matrices did not belong to fields then many of the conclusions we have reached would no longer hold.

13.2 ROW TRANSFORMATIONS

Although the formula for the inverse of a square matrix is very useful, there is a procedure which can often be used to obtain the inverse more quickly. This procedure is known as the method of row operations.

An operation on a matrix is called an **elementary row transformation** if and only if it is one of the following:

1 An interchange of two rows;
2 Multiplication of the elements in a row by a non-zero number;
3 Subtraction of the elements of one row from the corresponding elements of another row.

A sequence of elementary row transformations results in a **row transformation**. Of course the matrix will be changed as a result of a row transformation, but the matrix which results is said to be **row equivalent** to the original matrix.

An **elementary matrix** is a matrix obtained by performing an elementary row transformation on the identity matrix I. In fact, it is easy to see that an elementary row transformation can be effected by pre-multiplying the matrix by an elementary matrix.

□

$$E_1A = \begin{bmatrix} 0 & 0 & 1 \\ 0 & 1 & 0 \\ 1 & 0 & 0 \end{bmatrix} \begin{bmatrix} a & b & c \\ d & e & f \\ g & h & i \end{bmatrix} = \begin{bmatrix} g & h & i \\ d & e & f \\ a & b & c \end{bmatrix}$$

E_1 has been obtained from the identity matrix by interchanging row 1 and row 3. ■

□

$$E_2A = \begin{bmatrix} 1 & 0 & 0 \\ 0 & k & 0 \\ 0 & 0 & 1 \end{bmatrix} \begin{bmatrix} a & b & c \\ d & e & f \\ g & h & i \end{bmatrix} = \begin{bmatrix} a & b & c \\ kd & ke & kf \\ g & h & i \end{bmatrix}$$

E_2 has been obtained from the identity matrix by multiplying row 2 by k. ■

□

$$E_3A = \begin{bmatrix} 1 & 0 & 0 \\ 0 & 1 & -1 \\ 0 & 0 & 1 \end{bmatrix} \begin{bmatrix} a & b & c \\ d & e & f \\ g & h & i \end{bmatrix} = \begin{bmatrix} a & b & c \\ d-g & e-h & f-i \\ g & h & i \end{bmatrix}$$

E_3 has been obtained from the identity matrix by subtracting row 3 from row 2. ■

Suppose now it is possible, using a sequence of elementary row transformations, to reduce a matrix A to the identity matrix I. Then

$$(E_s \ldots E_2E_1)A = I$$

It follows that

$$E_s \ldots E_2E_1 = A^{-1}$$

Moreover,

$$A^{-1} = E_s \dots E_2E_1 = E_s \dots E_2E_1I$$

This provides a method for obtaining the inverse of a matrix A.

13.3 OBTAINING INVERSES

To obtain the inverse of a non-singular matrix using row transformations:

1 Write down an array consisting of the matrix A on the left-hand side and an identity matrix of the same order on the right-hand side.

2 Perform a sequence of elementary row transformations on the *entire array* with the object of converting the matrix A into an identity matrix.

3 As the matrix A on the left is transformed into the identity matrix I, so the identity matrix I on the right becomes transformed into the inverse of A.

As we carry out this procedure we shall observe at each stage that the matrix on the left gets to look more and more like an identity matrix, whereas the matrix on the right gets to look more and more like the inverse of A. It's rather like watching Dr Jekyll turn into Mr Hyde!

The best way of carrying out the procedure is to work systematically column by column, starting on the left. As a first step, we arrange things so that we obtain 1 in the (1, 1) position. We then subtract multiples of the first row from the other rows so that the first column becomes the first column of an identity matrix. When an element is fixed and its row is used in this way the element is called a **pivot**; it may help to encircle the pivot at each stage.

□ Use the method of elementary row operations to obtain the inverse of the matrix

$$\begin{bmatrix} 6 & 13 & 9 \\ 3 & 7 & 5 \\ 2 & 3 & 2 \end{bmatrix}$$

We begin with the array

$$\left[\begin{array}{ccc|ccc} 6 & 13 & 9 & 1 & 0 & 0 \\ 3 & 7 & 5 & 0 & 1 & 0 \\ 2 & 3 & 2 & 0 & 0 & 1 \end{array}\right]$$

Subtract row 3 from row 2 to obtain a leading element 1 in row 2:

$$\left[\begin{array}{ccc|ccc} 6 & 13 & 9 & 1 & 0 & 0 \\ 1 & 4 & 3 & 0 & 1 & -1 \\ 2 & 3 & 2 & 0 & 0 & 1 \end{array}\right]$$

Interchange row 1 and row 2 to obtain 1 in the correct position:

$$\left[\begin{array}{ccc|ccc} \textcircled{1} & 4 & 3 & 0 & 1 & -1 \\ 6 & 13 & 9 & 1 & 0 & 0 \\ 2 & 3 & 2 & 0 & 0 & 1 \end{array}\right]$$

Subtract multiples of row 1 from row 2 and row 3 to produce the zeros in the first column:

$$\left[\begin{array}{ccc|ccc} 1 & 4 & 3 & 0 & 1 & -1 \\ 0 & -11 & -9 & 1 & -6 & 6 \\ 0 & -5 & -4 & 0 & -2 & 3 \end{array}\right]$$

Subtract twice row 3 from row 2 to obtain -1 in the (2, 2) position:

$$\left[\begin{array}{ccc|ccc} 1 & 4 & 3 & 0 & 1 & -1 \\ 0 & -1 & -1 & 1 & -2 & 0 \\ 0 & -5 & -4 & 0 & -2 & 3 \end{array}\right]$$

Change the sign of row 2:

$$\left[\begin{array}{ccc|ccc} 1 & 4 & 3 & 0 & 1 & -1 \\ 0 & \textcircled{1} & 1 & -1 & 2 & 0 \\ 0 & -5 & -4 & 0 & -2 & 3 \end{array}\right]$$

Subtract four times row 2 from row 1 and add five times row 2 to row 3:

$$\left[\begin{array}{ccc|ccc} 1 & 0 & -1 & 4 & -7 & -1 \\ 0 & 1 & 1 & -1 & 2 & 0 \\ 0 & 0 & 1 & -5 & 8 & 3 \end{array}\right]$$

Finally, add row 3 to row 1 and subtract row 3 from row 2:

$$\left[\begin{array}{ccc|ccc} 1 & 0 & 0 & -1 & 1 & 2 \\ 0 & 1 & 0 & 4 & -6 & -3 \\ 0 & 0 & 1 & -5 & 8 & 3 \end{array}\right]$$

Check:

$$\begin{bmatrix} 6 & 13 & 9 \\ 3 & 7 & 5 \\ 2 & 3 & 2 \end{bmatrix}\begin{bmatrix} -1 & 1 & 2 \\ 4 & -6 & -3 \\ -5 & 8 & 3 \end{bmatrix} = \begin{bmatrix} 1 & 0 & 0 \\ 0 & 1 & 0 \\ 0 & 0 & 1 \end{bmatrix}$$

Notice how we move column by column through the matrix, working from the left. Only when the column has been reduced to the appropriate column of an identity matrix do we proceed to the next column. ■

Here is an example for you to try. Remember to make a note of the row operations at each stage so that, if you make a mistake, it can be corrected later. By the way, it is always worth checking by multiplication that you have the correct result. In practice it is only necessary to check the product one way round; so we check either $AB = I$ or $BA = I$ but not both.

□ Obtain the inverse of the matrix

$$\begin{bmatrix} 6 & 11 & 5 \\ 18 & 34 & 15 \\ 13 & 25 & 11 \end{bmatrix}$$

When you have completed this example, check below to see if you have the correct answer.

There are many ways of proceeding to reduce the array using row transformations. Here is one of the ways:

$$\left[\begin{array}{ccc|ccc} 6 & 11 & 5 & 1 & 0 & 0 \\ 18 & 34 & 15 & 0 & 1 & 0 \\ 13 & 25 & 11 & 0 & 0 & 1 \end{array}\right]$$

To obtain 1 in row 3 we transform r3 = r3 − 2r1, and to reduce the numbers we put r2 = r2 − 3r1:

$$\left[\begin{array}{ccc|ccc} 6 & 11 & 5 & 1 & 0 & 0 \\ 0 & 1 & 0 & -3 & 1 & 0 \\ 1 & 3 & 1 & -2 & 0 & 1 \end{array}\right]$$

Interchanging row 1 and row 3 produces 1 in the first row:

$$\left[\begin{array}{ccc|ccc} \textcircled{1} & 3 & 1 & -2 & 0 & 1 \\ 0 & 1 & 0 & -3 & 1 & 0 \\ 6 & 11 & 5 & 1 & 0 & 0 \end{array}\right]$$

To complete the first column, r3 = r3 − 6r1:

$$\left[\begin{array}{ccc|ccc} 1 & 3 & 1 & -2 & 0 & 1 \\ 0 & \textcircled{1} & 0 & -3 & 1 & 0 \\ 0 & -7 & -1 & 13 & 0 & -6 \end{array}\right]$$

The second column is completed by r1 = r1 − 3r2 and r3 = r3 + 7r2:

$$\left[\begin{array}{ccc|ccc} 1 & 0 & 1 & 7 & -3 & 1 \\ 0 & 1 & 0 & -3 & 1 & 0 \\ 0 & 0 & -1 & -8 & 7 & -6 \end{array}\right]$$

Add row 3 to row 1 and then change the sign of row 3 to produce

$$\left[\begin{array}{ccc|ccc} 1 & 0 & 0 & -1 & 4 & -5 \\ 0 & 1 & 0 & -3 & 1 & 0 \\ 0 & 0 & 1 & 8 & -7 & 6 \end{array}\right]$$

Check:

$$\begin{bmatrix} 6 & 11 & 5 \\ 18 & 34 & 15 \\ 13 & 25 & 11 \end{bmatrix}\begin{bmatrix} -1 & 4 & -5 \\ -3 & 1 & 0 \\ 8 & -7 & 6 \end{bmatrix} = \begin{bmatrix} 1 & 0 & 0 \\ 0 & 1 & 0 \\ 0 & 0 & 1 \end{bmatrix}$$ ■

The rules for simplifying determinants and the use of row operations have features in common. However, you must *never* use column operations when obtaining the inverse of a matrix by row operations. (As a matter of interest there is a parallel theory using column operations instead of row operations, but the two must be kept distinct.)

13.4 SYSTEMATIC ELIMINATION

Elementary row transformations can be used to solve a system of simultaneous equations directly. To do so we use a matrix known as the **augmented matrix**. This is a matrix consisting of the matrix of coefficients with an extra column for the constants on the right-hand side. Such a matrix can be written down once we are given a system of equations; conversely, given any matrix we can write down a system of equations for which it is the augmented matrix.

□

$ax + by = h$ $cx + dy = k$	$\left[\begin{array}{cc|c} a & b & h \\ c & d & k \end{array}\right]$
equations	augmented matrix

■

If we now perform row transformations on the augmented matrix we shall produce an equivalent system of equations. The method of **systematic elimination**, otherwise known as the **Gauss elimination method**, makes use of this fact.

We perform row operations on the augmented matrix with the object of reducing it to a state in which each row has at least as many leading zeros as the previous row. In other words, if the first three elements in a row are zero then at least the first three elements of each subsequent row will be zero.

When this procedure has been completed the equations can be reconstituted. There will then be one of three possibilities:

1 The equations have a unique solution;

2 The equations are inconsistent;
3 The equations have more than one solution.
It will be clear which of these possibilities holds. If it is either 1 or 3 then the solution can be determined by back substitution. In case 3, some of the unknowns can be chosen arbitrarily.

We shall illustrate all three cases.

□ Decide whether or not the following system of simultaneous equations is consistent and, if it is, obtain the solution:

$$\begin{aligned} x + 2y - 3z &= 1 \\ x - y + 4z &= 5 \\ 2x + y - 3z &= 2 \\ 4x - 2y - 2z &= 4 \\ x + y + 3z &= 6 \end{aligned}$$

Here we have five equations in three unknowns. The augmented matrix is

$$\left[\begin{array}{ccc|c} \textcircled{1} & 2 & -3 & 1 \\ 1 & -1 & 4 & 5 \\ 2 & 1 & -3 & 2 \\ 4 & -2 & -2 & 4 \\ 1 & 1 & 3 & 6 \end{array}\right]$$

To produce zeros in the first column we use r2 = r2 − r1, r3 = r3 − 2r1, r4 = r4 − 4r1 and r5 = r5 − r1:

$$\left[\begin{array}{ccc|c} 1 & 2 & -3 & 1 \\ 0 & -3 & 7 & 4 \\ 0 & -3 & 3 & 0 \\ 0 & -10 & 10 & 0 \\ 0 & -1 & 6 & 5 \end{array}\right]$$

Divide row 3 by −3 and interchange with row 2:

$$\left[\begin{array}{ccc|c} 1 & 2 & -3 & 1 \\ 0 & \textcircled{1} & -1 & 0 \\ 0 & -3 & 7 & 4 \\ 0 & -10 & 10 & 0 \\ 0 & -1 & 6 & 5 \end{array}\right]$$

Now r3 = r3 + 3r2, r4 = r4 + 10r2 and r5 = r5 + r2:

$$\left[\begin{array}{ccc|c} 1 & 2 & -3 & 1 \\ 0 & 1 & -1 & 0 \\ 0 & 0 & 4 & 4 \\ 0 & 0 & 0 & 0 \\ 0 & 0 & 5 & 5 \end{array}\right]$$

Lastly, r3 = r3/4 and r5 = r5 − 5r3:

$$\left[\begin{array}{ccc|c} 1 & 2 & -3 & 1 \\ 0 & 1 & -1 & 0 \\ 0 & 0 & 1 & 1 \\ 0 & 0 & 0 & 0 \\ 0 & 0 & 0 & 0 \end{array}\right]$$

So two of the equations are redundant, and the new equations are

$$\begin{aligned} x + 2y - 3z &= 1 \\ y - z &= 0 \\ z &= 1 \end{aligned}$$

From these, by back substitution, $z = 1$, $y = z = 1$ and

$$x = -2y + 3z + 1 = 2.$$ ■

You should always check the solution by substituting it directly into the *original* equations. It is only the work of a moment, and if a mistake has occurred you then have an opportunity to locate and correct it. It is not unknown for examiners to reserve a mark or two for evidence that the solution has been checked. It is usually easy to check it mentally, but always indicate at the end of your solution that it has been checked; 'solution checks' is often enough.

□ Given the equations

$$\begin{aligned} 2x - y + z &= 3 \\ x + 2y + z &= 5 \\ 3x - 4y + z &= 1 \\ 5x + 5y + 4z &= 18 \end{aligned}$$

decide whether or not they are consistent. If they are consistent, solve them.

The augmented matrix is

$$\left[\begin{array}{ccc|c} 2 & -1 & 1 & 3 \\ 1 & 2 & 1 & 5 \\ 3 & -4 & 1 & 1 \\ 5 & 5 & 4 & 18 \end{array}\right]$$

Interchanging rows 1 and 2 gives

$$\left[\begin{array}{ccc|c} \textcircled{1} & 2 & 1 & 5 \\ 2 & -1 & 1 & 3 \\ 3 & -4 & 1 & 1 \\ 5 & 5 & 4 & 18 \end{array}\right]$$

Then r2 = r2 − 2r1, r3 = r3 − 3r1 and r4 = r4 − 5r1 give

$$\left[\begin{array}{ccc|c} 1 & 2 & 1 & 5 \\ 0 & -5 & -1 & -7 \\ 0 & -10 & -2 & -14 \\ 0 & -5 & -1 & -7 \end{array}\right]$$

Since row 3 and row 4 are multiples of row 2, the system reduces to just two equations:

$$\begin{aligned} x + 2y + z &= 5 \\ -5y - z &= -7 \end{aligned}$$

This shows that the equations are consistent, and in fact it is possible to choose one of the variables arbitrarily. So if we put $z = t$, we can express the remaining variables in terms of t. Therefore if $z = t$ then $y = -(t - 7)/5$ and

$$\begin{aligned} x &= -2y - z + 5 \\ &= [2(t - 7)/5] - t + 5 \\ &= (-3t + 11)/5 \end{aligned}$$

■

□ Solve, if possible, the system of equations

$$\begin{aligned} w + 3x - y + z &= 4 \\ 2w - x + y - z &= 7 \\ 5w + x + y - z &= 20 \end{aligned}$$

The augmented matrix is

$$\left[\begin{array}{cccc|c} \textcircled{1} & 3 & -1 & 1 & 4 \\ 2 & -1 & 1 & -1 & 7 \\ 5 & 1 & 1 & -1 & 20 \end{array}\right]$$

First, r2 = r2 − 2r1 and r3 = r3 − 5r1 give

$$\left[\begin{array}{cccc|c} 1 & 3 & -1 & 1 & 4 \\ 0 & -7 & 3 & -3 & -1 \\ 0 & -14 & 6 & -6 & 0 \end{array}\right]$$

Next, r3 = r3 − 2r2 produces

$$\left[\begin{array}{cccc|c} 1 & 3 & -1 & 1 & 4 \\ 0 & -7 & 3 & -3 & -1 \\ 0 & 0 & 0 & 0 & 2 \end{array}\right]$$

The system is clearly inconsistent since the final 'equation' has become now $0 = 2$. ■

Here now are a few steps which will enable you to see how you are progressing.

13.5 Workshop

▷**Exercise** Use the formula to obtain the inverse of the matrix

$$\begin{bmatrix} 3 & 5 & 9 \\ 1 & 2 & 3 \\ 4 & 7 & 15 \end{bmatrix}$$

Check your answer by multiplication.

If everything works out right, proceed to step 4. Otherwise, go on to step 2.

---

2 If the product does not produce an identity matrix, check carefully all the stages:

1 the matrix of minors
2 the matrix of cofactors
3 the adjoint matrix
4 the determinant
5 the final multiplication.

Now see if you can get it right. If all is well move on to step 4. If not, go to step 3.

---

3 So we still have problems. If you have gone through all the stages in step 2 and checked for errors, there can be only one explanation. You must have 'simplified' the matrix in some way. Did you use row transformations, or the rules for simplifying determinants? If you did then you have made a *fatal* error. Remember, whenever we use row transformations we change the matrix. So you have found the inverse of a row equivalent matrix and not the one in which we are interested.

Award yourself a wooden spoon, solve the problem correctly and move on to step 4.

▷**Exercise** Use the method of row transformations to obtain the inverse of the matrix **4**

$$\begin{bmatrix} 3 & 1 & 2 \\ 7 & 3 & 4 \\ 9 & 3 & 8 \end{bmatrix}$$

You can check your answer by multiplication. If it is right, move on to step 7. If not, go to step 5.

If you have managed to convert the matrix on the left into an identity matrix but the matrix on the right didn't turn out to be the inverse, move on to step 6. **5**

If you have been unable to convert the matrix on the left into an identity matrix you have not been approaching the problem systematically. Remember to move across the matrix column by column:

1 Get the non-zero element into the correct position; then
2 Convert this element into a 1; then
3 Subtract multiples of this row from the others to produce the zeros.

Only when all this has been done do we move to the next column.

Try again, and if all is well now move on to step 7. If there are still problems, move to step 6.

There are a number of possible sources of error: **6**

1 You may have simply made an arithmetical slip. Errors have a habit of hiding away in the parts where you least expect to find them. Have a good look and see if you can spot one or two.

2 You may have carried out some illegal operation. For instance it is usually advisable to make sure that at each stage the row in which the pivot lies remains unaltered. Are you sure that you have performed each row operation on the entire row of six elements?

3 You may have used column transformations as well as row transformations. You must stick to row transformations; mixing transformations is strictly against the law!

Once you have located your error, make the necessary amendments and obtain the correct inverse.

7 **Exercise** Discuss the solution of the following system of equations:

$$\begin{aligned} x + 2y - z &= 2 \\ x - y + z &= 5 \\ 3x + 3y + az &= b + 8 \end{aligned}$$

in the three cases **a** $a = -1, b = 1$ **b** $a = -1, b = 2$ **c** $a = 1, b = 7$.

You may consider the cases separately if you wish, or you may choose to operate on the augmented matrix algebraically. Whichever method you decide on, complete the problem and take the final step to see if all is well.

8 The augmented matrix is

$$\left[\begin{array}{ccc|c} 1 & 2 & -1 & 2 \\ 1 & -1 & 1 & 5 \\ 3 & 3 & a & b+8 \end{array}\right]$$

This reduces to

$$\left[\begin{array}{ccc|c} 1 & 2 & -1 & 2 \\ 0 & -3 & 2 & 3 \\ 0 & 0 & a+1 & b-1 \end{array}\right]$$

So

a $a = -1, b = 1$ gives a redundant row, and the solution in terms of z is $y = (2z - 3)/3$ and $x = 3 - z/3$.

b $a = -1, b = 2$ produces a final equation $0 = 1$, and so the equations are inconsistent.

c $a = 1, b = 7$ gives a unique solution $z = 3, y = 1$ and $x = 3$.

13.6 PIVOTING

When we used the method of systematic elimination we avoided arithmetical complications by choosing the pivot with care. In particular we tried to perform our arithmetic with integers. On occasions we interchanged rows rather than divide through a row by a number which would have resulted in decimal representations. This is one way of keeping the arithmetic exact and so avoiding errors.

If we were programming a computer to perform this elimination method we should in each case reduce the non-zero element on the leading diagonal to 1 by dividing throughout the row. It would require some sophisticated programming to detect the presence of suitable rows which could be manipulated by the rules to produce, without division, a pivot 1.

The approach which the computer would take is shown in the following scheme:

$$\begin{bmatrix} * & * & * & * \\ * & * & * & * \\ * & * & * & * \end{bmatrix} \to \begin{bmatrix} 1 & * & * & * \\ * & * & * & * \\ * & * & * & * \end{bmatrix} \to$$

$$\begin{bmatrix} 1 & * & * & * \\ 0 & * & * & * \\ 0 & * & * & * \end{bmatrix} \to \begin{bmatrix} 1 & * & * & * \\ 0 & 1 & * & * \\ 0 & * & * & * \end{bmatrix} \to$$

$$\begin{bmatrix} 1 & * & * & * \\ 0 & 1 & * & * \\ 0 & 0 & * & * \end{bmatrix} \to \begin{bmatrix} 1 & * & * & * \\ 0 & 1 & * & * \\ 0 & 0 & 1 & * \end{bmatrix}$$

In the first transformation, row 1 is divided by the element in the (1, 1) position.

In the second transformation, multiples of row 1 are subtracted from the other rows to produce the zeros in column 1. Attention then turns to column 2.

In the third transformation, we divide row 2 by the element now in the (2, 2) position to produce 1 in the (2, 2) position. Then multiples of row 2 are subtracted from the other rows to produce the required zeros in column 2.

If a zero were to appear on the leading diagonal, a complication would arise. In such circumstances it would be necessary to locate a non-zero element in the same column below the leading diagonal and by means of a row interchange to bring it to the leading diagonal.

One of the problems which arises in practice with this method is due to round-off error in the numerical approximation.

One method of minimizing the effects of round-off error is to use **partial pivoting**. Partial pivoting consists of rearranging the equations in such a way that the numerically largest non-zero elements occur on the leading diagonal.

□ Here is a set of equations before partial pivoting:

$$\begin{aligned} x + 5y - z &= 8 \\ 4x - y + 2z &= 5 \\ 2x + 3y - 6z &= 1 \end{aligned}$$

In the first equation the largest coefficient is 5, in the second it is 4 and in the third it is -6. Therefore the system of equations after partial pivoting becomes

$$\begin{aligned} 4x - y + 2z &= 5 \\ x + 5y - z &= 8 \\ 2x + 3y - 6z &= 1 \end{aligned}$$

■

Partial pivoting ensures that when the rows are divided by the elements on the leading diagonal the numbers are reduced and errors are therefore controlled.

We shall now solve a practical problem involving binary arithmetic. In binary arithmetic there are just two numbers, 0 and 1. The following algebraic rules are satisfied:

$$\begin{aligned} 0 \times 0 &= 0 \qquad & 0 + 0 &= 0 \\ 0 \times 1 &= 0 & 1 + 0 &= 1 \\ 1 \times 0 &= 0 & 0 + 1 &= 1 \\ 1 \times 1 &= 1 & 1 + 1 &= 0 \end{aligned}$$

Most people have come across binary arithmetic; computers use it.

13.7 Practical

CODEWORDS

A seven-bit binary code consists of codewords $x_1x_2 \ldots x_7$ which satisfy the condition $H\mathbf{x}^{\mathrm{T}} = \mathrm{O}$, where $\mathbf{x} = [x_1, x_2, \ldots, x_7]$, O is the 7×1 zero vector and

$$H = \begin{bmatrix} 0 & 0 & 0 & 1 & 1 & 1 & 1 \\ 0 & 1 & 1 & 0 & 0 & 1 & 1 \\ 1 & 0 & 1 & 0 & 1 & 0 & 1 \end{bmatrix}$$

Obtain all the codewords.

See how this goes. We shall attack the problem one stage at a time.

We have the set of equations

$$\begin{aligned} x_4 + x_5 + x_6 + x_7 &= 0 \\ x_2 + x_3 + x_6 + x_7 &= 0 \\ x_1 + x_3 + x_5 + x_7 &= 0 \end{aligned}$$

If we remember that in binary arithmetic $1 + 1 = 0$ we see that these are equivalent to

$$x_4 = x_5 + x_6 + x_7$$

$$x_2 = x_3 + x_6 + x_7$$
$$x_1 = x_3 + x_5 + x_7$$

Can you see how many codewords there will be? When you have decided, move to the next stage and see if you are right.

We have expressed three of the unknowns x_4, x_2 and x_1 in terms of the other four. Each of these four has two possible values, 0 or 1. Consequently the total number of codewords is $2 \times 2 \times 2 \times 2 = 2^4 = 16$. Once we enumerate these 16 possibilities for x_3, x_5, x_6 and x_7 the equations will determine the other bits and so the codewords will be obtained.

We shall do this in two stages. First determine all the codewords which have $x_3 = 0$.

We enumerate (x_3, x_5, x_6, x_7) and use the equations to obtain x_4, x_2 and x_1 and thereby the codeword $x_1x_2 \ldots x_7$:

1 $(0, 0, 0, 0) \Rightarrow x_4 = 0, x_2 = 0, x_1 = 0 \Rightarrow 0000000$
2 $(0, 0, 0, 1) \Rightarrow x_4 = 1, x_2 = 1, x_1 = 1 \Rightarrow 1101001$
3 $(0, 0, 1, 0) \Rightarrow x_4 = 1, x_2 = 1, x_1 = 0 \Rightarrow 0101010$
4 $(0, 0, 1, 1) \Rightarrow x_4 = 0, x_2 = 0, x_1 = 1 \Rightarrow 1000011$
5 $(0, 1, 0, 0) \Rightarrow x_4 = 1, x_2 = 0, x_1 = 1 \Rightarrow 1001100$
6 $(0, 1, 0, 1) \Rightarrow x_4 = 0, x_2 = 1, x_1 = 0 \Rightarrow 0100101$
7 $(0, 1, 1, 0) \Rightarrow x_4 = 0, x_2 = 1, x_1 = 1 \Rightarrow 1100110$
8 $(0, 1, 1, 1) \Rightarrow x_4 = 1, x_2 = 0, x_1 = 0 \Rightarrow 0001111$

If something went wrong, check through things carefully and then obtain the remaining eight codewords.

We now put $x_3 = 1$:

9 $(1, 0, 0, 0) \Rightarrow x_4 = 0, x_2 = 1, x_1 = 1 \Rightarrow 1110000$
10 $(1, 0, 0, 1) \Rightarrow x_4 = 1, x_2 = 0, x_1 = 0 \Rightarrow 0011001$
11 $(1, 0, 1, 0) \Rightarrow x_4 = 1, x_2 = 0, x_1 = 1 \Rightarrow 1011010$
12 $(1, 0, 1, 1) \Rightarrow x_4 = 0, x_2 = 1, x_1 = 0 \Rightarrow 0110011$
13 $(1, 1, 0, 0) \Rightarrow x_4 = 1, x_2 = 1, x_1 = 0 \Rightarrow 0111100$
14 $(1, 1, 0, 1) \Rightarrow x_4 = 0, x_2 = 0, x_1 = 1 \Rightarrow 1010101$
15 $(1, 1, 1, 0) \Rightarrow x_4 = 0, x_2 = 0, x_1 = 0 \Rightarrow 0010110$
16 $(1, 1, 1, 1) \Rightarrow x_4 = 1, x_2 = 1, x_1 = 1 \Rightarrow 1111111$

It is interesting to note that each pair of codewords differs by at least three bits. This is an example of an error-correcting linear code.

13.8 CONCLUDING REMARKS

Before we end this chapter there are two points which should be made:

1 We have only touched on the theory of matrices. There is much more to them than this. For example we have solved systems of equations by using an elimination method, but we have not examined the theory behind the technique we have used. The key idea is that of the **rank** of a matrix, which is the number of linearly independent rows. It can be shown that

   a The rank of a matrix is unaltered when a row transformation is performed on it;

   b A set of equations is consistent if and only if the rank of the matrix of coefficients is the same as the rank of the augmented matrix.

2 When matrices are used in conjunction with differential equations some very powerful techniques become available. One of the most fruitful ideas arises from considering the equation

$$A\mathbf{x} = \lambda\mathbf{x}$$

where A is a square matrix and λ is a scalar. The non-zero vectors $\mathbf{x}$ which satisfy this equation are known as **eigenvectors** and the corresponding values of λ are then called the **eigenvalues**. One of the further exercises gives you an opportunity to examine the consequences of this equation. It is possible to express many differential and difference equations as eigenvalue problems.

SUMMARY

We now summarize what we have learnt in this chapter:

☐ The formula for calculating the inverse of a matrix:

$$A^{-1} = \frac{1}{|A|}\text{adj } A$$

☐ The meanings of the terms singular matrix and non-singular matrix

$$A \text{ is singular} \Leftrightarrow |A| = 0$$

☐ The operations on a matrix known as elementary row transformations

a an interchange of two rows;

b multiplication of the elements in a row by a non-zero number;

c subtraction of the elements of one row from the corresponding elements of another row.

☐ The method of obtaining the inverse of a matrix by using row transformations

$$A|I \rightarrow I|A^{-1}$$

☐ The use of row transformations to solve systems of simultaneous algebraic equations

$$\left[\begin{array}{ccc|c} * & * & * & * \\ * & * & * & * \\ * & * & * & * \end{array}\right] \rightarrow \left[\begin{array}{ccc|c} 1 & * & * & * \\ 0 & 1 & * & * \\ 0 & 0 & 1 & * \end{array}\right]$$

EXERCISES

1 Use the formula to obtain the inverses of each of the following matrices:

a $\begin{bmatrix} 3 & 8 & 2 \\ 4 & 10 & 3 \\ 2 & 5 & 1 \end{bmatrix}$ **b** $\begin{bmatrix} 5 & 3 & 8 \\ 7 & 5 & 11 \\ 3 & 2 & 5 \end{bmatrix}$

c $\begin{bmatrix} 11 & 7 & 18 \\ 4 & 3 & 6 \\ 14 & 9 & 23 \end{bmatrix}$ **d** $\begin{bmatrix} 8 & 5 & 13 \\ 10 & 7 & 16 \\ 13 & 9 & 21 \end{bmatrix}$

2 Obtain x for each of the following matrices if each one is singular:

a $\begin{bmatrix} x+1 & x+2 & x+5 \\ 5 & 9 & 4 \\ x+3 & x+7 & x+2 \end{bmatrix}$

b $\begin{bmatrix} x+2 & x+3 & 7 \\ 2 & 5 & -3 \\ x+4 & 9 & x+3 \end{bmatrix}$

c $\begin{bmatrix} x-1 & x+1 & x+4 \\ x-4 & x+2 & -3 \\ x-2 & 3x & x+1 \end{bmatrix}$

d $\begin{bmatrix} -x & x+1 & x+2 \\ x-3 & x-1 & -x+2 \\ -x+2 & 2x & x-1 \end{bmatrix}$

3 Use the method of row transformations to obtain the inverses of the matrices in question 1.

4 Use the method of row transformations to solve the following sets of simultaneous equations:

a $3x + 5y + 8z = 3$
$4x + 6y + 11z = 5$
$2x + 3y + 5z = 7$

b $4x + 2y + 3z = 5$
$6x + 3y + 5z = 8$
$11x + 5y + 8z = 7$

c $18x + 6y + 5z = -1$
$11x + 4y + 3z = 7$
$7x + 3y + 2z = 9$

d $7x + 3y + 8z = 4$
$5x + 2y + 5z = 5$
$11x + 5y + 13z = 6$

ASSIGNMENT

1 Obtain the inverse of the matrix

$$\begin{bmatrix} 2 & 4 & 7 \\ 3 & 7 & 9 \\ 1 & 2 & 3 \end{bmatrix}$$

2 An **orthogonal matrix** is a matrix which has its inverse equal to its transpose. Verify that

$$A = \frac{1}{3}\begin{bmatrix} 1 & -2 & 2 \\ 2 & -1 & -2 \\ -2 & -2 & -1 \end{bmatrix}$$

is an orthogonal matrix.

3 If

$$A = \begin{bmatrix} 3 & 2 & 4 \\ 2 & -2 & -6 \\ 4 & -6 & -1 \end{bmatrix} \qquad X = \begin{bmatrix} -2 & -2 & 1 \\ -2 & 1 & -2 \\ 1 & -2 & -2 \end{bmatrix}$$

show that $X^{\mathrm{T}}AX$ is a diagonal matrix.

4 Prove that if A and B are non-singular square matrices of the same order then AB is also non-singular and $(AB)^{-1} = B^{-1}A^{-1}$.

FURTHER EXERCISES

1 If A is a non-singular matrix, show in general that adj A is also non-singular and give an explicit formula for the inverse of adj A.

2 Obtain the relationship between the determinant of a square matrix A and the determinant of adj A, its adjoint matrix.

3 A company employs 45 people and there are three different grades of employee. The pay and profits are as follows:

Grade	*Pay/hour* (£)	*Profit/hour* (£)
1	2	4
2	4	−3
3	6	4

Obtain the numbers of employees in the various grades if the total wage bill is £200 per hour and the total profit is £75 per hour.

4 Show that if $AB = O$ and if either A or B is a non-singular square matrix then either $A = O$ or $B = O$.

5 Prove that, if A is a non-singular matrix, $(A^{\mathrm{T}})^{-1} = (A^{-1})^{\mathrm{T}}$.

6 Show that if the simultaneous equations

$$\begin{aligned} ax + by + h &= 0 \\ cx + dy + k &= 0 \\ ex + fy + l &= 0 \end{aligned}$$

are consistent, then

$$\begin{vmatrix} a & b & h \\ c & d & k \\ e & f & l \end{vmatrix} = 0$$

By means of an example, show that this condition is no guarantee of consistency.

7 If

$$A = \begin{bmatrix} 7 & 8 & 9 \\ 0 & 3 & 3 \\ -2 & -4 & -4 \end{bmatrix}$$

and the equation $AX = kX$ is satisfied for some non-zero matrix X, prove that det $(A - kI) = 0$. Hence or otherwise obtain the three possible values of k.

8

$$A = \begin{bmatrix} 1 & -1 & 2 \\ 3 & 4 & 0 \\ -2 & 2 & -4 \end{bmatrix}$$

show that **a** A is singular and **b** A satisfies the equation $A^3 - A^2 - 9A = 0$.

9 If the input and output of a system are denoted by **y** and **x** respectively and are related by an equation of the form $\mathbf{y} = M\mathbf{x}$ where M is a matrix, then M is called a transmission matrix. If the transmission matrix M for a waveguide below cutoff is

$$\begin{bmatrix} \cosh a & Z \sinh a \\ (1/Z) \sinh a & \cosh a \end{bmatrix}$$

then

a Show that M is non-singular and obtain its inverse;

b Show that for a cascade of n such waveguides the transmission matrix is

$$\begin{bmatrix} \cosh na & Z \sinh na \\ (1/Z) \sinh na & \cosh na \end{bmatrix}$$

10 If I is the 3×3 identity matrix and

$$S = \begin{bmatrix} 0 & 1 & 0 \\ -1 & 0 & 2 \\ 0 & -2 & 0 \end{bmatrix}$$

show that the matrix A given by $A = (I + S)(I - S)^{-1}$ is orthogonal. Hence, or otherwise, solve the matrix equation $AX = K$ where

$$X = \begin{bmatrix} x \\ y \\ z \end{bmatrix} \qquad K = \begin{bmatrix} 2 \\ 1 \\ 3 \end{bmatrix}$$

11 Show that if **x** is a column vector of order $n \times 1$ and A is a matrix of order $n \times n$ then the equation

$$A\mathbf{x} = \mathbf{y}$$

implies that **y** is a column vector of order $n \times 1$.

If $\mathbf{y} = \lambda\mathbf{x}$ where λ is a constant and $\mathbf{x} \neq \mathbf{0}$ deduce that

$$|A - \lambda I| = 0$$

where I is an identity matrix of order $n \times n$.

Determine the values of λ and the corresponding column vectors $\mathbf{x}_1$ and $\mathbf{x}_2$ with integer elements if

$$A = \begin{pmatrix} 2 & 3 \\ -1 & 6 \end{pmatrix}$$

Write down the matrix $P = (\mathbf{x}_1, \mathbf{x}_2)$ of order 2×2 and verify that $P^{-1}AP$ is a diagonal matrix.

Vectors 14

In Chapter 13 we used the term vector to mean an algebraic vector. The vectors which we describe in this chapter have rather more structure and are widely used in applications.

After completing this chapter you should be able to

- ☐ Apply the rules of vector addition and scalar multiplication;
- ☐ Obtain the scalar product of two vectors;
- ☐ Obtain the vector product of two vectors;
- ☐ Use triple scalar products and triple vector products;
- ☐ Use vector methods to solve simple problems;
- ☐ Differentiate vectors.

At the end of the chapter we look at a simple practical problem in particle dynamics.

14.1 DESCRIPTIONS

You probably have an intuitive idea of what is meant by magnitude and direction. **Magnitude** gives a measure of how large something is, and **direction** an indication of where it applies. For example, meteorologists talk of a north-easterly wind of force 6. There the magnitude is 6 on the Beaufort scale and the direction is from the north-east.

Some concepts which arise in practice are adequately described purely in terms of a magnitude. Mass, volume, height, speed and time are all examples of quantities of this kind, and we call them **scalar quantities**.

Other concepts need not only a magnitude to describe them but also a direction. Velocity and displacement are examples in which both a magnitude and a direction are involved; we call these **vector quantities**.

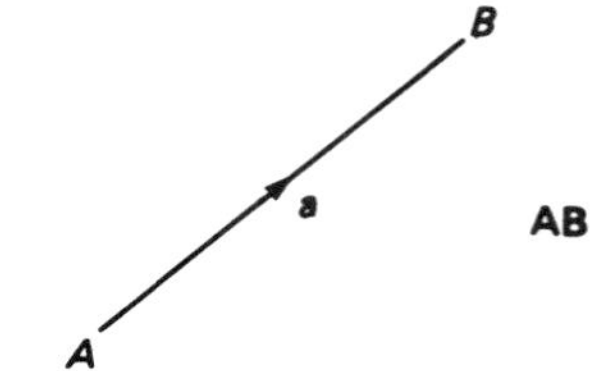

Fig. 14.1 Representation of a vector quantity.

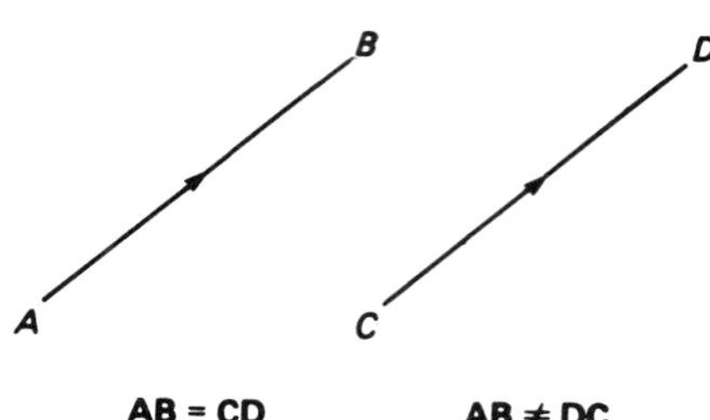

Fig. 14.2 Equivalent and non-equivalent vector quantities.

In brief, two scalar quantities of the same type can be compared by their magnitudes, whereas two vector quantities of the same type cannot be compared adequately in this way. We shall use real numbers to represent scalar quantities and directed line segments to represent vector quantities.

So to represent a vector quantity we choose an initial point A and construct a directed line segment **AB** with the same direction as the vector quantity and with length AB proportional to its magnitude (Fig. 14.1). It is sometimes convenient to represent the vector quantity **AB** by means of the notation **a**, but then we usually need to refer to a diagram showing AB with an arrowhead somewhere on it so that the direction of the vector quantity is unambiguous.

Two parallel directed line segments with the same direction will be regarded as equivalent vector quantities (Fig. 14.2).

Although anything with both magnitude and direction can be thought of as a vector quantity, it would be wrong to think that all such things are vectors. A set of **vectors** is a set of vector quantities with operations known as **vector addition** and **scalar multiplication**. We must describe how each of these operations is performed.

14.2 VECTOR ADDITION

Vector addition is performed by the **parallelogram law**. Suppose we take two vector quantities **OP** and **OQ** and construct the parallelogram $OPRQ$

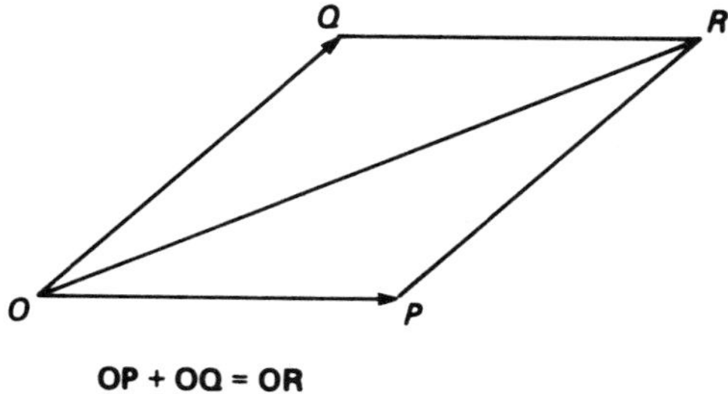

Fig. 14.3 The parallelogram law.

(Fig. 14.3). Then a necessary condition for **OP** and **OQ** to be vectors is

$$\mathbf{OP} + \mathbf{OQ} = \mathbf{OR}$$

Many vector quantities are in fact examples of vectors because they combine according to the parallelogram law. However, others do not satisfy this condition and so are not vectors.

□ A light aircraft travels the shortest route from London to Oxford and then from Oxford to York. The pilot keeps records of fuel consumed, travelling time and distance travelled.

For each leg, represent the journeys as vector quantities; the magnitude is the quantity recorded by the pilot, the direction is the direction of travel. Decide, in each case, which are vectors. You may neglect the curvature of the earth.

In Fig. 14.4 we may represent the journey from London to Oxford by **OP** and the journey from Oxford to York by **PR**. For these to constitute a set of vectors we require the journey from London to York to be represented by **OR**, where *OPRQ* is a parallelogram.

If the magnitudes are proportional to the fuel consumption, then in general

$$\mathbf{OP} + \mathbf{PR} \neq \mathbf{OR}$$

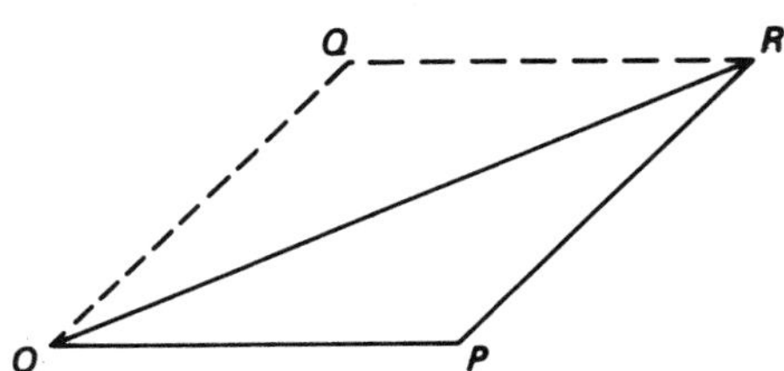

Fig. 14.4 The parallelogram.

To see why this is, we merely consider the situation in which there is a strong wind blowing in the direction from York to London. This would affect fuel consumption disproportionately. Therefore we do not have a set of vectors.

If the magnitudes are proportional to travelling time, then again we do not have a set of vectors for reasons similar to those given for fuel consumption.

If the magnitudes are proportional to the distances travelled then

$$\mathbf{OP} + \mathbf{PR} = \mathbf{OR}$$

These are a set of vectors, for if he had travelled from London, in the direction of **OR** and with distance OR, he would have arrived at York. ■

When no confusion is likely to occur we shall represent a vector by the notation $\mathbf{a}$ and its magnitude by $|\mathbf{a}|$.

When we employ this notation then it is often necessary to refer to a diagram.

Note that we shall avoid the temptation to write the magnitude of $\mathbf{a}$ simply as a. This is because when we consider differentiation of vectors a conflict of notation can lead to error.

□ In Fig. 14.5 the parallelogram law is $\mathbf{a} + \mathbf{b} = \mathbf{c}$. The word resultant is often used for the vector obtained by adding two vectors. The resultant is usually shown on the diagram by means of a double arrow. ■

We have said that parallel vectors with the same magnitude and direction are equivalent to one another. Therefore every vector has a representative in the form of a vector emanating from a single point O. The terminology **free vector** for a vector which can start anywhere, and **localized vector** for a vector which must start from a fixed point, is in common usage.

It is convenient for us to be able to represent vectors in terms of coordinates. Suppose $Oxyz$ is a rectangular three-dimensional coordinate system (Fig. 14.6). We introduce three **unit vectors** $\mathbf{i}$, $\mathbf{j}$ and $\mathbf{k}$ emanating from the origin O in the direction Ox, Oy and Oz respectively, each having a unit magnitude.

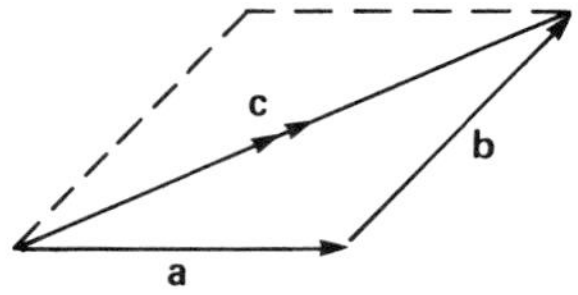

Fig. 14.5 The resultant.

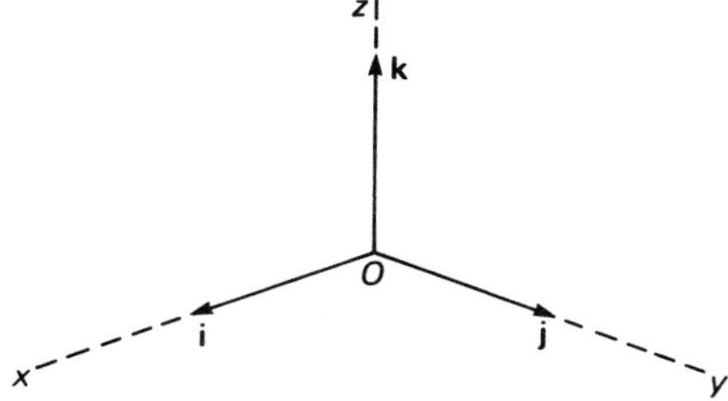

Fig. 14.6 The unit vectors **i**, **j** and **k**.

0 is a special vector called the **zero vector**; it has zero magnitude and arbitrary direction. Any vector which has magnitude 1 is called a **unit vector**.

BINARY OPERATIONS

We know how to add and multiply scalars, since these rules are the usual rules for dealing with real numbers.

We have also described the rule for adding two vectors **a** and **b** together to produce **a** + **b**. This is known as a **binary operation**:

$$(\mathbf{a}, \mathbf{b}) \rightarrow \mathbf{a} + \mathbf{b}$$

We think of this as the parallelogram law. Later we shall describe two further binary operations on vectors which produce a scalar known as the **scalar product**, and a vector known as the **vector product**.

Before this we need to describe how to multiply a scalar by a vector. We shall call this operation **scalar multiplication**. However, it must not be confused with the scalar product, which is an operation we shall describe later.

14.3 SCALAR MULTIPLICATION

If **a** is a vector and p is a real number then we define $p\mathbf{a}$ to be a vector with the following properties (Fig. 14.7):

1 If $p > 0$ then $p\mathbf{a}$ has the same direction as **a** and magnitude p times that of **a**.

2 If $p < 0$ then $p\mathbf{a}$ has the opposite direction to **a** and magnitude $(-p)$ times that of **a**.

3 If $p = 0$ then $p\mathbf{a}$ has arbitrary direction and magnitude 0.

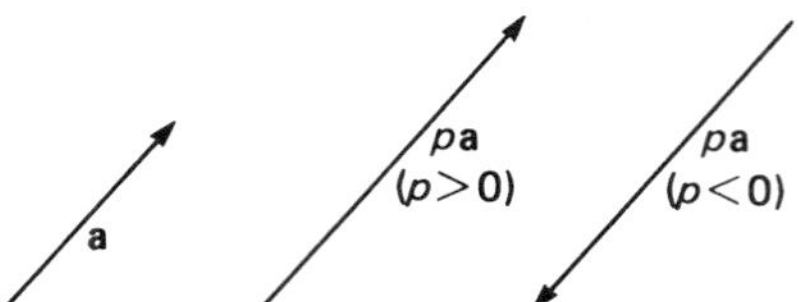

Fig. 14.7 Scalar multiplication.

Having dealt with vector addition and scalar multiplication, we can summarize the position so far in a set of rules. Suppose p and q are any real numbers (scalars) and **a**, **b** and **c** are arbitrary vectors. Then the following rules hold:

$$\mathbf{a} + \mathbf{b} = \mathbf{b} + \mathbf{a}$$
$$(\mathbf{a} + \mathbf{b}) + \mathbf{c} = \mathbf{a} + (\mathbf{b} + \mathbf{c})$$
$$\mathbf{a} + \mathbf{0} = \mathbf{0} + \mathbf{a} = \mathbf{a}$$
$$\mathbf{a} + (-\mathbf{a}) = (-\mathbf{a}) + \mathbf{a} = \mathbf{0}$$

$$p(\mathbf{a} + \mathbf{b}) = p\mathbf{a} + p\mathbf{b}$$
$$p(q\mathbf{a}) = (pq)\mathbf{a}$$
$$(p + q)\mathbf{a} = p\mathbf{a} + q\mathbf{a}$$

□ Show that there is only one zero vector **0**.

Suppose that there are two zero vectors $\mathbf{0}_1$ and $\mathbf{0}_2$. Then with $\mathbf{a} = \mathbf{0}_1$ and $\mathbf{0} = \mathbf{0}_2$ we can use

$$\mathbf{a} + \mathbf{0} = \mathbf{0} + \mathbf{a} = \mathbf{a}$$

to obtain

$$\mathbf{0}_1 + \mathbf{0}_2 = \mathbf{0}_2 + \mathbf{0}_1 = \mathbf{0}_1$$

Whereas with $\mathbf{a} = \mathbf{0}_2$ and $\mathbf{0} = \mathbf{0}_1$ we obtain

$$\mathbf{0}_2 + \mathbf{0}_1 = \mathbf{0}_1 + \mathbf{0}_2 = \mathbf{0}_2$$

Therefore $\mathbf{0}_1 = \mathbf{0}_2$. ■

This is quite an interesting result because originally we defined **0** to be a vector with zero magnitude and arbitrary direction. It might be thought therefore that there are an *infinity* of zero vectors. This example has shown that this is not the case. If you have nothing, it doesn't matter what you do with it!

14.4 COMPONENTS

Suppose we have a rectangular cartesian coordinate system $Oxyz$ and that, relative to this system, P is the point (a_1, a_2, a_3) (Fig. 14.8). We have

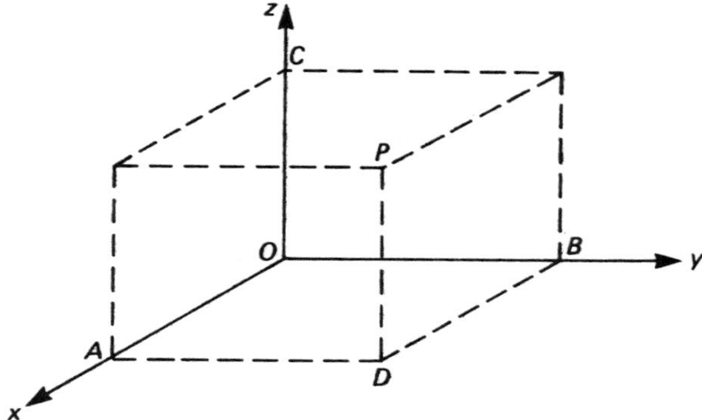

Fig. 14.8 Cartesian representation.

defined **i**, **j** and **k** to be unit vectors in the directions Ox, Oy and Oz respectively. Using the diagram,

$$\begin{aligned}\mathbf{r} = \mathbf{OP} &= \mathbf{OD} + \mathbf{DP}\\ &= \mathbf{OA} + \mathbf{AD} + \mathbf{DP}\\ &= \mathbf{OA} + \mathbf{OB} + \mathbf{OC}\\ &= a_1\mathbf{i} + a_2\mathbf{j} + a_3\mathbf{k}\end{aligned}$$

Also

$$\begin{aligned}|\mathbf{r}| = OP &= \sqrt{(OD^2 + DP^2)}\\ &= \sqrt{(OA^2 + AD^2 + DP^2)}\\ &= \sqrt{(a_1^2 + a_2^2 + a_3^2)}\end{aligned}$$

When a vector is expressed in this way we say it has been expressed in terms of **components**. (These are also known as resolutes or projections.)

If **r** is a vector it is customary to represent a unit vector in the same direction by $\hat{\mathbf{r}}$. We then have

$$\mathbf{r} = |\mathbf{r}|\,\hat{\mathbf{r}}$$

or equivalently

$$\hat{\mathbf{r}} = \frac{1}{|\mathbf{r}|}\,\mathbf{r}$$

□ If $\mathbf{a} = 3\mathbf{i} - 4\mathbf{j} + 5\mathbf{k}$ and $\mathbf{b} = \mathbf{i} - 5\mathbf{j} + 7\mathbf{k}$, obtain (a) $\mathbf{a} + \mathbf{b}$ (b) a unit vector in the direction of $\mathbf{a} - \mathbf{b}$ (c) a unit vector in the opposite direction to $\mathbf{a} - \mathbf{b}$.

a We have

$$\begin{aligned}\mathbf{a} + \mathbf{b} &= (3\mathbf{i} - 4\mathbf{j} + 5\mathbf{k}) + (\mathbf{i} - 5\mathbf{j} + 7\mathbf{k})\\ &= 4\mathbf{i} - 9\mathbf{j} + 12\mathbf{k}\end{aligned}$$

b First,

$$\begin{aligned}\mathbf{a} - \mathbf{b} &= (3\mathbf{i} - 4\mathbf{j} + 5\mathbf{k}) - (\mathbf{i} - 5\mathbf{j} + 7\mathbf{k}) \\ &= 2\mathbf{i} + \mathbf{j} - 2\mathbf{k}\end{aligned}$$

Therefore

$$|\mathbf{a} - \mathbf{b}| = \sqrt{[(2)^2 + (1)^2 + (-2)^2]} = \sqrt{9} = 3$$

So the required vector is

$$\frac{1}{3}(\mathbf{a} - \mathbf{b}) = \frac{2}{3}\mathbf{i} + \frac{1}{3}\mathbf{j} - \frac{2}{3}\mathbf{k}$$

c The required vector is

$$-\frac{1}{3}(\mathbf{a} - \mathbf{b}) = -\frac{2}{3}\mathbf{i} - \frac{1}{3}\mathbf{j} + \frac{2}{3}\mathbf{k}$$

■

14.5 THE SCALAR PRODUCT

Given any two vectors **a** and **b**, we define the **scalar product a · b** by the formula

$$\mathbf{a} \cdot \mathbf{b} = |\mathbf{a}||\mathbf{b}| \cos \theta$$

where θ is the angle between the two vectors (Fig. 14.9).

We observe that **a · b** is always a scalar, and this is why it is called the scalar product. Sometimes it is known as the **dot product** because of this notation, and to avoid confusion with scalar multiplication.

Note that since $\cos \theta = \cos (2\pi - \theta)$ it does not matter whether the included angle or the excluded angle between the two vectors is chosen. Moreover,

$$\mathbf{b} \cdot \mathbf{a} = |\mathbf{b}||\mathbf{a}| \cos \theta = \mathbf{a} \cdot \mathbf{b}$$

That is, **a · b = b · a** for all vectors **a** and **b**. This is known as the **commutative rule**, and we have shown that it holds for the scalar product. Later we shall see that for the other type of product, the vector product, the commutative law does *not* hold.

□ Show that if **a** and **b** are non-zero vectors then

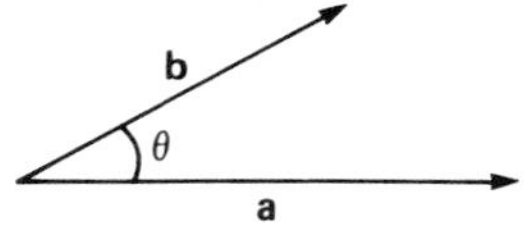

Fig. 14.9 The angle between two vectors.

$$\mathbf{a} \cdot \mathbf{b} = 0$$

if and only if **a** and **b** are mutually perpendicular.

Suppose first that $\mathbf{a} \cdot \mathbf{b} = 0$. Then

$$|\mathbf{a}|\,|\mathbf{b}| \cos \theta = 0$$

and since neither **a** nor **b** is the zero vector we conclude that $|\mathbf{a}| \neq 0$ and $|\mathbf{b}| \neq 0$. Consequently $\cos \theta = 0$ and so $\theta = \pi/2$ (or $3\pi/2$); therefore **a** and **b** are mutually perpendicular.

Conversely, if **a** and **b** are mutually perpendicular then $\theta = \pi/2$ (or $3\pi/2$) and it follows that $\mathbf{a} \cdot \mathbf{b} = 0$. ■

We have also

$$\mathbf{a} \cdot \mathbf{a} = |\mathbf{a}|\,|\mathbf{a}| \cos \theta = |\mathbf{a}|^2$$

so that

$$\mathbf{i} \cdot \mathbf{i} = \mathbf{j} \cdot \mathbf{j} = \mathbf{k} \cdot \mathbf{k} = 1$$
$$\mathbf{i} \cdot \mathbf{j} = \mathbf{j} \cdot \mathbf{k} = \mathbf{k} \cdot \mathbf{i} = 0$$

14.6 DIRECTION RATIOS AND DIRECTION COSINES

If we multiply a vector by a positive number we preserve the direction of the vector. Likewise if we multiply it by a negative number the direction is reversed. It follows that if

$$\mathbf{r} = l\mathbf{i} + m\mathbf{j} + n\mathbf{k}$$

where the components l, m and n are non-zero, then the ratio $l:m:n$ determines the direction (but not the sense) of the vector. Any three numbers which are in these proportions are known as **direction ratios**.

Another way of fixing the direction of a vector is by means of a unit vector in the same direction. Suppose that

$$\mathbf{u} = u_1\mathbf{i} + u_2\mathbf{j} + u_3\mathbf{k}$$

is a unit vector in the direction of the vector **r**. Then if α, β and γ are the angles between this vector and the axes Ox, Oy and Oz respectively (Fig. 14.10) we have

$$u_1 = \mathbf{u} \cdot \mathbf{i} = (1)(1) \cos \alpha = \cos \alpha$$
$$u_2 = \mathbf{u} \cdot \mathbf{j} = (1)(1) \cos \beta = \cos \beta$$
$$u_3 = \mathbf{u} \cdot \mathbf{k} = (1)(1) \cos \gamma = \cos \gamma$$

Therefore

$$\mathbf{u} = \cos \alpha\mathbf{i} + \cos \beta\mathbf{j} + \cos \gamma\mathbf{k}$$

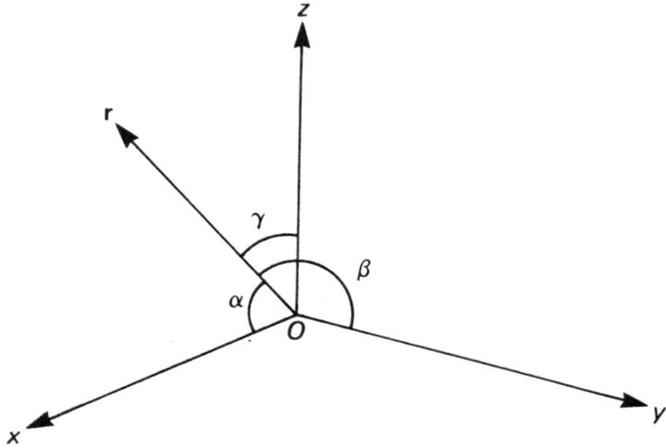

Fig. 14.10 Direction cosines.

These cosines are called the **direction cosines**.

Note that

$$|\mathbf{u}|^2 = 1 = \cos^2 \alpha + \cos^2 \beta + \cos^2 \gamma$$

Therefore the direction cosines are not independent of one another and cannot all be chosen arbitrarily.

14.7 ALGEBRAIC PROPERTIES

There are a number of algebraic properties which follow from the definition of the scalar product.

Suppose **a**, **b** and **c** are vectors and p is a scalar. Then

$$p(\mathbf{a} \cdot \mathbf{b}) = (p\mathbf{a}) \cdot \mathbf{b}$$
$$(\mathbf{a} + \mathbf{b}) \cdot \mathbf{c} = \mathbf{a} \cdot \mathbf{c} + \mathbf{b} \cdot \mathbf{c}$$

This is the **distributive rule** for the dot product. Note that we are using the convention of elementary algebra that multiplication takes precedence over addition. Without this convention we should need to include more brackets:

$$(\mathbf{a} + \mathbf{b}) \cdot \mathbf{c} = (\mathbf{a} \cdot \mathbf{c}) + (\mathbf{b} \cdot \mathbf{c})$$

□ Prove that if p is any scalar and **a** and **b** are arbitrary vectors then

$$p(\mathbf{a} \cdot \mathbf{b}) = (p\mathbf{a}) \cdot \mathbf{b}$$

We have

$$p(\mathbf{a} \cdot \mathbf{b}) = p|\mathbf{a}||\mathbf{b}| \cos \theta$$

First, if $p \geqslant 0$ then $|p\mathbf{a}| = p|\mathbf{a}|$. So

$$\begin{aligned} p(\mathbf{a} \cdot \mathbf{b}) &= p|\mathbf{a}||\mathbf{b}| \cos \theta \\ &= |p\mathbf{a}||\mathbf{b}| \cos \theta \\ &= (p\mathbf{a}) \cdot \mathbf{b} \end{aligned}$$

Next, if $p < 0$ then $|p\mathbf{a}| = (-p)|\mathbf{a}|$. So

$$\begin{aligned} p(\mathbf{a} \cdot \mathbf{b}) &= -[(-p)|\mathbf{a}||\mathbf{b}| \cos \theta] \\ &= -|p\mathbf{a}||\mathbf{b}| \cos \theta \\ &= |p\mathbf{a}||\mathbf{b}| \cos (\pi - \theta) \\ &= (p\mathbf{a}) \cdot \mathbf{b} \end{aligned}$$

■

Here is a very simple formula which enables us to determine the scalar product when the vectors are given in terms of coordinates. Suppose $\mathbf{a} = a_1\mathbf{i} + a_2\mathbf{j} + a_3\mathbf{k}$ and $\mathbf{b} = b_1\mathbf{i} + b_2\mathbf{j} + b_3\mathbf{k}$. Then

$$\mathbf{a} \cdot \mathbf{b} = a_1b_1 + a_2b_2 + a_3b_3$$

This is easy to establish, as you will see later in this chapter.

□ Obtain the acute angle between two diagonals which each pass through the centre of a cube. We remark that without the use of vectors this would be quite a difficult problem.

Consider a unit cube and position it in such a way that one corner is at the origin O and the axes Ox, Oy and Oz lie along the edges incident at O (Fig. 14.12). It does not matter which pair of the four principal diagonals we choose, as any pair is equivalent to any other. At first sight the diagonals EB and AD may appear to give a different angle from the pair EB and CF, but a few moments' consideration will show that this is not really so.

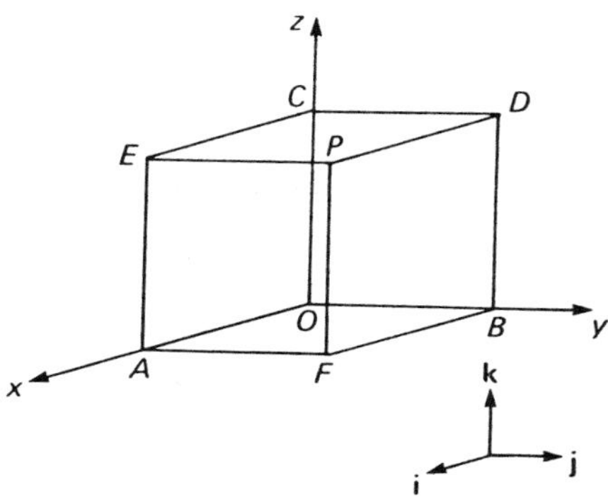

Fig. 14.12 A cube relative to $Oxyz$.

Using the diagram we require the angle between OP and AD:

$$\begin{aligned}\mathbf{OP} &= \mathbf{OA} + \mathbf{AF} + \mathbf{FP} = \mathbf{OA} + \mathbf{OB} + \mathbf{OC} = \mathbf{i} + \mathbf{j} + \mathbf{k}\\ \mathbf{AD} &= \mathbf{AF} + \mathbf{FP} + \mathbf{PD} = \mathbf{OB} + \mathbf{OC} - \mathbf{OA} = \mathbf{j} + \mathbf{k} - \mathbf{i}\end{aligned}$$

Now $\mathbf{OP} \cdot \mathbf{AD} = |\mathbf{OP}||\mathbf{AD}| \cos \theta$, where θ is the angle between OP and AD. Therefore

$$\begin{aligned}(\mathbf{i} + \mathbf{j} + \mathbf{k}) \cdot (-\mathbf{i} + \mathbf{j} + \mathbf{k}) &= |\mathbf{i} + \mathbf{j} + \mathbf{k}||-\mathbf{i} + \mathbf{j} + \mathbf{k}| \cos \theta\\ -1 + 1 + 1 &= \sqrt{3}\,\sqrt{3} \cos \theta\end{aligned}$$

and therefore $\theta = \cos^{-1} (1/3)$. ■

14.8 APPLICATIONS

We can use the scalar product to obtain the component of one vector in the direction of another. For example, suppose we have two vectors **a** and **b** and we require the component of **a** in the direction of **b**. We first obtain a unit vector $\mathbf{u} = \hat{\mathbf{b}}$, and then $\mathbf{a} \cdot \mathbf{u} = |\mathbf{a}|\,(1) \cos \theta$ is the required component.

Extending this idea, suppose we have a particle which is subject to a constant force **F**. The work done by the force is defined to be the magnitude of the force multiplied by the distance moved in the direction of the force. Equivalently the work done by the force is the displacement multiplied by the component of the force in the direction of the displacement. It follows that if the particle is displaced by **s** then the work done by **F** is simply $\mathbf{F} \cdot \mathbf{s}$ (Fig. 14.11).

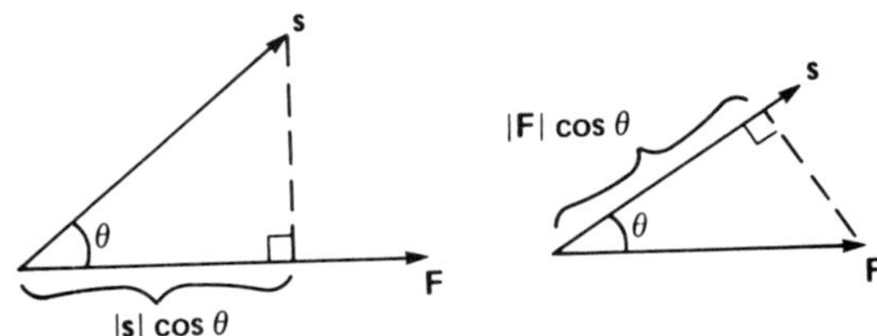

Fig. 14.11 The work done by a force.

□ A force of 3 newtons is applied from the origin to a particle placed at the point $P\,(1,2,2)$ and subject to a system of forces. The particle is displaced to $Q\,(3,4,5)$. If all distances are in metres, calculate the work W done by the force.

We have

$$\mathbf{F} = \mathbf{i} + 2\mathbf{j} + 2\mathbf{k}$$

and

$$\begin{aligned}\mathbf{PQ} &= \mathbf{PO} + \mathbf{OQ} = \mathbf{OQ} - \mathbf{OP}\\ &= (3\mathbf{i} + 4\mathbf{j} + 5\mathbf{k}) - (\mathbf{i} + 2\mathbf{j} + 2\mathbf{k}) = 2\mathbf{i} + 2\mathbf{j} + 3\mathbf{k}\end{aligned}$$

So

$$W = \mathbf{F} \cdot \mathbf{s} = (\mathbf{i} + 2\mathbf{j} + 2\mathbf{k}) \cdot (2\mathbf{i} + 2\mathbf{j} + 3\mathbf{k})$$
$$= 2 + 4 + 6 = 12 \text{ Nm}$$

■

14.9 THE VECTOR PRODUCT

We now introduce another binary operation known as the vector product. Remember that in general a binary operation takes two things and produces from it a third. We use binary operations all the time. For example, whenever we add two numbers together we are performing a binary operation.

Let us review the binary operations involving vectors which we have encountered so far:

1 scalar multiplication $(p, \mathbf{a}) \to p\mathbf{a}$
2 vector addition $(\mathbf{a}, \mathbf{b}) \to \mathbf{a} + \mathbf{b}$
3 scalar product $(\mathbf{a}, \mathbf{b}) \to \mathbf{a} \cdot \mathbf{b}$

We have seen that given two vectors **a** and **b** the scalar product results in a scalar $\mathbf{a} \cdot \mathbf{b}$. The new operation which we shall now consider produces from two vectors **a** and **b** another vector $\mathbf{a} \times \mathbf{b}$. The vector product is sometimes known as the **cross product** because of this notation. An alternative notation is $\mathbf{a} \wedge \mathbf{b}$.

Suppose **a** and **b** are any two non-zero vectors. Then we can choose representatives emanating from O (Fig. 14.13). Moreover, unless they are parallel to one another, they determine a plane in which each of them lies. There are two unit vectors perpendicular to this plane passing through O. It will be convenient to use the special symbols $\odot$ and $\oplus$; the first indicates a vector coming out of the plane towards you, and the second indicates a vector going into the plane away from you. The symbols are inspired by an arrow. The point shows it coming towards you, the tail feathers show it going away.

We shall use **n** for a unit vector normal to the plane. We define the **vector product** as

$$\mathbf{a} \times \mathbf{b} = |\mathbf{a}|\,|\mathbf{b}| \sin \theta \; \mathbf{n}$$

where the direction of **n** is obtained by the right-hand rule.

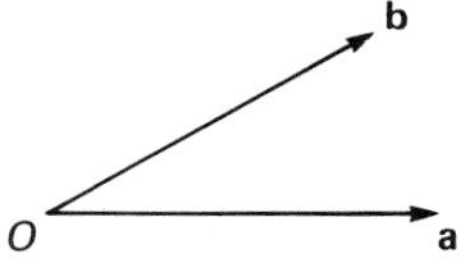

Fig. 14.13 Two vectors emanating from O.

The **right-hand rule** is sometimes known as the corkscrew rule. Ordinary wood screws, jar tops etc. have a right-hand thread. Imagine rotating a screw which is in a piece of wood from the direction **a** to the direction **b** (Fig. 14.14). Then the screw would tend to come out of the wood, and so the direction of **n** is out of the plane towards you. Conversely, if **a** and **b** are reversed then as we rotate the screw from **a** to **b** the screw goes into the wood (Fig. 14.15), and the direction of **n** is into the plane away from you.

For the sake of completeness, if either **a** or **b** is **0** then we define $\mathbf{a} \times \mathbf{b} = \mathbf{0}$ too.

We have in all cases

$$|\mathbf{a} \times \mathbf{b}| = |\mathbf{a}|\,|\mathbf{b}|\,|\sin \theta|$$

but of course $\mathbf{a} \times \mathbf{b} = -\mathbf{b} \times \mathbf{a}$ because the unit vectors concerned have opposite directions.

We can show that

$$\mathbf{a} \times \mathbf{b} = -\mathbf{b} \times \mathbf{a}$$

algebraically by using the definition (see also Fig. 14.16). We have

$$\mathbf{a} \times \mathbf{b} = |\mathbf{a}|\,|\mathbf{b}| \sin \theta \; \mathbf{n}$$

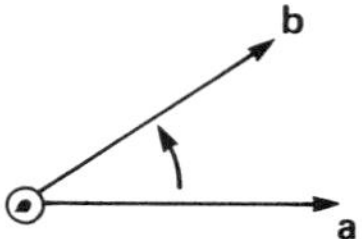

Fig. 14.14 $\mathbf{a} \times \mathbf{b}$ (out of the plane).

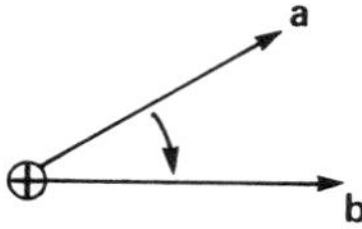

Fig. 14.15 $\mathbf{a} \times \mathbf{b}$ (into the plane).

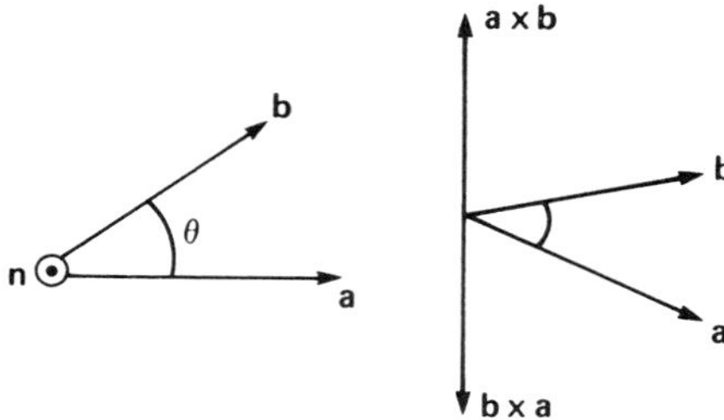

Fig. 14.16 The vector product.

So

$$\begin{aligned}\mathbf{b} \times \mathbf{a} &= |\mathbf{b}|\,|\mathbf{a}| \sin (2\pi - \theta)\, \mathbf{n} \\ &= -|\mathbf{a}|\,|\mathbf{b}| \sin \theta\, \mathbf{n} = -\mathbf{a} \times \mathbf{b}\end{aligned}$$

□ Show that if **a** and **b** are non-zero vectors then $\mathbf{a} \times \mathbf{b} = \mathbf{0}$ if and only if **a** and **b** are parallel.

Suppose $\mathbf{a} \times \mathbf{b} = \mathbf{0}$. Then $|\mathbf{a} \times \mathbf{b}| = 0$. So $|\mathbf{a}|\,|\mathbf{b}|\,|\sin \theta| = 0$, and since **a** and **b** are non-zero it follows that $\sin \theta = 0$. So $\theta = 0$ or $\theta = \pi$ and the vectors are parallel.

Conversely, suppose the vectors **a** and **b** are parallel. Then either $\theta = 0$ or $\theta = \pi$, and it follows that

$$\mathbf{a} \times \mathbf{b} = |\mathbf{a}|\,|\mathbf{b}| \sin \theta\, \mathbf{n} = 0\mathbf{n} = \mathbf{0}$$ ■

It follows of course that $\mathbf{a} \times \mathbf{a} = \mathbf{0}$, and in fact we can easily obtain the vector products of the vectors **i**, **j** and **k** (Fig. 14.17):

$$\begin{array}{ll} \mathbf{i} \times \mathbf{j} = \mathbf{k} & \mathbf{j} \times \mathbf{i} = -\mathbf{k} \\ \mathbf{j} \times \mathbf{k} = \mathbf{i} & \mathbf{k} \times \mathbf{j} = -\mathbf{i} \\ \mathbf{k} \times \mathbf{i} = \mathbf{j} & \mathbf{i} \times \mathbf{k} = -\mathbf{j} \end{array}$$

Corresponding to the algebraic properties we derived for the scalar product, there are similar properties for the vector product. Suppose **a** and **b** are arbitrary vectors and that p is a scalar. Then

$$\begin{aligned} p(\mathbf{a} \times \mathbf{b}) &= (p\mathbf{a}) \times \mathbf{b} \\ (\mathbf{a} + \mathbf{b}) \times \mathbf{c} &= \mathbf{a} \times \mathbf{c} + \mathbf{b} \times \mathbf{c} \end{aligned}$$

□ Prove that if p is a scalar and **a** and **b** are arbitrary vectors then

$$p(\mathbf{a} \times \mathbf{b}) = (p\mathbf{a}) \times \mathbf{b}$$

Fig. 14.18 illustrates the alternatives.

If $p > 0$ then

$$\begin{aligned}(p\mathbf{a}) \times \mathbf{b} &= |p\mathbf{a}|\,|\mathbf{b}| \sin \theta\, \mathbf{n} \\ &= p|\mathbf{a}|\,|\mathbf{b}| \sin \theta\, \mathbf{n} \\ &= p(\mathbf{a} \times \mathbf{b})\end{aligned}$$

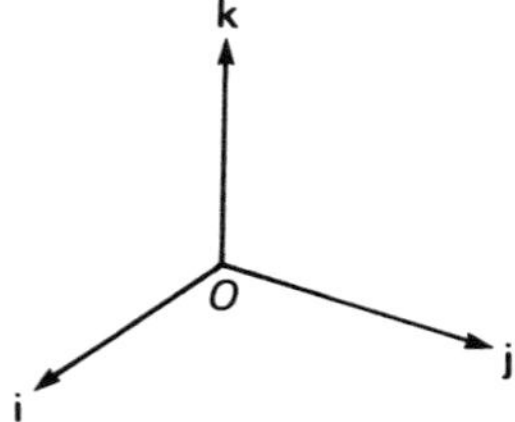

Fig. 14.17 The standard unit vectors.

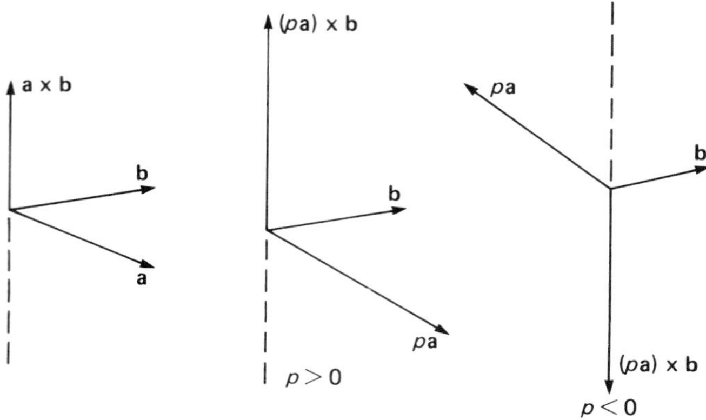

Fig. 14.18 Verification that $(p\mathbf{a}) \times \mathbf{b} = p(\mathbf{a} \times \mathbf{b})$.

If $p < 0$ then

$$\begin{aligned}(p\mathbf{a}) \times \mathbf{b} &= |p\mathbf{a}|\,|\mathbf{b}| \sin\theta\,(-\mathbf{n}) \\ &= (-p)\,|\mathbf{a}|\,|\mathbf{b}| \sin\theta\,(-\mathbf{n}) \\ &= p|\mathbf{a}|\,|\mathbf{b}| \sin\theta\,\mathbf{n} \\ &= p(\mathbf{a} \times \mathbf{b})\end{aligned}$$

If $p = 0$ the conclusion is immediate. ■

Here is a formula which enables us to determine the vector product when the vectors are given in terms of coordinates. Suppose $\mathbf{a} = a_1\mathbf{i} + a_2\mathbf{j} + a_3\mathbf{k}$ and $\mathbf{b} = b_1\mathbf{i} + b_2\mathbf{j} + b_3\mathbf{k}$. Then

$$\begin{aligned}\mathbf{a} \times \mathbf{b} &= (a_2b_3 - a_3b_2)\mathbf{i} - (a_1b_3 - a_3b_1)\mathbf{j} + (a_1b_2 - a_2b_1)\mathbf{k} \\ &= \begin{vmatrix} \mathbf{i} & \mathbf{j} & \mathbf{k} \\ a_1 & a_2 & a_3 \\ b_1 & b_2 & b_3 \end{vmatrix}\end{aligned}$$

Notice how useful the determinant notation is here, and how easy it is to remember this formula provided we know how to expand determinants (Chapter 12). The first row consists of the unit vectors **i**, **j** and **k**. The next two rows consist of the components of the vectors **a** and **b** respectively. It is interesting also to notice that, by the rules of determinants, if we interchange two rows the determinant changes sign. This gives another proof that $\mathbf{a} \times \mathbf{b} = -\mathbf{b} \times \mathbf{a}$. The formula is quite easy to obtain, as you will see later in the chapter.

Let's summarize what we have learnt so far:

□ We began with scalar quantities and vector quantities.

□ We added vectors together using the parallelogram law.
□ We multiplied vectors by scalars.
□ We expressed vectors in terms of components.
□ We defined the scalar product of two vectors.
□ We defined the vector product of two vectors.

Right! Are you ready for some steps? If necessary you can read through all we have covered once more.

14.10 Workshop

1

▷**Exercise** A man travels over the surface of the earth, first from Moscow to Paris and then from Paris to Cairo. Representing these journeys as vector quantities, where the magnitude of each is the shortest distance over the earth's surface, decide whether or not they are a set of vectors.

Think carefully about this before you decide.

2

They do *not* constitute a set of vectors because the parallelogram law does not in general apply where distances are measured over the curved surface of the earth.

If you were wrong, don't worry: we are only just getting warmed up! Try this exercise.

▷**Exercise** If $\mathbf{a} = a_1\mathbf{i} + a_2\mathbf{j} + a_3\mathbf{k}$, show that

$$\mathbf{a} \cdot \mathbf{i} = a_1 \qquad \mathbf{a} \cdot \mathbf{j} = a_2 \qquad \mathbf{a} \cdot \mathbf{k} = a_3$$

Try this using the rules carefully. Then step ahead.

3

First,

$$\begin{aligned}\mathbf{a} \cdot \mathbf{i} &= (a_1\mathbf{i} + a_2\mathbf{j} + a_3\mathbf{k}) \cdot \mathbf{i} \\ &= a_1(\mathbf{i} \cdot \mathbf{i}) + a_2(\mathbf{j} \cdot \mathbf{i}) + a_3(\mathbf{k} \cdot \mathbf{i}) \\ &= a_1 1 + a_2 0 + a_3 0 = a_1\end{aligned}$$

If you didn't manage that you should now be able to obtain $\mathbf{a} \cdot \mathbf{j}$ and $\mathbf{a} \cdot \mathbf{k}$.

Then go on to the next exercise.

▷**Exercise** If $\mathbf{a} = a_1\mathbf{i} + a_2\mathbf{j} + a_3\mathbf{k}$ and $\mathbf{b} = b_1\mathbf{i} + b_2\mathbf{j} + b_3\mathbf{k}$, show that

$$\mathbf{a} \cdot \mathbf{b} = a_1b_1 + a_2b_2 + a_3b_3$$

Try this using the results of the previous exercise.

4

The working is

$$\begin{aligned}\mathbf{a}\cdot\mathbf{b} &= \mathbf{a}\cdot(b_1\mathbf{i} + b_2\mathbf{j} + b_3\mathbf{k})\\ &= (\mathbf{a}\cdot\mathbf{i})b_1 + (\mathbf{a}\cdot\mathbf{j})b_2 + (\mathbf{a}\cdot\mathbf{k})b_3\\ &= a_1b_1 + a_2b_2 + a_3b_3\end{aligned}$$

This should have caused no difficulty. If it did then repeat the exercise, taking the vector **a** in components instead of **b**.

Then go on to the next exercise.

▷**Exercise** Without expressing the vectors in terms of components, show that

$$\mathbf{a}\cdot(\mathbf{b}+\mathbf{c}) = \mathbf{a}\cdot\mathbf{b} + \mathbf{a}\cdot\mathbf{c}$$

Don't assume any algebraic properties that we have not already discussed. Try it, then step forward.

---

5 We have

$$\mathbf{a}\cdot(\mathbf{b}+\mathbf{c}) = (\mathbf{b}+\mathbf{c})\cdot\mathbf{a}$$

because the scalar product is commutative. Then

$$= \mathbf{b}\cdot\mathbf{a} + \mathbf{c}\cdot\mathbf{a}$$

using the distributive rule. Finally

$$= \mathbf{a}\cdot\mathbf{b} + \mathbf{a}\cdot\mathbf{c}$$

again using the commutative property of the dot product.

That was a little algebraic perhaps, so let's solve a more practical problem.

▷**Exercise** A rectangular building has a square cross-section and is twice as long as it is wide. Two thin wires are to be inserted joining the corners at the top of one end of the building to the diagonally opposite corners at the bottom of the far end. Calculate the cosine of the angle between these two wires (Fig. 14.19).

This is very similar to the problem in the text involving the cube. If you can do that, you can do this.

---

6 We set up the building in a coordinate system (Fig. 14.20) just as we did for the cube (Fig. 14.12). Then in Fig. 14.20 we have

$$|\mathbf{OB}| = 2|\mathbf{OA}| = 2|\mathbf{OC}|$$

So, taking $|\mathbf{OA}| = 1$, we have

$$\begin{aligned}\mathbf{OP} &= \mathbf{OA} + \mathbf{AF} + \mathbf{FP}\\ &= \mathbf{OA} + \mathbf{OB} + \mathbf{OC} = \mathbf{i} + 2\mathbf{j} + \mathbf{k}\end{aligned}$$

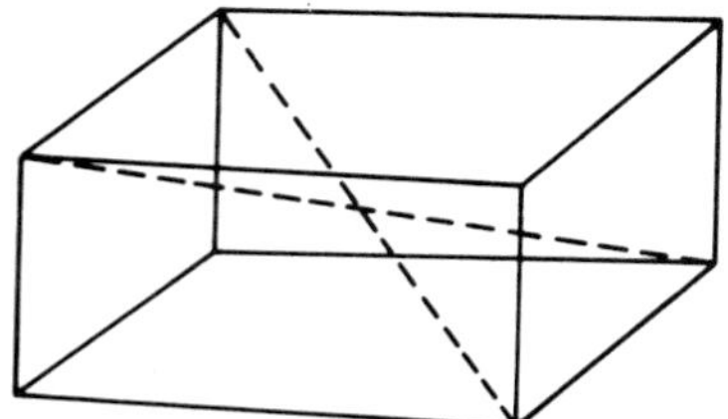

Fig. 14.19 Building with diagonals.

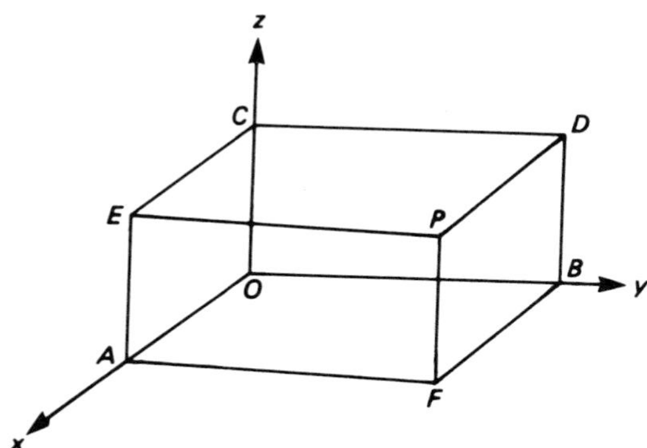

Fig. 14.20 Building relative to $Oxyz$.

$$\begin{aligned}\mathbf{AD} &= \mathbf{AF} + \mathbf{FP} + \mathbf{PD} \\ &= \mathbf{OB} + \mathbf{OC} - \mathbf{OA} = -\mathbf{i} + 2\mathbf{j} + \mathbf{k}\end{aligned}$$

Now

$$\mathbf{OP} \cdot \mathbf{AD} = |\mathbf{OP}|\,|\mathbf{AD}| \cos \theta$$

So

$$(\mathbf{i} + 2\mathbf{j} + \mathbf{k}) \cdot (-\mathbf{i} + 2\mathbf{j} + \mathbf{k}) = |\mathbf{i} + 2\mathbf{j} + \mathbf{k}|\,|-\mathbf{i} + 2\mathbf{j} + \mathbf{k}| \cos \theta$$

Therefore

$$\begin{aligned}-1 + 4 + 1 &= \sqrt{(1 + 4 + 1)}\sqrt{(1 + 4 + 1)} \cos \theta \\ 4 &= 6 \cos \theta\end{aligned}$$

so that $\cos \theta = 2/3$.

If you didn't get that right, try the cube again without looking at the solution. Then repeat the building exercise.

Now let's see if we can handle the vector product.

▷**Exercise** Expand

a $\mathbf{a} \times (\mathbf{b} + \mathbf{c})$

b $(\mathbf{a} + \mathbf{b}) \times (\mathbf{c} + \mathbf{d})$

To do this you need to use the same techniques as at step 5 for the scalar product, although the formulas differ a little. Try it and see how it goes.

7

a We have

$$\begin{aligned}\mathbf{a} \times (\mathbf{b} + \mathbf{c}) &= -(\mathbf{b} + \mathbf{c}) \times \mathbf{a} \\ &= -(\mathbf{b} \times \mathbf{a} + \mathbf{c} \times \mathbf{a}) \\ &= -\mathbf{b} \times \mathbf{a} - \mathbf{c} \times \mathbf{a} \\ &= \mathbf{a} \times \mathbf{b} + \mathbf{a} \times \mathbf{c}\end{aligned}$$

b First write $\mathbf{e} = \mathbf{c} + \mathbf{d}$. Then

$$\begin{aligned}(\mathbf{a} + \mathbf{b}) \times (\mathbf{c} + \mathbf{d}) &= (\mathbf{a} + \mathbf{b}) \times \mathbf{e} \\ &= \mathbf{a} \times \mathbf{e} + \mathbf{b} \times \mathbf{e} \\ &= \mathbf{a} \times (\mathbf{c} + \mathbf{d}) + \mathbf{b} \times (\mathbf{c} + \mathbf{d}) \\ &= \mathbf{a} \times \mathbf{c} + \mathbf{a} \times \mathbf{d} + \mathbf{b} \times \mathbf{c} + \mathbf{b} \times \mathbf{d}\end{aligned}$$

Did you manage that all right? If you didn't then you must be extra specially careful about the next problem.

▷**Exercise** Simplify $(\mathbf{a} + \mathbf{b}) \times (\mathbf{a} - \mathbf{b})$

Do be careful!

8

We have

$$\begin{aligned}(\mathbf{a} + \mathbf{b}) \times (\mathbf{a} - \mathbf{b}) &= \mathbf{a} \times \mathbf{a} + \mathbf{b} \times \mathbf{a} - \mathbf{a} \times \mathbf{b} - \mathbf{b} \times \mathbf{b} \\ &= \mathbf{0} + \mathbf{b} \times \mathbf{a} - \mathbf{a} \times \mathbf{b} - \mathbf{0} \\ &= 2(\mathbf{b} \times \mathbf{a})\end{aligned}$$

Did you fall into the trap of cancelling out $\mathbf{a} \times \mathbf{b}$ and deducing the incorrect answer $\mathbf{0}$? You must remember that for the vector product the commutative law does *not* hold.

Now try one last exercise.

▷**Exercise** Obtain the formula for the vector product of $\mathbf{a} = a_1\mathbf{i} + a_2\mathbf{j} + a_3\mathbf{k}$ and $\mathbf{b} = b_1\mathbf{i} + b_2\mathbf{j} + b_3\mathbf{k}$.

This is not too hard now that we know how to multiply out. Do this carefully, and remember to write down the vector products in the right order.

9

We have

$$\begin{aligned}\mathbf{a} \times \mathbf{b} &= (a_1\mathbf{i} + a_2\mathbf{j} + a_3\mathbf{k}) \times \mathbf{b} \\ &= a_1(\mathbf{i} \times \mathbf{b}) + a_2(\mathbf{j} \times \mathbf{b}) + a_3(\mathbf{k} \times \mathbf{b})\end{aligned}$$

$$\begin{aligned}
&= a_1\mathbf{i} \times (b_1\mathbf{i} + b_2\mathbf{j} + b_3\mathbf{k}) \\
&\quad + a_2\mathbf{j} \times (b_1\mathbf{i} + b_2\mathbf{j} + b_3\mathbf{k}) \\
&\quad + a_3\mathbf{k} \times (b_1\mathbf{i} + b_2\mathbf{j} + b_3\mathbf{k}) \\
&= a_1b_2(\mathbf{i} \times \mathbf{j}) + a_1b_3(\mathbf{i} \times \mathbf{k}) + a_2b_1(\mathbf{j} \times \mathbf{i}) \\
&\quad + a_2b_3(\mathbf{j} \times \mathbf{k}) + a_3b_1(\mathbf{k} \times \mathbf{i}) + a_3b_2(\mathbf{k} \times \mathbf{j}) \\
&= a_1b_2\mathbf{k} - a_1b_3\mathbf{j} - a_2b_1\mathbf{k} \\
&\quad + a_2b_3\mathbf{i} + a_3b_1\mathbf{j} - a_3b_2\mathbf{i}
\end{aligned}$$

Therefore we have

$$\begin{aligned}
\mathbf{a} \times \mathbf{b} &= (a_2b_3 - a_3b_2)\mathbf{i} - (a_1b_3 - a_3b_1)\mathbf{j} + (a_1b_2 - a_2b_1)\mathbf{k} \\
&= \begin{vmatrix} \mathbf{i} & \mathbf{j} & \mathbf{k} \\ a_1 & a_2 & a_3 \\ b_1 & b_2 & b_3 \end{vmatrix}
\end{aligned}$$

Well, now it's time to move on.

We have seen that if **a** and **b** are vectors then $\mathbf{a} \times \mathbf{b}$ is also a vector. We can therefore consider the effect of combining this vector with a third vector **c**. There are two operations that we need to examine: the triple scalar product and the triple vector product.

14.11 THE TRIPLE SCALAR PRODUCT

We now describe what is meant by the symbol

$$\mathbf{a} \cdot \mathbf{b} \times \mathbf{c}$$

What *can* it mean? There are only two possible ways of inserting brackets so that the operations can be performed consecutively, and these are

$$\mathbf{a} \cdot (\mathbf{b} \times \mathbf{c}) \quad \text{and} \quad (\mathbf{a} \cdot \mathbf{b}) \times \mathbf{c}$$

However, a few moments' thought shows that the expression $(\mathbf{a} \cdot \mathbf{b}) \times \mathbf{c}$ has no meaning. This is because it purports to calculate the vector product of a scalar $\mathbf{a} \cdot \mathbf{b}$ with a vector **c**. Now we defined the vector product as a binary operation in which two *vectors* were combined. Consequently the expression $(\mathbf{a} \cdot \mathbf{b}) \times \mathbf{c}$ is meaningless.

On the other hand the expression $\mathbf{a} \cdot (\mathbf{b} \times \mathbf{c})$ does have a meaning, and so it is this which we take as the definition of the symbol $\mathbf{a} \cdot \mathbf{b} \times \mathbf{c}$ when brackets are not inserted. Therefore

$$\mathbf{a} \cdot \mathbf{b} \times \mathbf{c} = \mathbf{a} \cdot (\mathbf{b} \times \mathbf{c})$$

Now **a** is a vector and $\mathbf{b} \times \mathbf{c}$ is also a vector; consequently $\mathbf{a} \cdot (\mathbf{b} \times \mathbf{c})$ is a scalar. For this reason this product of the three vectors is called a **triple scalar product**.

□ Obtain the triple scalar product $\mathbf{a} \cdot \mathbf{b} \times \mathbf{c}$ of three vectors in terms of components.

If you wish you can try this yourself. It is not difficult now that we know how to obtain the scalar product and the vector product in terms of components. Suppose

$$\begin{aligned}\mathbf{a} &= a_1\mathbf{i} + a_2\mathbf{j} + a_3\mathbf{k}\\ \mathbf{b} &= b_1\mathbf{i} + b_2\mathbf{j} + b_3\mathbf{k}\\ \mathbf{c} &= c_1\mathbf{i} + c_2\mathbf{j} + c_3\mathbf{k}\end{aligned}$$

Here is a *neat* way of deducing the result. Remember that

$$\begin{aligned}\mathbf{a} \cdot \mathbf{b} &= (a_1\mathbf{i} + a_2\mathbf{j} + a_3\mathbf{k}) \cdot (b_1\mathbf{i} + b_2\mathbf{j} + b_3\mathbf{k})\\ &= a_1b_1 + a_2b_2 + a_3b_3\end{aligned}$$

and this can be thought of as having been obtained in the following way. The unit vectors **i**, **j** and **k** in the second vector have each been replaced by the components a_1, a_2 and a_3 of the first vector. Using this idea we have

$$\begin{aligned}\mathbf{a} \cdot \mathbf{b} \times \mathbf{c} &= \mathbf{a} \cdot (\mathbf{b} \times \mathbf{c})\\ &= (a_1\mathbf{i} + a_2\mathbf{j} + a_3\mathbf{k}) \cdot \begin{vmatrix} \mathbf{i} & \mathbf{j} & \mathbf{k}\\ b_1 & b_2 & b_3\\ c_1 & c_2 & c_3\end{vmatrix}\\ &= \begin{vmatrix} a_1 & a_2 & a_3\\ b_1 & b_2 & b_3\\ c_1 & c_2 & c_3\end{vmatrix}\end{aligned}$$

■

One immediate consequence of this is that if two of the vectors have the same direction then the triple scalar product is *zero*. This is a consequence of the rule for determinants that if two rows are equal then the determinant is zero.

Using the work we did on determinants (Chapter 12) you will now be able to write down several equivalent expressions for $\mathbf{a} \cdot \mathbf{b} \times \mathbf{c}$. Remember, if we interchange the corresponding elements in two rows of a determinant then the determinant changes sign. So

$$\begin{aligned}\mathbf{a} \cdot \mathbf{b} \times \mathbf{c} &= -\begin{vmatrix} b_1 & b_2 & b_3\\ a_1 & a_2 & a_3\\ c_1 & c_2 & c_3\end{vmatrix} = -\mathbf{b} \cdot \mathbf{a} \times \mathbf{c}\\ &= \begin{vmatrix} b_1 & b_2 & b_3\\ c_1 & c_2 & c_3\\ a_1 & a_2 & a_3\end{vmatrix} = \mathbf{b} \cdot \mathbf{c} \times \mathbf{a}\end{aligned}$$

In fact the rules for determinants enable us to deduce that the important ingredient in a triple scalar product is not the relative positions of the signs × and ·, or the precise order of the vectors **a**, **b** and **c**, but the **cyclic order** of the vectors **a**, **b** and **c**.

If we preserve the cyclic order so that **b** follows **a**, **c** follows **b**, and **a** follows **c**, it does not matter where we put the dot and cross; we will obtain the same result. However, if we reverse the cyclic order so that **b** follows **c**, **c** follows **a**, and **a** follows **b**, then wherever we put the dot and cross the result will have the *opposite* sign. So

$$\mathbf{a} \cdot \mathbf{b} \times \mathbf{c} = \mathbf{c} \times \mathbf{a} \cdot \mathbf{b}$$
$$\mathbf{b} \times \mathbf{c} \cdot \mathbf{a} = -\mathbf{a} \cdot \mathbf{c} \times \mathbf{b}$$

For this reason a triple scalar product is sometimes written as $[\mathbf{a}, \mathbf{b}, \mathbf{c}]$; this defines the cyclic order, and is all that is needed. So

$$\begin{aligned}[\mathbf{a}, \mathbf{b}, \mathbf{c}] &= [\mathbf{b}, \mathbf{c}, \mathbf{a}] = [\mathbf{c}, \mathbf{a}, \mathbf{b}] \\ &= -[\mathbf{a}, \mathbf{c}, \mathbf{b}] = -[\mathbf{c}, \mathbf{b}, \mathbf{a}] = -[\mathbf{b}, \mathbf{a}, \mathbf{c}]\end{aligned}$$

PHYSICAL INTERPRETATIONS

If we consider the magnitude of $\mathbf{a} \times \mathbf{b}$, the vector product of two vectors **a** and **b**, we see that

$$|\mathbf{a} \times \mathbf{b}| = |\mathbf{a}|\,|\mathbf{b}|\,|\sin \theta|$$

This is equal to the area of the parallelogram formed by the two vectors (Fig. 14.21).

If now we consider $\mathbf{a} \cdot (\mathbf{b} \times \mathbf{c})$ we have

$$\mathbf{a} \cdot (\mathbf{b} \times \mathbf{c}) = |\mathbf{a}|\,|\mathbf{b} \times \mathbf{c}| \cos \phi$$

where ϕ is the angle between **a** and $\mathbf{b} \times \mathbf{c}$ (Fig. 14.22). Now $\mathbf{b} \times \mathbf{c}$ is perpendicular to the plane of **b** and **c**, and so it follows that $|\mathbf{a}| \cos \phi$ is the height of the parallelepiped Π formed by the vectors **a**, **b** and **c**. Therefore

$$\begin{aligned}\mathbf{a} \cdot (\mathbf{b} \times \mathbf{c}) &= \text{height of } \Pi \times \text{area of base of } \Pi \\ &= \text{volume of } \Pi\end{aligned}$$

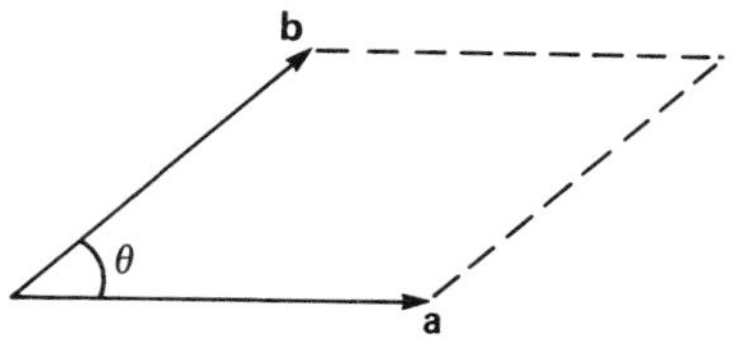

Fig. 14.21 The magnitude of $\mathbf{a} \times \mathbf{b}$.

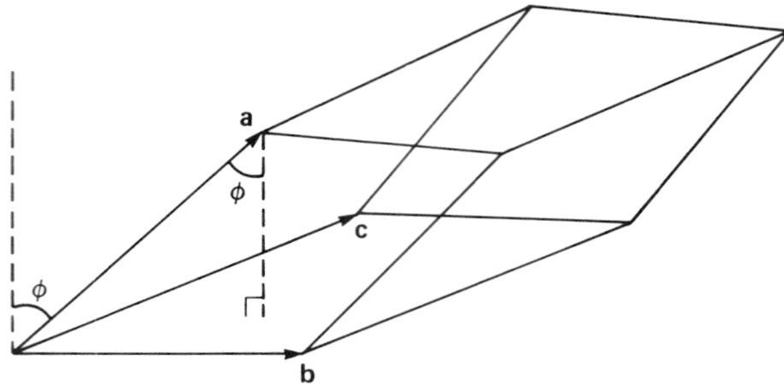

Fig. 14.22 The triple scalar product.

Strictly speaking we have relied a little too much on the diagram for, were **b** and **c** interchanged, cos θ would be negative. Nevertheless we have shown that the *magnitude* of $\mathbf{a} \cdot \mathbf{b} \times \mathbf{c}$ can be interpreted as the volume of the parallelepiped Π formed from the three vectors **a**, **b** and **c**.

Observe that this accords with several of the properties we have already discovered. In particular it shows that the triple scalar product depends only on the cyclic order of the three vectors concerned.

14.12 THE TRIPLE VECTOR PRODUCT

We have considered the scalar product of the vector **a** and the vector $\mathbf{b} \times \mathbf{c}$. We now turn our attention to the vector product of these vectors. The result $\mathbf{a} \times (\mathbf{b} \times \mathbf{c})$ will be a vector and so we call it a triple vector product.

At the outset it is important to realize that

$$\mathbf{a} \times (\mathbf{b} \times \mathbf{c}) \neq (\mathbf{a} \times \mathbf{b}) \times \mathbf{c}$$

Indeed there is no reason at all why these two triple vector products should be equal. However, the use of the product notation and the fact that the associative law usually holds for products leads the unwary to expect that it will hold. It certainly does not hold, and this means of course that we must be scrupulously careful to insert brackets correctly in any mathematical expression we write down. More marks are lost in examinations as a result of missing brackets than are lost through numerical inaccuracy.

What then is $\mathbf{a} \times (\mathbf{b} \times \mathbf{c})$? We know it is a vector because it is the result of taking the vector product of two vectors **a** and $\mathbf{b} \times \mathbf{c}$. Moreover it is perpendicular to both **a** and $\mathbf{b} \times \mathbf{c}$. Now $\mathbf{b} \times \mathbf{c}$ itself is perpendicular to the plane containing the vectors **b** and **c**, and since we have only three dimensions at our disposal we can make a deduction. Can you see what it is?

The implication is that $\mathbf{a} \times (\mathbf{b} \times \mathbf{c})$ must be in the *same plane* as $\mathbf{b}$ and $\mathbf{c}$. Now any vector in the plane of $\mathbf{b}$ and $\mathbf{c}$ can be written in the form $p\mathbf{b} + q\mathbf{c}$ where p and q are scalars. So we have deduced that

$$\mathbf{a} \times (\mathbf{b} \times \mathbf{c}) = p\mathbf{b} + q\mathbf{c}$$

It remains only to determine the scalars p and q.

If we take the dot product with $\mathbf{a}$ of each side of the last relation we obtain

$$0 = p(\mathbf{a} \cdot \mathbf{b}) + q(\mathbf{a} \cdot \mathbf{c})$$

So defining a new scalar t by

$$p = t(\mathbf{a} \cdot \mathbf{c})$$

we have

$$q = -t(\mathbf{a} \cdot \mathbf{b})$$

and therefore

$$\mathbf{a} \times (\mathbf{b} \times \mathbf{c}) = t[(\mathbf{a} \cdot \mathbf{c})\mathbf{b} - (\mathbf{a} \cdot \mathbf{b})\mathbf{c}]$$

It remains only to obtain the scalar t.

No loss of generality is obtained if we choose our coordinate system so that $\mathbf{a} = \lambda\mathbf{i}$, and since the expansion must hold for all $\mathbf{b}$ and $\mathbf{c}$ we can examine the component in the direction of $\mathbf{k}$. The right-hand side gives

$$t[\lambda c_1 b_3 - \lambda b_1 c_3]$$

The left-hand side gives, since $\mathbf{i} \times \mathbf{j} = \mathbf{k}$, the product of λ with the coefficient of $\mathbf{j}$ in $\mathbf{b} \times \mathbf{c}$. This is

$$-\lambda[b_1 c_3 - b_3 c_1] = \lambda[c_1 b_3 - b_1 c_3]$$

It follows that $t = 1$, and therefore the **triple vector product** is given by

$$\mathbf{a} \times (\mathbf{b} \times \mathbf{c}) = (\mathbf{a} \cdot \mathbf{c})\mathbf{b} - (\mathbf{a} \cdot \mathbf{b})\mathbf{c}$$

We are now in a position to calculate the other triple vector product $(\mathbf{a} \times \mathbf{b}) \times \mathbf{c}$. We have

$$\begin{aligned}(\mathbf{a} \times \mathbf{b}) \times \mathbf{c} &= -\,\mathbf{c} \times (\mathbf{a} \times \mathbf{b}) \\ &= -[(\mathbf{c} \cdot \mathbf{b})\mathbf{a} - (\mathbf{c} \cdot \mathbf{a})\mathbf{b}] \\ &= (\mathbf{c} \cdot \mathbf{a})\mathbf{b} - (\mathbf{c} \cdot \mathbf{b})\mathbf{a} \\ &= (\mathbf{a} \cdot \mathbf{c})\mathbf{b} - (\mathbf{b} \cdot \mathbf{c})\mathbf{a}\end{aligned}$$

So we have the two triple vector products

$$\begin{aligned}\mathbf{a} \times (\mathbf{b} \times \mathbf{c}) &= (\mathbf{a} \cdot \mathbf{c})\mathbf{b} - (\mathbf{a} \cdot \mathbf{b})\mathbf{c} \\ (\mathbf{a} \times \mathbf{b}) \times \mathbf{c} &= (\mathbf{a} \cdot \mathbf{c})\mathbf{b} - (\mathbf{b} \cdot \mathbf{c})\mathbf{a}\end{aligned}$$

These have a similar structure, which makes the expansion easy to remember if you are prepared to learn a little chant. To expand a triple vector product write down the middle vector and multiply it by the dot product of the others; then subtract the other *bracketed* vector multiplied by the dot product of the remaining two.

□ Write down the triple vector product $\mathbf{a} \times (\mathbf{b} \times \mathbf{c})$ of the vectors $\mathbf{a} = 2\mathbf{i} + \mathbf{j} - \mathbf{k}$, $\mathbf{b} = \mathbf{i} + \mathbf{j} - \mathbf{k}$ and $\mathbf{c} = \mathbf{i} + 2\mathbf{j} - \mathbf{k}$.

We have

$$\mathbf{a} \times (\mathbf{b} \times \mathbf{c}) = (\mathbf{a} \cdot \mathbf{c})\mathbf{b} - (\mathbf{a} \cdot \mathbf{b})\mathbf{c}$$

and

$$\begin{aligned} \mathbf{a} \cdot \mathbf{c} &= 2 \times 1 + 1 \times 2 + (-1) \times (-1) = 5 \\ \mathbf{a} \cdot \mathbf{b} &= 2 \times 1 + 1 \times 1 + (-1) \times (-1) = 4 \end{aligned}$$

So

$$\begin{aligned} \mathbf{a} \times (\mathbf{b} \times \mathbf{c}) &= 5(\mathbf{i} + \mathbf{j} - \mathbf{k}) - 4(\mathbf{i} + 2\mathbf{j} - \mathbf{k}) \\ &= \mathbf{i} - 3\mathbf{j} - \mathbf{k} \end{aligned}$$

Alternatively, we could determine $\mathbf{b} \times \mathbf{c}$ first and then $\mathbf{a} \times (\mathbf{b} \times \mathbf{c})$. Do this and check the answer. ■

14.13 DIFFERENTIATION OF VECTORS

All the vectors we have considered in the examples have been constant vectors. However, there is no reason why this should be so. Suppose O is a fixed point and P is a general point on a curve. As P moves along the curve the vector **OP** changes in both magnitude and direction (Fig. 14.23).

OP is called the **position vector** of the point P. We may write $\mathbf{OP} = \mathbf{r}(t)$ to represent this vector, where t is a parameter. It may be helpful to think

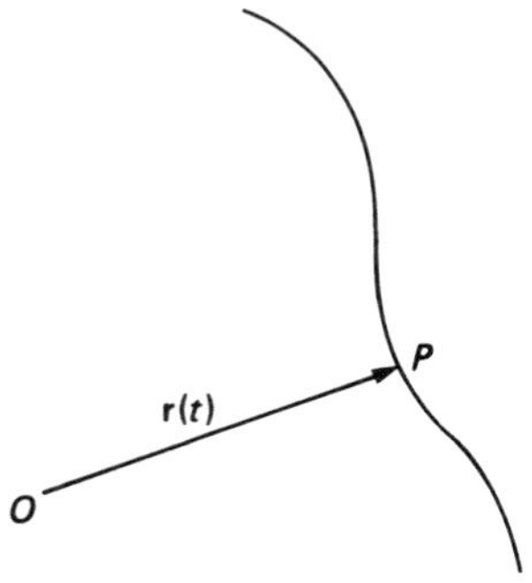

Fig. 14.23 A variable position vector.

of t as time, but this restriction is not essential. The variable t could be any scalar, for example temperature or angle. So as t changes, $\mathbf{r}(t)$ gives the position vector of a point which is moving along the curve.

Suppose δt is a non-zero change in t. We define

$$\dot{\mathbf{r}}(t) = \frac{\mathrm{d}}{\mathrm{d}t}\mathbf{r}(t) = \lim_{\delta t \to 0} \frac{\mathbf{r}(t + \delta t) - \mathbf{r}(t)}{\delta t}$$

This definition is a direct generalization to vectors of the idea of the derivative of a real function (Chapter 4).

Using Fig. 14.24 we have

$$\mathbf{r}(t + \delta t) - \mathbf{r}(t) = \mathbf{OQ} - \mathbf{OP} = \mathbf{PQ}$$

so that as $\delta t \to 0$ we obtain the direction of $\dot{\mathbf{r}}(t)$ as the direction of the tangent to the curve at P. We call $\dot{\mathbf{r}}(t)$ the **velocity vector** of P.

Similarly we define

$$\ddot{\mathbf{r}}(t) = \frac{\mathrm{d}}{\mathrm{d}t}\dot{\mathbf{r}}(t)$$

to be the **acceleration vector** of P. In general, the directions of $\mathbf{r}(t)$, $\dot{\mathbf{r}}(t)$ and $\ddot{\mathbf{r}}(t)$ are all different.

RULES FOR DIFFERENTIATION

Suppose $\mathbf{u}$, $\mathbf{v}$ and $\mathbf{w}$ are position vectors, and p is a scalar, each dependent on the variable t. Then provided the derivatives exist, the following rules hold:

$$\frac{\mathrm{d}}{\mathrm{d}t}(\mathbf{u} + \mathbf{v}) = \frac{\mathrm{d}\mathbf{u}}{\mathrm{d}t} + \frac{\mathrm{d}\mathbf{v}}{\mathrm{d}t}$$

$$\frac{\mathrm{d}}{\mathrm{d}t}(p\mathbf{u}) = p\frac{\mathrm{d}\mathbf{u}}{\mathrm{d}t} + \frac{\mathrm{d}p}{\mathrm{d}t}\mathbf{u}$$

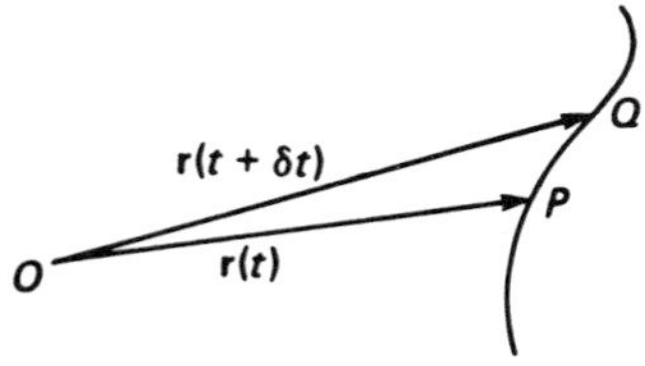

Fig. 14.24 Two neighbouring position vectors.

$$\frac{\mathrm{d}}{\mathrm{d}t}(\mathbf{u}\cdot\mathbf{v}) = \mathbf{u}\cdot\frac{\mathrm{d}\mathbf{v}}{\mathrm{d}t} + \frac{\mathrm{d}\mathbf{u}}{\mathrm{d}t}\cdot\mathbf{v}$$

$$\frac{\mathrm{d}}{\mathrm{d}t}(\mathbf{u}\times\mathbf{v}) = \mathbf{u}\times\frac{\mathrm{d}\mathbf{v}}{\mathrm{d}t} + \frac{\mathrm{d}\mathbf{u}}{\mathrm{d}t}\times\mathbf{v}$$

It is important to realize, in the last expansion, that the order in which we write down the vectors is crucial; in each term the vector **u** must precede the vector **v**.

The rules of differentiation enable us to differentiate in terms of components. Suppose

$$\mathbf{r}(t) = x(t)\mathbf{i} + y(t)\mathbf{j} + z(t)\mathbf{k}$$

where $x(t)$, $y(t)$ and $z(t)$ are scalars dependent on t. We have

$$\dot{\mathbf{r}}(t) = \dot{x}(t)\mathbf{i} + \dot{y}(t)\mathbf{j} + \dot{z}(t)\mathbf{k}$$
$$\ddot{\mathbf{r}}(t) = \ddot{x}(t)\mathbf{i} + \ddot{y}(t)\mathbf{j} + \ddot{z}(t)\mathbf{k}$$

□ A point P moves in space with its position vector given by

$$\mathbf{r}(t) = (\sin t)\mathbf{i} + (\cos t)\mathbf{j} + t\mathbf{k}$$

If the time t is given in seconds and the magnitude of $\mathbf{r}(t)$ is given in metres, obtain the velocity and speed of the point P.

We have, differentiating with respect to t,

$$\dot{\mathbf{r}}(t) = (\cos t)\mathbf{i} - (\sin t)\mathbf{j} + \mathbf{k}$$

and this is the velocity. Velocity is a vector quantity; it has a magnitude and a direction. The speed is the magnitude of the velocity, and so

$$\text{speed} = |\dot{\mathbf{r}}(t)| = \surd(\cos^2 t + \sin^2 t + 1) = \surd 2 \text{ m/s}$$

Don't make the mistake of obtaining $|\mathbf{r}(t)|$ and differentiating this with respect to time. If you do you will not obtain the speed. Here this procedure would give

$$|\mathbf{r}(t)| = \surd(\sin^2 t + \cos^2 t + t^2)$$
$$= \surd(1 + t^2)$$

which differentiating with respect to time gives

$$\frac{t}{\surd(1 + t^2)} \text{ m/s}$$

which is certainly not the speed of the point. ■

We shall now take a few more steps just to make sure we have the ideas of triple products and differentiation clear.

14.14 Workshop

▷ **Exercise** Obtain the equation satisfied by a point (x, y, z) which is in the same plane as the points $(1, -1, 2)$, $(3, 1, -1)$ and $(1, 1, -2)$. 1

Think carefully about this. Remember that the triple scalar product can be interpreted as the volume of the parallelepiped formed from the three vectors.

Suppose we let the three fixed points be A, B and C. Then the vectors **OA**, **OB** and **OC** are the position vectors of the points. If P is in the same plane as A, B and C then the vectors **AP**, **AB** and **AC** are coplanar. It therefore follows that their triple scalar product is zero. 2

If you didn't manage to argue this through then don't be concerned. Store the idea away in your mind for future use and see if you can complete the problem.

We have 3

$$\begin{aligned}
\mathbf{AP} = \mathbf{AO} + \mathbf{OP} = -\mathbf{OA} + \mathbf{OP} &= -(\mathbf{i} - \mathbf{j} + 2\mathbf{k}) + (x\mathbf{i} + y\mathbf{j} + z\mathbf{k}) \\
&= (x - 1)\mathbf{i} + (y + 1)\mathbf{j} + (z - 2)\mathbf{k} \\
\mathbf{AB} = \mathbf{AO} + \mathbf{OB} = -\mathbf{OA} + \mathbf{OB} &= -(\mathbf{i} - \mathbf{j} + 2\mathbf{k}) + (3\mathbf{i} + \mathbf{j} - \mathbf{k}) \\
&= 2\mathbf{i} + 2\mathbf{j} - 3\mathbf{k} \\
\mathbf{AC} = \mathbf{AO} + \mathbf{OC} = -\mathbf{OA} + \mathbf{OC} &= -(\mathbf{i} - \mathbf{j} + 2\mathbf{k}) + (\mathbf{i} + \mathbf{j} - 2\mathbf{k}) \\
&= 2\mathbf{j} - 4\mathbf{k}
\end{aligned}$$

The triple scalar product is zero, and so

$$\begin{vmatrix} x - 1 & y + 1 & z - 2 \\ 2 & 2 & -3 \\ 0 & 2 & -4 \end{vmatrix} = 0$$

This gives

$$\begin{aligned}
(x - 1)(-8 + 6) - 2[-4(y + 1) - 2(z - 2)] &= 0 \\
-2(x - 1) + 8(y + 1) + 4(z - 2) &= 0 \\
x - 1 - 4y - 4 - 2z + 4 &= 0 \\
x - 4y - 2z &= 1
\end{aligned}$$

Did you get there? Don't forget to check that A, B and C each satisfy this equation.

If all was well you can move on to step 5. If there were difficulties then try the next problem carefully. It is always better to spend a few seconds *planning* what you intend to do.

▷ **Exercise** Obtain the volume of the tetrahedron formed by the points O,

A, B and C, where O is the origin and A, B, C are the points $(1, -1, 2)$, $(3, 1, -1)$ and $(1, 1, -2)$.

You need to know that a tetrahedron has

$$\text{volume} = \frac{1}{3}\text{ area of base} \times \text{height}$$

---

4 The area of the triangle formed by two vectors is half the area of the parallelogram which they form. Therefore the required volume will be one-sixth of the volume of the parallelepiped formed by the three vectors **OA**, **OB** and **OC**.

The triple scalar product gives the volume of the parallelepiped.

$$\begin{aligned}\mathbf{OA} &= \mathbf{i} - \mathbf{j} + 2\mathbf{k}\\ \mathbf{OB} &= 3\mathbf{i} + \mathbf{j} - \mathbf{k}\\ \mathbf{OC} &= \mathbf{i} + \mathbf{j} - 2\mathbf{k}\end{aligned}$$

So we obtain

$$\begin{vmatrix} 1 & -1 & 2 \\ 3 & 1 & -1 \\ 1 & 1 & -2 \end{vmatrix} = (-2 + 1) + (-6 + 1) + 2(3 - 1)$$

$$= -1 - 5 + 4 = -2$$

expanding in terms of the first row. The magnitude of this is therefore 2 and the required volume is $\frac{2}{6} = \frac{1}{3}$.

---

5 **Exercise** Show that if $\mathbf{u} \cdot \mathbf{v} \neq -1$ then $\mathbf{x} = \mathbf{u}$ is the only solution of the equation

$$\mathbf{x} + (\mathbf{u} \wedge \mathbf{x}) \wedge \mathbf{v} = \mathbf{u}$$

Notice that this question uses the wedge notation for the vector product.

It is clear that $\mathbf{x} = \mathbf{u}$ is certainly a solution of the equation, but we must show it is the *only* solution.

---

6 Expanding the triple vector product we obtain

$$\mathbf{x} + (\mathbf{u} \cdot \mathbf{v})\mathbf{x} - (\mathbf{v} \cdot \mathbf{x})\mathbf{u} = \mathbf{u}$$

However, if we take the dot product with $\mathbf{v}$ of both sides of the equation

$$\mathbf{x} + (\mathbf{u} \wedge \mathbf{x}) \wedge \mathbf{v} = \mathbf{u}$$

we obtain a triple scalar product in which two of the vectors are the same:

$$(\mathbf{u} \wedge \mathbf{x}) \wedge \mathbf{v} \cdot \mathbf{v}$$

Consequently this product is zero and therefore

$$\mathbf{x} \cdot \mathbf{v} = \mathbf{u} \cdot \mathbf{v}$$

So

$$\mathbf{v} \cdot \mathbf{x} = \mathbf{u} \cdot \mathbf{v}$$

We now have

$$\mathbf{x} + (\mathbf{u} \cdot \mathbf{v})\mathbf{x} - (\mathbf{u} \cdot \mathbf{v})\mathbf{u} = \mathbf{u}$$
$$[1 + (\mathbf{u} \cdot \mathbf{v})]\mathbf{x} = [1 + (\mathbf{u} \cdot \mathbf{v})]\mathbf{u}$$

Since $1 + (\mathbf{u} \cdot \mathbf{v}) \neq 0$ we conclude that $\mathbf{x} = \mathbf{u}$.

If you managed that all right then move on to the practical. Otherwise, try a final exercise.

▷**Exercise** Obtain the general solution of the equation

$$\mathbf{u} \wedge \mathbf{x} + \mathbf{v} \wedge \mathbf{x} = \mathbf{u} \wedge \mathbf{v}$$

where **u** and **v** are non-parallel vectors.

Remember some of the tricks we have used before and see if you can sort this out. The argument is similar, in some ways, to the one we used to expand a triple vector product.

We have 7

$$[\mathbf{u} + \mathbf{v}] \wedge \mathbf{x} = \mathbf{u} \wedge \mathbf{v}$$

The right-hand side is a vector perpendicular to the plane of **u** and **v**, whereas the left-hand side is a vector perpendicular to the plane of $\mathbf{u} + \mathbf{v}$ and **x**. It follows that **x** is in the plane of **u** and **v**, so $\mathbf{x} = h\mathbf{u} + k\mathbf{v}$ where h and k are scalars. Substituting this expression for **x** into the vector equation gives

$$[\mathbf{u} + \mathbf{v}] \wedge [h\mathbf{u} + k\mathbf{v}] = \mathbf{u} \wedge \mathbf{v}$$
$$h(\mathbf{u} \wedge \mathbf{u}) + h(\mathbf{v} \wedge \mathbf{u}) + k(\mathbf{u} \wedge \mathbf{v}) + k(\mathbf{v} \wedge \mathbf{v}) = \mathbf{u} \wedge \mathbf{v}$$

Now $\mathbf{u} \wedge \mathbf{u} = 0 = \mathbf{v} \wedge \mathbf{v}$ and $\mathbf{v} \wedge \mathbf{u} = -\mathbf{u} \wedge \mathbf{v}$. Therefore

$$(-h + k)\mathbf{u} \wedge \mathbf{v} = \mathbf{u} \wedge \mathbf{v}$$

Now $\mathbf{u} \wedge \mathbf{v} \neq 0$, and so $-h + k = 1$. Therefore

$$\mathbf{x} = h\mathbf{u} + (h + 1)\mathbf{v} = h(\mathbf{u} + \mathbf{v}) + \mathbf{v}$$

where h is an arbitrary scalar.

14.15 Practical

PARTICLE DYNAMICS

A particle has position $P\ (t, t^2, t^3)$ at time t, relative to a rectangular cartesian coordinate system $Oxyz$. Obtain, for the particle at time t, the velocity, the speed and the magnitude of the acceleration. Distances are measured in metres, and time is measured in seconds.

All you have to do is put this in vector form; the rest is easy.

Let **OP** be the position vector of the particle. Then, in the usual notation, we have

$$\mathbf{OP} = \mathbf{r}(t) = t\mathbf{i} + t^2\mathbf{j} + t^3\mathbf{k}$$

The velocity is obtained by differentiating:

$$\dot{\mathbf{r}}(t) = \mathbf{i} + 2t\mathbf{j} + 3t^2\mathbf{k}$$

The magnitude of this gives the speed:

$$\text{speed} = \sqrt{(1 + 4t^2 + 9t^4)}\ \text{m/s}$$

For the acceleration we must differentiate again:

$$\ddot{\mathbf{r}}(t) = 2\mathbf{j} + 6t\mathbf{k}$$

and the magnitude of this is $\sqrt{(4 + 36t^2)}$ m/s^2.

Vectors are very useful indeed; they enable us to solve problems in three dimensions without having to try to visualize them. When they are used in conjunction with matrices and operators they become a very powerful mathematical instrument. For example Maxwell's equations can be expressed by this method, and these form the cornerstone of electromagnetics. We shall gain some insight into how each of these mathematical ingredients functions separately, but the powerful combination is outside the scope of our present studies.

SUMMARY

In this chapter we have covered the following topics:

☐ Vector algebra; addition and scalar multiplication of vectors

$$(\mathbf{a}, \mathbf{b}) \to \mathbf{a} + \mathbf{b}$$
$$(p, \mathbf{a}) \to p\mathbf{a}$$

☐ The scalar product and the vector product

$$\mathbf{a} \cdot \mathbf{b} = |\mathbf{a}|\,|\mathbf{b}| \cos \theta$$
$$\mathbf{a} \times \mathbf{b} = |\mathbf{a}|\,|\mathbf{b}| \sin \theta\, \mathbf{n}$$

where θ is the angle between the vectors and **n** is a unit vector.

☐ The use of position vectors.

☐ Triple scalar products

$$[\mathbf{a}, \mathbf{b}, \mathbf{c}] = \begin{vmatrix} a_1 & a_2 & a_3 \\ b_1 & b_2 & b_3 \\ c_1 & c_2 & c_3 \end{vmatrix}$$

☐ Triple vector products

$$\mathbf{a} \times (\mathbf{b} \times \mathbf{c}) = (\mathbf{a} \cdot \mathbf{c})\mathbf{b} - (\mathbf{a} \cdot \mathbf{b})\mathbf{c}$$
$$(\mathbf{a} \times \mathbf{b}) \times \mathbf{c} = (\mathbf{a} \cdot \mathbf{c})\mathbf{b} - (\mathbf{b} \cdot \mathbf{c})\mathbf{a}$$

☐ Differentiation of vectors.

EXERCISES

1 Determine $\mathbf{a} + \mathbf{b}$, $\mathbf{a} \cdot \mathbf{b}$, $\mathbf{a} \times \mathbf{b}$ for each of the following:

a $\mathbf{a} = 2\mathbf{i} + \mathbf{j} + 3\mathbf{k}$, $\quad \mathbf{b} = \mathbf{i} - 3\mathbf{j} + \mathbf{k}$

b $\mathbf{a} = \mathbf{i} + 4\mathbf{j} - 5\mathbf{k}$, $\quad \mathbf{b} = 2\mathbf{i} + 2\mathbf{j} + 3\mathbf{k}$

c $\mathbf{a} = -\mathbf{i} + 2\mathbf{j} - 3\mathbf{k}$, $\quad \mathbf{b} = \mathbf{i} + 3\mathbf{j} - 2\mathbf{k}$

d $\mathbf{a} = \mathbf{i} + \mathbf{j} + 2\mathbf{k}$, $\quad \mathbf{b} = -3\mathbf{i} + 4\mathbf{j} - \mathbf{k}$

2 Obtain two unit vectors perpendicular to each of the following:

a $\mathbf{a} = \mathbf{i} + 3\mathbf{j} - 5\mathbf{k}$, $\quad \mathbf{b} = \mathbf{i} + \mathbf{j}$

b $\mathbf{a} = \mathbf{i} + 2\mathbf{j} + 3\mathbf{k}$, $\quad \mathbf{b} = 3\mathbf{i} + 2\mathbf{j} - \mathbf{k}$

c $\mathbf{a} = -\mathbf{i} + \mathbf{j} - \mathbf{k}$, $\quad \mathbf{b} = \mathbf{i} - \mathbf{j} - \mathbf{k}$

d $\mathbf{a} = (\mathbf{i} + \mathbf{j}) \times (\mathbf{j} + \mathbf{k})$, $\quad \mathbf{b} = (\mathbf{i} - \mathbf{j}) \times (\mathbf{j} - \mathbf{k})$

3 Obtain the triple scalar product $[\mathbf{a}, \mathbf{b}, \mathbf{c}]$ for each of the following

a $\mathbf{a} = \mathbf{i} + \mathbf{j}$, $\quad \mathbf{b} = \mathbf{j} + \mathbf{k}$, $\quad \mathbf{c} = \mathbf{k} + \mathbf{i}$

b $\mathbf{a} = 2\mathbf{i} + \mathbf{j} + \mathbf{k}$, $\quad \mathbf{b} = \mathbf{i} + 2\mathbf{j} + \mathbf{k}$, $\quad \mathbf{c} = \mathbf{i} + \mathbf{j} + 2\mathbf{k}$

c $\mathbf{a} = 3\mathbf{i} - 2\mathbf{j} - \mathbf{k}$, $\quad \mathbf{b} = \mathbf{i} + 3\mathbf{j} - 2\mathbf{k}$, $\quad \mathbf{c} = 2\mathbf{i} + \mathbf{j} - 3\mathbf{k}$

d $\mathbf{a} = (\mathbf{i} + \mathbf{j}) \times \mathbf{k}$, $\quad \mathbf{b} = (\mathbf{j} + \mathbf{k}) \times \mathbf{i}$, $\quad \mathbf{c} = (\mathbf{k} + \mathbf{i}) \times \mathbf{j}$

4 Obtain the triple vector products $\mathbf{a} \times (\mathbf{b} \times \mathbf{c})$ and $(\mathbf{a} \times \mathbf{b}) \times \mathbf{c}$ for each of the sets of vectors given in exercise 3.

5 Obtain $\mathrm{d}\mathbf{r}/\mathrm{d}t$ for each of the vectors given below:
a $\mathbf{r} = \cos t\mathbf{i} - \sin t\mathbf{j} + t\mathbf{k}$
b $\mathbf{r} = \cos 3t\mathbf{i} + \sin 5t\mathbf{j} + t^2\mathbf{k}$
c $\mathbf{r} = (1 + t^2)^2\mathbf{i} + (1 + t^3)^3\mathbf{j} + (1 + t^4)^4\mathbf{k}$
d $\mathbf{r} = (t\mathbf{i} - t^2\mathbf{j} + t^3\mathbf{k}) \wedge (t^3\mathbf{i} + t^2\mathbf{j} + t\mathbf{k})$

ASSIGNMENT

1 For the vectors $\mathbf{a} = 2\mathbf{i} + \mathbf{j} - \mathbf{k}$ and $\mathbf{b} = \mathbf{i} + \mathbf{j} + 2\mathbf{k}$ obtain
(a) the scalar product $\mathbf{a} \cdot \mathbf{b}$
(b) the cosine of the angle between the vectors **a** and **b**
(c) the vector product $\mathbf{a} \times \mathbf{b}$
(d) the area of the parallelogram formed by the two vectors **a** and **b**.

2 Obtain a unit vector parallel to the Oyz plane and perpendicular to $\mathbf{i} + 4\mathbf{j} - 3\mathbf{k}$.

3 By putting $\mathbf{d} = \mathbf{b} \times \mathbf{c}$ initially or otherwise show that

$$(\mathbf{a} \times \mathbf{b}) \times (\mathbf{b} \times \mathbf{c}) = [\mathbf{a}, \mathbf{b}, \mathbf{c}]\mathbf{b}$$

4 Show that
(a) $[\mathbf{a} + \mathbf{b}, \mathbf{b} + \mathbf{c}, \mathbf{c} + \mathbf{a}] = 2[\mathbf{a}, \mathbf{b}, \mathbf{c}]$
(b) $[\mathbf{a} \times \mathbf{b}, \mathbf{b} \times \mathbf{c}, \mathbf{c} \times \mathbf{a}] = [\mathbf{a}, \mathbf{b}, \mathbf{c}]^2$

5 Solve the equation $\mathbf{a} \wedge \mathbf{x} + (\mathbf{b} \cdot \mathbf{x})\mathbf{a} = \mathbf{b}$, if $\mathbf{a} \cdot \mathbf{b} \neq 0$.

6 Obtain the constant t if the vectors $\mathbf{i} - \mathbf{j} + 2\mathbf{k}$, $5\mathbf{i} + t\mathbf{j} + 3\mathbf{k}$ and $-3\mathbf{i} + 2\mathbf{j} + \mathbf{k}$ are coplanar.

7 Show that if **u**, **v** and **w** are dependent on the parameter t and differentiable with respect to t then

$$\frac{\mathrm{d}}{\mathrm{d}t}[\mathbf{u}, \mathbf{v}, \mathbf{w}] = \left[\frac{\mathrm{d}\mathbf{u}}{\mathrm{d}t}, \mathbf{v}, \mathbf{w}\right] + \left[\mathbf{u}, \frac{\mathrm{d}\mathbf{v}}{\mathrm{d}t}, \mathbf{w}\right] + \left[\mathbf{u}, \mathbf{v}, \frac{\mathrm{d}\mathbf{w}}{\mathrm{d}t}\right]$$

FURTHER EXERCISES

1 ABC is a triangle; D is the midpoint of BC, and E is the midpoint of AC. Prove that $\mathbf{AB} = 2\,\mathbf{ED}$.

2 For the vector $\mathbf{a} = (\mathbf{i} - 2\mathbf{j} + 2\mathbf{k})/3$ and $\mathbf{b} = (-3\mathbf{i} - 5\mathbf{j} + 4\mathbf{k})/5$ determine
(a) the angle between **a** and **b**
(b) two unit vectors perpendicular to the plane of **a** and **b**.

3 By eliminating z and using differentiation, or otherwise, determine the vector

$$\mathbf{a} = x\mathbf{i} + y\mathbf{j} + z\mathbf{k}$$

which has all of the following properties:

(a) $\mathbf{a}$ is perpendicular to $\mathbf{i} + \mathbf{j} + \mathbf{k}$;
(b) $\mathbf{a}$ has twice the magnitude of $\mathbf{i} + \mathbf{j} + \mathbf{k}$;
(c) y is a minimum.

4 Suppose $\mathbf{a}$, $\mathbf{b}$, $\mathbf{c}$ and $\mathbf{d}$ are position vectors from the origin to the points A, B, C and D respectively.
(a) By expanding $(\mathbf{a} \wedge \mathbf{b}) \wedge (\mathbf{c} \wedge \mathbf{d})$ in two different ways, or otherwise, show that

$$[\mathbf{a}, \mathbf{b}, \mathbf{c}]\mathbf{d} + [\mathbf{a}, \mathbf{c}, \mathbf{d}]\mathbf{b} = [\mathbf{b}, \mathbf{c}, \mathbf{d}]\mathbf{a} + [\mathbf{a}, \mathbf{b}, \mathbf{d}]\mathbf{c}$$

(b) Show that

$$(\mathbf{b} \wedge \mathbf{c}) \cdot (\mathbf{a} \wedge \mathbf{d}) + (\mathbf{c} \wedge \mathbf{a}) \cdot (\mathbf{b} \wedge \mathbf{d}) + (\mathbf{a} \wedge \mathbf{b}) \cdot (\mathbf{c} \wedge \mathbf{d}) = 0$$

5 The path of a point is given vectorially by the equation

$$\mathbf{r} = (\cos^2 \theta)\mathbf{i} + (\cos \theta \sin \theta)\mathbf{j} + (\cos \theta)\mathbf{k}$$

Determine expressions for $\mathbf{r} \cdot \mathbf{r}$ and $\mathbf{r} \wedge \mathbf{r}$. Obtain the vector $\mathbf{r}$ which has (a) minimum magnitude (b) maximum magnitude.

6 The position vector of a moving point P is given by

$$\mathbf{r} = (\cos \omega t)\mathbf{i} + (\sin \omega t)\mathbf{j} + t\mathbf{k}$$

where t is time. Show that the direction of motion makes a constant angle α with the Oz axis, where $\omega = \tan \alpha$.

7 Show that $\mathbf{a} \cdot \mathbf{b} \leqslant |\mathbf{a}|\,|\mathbf{b}|$ for any two vectors $\mathbf{a}$ and $\mathbf{b}$. By using

$$|\mathbf{a} + \mathbf{b}|^2 = (\mathbf{a} + \mathbf{b}) \cdot (\mathbf{a} + \mathbf{b})$$

or otherwise, deduce the triangle inequality $|\mathbf{a} + \mathbf{b}| \leqslant |\mathbf{a}| + |\mathbf{b}|$.

8 Show that W, the work done by a constant force $\mathbf{F} = \alpha\mathbf{i} + \beta\mathbf{j} + \gamma\mathbf{k}$ in moving a mass from the point (a_1, b_1, c_1) to the point (a_2, b_2, c_2), is given by

$$W = \alpha(a_2 - a_1) + \beta(b_2 - b_1) + \gamma(c_2 - c_1)$$

9 An electron is constrained to move on a curve. Its position vector at time t is given by

$$\mathbf{r} = (2 \cos t)\mathbf{i} + (2 \sin t)\mathbf{j} + t\mathbf{k}$$

Show that its velocity and acceleration each have constant magnitude.

10 Suppose

$$\mathbf{u} = (\cos t)\mathbf{i} + (\sin t)\mathbf{j} + e^{-t}\mathbf{k}$$
$$\mathbf{v} = (-\sin t)\mathbf{i} + (\cos t)\mathbf{j} + e^{t}\mathbf{k}$$
$$\mathbf{r} = \mathbf{u} \times \mathbf{v}$$

Show that $\dot{\mathbf{r}}$ is always in the $\mathbf{i}$, $\mathbf{j}$ plane and that $\dot{\mathbf{r}}$ has magnitude $2\sqrt{(2)} \cosh t$.

11 The position vector of a point mass at time t is

$$\mathbf{r} = (\cos^2 t)\mathbf{i} + (\sin t \cos t)\mathbf{j} + (\sin t)\mathbf{k}$$

Determine the velocity $\dot{\mathbf{r}}$ and the acceleration $\ddot{\mathbf{r}}$. Show also that $|\mathbf{r}| = 1$ and that, at time t, $|\dot{\mathbf{r}}| = \sqrt{(1 + \cos^2 t)}$.

(a) The curvature $\varkappa$ is given by the formula

$$\varkappa = |\dot{\mathbf{r}} \times \ddot{\mathbf{r}}|/|\dot{\mathbf{r}}|^3$$

Show that the curvature at time t is $(3\cos^2 t + 5)^{1/2}/(1 + \cos^2 t)^{3/2}$.

(b) The torsion τ is given by the formula

$$1/\tau = (\dot{\mathbf{r}} \times \ddot{\mathbf{r}} \cdot \dddot{\mathbf{r}})/|\dot{\mathbf{r}} \times \ddot{\mathbf{r}}|^2$$

Show that the torsion at time t is $(3 + 5\sec^2 t)/6\sec t$.

12 An electron moves with constant angular velocity on a circle of radius 1. Show that when suitable axes are chosen its position vector can be expressed in the form

$$\mathbf{r} = (\cos \omega t)\mathbf{i} + (\sin \omega t)\mathbf{j}$$

Confirm, using vectors, that

(a) The velocity of the electron is perpendicular to the vector $\mathbf{r}$;

(b) The acceleration $\ddot{\mathbf{r}}$ is directed towards the centre of the circle.

13 A particle is constrained to move on the curve defined by $x = 2\cos t$, $y = 2\sin t$, $z = t\sqrt{5}$ relative to a rectangular cartesian coordinate system $Oxyz$. If distances are measured in metres and time t in seconds, show that the magnitude of the velocity is 3 m/s and the magnitude of the acceleration is 2 m/s^2.

14 The position vector of a helicopter at time t is given by

$$\mathbf{r} = t(\cos t\mathbf{i} + \sin t\mathbf{j} + \mathbf{k})$$

(a) obtain $\dot{\mathbf{r}}$ and show that

$$\ddot{\mathbf{r}} = -(2\sin t + t\cos t)\mathbf{i} + (2\cos t - t\sin t)\mathbf{j}$$

(b) Obtain the value of t if $\mathbf{r} \cdot \dot{\mathbf{r}} = 6$.

(c) Obtain $|\ddot{\mathbf{r}}|$ when $t = 3$.

(d) Obtain $\dot{\mathbf{r}} \times \ddot{\mathbf{r}}$ when $t = 0$.

15 Within a construction site for an aviary three points A, B and C are determined relative to a central reference point O and expressed using position vectors by

$$\begin{aligned} \mathbf{a} &= \mathbf{i} + \mathbf{j} + \mathbf{k} \\ \mathbf{b} &= \mathbf{i} + 2\mathbf{j} - 2\mathbf{k} \\ \mathbf{c} &= \mathbf{i} - \mathbf{j} + \mathbf{k} \end{aligned}$$

respectively.

(a) Obtain $\mathbf{r} = \mathbf{a} + \mathbf{b}$ and $\mathbf{s} = \mathbf{a} + 2\mathbf{c}$ and show that $\mathbf{r}$ and $\mathbf{s}$ are mutually perpendicular.

(b) Calculate $\mathbf{r} \times \mathbf{s}$ and obtain a unit vector which is at right angles to both $\mathbf{a}$ and also to $\mathbf{b}$.

(c) Determine the volume of the parallelepiped formed from the three vectors $\mathbf{a}$, $\mathbf{b}$ and $\mathbf{c}$ and verify that

$$\mathbf{r} \times \mathbf{s} = 2(\mathbf{r} \times \mathbf{c}) - \mathbf{a} \times \mathbf{b}$$

15 Integration 1

Now that our algebraic knowledge has been increased by studying complex numbers, matrices, determinants and vectors we return to the calculus to begin our work on integration.

After studying this chapter you should be able to

- ☐ Write down indefinite integrals of simple functions;
- ☐ Apply the four basic rules of integration correctly;
- ☐ Perform simple substitutions to determine integrals;
- ☐ Obtain integrals by putting simple rational functions into partial fractions.

At the end of the chapter we look at practical problems in gas compression and structural stress.

15.1 THE CONCEPT OF INTEGRATION

Integration is sometimes thought of as falling into two parts: indefinite integration and definite integration. We shall deal with integration in this way, and the link between the two will then become clear.

It is best to think of indefinite integration as the reverse procedure to that of differentiation. We know that given a differentiable real function F there is a unique real function f such that $F' = f$. We call f the derivative of F (Chapter 4).

Suppose we try to reverse this procedure. For any real function f:

1 Under what circumstances does there exist a real function F such that $F' = f$?

2 If there is a function F with this property, is it unique?

We shall not answer question 1 since it is beyond the scope of our work, but question 2 can be answered straight away: *no*! If F exists it is not the only function with this property.

□ We know that both f and g defined by

$$f(x) = x + 1 \quad \text{and} \quad g(x) = x \quad \text{when } x \in \mathbb{R}$$

have a derivative h defined by

$$h(x) = 1 \quad \text{when } x \in \mathbb{R}$$

In other symbols, if $y = x + 1$ and $z = x$ then

$$\frac{dy}{dx} = \frac{dz}{dx}$$

■

Suppose we have two real functions F and G which have the same derivative. What can we say about them? Suppose $F' = G'$: then $F' - G' = 0$ and therefore $(F - G)' = 0$. Consequently the function $F - G$ has the zero function as its derivative. Now it so happens that the only differentiable function, defined for all real x, which has zero derivative is a constant function.

In other symbols, if

$$\frac{dy}{dx} = \frac{dz}{dx}$$

then

$$\frac{dy}{dx} - \frac{dz}{dx} = 0$$

$$\frac{d}{dx}(y - z) = 0$$

Therefore

$$y - z = \text{constant}$$

Consequently to integrate a real function f we must

1 Determine any function F with derivative f;

2 Add an arbitrary constant.

This will then represent all those functions which have derivative f.

It is convenient to use a dummy variable such as x or t when describing functions, and the same is true when it comes to representing an indefinite integral.

Of course it is the presence of an arbitrary constant which gives rise to the name 'indefinite' integral. Suppose

$$\frac{\mathrm{d}}{\mathrm{d}x} F(x) = f(x)$$

Then we write

$$\int f(x)\,\mathrm{d}x = F(x) + C$$

where C is an arbitrary constant. We call this the **indefinite integral** of $f(x)$ with respect to x, and refer to $f(x)$ as the **integrand**. We shall also say that $f(x)$ is integrable, and we mean by this that the indefinite integral with respect to x exists.

Our work on differentiation will stand us in good stead, since we can write down a number of indefinite integrals. For example:

$$\int \mathrm{e}^x\,\mathrm{d}x = \mathrm{e}^x + C$$

$$\int \cos x\,\mathrm{d}x = \sin x + C$$

$$\int \sec^2 x\,\mathrm{d}x = \tan x + C$$

These are examples of **standard forms**. When we wish to perform an integration we shall attempt to reduce the integrand to a sum of standard forms. The techniques of integration may not at first seem quite as straightforward as the techniques of differentiation. However, practice will soon overcome this problem.

15.2 RULES FOR INTEGRATION

The rules for integration follow from the rules for differentiation. We shall suppose that $u = u(x)$ and $v = v(x)$ are real functions which are integrable and that c is a constant.

Sum rule

$$\int (u + v)\,\mathrm{d}x = \int u\,\mathrm{d}x + \int v\,\mathrm{d}x$$

This means that if we express the integrand as the sum of two parts we can then integrate each separately and add the results. You should not presume that this rule is self-evident, however reasonable it may appear; the corresponding property does not hold for products!

Factor rule

$$\int cu \, dx = c \int u \, dx$$

This means that we can divide the integrand by a constant and take this outside the integral sign. This rule only holds for constants, so that

$$\int 3 \tan x \, dx = 3 \int \tan x \, dx$$

but

$$\int x \tan x \, dx \neq x \int \tan x \, dx$$

Product rule

$$\int u \frac{dv}{dx} \, dx = uv - \int v \frac{du}{dx} \, dx$$

This is usually called the formula for **integration by parts** and it is useful for integrating certain products. It is an awkward rule to remember, and some students prefer to learn it in words. So here is the chant: 'The integral of a product of two functions is the first times the integral of the second, minus the integral of, the integral of the second times the derivative of the first.' You may well prefer to remember the formula!

Substitution rule

$$\int f(x) \, dx = \int f[g(u)]g'(u) \, du$$

where $x = g(u)$.

This is easy to apply because we make the substitution $x = g(u)$ and then

$$\frac{dx}{du} = g'(u)$$

which leads quite naturally to the substitution $dx = g'(u) \, du$.

There is one additional requirement. To every x there must correspond a value of u such that $x = g(u)$ and $g'(u)$ must remain bounded on any finite interval.

The substitution rule is perhaps the most useful rule for integration.

One little remark about arbitrary constants is in order. We do not allow arbitrary constants to proliferate each time we split an integral into two. This is because they can all be collected together in a sum which is itself an arbitrary constant. To avoid these constants appearing left, right and centre we shall suppose that the last integral to be determined is the guardian of the arbitrary constant. It's rather like a waiter watching customers at a table in a café. He doesn't mind if some of them leave, provided somebody is still sitting there to pay the bill. If the last customer gets up and walks out he says 'Oi! What about the bill?' or words to that effect. We must do the same; the last integral sign remaining owes a debt of the arbitrary constant.

□ Using the rules for differentiation and the definition of the indefinite integral deduce **a** the sum, **b** the factor, **c** the product and **d** the substitution rules.

We shall suppose that

$$\int u(x)\,\mathrm{d}x = U(x) + A$$

$$\int v(x)\,\mathrm{d}x = V(x) + B$$

where A and B are arbitrary constants. As usual we shall write $U = U(x)$ and $V = V(x)$.

a By definition, $U'(x) = u(x)$ and $V'(x) = v(x)$. Now

$$(U + V)' = U' + V' = u + v$$

Consequently

$$\begin{aligned}\int (u + v)\,\mathrm{d}x &= U + V + C \\ &= \int u\,\mathrm{d}x + \int v\,\mathrm{d}x\end{aligned}$$

b By definition, $(cU)' = cU' = cu$. Therefore

$$\int cu\,\mathrm{d}x = cU(x) + K$$

where K is an arbitrary constant. So

$$\int cu\,\mathrm{d}x = c\int u\,\mathrm{d}x$$

c The product rule for differentiation gives

$$\frac{\mathrm{d}}{\mathrm{d}x}(uv) = u\frac{\mathrm{d}v}{\mathrm{d}x} + v\frac{\mathrm{d}u}{\mathrm{d}x}$$

Therefore by the sum rule

$$\int \frac{\mathrm{d}}{\mathrm{d}x}(uv)\,\mathrm{d}x = \int u\frac{\mathrm{d}v}{\mathrm{d}x}\,\mathrm{d}x + \int v\frac{\mathrm{d}u}{\mathrm{d}x}\,\mathrm{d}x$$

So

$$uv = \int v\frac{\mathrm{d}u}{\mathrm{d}x} + \int u\frac{\mathrm{d}v}{\mathrm{d}x}$$

Rearranging,

$$\int u\frac{\mathrm{d}v}{\mathrm{d}x} = uv - \int v\frac{\mathrm{d}u}{\mathrm{d}x}$$

d We have

$$\frac{\mathrm{d}}{\mathrm{d}x}\left[\int f(x)\,\mathrm{d}x\right] = f(x) = f[g(u)]$$

So

$$\frac{\mathrm{d}}{\mathrm{d}x}\left[\int f(x)\,\mathrm{d}x\right]\frac{\mathrm{d}x}{\mathrm{d}u} = f[g(u)]\frac{\mathrm{d}x}{\mathrm{d}u}$$

Therefore by the chain rule for differentiation

$$\frac{\mathrm{d}}{\mathrm{d}u}\left[\int f(x)\,\mathrm{d}x\right] = f[g(u)]g'(u)$$

and so

$$\int f(x)\,\mathrm{d}x = \int f[g(u)]g'(u)\,\mathrm{d}u$$

■

Table 15.1 Standard derivatives

$f(x)$	$f'(x)$
x^n (n constant)	nx^{n-1}
e^x	e^x
$\ln x$ ($x > 0$)	x^{-1}
$\sin x$	$\cos x$
$\cos x$	$-\sin x$
$\tan x$	$\sec^2 x$
$\sec x$	$\sec x \tan x$
$\cot x$	$-\operatorname{cosec}^2 x$
$\operatorname{cosec} x$	$-\operatorname{cosec} x \cot x$
$\sinh x$	$\cosh x$
$\cosh x$	$\sinh x$
$\sin^{-1} x$	$1/\sqrt{(1 - x^2)}$
$\tan^{-1} x$	$1/(1 + x^2)$
$\sinh^{-1} x$	$1/\sqrt{(1 + x^2)}$
$\cosh^{-1} x$	$1/\sqrt{(x^2 - 1)}$

STANDARD FORMS

We shall now build up a table of standard forms which we shall use to integrate more complicated functions. To begin with we list some well-known derivatives in Table 15.1, as these will form the basis of the integrals table which we shall devise.

On the basis of these, and remembering that we can take any constant factor outside the integral sign, but not expressions which depend on x, it is possible to list the standard integrals in Table 15.2. We use the notation

$$\int f(x)\,\mathrm{d}x = F(x) + C$$

in the table.

Table 15.2 Standard integrals

$f(x)$	$F(x)$
x^n ($n \neq -1$, constant)	$x^{n+1}/(n+1)$
e^x	e^x
x^{-1} $(x > 0)$	$\ln x$
$\cos x$	$\sin x$
$\sin x$	$-\cos x$
$\sinh x$	$\cosh x$
$\cosh x$	$\sinh x$
$1/(1+x^2)$	$\tan^{-1} x$
$1/\sqrt{(1-x^2)}$	$\sin^{-1} x$
$1/\sqrt{(1+x^2)}$	$\sinh^{-1} x$
$1/\sqrt{(x^2-1)}$	$\cosh^{-1} x$
$\sec^2 x$	$\tan x$
$\sec x \tan x$	$\sec x$
$\operatorname{cosec}^2 x$	$-\cot x$
$\operatorname{cosec} x \cot x$	$-\operatorname{cosec} x$

□ We shall extend Table 15.2, using the rules of integration, to obtain

$$\int x^{-1}\,\mathrm{d}x \quad \text{when } x < 0$$

We use the substitution rule and put $x = -t$. Then $t > 0$ since $x < 0$. Also $\mathrm{d}x/\mathrm{d}t = -1$, and so we obtain

$$\begin{aligned}\int x^{-1}\,\mathrm{d}x &= \int \frac{1}{x}\frac{\mathrm{d}x}{\mathrm{d}t}\,\mathrm{d}t \\ &= \int \frac{1}{(-t)}(-1)\,\mathrm{d}t \\ &= \int \frac{1}{t}\,\mathrm{d}t\end{aligned}$$

But since $t > 0$ we know this integral is $\ln t + C$. So substituting back we have

$$\int x^{-1}\,dx = \ln(-x) + C \quad \text{when } x < 0$$

Of course we can combine both cases, $x > 0$ and $x < 0$, in one formula:

$$\int x^{-1}\,dx = \ln|x| + C \quad \text{when } x \neq 0$$

It is important to remember this, as occasionally it is possible to produce incorrect working by overlooking the possibility that $x < 0$. On the other hand, if x is clearly positive the modulus signs are entirely superfluous and should be omitted. ■

A word or two about the arbitrary constant is not out of place here. We know that given any real number y there exists a positive number x such that $y = \ln x$. This number x is equal to e^y, in fact. Therefore any arbitrary constant C can be written in the form $\ln k$ where $k > 0$.

One advantage of doing this is that we can then use the laws of logarithms to put the integral in a tidier form. One disadvantage is that it may take some algebraic work on your part to confirm that the answer you have obtained to an integration is equivalent to the one which it is stated you should obtain! For instance,

$$\ln(x^2 - 1) - \ln(x + 1) + C$$

is equivalent to

$$\ln k(x - 1)$$

where C and k (>0) are arbitrary constants.

Now let's do some integrations. In these we rearrange the integrand and then use standard forms together with the sum and factor rule.

15.3 Workshop

1 **Exercise** Obtain

$$I = \int e^x \cosh x\,dx$$

See if you can rearrange the integrand so that it becomes a sum of standard forms. When you have done so, take the next step.

2

If we use the definition of cosh x we see that we can split the integral into two. We have

$$\begin{aligned} I &= \int e^x \frac{e^x + e^{-x}}{2} \, dx \\ &= \frac{1}{2} \int (e^{2x} + 1) dx \\ &= \frac{1}{2} \int e^{2x} \, dx + \frac{1}{2} \int dx \end{aligned}$$

It is customary to use dx algebraically in this way and not to insist that it appears on the far right as in the standard notation

$$I = \frac{1}{2} \int e^{2x} \, dx + \frac{1}{2} \int 1 \, dx$$

We obtain

$$\begin{aligned} I &= \frac{1}{2} \frac{e^{2x}}{2} + \frac{1}{2} x + C \\ &= \frac{e^{2x}}{4} + \frac{x}{2} + C \end{aligned}$$

Now try this one. You have to think carefully.

▷ **Exercise** Obtain

$$\int \frac{dx}{1 + \sin x}$$

When you are ready, take the next step.

3

If you have split the integral into two by writing

$$\int dx + \int \frac{dx}{\sin x}$$

then you have made an algebraic blunder because

$$\frac{1}{A + B} \neq \frac{1}{A} + \frac{1}{B}$$

You can readily confirm this by putting $A = 1$ and $B = 1$. Instead, we multiply the numerator and denominator by $1 - \sin x$. On the one hand the integrand is unaltered, but on the other hand we can use the rules of elementary trigonometry to simplify it. So

$$\int \frac{dx}{1 + \sin x} = \int \frac{(1 - \sin x)}{(1 - \sin x)} \frac{dx}{(1 + \sin x)}$$

$$= \int \frac{1 - \sin x}{1 - \sin^2 x} dx$$

$$= \int \frac{1 - \sin x}{\cos^2 x} dx = \int (\sec^2 x - \sec x \tan x) dx$$

$$= \int \sec^2 x \, dx - \int \sec x \tan x \, dx$$

using the sum and factor rules.

Now each of these is a standard form which we included in Table 15.2. See then if you can write the answer down straight away. If not, you had better look back at the table and then write down the answer. Whichever way you proceed, as soon as you have finished move on to the final step.

4 This is what you should have written down:

$$\int \frac{dx}{1 + \sin x} = \int \sec^2 x \, dx - \int \sec x \tan x \, dx$$

$$= \tan x - \sec x + C$$

You may not have managed to tackle these steps, but do not worry at this stage. If we look at what we have done there are one or two features which we can remember for future use. First, we need to know our trigonometrical formulas forwards, backwards and inside out! Also, simple algebraic identities will often assist us.

We have said that we are content if we can resolve the integrand into a sum of standard forms, but in general two guidelines can be of help. They are

1 Remove denominators;
2 Resolve roots.

In the workshop we had a denominator which we wished to remove. How did we do it? We had to remember almost simultaneously that

$$1 - \sin^2 x = \cos^2 x$$
$$(1 - \sin x)(1 + \sin x) = 1 - \sin^2 x$$

This realization showed us that we needed to multiply numerator and denominator by $1 - \sin x$ to proceed effectively.

So we must train ourselves to inspect the integrand and to plan ahead. It's all a matter of practice.

15.4 INTEGRATION BY INSPECTION

Sometimes when we inspect an integrand it is possible to see straight away how to integrate it. One very common situation, where this is the case, is when the integral is of the form

$$\int \frac{f'(x)}{f(x)}\, dx$$

If we use the substitution rule and put $u = f(x)$ then we obtain, differentiating with respect to x, $u' = f'(x)$, and so

$$\begin{aligned}\int \frac{f'(x)}{f(x)}\, dx &= \int \frac{1}{u}\, du = \ln |u| + C \\ &= \ln |f(x)| + C\end{aligned}$$

□ Render the following integrals:

$$\int \frac{2x}{x^2 + 4}\, dx \qquad \int \frac{4x + 3}{x^2 + 1}\, dx$$

The first integral has the desired form: the numerator is the derivative of the denominator. So, without making a formal substitution, we can write down straight away

$$\begin{aligned}\int \frac{2x}{x^2 + 4}\, dx &= \ln (x^2 + 4) + C \\ &= \ln k(x^2 + 4)\end{aligned}$$

where k is constant. Note that we do not need to include modulus signs here since the argument is clearly *positive*.

The second integral needs to be manipulated slightly but is essentially the same. See if you can do it. When you have finished, read on and see if you are right.

You're not looking, are you? You really have tried your best? Well then, here it is:

$$\int \frac{4x + 3}{x^2 + 1}\, dx = 2 \int \frac{2x}{x^2 + 1}\, dx + 3 \int \frac{dx}{x^2 + 1}$$

Here we have used the sum rule and the factor rule to split things up. Notice particularly that only constants can appear to the left of each integral sign. The first integral is the logarithmic type which we have been discussing, and the second integral is a standard form. So we can write straight away

$$I = 2 \ln (x^2 + 1) + 3 \tan^{-1} x + C \qquad ■$$

The logarithmic type is a special case of a more general type, and we should be on the lookout for this too. This is an integral of the form

$$\int g(u)\, \mathrm{d}u$$

where $u = f(x)$ and g is a function which has a known integral.

□ Obtain each of the following integrals:

$$\int \frac{x^2}{\sqrt{(x^3 + 1)}}\, \mathrm{d}x \qquad \int x \sqrt{(1 + x^2)}\, \mathrm{d}x$$

Why not see if you can do something with these by yourself?

For the first integral, if we put $u = x^3 + 1$ then $\mathrm{d}u/\mathrm{d}x = 3x^2$ and we can use the factor rule to adjust the numerator to include the 3. Moreover, we know

$$\int \frac{\mathrm{d}u}{\sqrt{u}} = \int u^{-1/2}\, \mathrm{d}u = 2u^{1/2} + C$$

In words: raise the index by 1 and divide by the number so obtained. Therefore we have

$$\int \frac{x^2}{\sqrt{(x^3 + 1)}}\, \mathrm{d}x = \frac{1}{3} \int \frac{3x^3}{\sqrt{(x^3 + 1)}}\, \mathrm{d}x$$
$$= \frac{2}{3} \sqrt{(x^3 + 1)} + C$$

It is a common error to get this wrong by half recognizing the general form and ignoring the root sign by writing down a logarithm. Don't be one of those who makes that particular mistake! Now you have a go at the second integral.

Here we are then. It's a simple example of the general type, although an adjustment of the constant is necessary:

$$\int x \sqrt{(1 + x^2)}\, \mathrm{d}x = \frac{1}{2} \int 2x \sqrt{(1 + x^2)}\, \mathrm{d}x$$
$$= \frac{1}{2}\frac{2}{3} (1 + x^2)^{3/2} + C = \frac{1}{3} (1 + x^2)^{3/2} + C \qquad ■$$

15.5 INTEGRATION BY PARTIAL FRACTIONS

One type of integrand can be dealt with routinely. This is when the integrand is a quotient of two polynomials, for example

$$\frac{x^4 + 2x - 3}{x^3 + 2x^2 + x}$$

It is then possible to split the integrand into a sum of a number of partial fractions.

We shall use the example to illustrate the method. We first divide the denominator into the numerator so that, for the quotient which results, the degree of the numerator is less than the degree of the denominator. The rest of the integrand is then a polynomial, which we can integrate without difficulty.

We have

$$\frac{x^4 + 2x - 3}{x^3 + 2x^2 + x} = x - 2 + \frac{3x^2 + 4x - 3}{x^3 + 2x^2 + x}$$

The rational expression

$$Q = \frac{3x^2 + 4x - 3}{x^3 + 2x^2 + x}$$

remains to be resolved. We factorize the denominator of Q:

$$x^3 + 2x^2 + x = x(x^2 + 2x + 1) = x(x + 1)^2$$

Here the factors of the denominator are x and $x + 1$ (repeated). Therefore we obtain partial fractions with denominators x, $x + 1$ and $(x + 1)^2$ and *constant* numerators. So

$$Q = \frac{3x^2 + 4x - 3}{x^3 + 2x^2 + x} = \frac{A}{x} + \frac{B}{x + 1} + \frac{C}{(x + 1)^2}$$

If the partial fractions are recombined the two numerators must be identically equal. Now

$$\frac{A}{x} + \frac{B}{x + 1} + \frac{C}{(x + 1)^2} = \frac{A(x + 1)^2 + Bx(x + 1) + Cx}{x(x + 1)^2}$$

Therefore we require

$$3x^2 + 4x - 3 \equiv A(x + 1)^2 + Bx(x + 1) + Cx$$

The constants A, B and C can be obtained either by substituting values of x into the identity or by comparing the coefficients of powers of x on each side of it. In practice a mixture of the methods is usually the quickest.

Here if we put $x = 0$ we obtain $A = -3$. If we put $x = -1$ we obtain $3 - 4 - 3 = -C$, so $C = 4$. If we examine the coefficient of x^2 on each side of the identity we obtain $3 = A + B$ and so $B = 6$. Therefore

$$\frac{x^4 + 2x - 3}{x^3 + 2x^2 + x} = x - 2 - \frac{3}{x} + \frac{6}{x + 1} + \frac{4}{(x + 1)^2}$$

Each of these can be integrated without trouble. Try it and see how you get on.

We integrate term by term to obtain

$$\frac{x^2}{2} - 2x - 3 \ln |x| + 6 \ln |x + 1| - \frac{4}{x + 1} + C$$

Did you manage that? All we have done is to apply the ideas which we developed earlier and to integrate by sight. If you are having a few difficulties, then to begin with you can make the substitutions algebraically. However, as soon as possible you should train yourself to carry out these substitutions mentally. Otherwise you may lose time in examinations going through tedious algebraic routines that can be avoided.

STANDARD FORMS

Our table of standard integrals (Table 15.2) had two important omissions which we can now make good. They are the integrals of the tangent function and the secant function.

For the *tangent* function we have

$$\int \tan x \, dx = \int \frac{\sin x}{\cos x} \, dx$$

Now if we differentiate $\cos x$ with respect to x we obtain $-\sin x$. Therefore by the factor and substitution rules we have

$$\int \tan x \, dx = -\int \frac{-\sin x}{\cos x} \, dx = -\ln |\cos x| + C$$

$$= \ln |\sec x| + C$$

For the *secant* function we rearrange the integrand in a cunning way:

$$\int \sec x \, dx = \int \frac{\sec x \, (\sec x + \tan x)}{\sec x + \tan x} \, dx$$

The reason is that if we differentiate $\sec x + \tan x$ with respect to x we obtain

$$\sec x \tan x + \sec^2 x = \sec x \, (\sec x + \tan x)$$

So in fact the numerator is now the derivative of the denominator. (Of course you would not be expected to pluck that technique out of thin air on your own, but just savour for a moment the elegance of the move and remember it for future use.) Consequently

$$\int \sec x \,\mathrm{d}x = \ln |\sec x + \tan x| + C$$

Now you try to integrate the cotangent function and the cosecant function. When you have done them, check the answers below. The methods are very similar.

For the *cotangent* function we have

$$\int \cot x \,\mathrm{d}x = \ln |\sin x| + C$$

For the *cosecant* function we have

$$\int \operatorname{cosec} x \,\mathrm{d}x = \ln |\operatorname{cosec} x - \cot x| + C$$

Did you manage to sort out the sign in the second one?

There are some other obvious omissions in our table of integrals, for instance $\int \ln x \,\mathrm{d}x$. This can be rectified provided we use the product rule known as the formula for integration by parts. We now consider this rule in some detail.

15.6 INTEGRATION BY PARTS

If we have an integrand which is the product of two *different* types of function, for example

1 exponential × circular
2 polynomial × exponential
3 polynomial × circular

the formula for integration by parts can often resolve it. Here is the formula again:

$$\int u \frac{\mathrm{d}v}{\mathrm{d}x} \,\mathrm{d}x = uv - \int v \frac{\mathrm{d}u}{\mathrm{d}x} \,\mathrm{d}x$$

or equivalently

$$\int u \,\mathrm{d}v = uv - \int v \,\mathrm{d}u$$

Given a product, we shall need to decide which part to call u and which part $\mathrm{d}v$. There are four broad principles to adopt:

1 We must be able to write down v easily.
2 If there is a polynomial then the polynomial is usually u.
3 If having made a choice a more complicated integral results, then start again with the opposite choice.
4 If having decided on the correct choice for u and dv a further integration by parts is necessary, maintain the same type of functions for u and dv.

☐ Obtain

$$\int x e^{5x} \, dx$$

Here using principle 2 we have $u = x$ and $dv = e^{5x}\,dx$, so that $v = e^{5x}/5$. We do not need to include an arbitrary constant; if we include it, it will only cancel out later. Also $du = 1\,dx = dx$, so therefore

$$\begin{aligned}\int x e^{5x} \, dx &= \frac{x}{5} e^{5x} - \frac{1}{5} \int e^{5x} \, dx \\ &= \frac{x}{5} e^{5x} - \frac{1}{25} e^{5x} + C\end{aligned}$$ ■

Sometimes it doesn't matter which way we choose u and dv.

☐ Obtain the following integral:

$$I = \int e^x \cos x \, dx$$

Here whichever choice we make we are bound to succeed. For instance $u = \cos x$, $dv = e^x\,dx$ gives $du = -\sin x\,dx$ and $v = e^x$. So

$$I = \int e^x \cos x \, dx = e^x \cos x - \int e^x(-\sin x)\,dx$$

Now the integral on the right is no *worse* than the one we started with, so we continue integrating by parts. Maintaining our choice of types of function for u and v gives $u = \sin x$ and $dv = e^x$. If we were to choose them the other way round we should get back where we started!

So we have $du = \cos x$ and $v = e^x$, and so

$$\begin{aligned}I &= e^x \cos x + \int e^x \sin x \, dx \\ &= e^x \cos x + \left[e^x \sin x - \int e^x \cos x \, dx \right] \\ &= e^x \cos x + e^x \sin x - I\end{aligned}$$

So

$$2I = e^x \cos x + e^x \sin x$$

'Oi! What about the constant?' Oh yes, we forgot the waiter in the café, didn't we? We have removed the last integral sign, and we overlooked the fact that it is the guardian of some arbitrary constants. Therefore

$$2I = e^x \cos x + e^x \sin x + \text{constant}$$

That is,

$$I = \frac{1}{2}(e^x \cos x + e^x \sin x) + C$$

Now you try the same problem but with the opposite choice for u and dv. That is, you choose $u = e^x$ and $dv = \cos x$. It will still work out; keep calm, and keep a clear head! ■

Sometimes an integrand does not look like a product at all, but nevertheless can be determined using integration by parts. One such integral is $\int \ln x \, dx$. Suppose we put $u = \ln x$ and $dv = dx$; then we obtain $v = x$ and $du = x^{-1} \, dx$. So the integral of the *logarithm* function is

$$\begin{aligned} \int \ln x \, dx &= x \ln x - \int x \frac{1}{x} \, dx \\ &= x \ln x - x + C \end{aligned}$$

Integration by parts is always worth considering.

15.7 Workshop

It's time now to look at some integrals. We shall do this bearing in mind that all of them can be solved by the methods we have used in this chapter. It is important that you get used to looking at the integral and deciding *before* you start which approach you are going to adopt. Sometimes an integral looks quite fierce but on closer inspection we see that in reality it is very easy. The opposite is also true, unfortunately.

The integrals we are going to inspect also form the problems for this chapter. If you feel confident you can tackle them on your own and omit these steps altogether. However, you may prefer to take things a little more slowly and in that way build up confidence. These steps give ideas for the methods, but not the full solutions.

▷ **Exercise** Inspect these integrals: 1

1 $\displaystyle\int \frac{x+1}{x} \, dx$ **2** $\displaystyle\int \frac{x}{x+1} \, dx$ **3** $\displaystyle\int \tan^2 x \, dx$

When you have decided which approach to adopt, take the next step. Don't be afraid to try a few moves on paper.

2

1 We can split the integral into two, each a standard form.
2 It's best to put $u = x + 1$, or equivalently to rewrite the numerator as $(x + 1) - 1$.
3 We cannot integrate $\tan^2 x$ as it stands, but we can integrate $\sec^2 x$. Remember: $1 + \tan^2 x = \sec^2 x$.

If you have all those right, then move on to step 4. Otherwise, consider the next three integrals.

▷**Exercise** Inspect these integrals:

4 $\displaystyle\int \frac{x+1}{x-1}\,dx$ **5** $\displaystyle\int \frac{x+1}{x^2-1}\,dx$ **6** $\displaystyle\int x\,e^{-x^2}\,dx$

When you have decided what to do, take another step.

3

4 This is like problem 2. We can substitute $u = x - 1$.
5 There is a common factor which can be cancelled.
6 This is a case for a substitution: $u = x^2$ will get rid of the floating x.

Now move ahead to step 4.

4

Exercise Inspect these integrals:

7 $\displaystyle\int \frac{2x}{x^2+x}\,dx$ **8** $\displaystyle\int \frac{e^x}{e^x+1}\,dx$ **9** $\displaystyle\int \sin^2 x \cos x\,dx$

Move on when you are ready.

5

7 Cancel out the x.
8 Put $u = e^x + 1$.
9 Note that when we differentiate $\sin x$ with respect to x we obtain $\cos x$. Therefore substitute $u = \sin x$.

Now take a look at the next three.

▷**Exercise** Inspect these integrals:

10 $\displaystyle\int \frac{\sin x}{\cos^2 x + 1}\,dx$ **11** $\displaystyle\int \frac{e^x}{e^{2x}+1}\,dx$ **12** $\displaystyle\int \sin^{-1} x\,dx$

When you have made your decisions, take the next step.

6

Things are not quite so straightforward now.
10 We can reduce this to a standard form by putting $u = \cos x$.
11 A standard form is obtained by putting $u = e^x$.
12 A substitution isn't really going to help much here. This is one of those

cases where integration by parts is the best method; we can differentiate $\sin^{-1} x$.

If you managed all those then move ahead to step 8. Otherwise, try these before you step ahead.

▷**Exercise** Inspect these integrals:

13 $\int x \sqrt{(x^2 + 4)}\, dx$ **14** $\int \frac{\sin 2x}{\sqrt{(\cos^2 x + 9)}}\, dx$ **15** $\int x \ln x\, dx$

Only move on when you have made a clear decision in each case.

---

13 At first sight this might seem difficult, but note that if $u = x^2 + 4$ then $du = 2x\, dx$ and so the floating x can be removed. **7**

14 If we differentiate $\cos^2 x$ with respect to x we obtain $-2\cos x \sin x = -\sin 2x$. This is conveniently present in the numerator to make the substitution work smoothly.

15 This is a product of two different types of function and so is a clear candidate for integration by parts.

Now take the next step.

▷**Exercise** Inspect these integrals: **8**

16 $\int x \tan^{-1} x\, dx$ **17** $\int \frac{x + 2}{x^2 - 5x + 6}\, dx$ **18** $\int \tan^3 x\, dx$

As soon as you have decided for all three, take another step.

---

16 This is a clear candidate for integration by parts. **9**

17 The denominator factorizes and so we can use partial fractions.

18 This is certainly different from problem 3, although we get the hint from that one. We take $\tan^2 x$ and rewrite it as $\sec^2 x - 1$, so that $\tan^3 x = \tan x \sec^2 x - \tan x$. We can handle each of these.

Try the final three.

▷**Exercise** Inspect these integrals:

19 $\int \frac{dx}{x^4 - 1}$ **20** $\int \frac{x^3\, dx}{x^4 - 1}$ **21** $\int \frac{x^3\, dx}{\sqrt{(x^4 + 1)}}$

Then take the final step.

---

19 We can factorize $x^4 - 1$ and this will enable us to use partial fractions.

20 Substitute $u = x^4 - 1$ rather than use partial fractions.

21 The square root presents the difficulty. However, if you remember the guidelines there should be no serious problem. Put $u = x^4 + 1$; then $du = 4x^3\ dx$ and the substitution can be carried through successfully.

15.8 Practical

GAS COMPRESSION

The work done in compressing a gas from one volume to another can be expressed by

$$W = \int p\ dv$$

If the volume v_1 is compressed to the volume v_2 and if $pv^n = c$, where c is a constant and $n \neq 1$, show that

$$W = \frac{c}{1 - n}(v_2^{1-n} - v_1^{1-n})$$

Obtain W in the case $n = 1$.

See how you get on with this. We will offer the solution stage by stage.

---

1 There is one small point to note. W has been used in two different ways. First, it is the general symbol for the work done. Secondly, it is the amount of work done in compressing the gas from v_1 to v_2. To avoid confusion we shall refer to the second as W^*.

We have, if $n \neq 1$,

$$W = \int \frac{c}{v^n}\ dv = c \int v^{-n}\ dv$$
$$= \frac{cv^{1-n}}{1 - n} + A$$

where A is a constant.

If you did not get this, try now to calculate A.

---

2 When $v = v_1$ no work has been done, and so we have

$$0 = W_1 = \frac{cv_1^{1-n}}{1 - n} + A$$

So

$$A = -\frac{cv_1^{1-n}}{1 - n}$$

Therefore

$$W = \frac{cv^{1-n}}{1-n} - \frac{cv_1^{1-n}}{1-n}$$

when $v = v_2$ we have $W = W^*$, which is

$$W^* = \frac{cv_2^{1-n}}{1-n} - \frac{cv_1^{1-n}}{1-n} = \frac{c}{1-n}(v_2^{1-n} - v_1^{1-n})$$

Now you deal with the case $n = 1$.

---

3 If $n = 1$ then

$$W = \int \frac{c}{v}\,\mathrm{d}v = c \ln v + A$$

As before, when $v = v_1$, $W = 0$. Therefore

$$0 = c \ln v_1 + A$$

and so, replacing A,

$$W = c \ln v - c \ln v_1$$

So when $v = v_2$,

$$W^* = c \ln v_2 - c \ln v_1 = c \ln (v_2/v_1)$$

STRESS

In a thick cylinder, subject to internal pressure, the radial stress P at a distance r from the axis of the cylinder is given by

$$\int \frac{\mathrm{d}P}{a - P} = 2 \int \frac{\mathrm{d}r}{r}$$

where a is a constant. If the stress has magnitude P_0 at the inner wall ($r = r_0$) and if it may be neglected at the outer wall ($r = r_1$), show that

$$P = \frac{P_0 r_0^2}{r_1^2 - r_0^2}\left(\frac{r_1^2}{r^2} - 1\right)$$

Again we present the solution stage by stage.

---

1 Rearranging the integral we have

$$-\int \frac{(-1)}{a - P}\,\mathrm{d}P = 2 \int \frac{\mathrm{d}r}{r}$$

So

$$-\ln|a - P| = 2\ln|r| + A$$

That is,

$$-\ln|a - P| = 2\ln|r| + \ln k = \ln kr^2$$

$$\frac{1}{|a - P|} = kr^2$$

where $k > 0$ is the arbitrary constant.

There are two cases to consider: $a > P$ and $a < P$.

---

2 If $a > P$ then

$$\frac{1}{a - P} = kr^2$$

$$a - P = \frac{1}{kr^2}$$

When $P = P_0$, $r = r_0$ and so $a - P_0 = 1/kr_0^2$. When $P = 0$, $r = r_1$ and therefore $a = 1/kr_1^2$. Now $r_1 > r_0$ and $P_0 > 0$, so that

$$P_0 = a - \frac{1}{kr_0^2} = \frac{1}{kr_1^2} - \frac{1}{kr_0^2} < 0$$

which is a contradiction. Therefore the case $a > P$ cannot arise and there remains only $a < P$.

---

3 When $a < P$ we have

$$\frac{1}{a - P} = -kr^2$$

$$a - P = -\frac{1}{kr^2}$$

When $r = r_0$, $P = P_0$ and so $a - P_0 = -1/kr_0^2$. When $r = r_1$, $P = 0$ and so $a = -1/kr_1^2$. Consequently

$$P_0 = \frac{1}{kr_0^2} - \frac{1}{kr_1^2} = \frac{1}{k}\left(\frac{1}{r_0^2} - \frac{1}{r_1^2}\right)$$

from which

$$\frac{1}{k} = \frac{P_0 r_0^2 r_1^2}{r_1^2 - r_0^2}$$

and

$$a = -\frac{1}{kr_1^2} = \frac{-P_0 r_0^2}{r_1^2 - r_0^2}$$

Further,

$$\begin{aligned} P = a + \frac{1}{kr^2} &= -\frac{1}{kr_1^2} + \frac{1}{kr^2} \\ &= \frac{1}{kr_1^2}\left(-1 + \frac{r_1^2}{r^2}\right) \\ &= \frac{P_0 r_0^2}{r_1^2 - r_0^2}\left(\frac{r_1^2}{r^2} - 1\right) \end{aligned}$$

SUMMARY

To obtain an integral we first inspect the integrand. The following checklist can then help us to proceed:

- ☐ Is this an integral which can be done by sight?
- ☐ Is there a simple substitution which will reduce it to a standard form? (Don't forget to substitute for dx too!)
- ☐ Is it possible to rearrange the integrand so that it becomes a sum of standard forms?
- ☐ Can the integrand be rearranged using the theory of partial fractions?
- ☐ Can the integral be obtained by integration by parts?

Remember also that if you make a substitution you must always carry it through. Never allow two variables to appear together under the integral sign.

EXERCISES

1 Obtain each of the following integrals:

a $\displaystyle\int (1 + 2x + 3x^3)\,dx$

b $\displaystyle\int \{\exp x + 2\exp(-x)\}\,dx$

c $\displaystyle\int \left\{1 + \frac{1}{\sqrt{(1 + x^2)}}\right\} dx$

d $\displaystyle\int \sec x\,(\sec x + \tan x)\,dx$

2 Use a simple substitution to obtain each of the following integrals:

a $\int (1 + 2x)^7 \, \mathrm{d}x$

b $\int \cos 3x \, \mathrm{d}x$

c $\int x^2 \sin x^3 \, \mathrm{d}x$

d $\int \frac{1}{4 + x^2} \, \mathrm{d}x$

3 Resolve into partial fractions to obtain each of the following:

a $\int \frac{\mathrm{d}x}{(2 + x)(1 - 3x)}$

b $\int \frac{\mathrm{d}x}{x^3 + 3x}$

c $\int \frac{\mathrm{d}x}{x^3 - 4x}$

d $\int \frac{\mathrm{d}x}{(x^2 + 1)(x^2 + 4)}$

4 Use the method of integration by parts to obtain each of the following:

a $\int 2x \cos 3x \, \mathrm{d}x$

b $\int x \exp 3x \, \mathrm{d}x$

c $\int x^3 \exp x^2 \, \mathrm{d}x$

d $\int x^2 \ln (x^2 + 1) \, \mathrm{d}x$

5 Obtain the following integrals:

a $\int \left(\frac{x}{2} + \frac{2}{x} \right) \mathrm{d}x$

b $\int \frac{\mathrm{d}x}{1 + \cos x}$

c $\int x^2 \exp x^3 \, \mathrm{d}x$

d $\int x \sec^2 x^2 \, \mathrm{d}x$

ASSIGNMENT

Obtain each of the following integrals. If you are stuck at any stage, hints are given in section 15.7.

1 $\displaystyle\int \frac{x+1}{x}\,dx$ **2** $\displaystyle\int \frac{x}{x+1}\,dx$ **3** $\displaystyle\int \tan^2 x\,dx$

4 $\displaystyle\int \frac{x+1}{x-1}\,dx$ **5** $\displaystyle\int \frac{x+1}{x^2-1}\,dx$ **6** $\displaystyle\int x\,e^{-x^2}\,dx$

7 $\displaystyle\int \frac{2x}{x^2+x}\,dx$ **8** $\displaystyle\int \frac{e^x}{e^x+1}\,dx$ **9** $\displaystyle\int \sin^2 x \cos x\,dx$

10 $\displaystyle\int \frac{\sin x}{\cos^2 x+1}\,dx$ **11** $\displaystyle\int \frac{e^x}{e^{2x}+1}\,dx$ **12** $\displaystyle\int \sin^{-1} x\,dx$

13 $\displaystyle\int x\sqrt{(x^2+4)}\,dx$ **14** $\displaystyle\int \frac{\sin 2x}{\sqrt{(\cos^2 x+9)}}\,dx$ **15** $\displaystyle\int x \ln x\,dx$

16 $\displaystyle\int x\tan^{-1} x\,dx$ **17** $\displaystyle\int \frac{x+2}{x^2-5x+6}\,dx$ **18** $\displaystyle\int \tan^3 x\,dx$

19 $\displaystyle\int \frac{dx}{x^4-1}$ **20** $\displaystyle\int \frac{x^3\,dx}{x^4-1}$ **21** $\displaystyle\int \frac{x^3\,dx}{\sqrt{(x^4+1)}}$

FURTHER EXERCISES

1 Obtain each of the following integrals:

a $\displaystyle\int x\cos x^2\,dx$ **b** $\displaystyle\int \frac{1}{\sqrt{(x+3)}}\,dx$

c $\displaystyle\int \frac{e^t-1}{e^t+1}\,dt$ **d** $\displaystyle\int \frac{x^3}{1+x^4}\,dx$

2 Resolve each integral:

a $\displaystyle\int \frac{du}{u\ln u}$ **b** $\displaystyle\int (\sin^3\theta+\cos^3\theta)\,d\theta$

c $\displaystyle\int \sec^{-1} t\,dt$ **d** $\displaystyle\int \frac{\ln x}{x}\,dx$

3 Render each of the following:

a $\displaystyle\int \tan^{-1} u\,du$ **b** $\displaystyle\int \frac{dx}{1-\cos x}$

c $\displaystyle\int \frac{dx}{(x-1)^2(x+1)}$ **d** $\displaystyle\int \frac{\cos 2\theta\,d\theta}{(1-\sin\theta)(1-\cos\theta)}$

4 When a constant EMF E is applied to a coil with inductance L and resistance R the current i is given by the equation

$$\int \frac{L\,\mathrm{d}i}{E - Ri} = \int \mathrm{d}t$$

Determine the current i at time t if $i = 0$ when $t = 0$.

5 By Newton's law of cooling the surface temperature θ at time t of a sphere in isothermal surroundings of temperature θ_0 is given by the equation

$$\int \frac{\mathrm{d}\theta}{\theta - \theta_0} = -kt$$

When $t = 0$, $\theta = \theta_1$. Show that at time t

$$\theta = (\theta_1 - \theta_0)\mathrm{e}^{-kt} + \theta_0$$

6 When a uniform beam of length L is clamped horizontally at each end and carries a load of w per unit length, the deflection y at distance x from one end satisfies

$$\frac{\mathrm{d}^4 y}{\mathrm{d}x^4} = \frac{w}{EI}$$

where EI is constant and is the flexural rigidity of the beam. Use the information that $y = 0 = \mathrm{d}y/\mathrm{d}x$ at both $x = 0$ and $x = L$ to obtain an expression for the deflection y at a general point.

7 A spherical drop of liquid evaporates at a rate proportional to its surface area. Show that if r is the radius then $\mathrm{d}r/\mathrm{d}t$ is constant. Given that the volume halves in 30 minutes, determine how long it will take for the drop to evaporate completely.

8 If $c_p \equiv \cos p\theta$ and $s_p \equiv \sin p\theta$, verify that if m and n are real constants

$$\frac{\mathrm{d}}{\mathrm{d}\theta}\frac{c_n + s_n}{c_m + s_m} = \frac{(n - m)c_{n+m} - (n + m)s_{n-m}}{(c_m + s_m)^2}$$

Hence, or otherwise, obtain

$$\int \frac{\cos 3\theta - 3\sin\theta}{1 + \sin 2\theta}\,\mathrm{d}\theta$$

9 The acceleration of a missile is given by $f = \alpha\,\mathrm{e}^{-t}\cosh u$ where α is constant, t is time in seconds and u is its speed in kilometres per second. Show that this leads to the integral equation

$$\int \operatorname{sech} u\,\mathrm{d}u = \alpha \int \mathrm{e}^{-t}\,\mathrm{d}t$$

Obtain the speed at time t if initially it is zero, and show that as $t \to \infty$, $u \to \ln[\tan(\pi/4 + \alpha/2)]$.

Integration 2 — 16

In this chapter we shall extend our table of standard forms still further and learn how to deal with more difficult integrands.

After studying this chapter you should be able to

- ☐ Use trigonometrical substitutions to resolve algebraic integrands;
- ☐ Apply the standard algebraic substitutions to resolve trigonometrical integrands;
- ☐ Manipulate integrands and use tables of standard forms to resolve them;
- ☐ Assess an integrand to plan a suitable method of integration.

At the end of the chapter we consider practical problems in particle dynamics and ballistics.

16.1 INTEGRATION BY SUBSTITUTION

Although we have extended our table of standard forms slightly, there are still a few gaps. Three of these are the integrals

$$\int \sqrt{(1 + x^2)}\,\mathrm{d}x \qquad \int \sqrt{(1 - x^2)}\,\mathrm{d}x \qquad \int \sqrt{(x^2 - 1)}\,\mathrm{d}x$$

We shall now consider how to determine these. To do so we deal further with integration by substitution.

Sometimes it is not possible to reduce an integrand to standard forms either by inspection or by the use of partial fractions. In such circumstances a substitution is worth considering. Remember the broad guidelines:

1 Remove denominators;
2 Resolve roots.

A good facility with trigonometry and elementary algebra is essential for this. Before going any further it is vital that you are aware of the essential difference in kind between the two integrals $\int \sqrt{(1 + x)}\,dx$ and $\int \sqrt{(1 + x^2)}\,dx$. The first integral can be obtained by inspection. See if you can write the answer down.

If we differentiate $(1 + x)^{3/2}$ with respect to x we obtain

$$\frac{3}{2}(1 + x)^{1/2} = \frac{3}{2}\sqrt{(1 + x)}$$

so that

$$\int \sqrt{(1 + x)}\,dx = \frac{2}{3}(1 + x)^{3/2} + C$$

It is unfortunately a common error for students to write

$$\int \sqrt{(1 + x^2)}\,dx \text{ as } \frac{1}{3x}(1 + x^2)^{3/2} + C$$

using broadly the same style of argument. This is incorrect because $2x$ is not constant, so the factor rule cannot be used in this way. Indeed, were we to differentiate this result correctly we should not obtain the integrand.

So then we have a problem. How are we to determine $\int \sqrt{(1 + x^2)}\,dx$? The answer is that we must make a substitution.

It is no good putting $u^2 = 1 + x^2$ because, although at first this may appear to resolve the root, when we substitute for dx in terms of du another one appears. We must make a *trigonometrical* substitution, and for this purpose it helps to recall one of the following two identities:

$$1 + \tan^2 t = \sec^2 t$$
$$1 + \sinh^2 t = \cosh^2 t$$

For example, using the first we put $x = \tan t$. Then $dx = \sec^2 t\,dt$ and so

$$\sqrt{(1 + x^2)} = \sqrt{(1 + \tan^2 t)} = \sqrt{\sec^2 t} = \sec t$$

provided we choose t in the interval $(-\pi/2, \pi/2)$. This is certainly possible since for all real x there exists t in this interval such that $x = \tan t$. Now

$$\int \sqrt{(1 + x^2)}\,dx = \int \sec t \sec^2 t\,dt = \int \sec^3 t\,dt$$

How are we to determine $\int \sec^3 t\,dt$? If you think the answer is simply $(1/4)\sec^4 t + C$ you had better think again! For, if we differentiate $(1/4)\sec^4 t$ with respect to t we obtain $\sec^3 t \sec t \tan t$, which is not the integrand.

At first sight, then, this integral may seem difficult. However, suppose we rewrite $\sec^3 t$ as a product $\sec^2 t \sec t$. We can then integrate by parts. If $dv = \sec^2 t\, dt$ then $v = \tan t$, whereas if $u = \sec t$ then $du = \sec t \tan t\, dt$. Consequently

$$\int \sec t \sec^2 t\, dt = \sec t \tan t - \int \tan t \sec t \tan t\, dt$$
$$= \sec t \tan t - \int \sec t \tan^2 t\, dt$$

Now $\sec^2 t = 1 + \tan^2 t$, and so this becomes

$$\int \sec^3 t\, dt = \sec t \tan t - \int \sec t\,(\sec^2 t - 1)\, dt$$
$$= \sec t \tan t - \int (\sec^3 t - \sec t)\, dt$$

Consequently

$$2\int \sec^3 t\, dt = \sec t \tan t + \int \sec t\, dt$$
$$= \sec t \tan t + \ln|\sec t + \tan t| + \text{constant}$$

and so

$$\int \sec^3 t\, dt = \frac{1}{2}\sec t \tan t + \frac{1}{2}\ln|\sec t + \tan t| + C$$

It is worth remarking that for $t \in (-\pi/2, \pi/2)$, $\sec t + \tan t$ which equals $(1 + \sin t)/\cos t$ is positive, and so the modulus signs are superfluous.

It remains only to substitute back in terms of x. We had $x = \tan t$, so $\sec t = \sqrt{(1 + x^2)}$. Therefore finally

$$\int \sqrt{(1 + x^2)}\, dx = \frac{1}{2}x\sqrt{(1 + x^2)} + \frac{1}{2}\ln[x + \sqrt{(1 + x^2)}] + C$$

We will now use some steps to look at the other two integrals in the trio introduced at the start of the chapter.

16.2 Workshop

▷**Exercise** Determine the integrals 1

$$\int \sqrt{(1 - x^2)}\, dx \qquad \int \sqrt{(x^2 - 1)}\, dx$$

For $\int \sqrt{(1 - x^2)}\, dx$, decide what you should do and then take the next step.

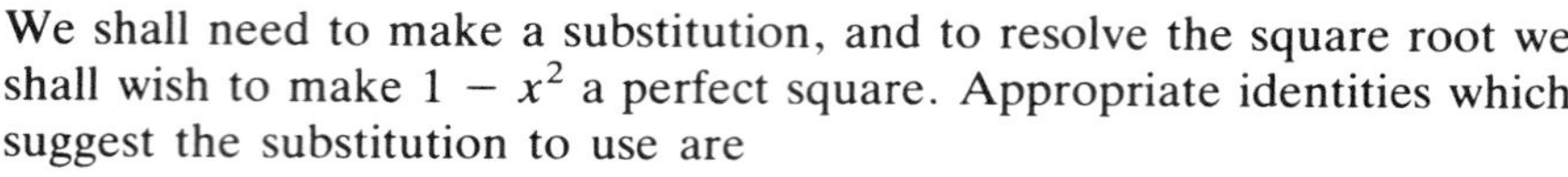

2 We shall need to make a substitution, and to resolve the square root we shall wish to make $1 - x^2$ a perfect square. Appropriate identities which suggest the substitution to use are

$$1 - \sin^2 t = \cos^2 t$$
$$1 - \tanh^2 t = \operatorname{sech}^2 t$$

Accordingly we should choose $x = \sin t$ (or $x = \cos t$) or $x = \tanh t$ (or $x = \operatorname{sech} t$). Which one we choose depends on mood and temperament! Let us be definite and select $x = \sin t$ so that we all do the same thing.

One small observation should be made before proceeding. If $x = \sin t$ then $|x| \leqslant 1$, but this is no problem since the integrand is not defined when $x^2 > 1$.

If you managed to sort out the integrand along these lines, then make the substitution and move ahead to step 4.

If you couldn't do it then look carefully at the identities and try your skill with the other integral. In other words, decide which substitution you would choose to obtain $\int \sqrt{(x^2 - 1)}\, dx$. As soon as you are ready take the next step.

---

3 The identities which are of use to us for $\int \sqrt{(x^2 - 1)}\, dx$ are

$$\sec^2 t - 1 = \tan^2 t$$
$$\cosh^2 t - 1 = \sinh^2 t$$

So you should substitute either $x = \sec t$ or $x = \cosh t$. The substitution $x = \cosh t$ presumes that $x \geqslant 1$. For the integral to exist we require $x^2 \geqslant 1$ so that either $x \geqslant 1$ or $x \leqslant -1$. The case $x \leqslant -1$ needs to be considered separately. The substitution $u = -x$ shows that when $x \leqslant -1$

$$\int \sqrt{(x^2 - 1)}\, dx = -\int \sqrt{(u^2 - 1)}\, du$$

where $u \geqslant 1$.

We shall return to this point once we have determined the integral in the case $x \geqslant 1$.

Of course it is always possible that you will think of a totally different approach which is nevertheless correct. The test will be whether or not you come up with the correct answer eventually.

Now return to $\int \sqrt{(1 - x^2)}\, dx$, substitute $x = \sin t$, simplify and take another step.

---

4 When $x = \sin t$ we have $dx = \cos t\, dt$ and $\sqrt{(1 - x^2)} = \cos t$. Once more we note that if we take $t \in (-\pi/2, \pi/2)$ then every $x \in (-1, 1)$ is attained. Also then $\cos t$ is positive, and so no modulus sign is needed when the value of the square root is calculated. We obtain

$$\int \sqrt{(1 - x^2)}\,\mathrm{d}x = \int \cos^2 t\,\mathrm{d}t$$

If you managed that then think how you are going to resolve this integral. When you have decided, try it, then move to step 6 and see if all is well.

If this did not work out correctly, look carefully to see where you made a mistake. Then try putting $x = \cosh t$ in the integral $\int \sqrt{(x^2 - 1)}\,\mathrm{d}x$. When you are satisfied with your answer, take a further step.

5

If we put $x = \cosh t$ then $\mathrm{d}x = \sinh t\,\mathrm{d}t$ and $\sqrt{(x^2 - 1)} = \sinh t$. So

$$\int \sqrt{(x^2 - 1)}\,\mathrm{d}x = \int \sinh^2 t\,\mathrm{d}t$$

All's well now!

Now return to the determination of $\int \sqrt{(1 - x^2)}\,\mathrm{d}x$, and consider what to do about obtaining $\int \cos^2 t\,\mathrm{d}t$. As soon as you have decided, move to step 6 to see if you are correct. It helps to remember a simple trigonometrical identity!

6

We have

$$\begin{aligned}\int \cos^2 t\,\mathrm{d}t &= \frac{1}{2}\int (1 + \cos 2t)\,\mathrm{d}t \\ &= \frac{t}{2} + \frac{\sin 2t}{4} + C \\ &= \tfrac{1}{2}t + \tfrac{1}{2}\sin t \cos t + C\end{aligned}$$

Did you get that? If so, substitute back in terms of x and then move on to step 8.

If something went wrong, try again this time with $\int \sinh^2 t\,\mathrm{d}t$ in the determination of $\int \sqrt{(x^2 - 1)}\,\mathrm{d}x$. When you have sorted things out, move to step 7.

7

We have

$$\begin{aligned}\int \sinh^2 t\,\mathrm{d}t &= \frac{1}{2}\int (\cosh 2t - 1)\,\mathrm{d}t \\ &= \frac{\sinh 2t}{4} - \frac{t}{2} + C \\ &= \tfrac{1}{2}\sinh t \cosh t - \tfrac{1}{2}t + C\end{aligned}$$

It's just a few trigonometrical identities and simple use of the substitution rule.

Now let's go back to the solution of $\int \sqrt{(1 - x^2)}\, dx$. We had $x = \sin t$, and we found that

$$\int \sqrt{(1 - x^2)}\, dx = \int \cos^2 t\, dt$$
$$= \tfrac{1}{2} t + \tfrac{1}{2} \sin t \cos t + C$$

Substitute back in terms of x, and when you are ready take a step.

---

8 We have $x = \sin t$ where $t \in (-\pi/2, \pi/2)$, and so $t = \sin^{-1} x$ and

$$\cos t = \sqrt{(1 - \sin^2 t)} = \sqrt{(1 - x^2)}$$

Consequently

$$\int \sqrt{(1 - x^2)}\, dx = \tfrac{1}{2} \sin^{-1} x + \tfrac{1}{2} x \sqrt{(1 - x^2)} + C$$

If you managed that, then all well and good. You may move on to the next section.

If things went wrong, then try the twin brother of the one we have just looked at. Given that, when $x = \cosh t$,

$$\int \sqrt{(x^2 - 1)}\, dx = \int \sinh^2 t\, dt$$
$$= \tfrac{1}{2} \sinh t \cosh t - \tfrac{1}{2} t + C$$

substitute back to express the integral in terms of x. When you have done that, take the last step.

---

9 If $x = \cosh t$ then

$$t = \cosh^{-1} x = \ln [x + \sqrt{(x^2 - 1)}]$$

Also $\sinh t = \sqrt{(x^2 - 1)}$, so that

$$\int \sqrt{(x^2 - 1)}\, dx = \frac{1}{2}x \sqrt{(x^2 - 1)} - \frac{1}{2} \cosh^{-1} x + C$$

Alternatively

$$\int \sqrt{(x^2 - 1)}\, dx = \frac{1}{2}x \sqrt{(x^2 - 1)} - \frac{1}{2} \ln [x + \sqrt{(x^2 - 1)}] + C$$

---

10 Remember that we have considered the case $x \geq 1$ only. You should use the substitution rule to confirm that when $x \leq -1$

$$\int \sqrt{(x^2 - 1)} dx = \frac{1}{2}x \sqrt{(x^2 - 1)} - \frac{1}{2} \ln [-x - \sqrt{(x^2 - 1)}] + C$$

So that in general, when $x^2 \geqslant 1$

$$\int \sqrt{(x^2 - 1)}\,dx = \frac{1}{2}x\sqrt{(x^2 - 1)} - \frac{1}{2}\ln|x + \sqrt{(x^2 - 1)}| + C$$

Let's list the three standard forms discussed so far in this chapter:

$$\int \sqrt{(1 + x^2)}\,dx = \frac{1}{2}x\sqrt{(1 + x^2)} + \frac{1}{2}\ln[x + \sqrt{(1 + x^2)}] + C$$

$$\int \sqrt{(1 - x^2)}\,dx = \frac{1}{2}x\sqrt{(1 - x^2)} + \frac{1}{2}\sin^{-1} x + C$$

$$\int \sqrt{(x^2 - 1)}\,dx = \frac{1}{2}x\sqrt{(x^2 - 1)} - \frac{1}{2}\ln|x + \sqrt{(x^2 - 1)}| + C$$

Notice the common pattern.

16.3 SPECIAL SUBSTITUTIONS

There are a number of integrals which are best dealt with by the use of special substitutions.

THE t SUBSTITUTION

Consider the two integrals

$$\int \frac{dx}{4 + 3\cos x} \qquad \int \frac{\sin x\,dx}{10 + \cos x}$$

These can be resolved by making the substitution $t = \tan(x/2)$, known universally as the ***t* substitution**. We shall need to remember the special formulas which are a consequence of this substitution. Here they are:

$$\sin x = \frac{2t}{1 + t^2} \qquad \cos x = \frac{1 - t^2}{1 + t^2}$$

$$\tan x = \frac{2t}{1 - t^2} \qquad \frac{dx}{dt} = \frac{2}{1 + t^2}$$

We must also remember of course that $t = \tan(x/2)$, and to express the result of the integration in terms of x. Surprisingly, some students remember the substitutions but forget what t is!

The derivation of these special formulas involves the use of the half-angle formulas of elementary trigonometry. See if you can derive the one for $\cos x$ yourself. Afterwards we shall see how to use these formulas to obtain integrals.

Here we go then!

$$\begin{aligned}\cos x &= 2\cos^2 (x/2) - 1\\ &= \frac{2}{\sec^2 (x/2)} - 1\\ &= \frac{2}{1 + \tan^2 (x/2)} - 1\\ &= \frac{2}{1+t^2} - 1\\ &= \frac{2 - (1 + t^2)}{1 + t^2}\\ &= \frac{1 - t^2}{1 + t^2}\end{aligned}$$

If you didn't manage that, then look at the working carefully. Observe the chain of thought: we express $\cos x$ in terms of $\cos^2 (x/2)$, which itself can be expressed in terms of $\sec^2 (x/2)$, which itself can be expressed in terms of $\tan^2 (x/2)$.

Now you try $\sin x$ and $\tan x$.

Now

$$\begin{aligned}\sin x &= 2\sin (x/2)\cos (x/2)\\ &= 2\tan (x/2)\cos^2 (x/2)\\ &= \frac{2\tan (x/2)}{\sec^2 (x/2)}\\ &= \frac{2\tan (x/2)}{1 + \tan^2 (x/2)}\\ &= \frac{2t}{1 + t^2}\end{aligned}$$

Also

$$\begin{aligned}\tan x &= \frac{2\tan (x/2)}{1 - \tan^2 (x/2)}\\ &= \frac{2t}{1 - t^2}\end{aligned}$$

Alternatively we can obtain $\tan x$ by dividing $\sin x$ by $\cos x$; at any rate we have a useful check that all is well.

Lastly obtain the substitution which enables us to put $\mathrm{d}x$ in terms of $\mathrm{d}t$. Don't move on until you have made an attempt!

Here is the correct working. $t = \tan(x/2)$, and so differentiating with respect to x using the chain rule:

$$\frac{dt}{dx} = \frac{1}{2}\sec^2(x/2)$$

$$= \frac{1}{2}[1 + \tan^2(x/2)]$$

$$= \frac{1 + t^2}{2}$$

$$\frac{dx}{dt} = \frac{2}{1 + t^2}$$

That wasn't too bad, was it? Let's hope you didn't forget the 1/2 when you applied the chain rule.

We shall now concern ourselves with the mechanics of the t substitution. When do we use it, and how does it work?

If the integrand

1 contains circular functions only

2 is free of powers or roots

3 is difficult to resolve by other means

then the t substitution *may* be of use.

The t substitution converts a trigonometrical integrand into an algebraic integrand consisting of a rational function. Unfortunately the process can be tedious because it is often necessary to use partial fractions to complete the integration. However, the good thing about integrals of this kind is that they are not inherently difficult. One thing, of course: we must remember to substitute back to eliminate t finally.

Well now, let's have a go at one. Let us look at the first of the two integrals introduced at the beginning of this section, namely

$$\int \frac{dx}{4 + 3\cos x}$$

You make the substitutions and see what you get.

We have

$$4 + 3\cos x = 4 + \frac{3(1 - t^2)}{1 + t^2}$$

$$= \frac{4(1 + t^2) + 3(1 - t^2)}{1 + t^2}$$

$$= \frac{7 + t^2}{1 + t^2}$$

So we have

$$\int \frac{dx}{4 + 3\cos x} = \int \frac{(1 + t^2)}{(7 + t^2)} \frac{2\,dt}{(1 + t^2)}$$

$$= \int \frac{2\,dt}{7 + t^2} = \frac{2}{7}\int \frac{dt}{1 + (t/\sqrt{7})^2}$$

$$= \frac{2}{\sqrt{7}}\tan^{-1}(t/\sqrt{7}) + C$$

So that

$$\int \frac{dx}{4 + 3\cos x} = \frac{2}{\sqrt{7}}\tan^{-1}\left[\frac{\tan(x/2)}{\sqrt{7}}\right] + C$$

Not very pretty, is it? However, we were lucky! We didn't need to use partial fractions.

Now you try the second integral, namely

$$\int \frac{\sin x\,dx}{10 + \cos x}$$

and don't forget your work on partial fractions. Only when you have completed it should you move on and see if you are correct.

Here we go then:

$$10 + \cos x = 10 + \frac{1 - t^2}{1 + t^2}$$

$$= \frac{10(1 + t^2) + 1 - t^2}{1 + t^2}$$

$$= \frac{11 + 9t^2}{1 + t^2}$$

So

$$\int \frac{\sin x\,dx}{10 + \cos x} = \int \frac{(1 + t^2)}{(11 + 9t^2)} \frac{2t}{(1 + t^2)} \frac{2\,dt}{(1 + t^2)}$$

$$= \int \frac{4t\,dt}{(11 + 9t^2)(1 + t^2)}$$

Now

$$\frac{4t}{(11 + 9t^2)(1 + t^2)} = \frac{A + Bt}{11 + 9t^2} + \frac{C + Dt}{1 + t^2}$$

So

$$4t \equiv (A + Bt)(1 + t^2) + (C + Dt)(11 + 9t^2)$$

First, $t = 0$ gives $0 = A + 11C$. The coefficient of t^2 gives $0 = A + 9C$. Therefore $A = 0$ and $C = 0$. The coefficient of t gives $4 = B + 11D$. The coefficient of t^3 gives $0 = B + 9D$. Consequently $D = 2$ and $B = -18$. You should check that this works by recombining the partial fractions into a rational expression.

So we have

$$\frac{4t}{(11 + 9t^2)(1 + t^2)} = \frac{-18t}{11 + 9t^2} + \frac{2t}{1 + t^2}$$

You could actually have done this by the cover-up method by replacing t^2 by u and ignoring the numerator initially. So we now have

$$\begin{aligned}\int \frac{\sin x \, dx}{10 + \cos x} &= \int \frac{4t \, dt}{(11 + 9t^2)(1 + t^2)} \\ &= \int \frac{-18t \, dt}{11 + 9t^2} + \int \frac{2t \, dt}{1 + t^2} \\ &= -\ln (11 + 9t^2) + \ln (1 + t^2) + \ln k \\ &= \ln k \left[\frac{1 + t^2}{11 + 9t^2}\right] \\ &= \ln k \left[\frac{1 + \tan^2(x/2)}{11 + 9 \tan^2(x/2)}\right]\end{aligned}$$

This simplifies even more. Can you sort it out?

Here is the working:

$$\begin{aligned}\int \frac{\sin x \, dx}{10 + \cos x} &= \ln k \left[\frac{\sec^2(x/2)}{11 + 9 \tan^2(x/2)}\right] \\ &= \ln k \left[\frac{1}{11 \cos^2(x/2) + 9 \sin^2(x/2)}\right] \\ &= \ln k \left[\frac{1}{2 \cos^2(x/2) + 9}\right] \\ &= \ln k \left[\frac{1}{\cos x + 10}\right] \\ &= -\ln (\cos x + 10) + C\end{aligned}$$

Now this is rather strange, isn't it? All that work: surely we could have done this quicker!

Could we? Yes, of course we could. We should have integrated by sight! Apart from a constant factor, the numerator of the integrand is the derivative of the denominator. The lesson to be learnt is quite plain: always look before you leap!

If you managed to spot this on your own, award yourself a special bonus. You can feel proud of yourself. If you didn't spot it, don't be alarmed; you were led into the longer method deliberately in order to make a point.

THE *s* SUBSTITUTION

If the integrand contains squares of circular functions then another substitution, which we shall call the *s* substitution, can be of use. As an example we shall consider the integral

$$\int \frac{dx}{9\sin^2 x + 4\cos^2 x}$$

The substitution is $s = \tan x$. Do not confuse this with the t substitution; here we do not have the half-angle.

We obtain at once the corresponding special formulas. First,

$$\frac{ds}{dx} = \sec^2 x = 1 + \tan^2 x = 1 + s^2$$

so that

$$\frac{dx}{ds} = \frac{1}{1 + s^2}$$

Also

$$\cos^2 x = \frac{1}{\sec^2 x} = \frac{1}{1 + s^2}$$

$$\sin^2 x = 1 - \cos^2 x$$

$$= 1 - \frac{1}{\sec^2 x} = 1 - \frac{1}{1 + s^2} = \frac{s^2}{1 + s^2}$$

You do not need to remember these formulas because you can, if you wish, obtain them directly using a right-angled triangle (Fig. 16.1).

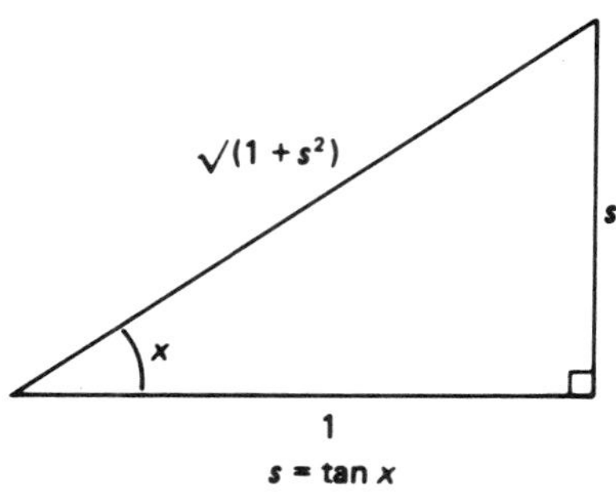

Fig. 16.1

Now you try the substitutions on the integral which we have taken as an example.

Here goes then. When $s = \tan x$,

$$9\sin^2 x + 4\cos^2 x = 9 - 5\cos^2 x = 9 - \frac{5}{1+s^2} = \frac{4+9s^2}{1+s^2}$$

So

$$\begin{aligned}\int \frac{dx}{9\sin^2 x + 4\cos^2 x} &= \int \frac{(1+s^2)}{(4+9s^2)}\frac{ds}{(1+s^2)} \\ &= \int \frac{ds}{(4+9s^2)} = \frac{1}{4}\int \frac{ds}{1+(3s/2)^2} \\ &= \frac{1}{4}\frac{2}{3}\tan^{-1}\left(\frac{3s}{2}\right) + C \\ &= \frac{1}{6}\tan^{-1}\left(\frac{3\tan x}{2}\right) + C\end{aligned}$$

We could have done this more quickly by dividing numerator and denominator by $\cos^2 x$ at the outset.

If you didn't get it right, then here is another integral to determine:

$$\int \frac{\tan x\, dx}{5\sin^2 x - 3\cos^2 x}$$

If you were all right, then read through the working just to check there are no surprises.

When $s = \tan x$ it follows that

$$5\sin^2 x - 3\cos^2 x = \frac{5s^2}{1+s^2} - \frac{3}{1+s^2} = \frac{5s^2-3}{1+s^2}$$

So

$$\begin{aligned}\int \frac{\tan x\, dx}{5\sin^2 x - 3\cos^2 x} &= \int \frac{(1+s^2)}{(5s^2-3)}\frac{s\, ds}{(1+s^2)} \\ &= \int \frac{s\, ds}{(5s^2-3)} = \frac{1}{10}\ln|5s^2-3| + C \\ &= \frac{1}{10}\ln|5\tan^2 x - 3| + C\end{aligned}$$

As in the previous example, we can reduce the integral a little more quickly if we divide numerator and denominator by $\cos^2 x$ first.

GUIDELINES FOR THE t AND s SUBSTITUTIONS

We have the following guidelines for these special substitutions. However, they should only be used if it is not possible to perform the integration by more direct means.

1 If the integrand contains circular functions without powers, consider putting $t = \tan(x/2)$.

2 If the integrand contains squares of circular functions, consider putting $s = \tan x$.

16.4 INTEGRATION USING STANDARD FORMS

Another technique of indefinite integration which we shall discuss briefly is the use of standard forms. We should always bear in mind that we already know the following standard forms:

1 $$\int \frac{dx}{1+x^2} = \tan^{-1} x + C$$

2 $$\int \frac{dx}{1-x^2} = \frac{1}{2}\ln\left|\frac{1+x}{1-x}\right| + C$$

3 $$\int \frac{dx}{\sqrt{(1-x^2)}} = \sin^{-1} x + C$$

4 $$\int \frac{dx}{\sqrt{(1+x^2)}} = \sinh^{-1} x + C = \ln[x + \sqrt{(1+x^2)}] + C$$

5 $$\int \frac{dx}{\sqrt{(x^2-1)}} = \cosh^{-1} |x| + C = \ln|x + \sqrt{(x^2-1)}| + C$$

6 $$\int \sqrt{(x^2-1)}\,dx = \{x\sqrt{(x^2-1)} - \ln|x + \sqrt{(x^2-1)}|\}/2 + C$$

7 $$\int \sqrt{(x^2+1)}\,dx = \{x\sqrt{(x^2+1)} + \ln[x + \sqrt{(x^2+1)}]\}/2 + C$$

8 $$\int \sqrt{(1-x^2)}\,dx = [x\sqrt{(1-x^2)} + \sin^{-1} x]/2 + C$$

We have derived most of these in the course of our work, but there remain two which we have not. Before going any further, and by way of revision, you should tackle them. Do you know which ones they are? They are numbers 2 and 5.

Consequently if the integrand is of the form $1/Q$, $1/\sqrt{Q}$ or $\sqrt{Q}$, where Q is a quadratic, then we can always determine the integral by a routine procedure. All we need do is remember the procedure for 'completing the square'. That is, given a quadratic of the form $x^2 + ax + b$ we must express it in the form $(x + h)^2 \pm k^2$.

Do you remember how to do this? It is simplicity itself! We take h as half the coefficient of x, calculate $(x + h)^2$ and then add or subtract a positive number k^2 so that the constant agrees with b. In algebraic terms we obtain

$$(x + \tfrac{1}{2}a)^2 + b - \tfrac{1}{4}a^2$$

which looks much more complicated than it is.

Here is a numerical example.

□ Express $x^2 - 4x + 9$ in the form $(x + h)^2 \pm k^2$.

The coefficient of x is -4, so h is -2. Now $(x - 2)^2 = x^2 - 4x + 4$, so that

$$x^2 - 4x + 9 = (x - 2)^2 + 5$$

So here $k^2 = 5$. ■

How does this help with integration? Well, when we see an integrand which is of the form $1/Q$, $1/\sqrt{Q}$ or $\sqrt{Q}$, where Q is a quadratic, we complete the square and substitute $X = x + h$. The integral will thereby be reduced to one of the standard forms. Be careful, however. Any stray x terms or other functions in the integrand change its nature completely and the method cannot be used.

So here is an example:

$$\int \frac{dx}{\sqrt{(x^2 + 6x + 18)}}$$

Completing the square:

$$x^2 + 6x + 18 = (x + 3)^2 + 9$$

So the integral becomes

$$\int \frac{dx}{\sqrt{[(x + 3)^2 + 3^2]}}$$

and the substitution $X = x + 3$ reduces it to a standard form. Now you complete this part of the calculation, and you will get the idea.

We have

$$\begin{aligned}\int \frac{dX}{\sqrt{(X^2 + 3^2)}} &= \frac{1}{3}\int \frac{dX}{\sqrt{[(X/3)^2 + 1]}}\\ &= \int \frac{d(X/3)}{\sqrt{[1 + (X/3)^2]}}\\ &= \ln\{(X/3) + \sqrt{[1 + (X/3)^2]}\} + C\end{aligned}$$

$$= \ln \{(x + 3)/3 + \sqrt{[1 + [(x + 3)/3]^2]}\} + C$$
$$= \ln k\,[x + 3 + \sqrt{(x^2 + 6x + 18)}]$$

Here we have absorbed some stray constants.

Now you try this one completely on your own:

$$\int \frac{\mathrm{d}x}{\sqrt{(15 - 2x - x^2)}}$$

Don't worry if things get in a mess. Help is at hand!

We complete the square but have to change the sign. So

$$x^2 + 2x - 15 = (x + 1)^2 - 16$$
$$15 - 2x - x^2 = 16 - (x + 1)^2$$

The integral is therefore

$$\int \frac{\mathrm{d}x}{\sqrt{[16 - (x + 1)^2]}}$$

Putting $X = x + 1$ we obtain

$$\int \frac{\mathrm{d}X}{\sqrt{(16 - X^2)}} = \frac{1}{4}\int \frac{\mathrm{d}X}{\sqrt{[1 - (X/4)^2]}}$$
$$= \int \frac{\mathrm{d}(X/4)}{\sqrt{[1 - (X/4)^2]}}$$
$$= \sin^{-1}(X/4) + C = \sin^{-1}[(x + 1)/4] + C$$

So there it is.

16.5 REDUCTION FORMULA

Sometimes it is possible to avoid unnecessary work when obtaining an integral by using a reduction formula. This is best illustrated using an example.

☐ Obtain

$$\int \tan^7 x \,\mathrm{d}x$$

The idea is to deal with a whole family of integrals at once. Suppose

$$I_n = \int \tan^n x \,\mathrm{d}x$$

so that $I_7 = \int \tan^7 x \, dx$ is the required integral. We try to obtain an equation which expresses I_n in terms of the same integral for smaller values of n. We have here

$$\begin{aligned} I_n &= \int \tan^n x \, dx = \int \tan^{n-2} x \tan^2 x \, dx \\ &= \int \tan^{n-2} x(\sec^2 x - 1) \, dx \\ &= \int \tan^{n-2} x \sec^2 x \, dx - \int \tan^{n-2} x \, dx \\ &= \int \tan^{n-2} x \sec^2 x \, dx - I_{n-2} \end{aligned}$$

Now we can write down the integral immediately – can't we? You do it and then move on!

We have

$$I_n = \frac{\tan^{n-1} x}{n-1} - I_{n-2} \qquad \text{when } n > 1$$

So that

$$\begin{aligned} I_7 &= \frac{\tan^6 x}{6} - I_5 \qquad \text{putting } n = 7 \\ I_5 &= \frac{\tan^4 x}{4} - I_3 \qquad \text{putting } n = 5 \\ I_3 &= \frac{\tan^2 x}{2} - I_1 \qquad \text{putting } n = 3 \end{aligned}$$

Furthermore, we know that

$$I_1 = \int \tan x \, dx = \ln |\sec x| + \text{constant}$$

So substituting we obtain

$$I_7 = \frac{1}{6}\tan^6 x - \frac{1}{4}\tan^4 x + \frac{1}{2}\tan^2 x - \ln |\sec x| + C$$ ■

By producing a reduction formula we have avoided repeating the same stage three times. Usually integration by parts is used in forming a reduction formula. However, do not rush into a reduction formula unnecessarily; it is often possible to make a simple substitution to resolve a difficult integral. Now here is one for you to try.

□ Obtain, by first deducing a reduction formula,

$$\int x^4 \, e^x \, dx$$

As soon as you have obtained the reduction formula, read on and see if you have it correct.

We put

$$I_n = \int x^n \, e^x \, dx$$

and integrate by parts:

$$I_n = x^n \, e^x - \int n \, x^{n-1} \, e^x \, dx$$

So

$$I_n = x^n \, e^x - nI_{n-1} \qquad n \text{ any integer}$$

Good! Now put in values of n so that we reduce things down to a very simple integral, and finish it off.

Putting $n = 4, 3, 2, 1, 0$ in turn we obtain:

$$\begin{aligned} I_4 &= x^4 \, e^x - 4I_3 \\ I_3 &= x^3 \, e^x - 3I_2 \\ I_2 &= x^2 \, e^x - 2I_1 \\ I_1 &= x^1 \, e^x - 1I_0 \end{aligned}$$

Now

$$I_0 = \int e^x \, dx = e^x + \text{constant}$$

Finally we have

$$\begin{aligned} I_4 &= x^4 \, e^x - 4x^3 \, e^x + 12x^2 \, e^x - 24x \, e^x + 24 \, e^x + C \\ &= (x^4 - 4x^3 + 12x^2 - 24x + 24)e^x + C \end{aligned}$$ ■

If you have followed everything so far in these integration chapters you should be able to make progress with all the usual integrals. However, it is important to know that not every integral can be obtained *analytically*. Sometimes two integrals may look superficially similar but are in fact totally different. To illustrate the point, $\int \sqrt{(\tan x)}\, dx$ and $\int \sqrt{(\sin x)}\, dx$ *look* alike, but the second cannot be obtained analytically using elementary functions whereas the first can. However, the integral of $\sqrt{\tan x}$, with respect to x, is a very hard nut to crack, and you would be best advised to leave it well alone! You have been warned.

16.6 Workshop

Now it's time to take steps. We are going to approach things in the same way as we did in the previous chapter. We are going to look at some integrals and decide the best approach. The integrals also form the assignment for this chapter. We are not at this stage going to obtain the integrals. Experience shows that the key to integration is not the ability to deal with technical detail but the ability to stand back and *plan ahead*. So out with the magnifying glass and on with the deerstalker:

We are going to inspect some more integrals. Examine each carefully and decide how you would proceed. You may like to make a few jottings to explore your ideas, but you should not at this stage complete the integration.

▷**Exercise** Inspect these integrals: **1**

$$\textbf{1}\ \int \frac{\cos x + 1}{\sin x - 1}\,dx \qquad \textbf{2}\ \int e^{\sin x} \cos x\, dx \qquad \textbf{3}\ \int \frac{x\, dx}{x^3 - 1}$$

When you have decided, take the next step.

2

1 At first sight you may think of going for the t substitution, but this may not be best. The integrand simplifies into a sum of standard forms if we multiply numerator and denominator by $\sin x + 1$.

2 Do not go for integration by parts; things will only get worse. In fact the integral can be obtained by sight, can it not ($u = \sin x$)?

3 This is a clear case for using partial fractions. The derivative of the denominator is not the numerator, and there is no obvious substitution to employ.

If all your answers agree with these, then move on to step 4. If some of your answers look as if they will not resolve the integral, then try these next three.

▷**Exercise** Inspect these integrals:

$$\textbf{4}\ \int \frac{\cos x}{\sin^2 x - 1}\,dx \qquad \textbf{5}\ \int e^{\sin^2 x} \sin 2x\, dx \qquad \textbf{6}\ \int \frac{x\, dx}{x^2 - 1}$$

When you are ready, read on!

3

4 If we simplify the integrand we obtain a standard form straight away. Alternatively we can put $u = \sin x$ and use a standard form. We could even use partial fractions following this substitution, but this is then rather long.

5 This is the sort of situation where it helps if we know our trigonometrical identities and can use them smoothly. When $u = \sin^2 x$ we obtain $du = 2 \sin x \cos x \, dx = \sin 2x \, dx$. Therefore this substitution resolves the integral.

6 There is obviously a method using partial fractions available to us. However, it is quite unnecessary to go to these lengths since, apart from a constant factor, the numerator is the derivative of the denominator.

Good! Now let's look at some more.

4

Exercise Inspect these integrals:

$$\mathbf{7} \int \sin^5 x \, dx \qquad \mathbf{8} \int \cos^4 x \, dx \qquad \mathbf{9} \int \cos 2x \sin x \, dx$$

As soon as you you have decided on a suitable method, look at the next step to see if you are right.

5

7 Odd powers of sine or cosine should cause us no difficulty. Here we keep one sine and convert the even power into cosines using $\cos^2 x + \sin^2 x = 1$. Finally we use the substitution rule, putting $u = \cos x$.

8 Even powers of sine or cosine must be converted into multiple angles. The key formulas to use here are $\cos^2 x = (1 + \cos 2x)/2$ and $\sin^2 x = (1 - \cos 2x)/2$. In this example we can convert the integrand into one containing a quadratic term in $\cos 2x$, which in turn, using the double-angle formulas again, can be converted into terms of $\cos 4x$. All terms are then integrable by sight.

9 There are several approaches here which work. For example, we could write $\cos 2x = 2 \cos^2 x - 1$ and then integrate term by term. Alternatively we could use the formula

$$2 \sin A \cos B = \sin (A + B) + \sin (A - B)$$

If you are all right with these, then move ahead to step 8. If not, look carefully at the integrands we have been considering and by making rough notes convince yourself that the approaches which have been suggested do in fact work. Then examine the next three integrals.

6

Exercise Inspect these integrals:

10 $\int \sec^6 x \, dx$ **11** $\int \sec^5 x \, dx$ **12** $\int \sec^5 x \tan x \, dx$

When you have decided what to do with these, take another step.

7

10 If we were to put $u = \tan x$ we should obtain $du = \sec^2 x \, dx$. So if we split $\sec^6 x$ into a product of $\sec^4 x$ with $\sec^2 x$, it remains only to express $\sec^4 x$ in terms of $\tan x$. The identity $\sec^2 x = 1 + \tan^2 x$ is all we need.

11 Although superficially this may seem the same as the previous integral, in fact the odd power makes all the difference. Here we need to think in terms of integration by parts, writing the integrand as $\sec^3 x \sec^2 x$.

12 This may look more involved than the others, but in reality it is much simpler. If $u = \sec x$ then $du = \sec x \tan x \, dx$, and the necessary $\tan x$ term is obligingly part of the integrand. A simple application of the substitution rule enables us to complete it.

Now try the last two exotic creatures.

8

▷**Exercise** Inspect these integrals:

13 $\int \frac{1}{x}\left(x + \frac{1}{x}\right)^{11}\left(x - \frac{1}{x}\right) dx$ **14** $\int (1 + \sin x)^7 \cos x \, dx$

How will you proceed? Make an attempt at **13** and take another of the steps.

9

13 You could multiply out and integrate term by term – all 24 terms! You may wonder if there is a better method. Yes, there is: the integral can be found by sight ($u = x + x^{-1}$).

If you are right then move on to the text following. Otherwise, try **14** and take the last step.

10

14 This is much the same as the previous integral. Of course it is possible to expand out the bracket so that the integrand snakes away all over the page, but we should never overlook a very simple substitution. Think carefully about which substitution to make!

That's it then. With judicious use of the rules of integration you should now be able to tackle any integral you are given. The key to success is experience, and this can only be obtained through practice. Therefore tackle as many integrals as possible; then it will have to be a very exceptional integral which catches you out!

We now consider two practical problems. One of the tasks of the next chapter will be to show how useful integration is when it comes to solving problems which arise in engineering and science.

16.7 Practical

PARTICLE ATTRACTION

A particle is attracted towards a fixed point O by a force inversely proportional to the square of its distance from O and directly proportional to its mass m. If it starts from rest at a distance a from O, its distance x from O at time t satisfies the equation

$$\frac{dx}{dt} = \pm\sqrt{(2k)}\left(\frac{1}{x} - \frac{1}{a}\right)^{1/2}$$

where k is a positive constant. Show that the time taken to reach O is

$$a^{3/2}\pi/2\sqrt{(2k)}$$

As a first stage, integrate the differential equation with respect to t by writing it as two integrals.

---

1 We have

$$\pm\sqrt{(2k)}\int dt = \int \frac{dx}{\sqrt{[(1/x) - (1/a)]}}$$

Now we have a difficulty. How are we to resolve the integral on the right? Remember the guidelines: remove denominators, resolve roots. Remember also $\sec^2 \theta - 1 = \tan^2 \theta$. Try something, then move on.

---

2 We shall put $1/x = (1/a) \sec^2 \theta$, so that $x = a \cos^2 \theta$. Initially $\theta = 0$ and $x = a$, and as θ increases, x decreases. We have

$$dx/d\theta = -2a \cos \theta \sin \theta$$

so that

$$\pm\sqrt{(2k)}\, t = \int \frac{-2a \cos \theta \sin \theta}{\sqrt{(1/a)} \tan \theta}\, d\theta$$

Now see if you can resolve this integral.

---

3 We have, using elementary trigonometry,

$$\pm\sqrt{(2k)}\, t = -2a\sqrt{a}\int \cos^2\theta\, d\theta$$
$$= -a\sqrt{a}\int 2\cos^2\theta\, d\theta$$
$$= -a\sqrt{a}\int (\cos 2\theta + 1)\, d\theta$$
$$= -a\sqrt{a}\left(\frac{1}{2}\sin 2\theta + \theta\right) + C$$

where C is the arbitrary constant.

Now determine C and complete the solution.

---

4 When $t = 0$, $x = a$ and so $\theta = 0$. Consequently $C = 0$.

As $x \to 0$, $\cos\theta \to 0$, and $\theta \to \pi/2$. Therefore if T is the time required to reach O we have

$$\pm\sqrt{(2k)}\, T = -a\sqrt{a}\left(0 + \frac{\pi}{2}\right)$$

Now $T > 0$, so

$$T = \frac{a\sqrt{(a)}\,\pi}{\sqrt{(2k)}} = \frac{\pi a^{3/2}}{2\sqrt{(2k)}}$$

BALL BEARING MOTION

A ball bearing of mass m is projected vertically upwards with speed u in a liquid which offers resistance of magnitude mkv, where v is the speed of the bearing and k is a constant. Given that

$$-m\frac{d^2x}{dt^2} = mg + mk\frac{dx}{dt}$$

where x is the height, t is the time and g is the acceleration due to gravity, show that the greatest height attained is

$$\frac{u}{k} - \frac{g}{k^2}\ln\left(1 + \frac{ku}{g}\right)$$

Determine also the maximum speed the ball bearing will attain when falling through the liquid.

The equation may be integrated with respect to t to express dx/dt in terms of x and t. At the maximum height, $dx/dt = 0$; it will therefore be useful if we can obtain the time T to reach the maximum height. If we substitute $v = dx/dt$ we can calculate the time t in terms of v. Make this substitution, then look ahead for confirmation.

---

1 If we put $v = \mathrm{d}x/\mathrm{d}t$ the equation becomes

$$-\mathrm{d}v/\mathrm{d}t = g + kv$$

Therefore

$$\int \frac{\mathrm{d}v}{g + kv} = -\int \mathrm{d}t$$

So

$$\frac{1}{k} \ln (g + kv) = -t + C$$

where C is the arbitrary constant.

Next determine C and the time T taken for the ball bearing to reach its greatest height.

---

2 When $t = 0$, $v = u$ and so

$$\frac{1}{k} \ln (g + ku) = C$$

Therefore

$$\frac{1}{k} \ln (g + kv) = -t + \frac{1}{k} \ln (g + ku)$$

$$kt = \ln \left(\frac{g + ku}{g + kv}\right)$$

The maximum height is when $v = 0$, so we have

$$kT = \ln \left(\frac{g + ku}{g}\right)$$

$$= \ln \left(1 + \frac{k}{g} u\right)$$

So

$$T = \frac{1}{k} \ln \left(1 + \frac{k}{g} u\right)$$

Now that we know how long it takes to reach its maximum height, we can return to the differential equation to obtain the greatest height. See if you can do it; you must integrate term by term. Do this and then move on.

---

3 Dividing out m we have

$$-\frac{d^2x}{dt^2} = g + k\frac{dx}{dt}$$

So that, integrating with respect to t,

$$-\frac{dx}{dt} = gt + kx + A$$

where A is the arbitrary constant. Now when $t = 0$, $x = 0$ and $dx/dt = u$ so that $A = -u$. So

$$-\frac{dx}{dt} = gt + ku - u$$

At the maximum height $dx/dt = 0$ and we know T. Try to finish the solution.

---

4 We obtain

$$gT - u + kX = 0$$

where X is the greatest height. Therefore

$$\begin{aligned} X &= \frac{1}{k}(u - gT) \\ &= \frac{1}{k}\left[u - \frac{g}{k}\ln\left(1 + \frac{ku}{g}\right)\right] \\ &= \frac{u}{k} - \frac{g}{k^2}\ln\left(1 + \frac{ku}{g}\right) \end{aligned}$$

Lastly we must obtain the terminal velocity: that is, the maximum speed which the ball bearing will attain when falling freely through the liquid.

The key thing to realize is that at the maximum speed there is no acceleration and so $d^2x/dt^2 = 0$. Therefore

$$g + k\frac{dx}{dt} = 0$$

So

$$\frac{dx}{dt} = -\frac{g}{k}$$

The negative sign indicates that the ball bearing is falling.

SUMMARY

We have seen how to tackle integrals using a number of techniques.

- ☐ Trigonometrical and algebraic substitutions. Remember:
  - **a** remove denominators
  - **b** resolve roots.
- ☐ Reduction of the integral to a sum of standard forms.
- ☐ Use of the special substitutions

$$t = \tan x/2$$

and

$$s = \tan x$$

- ☐ Resolution of a quadratic Q where the integral is one of three types:

$$1/Q,\ 1/\sqrt{Q},\ \sqrt{Q}$$

- ☐ Use of reduction formulas.

EXERCISES

1 Simplify, and thereby resolve, each of the following:

a $\displaystyle\int \frac{\exp 2x + 1}{\exp x + 1}\,dx$

b $\displaystyle\int \frac{\sin 2x + 2\cos x}{1 + \sin x}\,dx$

c $\displaystyle\int \ln(x \exp x)\,dx$

d $\displaystyle\int \frac{\sqrt{(1 + x^2)}}{(1 + x)^2 - 2x}\,dx$

2 Use appropriate substitutions, where necessary, to obtain

a $\displaystyle\int x \tan(1 + x^2)\,dx$

b $\displaystyle\int \frac{x \sin\sqrt{(1 + x^2)}}{\sqrt{(1 + x^2)}}\,dx$

c $\displaystyle\int \frac{\sin 2x}{(1 + \sin^2 x)}\,dx$

d $\displaystyle\int \sin 3x \cos 4x\,dx$

3 Obtain each of the following integrals by reducing it to a sum of standard forms:

a $\displaystyle\int \frac{\sin 2x}{2(1 + \sin x)}\,dx$

b $\displaystyle\int \frac{2 - x^2}{\sqrt{(1 - x^2)}}\,dx$

c $\displaystyle\int \frac{dx}{\{\sqrt{(x^2 + 1)} + \sqrt{(x^2 - 1)}\}}$

d $\displaystyle\int \frac{2x^2\,dx}{\sqrt{(1 + x^2)} - \sqrt{(1 - x^2)}}$

4 Use the t substitution to obtain the following integrals:

a $\displaystyle\int \frac{dx}{(4 - 3\tan x)}$

b $\displaystyle\int \frac{dx}{5\cos x - 12\sin x}$

5 Use a reduction formula to obtain each of the following integrals:

a $\displaystyle\int \cosh^7 x\,dx$

b $\displaystyle\int x^5 \cos x\,dx$

ASSIGNMENT

Obtain each of the following integrals. If you are stuck at any stage, hints are given in section 16.6.

1 $\displaystyle\int \frac{\cos x + 1}{\sin x - 1}\,dx$ **2** $\displaystyle\int e^{\sin x} \cos x\,dx$ **3** $\displaystyle\int \frac{x\,dx}{x^3 - 1}$

4 $\displaystyle\int \frac{\cos x}{\sin^2 x - 1}\,dx$ **5** $\displaystyle\int e^{\sin^2 x} \sin 2x\,dx$ **6** $\displaystyle\int \frac{x\,dx}{x^2 - 1}$

7 $\displaystyle\int \sin^5 x\,dx$ **8** $\displaystyle\int \cos^4 x\,dx$ **9** $\displaystyle\int \cos 2x \sin x\,dx$

10 $\displaystyle\int \sec^6 x\,dx$ **11** $\displaystyle\int \sec^5 x\,dx$ **12** $\displaystyle\int \sec^5 x \tan x\,dx$

13 $\displaystyle\int \frac{1}{x}\left(x + \frac{1}{x}\right)^{11}\left(x - \frac{1}{x}\right)dx$ **14** $\displaystyle\int (1 + \sin x)^7 \cos x\,dx$

FURTHER EXERCISES

1 Obtain each of the following integrals:

(a) $\int \operatorname{cosec}^3 x \, dx$

(b) $\int \tanh^3 x \, dx$

(c) $\int \frac{d\theta}{3 \cos \theta + 4 \sin \theta}$

2 Obtain each of the following integrals:

(a) $\int \frac{1}{x^2} \sqrt{\left(1 - \frac{1}{x}\right)} \, dx$

(b) $\int \tan^2 (\theta/2) \tan \theta \, d\theta$

(c) $\int \frac{1 + \sin \theta}{1 + \cos \theta} \exp \theta \, d\theta$

3 Determine:

(a) $\int \frac{1 - \tan^2 x}{1 + \tan x + \tan^2 x} \, dx$

(b) $\int \frac{d\theta}{5 + 12 \tan \theta}$

(c) $\int \frac{x + 2}{(x^2 + 2x + 2)^2} \, dx$

4 Obtain each of the following:

(a) $\int \frac{dx}{\cos x - \sin x}$

(b) $\int \frac{dx}{\sin^3 x + 3 \sin x}$

(c) $\int \cos 3x \cos 5x \, dx$

5 If $I_n = \int \sec^n x \, dx$, show that for $n \geqslant 2$

$$(n - 1)I_n = \sec^{n-2} x \tan x + (n - 2)I_{n-2}$$

6 If $I_n = \int \tan^n x \, dx$, show that for $n \geqslant 2$

$$I_n = \frac{1}{n - 1} \tan^{n-1} x - I_{n-2}$$

7 If $I_n = \int x^m (\ln x)^n \, dx$, show that if $n \in \mathbb{N}$

$$(m + 1)I_n = x^{m+1} (\ln x)^n - nI_{n-1}$$

8 A particle moves on a curve defined parametrically in the polar co-ordinate system by $r = \sec u$ and $\theta = \tan u - u$. Show that if the intrinsic coordinates s and ψ are each measured from the line $\theta = 0$ then

(a) $\left(\frac{ds}{d\theta}\right)^2 = \left(\frac{dr}{d\theta}\right)^2 + r^2$

(b) $\frac{dr}{d\theta} = r \cot (\psi - \theta)$

Hence or otherwise show that the particle moves on the spiral $s = \psi^2/2$.

9 Obtain each of the following integrals:

a $\int t \sin 2t \, dt$

b $\int (2t + 3)/(t^2 + 3t + 1) \, dt$

c $\int t\sqrt{2t + 1} \, dt$

10 Determine each of the following integrals:

a $\int_0^{\pi/2} 1/(5 + 4\cos x) \, dx$

b $\int_0^{\pi/2} \sin x/(5 + 4\cos x) \, dx$

c $\int_0^{\pi/2} \cos x/(5 + 4\cos x) \, dx$

11 If $I_n = \int_0^{\pi/2} \sin^n x \, dx$ show that

$$I_n = \frac{n-1}{n} I_{n-2} \qquad (n \geqslant 2)$$

Hence or otherwise evaluate

a $\int_0^{\pi/2} \sin^7 x \, dx$

b $\int_0^{\pi/2} \sin^5 x \cos^2 x \, dx$

Integration 3 — 17

In Chapters 15 and 16, we were learning the techniques of integration. This chapter will be concerned with applications.

After studying this chapter you should be able to
- ☐ Evaluate definite integrals;
- ☐ Examine simple improper integrals for convergence;
- ☐ Apply methods of integration to determine volumes of revolution, centres of mass, moments of inertia and other quantities.

At the end of this chapter we look at a practical problem concerning the radius of gyration of a body.

17.1 DEFINITE INTEGRATION

Suppose that $f(x)$ is integrable with respect to x for all $x \in [a, b]$ where a and b are real numbers, $a < b$. In other words this means that we can find

$$\int f(x)\, dx = F(x) + C$$

whenever $a \leqslant x \leqslant b$. In such circumstances we define the **definite integral** of $f(x)$ with respect to x, with upper limit of integration b and lower limit of integration a, by

$$\int_a^b f(x)\, dx = [F(x)]_a^b = F(b) - F(a)$$

So the procedure for finding a definite integral is to first find the indefinite integral, ignoring the arbitrary constant, and then subtract its value at the lower limit of integration from its value at the upper limit of integration.

The only point to watch is that if we use a substitution when we are performing the indefinite integral we must take care either to change the limits of integration or to substitute back in terms of the original variable before evaluating the definite integral.

□ Evaluate

$$\int_0^1 (1 + x)^2 \, dx$$

We have

$$\begin{aligned}\int_0^1 (1 + x)^2 \, dx &= [\tfrac{1}{3}(1 + x)^3]_0^1 \\ &= \tfrac{1}{3}[(1 + 1)^3 - (1 + 0)^3] = \tfrac{1}{3}(8 - 1) = 7/3\end{aligned}$$

Alternatively, if we make a substitution $u = 1 + x$ then we have $u = 2$ when $x = 1$ and $u = 1$ when $x = 0$. So

$$\int_0^1 (1 + x)^2 \, dx = \int_1^2 u^2 \, dt = [\tfrac{1}{3}u^3]_1^2 = 7/3$$ ■

We shall see later that a physical interpretation can be given for the definite integral.

17.2 IMPROPER INTEGRALS

Suppose that $f(x)$ is integrable with respect to x for all $x \in (a, b)$, where a and b are real numbers, $a < b$. In other words, this means that we can find

$$\int f(x) \, dx = F(x) + C$$

whenever $a < x < b$.

It may be that $f(x)$ is not defined when $x = a$ or $x = b$. We extend the definition of the definite integral under such circumstances by

$$\int_a^b f(x) \, dx = \lim_{x \to b-} F(x) - \lim_{x \to a+} F(x)$$

provided the limits exist.

Remember that $x \to b-$ means that x approaches b through numbers less than b, whereas $x \to a+$ means that x approaches a through numbers greater than a. You might like to imagine yourself imprisoned by the interval (a, b); then the left boundary is a and the right boundary is b, and all values in the interval are between these two extremes.

Of course if the integral exists throughout a *closed* interval $[a, b]$ it is not necessary to take limits. Equally if the integral of $f(x)$ with respect to x exists for $a \leqslant x < b$ then there is no need to take limits at a. We should then have

$$\int_a^b f(x)\,\mathrm{d}x = \lim_{x\to b-} F(x) - F(a)$$

It is possible to use these ideas to extend the definition further so that infinite integrals may be considered. An **infinite integral** occurs when either the upper limit of integration is ∞ or the lower limit of integration is $-\infty$, or both. For example, if the integral of $f(x)$ with respect to x exists throughout the interval $[0, \infty)$ then

$$\int_0^\infty f(x)\,\mathrm{d}x = \lim_{x\to\infty} F(x) - F(0)$$

provided the limit exists. If the limit exists the integral is said to **converge**; if it does not then it is said to **diverge**.

□ Evaluate each of the following integrals, if the integrals exist:

$$\int_0^\infty \mathrm{e}^{-t}\,\mathrm{d}t \qquad \int_0^1 x^{-1}\,\mathrm{d}x$$

For the first integral we have

$$\int \mathrm{e}^{-t}\,\mathrm{d}t = -\mathrm{e}^{-t} + C$$

So that

$$\begin{aligned}\int_0^\infty \mathrm{e}^{-t}\,\mathrm{d}t &= [-\mathrm{e}^{-t}]_0^\infty \\ &= \lim_{t\to\infty} (-\mathrm{e}^{-t}) - (-\mathrm{e}^0) \\ &= 0 - (-1) = 1\end{aligned}$$

In the second integral the integrand is not defined when $x = 0$, and since $x > 0$ throughout the interval we have

$$\int x^{-1}\,\mathrm{d}x = \ln x + C$$

So that

$$\int_0^1 x^{-1}\,\mathrm{d}x = \ln 1 - \lim_{x\to 0+} (\ln x)$$

However, a graph of the logarithmic function shows that $\ln x \to -\infty$ as $x \to 0+$, so that the limit does not *exist*. Consequently the improper integral does not exist. ■

17.3 AREA UNDER THE CURVE

At this stage an important question has to be considered. How can we tell when $f(x)$ is integrable with respect to x? We used the phrase '$f(x)$ is integrable with respect to x' in connection with the indefinite integral to mean that we could obtain a function F such that $F'(x) = f(x)$. However, this idea is too narrow when we come to the definite integral and we shall need to modify it. In particular we shall show that if a function $f:[a, b] \to \mathbb{R}$ is continuous then it has a definite integral over the interval.

To do this we show that we can physically identify the definite integral of a positive continuous function between the limits a and b with the area A of the region enclosed by the x-axis, the curve $y = f(x)$ and the lines $x = a$ and $x = b$.

Suppose that the function $f:[a, b] \to \mathbb{R}$ is continuous, and suppose $A(t)$ is the area enclosed by the x-axis, the curve $y = f(x)$ and the lines $x = a$ and $x = t$, so that $A(a) = 0$ and $A(b) = A$ (Fig. 17.1). If t changes by a small amount δt then the corresponding change in the area of the shaded region is $A(t + \delta t) - A(t)$. Furthermore, suppose that $f^*(t)$ is the *maximum* value of $f(x)$ when $x \in [t, t + \delta t]$, and that $f_*(t)$ is the *minimum* value of $f(x)$ when $x \in [t, t + \delta t]$. Then

$$f_*(t)\,\delta t \leqslant A(t + \delta t) - A(t) \leqslant f^*(t)\,\delta t$$

since $f_*(t)\,\delta t$ underestimates the value of $A(t + \delta t) - A(t)$ whereas $f^*(t)\,\delta t$ overestimates the value of $A(t + \delta t) - A(t)$. Now if $\delta t \neq 0$ we have

$$f_*(t) \leqslant \frac{A(t + \delta t) - A(t)}{\delta t} \leqslant f^*(t)$$

As $\delta t \to 0$ we notice that both $f^*(t) \to f(t)$ and $f_*(t) \to f(t)$. Also

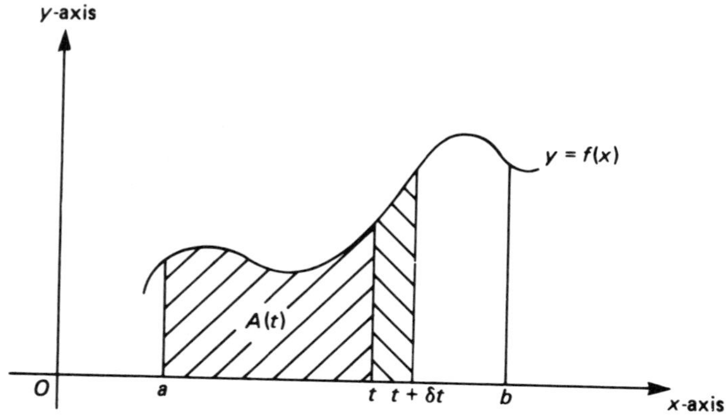

Fig. 17.1 The area under a curve.

$$\frac{A(t + \delta t) - A(t)}{\delta t} \to \frac{\mathrm{d}A}{\mathrm{d}t}$$

Therefore

$$f(t) \leqslant \frac{\mathrm{d}A}{\mathrm{d}t} \leqslant f(t)$$

so that $\mathrm{d}A/\mathrm{d}t = f(t)$. Consequently

$$\int f(t)\,\mathrm{d}t = A(t) + C$$

Now $A(a) = 0$ and $A(b) = A$, so that

$$A = A(b) - A(a) = \int_a^b f(x)\,\mathrm{d}x$$

This was what we wanted to show.

It should be stressed that in this argument we have tacitly used a number of properties of continuous functions without justification.

We can use the idea of a definite integral of a continuous function having a physical representation as the area 'under a curve' as a springboard to apply the calculus to a wide variety of different situations.

Suppose we look again at the area under the curve. We can imagine the interval $[a, b]$ divided up into n subintervals each of equal length δx. A typical subinterval can be represented as $[x, x + \delta x]$ (Fig. 17.2). Each subinterval corresponds to a strip of area δA which we can approximate by $f(x)\,\delta x$. This is the area of a rectangular region, and may be either an overestimate or an underestimate for the true area. However, $f(x)\,\delta x$ will

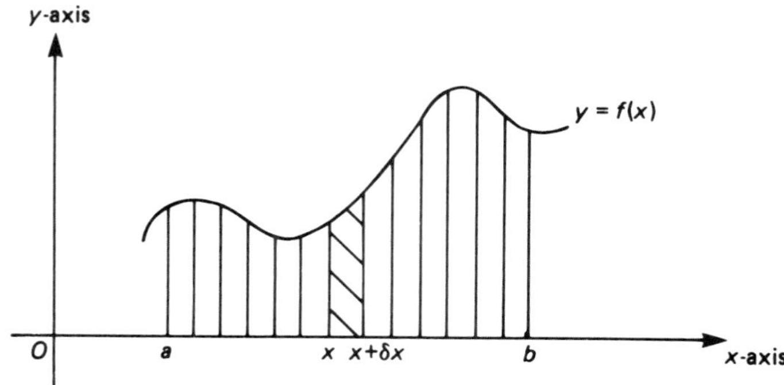

Fig. 17.2 Subdivision into strips.

approximate the true area when δx is very small. We may represent the total area in an informal way by

$$A = \sum \delta A \simeq \sum_{x=a}^{x=b} f(x)\, \delta x$$

where the summation sign Σ indicates that we are adding up the corresponding elements. In the second sum, a and b show that we are summing these elements from $x = a$ until $x = b$; in other words, over the interval $[a, b]$.

Now we already know what happens as $\delta x \to 0$ because we have already shown that

$$A = \int_a^b f(x)\, \mathrm{d}x$$

So a remarkable transformation occurs. As $\delta x \to 0$, the approximation becomes equality, the δ becomes d and the ugly duckling of a sigma sign becomes a beautiful swan of an integral sign!

We use this single example to infer a general method which we shall use to apply the calculus to a variety of problems. To fix the language for future use we shall refer to the idea of **partitioning** the interval $[a, b]$ into subintervals, and the corresponding portion of area δA which results will be termed an **element** of area. There are two conditions which are satisfied in this example and which must be satisfied in general:

1 The element which we choose and on which we base our approximation must be *typical*. That is, each element must be of this form and the approximation must be valid for each one.

2 By decreasing δx and so increasing the number of elements we must be certain that we could make the approximation arbitrarily close to the *true* result.

If and only if these two conditions are satisfied can we pronounce the magic words 'as δx tends to zero the approximation becomes good' and then carry out the following replacements:

$$\delta \to d \qquad \simeq \,\to\, = \qquad \sum \to \int$$

Let's try to visualize this in a more practical way.

17.4 VOLUME OF REVOLUTION

There are many ways of obtaining the volume of an egg, but one of them is closely related to the idea of integration and so we shall discuss it briefly. We can boil it, shell it and slice it up with an egg slicer. The egg will then have been converted into several small disc-like portions (Fig. 17.3). Then

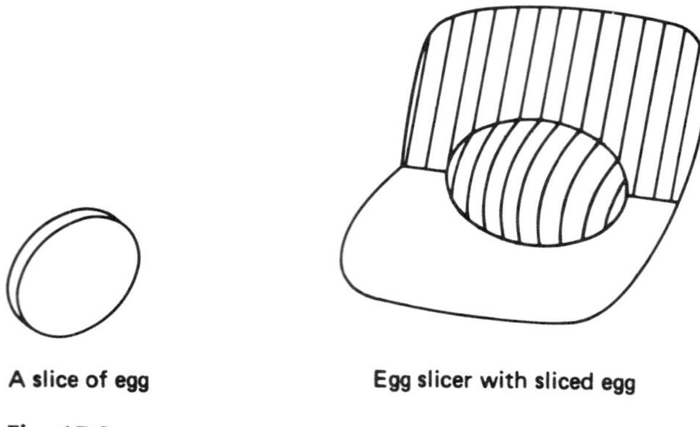

Fig. 17.3

we can measure the radius and thickness of each portion and calculate approximately its volume. We can then add up all the volumes corresponding to each of the slices and in that way obtain an approximation to the volume of the egg.

Some observations are worth making:

1 The smaller the gaps between the wires of the egg slicer, the closer the slices will be to discs and so the better the approximation.
2 If we choose a slice at random it is typical of the others; they can each be approximated by a disc.
3 We could obtain an approximation as close as we desired to the true volume of the egg just by making the subdivisions smaller and smaller.

Let us now look at this problem more systematically. The egg can be regarded as a solid of revolution. That is, we may suppose the region surrounded by the curve $y = f(x) \geqslant 0$, the x-axis and the lines $x = a$ and $x = b$ has been rotated through 2π degrees about the x-axis. In this way the egg is obtained (Fig. 17.4).

Now suppose we partition the interval $[a, b]$ into equal parts each of width δx. A typical element of volume will be a disc-like shape with its

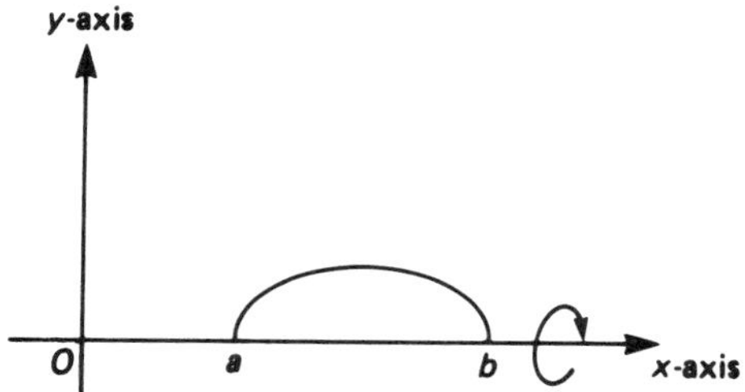

Fig. 17.4 Generating a solid of revolution.

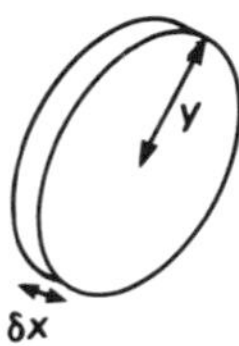

Fig. 17.5 A typical element.

centre a distance x from the origin. The radius will be the height $y = f(x)$ of the curve at x. So the volume of a typical element is approximately $\pi y^2\ \delta x$ (Fig. 17.5). We may represent the sum of all the elements by writing

$$V = \sum \delta V \simeq \sum_{x=a}^{x=b} \pi y^2\ \delta x$$

Now the two basic requirements are certainly satisfied and so we can pronounce the magic spell: 'as δx tends to zero the approximation becomes good'. Hey presto! We obtain

$$V = \int_{x=a}^{x=b} \pi y^2\ \mathrm{d}x$$

$$V = \int_{a}^{b} \pi y^2\ \mathrm{d}x$$

We should realize just how powerful this method is. Unfortunately it is easy to misuse it and thereby to obtain an incorrect result. For example, if we were to attempt to obtain the surface area of an egg by the same procedure we should still obtain discs as elements. It might be tempting to approximate the curved surface area of each disc by $2\pi y\ \delta x$, since $2\pi y$ is the perimeter and δx is the width of a typical disc. However,

$$A \neq \int_{x=a}^{x=b} 2\pi y\ \mathrm{d}x$$

What has gone wrong? Can you see?

The error here is that the element we have chosen does not typify the *extreme* case. For example, if $y = f(x)$ is particularly steep then the width of each element does not relate to the surface area (Fig. 17.6). Instead it is the length δs of the corresponding element of curve which is important. So, for a typical element, the curved surface area is approximately $2\pi y\ \delta s$. Hence the surface area required is

Fig. 17.6 An element.

$$A = \sum \delta A \simeq \sum_{x=a}^{x=b} 2\pi y \; \delta s$$

$$A = \int_{x=a}^{x=b} 2\pi y \; ds$$

We have a powerful method which can be used not only to calculate quantities which we already know about, but also to calculate quantities which may become of importance in the future and which have not even been considered at present.

Workshop

This chapter differs from the others in the book because there is no formal workshop; the workshop is dispersed among the text which follows. The reason for this is that we are about to derive a wide variety of formulas using, over and over again, the same basic principles of integral calculus. Therefore you can select for detailed study those which are particularly relevant to your branch of engineering. You will of course wish to use the others for practice and examples. In some ways it is like going on a 'field trip'!

17.5 LENGTH OF A CURVE

We begin by developing the formula for the length s of a curve $y = f(x)$ between the points where $x = a$ and $x = b$ (Fig. 17.7):

$$s = \int_a^b \left[1 + \left(\frac{dy}{dx}\right)^2\right]^{1/2} dx$$

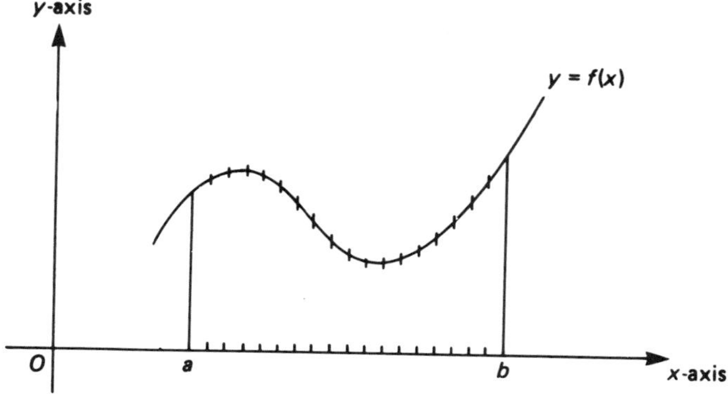

Fig. 17.7 Subdivision of a curve.

Suppose $y = f(x)$ is defined for $x \in [a, b]$. We wish to obtain the length of the curve over this interval. Dividing $[a, b]$ into subintervals each of length δx corresponds to a subdivision of the curve into portions of length δs. However, not all the subdivisions of the curve will necessarily have the same length even if δx becomes small (Fig. 17.8). We shall suppose that the curve is sufficiently smooth that when δx is small δs is given by the formula

$$(\delta s)^2 \simeq (\delta x)^2 + (\delta y)^2$$

and that this approximation becomes good as $\delta x \to 0$. So

$$\left(\frac{\delta s}{\delta x}\right)^2 \simeq 1 + \left(\frac{\delta y}{\delta x}\right)^2$$

and as $\delta x \to 0$

$$\left(\frac{\mathrm{d}s}{\mathrm{d}x}\right)^2 = 1 + \left(\frac{\mathrm{d}y}{\mathrm{d}x}\right)^2$$

so that

$$\frac{\mathrm{d}s}{\mathrm{d}x} = \left[1 + \left(\frac{\mathrm{d}y}{\mathrm{d}x}\right)^2\right]^{1/2}$$

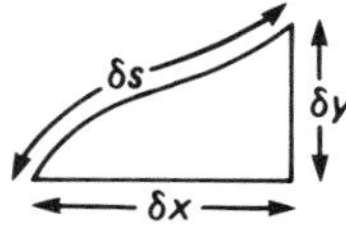

Fig. 17.8 Relating x, y and s.

assuming that s increases with x, so that $\mathrm{d}s/\mathrm{d}x \geqslant 0$. Now we have

$$s \simeq \sum_{x=a}^{x=b} \delta s$$

As $\delta x \to 0$ the approximation becomes good, and so

$$\begin{aligned} s &= \int_{x=a}^{x=b} \mathrm{d}s \\ &= \int_{x=a}^{x=b} \frac{\mathrm{d}s}{\mathrm{d}x}\,\mathrm{d}x \end{aligned}$$

Consequently we achieve the required formula for the length of a curve:

$$s = \int_a^b \left[1 + \left(\frac{\mathrm{d}y}{\mathrm{d}x}\right)^2\right]^{1/2} \mathrm{d}x$$

We remark that we are now in a position to give a formula for the surface area A produced by revolution of the curve around the x-axis:

$$A = 2\pi \int_a^b y \left[1 + \left(\frac{\mathrm{d}y}{\mathrm{d}x}\right)^2\right]^{1/2} \mathrm{d}x$$

Now for an example.

□ Obtain the length of the curve

$$y = \frac{x^2}{2} - \frac{1}{4}\ln x$$

between $x = 1$ and $x = 2$.

Try it yourself first.

We have

$$\frac{\mathrm{d}y}{\mathrm{d}x} = x - \frac{1}{4x}$$

Therefore

$$\begin{aligned} \left(\frac{\mathrm{d}s}{\mathrm{d}x}\right)^2 &= 1 + \left(x - \frac{1}{4x}\right)^2 \\ &= \left(x + \frac{1}{4x}\right)^2 \end{aligned}$$

$$\frac{\mathrm{d}s}{\mathrm{d}x} = x + \frac{1}{4x}$$

Consequently

$$s = \int_1^2 \left(x + \frac{1}{4x}\right) dx$$
$$= \left[\frac{x^2}{2} + \frac{1}{4} \ln x\right]_1^2$$
$$= \frac{3}{2} + \frac{1}{4} \ln 2$$

■

Sometimes a curve is described parametrically in the form $x = x(t)$, $y = y(t)$. We shall show that the length of the curve between the points $t = t_1$ and $t = t_2$ is given by

$$s = \int_{t_1}^{t_2} \sqrt{(\dot{x}^2 + \dot{y}^2)}\, dt$$

where $\dot{x} = x'(t)$ and $\dot{y} = y'(t)$.

Why not try this? It is not difficult.

If we partition the interval $[t_1, t_2]$ into equal parts each of length δt, this will produce corresponding elements δx and δy. We have

$$(\delta s)^2 \simeq (\delta x)^2 + (\delta y)^2$$

and as $\delta t \to 0$ the approximation becomes good. Now

$$\left(\frac{\delta s}{\delta t}\right)^2 \simeq \left(\frac{\delta x}{\delta t}\right)^2 + \left(\frac{\delta y}{\delta t}\right)^2$$

So as $\delta t \to 0$

$$\left(\frac{ds}{dt}\right)^2 = \left(\frac{dx}{dt}\right)^2 + \left(\frac{dy}{dt}\right)^2$$

Now assuming that s increases with t we have

$$s = \int_{t=t_1}^{t=t_2} ds$$
$$= \int_{t=t_1}^{t=t_2} \frac{ds}{dt} dt$$
$$= \int_{t_1}^{t_2} \left[\left(\frac{dx}{dt}\right)^2 + \left(\frac{dy}{dt}\right)^2\right]^{1/2} dt$$

Therefore the parametric formula for the length of a curve is

$$s = \int_{t_1}^{t_2} \sqrt{(\dot{x}^2 + \dot{y}^2)}\, dt$$

Here now is a problem using this formula.

□ Obtain the length of the curve $x = \theta + \sin\theta$, $y = 1 + \cos\theta$ between $\theta = 0$ and $\theta = \pi$.

When you have done this, move forward for the solution.

We need to convince ourselves that the curve doesn't do anything totally unexpected such as producing a figure of eight! One of the assumptions which we made was that, as the parameter increased, so too did the length of the curve. When $\theta = 0$ we have $x = 0$ and $y = 2$. Then as θ increases from 0 to π we see that x increases from 0 to π and y decreases from 2 to 0 (Fig. 17.9). As a matter of fact this is an interesting curve known as a cycloid. It is the curve described by a point on the rim of a car tyre as it moves along the road.

Now

$$x'(\theta) = 1 + \cos\theta$$
$$y'(\theta) = -\sin\theta$$

so that

$$\begin{aligned}\dot{x}^2 + \dot{y}^2 &= (1 + \cos\theta)^2 + (-\sin\theta)^2 \\ &= 1 + 2\cos\theta + \cos^2\theta + \sin^2\theta \\ &= 1 + 2\cos\theta + 1 = 2(1 + \cos\theta)\end{aligned}$$

Now $1 + \cos\theta = 2\cos^2(\theta/2)$, so that

$$\dot{x}^2 + \dot{y}^2 = 4\cos^2(\theta/2)$$

Finally

$$\begin{aligned}s &= \int_0^\pi \sqrt{(\dot{x}^2 + \dot{y}^2)}\,d\theta \\ &= \int_0^\pi 2\cos(\theta/2)\,d\theta\end{aligned}$$

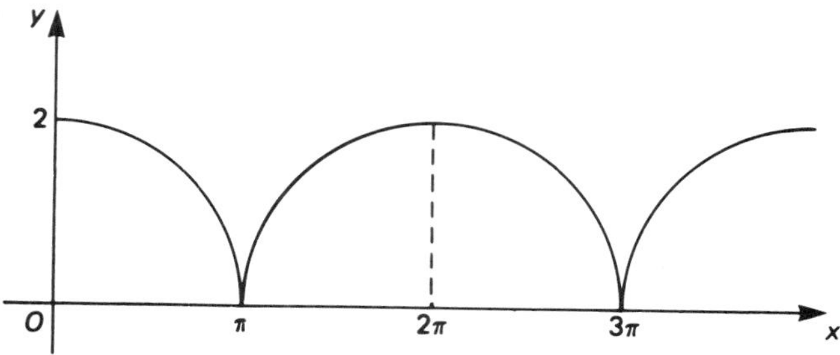

Fig. 17.9 A cycloid.

Note that cos (θ/2) is positive over the interval:

$$s = [4\sin(\theta/2)]_0^{\pi} = 4 - 0 = 4$$

■

17.6 CENTRES OF MASS

Suppose we have n particles of mass $m_1, m_2, \ldots, m_n$ positioned at points $(x_1, y_1, z_1), \ldots, (x_n, y_n, z_n)$ respectively relative to a rectangular cartesian coordinate system $Oxyz$. The centre of mass is the point $(\bar{x}, \bar{y}, \bar{z})$ where

$$M\bar{x} = \sum_{r=1}^{n} m_r x_r$$

$$M\bar{y} = \sum_{r=1}^{n} m_r y_r$$

$$M\bar{z} = \sum_{r=1}^{n} m_r z_r$$

and

$$M = \sum_{r=1}^{n} m_r$$

In many situations the system of particles behaves as if the mass M is concentrated at the centre of mass. If all the particles have equal mass then we have

$$\bar{x} = \frac{1}{n}\sum_{r=1}^{n} x_r$$

$$\bar{y} = \frac{1}{n}\sum_{r=1}^{n} y_r$$

$$\bar{z} = \frac{1}{n}\sum_{r=1}^{n} z_r$$

which is a purely geometrical property and is often called the **centroid** of the n points. We shall now use these concepts to obtain the position of centre of mass of a solid body and the position of a centroid of a uniform lamina.

□ Determine the position of the centre of mass of a uniform solid hemisphere of radius a.

Try this if you like.

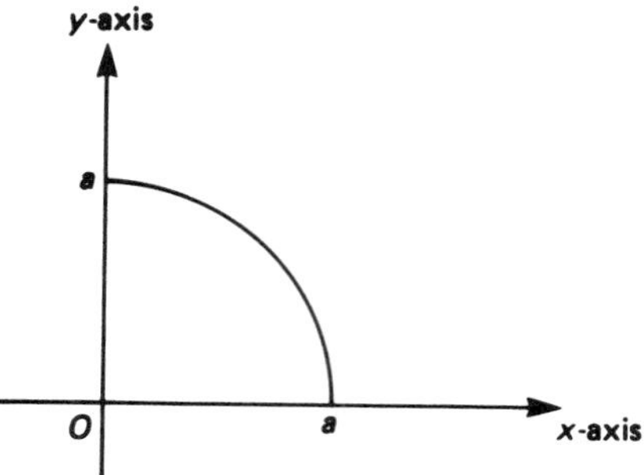

Fig. 17.10 Generating a hemisphere.

We first need to know the positions of the centroids of some simple objects. By symmetry, the centre of mass of a uniform rod is at the centre and the centre of mass of a uniform disc is also at the centre.

The hemisphere may be regarded as a solid of revolution obtained by rotating the portion of the circle $x^2 + y^2 = a^2$, $x \geqslant 0$ about the x-axis (Fig. 17.10). If we divide the interval $[0, a]$ into elements of length δx we divide up the hemisphere into discs each of radius y and width δx (Fig. 17.11).

Suppose the density of the hemisphere is ϱ. Then the mass of the elemental disc is $\varrho \pi y^2 \delta x$ approximately. So

$$M\bar{x} \simeq \sum_{x=0}^{x=a} \varrho\pi y^2 \, \delta x \, x$$

and

$$M \simeq \sum_{x=0}^{x=a} \varrho\pi y^2 \, \delta x$$

As $\delta x \to 0$ these approximations become good and so consequently

$$M\bar{x} = \int_0^a \varrho\pi x y^2 \, dx$$

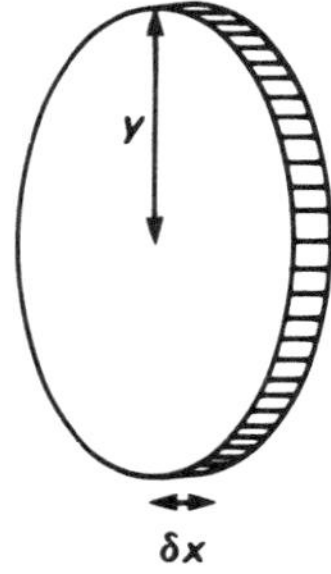

Fig. 17.11 A typical element.

$$M = \int_0^a \varrho\pi y^2 \, dx$$

Therefore

$$\begin{aligned} M &= \int_0^a \varrho\pi \, (a^2 - x^2) \, dx \\ &= \varrho\pi \left[a^2 x - \frac{x^3}{3} \right]_0^a \\ &= 2\varrho\pi a^3/3 \end{aligned}$$

$$\begin{aligned} M\bar{x} &= \int_0^a \varrho\pi \, (a^2 x - x^3) \, dx \\ &= \varrho\pi \left[\frac{a^2 x^2}{2} - \frac{x^4}{4} \right]_0^a \\ &= \varrho\pi a^4/4 \end{aligned}$$

Hence

$$\bar{x} = \varrho\pi \frac{a^4}{4} \frac{3}{2 \, \varrho\pi a^3} = \frac{3a}{8}$$

By symmetry, $\bar{y} = 0$ and $\bar{z} = 0$. ■

□ Obtain the position of the centroid of an arc of a circle, radius r, subtending an angle 2α at the centre.

Why not see if you can manage this on your own?

Using polar coordinates we can partition the arc into equal lengths each subtending an angle $\delta\theta$ at the centre (Fig. 17.12). So for each element of arc $\delta s = r \, \delta\theta$. Furthermore, referring to the diagram we can use symmetry to deduce that $\bar{y} = 0$, and it remains only to calculate $\bar{x}$. In order to determine the position of the centroid it may help to consider the circle as having a unit linear density. You may if you wish introduce a constant linear density ϱ, but this is unnecessary.

We have $r \, \delta\theta$ is the mass of a typical element and $r \cos \theta$ is the distance of the element from the y-axis. So

$$M\bar{x} \simeq \sum_{\theta=-\alpha}^{\theta=\alpha} r \, \delta\theta \, r \cos \theta$$

As $\delta\theta$ tends to zero the approximation becomes good and so

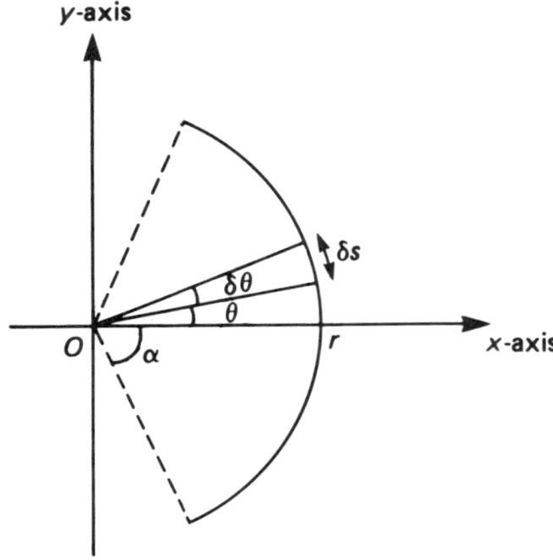

Fig. 17.12 Arc partitioned into equal lengths.

$$M\bar{x} = \int_{-\alpha}^{\alpha} r \, d\theta \; r \cos \theta$$

Moreover, M = length of arc × density = $2\alpha r$. So

$$\bar{x} = \frac{1}{2\alpha r} \int_{-\alpha}^{\alpha} r^2 \cos \theta \, d\theta$$

$$= \frac{r}{2\alpha} [\sin \theta]_{-\alpha}^{\alpha} = \frac{r \sin \alpha}{\alpha}$$

Of course α is expressed in terms of radians and not degrees. ■

□ Obtain the position of the centroid of a sector of a circle, radius a, subtending an angle 2α at the centre.

We divide the sector into elements each consisting of an arc subtending an angle 2α at the centre O (Fig. 17.13). The width of a typical arc with all its points distance r from O is δr. Now here we take the density per unit area as 1. We may consider each arc as having its mass concentrated at its centroid. In each case the centroid is at

$$\left(\frac{r \sin \alpha}{\alpha}, 0\right)$$

and the mass of an element is $2\alpha r \; \delta r$. Consequently

$$M\bar{x} \simeq \sum_{r=0}^{r=a} 2\alpha r \; \delta r \, \frac{r \sin \alpha}{\alpha}$$

As δr tends to zero the approximation becomes good and therefore

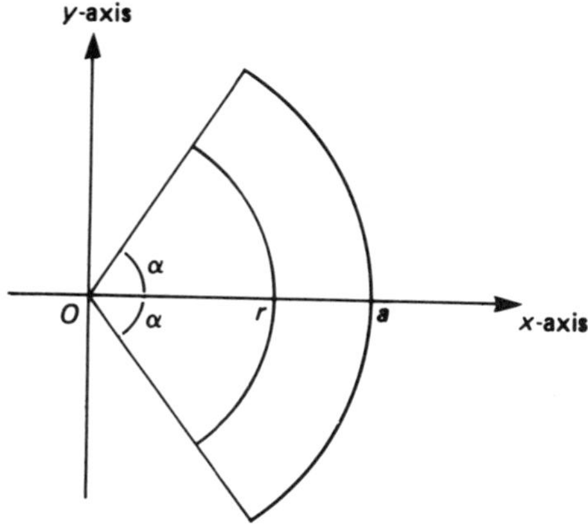

Fig. 17.13 Sector partitioned into concentric arcs.

$$\begin{aligned} M\bar{x} &= \int_0^a 2\alpha r \,\mathrm{d}r \,\frac{r \sin \alpha}{\alpha} \\ &= 2 \sin \alpha \int_0^a r^2 \,\mathrm{d}r \end{aligned}$$

Now M is the area of the sector, so $M = \alpha a^2$. So

$$\begin{aligned} \bar{x} &= \frac{2 \sin \alpha}{\alpha a^2} \int_0^a r^2 \,\mathrm{d}r \\ &= \frac{2 \sin \alpha}{\alpha a^2} \left[\frac{r^3}{3}\right]_0^a \\ &= \frac{2a \sin \alpha}{3\alpha} \end{aligned}$$

Once again we can appeal to symmetry to deduce that $\bar{y} = 0$. ■

17.7 THE THEOREMS OF PAPPUS

In the days before calculus, many techniques were employed to calculate volumes and surface areas. Two such techniques are attributed to Pappus, but their rediscovery 1300 years later by Guldin led to his name being linked with them also. These are the theorems.

Theorem 1 Suppose an arc rotates about an axis in its plane, which it does not cross. Then the curved surface area of the region which it describes is equal to the product of the length of arc and the distance travelled by the centroid of the arc.

To justify this we shall take the x-axis as the axis about which the curve is rotated and take $y = f(x)$, positive, as the curve itself (Fig. 17.14). If the arc is rotated through an angle θ then the surface area S swept out is easily obtained using calculus as

$$S = \int_{x=a}^{x=b} \theta y \, \mathrm{d}s$$

and of course the length of arc is given by

$$\int_{x=a}^{x=b} \mathrm{d}s$$

Now we need to obtain the position of the centroid. It will be sufficient for our purposes to obtain $\bar{y}$. As usual we partition the interval $[a, b]$ so that we obtain subintervals each of width δx. We have

$$\bar{y} \int_{x=a}^{x=b} \mathrm{d}s = \int_{x=a}^{x=b} y \, \mathrm{d}s$$

and the distance travelled by the centroid of the arc is $\theta\bar{y}$. Then

$$\begin{aligned} S &= \int_{x=a}^{x=b} \theta y \, \mathrm{d}s \\ &= \theta \int_{x=a}^{x=b} y \, \mathrm{d}s \\ &= \theta\bar{y} \int_{x=a}^{x=b} \mathrm{d}s \end{aligned}$$

that is, the distance travelled by the centroid times the length of the arc.

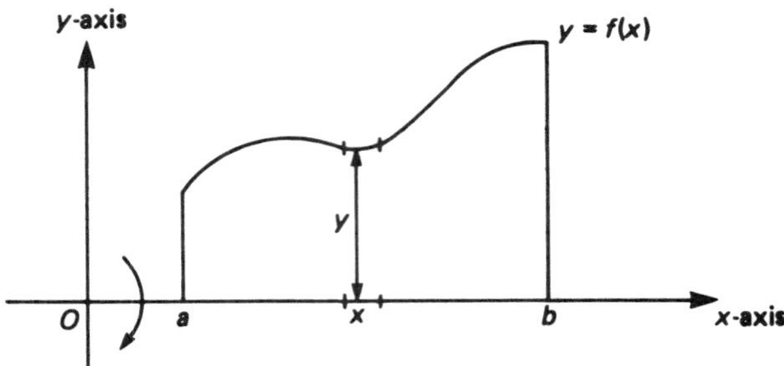

Fig. 17.14 Rotation of a curve.

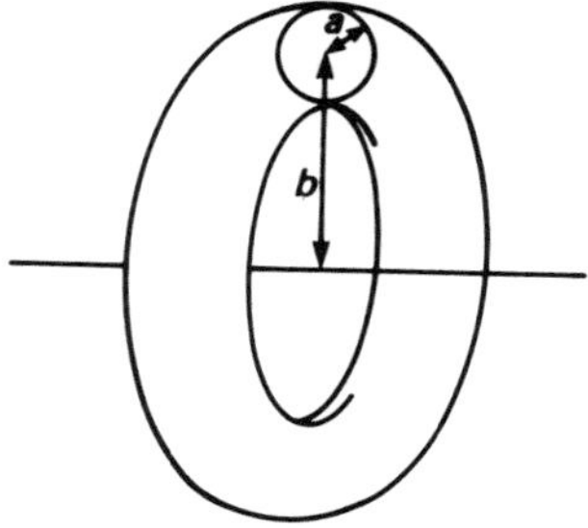

Fig. 17.15 Rotation of a circle.

☐ Obtain the surface area of a torus with inner radius $b - a$ and outer radius $b + a$.

A torus is sometimes known as an anchor ring, a tyre shape or a quoit. Try this example first, then look ahead. Calculus is not needed!

We can consider the torus as a circle of radius a rotated through 2π about an axis distance b ($>a$) from its centre (Fig. 17.15). The length of arc is the circumference of the circle $= 2\pi a$. The distance travelled by the centroid is $2\pi b$, since the centroid of a circle is at its centre. Therefore the surface area is

$$(2\pi a)(2\pi b) = 4\pi^2 ab$$

■

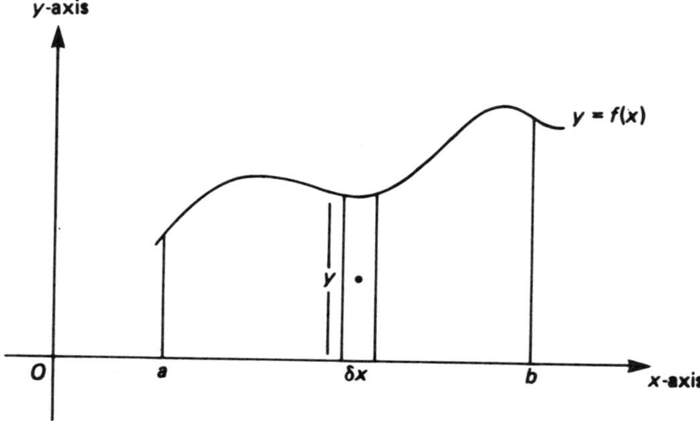

Fig. 17.16 Rotation of a region.

The second theorem of Pappus is very similar to the first. However, instead of rotating an arc we rotate a plane region.

Theorem 2 Suppose a region rotates about an axis in its plane, which it does not cross. Then the volume of the shape which it describes is equal to the product of the area of the region and the distance travelled by the centroid of the region.

See if you can justify this in the special case of the area enclosed by the curve $y = f(x)$, positive, the x-axis and the lines $x = a$ and $x = b$ rotating about the x-axis. The argument is very similar to the one we used for the first theorem.

If the region is rotated through an angle θ (Fig. 17.16) then the volume V swept out is given by

$$V = \int_{x=a}^{x=b} \frac{1}{2}y^2\theta \, dx$$

and the area of the region is given by

$$\int_{x=a}^{x=b} y \, dx$$

Now we need to obtain $\bar{y}$, the distance of the centroid of the region from the x-axis. We partition the interval $[a, b]$ so that we obtain subintervals each of width δx. Then, using the fact that the centroid of each strip is at its midpoint, we have

$$\bar{y} \int_{x=a}^{x=b} y \, dx = \int_{x=a}^{x=b} \frac{1}{2}y^2 \, dx$$

and the distance travelled by the centroid of the region is $\theta\bar{y}$. Then

$$\begin{aligned} V &= \int_{x=a}^{x=b} \frac{1}{2}y^2\theta \, dx \\ &= \theta \int_{x=a}^{x=b} \frac{1}{2}y^2 \, dx \\ &= \theta\bar{y} \int_{x=a}^{x=b} y \, dx \end{aligned}$$

that is, the distance travelled by the centroid times the area of the region. It is easy to adapt this argument to the more general case.

Pappus's theorems can be used the other way round to obtain the positions of centroids. See if you can do that with this example.

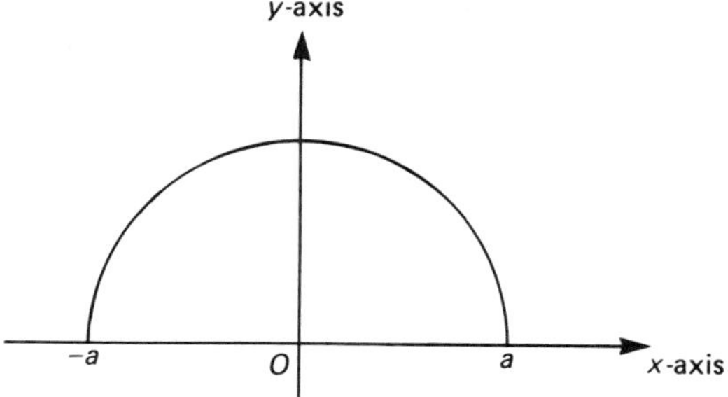

Fig. 17.17 Rotation of a semicircular region.

☐ Determine the position of the centroid of a semicircular region. Move on when you have tried this. You do not need calculus.

We can arrange the semicircle in a symmetric way as shown in Fig. 17.17. Then $\bar{x} = 0$ since the centroid must lie on the axis of symmetry. It remains only to determine $\bar{y}$.

Now if we rotate this region through 2π about the diameter, which is on the x-axis, we obtain a sphere as the solid of revolution. Using Pappus's second theorem we now have

$$\frac{4}{3}\pi a^3 = \frac{\pi a^2}{2} 2\pi\bar{y}$$

So

$$\bar{y} = \frac{4a}{3\pi}$$ ■

17.8 MOMENTS OF INERTIA

The product of the mass of a particle and its distance from some fixed axis is called the **first moment** of the particle about the axis. The product of the mass of a particle with the square of its distance from some fixed axis is called the **second moment** of the particle about the axis.

We have already used first moments implicitly when calculating the positions of centres of mass. Another name for the second moment is the **moment of inertia** of the particle about the axis. Moments of inertia are important in dynamics and so we shall consider the concept briefly.

Suppose we have a system of particles with masses $m_1, m_2, m_3, \ldots, m_n$ situated at the points $(x_1, y_1, z_1), (x_2, y_2, z_2), \ldots, (x_n, y_n, z_n)$ respectively relative to a rectangular cartesian coordinate system $Oxyz$ (Fig. 17.18). We denote by I_{Ox}, I_{Oy} and I_{Oz} the moments of inertia of the system about the axes Ox, Oy and Oz respectively. So

$$I_{Ox} = \sum_{i=1}^{n} m_i(y_i^2 + z_i^2)$$

$$I_{Oy} = \sum_{i=1}^{n} m_i(z_i^2 + x_i^2)$$

$$I_{Oz} = \sum_{i=1}^{n} m_i(x_i^2 + y_i^2)$$

If we sum these moments of inertia we obtain

$$I_{Ox} + I_{Oy} + I_{Oz} = 2 \Sigma\, m_i r_i^2$$

where r_i is the distance of the particle (x_i, y_i, z_i) from O. Note that the summation Σ is taken over all possible values of $i \in \{1, 2, \ldots, n\}$ and so we can simplify the notation by leaving out the limits.

□ Obtain the moment of inertia of a hollow spherical shell about a diameter.

We can partition the surface of the spherical shell into elements of area δA. So if the shell has uniform density ϱ each element has a mass $\varrho\, \delta A$. Then taking the origin at the centre of the shell we deduce by symmetry $I_{Ox} = I_{Oy} = I_{Oz}$. We also have

$$I_{Ox} + I_{Oy} + I_{Oz} = 3\, \Sigma\, \varrho\, \delta A\, r^2 = 2Mr^2$$

where r is the radius of the shell and M is its mass. Therefore

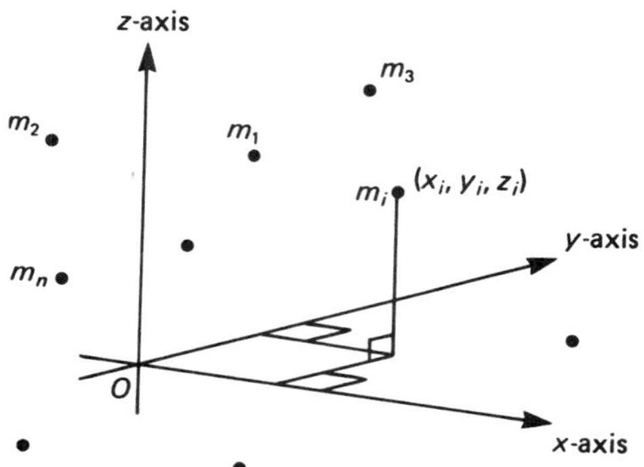

Fig. 17.18 System of particles.

$$3I_{Ox} = 2Mr^2$$

$$I_{Ox} = \frac{2}{3} Mr^2 \qquad \blacksquare$$

17.9 THE PERPENDICULAR AXIS THEOREM

An interesting relationship holds when all the particles in a system are in the same plane:

If a system of particles is coplanar then the moment of inertia of the system, about an axis perpendicular to its plane, is equal to the sum of the moments of inertia of the system about two mutually perpendicular axes, in the plane of the system, provided that all three axes are concurrent.

We may take the axis as Oz and the particles in the plane Oxy (Fig. 17.19). Then

$$I_{Ox} = \Sigma\, m_i y_i^2$$
$$I_{Oy} = \Sigma\, m_i x_i^2$$

So

$$I_{Ox} + I_{Oy} = \Sigma\, m_i(x_i^2 + y_i^2) = \Sigma\, m_i r_i^2 = I_{Oz}$$

This is a useful theorem, but it can only be used when the particles are coplanar. Naturally this extends to a plane lamina when we apply calculus, but it must *never* be misapplied to a solid body.

☐ Obtain the moments of inertia of a uniform solid disc of mass m and radius a about an axis through the centre perpendicular to its plane, and about a diameter.

Try this and see how it goes.

We begin by considering a uniform ring of radius r and uniform linear density ϱ (Fig. 17.20). Using polar coordinates, the perimeter can be split into elements each of length $r\,\delta\theta$. We therefore obtain an approximation for the moment of inertia about an axis Oz through the centre perpendicular to its plane:

$$I_{Oz} \simeq \sum_{\theta=0}^{2\pi} \varrho r\,\delta\theta\, r^2$$

The approximation becomes good as $\delta\theta$ tends to 0. Therefore

$$I_{Oz} = \int_0^{2\pi} \varrho r^3\,d\theta = \varrho r^3 2\pi$$

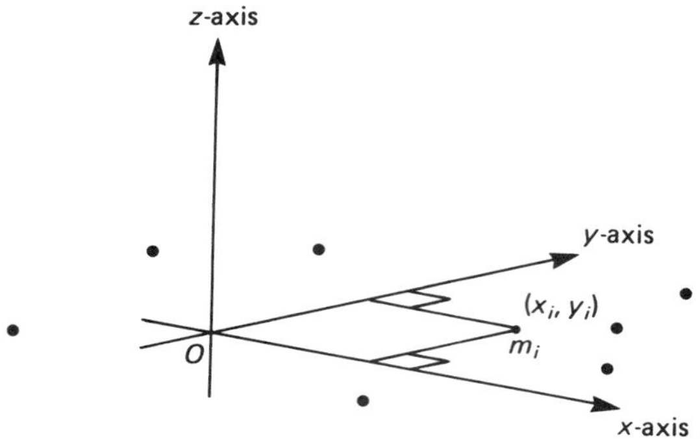

Fig. 17.19 System of coplanar particles.

But $M = 2\pi r\varrho$, so $I_{Oz} = Mr^2$.

Turning now to the disc, we split it into concentric rings as elements. A typical element has radius r and width δr. Using ϱ now for the area density, we have that the mass of an element is approximately $\varrho 2\pi r\ \delta r$; so it will contribute $\varrho 2\pi r\ \delta r\ r^2$ to the moment of inertia of the disc about Oz. Consequently

$$I_{Oz} \simeq \sum_{r=0}^{a} \varrho 2\pi r^3\ \delta r$$

and the approximation becomes good as δr tends to 0. Therefore

$$\begin{aligned} I_{Oz} &= \int_0^a 2\pi\varrho r^3\ \mathrm{d}r \\ &= 2\pi\varrho \left[\frac{r^4}{4}\right]_0^a \\ &= \pi\varrho a^4/2 \end{aligned}$$

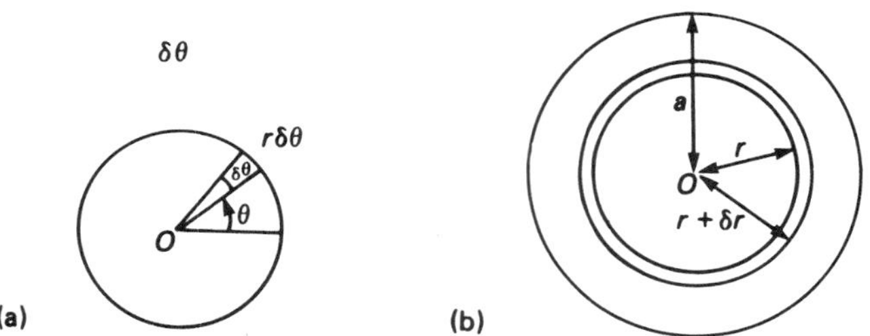

Fig. 17.20 (a) A uniform ring (b) Disc partitioned into concentric rings.

Now $m = \varrho\pi a^2$, and so the moment of inertia about an axis through the centre is $I_{Oz} = \frac{1}{2}ma^2$.

Lastly by symmetry $I_{Ox} = I_{Oy}$ and by the perpendicular axis theorem $I_{Ox} + I_{Oy} = I_{Oz}$. So the moment of inertia about a diameter is $I_{Ox} = I_{Oy} = \frac{1}{4}ma^2$. ■

Remember these results, because we often need them when calculating moments of inertia of other solids.

17.10 THE PARALLEL AXIS THEOREM

Another useful theorem which can be applied to solid bodies is known as the parallel axis theorem:

The moment of inertia of a system of particles about an axis is equal to the sum of the moment of inertia of the system about a parallel axis through the centre of mass and the product of the mass of the system with the square of the distance between the two axes.

Before we justify this we should note the reason for its importance. When calculating moments of inertia it is an error to assume that the mass of an element can be regarded as concentrated at the centre of mass. The parallel axis theorem must be used to obtain moments of inertia of elements about a given axis.

Suppose we are given a system of particles and a fixed axis (Fig. 17.21). We choose rectangular cartesian axes in such a way that

1 O is the centre of mass;
2 The fixed axis AB is parallel to the axis Oz;
3 The negative x-axis meets the fixed axis at A.

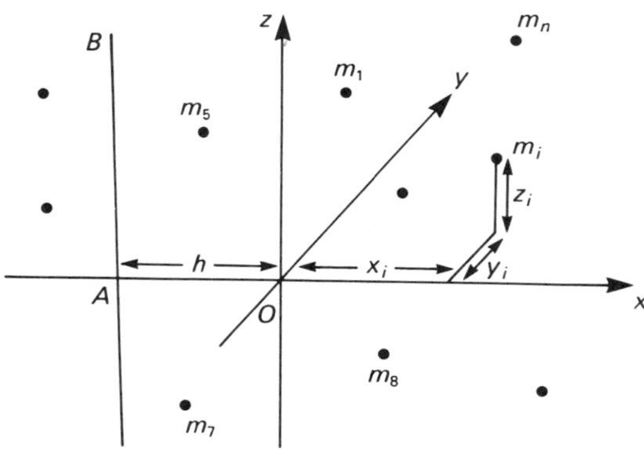

Fig. 17.21 System of particles and parallel axes.

We shall denote the fixed axis by AB and suppose that the distance between the two axes AB and Oz is h. Now

$$\begin{aligned} I_{AB} &= \Sigma\, m_i[(x_i + h)^2 + y_i^2] \\ &= \Sigma\, m_i(x_i^2 + y_i^2 + 2hx_i + h^2) \end{aligned}$$

But $\Sigma\, m_i x_i = 0$ because O is the centre of mass of the system. Consequently

$$\begin{aligned} I_{AB} &= \Sigma\, m_i(x_i^2 + y_i^2) + h^2\, \Sigma\, m_i \\ &= I_{Oz} + Mh^2 \end{aligned}$$

where M is the mass of the system.

□ Obtain the moment of inertia of a solid right circular cone about an axis through its vertex parallel to its base.

If you wish you can try this first on your own.

We take the height as h, the base radius as a and the mass as M. It will be convenient to take the axes as shown in Fig. 17.22 and the density as ϱ. So $M = \pi a^2 h\varrho/3$. We begin by slicing the cone into elements; a typical element is a disc with its centre at distance x from O, and with radius y and thickness δx. Regarding the cone as a solid of revolution we have

$$y = \frac{a}{h}x$$

Now the mass of an elemental disc is $\varrho\pi y^2\,\delta x$, and so the moment of inertia of the element about its diameter is $(1/4)\,(\varrho\pi y^2\,\delta x)\,y^2$. By the parallel axis theorem this element contributes

$$\tfrac{1}{4}(\varrho\pi y^2\,\delta x)\,y^2 + (\varrho\pi y^2\,\delta x)\,x^2$$

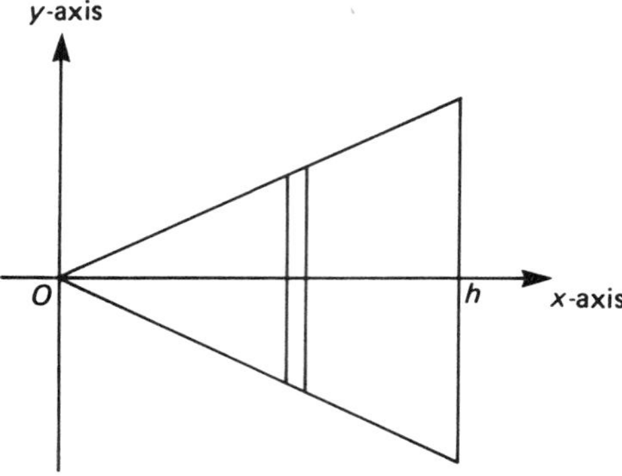

Fig. 17.22 Cross-section of solid right circular cone.

towards I_{Oy}. Therefore summing for all elements we have

$$I_{Oy} \simeq \sum_{x=0}^{h} \left(\frac{1}{4}y^2 + x^2\right) \varrho\pi y^2 \,\delta x$$

The approximation becomes good as δx tends to 0, and so

$$I_{Oy} = \int_0^h \varrho\pi \left(\frac{1}{4}y^2 + x^2\right) y^2 \,dx$$

Now

$$\begin{aligned}\left(\frac{y^2}{4} + x^2\right) y^2 &= \left(\frac{a^2x^2}{4h^2} + x^2\right)\left(\frac{a^2x^2}{h^2}\right) \\ &= Kx^4\end{aligned}$$

where

$$K = \frac{(a^2 + 4h^2)\,a^2}{4h^4}$$

So

$$\begin{aligned}I_{Oy} &= \varrho\pi K \int_0^h x^4 \,dx \\ &= \varrho\pi K \left[\frac{x^5}{5}\right]_0^h \\ &= \frac{\pi}{5}\varrho K h^5\end{aligned}$$

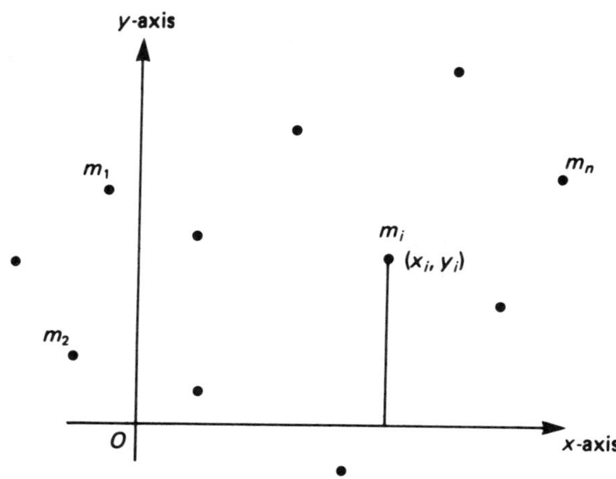

Fig. 17.23 System of particles with axes.

Now

$$\varrho = \frac{3M}{ha^2\pi}$$

so that

$$I_{Oy} = \frac{3M}{ha^2\pi} \frac{\pi h^5}{5} \frac{a^2 + 4h^2}{4h^4} a^2$$
$$= \frac{3M}{20}(a^2 + 4h^2)$$ ■

It is possible to define the product of inertia of a particle relative to two axes. For example, given the axes Ox and Oy and a system of particles (Fig. 17.23) then, in our usual notation, the **product of inertia** H_{xy} relative to these axes is given by

$$H_{xy} = \Sigma\, m_i x_i y_i$$

In order to generalize this to laminae and solid bodies we should need to extend the ideas of integration further to define double integrals and triple integrals. This is a simple matter but is beyond the scope of our present studies.

17.11 AVERAGE VALUES

We now look at two other quantities which are often calculated using integration. They are the mean value of a function over an interval, and the root mean square value of a function over an interval.

The **mean value** (MV) of f over the interval $[a, b]$ is given by

$$\text{MV} = \frac{1}{b - a} \int_a^b f(x)\, dx$$

We have already shown, if $f(x) \geqslant 0$ when $x \in [a, b]$, that the integral is the area enclosed by the curve, the x-axis and the lines $x = a$ and $x = b$ (Fig. 17.24). Therefore dividing by $b - a$ gives the average height of the curve.

If $f(x) \leqslant 0$, when $x \in [a, b]$, then a negative integral will be calculated. Consequently the mean value of the sine function, for instance, over the interval $[-\pi, \pi]$ is zero because $\sin(-x) = \sin x$ for all $x \in [-\pi, \pi]$. Therefore in this example as much of the area lies below the x-axis as lies above it.

In many applications this is an inadequate representation of the effect of the function over the interval. For instance, the effects of receiving

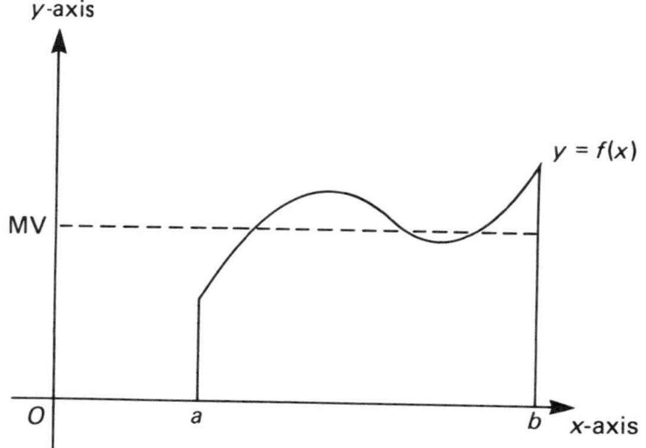

Fig. 17.24 The mean value of a function.

alternating current are certainly not zero! To obtain a more meaningful statistic, the RMS value is introduced.

The **root mean square value**, known as the RMS value, of a function f over the interval $[a, b]$ is the square root of the mean of the squares of the function:

$$\text{RMS} = \sqrt{\left\{\frac{1}{b-a}\int_a^b [f(x)]^2\,dx\right\}}$$

□ Obtain the mean value and the RMS value of $y = x^3 + 1$ over the interval $[-1, 1]$.

You can try this yourself if you wish. It is just a matter of substituting into the integrals and evaluating them. Why not have a go? We shall see who gets there first!

For the mean value we begin with

$$\int_{-1}^{1} (x^3 + 1)\,dx = \left[\frac{1}{4}x^4 + x\right]_{-1}^{1}$$

$$= \left(\frac{1}{4} + 1\right) - \left(\frac{1}{4} - 1\right) = 2$$

The length of the interval is $1 - (-1) = 2$ and so $\text{MV} = 2/2 = 1$.

For the RMS value we must first obtain

$$\begin{aligned}\int_{-1}^{1} (x^3 + 1)^2\,dx &= \int_{-1}^{1} (x^6 + 2x^3 + 1)\,dx \\ &= \left[\frac{1}{7}x^7 + \frac{2}{4}x^4 + x\right]_{-1}^{1} \\ &= \left(\frac{1}{7} + \frac{1}{2} + 1\right) - \left(-\frac{1}{7} + \frac{1}{2} - 1\right) \\ &= \frac{2}{7} + 2 = \frac{16}{7}\end{aligned}$$

We divide by the length of the interval and take the positive square root to obtain RMS = $\sqrt{[(16/7)/2]} = \sqrt{(8/7)}$. ■

17.12 RADIUS OF GYRATION

The moment of inertia I of a body of mass m about a given axis has dimensions ML^2; that is, it is the product of a mass with the square of a length. You possibly know that mass M, length L and time T are the basic building blocks in terms of which we can express many physical concepts. For example, acceleration has the dimensions LT^{-2}.

Indeed it is possible, by considering algebraic relationships between the ingredients in a physical problem, to derive the actual relationship by using dimensional analysis. In the case of a moment of inertia we see that $\sqrt{(I/m)}$ has the dimension of length ($\sqrt{(ML^2M^{-1})} = L$) and is called the **radius of gyration**.

The following practical problem involves the radius of gyration and uses polar coordinates.

17.13 Practical

METAL SPRING

□ A plane wire has the shape of the curve $r = f(\theta)$ where $\theta \in [\theta_1, \theta_2]$ in the polar coordinate system. Show that the radius of gyration k about an axis through the origin perpendicular to the plane of the curve satisfies

$$k^2 \int_{\theta_1}^{\theta_2} \sqrt{\left[\left(\frac{dr}{d\theta}\right)^2 + r^2\right]}\,d\theta = \int_{\theta_1}^{\theta_2} r^2 \sqrt{\left[\left(\frac{dr}{d\theta}\right)^2 + r^2\right]}\,d\theta$$

Hence or otherwise show that if $r = a \exp \theta$ and $\theta \in [0, \ln 10]$ then $k = a\sqrt{37}$.

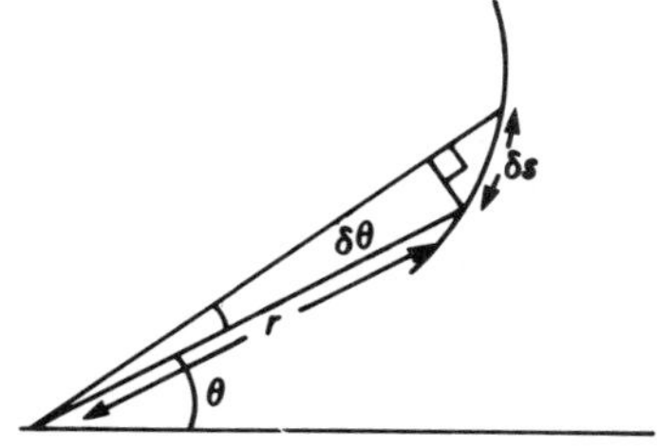

Fig. 17.25 Neighbouring points on $r = f(\theta)$.

It's a good idea to try this on your own first without looking at the solution. This is a problem in which you have to produce your own formula. We shall solve it stage by stage. Join in the solution when you feel you can.

---

1 We partition the spring into elements each of length δs (Fig. 17.25). Then if ϱ is linear density, we have that the mass of a typical element is $\varrho\,\delta s$. So the moment of inertia of this element about the required axis is given by

$$I_{\delta s} \simeq r^2 \varrho \; \delta s$$

The approximation becomes good as $\delta s \to 0$.

Use this to write down the mass m of the spring and the moment of inertia about the given axis.

---

2 We obtain immediately

$$m = \int_{\theta=\theta_1}^{\theta=\theta_2} \varrho \; ds$$

$$I = \int_{\theta=\theta_1}^{\theta=\theta_2} r^2 \varrho \; ds$$

Now use the definition of the radius of gyration to obtain an expression for k.

---

3 We have $k^2 m = I$, and so

$$k^2 \int_{\theta=\theta_1}^{\theta=\theta_2} \varrho \; ds = \int_{\theta=\theta_1}^{\theta=\theta_2} r^2 \varrho \; ds$$

Since ϱ is constant this becomes

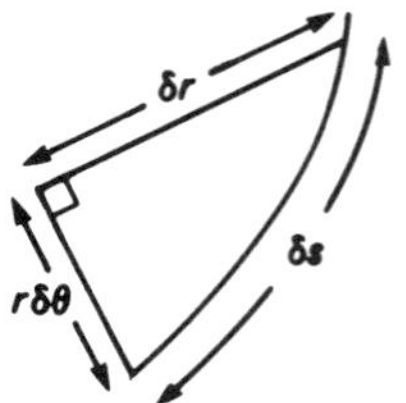

Fig. 17.26 Relating r, θ and s.

$$k^2 \int_{\theta=\theta_1}^{\theta=\theta_2} ds = \int_{\theta=\theta_1}^{\theta=\theta_2} r^2 \, ds$$

The next thing to do is to express $ds/d\theta$ in terms of $dr/d\theta$. See if you can do it.

---

4 We have, using Pythagoras's theorem (Fig. 17.26),

$$(\delta r)^2 + (r\,\delta\theta)^2 \simeq (\delta s)^2$$

So

$$\left(\frac{\delta r}{\delta\theta}\right)^2 + r^2 \simeq \left(\frac{\delta s}{\delta\theta}\right)^2$$

where the approximation becomes good as $\delta\theta \to 0$. Therefore

$$\left(\frac{dr}{d\theta}\right)^2 + r^2 = \left(\frac{ds}{d\theta}\right)^2$$

Consequently

$$k^2 \int_{\theta_1}^{\theta_2} \sqrt{\left[\left(\frac{dr}{d\theta}\right)^2 + r^2\right]} d\theta = \int_{\theta_1}^{\theta_2} r^2 \sqrt{\left[\left(\frac{dr}{d\theta}\right)^2 + r^2\right]} d\theta$$

Right! Now see if you can determine k for the portion of the spiral $r = a \exp \theta$ which has been specified.

---

5 We have $r = a \exp \theta$, $\theta_1 = 0$ and $\theta_2 = \ln 10$. So therefore $dr/d\theta = a \exp \theta$, and consequently

$$\left(\frac{dr}{d\theta}\right)^2 + r^2 = a^2 e^{2\theta} + a^2 e^{2\theta} = 2a^2 e^{2\theta}$$

Now from the previous relation we have

$$k^2\int_0^{\ln 10} (\surd 2)a\ \mathrm{e}^{\theta}\ \mathrm{d}\theta = \int_0^{\ln 10} (a\ \mathrm{e}^{\theta})^2\ (\surd 2)\,(a\ \mathrm{e}^{\theta})\,\mathrm{d}\theta$$

$$k^2a(\surd 2)\int_0^{\ln 10} \mathrm{e}^{\theta}\ \mathrm{d}\theta = a^3(\surd 2)\int_0^{\ln 10} \mathrm{e}^{3\theta}\ \mathrm{d}\theta$$

So

$$k^2[\mathrm{e}^{\theta}]_0^{\ln 10} = a^2\left[\frac{1}{3}\mathrm{e}^{3\theta}\right]_0^{\ln 10}$$

$$k^2(10 - 1) = \frac{a^2}{3}(1000 - 1)$$

from which $k^2 = 37a^2$, that is $k = a\surd 37$.

SUMMARY

We have seen how to

- ☐ Evaluate definite integrals.

$$\int_a^b f(x)\,\mathrm{d}x = F(b) - F(a)$$

if $F(x) = \int f(x)\,\mathrm{d}x$ exists for $x \in [a, b]$.

- ☐ Examine improper integrals for convergence.

$$\int_a^b f(x)\,\mathrm{d}x = \lim_{x\to b-} F(x) - \lim_{x\to a+} F(x)$$

if $F(x) = \int f(x)\,\mathrm{d}x$ exists for $x \in (a, b)$.

- ☐ Apply integral calculus to obtain formulas.

(a) Area under a curve $= \displaystyle\int_a^b y\,\mathrm{d}x$

(b) Volume of revolution $= \displaystyle\int_a^b \pi y^2\,\mathrm{d}x$

(c) Curved surface area of revolution $= \displaystyle\int_{x=a}^{x=b} 2\pi y\,\mathrm{d}s$

(d) Length of a curve $= \displaystyle\int_{x=a}^{x=b} \mathrm{d}s$

(e) Mean value $= \displaystyle\frac{1}{(b-a)}\int_a^b y\,\mathrm{d}x$

(f) Root mean square value $= \displaystyle\sqrt{\left\{\frac{1}{(b-a)}\int_a^b y^2\,\mathrm{d}x\right\}}$

if $y = f(x)$ is continuous for $x \in [a, b]$.

- ☐ Apply integral calculus to obtain centres of mass and moments of inertia of solid bodies.
- ☐ Use the theorems of Pappus to determine positions of centroids, lengths of curves, volumes and surface areas of revolution.

EXERCISES

1 Obtain the area enclosed by the curve $y = f(x)$, the x-axis and the ordinates at $x = a$ and $x = b$ for

a $y = \dfrac{1}{1 + x^2}$, $a = 0$, $b = 1$

b $y = x^2 + x$, $a = 0$, $b = 1$

c $y = \dfrac{x}{1 + x^2}$, $a = 0$, $b = 1$

d $y = \sin x + \sin 2x$, $a = 0$, $b = \pi$

2 Obtain the area enclosed by the curves

a $y = x^4$, $y = x$

b $y = 2x^2 + x$, $y = x^3 + 2x$

c $y = \dfrac{(x^3 + x^2 - 1)}{(x + 1)}$, $y = \dfrac{(x^2 + x - 1)}{(x + 1)}$

d $y = x \exp x^2$, $y = x \exp x$

3 Find the volume of revolution when $y = f(x) > 0$ is rotated through 2π about the x-axis

a $y = 1 - x^2$

b $y = x(1 - x) \exp x$

c $y = 2 - \cosh x$

d $y = (2 - x) \ln x$

4 Obtain the mean value of each of the following over the interval $[a, b]$:

a $y = \sin^2 x$, $a = 0$, $b = \pi$

b $y = \cosh x$, $a = 0$, $b = 1$

c $y = \ln x$, $a = 1$, $b = 2$

d $y = \tan x$, $a = 0$, $b = \pi/4$

5 Obtain the RMS values of each of the following over the interval [a, b]:

a $y = \cosh x$, $a = 0$, $b = 1$

b $y = \tan x$, $a = 0$, $b = \pi/4$

c $y = \sin x$, $a = 0$, $b = \pi$

d $y = \dfrac{x^2 + 1}{x^2}$, $a = 1$, $b = 2$

ASSIGNMENT

1 Obtain the length of the curve $e^y = \sec x$ between the points where $x = 0$ and $x = \pi/4$.

2 Determine the length of the curve $y = c \cosh(x/c)$ from $x = 0$ to $x = c \ln 2$.

3 Obtain the area enclosed by the curve $x = \cos^3 \theta$, $y = \sin^3 \theta$ where $\theta \in [0, 2\pi]$.

4 Sketch the polar curve $r = 2 \sin \theta$ and obtain the area of the region it encloses.

5 The portion of the curve

$$y = \tfrac{1}{2}x^2 - \tfrac{1}{4} \ln(1 + x) + x$$

between $x = 0$ and $x = 1$ is rotated about the x-axis through 2π radians. Obtain the area of the curved surface generated.

6 Obtain the position of the centre of mass of a uniform solid right circular cone of mass M, height h and base radius a.

7 Determine the moment of inertia of a uniform semicircular lamina of mass m and radius a about (a) its axis of symmetry (b) the bounding diameter. (Watch out for bounding diameters!)

8 Obtain the moment of inertia of a solid hemisphere of mass M and radius a about a tangent parallel to its plane face.

FURTHER EXERCISES

1 By putting $x = \tan\theta$ and then $\theta = \pi/4 - \phi$, or otherwise, show that

$$\int_0^1 \frac{\ln(1+x)}{1+x^2}\,dx = \frac{\pi}{8}\ln 2$$

2 Obtain the area of the region enclosed by the curves $y^2 = ax$ and $y^3 = ax^2$ and the position of its centroid.

3 Determine the length of the curve $4y = x^2 - \ln(x^2)$ between $x = 1$ and $x = 4$.

4 If

$$y = \tfrac{1}{4}x^2 + \tfrac{1}{4}x - \tfrac{1}{2}\ln(2x+1)$$

obtain the radius of curvature at the origin and the length of the curve between $x = -1/4$ and $x = 1/4$.

5 If

$$y = \tfrac{1}{2}\ln(1-x) - \tfrac{1}{2}\ln(1+x) + \tan^{-1}x$$

where $-1 < x < 1$, show that if the length of arc s is measured from the point where the curve crosses the x-axis then

$$s = -x - \tfrac{1}{2}\ln(1-x) + \tfrac{1}{2}\ln(1+x) + \tan^{-1}x$$

Hence, or otherwise, deduce that

$$y - s - x = \ln(1-x) - \ln(1+x)$$
$$y + s + x = 2\tan^{-1}x$$

6 Determine the moment of inertia of a solid right circular cone mass M, height h and base radius a about its axis of symmetry.

7 Evaluate where possible each of the following integrals:

(a) $\displaystyle\int_0^\infty x\ln x\,dx$

(b) $\displaystyle\int_0^\infty x\,e^{-x}\,dx$

(c) $\displaystyle\int_0^1 (1-x^2)^{-1/2}\,dx$

8 Prove that the area of the cardioid $r = a(1 + \cos\theta)$ is $3\pi a^2/2$ and that its perimeter is $8a$.

9 Show that the area of the ellipse

$$\frac{x^2}{a^2} + \frac{y^2}{b^2} = 1$$

is πab.

A solid circular ring of overall diameter 1.2 m is made from metal of elliptical cross-section. Each such ellipse has a major axis of length 0.4 m and a minor axis of length 0.2 m, and is such that its major axis is parallel to the axis of symmetry of the ring. Show that the volume of metal is $0.02\pi^2$ m^3.

10 A flat uniform metal plate PQR is in the shape of a triangle. Show that the moment of inertia of the plate about an axis through P parallel to QR is $Mh^2/2$, where M is the mass and h is the length of the perpendicular from P to QR. Without further integration prove that the moment of inertia about QR is $Mh^2/6$.

11 Show that when $3ay^2 = x(a - x)^2$ between $x = 0$ and $x = a$ is rotated about the x-axis through a complete revolution, the volume of the solid swept out is $\pi a^3/36$ and the surface area is $\pi a^2/3$.

12 If

$$I_n = \int_0^1 x^n \cos \pi x \, dx$$

and if n is a natural number, show that

$$\pi^2 I_n + n + n(n - 1)\, I_{n-2} = 0 \qquad (n > 1)$$

Hence, or otherwise, evaluate

$$\int_0^1 x^4 \cos \pi x \, dx$$

13 Determine area bounded by the curve

$$y = \frac{x(x-1)}{x-2}$$

and the x-axis between $x = 0$ and $x = 1$. This area is rotated through an angle of 2π about the x-axis. Calculate the volume of this solid of revolution.

18 Numerical techniques

In Chapters 14, 15 and 16 we investigated the technique of integration and saw how to apply it to a variety of situations. We noted that some integrals cannot be determined analytically using elementary functions. In the case of definite integrals a numerical method can often be employed. In this chapter we discuss some numerical techniques and include in this some methods for determining definite integrals.

After working through this chapter you should be able to

- ☐ Solve an equation of the form $f(x) = 0$ using one of four numerical techniques;
- ☐ Approximate derivatives of the first and second order and estimate the error involved;
- ☐ Apply the trapezoidal rule and Simpson's rule to evaluate definite integrals.

We shall then solve a practical problem concerning the approximation of the temperature in a heat-conducting fin.

18.1 THE SOLUTION OF THE EQUATION $f(x) = 0$

We know how to solve quadratic equations; there is a simple formula for doing this. It is even possible to write a complicated set of procedures and formulas which will enable us to solve cubic equations and quartic equations explicitly. However, in general it is impossible to do this for polynomial equations of degree greater than four, and it is impossible to solve many other types of algebraic equation explicitly.

In order to obtain solutions of algebraic equations of the form $f(x) = 0$ a number of numerical techniques have been developed, and we shall look at some of them. We shall not, however, submerge ourselves in a quagmire of detail but shall be content to see the overall method. With digital computers and programmable calculators readily available, much of the laborious and painful process of dealing with numerical techniques has been removed.

18.2 GRAPHICAL METHODS

If we cannot solve an equation of the form $f(x) = 0$ analytically, we can often obtain a solution by giving a rough sketch of the graph $y = f(x)$ and locating approximately those values of x at which the curve crosses the x-axis. Sometimes it is easier to rewrite the equation $f(x) = 0$ in the form $g(x) = h(x)$ and to determine the points at which the curves $y = g(x)$ and $y = h(x)$ intersect. In order to increase the accuracy of the approximation it may be necessary to draw the graphs in greater detail over a smaller interval, but provided we can calculate the values and have enough patience we should be able to obtain any degree of accuracy required.

However, there are obvious drawbacks with using a graphical method. Sketching graphs can be time consuming and liable to error even with a computer program, but perhaps more seriously it is difficult to estimate the accuracy of the solution which is obtained. A much more satisfactory method from many points of view is a numerical method, and we shall consider several of these in this chapter.

18.3 ITERATIVE METHODS

Many numerical methods use a recurrence relation of the form

$$x_{n+1} = F(x_n)$$

and we require $x_n \to a$ as $n \to \infty$, where a satisfies the equation $f(x) = 0$. We call x_n the nth **iterate** and x_0 the initial approximation or **starting value**. Therefore if h_n is the error in the nth iterate we have

$$x_n = a + h_n$$

Moreover, if the process is to converge then as $n \to \infty$ we have $x_n \to a$, so that

$$a = F(a) \quad \text{and} \quad f(a) = 0$$

Now $x_{n+1} = F(x_n)$, so

$$\begin{aligned} a + h_{n+1} &= F(a + h_n) \\ &= F(a) + h_n F'(a) + \tfrac{1}{2}h_n^2 F''(a + \theta h_n) \end{aligned}$$

where $\theta \in (0, 1)$. Here we are assuming that F has a Taylor expansion about the point a, and we are using the form of Taylor's theorem with the remainder after two terms (see Chapter 8).

For convergence we require $a = F(a)$, and so we obtain

$$h_{n+1} = h_n[F'(a) + \tfrac{1}{2}h_n F''(a + \theta h_n)]$$

Moreover, if the process is to converge we require for large n

$$|h_{n+1}| < |h_n|$$

That is, the error in the $(n + 1)$th iterate must eventually become less than the error in the nth iterate.

It can be shown that if the process is to converge and if F'' is bounded then

$$|F'(a)| < 1$$

□ Consider the equation

$$x^2 - 5x + 6 = 0$$

which we know to have roots at $x = 2$ and $x = 3$.

First, suppose we rewrite the equation as

$$x = 5 - \frac{6}{x} = F(x)$$

Then

$$F'(x) = \frac{6}{x^2}$$

So when $x = 2$, $F'(x) = 3/2 > 1$, and when $x = 3$, $F'(x) = 6/9 = 2/3 < 1$. This means that we expect the iteration

$$x_{n+1} = 5 - \frac{6}{x_n}$$

to converge near $x = 3$ but diverge near $x = 2$.

Secondly, suppose we rewrite the equation as

$$x = \frac{6}{5 - x} = G(x)$$

Then

$$G'(x) = -\frac{6}{(5 - x)^2}$$

So when $x = 2$, $G'(x) = -6/9$ and therefore $|G'(x)| < 1$; and when $x = 3$, $G'(x) = -6/4$ and therefore $|G'(x)| > 1$. This means that we expect the iteration

$$x_{n+1} = \frac{6}{5 - x_n}$$

to converge near $x = 2$ but diverge near $x = 3$. ■

The principal iterative method we shall consider is Newton's method. We begin, however, by considering some other numerical methods.

18.4 THE BISECTION METHOD

Suppose f is continuous and that we wish to solve the equation $f(x) = 0$. We first obtain two numbers a and b such that $f(a) < 0$ and $f(b) > 0$. We can then argue that somewhere in between a and b there is a solution of the equation (Fig. 18.1). This is a consequence of the intermediate value theorem, which is outside the scope of our work but is intuitively 'obvious'.

Let $c = (a + b)/2$. We then have one of three possibilities:

1 If $f(c) = 0$ we have found the required root.
2 If $f(c) < 0$ we can repeat the procedure with c replacing a.
3 If $f(c) > 0$ we can repeat the procedure with c replacing b.

After a bisection of the interval we either obtain the solution or we halve the length of the interval. So, if we repeat the process indefinitely, we must eventually arrive at the solution. The procedure stops when a and b agree to the required number of decimal places. In practice we rarely find that $f(c)$ is exactly 0 at any stage. However, we should always check that the root does satisfy the equation $f(x) = 0$ approximately.

We can make this more systematic by writing down the nth step:

1 Let a_n and b_n 'bracket' the root so that $f(a_n) < 0$ and $f(b_n) > 0$.
2 Put $c_n = (a_n + b_n)/2$.
3 If $f(c_n) = 0$ then the root is c_n.

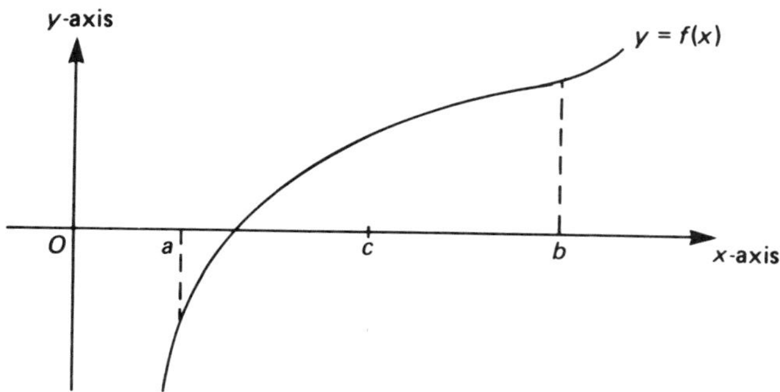

Fig. 18.1 The bisection method.

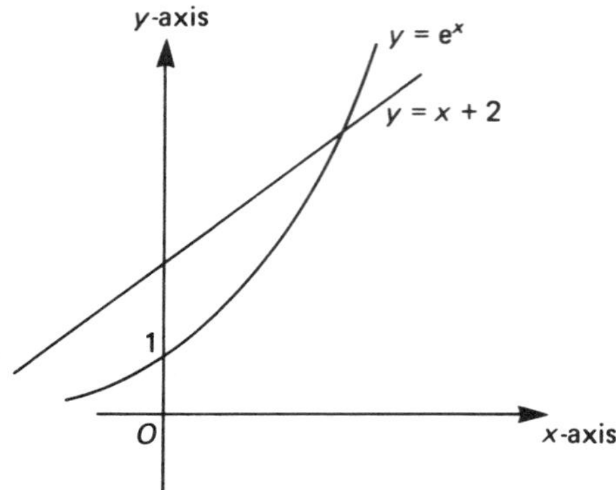

Fig. 18.2 The graphs of $y = e^x$ and $y = x + 2$.

4 If $f(c_n) < 0$ then $a_{n+1} = c_n$, $b_{n+1} = b_n$.
5 If $f(c_n) > 0$ then $a_{n+1} = a_n$, $b_{n+1} = c_n$.
We know that starting with $n = 0$ and $a_0 = a$, $b_0 = b$ we shall eventually obtain the root.

We now solve an equation using this method.

□ Obtain correct to three decimal places the positive root of $e^x = x + 2$.

A rough sketch of the equations $y = e^x$ and $y = x + 2$ reveals that there is indeed a positive root (Fig. 18.2).

Table 18.1

n	a_n	b_n	c_n	$f(c_n)$
0	1.000 00	2.000 00	1.500 00	0.981 69
1	1.000 00	1.500 00	1.250 00	0.240 34
2	1.000 00	1.250 00	1.125 00	−0.044 78
3	1.125 00	1.250 00	1.187 50	0.091 37
4	1.125 00	1.187 50	1.156 25	0.021 74
5	1.125 00	1.156 25	1.140 62	−0.011 91
6	1.140 63	1.156 25	1.148 44	0.004 83
7	1.140 63	1.148 44	1.144 54	−0.003 54
8	1.144 54	1.148 44	1.146 49	0.000 64
9	1.144 54	1.146 49	1.145 52	−0.001 44
10	1.145 52	1.146 49	1.146 01	−0.000 39

Put $f(x) = e^x - x - 2$. We begin by looking for two numbers which bracket the root: $f(1) = e - 3 < 0$ and $f(2) = e^2 - 5 > 0$. A table of values is the usual way to present the working (Table 18.1). As a general rule we must always work to two more places of decimals than that of the required accuracy, and so in this case we work to five. We can stop at $n = 10$ since no change to the third decimal place can now occur. So the required root is 1.146. ■

The main problem with the bisection method is that if the root is close to a or b the process will still take a long time to converge.

18.5 THE *REGULA FALSI* METHOD

One way of trying to compensate a little for the shortcomings of the bisection method is to attempt to use the curve itself in helping to locate the root.

In the *regula falsi* method we join the points $(a, f(a))$ and $(b, f(b))$ by a straight line and determine the point where it crosses the axis (Fig. 18.3). We know from our work on coordinate geometry (Chapter 3) that the equation of the straight line joining $(a, f(a))$ to $(b, f(b))$ is

$$\frac{y - f(a)}{f(b) - f(a)} = \frac{x - a}{b - a}$$

So when $y = 0$ we obtain

$$\begin{aligned} x &= a - \frac{(b - a)f(a)}{f(b) - f(a)} \\ &= \frac{a[f(b) - f(a)] - (b - a)f(a)}{f(b) - f(a)} \\ &= \frac{af(b) - bf(a)}{f(b) - f(a)} \end{aligned}$$

This formula gives an improved approximation:

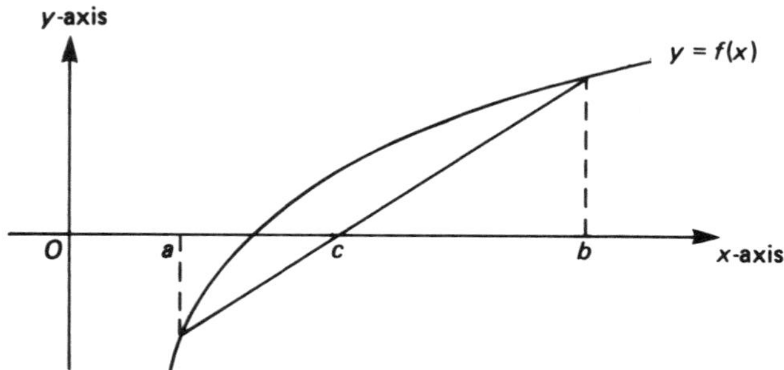

Fig. 18.3 The *regula falsi* method.

$$c_n = \frac{a_n f(b_n) - b_n f(a_n)}{f(b_n) - f(a_n)}$$

□ For the equation

$$f(x) = x^3 + 2x - 1$$

use the *regula falsi* method with $a = 0$ and $b = 1$ to obtain the first approximation c to the root.

We have $f(0) = -1$ and $f(1) = 1 + 2 - 1 = 2$, and so the interval $[0, 1]$ brackets the root. All we need to do now is to substitute $a = 0$ and $b = 1$ in

$$c_n = \frac{a_n f(b_n) - b_n f(a_n)}{f(b_n) - f(a_n)}$$

We obtain

$$c = \frac{0 \times 2 - 1 \times (-1)}{2 - (-1)} = \frac{1}{3}$$

The bisection method would have given $c = 1/2$. Furthermore $f(1/3) < 0$, and so if we were required to continue we should take $a_1 = 1/3$ and $b_1 = 1$. ■

18.6 THE SECANT METHOD

In the bisection method and the *regula falsi* method, *both* ends of the interval are liable to become modified as the method progresses. A technique similar in some ways to the *regula falsi* method, but which does not have this feature, is the secant method.

Unlike the bisection and *regula falsi* methods we do not require two initial approximations which bracket the root; nor is it necessary to check the sign of the value of the function at each stage. However, we do require two starting values x_0 and x_1. Suppose x_n is the nth approximation. Then we can join the points $(x_{n-1}, f(x_{n-1}))$, $(x_n, f(x_n))$ by a straight line and determine the point where this cuts the x-axis (Fig. 18.4).

In fact we have already determined this point for this line! We had

$$c_n = \frac{a_n f(b_n) - b_n f(a_n)}{f(b_n) - f(a_n)}$$

and so putting $a_n = x_{n-1}$, $b_n = x_n$ we obtain

$$x_{n+1} = \frac{x_{n-1} f(x_n) - x_n f(x_{n-1})}{f(x_n) - f(x_{n-1})}$$

□ If $f(x) = x^2 - 5$, obtain the formula corresponding to the secant method which gives x_{n+1} in terms of x_n and x_{n-1}.

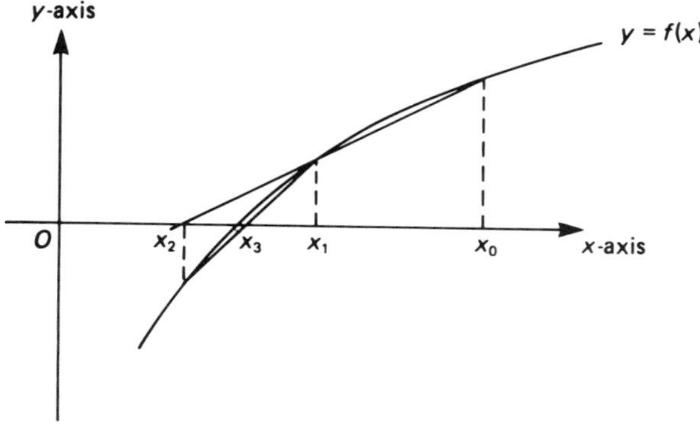

Fig. 18.4 The secant method.

We have

$$\begin{aligned} x_{n-1}f(x_n) - x_n f(x_{n-1}) &= x_{n-1}(x_n^2 - 5) - x_n(x_{n-1}^2 - 5) \\ &= x_n x_{n-1}(x_n - x_{n-1}) + 5(x_n - x_{n-1}) \\ &= (x_n x_{n-1} + 5)(x_n - x_{n-1}) \end{aligned}$$

and

$$\begin{aligned} f(x_n) - f(x_{n-1}) &= x_n^2 - 5 - (x_{n-1}^2 - 5) \\ &= x_n^2 - x_{n-1}^2 \\ &= (x_n - x_{n-1})(x_n + x_{n-1}) \end{aligned}$$

So

$$\begin{aligned} x_{n+1} &= \frac{(x_n x_{n-1} + 5)(x_n - x_{n-1})}{(x_n - x_{n-1})(x_n + x_{n-1})} \\ &= \frac{x_n x_{n-1} + 5}{x_n + x_{n-1}} \end{aligned}$$ ■

Two advantages of the secant method are that it is not necessary to choose starting values which bracket the root, and that we do not have to stop at each stage and check whether the function is positive or negative at the point.

The main disadvantage of this method is that convergence is no longer guaranteed. Although we shall not discuss the detailed circumstances in which convergence occurs, if

1 The starting values are chosen sensibly
2 The value of f' is non-zero at the root
3 The second-order derivative f'' is bounded then the method will work.

18.7 NEWTON'S METHOD

Yet another method, due originally to Newton, involves using the tangent to the curve (Fig. 18.5). Suppose x_n is an approximate root; then for many curves the tangent will cut the x-axis at a point which is closer to the true root. From the diagram

$$\text{slope of curve at } P = \frac{PR}{QR}$$

Therefore

$$f'(x_n) = \frac{f(x_n)}{x_n - x_{n+1}}$$

so that

$$x_{n+1} = x_n - \frac{f(x_n)}{f'(x_n)}$$

This is Newton's formula. It is an iterative formula because we obtain a new approximation each time n increases. We need just one starting value x_0.

□ Given the equation $f(x) = x^2 - 5$, obtain the iterative formula corresponding to Newton's method. Perform three iterations starting with $x = 2$ and $x = 3$. In each case work to three decimal places.

We have $f'(x) = 2x$, and so the formula gives

$$x_{n+1} = x_n - \frac{f(x_n)}{f'(x_n)}$$

$$x_{n+1} = x_n - \frac{x_n^2 - 5}{2x_n}$$

$$= \frac{2x_n^2 - (x_n^2 - 5)}{2x_n}$$

$$= \frac{x_n^2 + 5}{2x_n}$$

First we have $x_0 = 2$, and so

$$x_1 = \frac{x_0^2 + 5}{2x_0} = \frac{4 + 5}{4} = 2.25$$

$$x_2 = \frac{x_1^2 + 5}{2x_1} = \frac{(2.25)^2 + 5}{2(2.25)} = 2.236$$

$$x_3 = \frac{x_2^2 + 5}{2x_2} = \frac{(2.236)^2 + 5}{2(2.236)} = 2.236$$

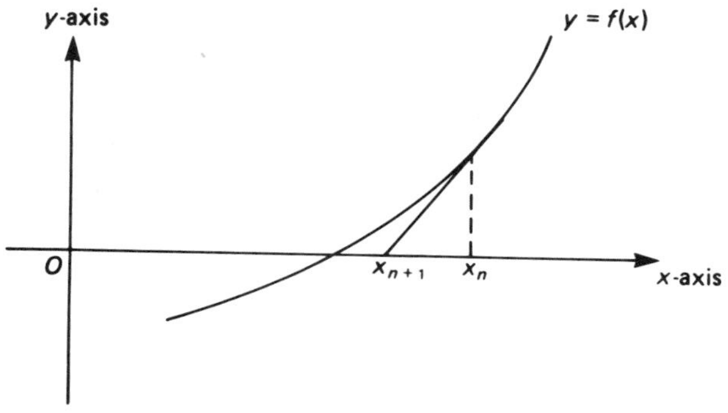

Fig. 18.5 Newton's method.

Secondly we have $x_0 = 3$, so

$$x_1 = \frac{x_0^2 + 5}{2x_0} = \frac{9 + 5}{6} = 2.333$$

$$x_2 = \frac{x_1^2 + 5}{2x_1} = \frac{(2.333)^2 + 5}{4.666} = 2.238$$

$$x_3 = \frac{x_2^2 + 5}{2x_2} = \frac{(2.238)^2 + 5}{2(2.238)} = 2.236$$

■

Newton's method is usually very good and can be applied to many problems. The number of correct decimal places is approximately doubled with each iteration.

The disadvantages with Newton's method are similar to those of the secant method. Principally we are not assured of convergence (Fig. 18.6), and certainly if f' is zero at a root then problems will occur. If f' is numerically small at any of the iterates then arithmetical difficulties such as rounding errors and arithmetic overflow will result.

There are two ways round this problem. One is to try it and see. That is, assume everything will be all right until shown otherwise. If a computer has been given the burden of calculation then either it will return an error message or it will calculate and calculate *ad nauseam*. A better method, but

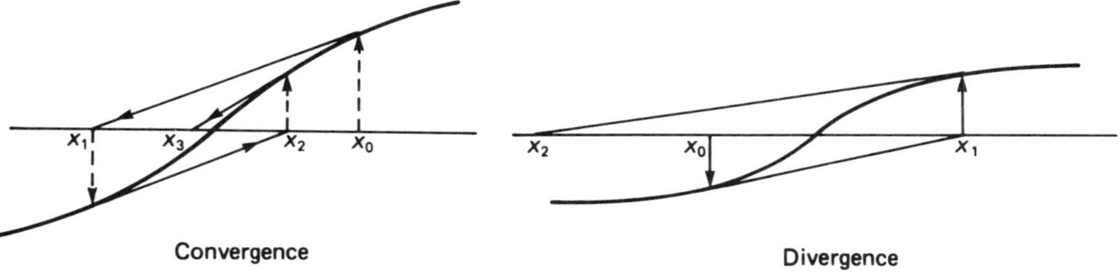

Fig. 18.6 Convergence and divergence of Newton's method.

one which requires some effort, is to try to anticipate any problems by making a rough sketch of $y = f(x)$.

Now we shall take a few steps.

18.8 Workshop

1

Exercise The bisection method is used to solve an equation $f(x) = 0$ and it is found that $f(0) = -1$ and $f(1) = 2$. If arithmetic is performed to seven decimal places, how many times will the method have to be applied to obtain a solution correct to five decimal places?

Give this some careful thought: it's not too difficult. Then step forward.

2

We need a_n and b_n to agree to five decimal places, and this will only be guaranteed if the difference between the two numbers is less than 0.000 000 1. For example, consider the numbers 0.435 674 9 and 0.435 675 0. The first rounds to 0.435 67 and the second to 0.435 68.

So we have in general

$$|a_n - b_n| < 0.000\,000\,1$$

Now

$$\begin{aligned}|a_0 - b_0| &= 1\\ |a_1 - b_1| &= 1/2\\ |a_2 - b_2| &= 1/4\\ |a_3 - b_3| &= 1/8\end{aligned}$$

In general, $|a_n - b_n| = 2^{-n}$. Therefore we must have $2^{-n} < 0.000\,000\,1$, which means we require $2^n > 10\,000\,000$. It follows at once that 24 applications are necessary to be certain.

If you got that right, move ahead to step 4.

If you went wrong then try this one.

▷ **Exercise** Arithmetic is performed to five decimal places, and after 12 applications of the bisection method a solution is stable to three decimal places. What was the maximum length of the interval initially?

Try this and then take a step. It may help to write down some numbers.

3

We need to ask the question: how different can the two numbers be if they agree to three decimal places? For example, 1.462 50 and 1.463 49 both agree to three decimal places. So we see that the difference between the two numbers is no more than 0.000 99.

Suppose $|a_0 - b_0| = r$. Then $|a_n - b_n| = 2^{-n}r$, and so we require $2^{-n}r < 0.000\,99$ when $n = 12$. Consequently $r < (0.000\,99)\,(4096) = 4.055\,04$, and this is the maximum length of the interval between a_0 and b_0.

Now for a rather different type of question.

▷**Exercise** A computer programmer wishes to apply Newton's method to solve an equation of the form $f(x) = 0$. He intends the computer to handle the derivative, and uses the approximation 4

$$f'(x_n) = \frac{f(x_n) - f(x_{n-1})}{x_n - x_{n-1}}$$

where x_n is the nth iterate. What is the iterative formula corresponding to this adaptation? How many starting values are needed?

Work your way through this and on to the next step.

We begin with the formula which Newton's method provides: 5

$$x_{n+1} = x_n - \frac{f(x_n)}{f'(x_n)}$$

So approximating the derivative in the prescribed fashion gives

$$\begin{aligned} x_{n+1} &= x_n - \frac{f(x_n)(x_n - x_{n-1})}{f(x_n) - f(x_{n-1})} \\ &= \frac{x_n[f(x_n) - f(x_{n-1})] - f(x_n)[x_n - x_{n-1}]}{f(x_n) - f(x_{n-1})} \\ &= \frac{f(x_n)x_{n-1} - f(x_{n-1})x_n}{f(x_n) - f(x_{n-1})} \end{aligned}$$

Clearly we shall need *two* starting values to set things going.

The computer programmer may *think* he is using Newton's method, but in fact he is using the secant method, isn't he?

If you made a mistake or would like some more practice, try the exercise below and take the following step; otherwise simply read it through.

▷**Exercise** Work out the iterative formula corresponding to Newton's method if it is intended to solve the equation $\sin x - x = 0$ by this method.

The key formula involves calculating 6

$$x - \frac{f(x)}{f'(x)}$$

Now $f(x) = \sin x - x$ and so $f'(x) = \cos x - 1$. Therefore

$$x - \frac{f(x)}{f'(x)} = x - \frac{\sin x - x}{\cos x - 1}$$

$$= \frac{x(\cos x - 1) - (\sin x - x)}{\cos x - 1}$$

$$= \frac{x \cos x - \sin x}{\cos x - 1}$$

So

$$x_{n+1} = \frac{x_n \cos x_n - \sin x_n}{\cos x_n - 1}$$

is the iterative formula which we require.

18.9 APPROXIMATIONS TO DERIVATIVES

We have seen that Newton's method uses the *derivative* of the function to produce an iterative formula. Unless we are going to tell the computer the derivative of the function when we supply it with the data (the function and the starting value), it will be necessary to produce some approximate formulas for derivatives. In fact you will need these formulas if ever you consider the numerical solutions of ordinary or partial differential equations.

Let us begin by recalling Taylor's expansion

$$f(x + h) = f(x) + hf'(x) + \frac{h^2}{2!} f''(c)$$

for some $c \in (x, x + h)$ (Chapter 8). This is the form of Taylor's expansion where the remainder is given after two terms. So therefore

$$\frac{f(x + h) - f(x)}{h} = f'(x) + \tfrac{1}{2}hf''(c)$$

This gives at once an approximate formula for the derivative together with an estimate of the error involved:

$$f'(x) \simeq \frac{f(x + h) - f(x)}{h}$$

If M is an upper bound for the modulus of f'' on the interval $(x, x + h)$ we can assert confidently that the error is at most $hM/2$. The fact that the error is no more than a constant multiple of h is expressed by saying that the error is of order h, which is written as $O(h)$. Of course as $h \to 0$ we have the error tends to 0 and the approximation becomes good.

A similar argument can be used to show that

$$f'(x) \simeq \frac{f(x) - f(x - h)}{h}$$

again with an error of $O(h)$.

We can obtain a better approximation if we take another term in Taylor's expansion:

$$f(x + h) = f(x) + hf'(x) + \frac{h^2}{2!} f''(x) + \frac{h^3}{3!} f^{(3)}(c_1)$$

for some $c_1 \in (x, x + h)$. Replacing h by $-h$ in this expansion gives

$$f(x - h) = f(x) - hf'(x) + \frac{h^2}{2!} f''(x) + \frac{h^3}{3!} f^{(3)}(c_2)$$

for some $c_2 \in (x - h, x)$. We shall use these to obtain an approximate formula of order h^2.

Subtracting the two expansions gives

$$f(x + h) - f(x - h) = 2hf'(x) + \frac{h^3}{3!} [f^{(3)}(c_1) + f^{(3)}(c_2)]$$

so that we obtain

$$\frac{f(x + h) - f(x - h)}{2h} = f'(x) + \frac{h^2}{12} [f^{(3)}(c_1) + f^{(3)}(c^2)]$$

Therefore if K is an upper bound for the modulus of $f^{(3)}$ on the interval $(x - h, x + h)$, we have

$$f'(x) \simeq \frac{f(x + h) - f(x - h)}{2h}$$

with an error no more than $h^2K/6$. So the error is of order h^2.

Yet one more term in the Taylor expansion provides an approximation for a second-order derivative which is also of order h^2. You might like to see if you can work it out for yourself.

We have

$$f(x + h) = f(x) + hf'(x) + \frac{h^2}{2!} f''(x) + \frac{h^3}{3!} f^{(3)}(x) + \frac{h^4}{4!} f^{(4)}(c_1)$$

for some $c_1 \in (x, x + h)$. Replacing h by $-h$ in this expansion gives

$$f(x - h) = f(x) - hf'(x) + \frac{h^2}{2!} f''(x) - \frac{h^3}{3!} f^{(3)}(x) + \frac{h^4}{4!} f^{(4)}(c_2)$$

for some $c_2 \in (x-h, x)$. (The c_1 and c_2 are not necessarily the same as the c_1 and c_2 when we considered the remainder after three terms.)

Adding the two expansions gives

$$f(x+h) + f(x-h) = 2f(x) + h^2 f''(x) + \frac{h^4}{4!}\left[f^{(4)}(c_1) + f^{(4)}(c_2)\right]$$

Therefore if K is an upper bound for the modulus of $f^{(4)}$ on the interval $(x-h, x+h)$ we obtain

$$f''(x) \simeq \frac{f(x+h) - 2f(x) + f(x-h)}{h^2}$$

with an error of no more than $h^2K/12$. So the error is of order h^2.

□ The deflection of a beam is believed to satisfy the equation

$$\frac{\mathrm{d}^2y}{\mathrm{d}x^2} = \mathrm{e}^{x^2}$$

The deflection is 0 at both $x = 0$ and $x = 1$. Estimate, using a second-order approximation for the derivative, the approximate deflections at 0.2, 0.4, 0.6 and 0.8.

We have approximately

$$\frac{f(x+h) - 2f(x) + f(x-h)}{h^2} = \mathrm{e}^{x^2}$$

Therefore taking $h = 0.2$ we have

$$\frac{f(x+0.2) - 2f(x) + f(x-0.2)}{0.04} = \mathrm{e}^{x^2}$$

Consequently,

$$f(x+0.2) - 2f(x) + f(x-0.2) = 0.04\,\mathrm{e}^{x^2}$$

Putting $x = 0.2$, 0.4, 0.6 and 0.8 in turn gives

$$\begin{aligned}
f(0.4) - 2f(0.2) + f(0) &= 0.041\,63\\
f(0.6) - 2f(0.4) + f(0.2) &= 0.046\,94\\
f(0.8) - 2f(0.6) + f(0.4) &= 0.057\,33\\
f(1) - 2f(0.8) + f(0.6) &= 0.075\,86
\end{aligned}$$

Therefore writing $f(0.2r) = f_r$ and using the information that $f_0 = f(0) = 0$ and $f_5 = f(1) = 0$, these equations become

$$\begin{aligned}
f_2 - 2f_1 + 0 &= 0.041\,63 && (1)\\
f_3 - 2f_2 + f_1 &= 0.046\,94 && (2)\\
f_4 - 2f_3 + f_2 &= 0.057\,33 && (3)\\
0 - 2f_4 + f_3 &= 0.075\,86 && (4)
\end{aligned}$$

Eliminating f_1 between (1) and (2) produces

$$2f_3 - 4f_2 + f_2 = 0.135\,51$$
$$2f_3 - 3f_2 = 0.135\,51 \qquad (5)$$

Eliminating f_4 between (3) and (4) produces

$$f_3 - 4f_3 + 2f_2 = 0.190\,52$$
$$-3f_3 + 2f_2 = 0.190\,52 \qquad (6)$$

Using (5) and (6) we can eliminate f_2:

$$4f_3 - 9f_3 = 0.842\,58$$

Therefore $-5f_3 = 0.842\,58$ and so $f_3 = -0.168\,52$. From (6),

$$2f_2 = 0.190\,52 + 3f_3 = -0.315\,04$$

Therefore $f_2 = -0.157\,52$. Equation (1) gives

$$2f_1 = f_2 - 0.041\,63 = -0.199\,15$$

So $f_1 = -0.099\,58$. Equation (4) gives

$$2f_4 = f_3 - 0.075\,86 = -0.244\,38$$

So $f_4 = -0.122\,19$.

We conclude that the deflections at 0.2, 0.4, 0.6 and 0.8 are approximately $-0.099\,58$, $-0.157\,52$, $-0.168\,52$ and $-0.122\,19$ respectively. If we were to analyse the approximation we should see that each of these could have an error not exceeding 0.0022. ■

Now for a few more steps.

18.10 Workshop

1

▷**Exercise** Obtain a finite difference approximation of order h^2 for the differential equation

$$\frac{d^2y}{dx^2} + \frac{dy}{dx} + y = \tan x$$

When you have tried this, step ahead.

2

We must use two approximations of order h^2, and we have just the right ones at our disposal. If we put $y = f(x)$ the equation becomes

$$f''(x) + f'(x) + f(x) = \tan x$$

If you got stuck then perhaps you would like to take over here. When you are ready, move on.

3 We have

$$f'(x) \simeq \frac{f(x+h) - f(x-h)}{2h}$$

$$f''(x) \simeq \frac{f(x+h) - 2f(x) + f(x-h)}{h^2}$$

So substituting into the equation:

$$\frac{f(x+h) - 2f(x) + f(x-h)}{h^2} + \frac{f(x+h) - f(x-h)}{2h} + f(x) \simeq \tan x$$

If we multiply up by $2h^2$ we obtain

$$2[f(x+h) - 2f(x) + f(x-h)] + h[f(x+h) - f(x-h)] + 2h^2 f(x) \simeq 2h^2 \tan x$$

$$(2+h)f(x+h) - 2(2-h^2)f(x) + (2-h)f(x-h) \simeq 2h^2 \tan x$$

If you managed that all right, then step out of the workshop and on to the next section. If there was a problem, then try this for size.

▷**Exercise** Derive the approximate formula for $f'(x)$ of order h

$$f'(x) \simeq \frac{f(x) - f(x-h)}{h}$$

and use it to obtain an approximate formula for

$$x^2 \frac{dy}{dx} + xy = \sin x$$

4 We use Taylor's expansion with the remainder after two terms:

$$f(x+h) = f(x) + hf'(x) + \frac{h^2}{2} f''(c)$$

where $c \in (x, x+h)$. Replacing h by $-h$ gives

$$f(x-h) = f(x) - hf'(x) + \frac{h^2}{2} f''(c)$$

where $c \in (x-h, x)$. Rearranging we obtain

$$f'(x) = \frac{f(x) - f(x-h)}{h} + \frac{h}{2} f''(c)$$

So

$$f'(x) \simeq \frac{f(x) - f(x-h)}{h}$$

where the approximation is of order h.

If you couldn't manage that, see now if you can do the last part and substitute into the differential equation.

5

We have

$$x^2 \frac{dy}{dx} + xy = \sin x$$

which becomes, on putting $y = f(x)$,

$$x^2 f'(x) + xf(x) = \sin x$$

Now

$$f'(x) \simeq \frac{f(x) - f(x - h)}{h}$$

and therefore

$$x^2 \frac{f(x) - f(x - h)}{h} + xf(x) \simeq \sin x$$

So

$$x^2[f(x) - f(x - h)] + hxf(x) \simeq h \sin x$$

that is,

$$x(x + h)f(x) - x^2 f(x - h) \simeq h \sin x$$

It is worth remarking that the effect of multiplying through by h, as we have, is to change the order of the error; so now the error is of order h^2.

18.11 NUMERICAL INTEGRATION

Although we can differentiate almost any function that is likely to arise, the situation is very different when it comes to integration. Unfortunately not only is a certain amount of skill needed to perform integration, but also there are some functions for which no indefinite integral exists using elementary functions. For example,

$$\int e^{x^2} dx$$

cannot be determined.

Consequently the problem we solved numerically, giving the deflection of a beam, cannot be solved analytically. Of course it is very unusual to have an indefinite integral like this. In practical situations we usually need to determine a *definite* integral, and so a numerical method is appropriate.

You will remember that we were able to show that any definite integral could be given a physical interpretation in terms of an area (Chapter 17). In the next two sections we use this fact to derive some formulas which can be used to determine integrals approximately.

18.12 THE TRAPEZOIDAL RULE

Consider the curve $y = f(x) \geqslant 0$ on the interval $[a, b]$ where $a < b$ (Fig. 18.7). Suppose we wish to determine the definite integral

$$\int_a^b f(x)\,\mathrm{d}x$$

We know that this is the area of the region enclosed by the curve, the x-axis and the lines $x = a$ and $x = b$.

In order to approximate this area we divide it up into n strips each parallel to the y-axis and each of equal width h. By doing this we partition the interval $[a, b]$ into n equal subintervals each of length $h = (b - a)/n$ formed by the points $x_k = a + kh$, where k increases from its initial value 0 by unit steps until the value $k = n$ is attained. There is a useful notation to represent this: $k = 0(1)n$.

Suppose that $y_k = f(x_k)$ for $k = 0(1)n$. Then (x_k, y_k) are the points where the edges of the strips cut the curve $y = f(x)$. If we join these points up we obtain a polygonal curve, and if h is small the area under the polygon will approximate the area under the curve.

The trapezoidal rule uses this approximation, and so each strip is approximated by a trapezium. You will recall that a trapezium is a quadrilateral with just one pair of parallel sides. If we look at a typical trapezium (Fig. 18.8) we have an area

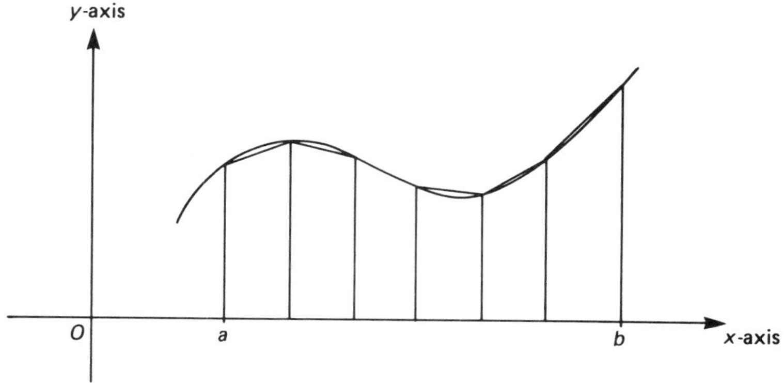

Fig. 18.7 The trapezoidal rule.

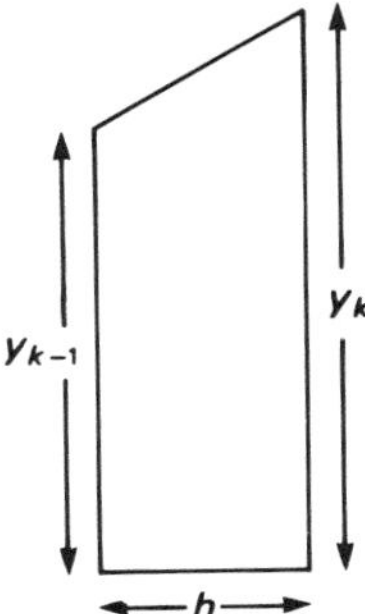

Fig. 18.8 A typical trapezium.

$$A_k = \tfrac{1}{2}h(y_{k-1} + y_k)$$

To obtain the total area A under the polygon we must add the values of A_k for $k = 1(1)n$:

$$\begin{aligned} A &= \sum_1^n A_k \\ &= \tfrac{1}{2}h[(y_0 + y_1) + (y_1 + y_2) + \ldots + (y_{n-1} + y_n)] \\ &= \tfrac{1}{2}h[y_0 + 2y_1 + 2y_2 + \ldots + 2y_{n-1} + y_n] \end{aligned}$$

Now $y_k = f(x_k) = f(a + kh)$ for $k = 0(1)n$ and $a + nh = b$, so

$$\begin{aligned} A &= \tfrac{1}{2}h[f(a) + 2f(a + h) + 2f(a + 2h) + \ldots \\ &\quad + 2f(a + \{n-1\}h) + f(b)] \\ &= \frac{h}{2}\left[f(a) + f(b) + 2\sum_{k=1}^{n-1} f(a + kh)\right] \end{aligned}$$

where $h = (b - a)/n$.

The formula written in this way looks rather complicated, but it is in fact very easy to apply. The trapezoidal method is as follows:

1. Partition the interval $[a, b]$ so that the area is divided up into equal strips of width h.
2. Calculate the corresponding y values.
3. Add the first value to the last value: call this P.
4. Add up all the other values: call this Q.
5. Calculate $P + 2Q$.
6. Multiply by h and divide by 2.

We have derived the **trapezoidal rule**:

$$\int_a^b f(x)\,dx \simeq \frac{h}{2}\left[f(a) + f(b) + 2\sum_{k=1}^{n-1} f(a + kh)\right]$$

If the maximum value of the modulus of f'' on the interval (a, b) is M then it can be shown that the error is no more than

$$\frac{h^2}{12} M(b - a)$$

For functions which have graphs of high curvature this error can be quite sizeable, and so the trapezoidal rule is quite limited. Theoretically one could argue that it is always possible to increase accuracy by taking more and more strips. Unfortunately this often results in error due to rounding off, and in problems arising from dealing with large numbers of very small numbers.

18.13 SIMPSON'S RULE

A better rule is known as Simpson's rule. This is named after the colourful seventeenth-century charlatan, rogue, plagiarist, astrologer and writer of mathematical textbooks, Thomas Simpson. How much he had to do with this rule will probably never be known, but he certainly published it.

In the trapezoidal rule we approximate the curve by a polygonal curve. Simpson's rule uses, instead of segments of straight lines $y = mx + c$, segments of parabolas $y = ax^2 + bx + c$. In order to do this it is necessary to divide up the area under the curve into an even number of strips. This means that there will be an odd number of points in the partition of the interval $[a, b]$.

To simplify matters we shall consider the area under two strips which we place symmetrically about the origin (Fig. 18.9). The x values are there-

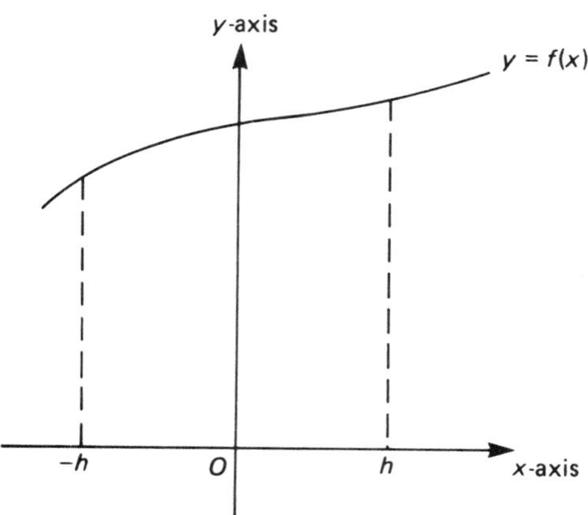

Fig. 18.9 A pair of strips.

fore $x = -h$, $x = 0$ and $x = h$, and the corresponding y values are $y = f(-h)$, $y = f(0)$ and $y = f(h)$ respectively. We are going to approximate the curve $y = f(x)$ by a parabola $y = ax^2 + bx + c$ passing through these three points.

The corresponding area is

$$\begin{aligned}\int_{-h}^{h} (ax^2 + bx + c)\,\mathrm{d}x &= [\tfrac{1}{3}ax^3 + \tfrac{1}{2}bx^2 + cx]_{-h}^{h}\\ &= (\tfrac{1}{3}ah^3 + \tfrac{1}{2}bh^2 + ch)\\ &\quad - [\tfrac{1}{3}a(-h)^3 + \tfrac{1}{2}b(-h)^2 + c(-h)]\\ &= \tfrac{2}{3}ah^3 + 2ch\end{aligned}$$

We must determine a and c.

The three points are to be on the parabola, and so they must satisfy its equation. Consequently

$$\begin{aligned}f(-h) &= a(-h)^2 + b(-h) + c\\ f(0) &= a0 + b0 + c\\ f(h) &= ah^2 + bh + c\end{aligned}$$

From the second $c = f(0)$, and then from the first and the third $f(-h) + f(h) = 2ah^2 + 2f(0)$. Using these, we have that the area under the parabola is

$$\begin{aligned}\tfrac{2}{3}ah^3 + 2ch &= \tfrac{1}{3}h[f(-h) + f(h) - 2f(0)] + 2hf(0)\\ &= \tfrac{1}{3}h[f(-h) + 4f(0) + f(h)]\end{aligned}$$

We can now return to the general situation. You will remember that we divided the area under the curve into an even number of strips. To emphasize this we shall let $2n$ be the number of strips, so that $2nh = (b - a)$ (Fig. 18.10). The x values corresponding to this partition are therefore $x_k = a + kh$, where $k = 0(1)2n$. To apply what we have just discovered we shall need to take the strips in pairs. A typical pair of strips is defined by the partition $\{x_{k-1}, x_k, x_{k+1}\}$.

We have seen that the corresponding area under a parabola is

$$\tfrac{1}{3}h[f(x_{k-1}) + 4f(x_k) + f(x_{k+1})]$$

If we approximate the area under the whole curve by parabolic segments in this way we must total them all to obtain the corresponding area A:

$$\begin{aligned}A &= \tfrac{1}{3}h\{[f(x_0) + 4f(x_1) + f(x_2)] + [f(x_2) + 4f(x_3) + f(x_4)] + \ldots\\ &\quad + [f(x_{2n-2}) + 4f(x_{2n-1}) + f(x_{2n})]\}\\ &= \tfrac{1}{3}h\{f(x_0) + f(x_{2n}) + 2[f(x_2) + f(x_4) + \ldots + f(x_{2n-2})]\\ &\quad + 4[f(x_1) + f(x_3) + \ldots + f(x_{2n-1})]\}\\ &= \frac{h}{3}\left[f(a) + f(b) + 2\sum_{r=1}^{n-1} f(a + 2rh) + 4\sum_{r=1}^{n} f(a + \{2r - 1\}h)\right]\end{aligned}$$

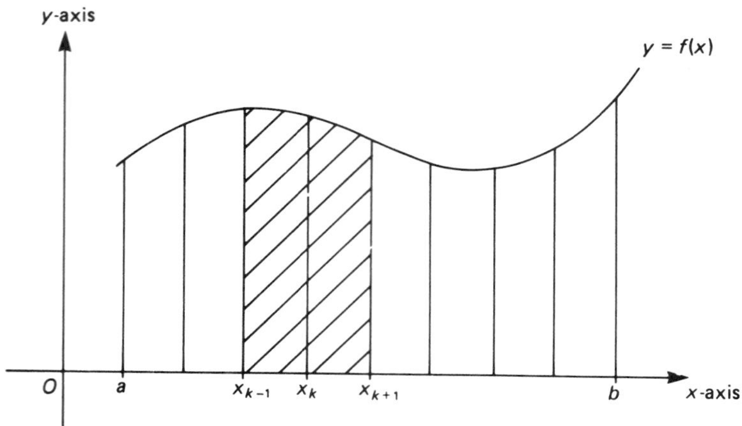

Fig. 18.10 Simpson's rule.

where $h = (b - a)/2n$.

Again we have quite a complicated formula, but it looks worse than it is. Here is Simpson's method:

1 Partition the interval $[a, b]$ into an even number of subintervals of equal length using points $x_0, x_1, \ldots, x_{2n}$.
2 Calculate the corresponding values of y.
3 Add the first value to the last value: call this P.
4 Add up all the values at the other even points: call this Q.
5 Add up all the values at the odd points: call this R.
6 Calculate $P + 2Q + 4R$.
7 Multiply by h and divide by 3.

We have derived **Simpson's rule**:

$$\int_a^b f(x)\,\mathrm{d}x \simeq \frac{h}{3}\left[f(a) + f(b) + 2\sum_{r=1}^{n-1} f(a + 2rh) + 4\sum_{r=1}^{n} f(a + \{2r - 1\}h)\right]$$

If the maximum value of the modulus of $f^{(4)}$ on the interval (a, b) is M, then it can be shown that the error is no more than

$$\frac{h^4}{180} M(b - a)$$

Of course if the curve happens to be a parabola then Simpson's rule will give an exact result. It may surprise you to learn that there are other integrals for which an exact result is obtained using Simpson's rule. For example, the integral of a polynomial of degree three is exactly determined using Simpson's rule.

Now it's time for some more steps.

18.14 Workshop

▷**Exercise** Use Simpson's rule to obtain 1

$$\int_0^{\pi/2} \sqrt{\sin x}\, dx$$

using six strips and performing arithmetic to five decimal places.

You will need to tabulate your work. Make sure you get the coefficients right. A table with the form shown below is probably the best way of proceeding. The totals can then be found and the appropriate factors used:

$f(x_0)$		
	$f(x_1)$	$f(x_2)$
	$f(x_3)$	$f(x_4)$
	$f(x_5)$	$f(x_6)$
	$\vdots$	$\vdots$
		$f(x_{2n-2})$
$f(x_{2n})$	$f(x_{2n-1})$	

We have six strips, so $2nh = 6h = (b - a) = \pi/2$. Therefore $h = \pi/12$. 2
There are seven points in the partition, and these are $x_k = k\pi/12$ where $k = 0(1)6$. The corresponding values of y_k are now calculated and arranged in the array:

0.000 00		
	0.508 74	0.707 11
	0.840 90	0.930 60
	0.982 82	
1.000 00		
1.000 00	2.332 46	1.637 71

The totals are given at the foot of each column.

Using 'four times the odd plus twice the even' we obtain

$$\begin{array}{r} 1.000\,00 \\ 9.329\,84 \\ 3.275\,42 \\ \hline 13.605\,26 \end{array}$$

from which

$$I \simeq \tfrac{1}{3}(\pi/12)\ (13.605\,26) = 1.187\,28$$

So the estimated value is 1.187.

This is an example where the estimate of error is very difficult to apply. To differentiate $\sqrt{\sin x}$ four times is bad enough, but that is not the least of our worries. It is not just that it is difficult to handle the trigonometry to examine the upper bound (which it is); more alarming, as $x \to 0+$ the fourth derivative increases without bound!

A good rule of thumb when dealing with numerical processes of this kind is to double the number of strips and recalculate the approximation. This should give an indication of how accurate the answer is.

If you made a mistake in the exercise, try the next one. If all was well, except possibly for a minor numerical slip, move on to the text following.

▷ **Exercise** Use Simpson's rule to obtain

$$\int_0^{\pi/2} \sqrt{\sin x}\ dx$$

using twelve strips and performing arithmetic to five decimal places.

This would be no problem to a computer with a suitable program, but without one you will have to press a few buttons. After all, you have to pay a penalty for getting the last problem wrong, and you may as well make sure your calculator earns its keep. Don't duck out of this.

---

3 We have twelve strips, so $2nh = 12h = (b - a) = \pi/2$. Therefore $h = \pi/24$. There are thirteen points in the partition, and these are $x_k = k\pi/24$ where $k = 0(1)12$. The corresponding values of y_k are now calculated and arranged in the array:

0.000 00		
	0.361 28	
		0.508 74
	0.618 61	
		0.707 11
	0.780 23	
		0.840 90
	0.890 70	
		0.930 60
	0.961 19	
		0.982 82
1.000 00	0.995 71	
1.000 00	4.607 72	3.970 17

Using 'four times the odd plus twice the even' we obtain

$$\begin{array}{r} 1.000\,00 \\ 18.430\,88 \\ 7.940\,34 \\ \hline 27.371\,22 \end{array}$$

from which

$$I \simeq \tfrac{1}{3}(\pi/24)\ (27.371\,22) = 1.194\,29$$

So we estimate the value as 1.194.

Comparing this result with the one obtained previously, we see that they agree to two decimal places that the integral is 1.19. We might feel there is a possibility that still more strips could change the last decimal place, and so we will opt for 1.2 as an approximate value.

Well, that's almost all there is for this chapter. We have examined a number of numerical techniques, and some of these you may be able to employ when solving equations which cannot be handled analytically. Numerical analysis is a vast subject, and there are many different techniques which have been developed to deal with almost any problem which can arise. High-speed computers have revolutionized the approach to many of them, but you must remember that a computer does not always give the whole picture. The program has been written by a human being and things might have been overlooked. An answer given by a computer is not always right, any more than something printed in a book is always right. Forewarned is forearmed.

Now it's time to take your forearms off the desk and tackle a practical problem.

18.15 Practical

CONDUCTION

If $\theta = \theta(x)$ is the temperature at any point distance x from one end of a heat-conducting fin (Fig. 18.11), then θ satisfies the equation

$$\frac{d^2\theta}{dx^2} + \frac{1}{x}\frac{d\theta}{dx} - \frac{r^2}{x}\theta = 0$$

where

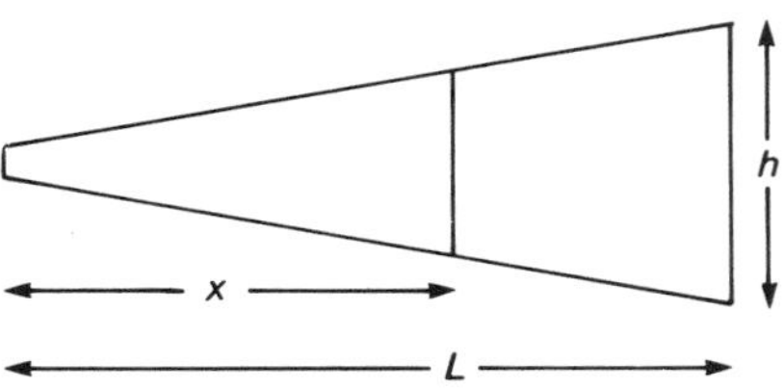

Fig. 18.11 Section of fin.

$$r^2 = \frac{2\sigma L}{kh}\left[1 + \left(\frac{h}{2L}\right)^2\right]^{1/2}$$

Here h is the width of the fin, L is the length, σ is the coefficient of heat transfer and k is the thermal conductivity. It is given that $\theta(L) = \theta_1$ and $\theta(0) = \theta_0$, and the units are standardized so that $L = 1$, $\theta_0 = 20$, $\theta_1 = 100$ and $r = 2$. Approximate the equation by means of approximations of order h^2 where $h = 0.25$, and thereby estimate θ at $x = 0.25$, 0.5 and 0.75.

It is a good idea to see if you can do this by yourself. We have already solved a very similar problem, and so you have a blueprint with which to work. As usual we shall solve it stage by stage so that you can join in with the solution as soon as possible.

The first thing to do is to use the approximations in the equation.

---

1 We have

$$x\frac{\mathrm{d}^2\theta}{\mathrm{d}x^2} + \frac{\mathrm{d}\theta}{\mathrm{d}x} - r^2\theta = 0$$

So

$$x\frac{\theta(x+h) - 2\theta(x) + \theta(x-h)}{h^2} + \frac{\theta(x+h) - \theta(x-h)}{2h} - r^2\theta(x) = 0$$

$$2x[\theta(x+h) - 2\theta(x) + \theta(x-h)] + h[\theta(x+h) - \theta(x-h)] - 2h^2r^2\theta(x) = 0$$

$$2x[\theta(x+0.25) - 2\theta(x) + \theta(x-0.25)] + 0.25[\theta(x+0.25) - \theta(x-0.25)] - 0.5\theta(x) = 0$$

$$[2x + 0.25]\theta(x+0.25) - [4x + 0.5]\theta(x) + [2x - 0.25]\theta(x - 0.25) = 0$$

Now we have the basic equation we need to estimate the required temperatures. To ease the notation we shall put $a = \theta(0.25)$, $b = \theta(0.5)$ and $c = \theta(0.75)$. Try it and see how it goes.

---

2 We use the boundary conditions and put $x = 0.25$, $x = 0.5$ and $x = 0.75$ in turn to obtain

$$0.75b - 1.5a + 0.25(20) = 0$$
$$1.25c - 2.5b + 0.75a = 0$$
$$1.75(100) - 3.5c + 1.25b = 0$$

Finally, solve these to obtain estimates of the temperatures that we require.

3 The equations become

$$3b - 6a + 20 = 0 \quad (1)$$
$$5c - 10b + 3a = 0 \quad (2)$$
$$700 - 14c + 5b = 0 \quad (3)$$

Equation (1) gives $a = (3b + 20)/6$, and equation (3) gives $c = (700 + 5b)/14$. Substituting these into (2) gives

$$3500 + 25b - 140b + 21b + 140 = 0$$

Therefore $94b = 3640$ and so $b = 38.723$. Lastly, substituting back we obtain

$$a = (3 \times 38.72 + 20)/6 = 22.695$$
$$c = (700 + 5 \times 38.72)/14 = 63.830$$

Therefore the temperatures where $x = 0.25$, $x = 0.5$ and $x = 0.75$ are 22.7, 38.7 and 63.8 approximately.

SUMMARY

Here is a list of the topics we have studied in this chapter:

☐ Solutions of the equation $f(x) = 0$ by
  a bisection method
  b *regula falsi*
  c secant method
  d Newton's method.

☐ Approximations for derivatives
  a approximations of order h

$$f'(x) \simeq \frac{f(x+h) - f(x)}{h}$$

$$f'(x) \simeq \frac{f(x) - f(x-h)}{h}$$

  b approximations of order h^2

$$f'(x) \simeq \frac{f(x+h) - f(x-h)}{2h}$$

$$f''(x) \simeq \frac{f(x+h) - 2f(x) + f(x-h)}{h^2}$$

☐ Numerical integration
  a trapezoidal rule
  b Simpson's rule.

EXERCISES

1 Solve the following equations by the bisection method giving your answers correct to four significant figures:
a $x = 2\sin^2 x$
b $\exp x = 1 + \cos x^2$

2 Using the *regula falsi* method and the equation

$$f(x) = x^3 \exp x - 1 = 0$$

obtain a formula to give a better approximation c in terms of a and b where $f(a) < 0$ and $f(b) > 0$. Obtain integer values of a and b.

3 Using the secant method and the equation

$$f(x) = x^3 + x - 5 = 0$$

obtain an iterative formula which gives x_{n+1} in terms of x_n and x_{n-1}. Calculate suitable integer starting values x_0 and x_1.

4 Use Newton's method to solve each of the following equations correct to five decimal places:
a $x^3 + x^2 = 4x - 1$
b $x - 2 = \ln x$

5 Use the trapezoidal method with six strips and working to five decimal places to estimate

a $\displaystyle\int_0^{\pi/2} \sin\sqrt{(x^2 + 1)}\,dx$

b $\displaystyle\int_0^{\pi/2} \ln(1 + \sin x)\,dx$

6 Use Simpson's rule, with six strips, working to five decimal places, to estimate

a $\displaystyle\int_0^{1} \exp(1 - x^2)\,dx$

b $\displaystyle\int_0^{\pi/2} \cos\sqrt{x}\,dx$

ASSIGNMENT

1 The bisection method is used to solve the equation $e^x = 3x$ using $[0, 1]$ as the initial interval and working to five decimal places. How many steps are needed to guarantee accuracy to three decimal places? Solve the equation using this method to achieve this accuracy.

2 Calculate the iterative formula corresponding to Newton's method for solving the equation $f(x) = e^x - 3x = 0$. Solve this equation using 0 as a starting value to obtain a solution accurate to three decimal places.

3 Determine the iterative formula corresponding to the secant method for solving $f(x) = e^x - 3x = 0$. Work through three iterations with starting values $x_0 = 0$ and $x_1 = 1$.

4 By writing $y = f(x)$, and using approximations of order h^2 for the derivatives, obtain a finite difference approximation for the differential equation

$$\frac{d^2y}{dx^2} + 2\frac{dy}{dx} + y = \sec x$$

5 Using six strips and working to seven decimal places, estimate the value of

$$\int_0^1 x e^{x^2} dx$$

by Simpson's rule. Check the accuracy of your solution by (a) doubling the number of strips (b) obtaining the exact integral.

FURTHER EXERCISES

1 Use Simpson's rule to evaluate

$$\int_0^{\pi} \frac{dt}{\cos t + 2}$$

using five ordinates. Determine the integral directly and thereby show that Simpson's rule underestimates the integral by less than 0.28%.

2 Show graphically that the curves $y = e^x - 1$ and $y = \ln(x + 2)$ intersect at two points. Use Newton's method to obtain the larger of the roots of the equation $e^x = 1 + \ln(x + 2)$. Give your answer correct to two decimal places.

3 Use a graph to show that the equation $x^2 - 4 = \ln x$ has just two real roots. Obtain, using Newton's method, the larger of the roots correct to three decimal places.

4 Use any numerical method to obtain the roots of the equation $x = 2 - e^x$ correct to three decimal places.

5 Starting with the approximation $x = 0.6$, obtain the real root of the equation $x = e^{-x}$ correct to four decimal places.

6 Show that if α is an approximate root of the equation $x \ln x - x = a$ (where a is real) and if $p = \ln \alpha - \alpha$, then a better approximation is given by β where $\beta = (a + \alpha)/(p + \alpha)$.

7 Evaluate approximately

$$\int_0^1 \frac{dx}{1 + x^2}$$

using eight strips (seven ordinates) by (a) the trapezoidal rule (b) Simpson's rule. Work to five decimal places and show that the error in the result obtained by the trapezoidal rule is less than 0.0026. Determine the integral exactly and thereby show that Simpson's rule is accurate in this instance.

8 Simpson's rule gives the exact result if the function to be integrated is a polynomial of degree three or less. Use this fact and two strips to derive the formula $V = \pi h(3a^2 + 3b^2 + h^2)/6$ for the volume of a segment of a sphere of height h and base radii a and b.

9 The response u of a system is given at time t by the equation $\mathrm{d}u/\mathrm{d}t = -u$ where the units have been standardized. Put $u = u(t)$ and approximate the derivative with an approximation of order h.

a Use the initial value $u(0) = 1$ with the step length $h = 0.2$ and five iterations to estimate $u(1)$.

b Use integration to obtain the exact solution and thereby calculate the percentage error in **a**.

(The numerical method outlined in this problem is usually known as Euler's method for solving a first-order differential equation.)

10 Write down Maclaurin's expansion (Taylor's expansion about the point 0) for the function $y = y(x)$ which has derivatives of all orders, and show that if $y' = 1 - xy$ and $y(0) = 0$ then

$$y(h) = h - h^3/3 + h^5/15 - h^7/105 + \dots$$

Estimate the value of $y(0.5)$ to four decimal places.

11 Show that the equation $2x^2 - 4\mathrm{e}^{-x} - 7 = 0$

has a root near 1.4 and use Newton's method to obtain an approximation which is correct to 3 significant figures.

12 Show that when Newton's method is applied to the equation

$$\mathrm{e}^{-x} - \sin x = 0$$

the iterative formula for x_{n+1} becomes

$$x_{n+1} = x_n + \frac{\mathrm{e}^{-x_n} - \sin x_n}{\mathrm{e}^{-x_n} + \cos x_n}$$

Hence, or otherwise, obtain the smallest positive root correct to 3 decimal places.

13 A semi-circular plot of land is to be divided into two equal areas by means of a fence which extends on a chord from one end of the diameter to the perimeter. The angle between the chord and the diameter is θ (radians). Show that

$$2\theta + \sin 2\theta = \frac{\pi}{2}$$

Solve this equation, by means of any suitable numerical method, to determine θ to three decimal places. To enable the construction to be completed, convert this angle into degrees.

19 First-order differential equations

Although we have studied integration as the reverse process of differentiation, integrals do not always present themselves explicitly in applications. Instead an equation involving derivatives is obtained and some method of eliminating these derivatives is needed. This is the subject of our next section of work, differential equations.

After studying this chapter you should be able

- □ Identify a differential equation and be able to state its order and degree;
- □ Recognize the standard form of three basic types of first-order differential equation;
- □ Solve variables separable, linear and homogeneous equations;
- □ Apply simple substitutions to convert equations into one of these three types.

At the end of this chapter we shall solve practical problems in circuits and vehicle braking.

19.1 TERMINOLOGY

You have already solved some first-order differential equations, although you may not be aware of the fact! The process of integration is equivalent to the solution of a very simple first-order differential equation. If you don't forget to put in the arbitrary constant you will obtain the general solution.

What then is a differential equation, and what do we mean by saying that we have solved it? An ordinary **differential equation** is an equation involving two variables (say x and y) and ordinary derivatives. The highest-order derivative which is present determines the **order** of the differential equation. So for instance

$$\left(\frac{d^2y}{dx^2}\right)^3 + \left(\frac{dy}{dx}\right)^4 + y = 0$$

is a differential equation of the second order. Notice here that the second-order derivative occurs raised to the power of 3, and so this is a differential equation of the third **degree**. The first-order derivative occurs raised to the power of 4 but, since this is not the derivative which determines the order of the differential equation, this does not affect the degree of the equation. To avoid complications, and because differential equations of high degree do not arise often in applications, we shall confine our attention to equations of the first degree.

□ Identify the degrees and orders of the following differential equations:

a $\left(\frac{d^3y}{dx^3}\right)^2 - x\left(\frac{d^2y}{dx^2}\right)^4 + y = 0$

b $y\left(\frac{d^2y}{dx^2} + x\right) + \left(\frac{dy}{dx}\right)^3 + x = 0$

c $\left(\frac{d^2y}{dx^2}\right)^3 + \left(\frac{dy}{dx} + y\right)^2 = 0$

when you have made your decision, read on and check you are right.

Here are the results:

a This is a third-order equation of degree 2.
b This is a second-order equation of degree 1.
c This is a second-order equation of degree 3. ■

Now that we have described what a differential equation is, we need to say what we mean by saying we have solved the equation. Usually we are given not only the differential equation itself but also some initial or boundary conditions which have to be satisfied too. For the moment, though, let us confine our attention to the differential equation itself.

A solution of the differential equation is an equation between the two variables concerned, x and y, which is

1 Free of derivatives;
2 Consistent with the differential equation.

Differential equations have many solutions, but we shall be interested in the **general solution**. Consider for example the equation

$$\frac{dy}{dx} = 1$$

By inspection we see that $y = x$ is a solution of this equation, and that so too is $y = x - 5$ or indeed $y = x + C$ where C is any constant. Moreover, any solution of this differential equation can be expressed in the form $y = x + C$ where C is some arbitrary constant.

This should come as no surprise to us, of course. We began with a first-order differential equation and have obtained a solution which is necessarily free of the derivative. Therefore a process equivalent to a single integration must have occurred, and this is bound to result in the presence of an arbitrary constant.

Continuing this line of thought, we shall expect the general solution of a differential equation of order n and degree 1 to contain n independent arbitrary constants.

So we already know how to solve one very important type of first-order differential equation:

$$\frac{dy}{dx} = f(x)$$

To solve this equation we merely integrate both sides with respect to x; remembering not to omit the arbitrary constant.

In fact we can generalize this very slightly to deal with a whole class of differential equations. These are equations which can be expressed in the form

$$\frac{dy}{dx} = \frac{f(x)}{g(y)}$$

where f and g are real functions.

19.2 VARIABLES SEPARABLE EQUATIONS

Any differential equation which can be expressed in the form

$$\frac{dy}{dx} = \frac{f(x)}{g(y)}$$

where f and g are real functions, is known as a **variables separable** differential equation. It can be solved easily by writing it in the form

$$g(y)\,\frac{dy}{dx} = f(x)$$

and integrating both sides with respect to x. We obtain then

$$\int g(y)\,dy = \int f(x)\,dx$$

Naturally we do not obtain two independent arbitrary constants because the equation

$$G(y) + A = F(x) + B$$

where A and B are arbitrary constants can be replaced by

$$G(y) = F(x) + C$$

where $C = B - A$.

The problem of solving variables separable differential equations therefore is twofold:

1 Identifying which equations are variables separable;
2 Performing the necessary integrations.

□ Solve, for $y > 0$, the equation

$$\frac{\mathrm{d}y}{\mathrm{d}x} - xy = y$$

We may rewrite this equation in the form

$$\frac{\mathrm{d}y}{\mathrm{d}x} = y + xy = y(1 + x)$$

So that, since $y \neq 0$,

$$\frac{1}{y}\frac{\mathrm{d}y}{\mathrm{d}x} = 1 + x$$

Consequently

$$\int \frac{1}{y}\,\mathrm{d}y = \int (1 + x)\,\mathrm{d}x$$

And so

$$\ln y = x + \tfrac{1}{2}x^2 + C$$

Note that, strictly speaking, without the information that $y > 0$ we should have to represent the solution as

$$\ln |y| = x + \tfrac{1}{2}x^2 + C \qquad (y \neq 0)$$

and also include the solution $y = 0$. In fact we shall limit discussions of this kind in this chapter because to do so could obscure the methods. ■

□ Solve the differential equation

$$\frac{\mathrm{d}y}{\mathrm{d}x} = \mathrm{e}^{x-y}$$

At first sight this may not look like a variables separable equation, but we need to remember from the laws of indices that

$$e^{x-y} = e^x/e^y$$

Now it is clear that we can put all the terms in x on the right and all the terms in y on the left and proceed to integrate throughout with respect to x. This produces

$$\int e^y \, dy = \int e^x \, dx$$

From which

$$e^y = e^x + C$$

where C as usual is the arbitrary constant. If we wish, we can make y the subject of the equation. Then

$$y = \ln (e^x + C)$$

By the way, don't make the mistake of delaying to put in the arbitrary constant. It must go in at the moment of integration; it cannot be added on as an afterthought. For instance, here we would obtain (ignoring the arbitrary constant)

$$e^y = e^x$$

From which

$$y = x$$

Whoops, forgot the arbitrary constant. So

$$y = x + C$$

No! This will not do. Watch out for this and similar errors. ■

Now let's take some steps to make sure we know how to separate the variables.

19.3 Workshop

▷ **Exercise** Obtain the general solution of the following differential equation: 1

$$\frac{dy}{dx} = \sin (x + y) + \sin (x - y)$$

If after you have given the matter some thought you can't see how the variables separate, then take another step for a clue. Otherwise try to solve the equation and move ahead to step 3.

2 The expansion formulas for sin $(x + y)$ and sin $(x - y)$ when added together reduce the left to the product of a sine and a cosine. Alternatively you can combine them using the addition formulas. The result is the same either way:

$$\frac{dy}{dx} = 2 \sin x \cos y$$

Now solve the equation, and when you have finished take the next step.

3 We have

$$\frac{dy}{dx} = 2 \sin x \cos y$$

from which, since $y \neq 0$,

$$\int \sec y \, dy = \int 2 \sin x \, dx$$

so that

$$\ln |\sec y + \tan y| = -2 \cos x + C$$

Here you would not be expected to rearrange the equation in order to make y the subject, as this is not particularly easy to do. Nevertheless, as a slight diversion and to help brush up our algebraic and trigonometrical skill, we shall have a go. Why not have a go at it yourself first? If you don't feel you could manage it, at least follow through the working carefully. You should be able to pick up a few useful tips.

4 We must take great care when dealing with exponentials and logarithms. It may be very tempting to expand out a logarithm, but if we replace the left-hand side by

$$\ln |\sec y| + \ln |\tan y|$$

we shall have made a bad mistake. Instead we remove the logarithm by using the definition to obtain

$$|\sec y + \tan y| = e^{-2 \cos x + C}$$

From which

$$|\sec y + \tan y| = A \, e^{-2 \cos x}$$

where A is another arbitrary constant (in fact positive since $C = \ln A$).

Let us write

$$k = A\ e^{-2\cos x}$$

so that we now have

$$\sec y + \tan y = \pm k \qquad (1)$$

Multiplying through by cos y gives

$$1 + \sin y = \pm k \cos y$$

Squaring produces

$$\begin{aligned}(1 + \sin y)^2 = k^2 \cos^2 y &= k^2 (1 - \sin^2 y)\\ \sin^2 y + 2 \sin y + 1 &= k^2 - k^2 \sin^2 y\\ (1 + k^2) \sin^2 y + 2 \sin y + (1 - k^2) &= 0\end{aligned}$$

This factorizes to give

$$(\sin y + 1)[(1 + k^2) \sin y + (1 - k^2)] = 0$$

Now $\sin y \neq -1$, as can be seen by considering the original equation (1), and consequently

$$\sin y = -\frac{1 - k^2}{1 + k^2}$$

Now

$$k^2 = A^2\ e^{-4\cos x}$$

Therefore finally we obtain

$$\sin y = \frac{A^2\ e^{-4\cos x} - 1}{A^2\ e^{-4\cos x} + 1}$$

Here now is another equation which, although not variables separable as it stands, can be transformed into a variables separable equation easily by means of a substitution.

▷ **Exercise** Solve

$$\frac{dy}{dx} = \sin (x + y)$$

Have a look at this and see if you can choose the substitution. Then step ahead.

Suppose we put $z = x + y$. Then we obtain, differentiating with respect to x, 5

$$\frac{dz}{dx} = 1 + \frac{dy}{dx}$$

So the equation becomes, on substituting,

$$\frac{dz}{dx} = 1 + \sin z$$

When we separate the variables we are left with two integrals to find – one easy, the other not so easy:

$$\int \frac{dz}{1 + \sin z} = \int dx$$

We shall concentrate on the integral on the left for the moment:

$$\int \frac{dz}{1 + \sin z} = \int \frac{1 - \sin z}{\cos^2 z}\, dz$$

$$= \int (\sec^2 z - \sec z \tan z)\, dz$$

$$= \tan z - \sec z + C$$

The solution is therefore

$$x = \tan z - \sec z + C$$

So that substituting back we obtain

$$x = \tan (x + y) - \sec (x + y) + C$$

19.4 LINEAR EQUATIONS

We are now ready to look at the next type of first-order differential equation. Any equation which can be written in the form

$$\frac{dy}{dx} + Py = Q$$

where P and Q depend on x only, is known as a **linear** first-order differential equation.

Here is an example:

$$x \frac{dy}{dx} = x^3 y + 1 + x^4$$

In order to arrange this in standard form we must divide by x and take the term in y to the left. When this is done we have

$$\frac{dy}{dx} - x^2 y = \frac{1}{x} + x^3$$

From which

$$P = -x^2 \qquad Q = \frac{1}{x} + x^3$$

Sometimes an equation is both variables separable and linear! See if you can construct an example which is both. It's not too difficult if you bear in mind the special features which each one has to have. Here is one example:

$$\frac{dy}{dx} = xy + x$$

Now let us see how to go about solving a linear differential equation. We shall first describe the process by which we obtain the solution and pick out the crucial steps later to obtain a direct method. Don't be too concerned if things seem a little complicated at first; we won't have to go through all this work whenever we want to solve an equation! To have a good understanding, however, it is best to have a peep behind the scenes.

The linear equation is an example of a general class of differential equations which can be solved by means of a device known as an integrating factor. An **integrating factor** is an expression which when multiplied through the equation makes it easy to integrate. Suppose in the case which we are considering ($(dy/dx) + Py = Q$) it is possible to multiply throughout by some expression I and thereby express the equation in the form

$$\frac{d}{dx}(Iy) = IQ$$

Of course you may object to this on the grounds that I may not exist. However, suspend disbelief for a little longer to discover the properties which I would have to possess.

Using the product rule we obtain

$$I\frac{dy}{dx} + y\frac{dI}{dx} = IQ$$

So that comparing with

$$I\frac{dy}{dx} + IPy = IQ$$

we deduce that

$$y\frac{dI}{dx} = IPy$$

$$\frac{dI}{dx} = IP$$

So there it is: we are looking for I satisfying the equation

$$\frac{dI}{dx} = IP$$

Any I which satisfies this will do, so that here (unusually) we can forget about an arbitrary constant!

Now this is a variables separable differential equation for I, since P depends solely on x. So separating the variables we obtain

$$\int \frac{dI}{I} = \int P \, dx$$

from which

$$\ln I = \int P \, dx$$

The expression on the right can be obtained easily by direct integration, and so we have

$$I = e^{\int P \, dx}$$

Remarkably, then, it is possible to solve the linear type of differential equation by means of an integrating factor, and the equation then becomes

$$\frac{d}{dx}(Iy) = IQ$$

From this it follows immediately that

$$Iy = \int IQ \, dx$$

Before we do an example we shall summarize this method for solving a linear differential equation:

1 Express the equation in the form

$$\frac{dy}{dx} + Py = Q \qquad (1)$$

where P and Q depend solely on x.

2 Identify P and Q and calculate the integrating factor

$$I = e^{\int P \, dx}$$

3 Multiply equation (1) by I to reduce it to the form

$$\frac{\mathrm{d}}{\mathrm{d}x}(Iy) = IQ$$

4 Integrate throughout to obtain the solution

$$Iy = \int IQ\,\mathrm{d}x$$

There are just three other points to remember. First, we can ignore any arbitrary constant when calculating I. Secondly, we do not need to worry about how the integrating factor works. When we multiply through by I the left-hand side of the equation automatically becomes

$$\frac{\mathrm{d}}{\mathrm{d}x}(\text{integrating factor} \times y)$$

Thirdly, in the final integration we must include an arbitrary constant as usual.

Well now! It's time to work through an example.

☐ Solve the differential equation

$$\cos x\,\frac{\mathrm{d}y}{\mathrm{d}x} + y \sin x = \sin 2x$$

We arrange it in standard form by dividing throughout by $\cos x$. It helps if we remember the trigonometrical identity $\sin 2x = 2 \sin x \cos x$. So we now have

$$\frac{\mathrm{d}y}{\mathrm{d}x} + y \tan x = 2 \sin x$$

Of course we have now also assumed that $\cos x$ is non-zero, and we should ensure that we represent that fact in the solution.

The equation is now in standard form, and so we can identify P and Q:

$$P = \tan x \qquad Q = 2 \sin x$$

Now we calculate the integrating factor I. We have

$$\int P\,\mathrm{d}x = \int \tan x\,\mathrm{d}x = -\ln(\cos x) = \ln(\sec x)$$

So

$$I = \mathrm{e}^{\ln(\sec x)} = \sec x$$

Strictly speaking we have assumed $\cos x > 0$, but this is also an integrating factor if $\cos x < 0$.

Multiplying through the equation by I produces

$$\frac{d}{dx}[(\sec x)y] = 2 \sin x \sec x = 2 \tan x$$

Integrating throughout with respect to x gives

$$(\sec x)y = \int 2 \tan x \, dx = -2 \ln (\cos x) + C$$

$$= \ln (\sec^2 x) + C = \ln (A \sec^2 x)$$

where C and A are arbitrary constants. Finally,

$$y = \cos x \ln (A \sec^2 x)$$

where A is a positive arbitrary constant. Note that the solution is not defined when $\cos x = 0$, and this is consistent with the remark we made earlier. ■

Now it is time for you to try to solve one of these equations. Remember the four stages in the solution and you should have no difficulty.

19.5 Workshop

1 **Exercise** Solve the differential equation

$$x \frac{dy}{dx} = y + x^3$$

First put the equation into standard form and identify P and Q. As soon as you have done this, check ahead that you have got things right.

---

2 You must divide through by x to put the equation into the standard form. Observe that if $x = 0$ then $y = 0$; therefore our solution must include this possibility. If you did the rearrangement correctly you should obtain

$$\frac{dy}{dx} - \frac{y}{x} = x^2$$

from which $P = -x^{-1}$ and $Q = x^2$.

Did you include the minus sign? It is a common error to overlook it! If you made a slip then notice where things went wrong to avoid the mistake next time.

Now calculate the integrating factor. Only when you have done this should you read on!

---

3 You should have written down

$$\int P \, dx = \int -x^{-1} \, dx = -\ln x$$

If the integral caused problems, perhaps you differentiated instead of integrated. Remember: differentiation is not integration, and integration is not differentiation!

Next the integrating factor I is obtained from

$$I = \exp(-\ln x) = \exp(\ln x^{-1}) = x^{-1}$$

Maybe a slip or two has occurred here. If you wrote down $-x$ then you have misused the laws of logarithms. You should not include the arbitrary constant as it is superfluous at this stage.

Once you have corrected any errors, use the integrating factor to solve the equation. As soon as you have done this, take the next step to see the result.

4

Multiplying the equation by I produces straight away

$$\frac{d}{dx}\left(\frac{y}{x}\right) = \frac{x^2}{x} = x$$

There is no need to go back and wrestle with the equation. The chances are that if you have done so you will have made some sort of mistake.

Lastly, integrating with respect to x gives the solution

$$\frac{y}{x} = \frac{x^2}{2} + C$$

so that on multiplying up by x we obtain

$$y = \frac{x^3}{2} + Cx$$

Note that in this form we include the solution at $x = 0$.

That was quite straightforward, wasn't it? Notice the four stages to the solution, and how we followed each one through to the end.

Of course not all problems are quite as easy as this one. Often a rather unpleasant integral is produced and has to be sorted out. In the worst situations it is not possible to complete the second stage. In such circumstances we must either find another method or resort to some numerical technique to obtain the solution. However, the numerical option is only available if we have a boundary condition and therefore do not require the general solution.

Here is an exercise which is not quite as straightforward as the previous one but nevertheless can be solved by proceeding through all the stages.

▷ **Exercise** Solve the equation

$$\exp x \frac{dy}{dx} + y \exp 2x = \exp 3x$$

If you managed the previous one without mistakes you might like to have a go at this completely on your own. Make a real effort and see if you can manage it. When you have finished, look ahead at the answer and see if you are right. Even if your answer looks wrong it may nevertheless be correct because it is often possible to write solutions to differential equations in different forms. Therefore, if you are in doubt, differentiate your answer and see if it satisfies the differential equation; remember the chain rule!

If you made mistakes with the previous example then a few more steps are necessary. First, put the equation into standard form so that it is recognizable as a linear type and identify P and Q. Then step ahead.

---

5 To put it into standard form so that it is recognizable as a linear type it is necessary to divide throughout by $\exp x$. This gives

$$\frac{dy}{dx} + y \exp x = \exp 2x$$

from which it follows that

$$P = \exp x \qquad Q = \exp 2x$$

If you didn't manage that then you have forgotten the laws of indices and you would be well advised to brush up on them (Chapter 1).

Now obtain the integrating factor. When you have done so, take the next step.

---

6 Now $\int \exp x \, dx = \exp x$, so that

$$I = \exp e^x$$

You didn't make an incorrect simplification, did you? The answer is certainly not x.

Lastly, complete the solution and take the final steps.

---

7 Multiplying through the equation by I gives

$$\frac{d}{dx}[(\exp e^x)y] = \exp e^x \exp 2x$$

so that

$$(\exp e^x)y = \int \exp e^x \exp 2x \, dx$$

If you've made it to here, all that remains is to determine the rather nasty looking integral on the right. A substitution is called for; can you see what it is?

It's best to put $u = \exp x$. Then $\mathrm{d}u/\mathrm{d}x = \exp x = u$, so the integral reduces to 8

$$\int u \exp u \, \mathrm{d}u$$

and this can be obtained by integration by parts:

$$\int u \exp u \, \mathrm{d}u = u \exp u - \int \exp u \, \mathrm{d}u = u \exp u - \exp u + C$$

So we have

$$(\exp \mathrm{e}^x)y = \exp x \exp \mathrm{e}^x - \exp \mathrm{e}^x + C$$

and therefore

$$y = \exp x - 1 + C \exp -\mathrm{e}^x$$

where C is an arbitrary constant.

19.6 BERNOULLI'S EQUATION

You will remember that when we were considering variables separable equations there were some equations which, although not variables separable as they stood, could be made so by means of a substitution. In a similar way it is sometimes possible to reduce an equation to a linear type by making a substitution. Although there are many different situations where this is true, one type of equation – known as Bernoulli's equation – is worthy of special note.

Bernoulli's equation is any equation which can be expressed in the form

$$\frac{\mathrm{d}y}{\mathrm{d}x} + Py = Qy^n$$

where P and Q depend solely on x, and n is constant ($n \neq 1$).

You will notice how similar this equation is to the standard form of the linear equation. In fact the only difference is the term in y on the right. To reduce the equation to a linear equation we make the substitution

$$z = \frac{1}{y^{n-1}}$$

It then follows that

$$\frac{dz}{dx} = \frac{-(n-1)}{y^n}\frac{dy}{dx}$$

Now dividing through Bernoulli's equation by y^n gives

$$\frac{1}{y_n}\frac{dy}{dx} + \frac{P}{y^{n-1}} = Q$$

so that when the substitutions are made,

$$\frac{1}{-(n-1)}\frac{dz}{dx} + Pz = Q$$

This is now a linear type where the variables are z and x, so it can be solved by the standard technique. Finally, the solution of the original equation can be obtained by substituting back for z.

Let us do an example.

□ Obtain the general solution to the equation

$$\frac{dy}{dx} + \frac{y}{x} = xy^2$$

This is Bernoulli's equation in the case $n = 2$, so we make the substitution $z = y^{-1}$. Then

$$\frac{dz}{dx} = \frac{-1}{y^2}\frac{dy}{dx}$$

so that dividing the equation through by y^2 yields

$$\frac{1}{y^2}\frac{dy}{dx} + \frac{1}{xy} = x$$

from which

$$-\frac{dz}{dx} + \frac{z}{x} = x$$

$$\frac{dz}{dx} - \frac{z}{x} = -x$$

This is a linear type in standard form and so we solve it in the usual way. Why not have a go on your own?

$P = -x^{-1}$ and $Q = -x$, so that

$$\int P\,dx = \int -x^{-1}\,dx = -\ln x$$

Consequently

$$I = e^{-\ln x} = x^{-1}$$

Therefore

$$\frac{d}{dx}(x^{-1}z) = -xx^{-1} = -1$$

From which

$$\frac{z}{x} = -x + C$$

so that $z = -x^2 + Cx$. Now $z = y^{-1}$; therefore substituting gives

$$y = (-x^2 + Cx)^{-1}$$

where C is an arbitrary constant. ■

19.7 HOMOGENEOUS EQUATIONS

Finally we shall consider a type of differential equation known as homogeneous. Before describing how to recognize an equation of this kind we need to say what is meant by a homogeneous function.

Consider the two expressions

$$\begin{aligned} f(x, y) &= x^2 + 2xy - y^2 \\ g(x, y) &= x^3 + 3xy - y^3 \end{aligned}$$

Suppose we replace x and y by tx and ty respectively. Then we obtain

$$\begin{aligned} f(tx, ty) &= (tx)^2 + 2(tx)(ty) - (ty)^2 \\ &= t^2(x^2 + 2xy - y^2) = t^2 f(x, y) \end{aligned}$$

We say that f is homogeneous of degree 2.

On the other hand,

$$g(tx, ty) = (tx)^3 + 3(tx)(ty) - (ty)^3$$

and it is not possible to extract all the 't's as factors.

In general, if f is a function of two variables x and y, we say that f is a **homogeneous function** of degree n if and only if

$$f(tx, ty) = t^n f(x, y)$$

Let us see how this applies to differential equations.

A first-order differential equation is said to be a **homogeneous equation** if and only if it can be expressed in the form

$$\frac{\mathrm{d}y}{\mathrm{d}x} = \frac{f(x, y)}{g(x, y)}$$

where f and g are both homogeneous functions of the same degree.

If the expression on the right is a quotient of two homogeneous functions of the same degree, then if we substitute $y = vx$ into it, all the 'x's cancel out and it reduces to terms in v. It follows that the substitution $y = vx$ reduces any homogeneous equation to a variables separable type.

Let us illustrate the method by means of an example.

□ Solve the equation

$$\frac{\mathrm{d}y}{\mathrm{d}x} = \frac{xy - y^2}{x^2 + xy}$$

We apply our simple test. We put $y = vx$ into the right of the equation and see what results. This gives

$$\frac{xvx - (vx)^2}{x^2 + xvx} = \frac{v - v^2}{1 + v}$$

which is an expression depending solely on v.

Now we use the same substitution $y = vx$ to reduce the equation to a variables separable type. Let us see how this works. If $y = vx$ then, differentiating with respect to x,

$$\frac{\mathrm{d}y}{\mathrm{d}x} = v + x\frac{\mathrm{d}v}{\mathrm{d}x}$$

so the equation becomes

$$v + x\frac{\mathrm{d}v}{\mathrm{d}x} = \frac{v - v^2}{1 + v}$$

Therefore

$$\begin{aligned} x\frac{\mathrm{d}v}{\mathrm{d}x} &= \frac{v - v^2}{1 + v} - v \\ &= \frac{v - v^2 - v(1 + v)}{1 + v} \\ &= \frac{-2v^2}{1 + v} \end{aligned}$$

This is now variables separable, and consequently

$$\int \frac{1 + v}{-2v^2}\,\mathrm{d}v = \int \frac{1}{x}\,\mathrm{d}x$$

Thus

$$\frac{1}{2v} - \frac{1}{2}\ln|v| = \ln|x| + C$$

where C is an arbitrary constant.

Finally, we substitute back to remove v from the equation since v was a term which we introduced ourselves:

$$\frac{x}{2y} - \frac{1}{2}\ln\left|\frac{y}{x}\right| = \ln|x| + C$$

which is

$$\begin{aligned}\frac{x}{2y} &= \frac{1}{2}\ln\left|\frac{y}{x}\right| + \ln|x| + C\\ &= \frac{1}{2}\ln|xy| + C\end{aligned}$$

■

Ready to take a few steps?

19.8 Workshop

▷ **Exercise** Consider the differential equation 1

$$\frac{dy}{dx} = \frac{y^2 - xy}{xy + x^2}$$

Make a substitution to reduce the right-hand side to terms in v. When you have done so, take the next step.

Putting $y = vx$ into the right-hand side of the equation gives 2

$$\frac{(vx)^2 - x(vx)}{x(vx) + x^2}$$

Cancelling x^2 top and bottom results in

$$\frac{v^2 - v}{v + 1}$$

which depends solely on v.

All right so far? If not then make sure you know where you went wrong. Now complete the reduction to variables separable type and take another step.

Substituting into the equation produces 3

$$x\frac{dv}{dx} + v = \frac{v^2 - v}{v + 1}$$

from which

$$x\frac{dv}{dx} = \frac{-2v}{v + 1}$$

That should have caused no problems. Now solve the equation in terms of v and x and take another step.

---

4 Separating the variables and integrating gives

$$\int \frac{1}{x}\,dx = \int \frac{v + 1}{-2v}\,dv$$

So

$$\ln |x| = -\frac{v}{2} - \frac{1}{2}\ln |v| + \ln k$$

where k is an arbitrary constant.

Check carefully if you have a different answer. Of course the constant may be expressed differently. Lastly, substitute back to eliminate v and take the final step.

---

5 You should obtain

$$\ln |x| = -\frac{y}{2x} - \frac{1}{2}\ln\left|\frac{y}{x}\right| + \ln k$$

Multiplying through by 2 and rearranging gives

$$2\ln |x| + \ln\left|\frac{y}{x}\right| - 2\ln k + \frac{y}{x} = 0$$

Using the laws of logarithms, in particular $2 \ln a = \ln a^2$ and $\ln (ab) = \ln a + \ln b$, produces the solution

$$x \ln (A|xy|) + y = 0$$

where A is an arbitrary constant ($A = k^{-2}$).

19.9 REDUCIBLE EQUATIONS

As with the other types of first-order equation we have discussed, there are some equations which although not homogeneous can be made so by means of a change of variable. For instance, consider an equation of the form

$$\frac{dy}{dx} = \frac{a_1x + b_1y + c_1}{a_2x + b_2y + c_2}$$

We wish to transform the variables x and y to obtain new variables X and Y. We take $X = x - A$ and $Y = y - B$. So the equation becomes

$$\frac{dY}{dX} = \frac{a_1X + b_1Y}{a_2X + b_2Y}$$

In order to achieve this objective we require

$$\begin{aligned} a_1A + b_1B + c_1 &= 0 \\ a_2A + b_2B + c_2 &= 0 \end{aligned}$$

This means that we can think of (A, B) as the point at which the pair of straight lines

$$\begin{aligned} a_1x + b_1y + c_1 &= 0 \\ a_2x + b_2y + c_2 &= 0 \end{aligned}$$

intersects. (We shall consider later the case of parallel lines, where there is no point of intersection.)

So now we have a simple method. Let us state precisely what we have to do. To solve the **reducible equation**

$$\frac{dy}{dx} = \frac{a_1x + b_1y + c_1}{a_2x + b_2y + c_2}$$

1 Obtain the point (A, B) where the straight lines

$$\begin{aligned} a_1x + b_1y + c_1 &= 0 \\ a_2x + b_2y + c_2 &= 0 \end{aligned}$$

intersect.

2 Change variables by writing $X = x - A$ and $Y = y - B$ so that

$$\frac{dy}{dx} = \frac{dY}{dx} = \frac{dY}{dX}\frac{dX}{dx} = \frac{dY}{dX}1 = \frac{dY}{dX}$$

3 The equation then becomes

$$\frac{dY}{dX} = \frac{a_1X + b_1Y}{a_2X + b_2Y}$$

which is homogeneous and can be solved in the usual way.

4 Replace X and Y by $x - A$ and $y - B$ respectively to obtain the solution to the original equation.

Time now for an example. We shall see that sometimes a nasty integral results. However, if we remember our work on integration (Chapters 15–17) and a few basic standard forms this task should present no great difficulty.

□ Solve the equation

$$\frac{dy}{dx} = \frac{x + y - 3}{2x - y - 3}$$

First solve the pair of simultaneous equations

$$\begin{aligned} x + y - 3 &= 0 \\ 2x - y - 3 &= 0 \end{aligned}$$

Subtracting gives at once

$$-x + 2y = 0$$

So $x = 2y$. Substituting for x in the first equation we obtain $3y - 3 = 0$, and therefore $y = 1$ and $x = 2$.

Now make the substitutions $X = x - A$ and $Y = y - B$; in this case, $X = x - 2$ and $Y = y - 1$. This produces

$$\frac{dY}{dX} = \frac{X + Y}{2X - Y}$$

Notice that we do not have to think very hard at this stage. First, the coefficients of X and Y are the same as the coefficients of x and y respectively in the original equation. Secondly, we can replace the derivative dy/dx by dY/dX since we have already considered the change of variable in detail.

Now solve the homogeneous equation

$$\frac{dY}{dX} = \frac{X + Y}{2X - Y}$$

As usual we make the substitution $Y = VX$ (it is neater to use capital letters throughout). This gives

$$X\frac{dV}{dX} + V = \frac{X + VX}{2X - VX} = \frac{1 + V}{2 - V}$$

Consequently

$$X\frac{dV}{dX} = \frac{1 + V}{2 - V} - V = \frac{1 - V + V^2}{2 - V}$$

As a result we obtain

$$\int \frac{dX}{X} = \int \frac{2 - V}{1 - V + V^2}\, dV$$

$$= \int \frac{2 - V}{(V - 1/2)^2 + 3/4} \, dV$$
$$= -\frac{1}{2} \int \frac{(2V - 1) - 3}{(V - 1/2)^2 + 3/4} \, dV$$

The numerator has been split up so that one part is the derivative of the denominator and the other is constant. In this way the integral can be seen to be the sum of two, one a logarithm and the other an inverse tangent:

$$-\tfrac{1}{2} \ln [(V - 1/2)^2 + 3/4] + (3/2)(2/\sqrt{3}) \tan^{-1}[(V - 1/2)/(\sqrt{3}/2)] + \text{constant}$$

So the solution is

$$\ln X = -\tfrac{1}{2} \ln (V^2 - V + 1) + (\sqrt{3}) \tan^{-1} [(2V - 1)/\sqrt{3}] + \text{constant}$$

Now $Y = VX$, where $X = x - 2$ and $Y = y - 1$. So multiplying by 2 we obtain

$$2 \ln X + \ln (V^2 - V + 1) = 2(\sqrt{3}) \tan^{-1} [(2V - 1)/\sqrt{3}] + \text{constant}$$

Eliminating V,

$$\ln (Y^2 - YX + X^2) = 2(\sqrt{3}) \tan^{-1} [(2Y - X)/(\sqrt{3})X] + \text{constant}$$

Lastly, substituting in terms of x and y,

$$\begin{aligned} &\ln [(y - 1)^2 - (y - 1)(x - 2) + (x - 2)^2] \\ &\qquad = 2(\sqrt{3}) \tan^{-1} \{[2(y - 1) - (x - 2)]/(\sqrt{3})(x - 2)\} + C \end{aligned}$$

Therefore

$$\begin{aligned} &\ln[y^2 - 2y + 1 - (xy - x - 2y + 2) + x^2 - 4x + 4] \\ &\qquad = 2(\sqrt{3}) \tan^{-1} [(2y - x)/(\sqrt{3})(x - 2)] + C \end{aligned}$$

Finally,

$$\ln (x^2 + y^2 - xy - 3x + 3) = 2(\sqrt{3}) \tan^{-1} [(2y - x)/(\sqrt{3})(x - 2)] + C$$

where C is an arbitrary constant. ■

Whew!

We now dispose of the problem of **parallel lines** which we mentioned earlier. If the simultaneous equations

$$\begin{aligned} a_1x + b_1y + c_1 &= 0 \\ a_2x + b_2y + c_2 &= 0 \end{aligned}$$

have no point of intersection, then the differential equation can be converted into a variables separable equation quite easily by substituting

$$z = a_2x + b_2y + c_2$$

In fact there are many substitutions which will have the same effect. For instance,

$$z = a_2x + b_2y$$
$$z = a_1x + b_1y$$
$$z = a_1x + b_1y + c_1$$

Here is an example of this kind.

☐ Solve the differential equation

$$\frac{dy}{dx} = \frac{2x - 4y - 3}{x - 2y + 1}$$

We observe that the equations

$$2x - 4y - 3 = 0$$
$$x - 2y + 1 = 0$$

have no common solution, and so we put

$$z = x - 2y + 1$$

Then

$$\frac{dz}{dx} = 1 - 2\frac{dy}{dx} = 1 - 2\frac{2z - 5}{z}$$

Therefore

$$\frac{dz}{dx} = \frac{z - 4z + 10}{z} = \frac{-3z + 10}{z}$$

Consequently,

$$x = \int \frac{z}{-3z + 10}\, dz$$
$$= -\frac{1}{3}\int \frac{(-3z + 10) - 10}{-3z + 10}\, dz$$

Observe how we can rearrange the numerator so that the denominator divides into it, leaving a simple integral. So

$$x = -\frac{1}{3}\int dz - \frac{10}{9}\int \frac{-3}{-3z + 10}\, dz$$

Here we have arranged things so that the numerator in the second integral is the derivative of the denominator. Then

$$x = -\frac{z}{3} - \frac{10}{9}\ln(-3z + 10) + C$$

To be strict we should take the modulus of $-3z + 10$ when finding the logarithm. One way to avoid any problems (which are unlikely to occur) is to use the laws of logarithms and rewrite the solution as

$$x = -\frac{z}{3} - \frac{5}{9}\ln(-3z + 10)^2 + C$$

Eliminating z produces

$$x = -\frac{x - 2y + 1}{3} - \frac{5}{9}\ln(-3x + 6y + 7)^2 + C$$ ■

That wasn't too bad, was it? Now you do one – and be careful with the integral.

□ Obtain the general solution to the differential equation

$$\frac{dy}{dx} = \frac{x - 2y + 3}{-2x + 4y + 7}$$

When you have completed your working, follow through the solution and see how things differ.

First we examine the two equations

$$\begin{aligned} x - 2y + 3 &= 0 \\ -2x + 4y + 7 &= 0 \end{aligned}$$

and see at once that they have no common solution.

Next make a substitution. $z = -2x + 4y + 7$ is suitable, but you may prefer $z = x - 2y$. Even if you have solved this problem successfully it is a good idea to follow through the substitution $z = x - 2y$ to see how it all comes eventually to the same thing. When $z = -2x + 4y + 7$ we have

$$\frac{dz}{dx} = -2 + 4\frac{dy}{dx} = -2 + 4\,\frac{-(z - 7)/2 + 3}{z}$$

So

$$\frac{dz}{dx} = \frac{-2z - 2(z - 7) + 12}{z} = \frac{-4z + 26}{z}$$

Therefore

$$\int \frac{z\,dz}{-4z + 26} = \int dx$$

It follows that

$$x = -\frac{1}{4}\int \frac{z - (13/2) + (13/2)}{z - (13/2)}\,dz$$

$$= -\frac{z}{4} - \frac{13}{8}\int \frac{dz}{z - (13/2)}$$

$$= -\frac{z}{4} - \frac{13}{8}\ln|z - (13/2)| + C$$

So that, substituting back, the solution is

$$x = -\frac{-2x + 4y + 7}{4} - \frac{13}{8}\ln|-2x + 4y + (1/2)| + C$$

So

$$2x + 4y - 7 = -\frac{13}{2}\ln|-2x + 4y + (1/2)| + 4C$$

$$4x + 8y - 14 + 13\ln|-2x + 4y + (1/2)| = A$$

where A is the arbitrary constant.

Although your answer may not look like this, it may be equivalent. Check to see if it is. For example, using the laws of logarithms, it is possible to express this answer in the form

$$4x + 8y - 14 + 13\ln|-4x + 8y + 1| = B$$

where B is an arbitrary constant. One way to check if two answers are equivalent is in each case to put the constant on the right and the rest of the solution on the left. Then subtract the two left-hand sides; if they differ by a constant, all is well. ■

19.10 BOUNDARY CONDITIONS

We have not mentioned the situation where we are given some boundary condition which the differential equation and therefore its solution must satisfy. In these circumstances we are not greatly interested in the general solution.

There are many methods for solving differential equations, and some of these make use of the boundary condition at the outset. However, for first-order differential equations we shall continue to solve them analytically by first obtaining the general solution and then determining the arbitrary constant. A single example will suffice to show how this is done; it really is very easy indeed!

□ Solve the equation

$$\frac{dy}{dx} = \tan(2x + y) - 2$$

given that $y = 3\pi/2$ when $x = 0$. This initial condition is sometimes written as $y(0) = 3\pi/2$, a useful shorthand notation.

Before reading any further, see if you can solve the equation on your own. You never know your own strength until you test it!

If we attempt to expand out the tangent we shall soon find things difficult. It is clear also that the equation does not obviously fall into any one of the categories which we have discussed. So we shall need to adapt it. We note that if $z = 2x + y$ then

$$\frac{dz}{dx} = 2 + \frac{dy}{dx} = 2 + \tan z - 2 = \tan z$$

so that the equation becomes variables separable. We have therefore

$$\int dx = \int \cot z \, dz = \int \frac{\cos z}{\sin z} \, dz$$

so that

$$x = \ln |\sin z| + C = \ln |\sin (2x + y)| + C$$

Now we use the initial condition and substitute into the equation the pair of values for x and y to determine C. We obtain

$$\begin{aligned} 0 &= \ln |\sin [0 + (3\pi/2)]| + C \\ &= \ln |-1| + C = \ln 1 + C = 0 + C \end{aligned}$$

So $C = 0$, and consequently

$$x = \ln |\sin (2x + y)|$$

or

$$\sin^2 (2x + y) = e^{2x}$$ ■

Now it's time to consider some applications. We shall solve two problems, one electrical and one mechanical. You can either take your pick or solve them both. We shall tackle them stage by stage so that you can participate in the solution whenever you wish.

19.11 Practical

CIRCUIT CURRENT

An RL series circuit has an EMF $E \sin \omega t$ where E is constant. The current i satisfies, at time t,

$$L \frac{di}{dt} + Ri = E \sin \omega t$$

Obtain the current at time t if initially it is zero.
Try it and see how it goes.

This is clearly a first-order linear equation:

$$\frac{di}{dt} + \frac{R}{L} i = \frac{E}{L} \sin \omega t$$

We have $P = R/L$, so

$$\int P \, dt = \int \frac{R}{L} \, dt = \frac{Rt}{L}$$

Therefore the integrating factor is $e^{Rt/L}$.
Now solve the equation.

We have

$$\frac{d}{dt} (e^{Rt/L} \, i) = \frac{E}{L} \sin \omega t \, e^{Rt/L}$$

Consequently

$$e^{Rt/L} \, i = \frac{E}{L} \int e^{Rt/L} \sin \omega t \, dt$$

Now

$$\int e^{ax} \sin bx \, dx = \frac{e^{ax}}{a^2 + b^2} (a \sin bx - b \cos bx) + C$$

Write down the solution and move on to the next stage.

We have

$$e^{Rt/L} \, i = \frac{(E/L)e^{Rt/L}}{(R^2/L^2) + \omega^2} \left(\frac{R}{L} \sin \omega t - \omega \cos \omega t \right) + C$$

where C is the arbitrary constant. Determine C and complete the solution.

We have

$$i = \frac{E}{R^2 + \omega^2 L^2} (R \sin \omega t - \omega L \cos \omega t) + C \, e^{-Rt/L}$$

When $t = 0$, $i = 0$, and therefore

$$C = \frac{E\omega L}{R^2 + \omega^2 L^2}$$

Consequently

$$i = \frac{E}{R^2 + \omega^2 L^2} (R \sin \omega t - \omega L \cos \omega t + \omega L \, \mathrm{e}^{-Rt/L})$$

VEHICLE BRAKING

The brakes of a vehicle are applied when the speed is u. Subsequently its speed v satisfies the equation

$$v \frac{\mathrm{d}v}{\mathrm{d}x} = \left(-2 - \frac{v}{3u}\right) a$$

where a is a constant and x is the distance travelled after braking. Obtain the distance the vehicle travels after the brakes are applied before it comes to rest.

Move on when you have sorted out the equation.

The equation is variables separable:

$$v \frac{\mathrm{d}v}{\mathrm{d}x} = \frac{-6u - v}{3u} a$$

So

$$\int \frac{3uv \, \mathrm{d}v}{-6u - v} = a \int \mathrm{d}x$$

Obtain these integrals and thereby the general solution.

We have

$$\begin{aligned} ax &= -3u \int \frac{v \, \mathrm{d}v}{v + 6u} \\ &= -3u \int \frac{v + 6u - 6u}{v + 6u} \, \mathrm{d}v \\ &= -3u \int \mathrm{d}v + 18u^2 \int \frac{\mathrm{d}v}{v + 6u} \\ &= -3uv + 18u^2 \ln (v + 6u) + C \end{aligned}$$

where C is an arbitrary constant.

Determine C and complete the solution.

When $x = 0$, $v = u$, and so

$$0 = -3u^2 + 18u^2 \ln 7u + C$$

Therefore

$$C = 3u^2 - 18u^2 \ln 7u$$

Consequently

$$ax = -3uv + 18u^2 \ln (v + 6u) + 3u^2 - 18u^2 \ln 7u$$

When $v = 0$, $x = d$ the stopping distance, so

$$\begin{aligned} ad &= 0 + 18u^2 \ln 6u + 3u^2 - 18u^2 \ln 7u \\ &= 18u^2 \ln (6/7) + 3u^2 \end{aligned}$$

Therefore

$$d = (1/a)[18u^2 \ln (6/7) + 3u^2]$$

SUMMARY

We have solved first-order ordinary differential equations which are of the following types:

☐ Variables separable

$$\frac{dy}{dx} = \frac{f(x)}{g(y)}$$

☐ Linear

$$\frac{dy}{dx} + Py = Q$$

☐ Homogeneous

$$\frac{dy}{dx} = \frac{f(x, y)}{g(x, y)}$$

We have seen that other equations can be reduced to these by means of substitutions. In particular we considered:

☐ Bernoulli's equation

$$\frac{dy}{dx} + Py = Qy^n$$

☐ Reducible

$$\frac{dy}{dx} = \frac{a_1x + b_1y + c_1}{a_2x + b_2y + c_2}$$

EXERCISES

1 Solve, by separating the variables,

a $x\frac{dy}{dx} = xy + y$

b $x^2(y + 1)\frac{dy}{dx} = y(x + 1)$

c $\exp(x + y)\frac{dy}{dx} = \frac{x}{y}$

2 Solve the linear equations

a $\frac{y}{x} + \frac{dy}{dx} = \exp x$

b $\frac{y}{x} - \frac{dy}{dx} + x^2 \exp x = 0$

c $\frac{dy}{dx} + y \cot x = \cos x$

3 Solve the homogeneous equations

a $\frac{dy}{dx} = \frac{2x + y}{x - y}$

b $(x + 4y)\frac{dy}{dx} = x - 2y$

c $(x^2 + xy)\frac{dy}{dx} = y^2 - xy$

4 Solve each of the following equations:

a $\frac{dy}{dx} + 2xy = y^2 \exp x^2$

b $x\frac{dy}{dx} - y = x^2y^2$

c $(x + y - 3)\frac{dy}{dx} = 2x - y$

ASSIGNMENT

Solve each of the following differential equations:

1 $e^y \frac{dy}{dx} = \frac{2x}{x^2 + 1}$

2 $\frac{dx}{dt} = x \sec^2 t$

3 $2 \sec 2u \frac{dv}{du} = \left(y + \frac{1}{y}\right)$

4 $\dfrac{dy}{dx} = \dfrac{y^2 - x + xy^2 - 1}{2xy}$

5 $\dfrac{dy}{d\theta} + y \cot \theta + 1 = 0$

6 $x \ln x \dfrac{dy}{dx} + y = \dfrac{x^2}{\sqrt{(x^2 + 1)}}$

7 $x + \dfrac{dx}{dt} = 2e^t$

8 $x \dfrac{dy}{dx} + y = 2x$

9 $\dfrac{dy}{dx} = \dfrac{x + y}{x}$

10 $\dfrac{ds}{dt} = \dfrac{2s^2 + t^2}{2st}$

11 $\dfrac{dy}{dx} = \dfrac{2y^2 - xy + x^2}{2xy - x^2}$

12 $\dfrac{dy}{dx} = \dfrac{y}{x} + \cos\left(\dfrac{y}{x}\right)$

FURTHER EXERCISES

Solve each of the differential equations 1 to 10:

1 $\dfrac{dy}{dx} = \dfrac{x^2}{y^2 - x^2} + \dfrac{y}{x}$

2 $\dfrac{x}{x + y} \dfrac{dy}{dx} = \dfrac{y}{x}$

3 $\sin 2t \dfrac{dx}{dt} = 2(\sin t - x)$

4 $x + y \dfrac{dy}{dx} = (x^2 + y^2) \cot x$

5 $\dfrac{du}{dv} = \dfrac{2u - 5}{v - 3}$

6 $(6x + y) \dfrac{dy}{dx} = x - 6y$

7 $\dfrac{dp}{dq} = \dfrac{q + 3p - 9}{3q - p - 7}$

8 $x \dfrac{dy}{dx} = y + x \tan\left(\dfrac{y}{x}\right)$

9 $x^2 \dfrac{dy}{dx} = (x - y)^2 + xy$ where $y(1) = 0$

10 $\cos^2 \theta \dfrac{dr}{d\theta} + r \cos \theta = 1 + \sin \theta$ given $r = 1$ when $\theta = 0$

11 A circuit consists of two branches in parallel. One branch consists of a resistance R ohms and a capacitance C farads. The other branch consists of another resistance R ohms and an inductance L henries. When an EMF $E \sin \omega t$ is applied, the branch currents i_1 and i_2 satisfy the two relations

$$L \frac{di_1}{dt} + Ri_1 = E \sin \omega t$$

$$R \frac{di_2}{dt} + \frac{1}{C} i_2 = \omega E \cos \omega t$$

Show that if the circuit is initially quiescent and is tuned so that $CR^2 = L$ then the total current $i_1 + i_2$ will be $(E/R) \sin \omega t$.

12 An EMF $E \sin \omega t$ is applied to an RC series circuit. Show that the current i is given by

$$RC \frac{di}{dt} + i = \omega EC \cos \omega t$$

Initially the circuit was quiescent. Obtain an expression for the charge on the capacitor at time t.

13 The rate of decay of a radioactive substance is proportional to the quantity Q which remains. Initially $Q = Q_0$. Show that if it takes T hours for the quantity to reduce by 50% it will take $T(\log_2 Q_0 - \log_2 Q_1)$ hours for the quantity to reduce from Q_0 to Q_1.

14 Newton's law of cooling states that the rate of fall of temperature of a body is approximately proportional to the excess temperature over that of its surroundings. If θ_1 is the temperature initially and θ_0 is the surrounding temperature ($\theta_1 > \theta_0$) and if it takes T minutes to cool to $(\theta_1 + \theta_0)/2$, show that the time taken to cool to $(\theta_1 + n\theta_0)/(n + 1)$, where n is a positive integer, is $T \log_2 (n + 1)$.

15 The equation of motion of a particle, which is attracted to the origin O is

$$v\frac{dv}{dx} + v^2 + 2x = 0$$

where v is the speed of the particle.

By putting $y = v^2$, or otherwise, solve the equation for v given that $v = 0$ when $x = a$.

Show that when $x = 0$ $\qquad v^2 = 1 - e^{2a} + 2a\, e^{2a}$

16 A small missile of mass m is projected vertically with initial speed u into the air. Air resistance at speed v may be presumed to be mkv^2, where k is constant. Show that when the missile returns it will have a speed of $u[1 + ku^2/g]^{-1/2}$, where g is the acceleration due to gravity.

20 Second-order differential equations

In Chapter 19 we solved some of the first-order differential equations which tend to arise in applications. We now turn our attention to second-order differential equations.

After studying this chapter you should be able to

- ☐ Recognize a second-order linear differential equation;
- ☐ Write down the general solution in the homogeneous case;
- ☐ Use the method of trial solutions to obtain a particular solution in the non-homogeneous case;
- ☐ Anticipate the breakdown case and remedy the situation;
- ☐ Solve a general linear second-order differential equation with constant coefficients.

At the end of the chapter we solve practical problems in filtering, circuits and mechanical oscillations.

20.1 LINEAR DIFFERENTIAL EQUATIONS

There are many types of second-order differential equation, but one in particular arises frequently in applications. This is known as a **linear** differential equation with **constant coefficients**. It can be expressed in the form

$$a\,\frac{d^2y}{dx^2} + b\,\frac{dy}{dx} + cy = f(x)$$

where a, b and c are real constants ($a \neq 0$) and $f(x)$ depends solely on x.

The special case where $f(x)$ is identically zero is known as the homogeneous case. Since this equation is not only easy to solve but also of relevance in the solution of the general equation, the non-homogeneous case, we shall consider it first.

20.2 THE HOMOGENEOUS CASE

We are concerned with the differential equation

$$a\frac{d^2y}{dx^2} + b\frac{dy}{dx} + cy = 0$$

where a, b and c are real constants and a is non-zero.

This is a second-order differential equation and so its general solution will contain two independent arbitrary constants. Later we shall be able to reduce the solution of this equation to a simple routine, but first we see how the routine arises.

We show firstly that if $y = u$ and $y = v$ are solutions of the equation then so also is $y = Au + Bv$, where A and B are arbitrary constants. This is important because it implies that in order to obtain the general solution it is sufficient to obtain any two linearly independent solutions.

Note that if the identity $Au + Bv \equiv 0$, where A and B are constants, is satisfied only when both $A = 0$ and $B = 0$, then u and v are said to be **linearly independent**.

Suppose then

$$a\frac{d^2u}{dx^2} + b\frac{du}{dx} + cu = 0$$

and

$$a\frac{d^2v}{dx^2} + b\frac{dv}{dx} + cv = 0$$

To show that $y = Au + Bv$ is also a solution we shall substitute this value for y into the left-hand side of the differential equation and deduce the result is zero. Now from $y = Au + Bv$ we deduce, by differentiating, that

$$\frac{dy}{dx} = A\frac{du}{dx} + B\frac{dv}{dx}$$

$$\frac{d^2y}{dx^2} = A\frac{d^2u}{dx^2} + B\frac{d^2v}{dx^2}$$

So

$$a\frac{d^2y}{dx^2} + b\frac{dy}{dx} + cy$$

$$= a\left(A\frac{d^2u}{dx^2} + B\frac{d^2v}{dx^2}\right) + b\left(A\frac{du}{dx} + B\frac{dv}{dx}\right) + c(Au + Bv)$$

$$= A\left(a\frac{d^2u}{dx^2} + b\frac{du}{dx} + cu\right) + B\left(a\frac{d^2v}{dx^2} + b\frac{dv}{dx} + cv\right)$$

$$= 0 + 0 = 0$$

So we have shown that if $y = u$ and $y = v$ are any two solutions of the differential equation

$$a\frac{d^2y}{dx^2} + b\frac{dy}{dx} + cy = 0$$

then $y = Au + Bv$ is also a solution, where A and B are constants which may be arbitrarily chosen.

The outcome of all this is that if we can find two linearly independent solutions of this differential equation we can find the general solution. How are we to find these solutions? Well it so happens that it is fairly easy to spot one. Remember that when we differentiate e^x with respect to x the answer remains e^x. We adapt this observation very slightly and look for a solution of the form e^{mx}, where m is a constant; we shall wish to determine m.

Now if $y = e^{mx}$ is a solution it follows that

$$\frac{dy}{dx} = m\,e^{mx} \qquad \frac{d^2y}{dx^2} = m^2\,e^{mx}$$

Consequently, substituting these expressions into the differential equation,

$$am^2\,e^{mx} + bm\,e^{mx} + c\,e^{mx} = 0$$

and since e^{mx} is never zero we can divide through by it to obtain

$$am^2 + bm + c = 0$$

This is a very familiar equation, which you probably recognize straight away: it is a quadratic equation. Because of its importance in the solution of this differential equation it is given a special name: the **auxiliary equation**.

Notice the pattern, and see how easy it is to write down the auxiliary equation straight away from the differential equation. The second-order derivative is replaced by m^2, the first-order derivative is replaced by m, and y is replaced by 1. There is no need to think!

Given the auxiliary equation

$$am^2 + bm + c = 0$$

there are three situations which can occur:

1 The equation has two distinct real roots m_1 and m_2;
2 The equation has two equal roots m;
3 The equation has complex roots $m = \alpha \pm i\beta$.

We shall deal with each of these cases in turn.

Case 1

The auxiliary equation $am^2 + bm + c = 0$ has distinct real roots m_1 and m_2.

Here we have now two distinct linearly independent solutions of the differential equation, namely $u = e^{m_1 x}$ and $v = e^{m_2 x}$. So the general solution is

$$y = A\ e^{m_1 x} + B\ e^{m_2 x}$$

where A and B are arbitrary constants.

Case 2

The auxiliary equation $am^2 + bm + c = 0$ has two equal roots m, necessarily real since $b^2 = 4ac$.

At first sight we may seem to be in difficulties since we have only one solution. However, in these circumstances it is easy to verify that $y = x\ e^{mx}$ is another solution. To see this we simply differentiate, substitute the results into the left-hand side of the differential equation and check that the outcome is zero. You may like to try this for yourself, but in either event here is the working in full.

If $y = x\ e^{mx}$ then

$$\frac{dy}{dx} = e^{mx} + mx\ e^{mx}$$

$$\frac{d^2y}{dx^2} = m\ e^{mx} + m\ e^{mx} + m^2x\ e^{mx} = 2m\ e^{mx} + m^2x\ e^{mx}$$

So substituting into the left-hand side of the auxiliary equation gives

$$\begin{aligned} &a(2m\ e^{mx} + m^2x\ e^{mx}) + b(e^{mx} + mx\ e^{mx}) + cx\ e^{mx} \\ &\quad = (am^2 + bm + c)\ x\ e^{mx} + (2am + b)\ e^{mx} \end{aligned}$$

Now $am^2 + bm + c = 0$ because the auxiliary equation is satisfied by m, and $2am + b = 0$ because the auxiliary equation has equal roots $m = -b/(2a)$. So we now have two linearly independent solutions of the differential equation, $u = e^{mx}$ and $v = x\ e^{mx}$. The general solution is consequently

$$y = A\ e^{mx} + Bx\ e^{mx} = (A + Bx)\ e^{mx}$$

where A and B are arbitrary constants.

Case 3

The auxiliary equation $am^2 + bm + c = 0$ has complex roots $m = \alpha \pm i\beta$.

This is similar to case 1. In fact if we were content to have a solution containing complex numbers we need go no further. However, the differential equation itself did not have any complex numbers in it and there is no reason why the solution should contain any; such equations often arise from practical situations where complex numbers would seem very out of place. Luckily we can express the solution in a form which is entirely free of complex numbers, and this we now do.

Following case 1 we have the general solution

$$y = P\,e^{(\alpha+i\beta)x} + Q\,e^{(\alpha-i\beta)x}$$

where P and Q are arbitrary constants. So

$$\begin{aligned} y &= e^{\alpha x}(P\,e^{i\beta x} + Q\,e^{-i\beta x}) \\ &= e^{\alpha x}[P(\cos\beta x + i\sin\beta x) + Q(\cos\beta x - i\sin\beta x)] \\ &= e^{\alpha x}[(P+Q)\cos\beta x + (Pi - Qi)\sin\beta x] \\ y &= e^{\alpha x}(A\cos\beta x + B\sin\beta x) \end{aligned}$$

where A and B are arbitrary constants.

The solution is now free of complex numbers. However, there are several different ways of expressing this. For example, another is

$$y = e^{\alpha x}R\cos(\beta x - \theta)$$

where R and θ are arbitrary constants. This follows immediately from elementary trigonometry, since we can always express $a\cos\theta + b\sin\theta$ as $r\cos(\theta - \alpha)$.

In summary, to obtain the general solution of the equation

$$a\frac{d^2y}{dx^2} + b\frac{dy}{dx} + cy = 0$$

where a, b and c are real constants ($a \neq 0$):

1 Write down the auxiliary equation $am^2 + bm + c = 0$.

2 Solve this quadratic equation to obtain the roots m_1 and m_2.

3 Select from three cases:

a If the roots m_1 and m_2 are both real and distinct,

$$y = A\,e^{m_1 x} + B\,e^{m_2 x}$$

b If the roots m_1 and m_2 are equal, so $m_1 = m_2 = m$,

$$y = (A + Bx)\,e^{mx}$$

c If the roots m_1 and m_2 are complex, so $m = \alpha \pm i\beta$,

$$y = e^{\alpha x}(A\cos\beta x + B\sin\beta x)$$

where A and B are arbitrary constants.

It really is very easy. We shall see how simple it all is by taking some steps.

20.3 Workshop

1

Exercise Obtain the general solutions of the following differential equations:

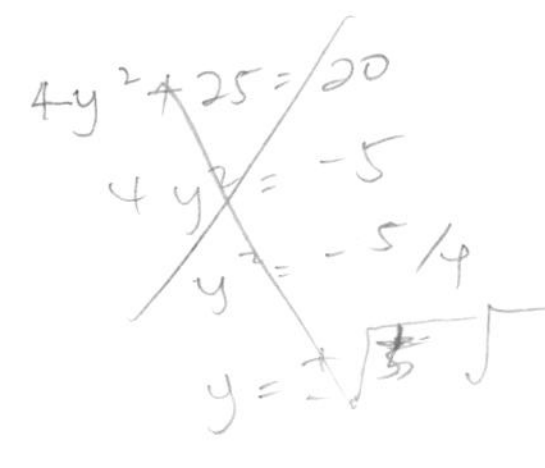

a $\dfrac{d^2y}{dx^2} - 6\dfrac{dy}{dx} + 5y = 0$

b $4\dfrac{d^2y}{dx^2} + 25y = 20\dfrac{dy}{dx}$

c $\dfrac{d^2y}{dx^2} + 25y = 0$

d $\dfrac{d^2y}{dx^2} + 25\dfrac{dy}{dx} = 0$

Four examples and only three cases; at least one case must occur more than once!

First write down the auxiliary equations for **a** and **b**. Remember that there is no need to do any mathematics at this stage: no differentiating, and no substituting into the differential equation. We have dealt with all that once and for all. We simply write down the auxiliary equation.

Done it? Step ahead.

Well then, here are the results you should obtain: **2**

a $m^2 - 6m + 5 = 0$

b $4m^2 - 20m + 25 = 0$

If all is well, write down the auxiliary equations for **c** and **d** and move ahead to step 3.

If you have made an error, look back carefully through what we have done and see where you went wrong. When you are satisfied that you can write down an auxiliary equation correctly, taking care about signs, try doing so for **c** and **d**. If you are confident that you have done it correctly, then read on to check that all is well. If there are still problems you had better go back to the main text and read things through slowly and carefully so that you understand it properly.

Here then are the other two auxiliary equations: **3**

c $m^2 + 25 = 0$

d $m^2 + 25m = 0$

Now the time has come to solve each of the four quadratic equations. Of course this is very elementary work, but it is surprising how many mistakes creep in at this stage. See if you can solve them correctly.

Here are the roots: **4**

a $m = 1$ or $m = 5$

b $m = 5/2$ (repeated)

c $m = \pm 5\text{i}$

d $m = 0$ or $m = -25$

Is all well? We must decide for each one which case it is, and then write down the corresponding solution. To begin with, try **a** and **b**. When you have finished, move to the next step to check they are correct.

5

Here are the answers:

a $y = A\ \mathrm{e}^{x} + B\ \mathrm{e}^{5x}$

b $y = (A + Bx)\ \mathrm{e}^{5x/2}$

where A and B are arbitrary constants. Of course it does not matter if you have A and B the other way round or have used some other letters.

If all is well you can now see if you can deal properly with **c** and **d**, and then move ahead to step 7.

If not, check carefully to see what went wrong. Look back at the summary of the method; possibly you identified the cases incorrectly. When you are confident that you know what went wrong, try these equations and see how it goes.

▷**Exercise** Obtain the general solutions of the following differential equations:

e $\dfrac{\mathrm{d}^2y}{\mathrm{d}x^2} - 14\dfrac{\mathrm{d}y}{\mathrm{d}x} + 49y = 0$

f $\dfrac{\mathrm{d}^2y}{\mathrm{d}x^2} + 6y = 5\dfrac{\mathrm{d}y}{\mathrm{d}x}$

First obtain the auxiliary equation, then the values for m and finally the correct form of the solution. Then step ahead.

6

The auxiliary equations are:

e $m^2 - 14m + 49 = 0$

f $m^2 - 5m + 6 = 0$

The roots are:

e $m = 7$ (repeated)

f $m = 2$ and $m = 3$

The solutions are:

e $y = (A + Bx)\ \mathrm{e}^{7x}$

f $y = A\ \mathrm{e}^{2x} + B\ \mathrm{e}^{3x}$

where A and B are arbitrary constants.

If things are still going wrong it is best to read through the chapter again and see if you can get things straight. Otherwise, see if you can now deal properly with **c** and **d** and then take a further step to see how things worked out.

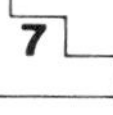

These are the solutions. For **c**, $m = \pm 5\mathrm{i}$ so that $\alpha = 0$ and $\beta = 5$. Consequently,

c $y = e^0(A \cos 5x + B \sin 5x)$
$= A \cos 5x + B \sin 5x$
d $y = A\,e^0 + B\,e^{-25x} = A + B\,e^{-25x}$

where A and B are arbitrary constants.

Here then are the general solutions **a–d** again:
a $y = A\,e^x + B\,e^{5x}$
b $y = (A + Bx)\,e^{5x/2}$
c $y = A \cos 5x + B \sin 5x$
d $y = A + B\,e^{-25x}$
where A and B are arbitrary constants.

All should be well with **d**, but **c** may have caused some difficulty. If you have a clean bill of health, you may move on to the next section of work.

Otherwise, here are two more equations where the roots of the auxiliary equation turn out to be complex numbers. Try these so that you can become confident that you can solve such equations.

▷**Exercise** Obtain the general solutions of the following differential equations:

g $\dfrac{d^2y}{dx^2} - 2\dfrac{dy}{dx} + 2y = 0$

h $\dfrac{d^2y}{dx^2} + 36y = 0$

When you have finished, move ahead to step 8 to see the results.

8

The auxiliary equations are:
g $m^2 - 2m + 2 = 0$
h $m^2 + 36 = 0$
The roots are:
g $m = 1 \pm i$, so $\alpha = 1$ and $\beta = 1$
h $m = \pm 6i$, so $\alpha = 0$ and $\beta = 6$
The solutions are:
g $y = e^x(A \cos x + B \sin x)$
h $y = e^0(A \cos 6x + B \sin 6x)$
$= A \cos 6x + B \sin 6x$
where A and B are arbitrary constants.

If there are still problems then it is best to look back through the material of this chapter to sort things out.

20.4 THE NON-HOMOGENEOUS CASE

We now turn our attention once more to the solution of second-order linear differential equations with constant coefficients. As we said before, such an equation can be expressed in the form

$$a\frac{d^2y}{dx^2} + b\frac{dy}{dx} + cy = f(x) \tag{1}$$

where a, b and c are real constants ($a \neq 0$) and $f(x)$ depends solely on x.

We have disposed completely of the homogeneous case $f(x) \equiv 0$. However it turns out, as we shall see in a moment, that the solution of the homogeneous case is part and parcel of the solution of the non-homogeneous case.

Suppose for the moment that we know how to obtain a solution $y = v$ to the differential equation. This solution is a particular solution and is not likely to contain any arbitrary constants. Then substituting into (1) gives

$$a\frac{d^2v}{dx^2} + b\frac{dv}{dx} + cv = f(x) \tag{2}$$

Subtracting (2) from (1) and simplifying gives

$$a\frac{d^2}{dx^2}(y - v) + b\frac{d}{dx}(y - v) + c(y - v) = f(x) - f(x) = 0$$

so that putting $u = y - v$ we have

$$a\frac{d^2u}{dx^2} + b\frac{du}{dx} + cu = 0 \tag{3}$$

Now this is very significant, although its importance may not occur to you straight away. Just think. We know how to solve (3), so we can obtain u containing two arbitrary constants. Moreover, $y = u + v$ and so we can obtain the general solution to (1) provided we can obtain *any* solution at all to it.

We call u the **complementary function** (or complementary part) and v a **particular integral** (or particular solution). The problem of solving the **non-homogeneous** case has therefore essentially become reduced to that of obtaining a particular solution of the differential equation:

general solution (y) = complementary part (u) + particular solution (v)

20.5 THE PARTICULAR SOLUTION

We have already seen how to find the complementary part, so we now concentrate our attention on finding a particular solution. There are two principal methods which can be used to do this, and each has something to be said for it. The methods are known as

1 the method of the operator D
2 the method of trial solution.

The method of the operator D is a formal method using the linear operator D (differentiation) in an algebraic way to derive a particular solution.

We shall be discussing this method in the more general context of linear operators (Chapter 22) and so we shall not consider it here.

Instead we shall consider the method of **trial solution**. Basically what we do is we examine $f(x)$ and attempt to find a solution of the same form.

□ Consider the equation

$$\frac{d^2y}{dx^2} + 5\frac{dy}{dx} + 6y = 24$$

Here $f(x) = 24$, and so we might wonder if there is a solution of the form $y = k$ where k is a constant.

To see if this is possible, we tentatively suppose that $y = k$ is a particular solution and substitute into the equation to see if we can find k. We have, if $y = k$,

$$\frac{dy}{dx} = 0 \qquad \frac{d^2y}{dx^2} = 0$$

So substituting,

$$0 + 0 + 6k = 24$$

So $k = 4$ and consequently $y = 4$ is a particular solution.

You can easily check that the complementary part is

$$u = A\,e^{-2x} + B\,e^{-3x}$$

So substituting,

$$y = A\,e^{-2x} + B\,e^{-3x} + 4$$ ■

There are two points to be careful about here:

1 Do *not* call the complementary part y. It is only part of the solution; by itself it does not even satisfy the equation. It is better to call it u or CP.

2 If there are initial conditions such as $y(0) = 1$ and $y'(0) = 2$ then we must obtain the general solution to the equation before we make any attempt to use them. We must *never* substitute these values into the complementary part in an attempt to determine the constants A and B.

How, then, are we to decide which trial solutions to use? Well, it is important to realize that it is not always possible to obtain an analytic solution to the differential equation by this method. In fact there are relatively few functions f for which particular solutions exist. However, we can construct a table and the recommended trial solution in some simple cases. There is a set of circumstances in which the trial solution will not work; we shall consider this later.

Suitable trial solutions for selected functions are shown in Table 20.1. In this table k is supposed constant, and the constants a, b, c and d are to be determined by trial solution.

Table 20.1

$f(x)$	Trial solution
Constant	Constant $y = k$
Polynomial, e.g. $x^2 + 1$	Polynomial of same degree $y = ax^2 + bx + c$
e^{kx}	$y = a\,e^{kx}$
$\sin kx$ or $\cos kx$	$y = a\cos kx + b\sin kx$
$\sinh kx$ or $\cosh kx$	$y = a\,e^{kx} + b\,e^{-kx}$ or $y = c\cosh kx + d\sinh kx$

Note that if $f(x)$ is a sum of several functions then the corresponding trial solution can be obtained by using an appropriate sum of trial solutions. A similar rule holds for products, provided we interpret the product of trial solutions in the widest sense. We shall later consider an example which illustrates this point.

□ Consider the equation

$$\frac{d^2y}{dx^2} - 7\frac{dy}{dx} + 10y = 2\,e^{-x}$$

From Table 20.1 we see that a suitable trial solution is $y = a\,e^{-x}$, where a is a constant which we shall need to obtain. (The presence of the factor 2 has no influence on the choice of trial solution.) Now differentiating we obtain

$$\frac{dy}{dx} = -a\,e^{-x} \qquad \frac{d^2y}{dx^2} = a\,e^{-x}$$

Therefore substituting,

$$a\,e^{-x} - 7(-a\,e^{-x}) + 10a\,e^{-x} = 2e^{-x}$$

Since e^{-x} is never zero we can divide out to obtain

$$a + 7a + 10a = 2$$

from which $a = 1/9$. So a particular solution (PS) is $y = e^{-x}/9$.

By way of revision, write down the complementary part (CP) and thereby the general solution (GS). It shouldn't take more than three minutes. When you have done it, move ahead to check the result.

Here it is then:

$$\begin{aligned} \text{CP} &= A\,e^{2x} + B\,e^{5x} \\ \text{PS} &= e^{-x}/9 \\ \text{GS} &= \text{CP} + \text{PS} \end{aligned}$$

so that

$$y = A\,e^{2x} + B\,e^{5x} + e^{-x}/9$$

is the general solution. ■

Now it's time for you to take a few steps on your own.

20.6 Workshop

1

▷**Exercise** Solve the differential equations

a $\dfrac{d^2y}{dx^2} - 6\dfrac{dy}{dx} + 8y = 2x^2$

b $\dfrac{d^2y}{dx^2} + 2\dfrac{dy}{dx} + 5y = \cos 2x$

Write down in each case a suitable trial solution. Only when you have done this should you read on.

The trial solution for equation **a** is 2

$$y = ax^2 + bx + c$$

Don't forget to include $bx + c$. We must allow for the possibility of a *general* polynomial of degree 2, and this would include a term in x and a constant.

If you didn't get that right then check your trial solution for equation **b** before taking the next step.

The trial solution for equation **b** is 3

$$y = a\cos 2x + b\sin 2x$$

It is worth remarking that we should use the same trial solution in the case $f(x) = \cos 2x + \sin 2x$.

Good, now we can proceed to obtain particular solutions. Let's concentrate on equation **a** for the moment. Have a go!

From $y = ax^2 + bx + c$ it follows that 4

$$\frac{dy}{dx} = 2ax + b \qquad \frac{d^2y}{dx^2} = 2a$$

So on substituting into equation **a** we require the following equation to hold for all x:

$$2a - 6(2ax + b) + 8(ax^2 + bx + c) = 2x^2$$

Therefore

$$(8a - 2)x^2 + (8b - 12a)x + (8c - 6b + 2a) \equiv 0$$

Consequently $8a = 2$, from which $a = 1/4$. Next $8b - 12a = 0$, from which $b = 3/8$. Finally $8c - 6b + 2a = 0$, from which $4c = 3b - a = 9/8 - 1/4 = 7/8$ and $c = 7/32$.

Therefore a particular solution for **a** is

$$y = \frac{x^2}{4} + \frac{3x}{8} + \frac{7}{32}$$

If that didn't quite work out in the way it should, see where you went wrong and try extra carefully to find a particular solution for equation **b**. When it has been done, move to the next step.

---

5 Using $y = a \cos 2x + b \sin 2x$, we have

$$\frac{dy}{dx} = -2a \sin 2x + 2b \cos 2x$$

$$\frac{d^2y}{dx^2} = -4a \cos 2x - 4b \sin 2x$$

So substituting these into the differential equation, we are seeking to satisfy the identity

$$(-4a \cos 2x - 4b \sin 2x) + 2(-2a \sin 2x + 2b \cos 2x) + 5(a \cos 2x + b \sin 2x) \equiv \cos 2x$$

So we require

$$(-4a + 4b + 5a) \cos 2x + (-4b - 4a + 5b) \sin 2x \equiv \cos 2x$$

It follows that $a + 4b = 1$ and $b - 4a = 0$. So $17a = 1$, and consequently $a = 1/17$ and $b = 4/17$.

A particular solution for **b** is therefore

$$y = \frac{\cos 2x + 4 \sin 2x}{17}$$

Lastly, write down the general solutions and take the final step.

---

6 Here are the answers. You should not have had any difficulty here.

a $y = A\,e^{2x} + B\,e^{4x} + \dfrac{x^2}{4} + \dfrac{3x}{8} + \dfrac{7}{32}$

b $y = e^{-x}(A\cos 2x + B\sin 2x) + \dfrac{\cos 2x + 4\sin 2x}{17}$

where A and B are arbitrary constants.

20.7 THE BREAKDOWN CASE

As we mentioned earlier, there is one situation in which it is possible to anticipate that the trial solution will not work. This is known as the breakdown case – not because of its effect on a hard-working student, but because the standard trial solution does not produce a particular solution. To anticipate when this is going to arise it is essential that we find the complementary part of the solution first. There is much to be said in favour of doing this anyhow, since it is a routine procedure and in an examination represents easy marks.

Suppose the trial solution y which is suggested by Table 20.1 is already present, with some suitable choice of A and B, in the complementary part. This means that y satisfies the homogeneous equation; that is, the equation when $f(x) \equiv 0$. Consequently it cannot possibly satisfy the non-homogeneous equation: that is, the equation when $f(x) \neq 0$. So then it's a dead duck!

What are we to do about it? Luckily there is a simple remedy:

1 Locate the part of the trial solution which corresponds to the complementary part;
2 Multiply it by x and construct a new trial solution;
3 Check again with the complementary part;
4 Repeat this procedure, if necessary, to ensure that the trial solution contains no terms in the complementary part.

An example will illustrate the procedure adequately.

□ Solve the equation

$$\frac{d^2y}{dx^2} - 6\frac{dy}{dx} + 9y = e^{3x}$$

First we find the complementary part. The auxiliary equation is

$$m^2 - 6m + 9 = 0$$

from which $(m - 3)^2 = 0$ and so $m = 3$ (repeated). Consequently

$$\text{CP} = (A + Bx)e^{3x}$$

Now we seek a particular solution. Here $f(x) = e^{3x}$, and so the standard trial solution is $y = a\,e^{3x}$. However, this is already part of the complementary part ($A = a$ and $B = 0$). So we try instead $y = ax\,e^{3x}$ and check if this is all right. Is it?

No it isn't, is it? If we choose $A = 0$ and $B = a$ then we see it is still part of the complementary part. We therefore repeat the prescription, and this time all is well: $y = ax^2 \mathrm{e}^{3x}$ is suitable.

The main advantage in anticipating the breakdown case is that we avoid waste of time and effort, for the standard trial solution will fail anyway and we will find ourselves back at square one. ■

A few steps will convince you how easy it is to anticipate the breakdown case and take appropriate action.

20.8 Workshop

1 **Exercise** Suppose

$$\mathrm{CP} = A\,\mathrm{e}^{4x} + B\,\mathrm{e}^{-2x}$$

and

$$f(x) = \mathrm{e}^{-2x}$$

What would be an appropriate trial solution?

When you have completed your answer, take the next step and see if you were correct.

---

2 Our initial trial solution would be

$$y = a\,\mathrm{e}^{-2x}$$

However, this is already present in the complementary part when $A = 0$ and $B = a$, and so we have the breakdown case. Consequently we select

$$y = ax\,\mathrm{e}^{-2x}$$

and this is fine.

Did you manage that? If you did, then move to step 4. If you made an error, follow through the argument carefully and then do this one.

▷ **Exercise** Find an appropriate trial solution for

$$\mathrm{CP} = (A + Bx)\,\mathrm{e}^{-2x}$$
$$f(x) = \mathrm{e}^{-2x}$$

Try it, then step ahead.

---

3 Our initial trial solution would be

$$y = a\,\mathrm{e}^{-2x}$$

However, this is already present in the complementary part when $A = a$ and $B = 0$ and so we have the breakdown case. Consequently

$$y = ax\,e^{-2x}$$

but this too is the breakdown case. We see this by putting $A = 0$ and $B = a$. Therefore

$$y = ax^2\,e^{-2x}$$

and this will certainly do.

Got it now? Step forward.

4

▷**Exercise** Find an appropriate trial solution for

$$\begin{aligned} \text{CP} &= (A\,e^{4x} + B\,e^{-2x}) \\ f(x) &= \cosh 2x + 1 \end{aligned}$$

Try it, then step ahead.

5

We can easily make a mistake here. If we use the standard trial solution in the form

$$y = a \cosh 2x + b \sinh 2x + c$$

we shall have failed to appreciate the difficulty. However, if we first express $f(x)$ in exponential form then the light will begin to dawn:

$$f(x) = \tfrac{1}{2}(e^{2x} + e^{-2x}) + 1$$

The e^{2x} term and the constant term are no problem, but the term in e^{-2x} is another matter altogether. If it had appeared on its own we should have the standard trial solution

$$y = a\,e^{-2x}$$

which is the breakdown case ($A = 0$ and $B = a$). So we should modify our trial solution and try instead

$$y = ax\,e^{-2x}$$

Consequently our trial solution should be, in the problem we are considering,

$$y = ax\,e^{-2x} + b\,e^{2x} + c$$

If you couldn't get that, try the next exercise. If you were successful, move to step 7.

▷**Exercise** Find an appropriate trial solution for

$$\text{CP} = (A + Bx)\,\mathrm{e}^{-2x}$$
$$f(x) = \cosh 2x + 1$$

Then take another step.

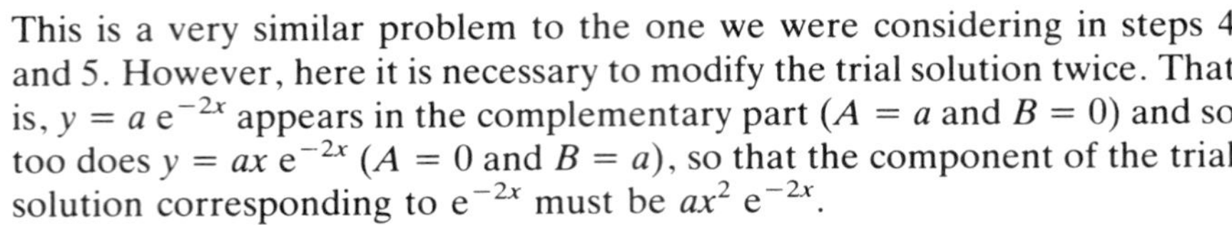

6 This is a very similar problem to the one we were considering in steps 4 and 5. However, here it is necessary to modify the trial solution twice. That is, $y = a\,\mathrm{e}^{-2x}$ appears in the complementary part ($A = a$ and $B = 0$) and so too does $y = ax\,\mathrm{e}^{-2x}$ ($A = 0$ and $B = a$), so that the component of the trial solution corresponding to e^{-2x} must be $ax^2\,\mathrm{e}^{-2x}$.

Therefore our trial solution is

$$y = ax^2\,\mathrm{e}^{-2x} + b\,\mathrm{e}^{2x} + c$$

If there are still problems, read through the text carefully and try the exercises again. Then move on.

---

7 **Exercise** Find an appropriate trial solution for

$$\text{CP} = \mathrm{e}^{-2x}(A\cos x + B\sin x)$$
$$f(x) = \mathrm{e}^{-2x}\cos x$$

Then take the final step.

---

8 If we had $f(x) = \cos x$ we should try

$$y = a\cos x + b\sin x$$

On the other hand, if we had $f(x) = \mathrm{e}^{-2x}$ we should try

$$y = c\,\mathrm{e}^{-2x}$$

For the product we can generalize and try

$$y = \mathrm{e}^{-2x}(a\cos x + b\sin x)$$

where the constant c has been absorbed by a and b. However, this appears in the complementary part ($A = a$ and $B = b$), and so finally we try instead

$$y = x\,\mathrm{e}^{-2x}(a\cos x + b\sin x)$$

There is a subtle point which is worth a remark. Suppose $f(x) = x\cos x$. Then corresponding to x we should normally try $ax + b$, and corresponding to $\cos x$ we should normally try $c\cos x + d\sin x$. We might therefore think that we should try

$$y = (ax + b)(c\cos x + d\sin x)$$

or, absorbing one of the constants,

$$y = (x + c)(a \cos x + b \sin x)$$

(a, b and c are different here, of course.) However, this presumes relationships between the coefficients which may not hold. Instead we must consider the generalized product and try

$$y = ax \cos x + bx \sin x + c \cos x + d \sin x$$

In summary, to obtain the general solution of a non-homogeneous second-order linear differential equation:

$$a \frac{d^2y}{dx^2} + b \frac{dy}{dx} + cy = f(x)$$

where a, b and c are real constants ($a \neq 0$):

1 Obtain u, the complementary part. This is the general solution to the equation

$$a \frac{d^2y}{dx^2} + b \frac{dy}{dx} + cy = 0$$

2 Obtain v, a particular solution of the equation

$$a \frac{d^2y}{dx^2} + b \frac{dy}{dx} + cy = f(x)$$

3 Then the general solution is given by $y = u + v$. That is, general solution = complementary part + particular solution.

4 If initial conditions are given then A and B, the two arbitrary constants generated by the complementary part, can now be determined.

20.9 HIGHER-ORDER EQUATIONS

The methods which we have developed can be generalized to higher-order linear differential equations with real constant coefficients. The generalization holds no surprises.

We begin by writing down the auxiliary equation and obtaining its roots. For example,

$$am^3 + bm^2 + cm + d = 0$$

where a, b, c and d are real constants.

The complementary part is constituted in the following way:

1 A distinct root m contributes

$$A\,e^{mx}$$

to the complementary part.

2 Equal roots $m_1 = m_2 = m_3$ $(=m)$ contribute

$$(A + Bx + Cx^2)\ \mathrm{e}^{mx}$$

to the complementary part.

3 Complex roots always occur in conjugate pairs $\alpha \pm \mathrm{i}\beta$, and so these contribute

$$\exp(\alpha x)\{A \cos \beta x + B \sin \beta x\}$$

to the complementary part.

In this description A, B and C are of course arbitrary constants.

20.10 DAMPING

Suppose we consider the equation

$$a\frac{\mathrm{d}^2x}{\mathrm{d}t^2} + b\frac{\mathrm{d}x}{\mathrm{d}t} + cx = f(t)$$

Then

1 If $b^2 - 4ac < 0$ and if the roots of the auxiliary equation are $\alpha \pm \mathrm{i}\beta$ we have

$$\alpha = -\frac{b}{2a} \qquad \beta = \frac{\sqrt{(4ac - b^2)}}{2a}$$

The complementary part is then

$$\exp(\alpha t)(A \cos \beta t + B \sin \beta t)$$

α is called the **damping factor**. If $\alpha < 0$ then as $t \to \infty$ the complementary part will decay. This means that the complementary part will tend to 0 as t tends to ∞. The angular frequency β is known as the **natural frequency** of the equation.

2 If $b^2 - 4ac = 0$ the system (the physical system which gives rise to the equation) is said to be **critically damped**, for then $\alpha = -b/(2a)$ and $\beta = 0$.

3 If $b^2 - 4ac > 0$ the system is said to be **overdamped**.

□ The equation of simple harmonic motion is

$$\frac{\mathrm{d}^2x}{\mathrm{d}t^2} + \omega^2 x = 0$$

Here $\alpha = 0$ and $\beta = \omega$, so that $x = A \cos \omega t + B \sin \omega t$. The natural frequency is ω and there is no damping. ■

20.11 RESONANCE

As an example, consider an LC series circuit to which an EMF $E \sin pt$ is applied; L, C, E and p are positive real constants. The charge q on the capacitor is given by

$$\frac{d^2q}{dt^2} + \frac{q}{LC} = \frac{E}{L} \sin pt$$

The auxiliary equation is

$$m^2 + \frac{1}{LC} = 0$$

and so $m = \pm i\omega$ where $\omega = 1/\sqrt{(LC)}$ is the natural frequency.

If p is set equal to ω then we have the breakdown case and consequently

$$q = A \cos \omega t + B \sin \omega t - \frac{Et}{2\omega L} \cos \omega t$$

The significance of this is that q is unbounded, so in practice the charge will increase until the capacitor fails. This contrasts sharply with the case where $p \neq \omega$:

$$q = A \cos \omega t + B \sin \omega t + \frac{E}{L(\omega^2 - p^2)} \sin pt$$

Here q remains bounded.

The frequency ω is called the **resonant frequency**. Resonance occurs when the frequency of f, the forcing function, is tuned to that of the natural frequency. Resonance occurs in a wide variety of situations. For instance, platoons of soldiers break step when marching over a bridge so that there is no danger of resonance undermining the structure.

20.12 TRANSIENT AND STEADY STATE

Any part of the solution x of a differential equation which tends to zero as the independent variable t tends to infinity is known as a **transient**. When t is large enough for the transients to be neglected, that which remains is known as the **steady state**. In this way we obtain the equation

general solution = transient + steady state

It is a mistake, however, to assume that the complementary part is necessarily the transient and that the particular solution is the steady state, although in some cases this is true.

□ Solve the differential equation

$$\frac{d^2x}{dt^2} + \frac{dx}{dt} - 6x = e^{-t}$$

Identify the transient and steady state.

The auxiliary equation is $m^2 + m - 6 = 0$, from which $m = -3$ or $m = 2$. The complementary part is therefore $A\,e^{-3t} + B\,e^{2t}$.

For a particular solution we try $x = a\,e^{-t}$, from which $x' = -a\,e^{-t}$ and $x'' = a\,e^{-t}$. Therefore $a\,e^{-t} - a\,e^{-t} - 6a\,e^{-t} = e^{-t}$. Consequently $a = -1/6$ and a particular solution is $x = -e^{-t}/6$.

The general solution is now

$$x = A\,e^{-3t} + B\,e^{2t} - e^{-t}/6$$

Here the transient is $A\,e^{-3t} - e^{-t}/6$ and the steady state is $B\,e^{2t}$. ■

□ Obtain the transient and steady state for the equation

$$4\frac{d^2x}{dt^2} + 9x = e^{-2t}$$

Do this before you read any more.

You will have obtained the complementary part $A\cos(3t/2) + B\sin(3t/2)$ and a particular solution $e^{-2t}/25$. So the general solution is

$$x = A\cos(3t/2) + B\sin(3t/2) + e^{-2t}/25$$

Here the transient is the particular solution $e^{-2t}/25$, and the steady state is the complementary part $A\cos(3t/2) + B\sin(3t/2)$. ■

We now work through examples which include some initial conditions.

20.13 Practical

PRESSURE FILTER

The transpose displacement x of a circular pressure filter at time t is known to satisfy the equation

$$\frac{d^2x}{dt^2} + 2p\frac{dx}{dt} + p^2 = 0$$

where p is a constant. If initially there was no displacement and the speed of displacement x' was a constant q, obtain the displacement x at time t.

There is one nasty trap into which the unwary are likely to step. The equation is not a homogeneous linear equation, for there is no term in x. Let's rearrange it in standard form

$$\frac{d^2x}{dt^2} + 2p\frac{dx}{dt} = -p^2$$

Now we can proceed.

First we seek the complementary part. The auxiliary equation is

$$m^2 + 2mp = 0$$

from which $m(m + 2p) = 0$, so $m = 0$ or $m = -2p$. Therefore

$$\text{CP} = A\,e^0 + B\,e^{-2pt} = A + B\,e^{-2pt}$$

Note that the variables are x and t and not y and x respectively.

Now we want a particular solution. Here $f(t) = -p^2$, a constant, so we try $x = a$, a constant. This is the breakdown case; $A = a$ and $B = 0$.

Therefore we modify the trial solution and try $x = at$. With this choice of x we have $x' = a$ and $x'' = 0$, so that substituting we require $0 + 2ap = -p^2$ from which $a = -p/2$. So

$$\text{PS} = -\tfrac{1}{2}pt$$

Therefore the general solution is given by

$$x = A + B\,e^{-2pt} - \tfrac{1}{2}pt \qquad (1)$$

Now we use the initial conditions to determine A and B. Differentiating throughout with respect to t we obtain

$$x' = -2Bp\,e^{-2pt} - \tfrac{1}{2}p \qquad (2)$$

When $t = 0$ we obtain from (1) and (2)

$$0 = A + B$$
$$q = -2Bp - \tfrac{1}{2}p$$

So $B = -(p + 2q)/4p$, and $A = (p + 2q)/4p$.

Finally the solution is

$$x = \frac{p + 2q}{4p}\left(1 - e^{-2pt}\right) - \frac{pt}{2}$$

Here are two problems for *you* to try. The first is an electrical problem, the second a mechanical problem. You may choose which you wish to do.

LC CIRCUIT

An alternating EMF $E \sin nt$ is applied to a quiescent circuit consisting of an inductance L and a capacitance C in series. Obtain the current at time $t > 0$, if $\omega^2 = 1/(LC) \neq n^2$.

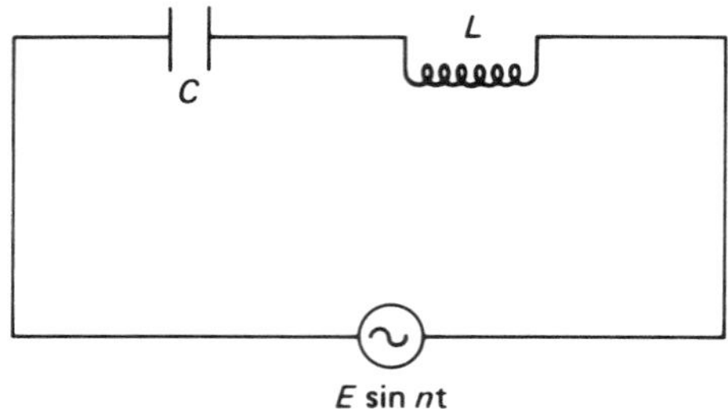

Fig. 20.1 An LC series circuit.

If you cannot cope with the electrical side of this problem, read through the first stage and take over the solution then.

The circuit is illustrated in Fig. 20.1. We have

$$L\frac{\mathrm{d}i}{\mathrm{d}t} + \frac{q}{C} = E \sin nt$$

where i is the current and q is the charge on the capacitor. Now $i = \mathrm{d}q/\mathrm{d}t$, and so

$$\frac{\mathrm{d}i}{\mathrm{d}t} = \frac{\mathrm{d}^2q}{\mathrm{d}t^2}$$

Therefore

$$\frac{\mathrm{d}^2q}{\mathrm{d}t^2} + \frac{1}{LC}q = \frac{E}{L}\sin nt$$

$$\frac{\mathrm{d}^2q}{\mathrm{d}t^2} + \omega^2 q = \frac{E}{L}\sin nt$$

so that ω is the natural frequency of the circuit. Next we must solve this differential equation.

We begin with the complementary part – a standard routine procedure. The auxiliary equation is

$$m^2 + \omega^2 = 0$$

so that $m = \pm \mathrm{j}\omega$. (Notice that here because i denotes current we are adopting the usual practice of writing j instead of the complex number i.)

With m in the form $\alpha \pm \mathrm{j}\beta$ we see that $\alpha = 0$ and $\beta = \omega$. Consequently

$$\begin{aligned}\mathrm{CP} &= \mathrm{e}^0(A\cos \omega t + B \sin \omega t)\\ &= A \cos \omega t + B \sin \omega t\end{aligned}$$

where A and B are arbitrary constants.

The next step is to find a particular solution.

A glance at the right-hand side of the equation enables us to infer the form of a particular solution. We try $q = a \sin nt + b \cos nt$ and differentiate twice with respect to t to obtain

$$\dot{q} = an \cos nt - bn \sin nt$$
$$\ddot{q} = -an^2 \sin nt - bn^2 \cos nt - n^2(a \sin nt + b \cos nt)$$

So substituting,

$$-n^2(a \sin nt + b \cos nt) + \omega^2(a \sin nt + b \cos nt) = (E/L) \sin nt$$

from which $a(\omega^2 - n^2) = E/L$ and $b(\omega^2 - n^2) = 0$. Since $\omega \neq n$ we can deduce

$$a = \frac{E}{L(\omega^2 - n^2)} \qquad b = 0$$

Therefore a particular solution is

$$q = \frac{E \sin nt}{L(\omega^2 - n^2)}$$

The general solution is then

$$q = A \cos \omega t + B \sin \omega t + \frac{E \sin nt}{L(\omega^2 - n^2)}$$

Initially the circuit is quiescent. This means there is no charge on the capacitor and there is no current. Use this information to obtain the arbitrary constants A and B.

When $t = 0$, $q = 0$ and so $A = 0$. Therefore

$$q = B \sin \omega t + \frac{E \sin nt}{L(\omega^2 - n^2)}$$

So

$$i = \frac{\mathrm{d}q}{\mathrm{d}t} = B\omega \cos \omega t + \frac{En \cos nt}{L(\omega^2 - n^2)}$$

When $t = 0$, $i = 0$ and so

$$B\omega + \frac{En}{L(\omega^2 - n^2)} = 0$$

$$B\omega = -\frac{En}{L(\omega^2 - n^2)}$$

Finally

$$i = -\frac{En \cos \omega t}{L(\omega^2 - n^2)} + \frac{En \cos nt}{L(\omega^2 - n^2)}$$

$$= \frac{En}{L(\omega^2 - n^2)} (\cos nt - \cos \omega t)$$

You may remember in the further exercises of Chapter 8 using l'Hospital's rule to obtain i when $\omega = n$. This of course corresponds to the breakdown case.

OSCILLATING BODY

A small body of mass m performs oscillations controlled by a spring of stiffness λ and subject to a frictional force of constant magnitude F (Fig. 20.2). The equation which describes the motion is

$$m\ddot{x} = -\lambda x + F$$

where x is the displacement from the position in which the spring has zero tension. The body is released from rest with a displacement a. Obtain the displacement when it next comes to rest.

Try this and see how it goes. We have one slight difficulty: some of the notation which we usually employ has been used here in a different way. We must be nimble in mind and prepared to use other symbols.

We begin as usual by obtaining the complementary part. Let us use u for the variable in the auxiliary equation. We then have

$$mu^2 + \lambda = 0$$

so that putting $\omega^2 = \lambda/m$ (positive) we obtain

$$u^2 + \omega^2 = 0$$

from which $u = \pm i\omega$. Consequently

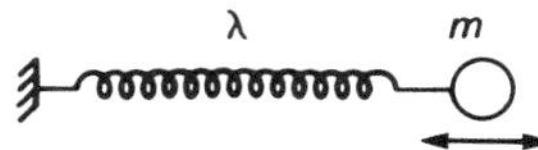

Fig. 20.2 Spring and mass.

$$CP = A \cos \omega t + B \sin \omega t$$

where A and B are arbitrary constants.
Now find a particular solution.

Here the forcing function is F, a constant. Therefore we look for a constant solution. Suppose $x = c$, a constant (we cannot use a). Then substituting, $\lambda c = F$ and so $c = F/\lambda$. A particular solution is therefore obtained:

$$PS = F/\lambda$$

The general solution is then

$$x = A \cos \omega t + B \sin \omega t + F/\lambda$$

Now complete the solution by first determining A and B.

When $t = 0$, $x = a$, so

$$a = A + F/\lambda$$
$$A = a - F/\lambda$$

Also when $t = 0$, $\dot{x} = 0$. Now

$$\dot{x} = -A\omega \sin \omega t + B\omega \cos \omega t$$

so that $0 = B$.
We have

$$x = (a - F/\lambda) \cos \omega t + F/\lambda$$

and also

$$\dot{x} = -(a - F/\lambda)\, \omega \sin \omega t$$

When the body is next at rest, $\dot{x} = 0$ and so we have $\sin \omega t = 0$. This first occurs when $\omega t = \pi$, and at this time $\cos \omega t = -1$. At this moment the displacement is

$$d = (a - F/\lambda)\,(-1) + F/\lambda = 2F/\lambda - a$$

SUMMARY

To obtain the general solution of a non-homogeneous second-order linear differential equation:

$$a\frac{d^2y}{dx^2} + b\frac{dy}{dx} + cy = f(x)$$

where a, b and c are real constants ($a \neq 0$):

☐ The complementary part u is the general solution of the equation

$$a\frac{d^2y}{dx^2} + b\frac{dy}{dx} + cy = 0$$

To obtain this, write down and solve the auxiliary equation

$$am^2 + bm + c = 0$$

and obtain the roots m_1 and m_2. There are three cases:

a If the roots m_1 and m_2 are both real and distinct,

$$u = A\,e^{m_1x} + B\,e^{m_2x}$$

b If the roots m_1 and m_2 are equal, so $m_1 = m_2 = m$,

$$u = (A + Bx)\,e^{mx}$$

c If the roots m_1 and m_2 are complex, so $m = \alpha \pm i\beta$,

$$u = e^{\alpha x}(A\cos\beta x + B\sin\beta x)$$

A and B are arbitrary constants.

☐ Examine u carefully to see whether $f(x)$ corresponds to the breakdown case. Then obtain v, a particular solution of the equation

$$a\frac{d^2y}{dx^2} + b\frac{dy}{dx} + cy = f(x)$$

using a trial solution.

☐ Then

$$y = u + v$$

general solution = complementary part + particular solution

☐ If boundary conditions are given then the constants A and B can be determined.

EXERCISES

1 Obtain the general solution of

a $2\dfrac{d^2x}{dt^2} - 7\dfrac{dx}{dt} + 3x = 0$

b $\dfrac{d^2x}{dt^2} - 2\dfrac{dx}{dt} + 10x = 0$

c $9\dfrac{d^2y}{dx^2} - 24\dfrac{dy}{dx} + 16y = 0$

2 Obtain the general solution of

a $2\dfrac{d^2x}{dt^2} - 9\dfrac{dx}{dt} - 5x = t$

b $3\dfrac{d^2y}{dx^2} - 8\dfrac{dy}{dx} + 4y = e^{2x}$

c $9\dfrac{d^2u}{dx^2} - 9\dfrac{du}{dx} + 2u = e^x$

d $5\dfrac{d^2y}{dx^2} - 4\dfrac{dy}{dx} + y = \cos x$

3 Obtain the solution which satisfies the conditions that when $t = 0, x = 0$ and $dx/dt = 0$ for

a $\dfrac{d^2x}{dt^2} - 6\dfrac{dx}{dt} + 10x = \sin t$

b $3\dfrac{d^2x}{dt^2} - 16\dfrac{dx}{dt} + 5x = e^{5t}$

c $25\dfrac{d^2x}{dt^2} - 30\dfrac{dx}{dt} + 9x = t\,e^{3t/5}$

ASSIGNMENT

Obtain the general solutions of each of the following differential equations:

1 $4\dfrac{d^2y}{dx^2} - 4\dfrac{dy}{dx} + y = e^x$

2 $\dfrac{d^2y}{dt^2} + 4\dfrac{dy}{dt} + 8y = \cos 2t$

3 $\dfrac{d^2x}{dt^2} + 2\dfrac{dx}{dt} - 3x = e^{2t}$

4 $\dfrac{d^2u}{dv^2} - 8\dfrac{du}{dv} + 16u = v^2$

5 $\dfrac{d^2s}{dt^2} + 6\dfrac{ds}{dt} + 10s = \cos t$

6 $\dfrac{d^2u}{dt^2} - 7\dfrac{du}{dt} + 10u = 1 + e^{5t}$

7 $\dfrac{d^2y}{dx^2} + 6\dfrac{dy}{dx} + 9y = e^{-3x} + e^x$

8 $\dfrac{d^2y}{du^2} - 3\dfrac{dy}{du} - 10y = \cosh 2u$

9 $\dfrac{d^2y}{dx^2} + 2\dfrac{dy}{dx} + 10y = e^{-x} \cos 3x$

10 $\dfrac{d^2y}{dw^2} + \dfrac{dy}{dw} - 2y = w \cos w$

FURTHER EXERCISES

1 A constant EMF E is applied to a series circuit with resistance R, capacitance C and inductance L. Given that

$$L\frac{di}{dt} + Ri + \frac{q}{C} = E$$

where, at time t, q is the charge on the capacitor and i is the current. Show that the system will oscillate if $4L > CR^2$.

2 The differential equation representing the simple harmonic motion (SHM) of a particle of unit mass is

$$\ddot{x} = -\lambda^2 x$$

where λ is a constant and the dots denote differentiation with respect to time. Solve this equation and express x in terms of t, given that x is zero when $t = 0$ and that the speed is u at $x = a$.

3 A capacitor of capacitance C discharges through a circuit of resistance R and inductance L. Show that if $CR^2 = 4L$ the discharge is just non-oscillatory. The initial voltage is E and $CR^2 = 4L$. Show that the charge q on the capacitor and the current i are given by

$$q = \frac{2E}{R}\left(\frac{2L}{R} + t\right) \exp(-Rt/2L)$$

$$i = -\frac{Et}{L} \exp(-Rt/2L)$$

4 The differential equation for the deflection y of a light cantilever of length c clamped horizontally at one end and with a concentrated load W at the other satisfies the equation

$$EI\frac{d^2y}{dx^2} = W(c - x)$$

where EI is the flexural rigidity and is constant. Show that the deflection at the free end is $Wc^3/3EI$.

5 The displacement x in metres at time t in seconds of a vibrating governor is given by the differential equation

$$\ddot{x} + x = \sin 2t$$

where dots denote differentiation. Initially the displacement and the speed are zero. Show that the next time the speed is instantaneously zero is when $t = 2\pi/3$ seconds.

6 An EMF $E \sin \omega t$ (where E and ω are constant) is applied to an RLC series circuit. The charge q on the capacitor and the current i are both initially zero. Show that if $CR^2 = 4L$ and $\omega^2 = 1/LC$ then at time t

$$i = (E/R)\,[\sin \omega t - \omega t \exp(-\omega t)]$$

7 The components of acceleration for a model which simulates the movement of a particle in a plane are

$$\ddot{x} = \omega\dot{y}$$
$$\ddot{y} = a\omega^2 - \omega\dot{x}$$

where a and ω are constant. When $t = 0$ the particle is stationary at the origin. Show that subsequently it describes the curve defined parametrically by

$$x = a(\theta - \sin\theta)$$
$$y = a(1 - \cos\theta)$$

where $\theta = \omega t$.

8 A light horizontal strut of length L and flexural rigidity EI carries a concentrated load W at its midpoint. It is supported at each end and subjected to a compressive force P. The deflection y at a point distance x from one end is given by

$$\frac{d^2y}{dx^2} + n^2y = -\frac{Wn^2x}{2P} \qquad \left(0 \leqslant x \leqslant \frac{L}{2}\right)$$

where $n^2 = P/EI$. Solve this equation to show that the greatest deflection of the strut which occurs at its midpoint is

$$\frac{WL}{4P}\left[\frac{\tan(nL/2)}{nL/2} - 1\right]$$

9 The current i in an LRC series circuit satisfies

$$L\frac{d^2i}{dt^2} + R\frac{di}{dt} + \frac{1}{C}i = E\cos nt$$

where L, R, C, E and n are constant and t denotes time. Given that R is positive, show that the exponential terms in the solution of this equation are transient. Show further that when the transient terms are ignored,

$$i = E\frac{nR\sin nt + (1/C - Ln^2)\cos nt}{R^2n^2 + (1/C - Ln^2)^2}$$

10 A beam of length L and of weight w per unit length is clamped horizontally at both ends. The beam is subject to an axial compressive load P. The deflection y is related to the distance x from one end by the equation

$$EIy'' + Py = G - \tfrac{1}{2}wLx + \tfrac{1}{2}wx^2 \qquad \left(0 \leqslant x \leqslant \frac{L}{2}\right)$$

where G is the clamping couple, E is Young's modulus and I is the moment of inertia, and the dashes indicate differentiation with respect to x. Show that

$$y = \frac{1}{P}\left[\left(\frac{w}{n^2} - G\right)\cos nx + \frac{wL}{2n}\sin nx + G - \frac{w}{n^2} - \frac{1}{2}wLx + \frac{1}{2}wx^2\right]$$

where $n^2 = P/EI$.

11 The displacement x metres at time t seconds of a vibrating membrane is given by the differential equation

$$\frac{\mathrm{d}^2x}{\mathrm{d}t^2} + x = \sin 2t$$

Initially, when $t = 0$ both x and $\mathrm{d}x/\mathrm{d}t$ are zero. Show that the next time that $\mathrm{d}x/\mathrm{d}t = 0$ is when $t = 2\pi/3$.

12 The equation of motion of a bead executing damped oscillations on a straight line is given by

$$\frac{\mathrm{d}^2x}{\mathrm{d}t^2} + 4\frac{\mathrm{d}x}{\mathrm{d}t} + 5x = 2a\sin t$$

Initially the distance $x = a$ and the speed $\mathrm{d}x/\mathrm{d}t = 0$. Show that

$$x = \frac{a}{4}\mathrm{e}^{-2t}[9\sin t + 5\cos t] + \frac{a}{4}[\sin t - \cos t]$$

Show further that under these circumstances the bead eventually executes simple harmonic motion of amplitude $a\sqrt{2}/4$.

Fourier series 21

We have already seen how certain functions can be expressed by means of Taylor series. Although this is a powerful technique, and is one of the cornerstones of numerical methods, there is one major disadvantage; relatively few functions can be represented by power series. In this chapter we shall see that Fourier series can be used for a much wider class of functions.

After completing this chapter you should be able to

- ☐ Obtain the Fourier series of functions defined on the interval $(-\pi, \pi)$;
- ☐ Distinguish between an odd function and an even function;
- ☐ Obtain Fourier sine series and Fourier cosine series;
- ☐ Apply Dirichlet's conditions to determine the convergence of a general Fourier series;
- ☐ Represent a wide class of functions by means of Fourier series.

At the end of this chapter we shall consider a practical problem involving a full-wave rectifier.

21.1 FOURIER SERIES

In the early nineteenth century it was a Frenchman Joseph Fourier who first had the idea of using a trigonometrical series rather than a power series to represent more general functions. Fourier was an applied mathematician working on the theory of heat and it was some time before his ideas caught the imagination of those studying the theory of functions. Power series are relatively easy to use and a considerable amount was known about their convergence. By contrast almost nothing was known about the convergence of trigonometrical series.

Fourier's belief was that virtually any function could be represented as a trigonometrical series using sines and cosines. This idea met considerable scepticism at first but broadly speaking Fourier was shown to be right. Where did he get the idea from? The answer may lie in the fact that he could sense it in the work he was doing on heat. In a similar manner, any sound consists of an accumulation of vibrations at different pitches, and yet we know that when these are all put together, a whole kaleidoscope of different sounds can result. Fourier's instinctive idea was a triumph of 'lateral thinking' and we shall now explore the basic concepts.

To start with let us suppose that $f(x)$ is defined on the interval $-\pi \leqslant x \leqslant \pi$ and that on this interval $f(x)$ can be represented by a series of cosine and sine functions. Specifically we shall suppose that

$$\begin{aligned} f(x) &= \frac{a_0}{2} + a_1 \cos x + a_2 \cos 2x + a_3 \cos 3x + \cdots \\ &\quad + b_1 \sin x + b_2 \sin 2x + b_3 \sin 3x + \cdots \\ &= \frac{a_0}{2} + \sum_{r=1}^{\infty} \{a_r \cos rx + b_r \sin rx\} \end{aligned}$$

A coefficient $a_0/2$ (rather than the more obvious constant a_0) has been introduced so that later we can include it in a general formula for the coefficients a_n. If we didn't include the '2' here we would have to include it later in a special formula for a_0. The notation which we are using is the one which is employed almost universally. Note also that we need to include a constant term for the cosines because $\cos rx = 1$ when $r = 0$ whereas $\sin rx = 0$ when $r = 0$.

To derive a formula for a_n we shall multiply through the equation by $\cos nx$ and integrate over the interval $(-\pi, \pi)$. Similarly, to derive a formula for b_n we shall multiply through the equation by $\sin nx$ and integrate over the interval $(-\pi, \pi)$. We shall need one or two results which you may care to derive as a revision exercise on integration. Here they are:

$$\begin{aligned} \int_{-\pi}^{\pi} \cos mx \cos nx \; dx &= 0 \text{ if } m \neq n \\ \int_{-\pi}^{\pi} \cos mx \cos nx \; dx &= \pi \text{ if } m = n \\ \int_{-\pi}^{\pi} \sin mx \sin nx \; dx &= 0 \text{ if } m \neq n \\ \int_{-\pi}^{\pi} \sin mx \sin nx \; dx &= \pi \text{ if } m = n \\ \int_{-\pi}^{\pi} \sin mx \cos nx \; dx &= 0 \end{aligned}$$

Let us assume therefore that $f(x)$ can be represented by the trigonometrical series

$$f(x) = \frac{a_0}{2} + \sum_{r=1}^{\infty}\{a_r \cos rx + b_r \sin rx\}$$

We shall obtain a_n and b_n separately.

1 We multiply through by $\cos nx$ and integrate over the interval $(-\pi, \pi)$, observing that every term on the right-hand side will produce zero except the nth term. Consequently we obtain

$$\int_{-\pi}^{\pi} f(x) \cos nx \; dx = a_n \pi$$

Therefore

$$a_n = \frac{1}{\pi}\int_{-\pi}^{\pi} f(x) \cos nx \; dx$$

The question of the constant $a_0/2$ needs to be considered. If we integrate $f(x)$ over the interval $(-\pi, \pi)$ we obtain

$$\begin{aligned}
\int_{-\pi}^{\pi} f(x) \; dx &= \frac{a_0}{2}\int_{-\pi}^{\pi} dx \\
&= \frac{a_0}{2}\; 2\pi \\
&= a_0\; \pi \\
\text{so } a_0 &= \frac{1}{\pi}\int_{-\pi}^{\pi} f(x) \; dx
\end{aligned}$$

which is consistent with the formula for a_n which we derived for $n \neq 0$.

2 We multiply through by $\sin nx$ and integrate over the interval $(-\pi, \pi)$, again observing that every term on the right-hand side will produce zero except the nth term. Therefore

$$\int_{-\pi}^{\pi} f(x) \sin nx \; dx = b_n \pi$$

So

$$b_n = \frac{1}{\pi}\int_{-\pi}^{\pi} f(x) \sin nx \; dx$$

Naturally we have presumed quite a lot in all this. For instance, we have presumed that the series does in fact converge to $f(x)$, that we are justified in multiplying through by $\cos nx$ and $\sin nx$, integrating term by term, and that the result which we have obtained is meaningful. We shall consider questions of this kind a little later but presuming this, we have now obtained formulae which enable us to calculate the coefficients a_n and b_n. Here they are again:

$$a_n = \frac{1}{\pi}\int_{-\pi}^{\pi} f(x)\cos nx \; dx$$
$$b_n = \frac{1}{\pi}\int_{-\pi}^{\pi} f(x)\sin nx \; dx$$

It will perhaps come as no surprise to learn that the coefficients in the trigonometrical series are known as Fourier coefficients.

☐ Suppose $f(x) = x$ throughout the interval $(-\pi, \pi)$ and that

$$f(x) = \frac{a_0}{2} + \sum_{r=1}^{\infty}\{a_r \cos rx + b_r \sin rx\}$$

Obtain the trigonometrical series explicitly.

We have straight away

$$a_n = \frac{1}{\pi}\int_{-\pi}^{\pi} f(x)\cos nx \; dx$$

Here

$$a_n = \frac{1}{\pi}\int_{-\pi}^{\pi} x\cos nx \; dx$$

Integrate by parts, when $n \neq 0$

$$\begin{aligned}
a_n &= \frac{1}{\pi}\left[x\left(\frac{1}{n}\sin nx\right) - \int\left(\frac{1}{n}\sin nx\right)\; dx\right]_{-\pi}^{\pi} \\
&= \frac{1}{\pi}\left[0 - \frac{1}{n}\int_{-\pi}^{\pi} \sin nx \; dx\right] \\
&= -\frac{1}{n\pi}\int_{-\pi}^{\pi} \sin nx \; dx \\
&= -\frac{1}{n\pi}\left[-\frac{1}{n}\cos nx\right]_{-\pi}^{\pi} \\
&= \frac{1}{n^2\pi}[\cos n\pi - \cos(-n\pi)] \\
&= 0
\end{aligned}$$

Whereas for a_0

$$
\begin{aligned}
a_0 &= \frac{1}{\pi}\int_{-\pi}^{\pi} x \, \mathrm{d}x \\
&= \frac{1}{\pi}\left[\frac{x^2}{2}\right]_{-\pi}^{\pi} \\
&= \frac{1}{\pi}[0] \\
&= 0
\end{aligned}
$$

We now turn our attention to b_n:

$$b_n = \frac{1}{\pi}\int_{-\pi}^{\pi} x \sin nx \, \mathrm{d}x$$

Again, integrate by parts,

$$
\begin{aligned}
b_n &= \frac{1}{\pi}\left[x\left(\frac{-\cos nx}{n}\right) - \int\left(\frac{-\cos nx}{n}\right) \, \mathrm{d}x\right]_{-\pi}^{\pi} \\
&= \frac{1}{\pi}\left[\left\{\frac{-\pi\cos n\pi}{n} - \frac{\pi\cos(-n\pi)}{n}\right\} + \frac{1}{n}\int_{-\pi}^{\pi}\cos nx \, \mathrm{d}x\right]
\end{aligned}
$$

Now $\cos(-n\pi) = \cos n\pi = (-1)^n$ and so

$$
\begin{aligned}
b_n &= \frac{-2\cos n\pi}{n} + \frac{1}{n^2\pi}[\sin nx]_{-\pi}^{\pi} \\
&= \frac{-2(-1)^n}{n} + 0 \\
&= \frac{2}{n}(-1)^{n+1}
\end{aligned}
$$

We have therefore obtained the trigonometrical series

$$
\begin{aligned}
\sum_{n=1}^{\infty}\left\{\frac{2}{n}(-1)^{n+1}\sin nx\right\} &= 2\sum_{n=1}^{\infty}\left\{(-1)^{n+1}\frac{\sin nx}{n}\right\} \\
&= 2\left\{\sin x - \frac{\sin 2x}{2} + \frac{\sin 3x}{3} - \cdots\right\}
\end{aligned}
$$

to represent $f(x) = x$ in the assumption that equality holds on the interval $(-\pi, \pi)$ and that we can multiply the series term by term and then integrate without disturbing or distorting the convergence. ■

Of course $f(x) = x$ is a rather strange example to use to obtain a trigonometrical series because we have no difficulty whatever in dealing with polynomial functions. It is the more obscure functions that occur in practical applications which concern us. However there are two principal reasons why we have worked through this example. First it provides a relatively simple exercise for us to use to illustrate how to calculate the Fourier coefficients, and secondly it leads us into a discussion of a special feature which some functions possess and which enables us to reduce the work in finding their Fourier coefficients.

21.2 ODD AND EVEN FUNCTIONS

Suppose $f(x)$ is defined on the interval $-\pi \leqslant x \leqslant \pi$, then f is said to be an **odd** function if

$$f(-x) = -f(x) \quad \text{whenever} \quad -\pi \leqslant x \leqslant \pi$$

You already know of many functions which are odd functions. Here are a few examples: $x, x^3, \sin x, \sinh x$. Odd functions are easily recog-

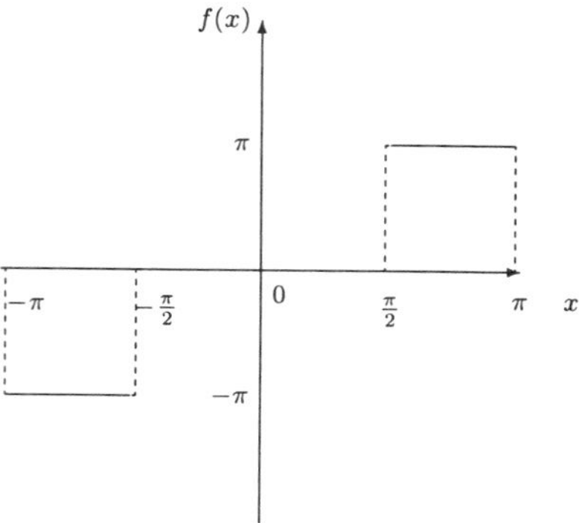

Fig. 21.1: An odd function.

nized by their graphs; they are symmetrical with respect to the origin (Fig. 21.1).
Suppose $f(x)$ is defined on the interval $-\pi \leqslant x \leqslant \pi$, then f is said to be an **even** function if

$$f(-x) = f(x) \quad \text{whenever} \quad -\pi \leqslant x \leqslant \pi$$

Here are a few examples of even functions: $1, x^2, \cos x, \cosh x$. Even functions are easily recognized by their graphs; they are symmetrical about the y-axis (Fig. 21.2).

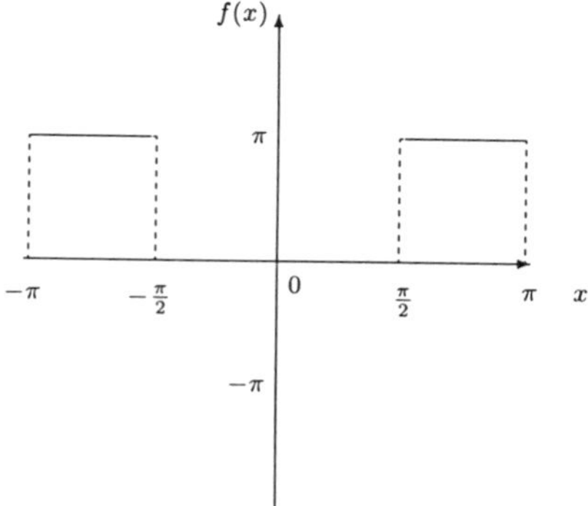

Fig. 21.2: An even function.

In fact the identity

$$f(x) \equiv \left[\frac{f(x)+f(-x)}{2}\right] + \left[\frac{f(x)-f(-x)}{2}\right]$$

shows that every function f defined on the interval $(-\pi,\ \pi)$ can be regarded as the sum of an even function and an odd function.

In the previous example we found a trigonometrical series corresponding to an odd function and it had one curious feature. Did you notice what it was? There were no cosine terms whatever. We went through the motions of calculating the Fourier coefficients a_n only to find they were all zero. Was this a coincidence or is there something deeper here?

Let us suppose that

$$f(x) = \frac{a_0}{2} + \sum_{r=1}^{\infty}\{a_r \cos rx + b_r \sin rx\}$$

and that f is an odd function on the interval $(-\pi,\ \pi)$.

We have already derived formulae for the Fourier coefficients and we have

$$\begin{aligned} a_n &= \frac{1}{\pi}\int_{-\pi}^{\pi} f(x)\cos nx \ \mathrm{d}x \\ b_n &= \frac{1}{\pi}\int_{-\pi}^{\pi} f(x)\sin nx \ \mathrm{d}x \end{aligned}$$

Therefore

$$a_n = \frac{1}{\pi}\int_{-\pi}^{\pi} f(x)\cos nx \ \mathrm{d}x$$

$$= \frac{1}{\pi}\left\{\int_{-\pi}^{0} f(x)\cos nx \,\mathrm{d}x + \int_{0}^{\pi} f(x)\cos nx \,\mathrm{d}x\right\}$$

Putting $t = -x$ in the first integral and using $f(-t) = -f(t)$ we obtain

$$\begin{aligned} a_n &= \frac{1}{\pi}\left\{\int_{\pi}^{0} f(-t)\cos(-nt)\,(-\mathrm{d}t) + \int_{0}^{\pi} f(x)\cos nx \,\mathrm{d}x\right\} \\ &= \frac{1}{\pi}\left\{\int_{\pi}^{0} f(t)\cos nt \,\mathrm{d}t + \int_{0}^{\pi} f(x)\cos nx \,\mathrm{d}x\right\} \\ &= \frac{1}{\pi}\left\{-\int_{0}^{\pi} f(t)\cos nt \,\mathrm{d}t + \int_{0}^{\pi} f(x)\cos nx \,\mathrm{d}x\right\} \end{aligned}$$

Observe that t is a dummy variable in the first integral just as x is a dummy variable in the second. Consequently these two integrals are equal and so cancel one another out.

Therefore we have shown that for an odd function $a_n = 0$. Moreover we can use the same idea to simplify slightly the formula for b_n in the case of an odd function. We obtain

$$\begin{aligned} b_n &= \frac{1}{\pi}\int_{-\pi}^{\pi} f(x)\sin nx \,\mathrm{d}x \\ &= \frac{1}{\pi}\left\{\int_{-\pi}^{0} f(x)\sin nx \,\mathrm{d}x + \int_{0}^{\pi} f(x)\sin nx \,\mathrm{d}x\right\} \end{aligned}$$

Put $t = -x$ in the first integral:

$$\begin{aligned} &= \frac{1}{\pi}\left\{\int_{\pi}^{0} f(-t)\sin(-nt)\,(-\mathrm{d}t) + \int_{0}^{\pi} f(x)\sin nx \,\mathrm{d}x\right\} \\ &= \frac{1}{\pi}\left\{-\int_{\pi}^{0} f(t)\sin nt \,\mathrm{d}t + \int_{0}^{\pi} f(x)\sin nx \,\mathrm{d}x\right\} \\ &= \frac{1}{\pi}\left\{\int_{0}^{\pi} f(t)\sin nt \,\mathrm{d}t + \int_{0}^{\pi} f(x)\sin nx \,\mathrm{d}x\right\} \\ &= \frac{2}{\pi}\int_{0}^{\pi} f(x)\sin nx \,\mathrm{d}x \end{aligned}$$

We are able to conclude therefore that if f is an **odd** function defined on the interval $(-\pi,\ \pi)$ and if

$$f(x) = \frac{a_0}{2} + \sum_{r=1}^{\infty}\{a_r \cos rx + b_r \sin rx\}$$

then

$$\begin{aligned} b_n &= \frac{2}{\pi}\int_0^{\pi} f(x)\sin nx \, dx \\ \text{and } a_n &= 0 \text{ for any integer } n \geqslant 0 \end{aligned}$$

In a similar way it is easy to deduce that if f is an **even** function defined on the interval $(-\pi, \pi)$ and if

$$f(x) = \frac{a_0}{2} + \sum_{r=1}^{\infty}\{a_r \cos rx + b_r \sin rx\}$$

then

$$\begin{aligned} a_n &= \frac{2}{\pi}\int_0^{\pi} f(x)\cos nx \, dx \\ \text{and } b_n &= 0 \text{ for any integer } n \geqslant 0 \end{aligned}$$

It is a good idea to see if you can deduce this on your own. If you get into difficulties then you can always use the derivation we obtained for an odd function as a model. The algebra is very similar indeed.

21.3 SINE SERIES AND COSINE SERIES

Before we move ahead it will pay us just to stand back a moment and think a little about the two results we have obtained in the case of odd functions and even functions. Notice that the values of a_n and b_n can be obtained once f is known on the interval $(0, \pi)$. Of course this is because f was either an odd function or an even function and so the values of f on the interval $(-\pi, 0)$ are determined.

However, suppose we have a function defined on the interval $(0, \pi)$, we could use these formulae to obtain a_n and b_n. Now this would imply that we had an understanding that the function f was either an odd function or an even function. With this understanding we should obtain a series containing sines only or a series containing cosines only, accordingly.

Therefore given a function defined on the interval $(0, \pi)$ we can obtain either a cosine series or a sine series to represent it on this interval. Implicitly we have extended the definition of our function to the interval $(-\pi, \pi)$. For a sine series the extended function is an odd function. For a cosine series the extended function is an even function. Of course these extended functions differ outside the interval.

For the sine series

$$f(x) = -f(-x) \text{ whenever } -\pi \leqslant x < 0$$

For the cosine series

$$f(x) = f(-x) \text{ whenever } -\pi \leqslant x < 0$$

However we do not need to bother ourselves too much with extending the definitions formally because the extensions are implicit in the formulae for a_n and b_n.

☐ Obtain a cosine series for the function defined by $f(x) = x$ on the interval $(0, \pi)$.

Our first example concerned the function defined by $f(x) = x$ on the interval $(-\pi,\ \pi)$. We now know the function to be an odd function and so it can be represented by a sine series which we determined. In this example we are concerned with an *even* function defined on the interval $(-\pi,\ \pi)$. We could if we wish make the definition of such a function explicit. It can be defined by $f(x) = |x|$ when $-\pi \leqslant x \leqslant \pi$. As we know it is symmetrical about the y-axis.

For this example we do not need to concern ourselves with these details but can instead proceed to obtain the cosine series by evaluating the formula for a_n.

We have

$$f(x) = \frac{a_0}{2} + \sum_{n=1}^{\infty} a_n \cos nx$$

where

$$a_n = \frac{2}{\pi}\int_0^{\pi} f(x)\cos nx \, dx$$

We shall need to obtain a_0 separately but for the moment we shall calculate a_n when $n \neq 0$. Integrating by parts we have

$$\begin{aligned}
a_n &= \frac{2}{\pi}\int_0^{\pi} x\cos nx \, dx \\
&= \frac{2}{\pi}\left[x\left(\frac{\sin nx}{n}\right) - \int \frac{\sin nx}{n}\, dx\right]_0^{\pi} \\
&= -\frac{2}{n\pi}\int_0^{\pi} \sin nx \, dx \\
&= -\frac{2}{n\pi}\left[-\frac{\cos nx}{n}\right]_0^{\pi} \\
&= \frac{2}{n^2\pi}[\cos n\pi - \cos 0] \\
&= \frac{2}{n^2\pi}[(-1)^n - 1]
\end{aligned}$$

Now we must obtain a_0. We have

$$\begin{aligned} a_0 &= \frac{2}{\pi}\int_0^{\pi} x \, dx \\ &= \frac{2}{\pi}\left[\frac{x^2}{2}\right]_0^{\pi} \\ &= \frac{2}{\pi}\left[\frac{\pi^2}{2}\right] \\ &= \pi \end{aligned}$$

We observe that $a_n = 0$ when n is even but non-zero and $a_n = -4/n^2\pi$ when n is odd. So that writing $n = 2r + 1$ we have

$$f(x) = \frac{\pi}{2} - \frac{4}{\pi}\sum_{r=0}^{\infty}\frac{\cos(2r+1)\pi}{(2r+1)^2}$$

21.4 DIRICHLET'S CONDITIONS

The time has now come for us to discuss briefly the question of convergence of the trigonometrical series we have been discussing. We shall state some conditions due to Dirichlet which enable us to proceed meaningfully.

We first introduce a useful notation. Suppose we have a function f defined on the interval $(-\pi, \pi)$. We shall write

$$f(x) \sim \frac{a_0}{2} + \sum_{r=1}^{\infty}\{a_r \cos rx + b_r \sin rx\}$$

if and only if

$$\begin{aligned} a_n &= \frac{1}{\pi}\int_{-\pi}^{\pi} f(x)\cos nx \, dx \text{ and} \\ b_n &= \frac{1}{\pi}\int_{-\pi}^{\pi} f(x)\sin nx \, dx \end{aligned}$$

Notice that we are not saying that the series converges or that, if it does, it converges to $f(x)$. All we are saying is that the coefficients a_n and b_n are determined by the integrals. The trigonometrical series on the right-hand side obtained by this procedure is called a Fourier series and it should be stressed that no claims whatever are made about its convergence at this stage.

So given a function f defined on the interval $(-\pi, \pi)$ we can write

$$f(x) \sim \frac{a_0}{2} + \sum_{r=1}^{\infty}\{a_r \cos rx + b_r \sin rx\}$$

The series on the right, periodic with period 2π, is called a Fourier series for f. The Fourier coefficients are determined by

$$a_n = \frac{1}{\pi}\int_{-\pi}^{\pi} f(x)\cos nx \, dx$$
$$b_n = \frac{1}{\pi}\int_{-\pi}^{\pi} f(x)\sin nx \, dx$$

Dirichlet's conditions Suppose the function f is defined on the interval $(-\pi, \pi)$ in such a way that

1 f has only a finite number of maxima and minima on the interval.

2 f has only a finite number of discontinuities on the interval.

3 f has only finite discontinuities on the interval.

Then when $-\pi < x < \pi$ the Fourier series converges to

$$\frac{1}{2}\left[\lim_{t\to 0+} f(x+t) + \lim_{t\to 0+} f(x-t)\right]$$

Note that at points where f is continuous

$$\lim_{t\to 0+} f(x+t) = f(x) = \lim_{t\to 0+} f(x-t)$$

and so the series does indeed converge to $f(x)$ at points where f is continuous.

At any point x of discontinuity we have

$\lim_{t\to 0+} f(x+t)$ is the right-hand limit of f at x

$\lim_{t\to 0+} f(x-t)$ is the left-hand limit of f at x

So that

$$\frac{1}{2}\left[\lim_{t\to 0+} f(x+t) + \lim_{t\to 0+} f(x-t)\right]$$

is the mean value of the right-hand and left-hand limit of f at x.

21.5 HARMONICS

In the Fourier series

$$\frac{a_0}{2} + \sum_{n=1}^{\infty} \{a_n \cos nx + b_n \sin nx\}$$

The term $a_n \cos nx + b_n \sin nx$ is known as the nth harmonic, so that a Fourier series consists of a constant term and a sum of harmonics. When we obtained a Fourier cosine series for $f(x) = x$ on the interval $0 \leqslant x \leqslant \pi$ we observed that there were no even harmonics although the constant term was non-zero. We can save ourselves effort if we can foresee when odd harmonics or even harmonics will be absent.

□ Show that if f satisfies

$$f(x) = f(x + \pi) \text{ whenever } -\pi \leqslant x \leqslant 0$$

then there are no odd harmonics present in the corresponding Fourier series of period 2π.

Notice that by specifying the period we in fact fix the series. We wish to show that $a_n = 0$ and $b_n = 0$ when n is odd. We use the definitions of a_n and b_n to obtain this result directly from the Fourier integrals themselves.

$$\begin{aligned} a_n &= \frac{1}{\pi} \int_{-\pi}^{\pi} f(x) \cos nx \, dx \\ b_n &= \frac{1}{\pi} \int_{-\pi}^{\pi} f(x) \sin nx \, dx \end{aligned}$$

We shall deal with a_n and leave the case of b_n as an exercise.

Our task is to show that when n is odd, $a_n = 0$. Now

$$\begin{aligned} a_n &= \frac{1}{\pi} \int_{-\pi}^{\pi} f(x) \cos nx \, dx \\ &= \frac{1}{\pi} \left\{ \int_{-\pi}^{0} f(x) \cos nx \, dx + \int_{0}^{\pi} f(x) \cos nx \, dx \right\} \end{aligned}$$

Put $t = x + \pi$ in the first integral and observe that since $-\pi \leqslant x \leqslant 0$, $0 \leqslant t \leqslant \pi$. Also

$$\begin{aligned} f(x) &= f(x + \pi) = f(t) \\ \cos n(t - \pi) &= \cos nt \cos n\pi + \sin nt \sin n\pi \end{aligned}$$

$$= (-1)^n \cos nt$$

$$\text{Likewise } \sin n(t-\pi) = \sin nt \cos n\pi - \cos nt \sin n\pi$$

$$= (-1)^n \sin nt$$

Therefore

$$\begin{aligned}
a_n &= \frac{1}{\pi}\left\{\int_0^\pi f(t)\cos n(t-\pi)\ \mathrm{d}t + \int_0^\pi f(x)\cos nx\ \mathrm{d}x\right\} \\
&= \frac{1}{\pi}\left\{\int_0^\pi f(t)\cos nt \cos n\pi\ \mathrm{d}t + \int_0^\pi f(x)\cos nx\ \mathrm{d}x\right\} \\
&= \frac{1}{\pi}\left\{(-1)^n\int_0^\pi f(t)\cos nt\ \mathrm{d}t + \int_0^\pi f(x)\cos nx\ \mathrm{d}x\right\} \\
&= \frac{1}{\pi}\left\{(-1)^n\int_0^\pi f(x)\cos nx\ \mathrm{d}x + \int_0^\pi f(x)\cos nx\ \mathrm{d}x\right\} \\
&= \frac{1}{\pi}\left\{[1+(-1)^n]\int_0^\pi f(x)\cos nx\ \mathrm{d}x\right\}
\end{aligned}$$

As anticipated, we observe that $a_n = 0$ when n is odd. ■

▷ **Exercise** Show that if f satisfies

$$f(x) = -f(x+\pi) \text{ whenever } -\pi \leqslant x \leqslant 0$$

then there are no even harmonics present in the corresponding Fourier series of period 2π.

21.6 FOURIER SERIES OVER ANY FINITE INTERVAL

We have only considered functions which are defined on the interval $(-\pi,\ \pi)$ but it is a simple matter to extend what we have been doing to the more general interval $(-l,\ l)$. We could even extend one step further to an arbitrary interval $(a,\ b)$ but in doing so we would lose the poetry of the formulae. There is in fact little to be said for too much generality since we can always perform a transformation to deal with an arbitrary interval. Nevertheless we shall deal with the interval $(-l,\ l)$ since this will show us how to handle transformations and as it happens the symmetry of the formulae is preserved.

Suppose that f is defined on the interval $(-l,\ l)$ and satisfies Dirichlet's conditions on the interval. We have $f(x)$ is defined whenever

$-l \leqslant x \leqslant l$ so if we substitute $t = \pi x/l$ and write $f(x) = g(t)$ we see that $g(t)$ is defined for $-\pi \leqslant t \leqslant \pi$ and so

$$g(t) \sim \frac{a_0}{2} + \sum_{n=1}^{\infty}\{a_n \cos nt + b_n \sin nt\}$$

where

$$\begin{aligned} a_n &= \frac{1}{\pi}\int_{-\pi}^{\pi} g(t)\cos nt \, \mathrm{d}t \\ b_n &= \frac{1}{\pi}\int_{-\pi}^{\pi} g(t)\sin nt \, \mathrm{d}t \end{aligned}$$

All we now need to do is to substitute back in terms of x to obtain the required formulae. In fact the algebra for a_n and b_n is so similar that we shall work through only one, b_n. The other one, a_n, is left as an exercise.

Here we go then.

$$\begin{aligned} t &= \frac{\pi x}{l} \\ \frac{\mathrm{d}t}{\mathrm{d}x} &= \frac{\pi}{l} \\ \sin nt &= \sin\left(\frac{n\pi x}{l}\right) \end{aligned}$$

So

$$\begin{aligned} b_n &= \frac{1}{\pi}\int_{-l}^{l} f(x)\sin\left(\frac{n\pi x}{l}\right)\frac{\pi}{l}\,\mathrm{d}x \\ &= \frac{1}{l}\int_{-l}^{l} f(x)\sin\left(\frac{n\pi x}{l}\right)\mathrm{d}x \end{aligned}$$

We therefore have

$$f(x) \sim \frac{a_0}{2} + \sum_{n=1}^{\infty}\left\{a_n \cos\left(\frac{n\pi x}{l}\right) + b_n \sin\left(\frac{n\pi x}{l}\right)\right\}$$

where

$$\begin{aligned} a_n &= \frac{1}{l}\int_{-l}^{l} f(x)\cos\left(\frac{n\pi x}{l}\right)\mathrm{d}x \\ b_n &= \frac{1}{l}\int_{-l}^{l} f(x)\sin\left(\frac{n\pi x}{l}\right)\mathrm{d}x \end{aligned}$$

Note that we can test if this looks right by putting $l = \pi$ to see whether or not these formulae reduce to the ones we had before. They do, so all seems well.

It is now time for you to test your skill at working with Fourier series. We shall tackle several fairly easy problems so that we get the ideas fixed.

21.7 Workshop

1 The function f is defined on the interval $(0, \pi)$ by

$$\begin{aligned} f(x) &= \pi \qquad 0 \leqslant x \leqslant \frac{\pi}{2} \\ &= 0 \qquad \frac{\pi}{2} < x \leqslant \pi \end{aligned}$$

Obtain a Fourier cosine series for f with period 2π on the interval $(0, \pi)$.
Try this carefully and then move ahead to check things are all right.

2 We have

$$f(x) \sim \frac{a_0}{2} + \sum_{n=1}^{\infty} a_n \cos nx$$

Where, for all integers $n \geqslant 0$

$$a_n = \frac{2}{\pi} \int_0^{\pi} f(x) \cos nx \, \mathrm{d}x$$

Notice that f is defined by a different formula on each of the two subintervals $(0, \pi/2)$ and $(\pi/2, \pi)$ so it is necessary for us to split the integral in each case.

To calculate a_n we consider a_0 separately:

$$\begin{aligned} a_0 &= \frac{2}{\pi} \int_0^{\pi} f(x) \, \mathrm{d}x \\ &= \frac{2}{\pi} \left\{ \int_0^{\pi/2} \pi \, \mathrm{d}x + \int_{\pi/2}^{\pi} 0 \, \mathrm{d}x \right\} \\ &= 2 \int_0^{\pi/2} \mathrm{d}x \\ &= 2[x]_0^{\pi/2} = \pi \end{aligned}$$

Now for a_n when $n \neq 0$

$$a_n = \frac{2}{\pi} \int_0^{\pi} f(x) \cos nx \, \mathrm{d}x$$

$$= \frac{2}{\pi}\left\{\int_0^{\pi/2} \pi \cos nx \, dx + \int_{\pi/2}^{\pi} 0 \cos nx \, dx\right\}$$

$$= 2\int_0^{\pi/2} \cos nx \, dx$$

$$= 2\left[\frac{\sin nx}{n}\right]_0^{\pi/2}$$

$$= 2\frac{\sin(\frac{n\pi}{2})}{n}$$

Now $\sin(n\pi/2)$ is zero whenever n is even. Suppose n is odd, so $n = 2r+1$, then $\sin(n\pi/2) = (-1)^r$. We therefore have

$$f(x) \sim \frac{\pi}{2} + 2\sum_{r=0}^{\infty}(-1)^r \frac{\cos(2r+1)x}{2r+1}$$

You may have decided not to simplify your answer as far as this and instead have left it in the form

$$f(x) \sim \frac{\pi}{2} + 2\sum_{n=1}^{\infty} \frac{\sin(\frac{n\pi}{2})}{n} \cos nx$$

This is a perfectly valid alternative form.

Now let us look at a follow-up question.

▷ **Exercise** Use Dirichlet's conditions to discuss the convergence of the series obtained in the previous exercise at the specific points $x = 0$, $x = \pi/2$ and $x = \pi$.

We see that f is continuous at all points except possibly 0, $\pi/2$ and π. In fact we certainly have a finite discontinuity at $\pi/2$. Dirichlet's conditions show that the Fourier series will converge to the average of the left-hand limit and the right-hand limit there. Now the right-hand limit is 0 and the left-hand limit is π so we deduce that the series converges to $\pi/2$ at $x = \pi/2$. **3**

Now let us turn our attention to what happens at $x = 0$. Remember we were asked to obtain a cosine series so we have in effect extended f to the interval $(-\pi, \pi)$ in such a way that it becomes an even function. Consequently at 0 both the right-hand limits and the left-hand limits are π and indeed the extended function is continuous at 0. We conclude that when $x = 0$, the series converges to π.

Lastly we consider $x = \pi$. Remember that the series is periodic with period 2π. It follows that the right-hand and left-hand limits are both in effect zero at $x = \pi$ and that therefore we can conclude that the series converges to 0 at $x = \pi$.

▷ **Exercise** Substitute $x = 0$, $x = \pi/2$ and $x = \pi$ into the Fourier cosine series to obtain a series expansion for $\pi/4$ in two different ways.

4

We have

$$f(x) \sim \frac{\pi}{2} + 2\sum_{r=0}^{\infty}(-1)^r \frac{\cos(2r+1)x}{2r+1}$$

Now when $x = 0$ we have deduced that the series converges to π and so we obtain

$$\begin{aligned} \pi &= \frac{\pi}{2} + 2\sum_{r=0}^{\infty}(-1)^r \frac{\cos(2r+1)0}{2r+1} \\ \frac{\pi}{2} &= 2\sum_{r=0}^{\infty} \frac{(-1)^r}{2r+1} \\ \frac{\pi}{4} &= \sum_{r=0}^{\infty} \frac{(-1)^r}{2r+1} \\ &= 1 - \frac{1}{3} + \frac{1}{5} - \frac{1}{7} + \cdots \end{aligned}$$

When $x = \pi/2$, the series converges to $\pi/2$ and therefore

$$\frac{\pi}{2} = \frac{\pi}{2} + 2\sum_{r=0}^{\infty}(-1)^r \frac{\cos\frac{(2r+1)\pi}{2}}{2r+1}$$

Now $\cos(2r+1)\pi/2 = 0$ for every integer r and so we obtain no new information from this series. We do however notice that we have consistency, which might reassure us if we feared we had made an error somewhere.

When $x = \pi$, the series converges to 0 and so we obtain

$$0 = \frac{\pi}{2} + 2\sum_{r=0}^{\infty}(-1)^r \frac{\cos(2r+1)\pi}{2r+1}$$

Now $\cos(2r+1)\pi = (-1)^{2r+1} = -1$ and so we obtain

$$\begin{aligned} -\frac{\pi}{2} &= -2\sum_{r=0}^{\infty} \frac{(-1)^r}{2r+1} \\ \frac{\pi}{4} &= \sum_{r=0}^{\infty} \frac{(-1)^r}{2r+1} \\ &= 1 - \frac{1}{3} + \frac{1}{5} - \frac{1}{7} + \cdots \text{ as before.} \end{aligned}$$

▷ **Exercise** Suppose that instead of choosing a Fourier cosine series for f on the interval $(0, \pi)$ defined by

$$\begin{aligned} f(x) &= \pi \qquad 0 \leqslant x \leqslant \frac{\pi}{2} \\ &= 0 \qquad \frac{\pi}{2} < x \leqslant \pi \end{aligned}$$

we had chosen a Fourier sine series. Without obtaining the series state how it would have converged at $x = 0$, $x = \pi/2$ and $x = \pi$.

We need to remember that a Fourier sine series for a function defined on an interval $(0, \pi)$ is the same as that of an odd function defined on the interval $(-\pi, \pi)$. We can therefore apply Dirichlet's conditions to such a function which we can extend with period 2π outside the interval $(-\pi, \pi)$. **5**

We note that at $x = 0$ the right-hand limit is π and the left-hand limit is $-\pi$ so that the series will converge to 0 at $x = 0$.

The point at which $x = \pi/2$ is within the interval $(0, \pi)$ and so the convergence behaviour of the series will be the same as it was for the Fourier cosine series. We found at step **3** that the series converges to $\pi/2$ when $x = \pi/2$. However although the Fourier sine series and the Fourier cosine series each converge to the same values *within* the interval $(0, \pi)$ the two series are very different.

When $x = \pi$ we observe that both the right-hand limit and the left-hand limit of f are zero. Consequently the Fourier sine series converges to zero too.

For the sake of completeness we shall conclude the workshop by determining the Fourier sine series we have been discussing.

▷ **Exercise** Obtain the Fourier sine series of period 2π which represents f on the interval $(0, \pi)$ where

$$\begin{aligned} f(x) &= \pi \qquad 0 \leqslant x \leqslant \frac{\pi}{2} \\ &= 0 \qquad \frac{\pi}{2} < x \leqslant \pi \end{aligned}$$

As we know, f is defined differently on each of the two subintervals $(0, \pi/2)$ and $(\pi/2, \pi)$ and so **6**

$$\begin{aligned} b_n &= \frac{2}{\pi}\int_0^{\pi} f(x)\sin nx \, dx \\ &= \frac{2}{\pi}\left\{\int_0^{\pi/2} \pi \sin nx \, dx + \int_{\pi/2}^{\pi} 0 \sin nx \, dx\right\} \end{aligned}$$

$$= 2\int_0^{\pi/2} \sin nx \, dx$$

$$= 2\left[\frac{-\cos nx}{n}\right]_0^{\pi/2}$$

$$= 2\left(\frac{1-\cos(\frac{n\pi}{2})}{n}\right)$$

Now $\cos(n\pi/2)$ is zero whenever n is odd. Therefore putting $n = 2r+1$ we obtain $b_{2r+1} = 2/(2r+1)$. On the other hand when n is even, we may write $n = 2r$, then $\cos(n\pi/2) = \cos r\pi = (-1)^r$. We therefore have

$$f(x) \sim 2\sum_{r=0}^{\infty}\left\{\frac{\sin(2r+1)x}{2r+1} + [1-(-1)^r]\frac{\sin 2rx}{2r}\right\}$$

$$= 2\sum_{r=0}^{\infty}\left\{\frac{1}{2r+1}\sin(2r+1)x + \left[\frac{1-(-1)^r}{2r}\right]\sin 2rx\right\}$$

Finally putting $x = \pi/2$ we have $\sin 2rx = \sin\pi = 0$ and $\sin(2r+1)x = \sin(2r+1)\pi = (-1)^r$.

$$\frac{\pi}{2} = 2\sum_{r=0}^{\infty}\left\{\frac{(-1)^r}{2r+1} + [1-(-1)^r]\frac{0}{2r}\right\}$$

$$= 2\sum_{r=0}^{\infty}\frac{(-1)^r}{2r+1}$$

$$\frac{\pi}{4} = \sum_{r=0}^{\infty}\frac{(-1)^r}{2r+1} \quad \text{as before.}$$

21.8 FURTHER DEVELOPMENTS

It is possible to obtain a general Fourier series of period $2l$ for a function f defined on any finite interval $(h, h+2l)$ and which satisfies Dirichlet's conditions on the interval. The formulae involved are very similar to the ones which we have investigated and it is a simple matter to derive them:

$$f(x) \sim \frac{a_0}{2} + \sum_{n=1}^{\infty} \left\{ a_n \cos\left(\frac{n\pi x}{l}\right) + b_n \sin\left(\frac{n\pi x}{l}\right) \right\}$$

where

$$a_n = \frac{1}{l} \int_h^{h+2l} f(x) \cos\left(\frac{n\pi x}{l}\right) \, dx$$

$$b_n = \frac{1}{l} \int_h^{h+2l} f(x) \sin\left(\frac{n\pi x}{l}\right) \, dx$$

One point is well worth bearing in mind. Given a function f defined on an interval $(0, k)$ and which satisfies Dirichlet's conditions on the interval, we have three obvious choices.

1 We could obtain a general Fourier series, of period k, to represent f on the interval $(0, k)$.

2 We could obtain a Fourier cosine series, of period $2k$, to represent f on the interval $(0, k)$.

3 We could obtain a Fourier sine series, of period $2k$, to represent f on the interval $(0, k)$.

It is important to appreciate that each of these series will converge to f in the interval $(0, k)$ but that outside the interval they will differ considerably.

The advent of modern computers has increased the demand for Fourier analysis. Harmonic analysis involves using approximate formulae, sums instead of integrals, for the Fourier coefficients a_n and b_n. This is of practical advantage because it is quite common for a function to be defined only in terms of a finite collection of data values. The underlying assumption, of course, is that all but a finite number of the intermediate values could be inferred by continuity considerations. One of the most well-known methods for dealing with harmonic analysis is an algorithm known as the method of 'fast Fourier transforms'.

In general, Fourier series tend to converge quickly, in the sense that with only a few harmonics the 'shape' of the function becomes evident and the approximate values are good. However there are many examples where the convergence behaviour is not quite as expected. One of the most celebrated examples of these is known as 'Gibb's phenomenon' where it was found that the numerical error near a point of discontinuity can be significantly greater than at points of continuity.

21.9 Practical

We now consider a practical application of Fourier series which involves a rectifier in an electrical circuit.

□ When an electromotive force E acts on a full-wave rectifer the output voltage is $|E|$. Show that if an EMF, $E_0 \sin wt$ acts on such a rectifier, where $E_0 > 0$, then it may be represented on the half-range by a Fourier cosine series

$$E_0 \left\{ \frac{2}{\pi} - \frac{4}{\pi} \sum_{n=1}^{\infty} \frac{\cos 2nwt}{4n^2 - 1} \right\}$$

Since $E_0 > 0$ we deduce that the output voltage corresponding to $E_0 \sin wt$ is $E_0 |\sin wt|$. The imposed EMF $E_0 \sin wt$ is clearly periodic and the half period T is obtained from the equation $wT = \pi$, since this is the smallest value of $T > 0$ at which $\sin wT = 0$.

We therefore require a Fourier cosine series on the interval $(0, T)$ where $T = \pi/w$ to represent the output voltage $e(t)$.

We note also that on the interval $(0, \pi/w)$, $\sin wt > 0$. We now have

$$\begin{aligned} e(t) &\sim \frac{a_0}{2} + \sum_{n=1}^{\infty} a_n \cos\left(\frac{n\pi\, t}{T}\right) \\ &= \frac{a_0}{2} + \sum_{n=1}^{\infty} a_n \cos nwt \end{aligned}$$

where, for all integers n,

$$\begin{aligned} a_n &= \frac{2}{T} \int_0^T e(t) \cos\left(\frac{n\pi t}{T}\right)\, dt \\ &= \frac{2wE_0}{\pi} \int_0^{\pi/w} \sin wt \cos nwt\, dt \end{aligned}$$

Now

$$\sin(A + B) + \sin(A - B) = 2 \sin A \cos B$$

so with $A = wt$ and $B = nwt$ we have

$$\begin{aligned} a_n &= \frac{wE_0}{\pi} \int_0^{\pi/w} [\sin(n+1)wt + \sin(1-n)wt]\, dt \\ &= \frac{wE_0}{\pi} \int_0^{\pi/w} [\sin(n+1)wt - \sin(n-1)wt]\, dt \\ &= \frac{wE_0}{\pi} \left[-\frac{\cos(n+1)wt}{(n+1)w} + \frac{\cos(n-1)wt}{(n-1)w} \right]_0^{\pi/w} \end{aligned}$$

except when $n = 1$.

We can easily obtain a_1 separately for we have

$$a_1 = \frac{wE_0}{\pi}\int_0^{\pi/w} \sin 2wt \, dt = \frac{wE_0}{\pi}\left[-\frac{\cos 2wt}{2w}\right]_0^{\pi/w}$$

$$= -\frac{E_0}{2\pi}[\cos 2\pi - \cos 0] = 0$$

So that for $n \neq 1$ we have

$$\begin{aligned} a_n &= \frac{E_0}{\pi}\left[-\frac{\cos(n+1)\pi}{n+1} + \frac{1}{n+1} \right. \\ &\qquad \left. + \frac{\cos(n-1)\pi}{n-1} - \frac{1}{n-1}\right] \\ &= \frac{E_0}{\pi}\left[-\frac{(-1)^{n+1}}{n+1} + \frac{1}{n+1} + \frac{(-1)^{n-1}}{n-1} - \frac{1}{n-1}\right] \\ &= \frac{(-1)^{n-1}E_0}{\pi}\left[-\frac{1}{n+1} + \frac{1}{n-1}\right] - \frac{2E_0}{\pi(n^2-1)} \\ &= \frac{2(-1)^{n-1}E_0}{\pi(n^2-1)} - \frac{2E_0}{\pi(n^2-1)} \end{aligned}$$

We now note that $(-1)^{n-1} = 1$ when n is odd and that consequently, when n is odd, $a_n = 0$. Therefore the only non-zero terms which occur are when n is even and so we may replace n by $2m$ to simplify a_n:

$$a_n = a_{2m} = \frac{-2E_0}{\pi(n^2-1)} - \frac{2E_0}{\pi(n^2-1)}$$

$$= -\frac{4E_0}{\pi(n^2-1)} = -\frac{4E_0}{\pi(4m^2-1)}$$

Note also that

$$a_0 = \frac{4E_0}{\pi} \text{ so that } \frac{a_0}{2} = \frac{2E_0}{\pi}$$

Substituting for a_n in the Fourier series we obtain

$$\begin{aligned} e(t) &\sim \frac{a_0}{2} + \sum_{n=1}^{\infty} a_n \cos\left(\frac{n\pi\, t}{T}\right) \\ &= \frac{a_0}{2} + \sum_{n=1}^{\infty} a_n \cos nwt \\ &= \frac{a_0}{2} + \sum_{m=1}^{\infty} a_{2m} \cos 2mwt \end{aligned}$$

$$= \frac{2E_0}{\pi} + \sum_{m=1}^{\infty}\left[-\frac{4E_0}{\pi(4m^2-1)}\right]\cos 2mwt$$

$$= \frac{2E_0}{\pi} - \frac{4E_0}{\pi}\sum_{m=1}^{\infty}\left[\frac{\cos 2mwt}{4m^2-1}\right]$$

$$= E_0\left[\frac{2}{\pi} - \frac{4}{\pi}\sum_{n=1}^{\infty}\frac{\cos 2nwt}{4n^2-1}\right]$$

Note that at the final stage we have replaced the dummy variable m by the dummy variable n in order to obtain the required form. ■

It is worth remarking that the Fourier series we have obtained corresponds to that of an even function over the interval $(-\pi/w,\ \pi/w)$ and that $\sin wt$, $0 \leqslant t \leqslant \pi/w$, is symmetrical about the line $t = \pi/2w$. Moreover, the Fourier series we have obtained is periodic outside the interval $(0,\ \pi/w)$ with period π/w.

Observe too that Dirichlet's conditions are satisfied and that at the end-points of the interval $(0,\ \pi/w)$ the Fourier series converges to zero.

Now this is precisely the response $E_0|\sin wt|$ from this full-wave rectifier and so, in this particular case, the Fourier series does indeed provide a faithful representation of the response for all t.

SUMMARY

□ Given a function f defined on the interval $(-l,\ l)$ we can obtain a Fourier series for f of period $2l$:

$$f(x) \sim \frac{a_0}{2} + \sum_{n=1}^{\infty} \left\{ a_n \cos\left(\frac{n\pi x}{l}\right) + b_n \sin\left(\frac{n\pi x}{1}\right) \right\}$$

where, for integers $n \geqslant 0$,

$$a_n = \frac{1}{2l} \int_{-l}^{l} f(x) \cos\left(\frac{n\pi x}{l}\right)\, \mathrm{d}x$$

$$b_n = \frac{1}{2l} \int_{-l}^{l} f(x) \sin\left(\frac{n\pi x}{l}\right)\, \mathrm{d}x$$

□ Given a function f defined on the interval $(0,\ l)$ we can obtain a Fourier cosine series for f of period $2l$:

$$f(x) \sim \frac{a_0}{2} + \sum_{n=1}^{\infty} a_n \cos\left(\frac{n\pi x}{l}\right)$$

where, for integers $n \geqslant 0$,

$$a_n = \frac{1}{l} \int_{0}^{l} f(x) \cos\left(\frac{n\pi x}{l}\right)\, \mathrm{d}x$$

□ Given a function f defined on the interval $(0,\ l)$ we can obtain a Fourier sine series for f of period $2l$:

$$f(x) \sim \sum_{n=1}^{\infty} b_n \sin\left(\frac{n\pi x}{l}\right)$$

where, for integers $n \geqslant 1$,

$$b_n = \frac{1}{l} \int_{0}^{l} f(x) \sin\left(\frac{n\pi x}{l}\right)\, \mathrm{d}x$$

EXERCISES

1 Obtain a Fourier cosine series of period 2π for each of the following functions:

a $$\begin{aligned} f(x) &= 0, & 0 \leqslant x < \pi/2 \\ f(x) &= x - \pi/2, & \pi/2 < x \leqslant \pi \end{aligned}$$

b $$\begin{aligned} f(x) &= \pi/2, & 0 \leqslant x < \pi/2 \\ f(x) &= \pi - x, & \pi/2 < x \leqslant \pi \end{aligned}$$

2 Obtain a Fourier sine series of period 2π for each of the following functions:

a $$\begin{aligned} f(x) &= 0, & 0 \leqslant x < \pi/2 \\ f(x) &= \pi - x, & \pi/2 < x \leqslant \pi \end{aligned}$$

b $$\begin{aligned} f(x) &= x, & 0 \leqslant x < \pi/2 \\ f(x) &= \pi - x, & \pi/2 < x \leqslant \pi \end{aligned}$$

3 Obtain a Fourier cosine series, of period 2π, which converges to $\sin x$ on the interval $0 < x < \pi$.

4 A function f is defined by the equations:

$$\begin{aligned} f(x) &= \pi/3 & 0 \leqslant x \leqslant \pi/3 \\ &= x & \pi/3 \leqslant x \leqslant 2\pi/3 \\ &= 0 & 2\pi/3 < x \leqslant \pi \end{aligned}$$

Obtain a Fourier cosine series for f which has period 2π and which converges to f on the interval $(0,\ \pi)$. State the numbers to which the series converges when $x = 0,\ \pi/3,\ 2\pi/3,\ \pi,\ 2\pi$.

ASSIGNMENT

1 Obtain a Fourier cosine series of period $2l$ for each of the following functions:

a $$\begin{aligned} f(x) &= 0, & 0 \leqslant x < l/2 \\ f(x) &= l, & l/2 < x \leqslant l \end{aligned}$$

b $$\begin{aligned} f(x) &= 0, & 0 \leqslant x < l/3 \\ f(x) &= l, & l/3 < x \leqslant 2l/3 \\ f(x) &= 0, & 2l/3 < x \leqslant l \end{aligned}$$

2 Obtain a general Fourier series of period $2l$ for each of the following functions:

a $$\begin{aligned} f(x) &= l, & 0 \leqslant x < l \\ f(x) &= 0, & l < x \leqslant 2l \end{aligned}$$

b $$\begin{aligned} f(x) &= l, & 0 \leqslant x < 2l/3 \\ f(x) &= 0, & 2l/3 < x \leqslant 4l/3 \\ f(x) &= l, & 4l/3 < x \leqslant 2l \end{aligned}$$

3 Obtain a Fourier sine series, of period $2l$, which converges to $\cos(\pi x/l)$ on the interval $0 < x < l$.

4 The function f defined by $f(t) = \pi - t$ when $0 \leqslant t \leqslant \pi$ is to be represented by means of a Fourier cosine series and also by a

Fourier sine series on the interval $(0,\ \pi)$ each of period 2π. Show that these are

$$\frac{\pi}{2}+\frac{4}{\pi}\sum_{n=1}^{\infty}\frac{\cos(2n-1)t}{(2n-1)^2}$$

and

$$2\sum_{n=1}^{\infty}\frac{\sin nt}{n}$$

respectively. Explain how the convergences of these series differ on the interval $(0,\ 2\pi)$.

FURTHER EXERCISES

1 Obtain a Fourier series of period π for each of the following functions:

a
$$\begin{aligned} f(x) &= 0, & 0 \leqslant x < \pi/2 \\ f(x) &= x-\pi/2, & \pi/2 < x \leqslant \pi \end{aligned}$$

b
$$\begin{aligned} f(x) &= \pi/2, & 0 \leqslant x < \pi/2 \\ f(x) &= \pi - x, & \pi/2 < x \leqslant \pi \end{aligned}$$

c
$$\begin{aligned} f(x) &= 0, & 0 \leqslant x < \pi/2 \\ f(x) &= \pi - x, & \pi/2 < x \leqslant \pi \end{aligned}$$

d
$$\begin{aligned} f(x) &= x, & 0 \leqslant x < \pi/2 \\ f(x) &= \pi - x, & \pi/2 < x \leqslant \pi \end{aligned}$$

2 Show that the Fourier series of period 2π corresponding to $f(x) = x^2$ where $-\pi \leqslant x \leqslant \pi$ is

$$\frac{\pi^2}{3}+4\sum_{r=1}^{\infty}(-1)^r\frac{\cos rx}{r^2}$$

Use the series to show that

$$\frac{\pi^2}{6}=\sum_{r=1}^{\infty}\frac{1}{r^2}$$

3 A function f of period 10 is defined by

$$\begin{aligned} f(x) &= 0 \quad \text{for} \quad -5 < x < 0 \\ &= 4 \quad \text{for} \quad 0 < x < 5 \end{aligned}$$

Show that the Fourier series, of period 10 corresponding to this function is

$$2+\frac{8}{\pi}\sum_{r=0}^{\infty}\frac{1}{2r+1}\sin\left(\frac{(2r+1)\pi x}{5}\right)$$

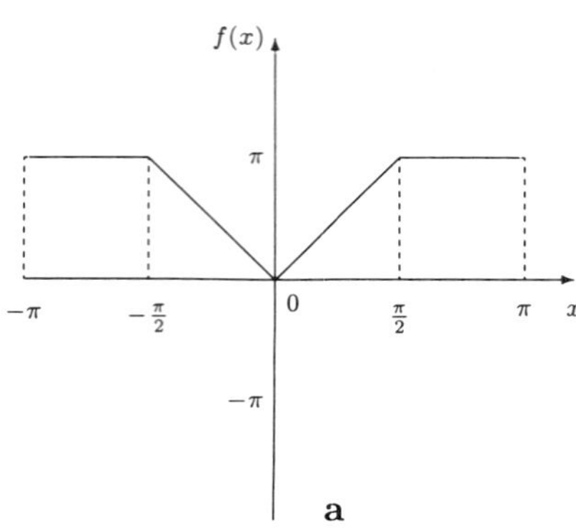

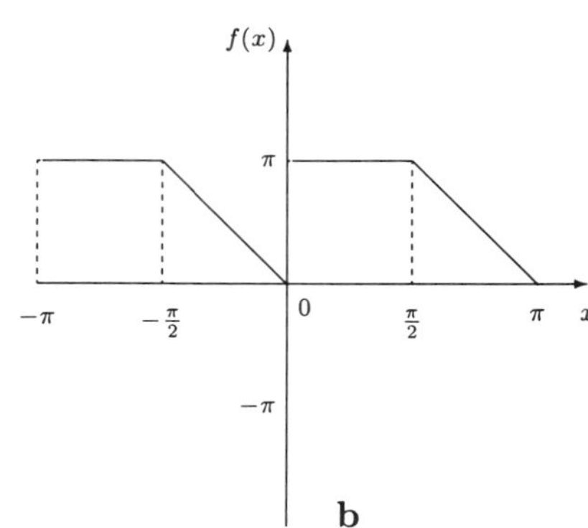

4 Obtain Fourier series corresponding to the each of the functions f illustrated in diagrams **a** and **b** and which converge to $f(x)$, at each point x where f is continuous, on the interval $-\pi < x < \pi$. At the end points, and at any points of discontinuity, state the numbers to which the series converge.

5 Explain how x can be represented on the interval $(0, \pi)$ by either a Fourier sine series or a Fourier cosine series. Hence or otherwise show

$$x \sim \frac{\pi}{2} - \frac{4}{\pi}\sum_{r=0}^{\infty}\frac{1}{(2r+1)^2}\cos(2r+1)x$$

Hence or otherwise deduce that

$$\frac{\pi^2}{8} = 1 + \frac{1}{3^2} + \frac{1}{5^2} + \cdots + \frac{1}{(2r+1)^2} + \cdots$$

6 Show that a general Fourier series of period 2π for the function defined by $f(x) = \sin x$ when $0 \leqslant x \leqslant \pi$ and $f(x) = 0$ when $\pi \leqslant x \leqslant 2\pi$ is

$$\frac{1}{\pi} + \frac{1}{2}\sin x - \frac{2}{\pi}\sum_{n=1}^{\infty}\frac{\cos 2nx}{4n^2-1}$$

Hence or otherwise deduce the sum of the series

$$\frac{1}{1\times 3} + \frac{1}{3\times 5} + \frac{1}{5\times 7} + \cdots$$

7 The rectified half-wave sine current is defined by

$$\begin{aligned} f(t) &= \sin w\pi t & 0 \leqslant t \leqslant 1/w \\ &= 0 & 1/w < t \leqslant 2/w \end{aligned}$$

Obtain a Fourier series for f of period $2/w$ and use Dirichlet's conditions to deduce the numbers to which it converges when $t = k/w$, where k is any natural number.

8 Obtain **a** a cosine series and **b** a sine series of period 4 for the function defined by

$$\begin{aligned} f(x) &= 2(1-x) && 0 < x \leqslant 1 \\ &= 0 && 0 < x \leqslant 2 \end{aligned}$$

9 An even function f is defined by the following equations

$$\begin{aligned} f(x) &= f(x-4) && \text{all real } x \\ &= 1 && 0 \leqslant x < 1/2 \\ &= 2(1-x) && 1/2 \leqslant x < 3/2 \\ &= -1 && 3/2 \leqslant x < 2 \end{aligned}$$

Show that

$$f(x) \sim \frac{16}{\pi^2} \sum_{n=1}^{\infty} \frac{1}{n^2} \sin\frac{n\pi}{2} \sin\frac{n\pi}{4} \cos\frac{n\pi x}{2}$$

10 The function f is defined by

$$\begin{aligned} f(t) &= t && 0 \leqslant t < \pi/2 \\ &= \pi/2 && \pi/2 \leqslant t < \pi \end{aligned}$$

Obtain the following representation for f as a Fourier cosine series

$$\frac{3\pi}{8} + \frac{2}{\pi} \sum_{n=1}^{\infty} \left(\cos\frac{n\pi}{2} - 1\right) \frac{\cos nt}{n^2}$$

Sketch the graph of the series on the interval $(-2\pi,\ 2\pi)$ and show further that

$$\frac{3\pi^2}{16} = 1 + \frac{2}{2^2} + \frac{1}{3^2} + \frac{1}{5^2} + \frac{2}{6^2} + \frac{1}{7^2} + \frac{1}{9^2} + \frac{2}{10^2} + \frac{1}{11^2} + \cdots$$

22 Laplace transforms

In Chapters 19 and 20 we solved differential equations by first obtaining the general solution and then using initial conditions to obtain the solution which was relevant. In this chapter we shall introduce a different approach in which we make use of the initial conditions at the outset and so do not obtain the general solution at all.

After completing this chapter you should be able to

- ☐ Recognize and distinguish between Fourier transforms and Laplace transforms;
- ☐ Obtain the Laplace transform and inverse Laplace transform of simple functions;
- ☐ Solve linear ordinary differential equations using Laplace transform methods;
- ☐ Use the Dirac 'delta function' to represent impulses and solve equations in which these arise.

At the end of this chapter we shall consider a practical problem involving a 'spike function' and an electrical circuit.

22.1 INTEGRAL TRANSFORMS

The underlying idea behind the use of integral transforms is that they give a method of replacing a differential equation by an algebraic equation. Intuitively algebraic equations are easier to solve than differential equations. What happens is that we normally have a differential equation together with a set of initial conditions and we wish to obtain the solution. To solve it we transform the differential equation into an algebraic one, solve that, and then reverse the transformation process to obtain the solution which we require. The only difficulty with this simple idea therefore rests in the transformations themselves. We shall devise a set of rules to help us with this but we must be prepared to face some difficulties somewhere along the line. These are usually to be found in obtaining the inverse transform. However these days things are much better than they used to be because computer algebra packages can take the drudgery out of solving routine calculations.

One word of warning. The sight of an integral transform might be rather daunting. Don't be put off, try to get the overview as to what is going on and don't be alarmed by the detail.

Suppose that $f(t)$ is defined whenever $a < t < b$ and that $K(s,t)$ is some expression in both s and t. We define the integral transform $T[f(t)]$, with kernel $K(s,t)$, on the interval (a,b), by:

$$T[f(t)] = \int_a^b K(t,s)\ f(t)\ \mathrm{d}t$$

whenever the integral exists.

Observe that we started with $f(t)$, which depended on an unknown t, and have produced something which depends on another unknown s; s is, of course, a parameter.

Now all this is very general but as soon as we specify the kernel $K(s,t)$ and the interval (a,b) we obtain a specific transform. The choice of the kernel and the interval is determined by what we wish the transform to achieve.

□ $K(s,t) = \mathrm{e}^{-st}$ and $(a,b) = (0,\infty)$ produces the **Laplace** transform.
$K(s,t) = \cos st$ or $K(s,t) = \sin st$ and $(a,b) = (0,\infty)$ produces **Fourier** transforms.

$K(s,t) = \cos nt$ or $K(s,t) = \sin nt$ and $(a,b) = (0,\pi)$, where $n \geqslant 0$ is an integer, produces **finite** Fourier transforms.

■

In this chapter we shall concentrate our attention on Laplace transforms because it is these that are used to solve ordinary differential equations. Were we to be solving partial differential equations then we could apply Fourier transforms of one type or another. You will probably find it necessary to use these transforms in due course but, once you have understood and become familiar with the techniques required to use Laplace transforms effectively, this should present you with no great difficulty.

We shall use the notation

$$F(s) = \mathcal{L}[f(t)] = \int_0^\infty \mathrm{e}^{-st}\ f(t)\ \mathrm{d}t$$

to represent the Laplace transform of $f(t)$. We shall hope to reverse the process to recover the expression in t, $f(t)$ which has some given expression in s, $F(s)$, as its Laplace transform.

In fact we are being just a little too optimistic but it turns out that for all practical purposes we can reverse the process of finding the Laplace transform.

□ T is an integral transform. Show that if $f(t)$ and $g(t)$ are expressions in t, and h and k are constants, then:

$$T[hf(t) + kg(t)] = hT[f(t)] + kT[g(t)]$$

This rule is a linearity rule. We shall show it is true in general so that we then have the property established for *any* integral transform. It is in fact a simple exercise on integration.

$$\begin{aligned}
T[hf(t) + kg(t)] &= \int_a^b [hf(t) + kg(t)]K(s,t)\ \mathrm{d}t \\
&= \int_a^b [hf(t)K(s,t) + kg(t)K(s,t)]\ \mathrm{d}t \\
&= \int_a^b hf(t)K(s,t)\ \mathrm{d}t + \int_a^b kg(t)K(s,t)\ \mathrm{d}t \\
&= h\int_a^b f(t)K(s,t)\ \mathrm{d}t + k\int_a^b g(t)K(s,t)\ \mathrm{d}t \\
&= hT[f(t)] + kT[g(t)]
\end{aligned}$$

■

There is one question which needs to be considered before long and that is the question of existence. When is it possible to obtain the integral transform of a function? Clearly this will depend not only on the function itself but also on the kernel and the interval (a, b). If we are more specific and ask when the *Laplace* transform of a function exists then we need only concern ourselves with $f(t)$ because the interval $(0, \infty)$ and the kernel e^{-st} are already fixed. This all comes down to whether or not the integral

$$\int_0^\infty f(t)\mathrm{e}^{-st}\ \mathrm{d}t$$

exists. Considerations of this kind are rather advanced but suffice it to say that if $f(t)$ is piecewise continuous on the closed interval $[0, \infty]$

then we shall have a fighting chance of settling the matter. A function is piecewise continuous on an interval if it is possible to partition the interval into a finite number of subintervals in such a way that (1) the function is continuous on each subinterval and (2) the function has a finite right-hand and left-hand limit at each point of discontinuity. However we shall not always restrict our attention to these functions but, occasionally in a somewhat cavalier fashion, we shall allow ourselves more freedom; the only proviso being that the Laplace transform will be presumed to exist.

One other point needs to be made. The parameter s is undefined and so we can take it to be almost anything we like. For instance we can assume it is a very large, but unspecified, positive number; or even a complex number, the real part of which is very large.

There is a whole host of rules for obtaining Laplace transforms and we shall consider these shortly but first we shall find our feet by obtaining a few of the easier transforms by direct integration.

☐ Obtain the Laplace transform of each of the following functions: **a** 1 **b** e^{at} **c** t **d** $1 + 2t + 3e^t$.

a We have

$$\begin{aligned}\mathcal{L}[1] &= \int_0^\infty \mathrm{e}^{-st}\,\mathrm{d}t \\ &= \left[\frac{\mathrm{e}^{-st}}{-s}\right]_0^\infty \\ &= \frac{1}{s}\end{aligned}$$

Here we are supposing that $s > 0$, so that $\lim_{t\to\infty} \mathrm{e}^{-st} = 0$.
b The integral gives

$$\begin{aligned}\mathcal{L}[\mathrm{e}^{at}] &= \int_0^\infty \mathrm{e}^{at}\,\mathrm{e}^{-st}\,\mathrm{d}t \\ &= \int_0^\infty \mathrm{e}^{-(s-a)}\,\mathrm{d}t \\ &= \frac{1}{s-a}\end{aligned}$$

Here we are supposing that $s > a$, so that $\lim_{t\to\infty} \mathrm{e}^{-(s-a)} = 0$.
c Here

$$\mathcal{L}[t] = \int_0^\infty t\,\mathrm{e}^{-st}\,\mathrm{d}t \quad \text{integrate by parts}$$

$$= \left[\frac{t\ \mathrm{e}^{-st}}{(-s)} - \int \frac{\mathrm{e}^{-st}}{(-s)}\ \mathrm{d}t\right]_0^\infty$$

$$= \frac{1}{s}\int_0^\infty \mathrm{e}^{-st}\ \mathrm{d}t$$

$$= \frac{1}{s^2}$$

There are two points worth noting here. The first is that we have made use of example **a**. We could have emphasized this by the inclusion of an extra step:

$$\frac{1}{s}\int_0^\infty \mathrm{e}^{-st}\ \mathrm{d}t = \frac{1}{s}\mathcal{L}[1] = \frac{1}{s^2}$$

The second is that we have made use of a property of limits

$$\lim_{t\to\infty}(t\mathrm{e}^{-st}) = 0$$

This is a property which we shall require for every $f(t)$ that we wish to transform. It is sometimes expressed by saying the $f(t)$ is of **exponential order**. We shall presume that for all the functions which we use

$$\lim_{t\to\infty}(f(t)\mathrm{e}^{-st}) = 0$$

d We use the linearity property which we derived for any integral transform

$$\begin{aligned}
\mathcal{L}[1+2t+3t^2] &= \mathcal{L}[1] + 2\mathcal{L}[t] + 3\mathcal{L}[\mathrm{e}^t] \\
&= \frac{1}{s} + 2\frac{1}{s^2} + 3\frac{1}{s-1} \\
&= \frac{s+2}{s^2} + \frac{3}{s-1} \\
&= \frac{(s+2)(s-1)+3s^2}{s^2(s-1)} \\
&= \frac{4s^2+s-2}{s^2(s-1)}
\end{aligned}$$

■

The first two of these examples illustrate a more general property which we shall find useful a little later. In fact the result which we shall now obtain is often known as a shifting theorem.

The first shifting theorem Suppose that $F(s) = \mathcal{L}[f(t)]$, then

$$F(s+a) = \mathcal{L}[\mathrm{e}^{-at}\ f(t)]$$

Proof Given that $F(s) = \mathcal{L}[f(t)]$ we have

$$\begin{aligned} F(s) &= \int_0^\infty \mathrm{e}^{-st} f(t)\ \mathrm{d}t \\ \text{So } F(s+a) &= \int_0^\infty \mathrm{e}^{-(s+a)t} f(t)\ \mathrm{d}t \\ &= \int_0^\infty \mathrm{e}^{-st}\ [\mathrm{e}^{-at} f(t)]\ \mathrm{d}t \\ &= \mathcal{L}[\mathrm{e}^{-at}\ f(t)] \end{aligned}$$

■

The time has come to develop some more general formulae which will enable us to determine other Laplace transforms. We should not lose sight of the fact that ultimately we shall wish to reverse the process and the work we did on partial fractions will be needed.

Property 1 Suppose $f'(t)$ exists, then under the usual conditions

$$\mathcal{L}[f'(t)] = s\ \mathcal{L}[f(t)] - \lim_{t\to 0+} [f(t)]$$

Proof We shall perform the integration, commenting later on some of the difficulties which arise.

$$\begin{aligned} \mathcal{L}[f'(t)] &= \int_0^\infty f'(t)\mathrm{e}^{-st}\ \mathrm{d}t \\ &= \int_0^\infty \mathrm{e}^{-st} f'(t)\ \mathrm{d}t \quad \text{integrate by parts} \\ &= [\mathrm{e}^{-st} f(t)]_0^\infty - \int_0^\infty (-s)\mathrm{e}^{-st} f(t)\ \mathrm{d}t \\ &= [\mathrm{e}^{-st} f(t)]_0^\infty + s\int_0^\infty \mathrm{e}^{-st} f(t)\ \mathrm{d}t \\ &= \lim_{t\to\infty} [\mathrm{e}^{-st} f(t)] - \lim_{t\to 0+} [\mathrm{e}^{-st} f(t)] + s\mathcal{L}[f(t)] \\ &= -\lim_{t\to 0+} [e^{-st} f(t)] + s\mathcal{L}[f(t)] \\ &= s\mathcal{L}[f(t)] - \lim_{t\to 0+} [f(t)] \end{aligned}$$

We have certainly applied the product rule for limits to deduce that

$$\lim_{t\to 0+} [\mathrm{e}^{-st} f(t)] = \lim_{t\to 0+} [\mathrm{e}^{-st}] \times \lim_{t\to 0+} [f(t)]$$

and the fact that

$$\lim_{t\to 0+} [\mathrm{e}^{-st}] = 1$$

to obtain this result. We have already mentioned that we shall be supposing that all the functions under consideration possess the property that

$$\lim_{t\to\infty} [\mathrm{e}^{-st} f(t)] = 0$$

The significance of this result is easy to miss but we will do well to look at it carefully because it is of great importance. To see it we shall change the notation to write

$$x = f(t) \text{ so that } \frac{\mathrm{d}x}{\mathrm{d}t} = f'(t)$$

We then have

$$\mathcal{L}\left[\frac{\mathrm{d}x}{\mathrm{d}t}\right] = s\mathcal{L}[x] - \lim_{t\to 0+} x$$

The key thing to note is that the derivative $\mathrm{d}x/\mathrm{d}t$ has been transformed in terms of the transform of x itself and also the 'initial' value of x. The effect of taking the limit of x as t tends to $0+$ is to track back to the value of x when t is 0. You will observe that we do not need to know the value of x when t is 0, we are more interested in its value immediately afterwards. This is a very common requirement in Science and Engineering and this initial value can often be inferred even if it is not known beforehand.

Repeated application of this property will enable us to take any linear differential equation with constant coefficients and transform it into an algebraic equation containing $X = \mathcal{L}[x]$ and the initial values of x and its derivatives. We therefore are able to take a differential equation of this form and transform it into an algebraic equation for X. Our ability to extract X explicitly from the equation is not in doubt and it only remains to reverse the Laplace transform to obtain the solution x.

Note that for instance:

$$\begin{aligned}
\mathcal{L}\left[\frac{\mathrm{d}^2x}{\mathrm{d}t^2}\right] &= s\mathcal{L}\left[\frac{\mathrm{d}x}{\mathrm{d}t}\right] - \lim_{t\to 0+} \frac{\mathrm{d}x}{\mathrm{d}t} \\
&= s\left\{s\mathcal{L}[x] - \lim_{t\to 0+} x\right\} - \lim_{t\to 0+} \frac{\mathrm{d}x}{\mathrm{d}t} \\
&= s^2\mathcal{L}[x] - s\lim_{t\to 0+} x - \lim_{t\to 0+} \frac{\mathrm{d}x}{\mathrm{d}t}
\end{aligned}$$

☐ Obtain the Laplace transforms of $\sin at$ and $\cos at$.

We can use the property which we have just been discussing to obtain each of these transforms. We observe that

$$
\begin{aligned}
\frac{\mathrm{d}}{\mathrm{d}t}[\sin at] &= a\cos at \\
\frac{\mathrm{d}}{\mathrm{d}t}[\cos at] &= -a\sin at \\
\text{so } \mathcal{L}\left[\frac{\mathrm{d}}{\mathrm{d}t}\sin at\right] &= a\mathcal{L}[\cos at] \\
\text{and } \mathcal{L}\left[\frac{\mathrm{d}}{\mathrm{d}t}\cos at\right] &= -a\mathcal{L}[\sin at] \\
\text{but } \mathcal{L}\left[\frac{\mathrm{d}}{\mathrm{d}t}\sin at\right] &= s\mathcal{L}[\sin at] - \lim_{t\to 0+}[\sin at] \\
&= s\mathcal{L}[\sin at] - \sin[0] = s\mathcal{L}[\sin at] \\
\text{likewise } \mathcal{L}\left[\frac{\mathrm{d}}{\mathrm{d}t}\cos at\right] &= s\mathcal{L}[\cos at] - \lim_{t\to 0+}[\cos at] \\
&= s\mathcal{L}[\cos at] - \cos[0] = s\mathcal{L}[\cos at] - 1
\end{aligned}
$$

Writing $X = \mathcal{L}[\sin at]$ and $Y = \mathcal{L}[\cos at]$ we have

$$
\begin{aligned}
sX &= aY \\
sY - 1 &= -aX \\
\text{Therefore } s^2X &= s(aY) \\
&= a(sY) \\
&= a(-aX + 1) \\
&= -a^2X + a \\
\text{from which } (s^2 + a^2)X &= a \\
X &= \frac{a}{s^2 + a^2} \\
Y &= \frac{s}{s^2 + a^2} \\
\mathcal{L}[\sin at] &= \frac{a}{s^2 + a^2}
\end{aligned}
$$

$$\mathcal{L}[\cos at] = \frac{s}{s^2 + a^2}$$

■

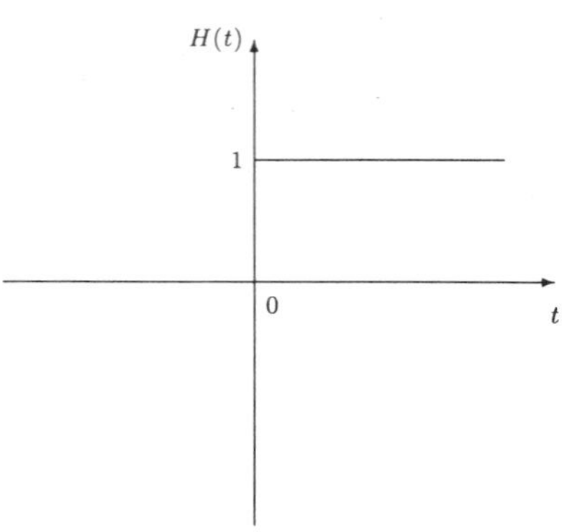

Fig. 22.1: The Heaviside unit function.

22.2 THE HEAVISIDE UNIT FUNCTION

In many physical situations things change suddenly; a switch is thrown, brakes are applied, collisions occur. The Heaviside unit function, H, is a very useful function for representing sudden change (Fig. 22.1).

We define $H(t)$ at all numbers t except $t = 0$ by

$$H(t) = \begin{cases} 1 & t > 0 \\ 0 & t < 0 \end{cases}$$

Consequently $H(t - a) = 1$ when $t > a$ and $H(t - a) = 0$ when $t < a$. This simple observation enables us to represent quite complicated events by means of a single equation.

□ When a machine is switched on at time $t = 0$ the response R is given for time $t \leqslant t_1$ by $R(t) = Pt/t_1$, where P is a constant. From time t_1 the response remains constant until time t_2 when the machine is switched off. Thereafter the response is given by $R(t) = P \exp -(t-t_2)$. Use the Heaviside unit function to represent the response by a single equation.

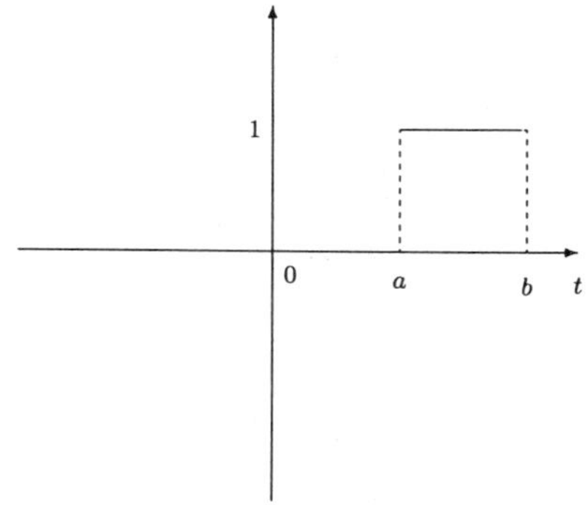

Fig. 22.2: The graph of $H(t-a) - H(t-b)$.

Before solving this problem we shall make a simple observation. Suppose $0 \leqslant a \leqslant b$ then

$$H(t-a) - H(t-b)$$

has the value 0 before $t = a$, the value 0 after $t = b$ but the value 1 between $t = a$ and $t = b$ (Fig. 22.2). We can use expressions such as these as units to build up quite complicated formulae. For if we multiply a term by $H(t-a) - H(t-b)$ then it will only be effective between the times $t = a$ and $t = b$.

In this case, we have the response Pt/t_1 between time $t = 0$ and $t = t_1$ so we represent this by

$$\frac{Pt}{t_1}[H(t) - H(t-t_1)]$$

Subsequently, the response is $P\exp -(t-t_2)$ and there is no prescribed time at which this response ceases; the response attenuates naturally. We may therefore represent this by

$$Pe^{-(t-t_2)}H(t-t_2)$$

Finally we put these together into one equation and obtain

$$\begin{aligned} R(t) &= \frac{Pt}{t_1}[H(t) - H(t-t_1)] + Pe^{-(t-t_2)}H(t-t_2) \\ &= P\left\{\frac{t}{t_1}[H(t) - H(t-t_1)] + e^{-(t-t_2)}H(t-t_2)\right\} \end{aligned}$$

■

□ Obtain the Laplace transform of $H(t-a)$ where $a > 0$.

We apply the definition and calculate the integral.

$$\begin{aligned}
\mathcal{L}[H(t-a)] &= \int_0^\infty H(t-a)\mathrm{e}^{-st}\,\mathrm{d}t \\
&= \left\{\int_0^a H(t-a)\mathrm{e}^{-st}\,\mathrm{d}t + \int_a^\infty H(t-a)\mathrm{e}^{-st}\,\mathrm{d}t\right\} \\
&= \left\{\int_0^a 0\,\mathrm{e}^{-st}\,\mathrm{d}t + \int_a^\infty 1\,\mathrm{e}^{-st}\,\mathrm{d}t\right\} \\
&= \int_a^\infty \mathrm{e}^{-st}\,\mathrm{d}t \text{ Put } x = t-a \\
&= \int_0^\infty \mathrm{e}^{-s(x+a)}\,\mathrm{d}x \\
&= e^{-as}\int_0^\infty \mathrm{e}^{-sx}\,\mathrm{d}x \\
&= e^{-as}\mathcal{L}[1] = \frac{\mathrm{e}^{-as}}{s}
\end{aligned}$$

■

When we generalize this we obtain a very useful property of the Laplace transform known as the second shifting theorem.

The second shifting theorem If $F(s) = \mathcal{L}[f(t)]$ then

$$\mathcal{L}[H(t-a)f(t-a)] = \mathrm{e}^{-as}F(s)$$

See if you can derive it yourself; it is quite straightforward.

Proof $\mathcal{L}[H(t-a)f(t-a)]$

$$\begin{aligned}
&= \int_0^\infty H(t-a)f(t-a)\mathrm{e}^{-st}\,\mathrm{d}t \\
&= \left\{\int_0^a H(t-a)f(t-a)\mathrm{e}^{-st}\,\mathrm{d}t + \int_a^\infty H(t-a)f(t-a)\mathrm{e}^{-st}\,\mathrm{d}t\right\} \\
&= \left\{\int_0^a 0\,\mathrm{e}^{-st}\,\mathrm{d}t + \int_a^\infty f(t-a)\mathrm{e}^{-st}\,\mathrm{d}t\right\} \\
&= \int_a^\infty f(t-a)\mathrm{e}^{-st}\,\mathrm{d}t \quad \text{Put } x = t-a \\
&= \int_0^\infty f(x)\mathrm{e}^{-s(x+a)}\,\mathrm{d}x \\
&= e^{-as}\int_0^\infty f(x)\mathrm{e}^{-sx}\,\mathrm{d}x
\end{aligned}$$

$$= e^{-as} \int_0^\infty f(t)e^{-st}\ dt$$

$$= e^{-as}\mathcal{L}[f(t)] = e^{-as}\ F(s)$$

■

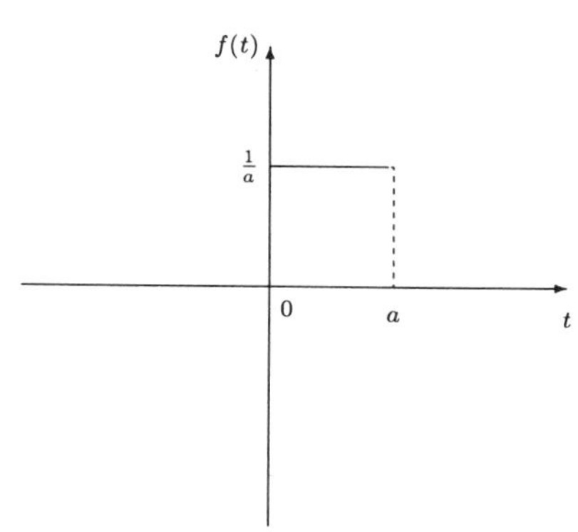

Fig. 22.3: As $a \to 0+$, $f(t) \to \delta(t)$.

22.3 THE DIRAC DELTA FUNCTION

The Heaviside unit function provides a mechanism by which to discuss a very interesting concept which is known as the **Dirac delta function**. We lead into this by considering the following function (Fig. 22.3):

$$f(t) = \begin{cases} 0 & t < 0 \\ 1/a & 0 < t < a \\ 0 & t > a \end{cases}$$

We can represent this as a single equation by writing

$$f(t) = \frac{1}{a}[H(t) - H(t-a)]$$

For each number $a > 0$ we obtain a function of this kind but for each of them the area under the curve is always 1. In terms of an integral we can express this by writing

$$\int_{-\infty}^{\infty} f(t)\ dt = 1$$

Moreover we can determine the Laplace transform of f easily, since

$$\begin{aligned}
\mathcal{L}[f(t)] &= \mathcal{L}\left[\frac{1}{a}[H(t) - H(t-a)]\right] \\
&= \frac{1}{a}\,\mathcal{L}[H(t) - H(t-a)] \\
&= \frac{1}{a}\,[\mathcal{L}[H(t)] - [H(t-a)]] \\
&= \frac{1}{a}\left[\frac{1}{s} - \frac{\mathrm{e}^{-as}}{s}\right] \\
&= \frac{1}{as}\,[1 - \mathrm{e}^{-as}] \\
&= \frac{1 - \mathrm{e}^{-as}}{as}
\end{aligned}$$

We now consider what happens as $a \to 0+$. We define

$$\delta(t) = \lim_{a \to 0+} f(t)$$

The area under the curve is 1 whatever the value of a and so the limit too will be 1. So we define

$$\int_{-\infty}^{\infty} \delta(t)\ \mathrm{d}t = \lim_{a \to 0+}\left[\int_{-\infty}^{\infty} f(t)\ \mathrm{d}t\right] = 1$$

Also, in a similar way, we can determine what happens to the Laplace transform

$$\begin{aligned}
\mathcal{L}[\delta(t)] &= \lim_{a \to 0+} \mathcal{L}[f(t)] \\
&= \lim_{a \to 0+} [\frac{1 - \mathrm{e}^{-as}}{as}] \\
&= 1 \text{ using l'Hospital's rule}
\end{aligned}$$

We have one other property of $\delta(t)$ which can be expressed by writing

$$\delta(t) = 0 \qquad t \neq 0$$

One thing is certain, $\delta(t)$, the Dirac delta function, is not a function in the technical sense in which we have used the word but nevertheless is very useful for representing impulsive motion. $\delta(t)$ is sometimes called a **generalized** function; electrical engineers like to call it a 'spike' function. Something measurable occurs instantaneously at time $t = 0$ but at no other time.

The basic features of the Dirac delta function are contained in the following three properties which can be derived:

$$\int_{-\infty}^{\infty} \delta(t)\,\mathrm{d}t = 1$$

$$\mathcal{L}[\delta(t-a)] = e^{-as}$$

$$\delta(t) = 0 \qquad t \neq 0$$

These, together with the other general properties of Laplace transforms which we shall develop, will be sufficient for our purposes.

□ An impulsive blow of 5 newtons is applied to a structure ten seconds after an experiment begins. Thirty seconds later a further impulsive blow of 10 newtons is applied. Represent the applied force, $f(t)$ after t seconds, in newtons using the Dirac delta function and thereby obtain its Laplace transform.

We have impulsive blows applied at time $t = 10$ and time $t = 40$. We can represent unit blows at these times by $\delta(t-10)$ and $\delta(t-40)$ respectively. Therefore

$$f(t) = 5\delta(t-10) + 10\delta(t-40)\ \mathrm{N}$$

Now

$$\begin{aligned}\mathcal{L}[f(t)] &= \mathcal{L}[5\delta(t-10) + 10\delta(t-40)] \\ &= 5\,\mathcal{L}[\delta(t-10)] + 10\mathcal{L}[\delta(t-40)] \\ &= 5\mathrm{e}^{-10s}\mathcal{L}[\delta(t)] + 10\mathrm{e}^{-40s}\mathcal{L}[\delta(t)] \\ &= 5\mathrm{e}^{-10s} + 10\mathrm{e}^{-40s}\end{aligned}$$

■

22.4 THE INVERSE LAPLACE TRANSFORM

We have described how to obtain the Laplace transform of a function either by first principles, that is by evaluating an improper integral, or by using some of the rules which have been determined. The time has now come to settle the important question of whether or not two different functions can have the same Laplace transform.

Lerch's theorem If two functions f and g, defined for positive real numbers, have the same Laplace transform then they differ by a **Null** function.

A null function N is a function such that

$$\int_0^T N(t)\,dt = 0 \qquad \text{for all } T > 0$$

In other words there is no area whatever under the curve $x = N(t)$. In one sense this is very reassuring because we now know that for all practical purposes we can regard the inverse of a Laplace transform as unique. It must be remembered however that this also means we could change the value of a function f at an arbitrary number of points without affecting the value of its Laplace transform F. Nonetheless for all practical purposes once $F(s)$ is known we can obtain $f(t)$ and for this reason we write $f(t) = \mathcal{L}^{-1}[F(s)]$ and call $f(t)$ the **inverse** Laplace transform of $F(s)$.

The process by which we determine an inverse Laplace transform requires us to express $F(s)$ in a form from which it is possible to recognize standard Laplace transforms. Before we can do so, therefore, we need to have a small table (Table 22.1) of Laplace transforms available as a reference guide. You will observe that in this table not only have we included the Laplace transforms of a few standard functions but also the two shifting theorems. We shall find that these are very useful indeed.

The convolution theorem We know from bitter experience that 'the integral of a product is *not* the product of the respective integrals'. A similar difficulty occurs with inverse Laplace transforms. In other words 'the inverse Laplace transform of a product is *not* the product of the respective inverse Laplace transforms'. However, there is a useful rule for inverse Laplace transforms and it is known as the convolution theorem.

Suppose that $F(s) = \mathcal{L}[f(t)]$ and $G(s) = \mathcal{L}[g(t)]$, then

$$F(s)G(s) = \mathcal{L}\left[\int_0^t f(t-u)g(u)\,du\right] = \mathcal{L}\left[\int_0^t f(u)g(t-u)\,du\right]$$

We shall occasionally use this property to determine inverse Laplace transforms and indeed in the workshop we shall apply this rule.

It would be wrong to give the impression that we can choose a Laplace transform $F(s)$ arbitrarily and then obtain $\mathcal{L}^{-1}[F(s)]$. There are restrictions on the choice of $F(s)$; for example $\lim_{s\to\infty} F(s) = 0$. We shall explore some of these restrictions on $F(s)$ in the 'Further exercises' at the end of this chapter.

For the moment, though, we have essentially two skills to acquire. First the skill of obtaining a Laplace transform and secondly the skill of obtaining the inverse Laplace transform. We have already observed that taking the Laplace transform results in rational expressions in s sometimes multiplied by an exponential function involving s. It may therefore be necessary for us to resolve the rational factor into partial fractions in order to recognize it as a sum of standard forms. We should ignore the presence of the exponential factor until the last stage since its presence merely indicates that we need to use the second shifting theorem.

Table 22.1: A short table of Laplace transforms

$f(t)$	$F(s)$	$f(t)$	$F(s)$
1	$\frac{1}{s}$	$t \sin at$	$\frac{2as}{(s^2+a^2)^2}$
e^{-at}	$\frac{1}{(s+a)}$	$t \cos at$	$\frac{s^2-a^2}{(s^2+a^2)^2}$
$\sin at$	$\frac{a}{(s^2+a^2)}$	$e^{-at} f(t)$	$F(s+a)$
$\cos at$	$\frac{s}{(s^2+a^2)}$	$f(t-a)H(t-a)$	$e^{-as}F(s)$
t	$\frac{1}{s^2}$	$H(t)$	$\frac{1}{s}$
$t^n, n \in \mathbb{N}$	$\frac{n!}{s^{n+1}}$	$\delta(t-a)$	e^{-as}
$\sinh at$	$\frac{a}{(s^2-a^2)}$	$t\ f(t)$	$-F'(s)$
$\cosh at$	$\frac{s}{(s^2-a^2)}$	$\int_0^t f(x)\ dx$	$\frac{F(s)}{s}$

By the same token, if we can express $F(s)$ in the form $F(S)$, where $S = s + a$ and a is constant, we can apply the first shifting theorem in order to obtain the inverse transform.

Practise makes perfect and so with no more ado we shall roll up our sleeves and concentrate our efforts on the workshop.

22.5 Workshop

Here is a straightforward problem just to get things started. Don't forget to use Table 22.1, when necessary. 1

▷ **Exercise** Obtain the Laplace transforms of the following:

1 $4t - 7$

2 $4\sin 3t - 5\cos 2t$

3 $(t-2)^2$

Have a good go at these before you move on.

2

1 We use the linearity of the Laplace transform

$$\begin{aligned}\mathcal{L}[4t-7] &= 4\mathcal{L}[t] - 7\mathcal{L}[1] \\ &= 4\left[\frac{1}{s^2}\right] - 7\left[\frac{1}{s}\right] \\ &= \frac{4}{s^2} - \frac{7}{s} \\ &= \frac{4-7s}{s^2}\end{aligned}$$

2 In this exercise we again use linearity and properties of the Laplace transform of the circular functions, which we derived earlier.

$$\begin{aligned}\mathcal{L}[4\sin 3t - 5\cos 2t] &= 4\mathcal{L}[\sin 3t] - 5\mathcal{L}[\cos 2t] \\ &= 4\left[\frac{3}{s^2+3^2}\right] - 5\left[\frac{s}{s^2+2^2}\right] \\ &= \frac{12}{s^2+9} - \frac{5s}{s^2+4}\end{aligned}$$

Note that we do not need to express the Laplace transform as a single rational expression. It doesn't simplify it in any way.

3 Again we multiply out the brackets and use the linearity rule.

$$\begin{aligned}\mathcal{L}[(t-2)^2] &= \mathcal{L}[t^2 - 4t + 4] \\ &= \mathcal{L}[t^2] - 4\,[\mathcal{L}[t]] + 4\,[\mathcal{L}[1]] \\ &= \frac{2}{s^3} - 4\left[\frac{1}{s^2}\right] + 4\left[\frac{1}{s}\right] \\ &= 2\left[\frac{1-2s+2s^2}{s^3}\right]\end{aligned}$$

Did you manage those? Here is a problem involving inverse Laplace transforms.

▷ **Exercise** Obtain the inverse Laplace transforms of each of the following:

1 $s/(s^2+9)$
2 $s/(s^2+2s+2)$
3 $s/(s^2+1)^2$

---

1 If we look carefully at Table 22.1 we shall notice that this transform is a standard form. 3

$$\begin{aligned}\frac{s}{s^2+9} &= \frac{s}{s^2+a^2} \text{ with } a=3\\ &= \mathcal{L}[\cos at]\\ &= \mathcal{L}[\cos 3t]\\ \text{So } \mathcal{L}^{-1}\left[\frac{s}{s^2+9}\right] &= \cos 3t\end{aligned}$$

2 Although $s/(s^2+2s+2)$ is not one of our standard forms it is a simple matter to express it in terms of them:

$$\begin{aligned}\frac{s}{s^2+2s+2} &= \frac{s}{(s+1)^2+1}\\ &= \frac{s+1}{(s+1)^2+1} - \frac{1}{(s+1)^2+1}\end{aligned}$$

Now

$$\mathcal{L}[\cos t] = \frac{s}{s^2+1} \text{ and } \mathcal{L}[\sin t] = \frac{1}{s^2+1}$$

and so using the first shifting theorem

$$\frac{s+1}{(s+1)^2+1} = \mathcal{L}[\mathrm{e}^{-t}\ \cos t] \text{ and } \frac{1}{(s+1)^2+1} = \mathcal{L}[\mathrm{e}^{-t}\ \sin t]$$

Consequently

$$\begin{aligned}\mathcal{L}^{-1}\left[\frac{s}{s^2+2s+2}\right] &= e^{-t}\ \cos t - \mathrm{e}^{-t}\ \sin t\\ &= e^{-t}\ [\cos t - \sin t]\end{aligned}$$

3 We have $s/(s^2+1)^2$ and if we look at Table 22.1 we see that

$$\mathcal{L}^{-1}[s/(s^2+1)] = \cos t \text{ and } \mathcal{L}^{-1}[1/(s^2+1)] = \sin t$$

Here we have the product of these transforms and so we can apply the convolution theorem.

$$\begin{aligned}
\mathcal{L}^{-1}\left[\frac{s}{(s^2+1)^2}\right] &= \int_0^t \sin(t-u)\cos u \; \mathrm{d}u \\
&= \frac{1}{2}\int_0^t \{\sin[(t-u)+u] + \sin[(t-u)-u]\} \; \mathrm{d}u \\
&= \frac{1}{2}\int_0^t \{\sin t + \sin(t-2u)\} \; \mathrm{d}u \\
&= \frac{1}{2}\sin t \int_0^t \mathrm{d}u + \frac{1}{2}\int_0^t \sin(t-2u) \; \mathrm{d}u \\
&= \left(\frac{1}{2}\sin t\right)[u]_0^t + \left(-\frac{1}{4}\right)[-\cos(t-2u)]_0^t \\
&= \frac{1}{2}\;[t\sin t] + \frac{1}{4}[\cos(-t) - \cos t] \\
&= \frac{1}{2}\;[t\sin t]
\end{aligned}$$

Lastly we shall solve a differential equation of the sort that often arises in practice. It will involve us with the entire process: taking the Laplace transform, solving the algebraic equation and finally obtaining the inverse Laplace transform to solve the equation.

Exercise Solve the equation

$$\frac{\mathrm{d}^2x}{\mathrm{d}t^2} - 4\frac{\mathrm{d}x}{\mathrm{d}t} + x = 2t$$

given that initially $x = 0$ and $\dot{x} = \mathrm{d}x/\mathrm{d}t = 1$.

See how you get on first of all and only check your working afterwards or if you come to a grinding halt.

4

We shall use the notation $X = \mathcal{L}[x]$ and begin by taking the Laplace transform of the equation. Remember that the Laplace transform is a linear transformation and therefore we can transform the equation term by term. The algebraic equation which results is often called the **subsidiary** equation.

$$\mathcal{L}\left[\frac{\mathrm{d}^2x}{\mathrm{d}t^2}\right] - 4\,\mathcal{L}\left[\frac{\mathrm{d}x}{\mathrm{d}t}\right] + \mathcal{L}[x] = 2\mathcal{L}[t]$$

Now we know that

$$\mathcal{L}\left[\frac{\mathrm{d}^2x}{\mathrm{d}t^2}\right] = s^2X - s\lim_{t\to 0+} x - \lim_{t\to 0+}\dot{x}$$

and also that

$$\mathcal{L}\left[\frac{dx}{dt}\right] = sX - \lim_{t\to 0+} x$$

But we know

$$\lim_{t\to 0+} x = 0 \text{ and } \lim_{t\to 0+} \dot{x} = 1$$

Therefore the subsidiary equation becomes

$$[s^2X - 1] - 4\,[sX - 0] + X = \frac{2}{s^2}$$

This is a good moment to pause. If you didn't manage to get to this step correctly then see if you can take it over from here. Remember we need to solve for X and then find the inverse Laplace transform to obtain x in terms of t.

Collecting the terms in X on one side of the equation together we have 5

$$(s^2 - 4s + 1)X = 1 + \frac{2}{s^2}$$

Therefore

$$X = \frac{1}{s^2 - 4s + 1} + \frac{2}{s^2(s^2 - 4s + 1)}$$

Now it is a simple, but tedious, exercise on partial fractions to obtain

$$\frac{1}{s^2(s^2 - 4s + 1)} = \frac{4s + 1}{s^2} + \frac{-4s + 15}{s^2 - 4s + 1}$$

So we have

$$\begin{aligned}
X &= \frac{1}{s^2 - 4s + 1} + 2\left[\frac{4s+1}{s^2} + \frac{-4s+15}{s^2 - 4s + 1}\right] \\
&= \frac{31}{s^2 - 4s + 1} + \frac{8s + 2}{s^2} + \left[\frac{-8s}{s^2 - 4s + 1}\right] \\
&= \frac{31}{(s-2)^2 - 3} + 8\left[\frac{1}{s}\right] + 2\left[\frac{1}{s^2}\right] + \left[\frac{-8s}{(s-2)^2 - 3}\right] \\
&= \frac{15}{(s-2)^2 - (\sqrt{3})^2} + 8\left[\frac{1}{s}\right] + 2\left[\frac{1}{s^2}\right] + \left[\frac{-8(s-2)}{(s-2)^2 - (\sqrt{3})^2}\right] \\
&= \frac{5(\sqrt{3})^2}{(s-2)^2 - (\sqrt{3})^2} + 8\left[\frac{1}{s}\right] + 2\left[\frac{1}{s^2}\right] - 8\left[\frac{(s-2)}{(s-2)^2 - (\sqrt{3})^2}\right]
\end{aligned}$$

We have carefully arranged things here so that we can recognize standard forms. We can see cosh and sinh here, together with an application of the first shifting theorem.

For example

$$\mathcal{L}[\sinh t\sqrt{3}] = \frac{\sqrt{3}}{s^2 - (\sqrt{3})^2}$$

and therefore

$$\mathcal{L}[e^{2t}\sinh t\sqrt{3}] = \frac{\sqrt{3}}{(s-2)^2 - (\sqrt{3})^2}$$

Similarly

$$\mathcal{L}[\cosh t\sqrt{3}] = \frac{s}{s^2 - (\sqrt{3})^2}$$

and therefore

$$\mathcal{L}[e^{2t}\cosh t\sqrt{3}] = \frac{s-2}{(s-2)^2 - (\sqrt{3})^2}$$

Consequently we can now obtain x, the inverse Laplace transform of X

$$\begin{aligned} x &= 5\sqrt{3}[\mathrm{e}^{2t}\sinh t\sqrt{3}] + 8\ H(t) + 2t - 8\ [\mathrm{e}^{2t}\cosh t\sqrt{3}] \\ &= e^{2t}[5\sqrt{3}\sinh t\sqrt{3} - 8\ \cosh t\sqrt{3}] + 8H(t) + 2t \end{aligned}$$

Unfortunately, it is quite usual for unattractive numerical expressions to arise when solving equations involving Laplace transforms.

22.6 Practical

We now solve an example which arises in a practical setting and which cannot easily be solved without the techniques which we have developed in this chapter.

□ An initially quiescent (L, R) circuit consists of an inductance L henries and a resistance R ohms in series. An electromotive force E volts is applied between time $t = 1$ and time $t = 2$. A power surge of $10E$ volts occurs instantaneously at time $t = 3$. Determine the current in the circuit at time t. Obtain further the change in power in the resistor at time $t = 3$. Time is measured in hours.

Electrical Engineers will be able to write down the equation for the potential drop across the circuit straight away. Others may have to accept it on trust. The potential difference across the (L, R) series circuit is

$$L\ \frac{\mathrm{d}i}{\mathrm{d}t} + iR$$

where i is the current in amps.

We now turn our attention to the EMF. This we do by using both the Heaviside unit function and the Dirac delta function.

$$E\ [H(t-1) - H(t-2)] + 10E\ \delta(t-3)$$

This shows we have a step of magnitude E between time $t = 1$ and $t = 2$ and an impulse of magnitude $10E$ at $t = 3$.

We therefore obtain the equation

$$L\ \frac{\mathrm{d}i}{\mathrm{d}t} + iR = E\ [H(t-1) - H(t-2)] + 10E\ \delta(t-3)$$

We know that initially the circuit is quiescent and this means that $i = 0$ when $t = 0$.

Now we take the Laplace transform, term by term, and write $I = \mathcal{L}[i]$. We therefore obtain the subsidiary equation

$$L\ \{sI - 0\} + RI = E\ \left\{\frac{\mathrm{e}^{-s}}{s} - \frac{\mathrm{e}^{-2s}}{s}\right\} + 10E\ \mathrm{e}^{-3s}$$

Our next task is to solve this equation for I. Bearing in mind that we shall wish to obtain the inverse Laplace transform we rearrange each of the terms in such a way that it becomes recognizable as a standard form.

$$[Ls + R]I = \left(\frac{E}{s}\right)\mathrm{e}^{-s} - \left(\frac{E}{s}\right)\mathrm{e}^{-2s} + 10E\ \mathrm{e}^{-3s}$$

Now using the cover-up rule for partial fractions we have

$$\frac{1}{s(Ls+R)} = \frac{\left(\frac{1}{R}\right)}{s} - \frac{\left(\frac{L}{R}\right)}{Ls+R}$$

Using this we obtain

$$\begin{aligned}
I &= \left[\frac{E}{s(Ls+R)}\right]\ \mathrm{e}^{-s} - \left[\frac{E}{s(Ls+R)}\right]\ \mathrm{e}^{-2s} + 10\left[\frac{E}{Ls+R}\right]\ \mathrm{e}^{-3s} \\
&= \left[\frac{\left(\frac{E}{R}\right)}{s} - \frac{\left(\frac{EL}{R}\right)}{Ls+R}\right]\mathrm{e}^{-s} - \left[\frac{\left(\frac{E}{R}\right)}{s} - \frac{\left(\frac{EL}{R}\right)}{Ls+R}\right]\mathrm{e}^{-2s} + 10\left(\frac{E}{Ls+R}\right)\mathrm{e}^{-3s} \\
&= \left[\frac{\left(\frac{E}{R}\right)}{s} - \frac{\left(\frac{E}{R}\right)}{(s+\frac{R}{L})}\right]\mathrm{e}^{-s} - \left[\frac{\left(\frac{E}{R}\right)}{s} - \frac{\left(\frac{E}{R}\right)}{(s+\frac{R}{L})}\right]\mathrm{e}^{-2s} + 10\left[\frac{\frac{E}{L}}{(s+\frac{R}{L})}\right]\mathrm{e}^{-3s} \\
&= \left[\frac{\left(\frac{E}{R}\right)}{s}\right]\ \mathrm{e}^{-s} - \left[\frac{\left(\frac{E}{R}\right)}{(s+\frac{R}{L})}\right]\ \mathrm{e}^{-s} - \left[\frac{\left(\frac{E}{R}\right)}{s}\right]\ \mathrm{e}^{-2s} + \left[\frac{\left(\frac{E}{R}\right)}{(s+\frac{R}{L})}\right]\ \mathrm{e}^{-2s} \\
&\qquad + 10\left[\frac{\frac{E}{L}}{(s+\frac{R}{L})}\right]\ \mathrm{e}^{-3s}
\end{aligned}$$

We can now obtain i, the inverse Laplace transform of I. Here are the key properties which we shall use:

$$\begin{aligned}
\mathcal{L}^{-1}\left[\frac{1}{s+b}\right] &= e^{-bt} \\
\mathcal{L}^{-1}\left[\frac{\mathrm{e}^{-as}}{s}\right] &= H(t-a) \\
\mathcal{L}^{-1}\left[\frac{\mathrm{e}^{-as}}{s+b}\right] &= e^{-b[t-a]}H(t-a)
\end{aligned}$$

Therefore we obtain

$$i = \left(\frac{E}{R}\right) H(t-1) - \left(\frac{E}{R}\right) \mathrm{e}^{-\frac{R(t-1)}{L}} H(t-1) - \left(\frac{E}{R}\right) H(t-2)$$
$$+ \left(\frac{E}{R}\right) \mathrm{e}^{-\frac{R(t-2)}{L}} H(t-2) + 10\left(\frac{E}{L}\right) \mathrm{e}^{-\frac{R(t-3)}{L}} H(t-3)$$

To complete the problem we need to see how the current changes at the critical times $t = 1$, $t = 2$ and $t = 3$. We have

$$t < 1 \qquad i = 0$$

$$\begin{aligned}
1 < t < 2 \qquad i &= \left(\frac{E}{R}\right) - \left(\frac{E}{R}\right) \mathrm{e}^{-\frac{R(t-1)}{L}} \\
&= \frac{E}{R}\left[1 - \mathrm{e}^{-\frac{R(t-1)}{L}}\right]
\end{aligned}$$

$$\begin{aligned}
2 < t < 3 \qquad i &= \left(\frac{E}{R}\right) - \left(\frac{E}{R}\right) \mathrm{e}^{-\frac{R(t-1)}{L}} - \left(\frac{E}{R}\right) + \left(\frac{E}{R}\right) \mathrm{e}^{-\frac{R(t-2)}{L}} \\
&= \left(\frac{E}{R}\right) \mathrm{e}^{-\frac{R(t-2)}{L}} - \left(\frac{E}{R}\right) \mathrm{e}^{-\frac{R(t-1)}{L}} \\
&= \frac{E}{R}\left[\mathrm{e}^{-\frac{R(t-2)}{L}} - \mathrm{e}^{-\frac{R(t-1)}{L}}\right] \\
&= \frac{E}{R}[\mathrm{e}^{\frac{R}{L}} - 1]\mathrm{e}^{-\frac{R(t-1)}{L}}
\end{aligned}$$

$t > 3$

$$i = \left(\frac{E}{R}\right) - \left(\frac{E}{R}\right) \mathrm{e}^{-\frac{R(t-1)}{L}} - \left(\frac{E}{R}\right) + \left(\frac{E}{R}\right) \mathrm{e}^{-\frac{R(t-2)}{L}} + 10\left(\frac{E}{L}\right) \mathrm{e}^{-\frac{R(t-3)}{L}}$$

$$= -\left(\frac{E}{R}\right) \mathrm{e}^{-\frac{R(t-1)}{L}} + \left(\frac{E}{R}\right) \mathrm{e}^{-\frac{R(t-2)}{L}} + 10\left(\frac{E}{L}\right) \mathrm{e}^{-\frac{R(t-3)}{L}}$$

$$= 10\left(\frac{E}{L}\right) \mathrm{e}^{-\frac{R(t-3)}{L}} + \left(\frac{E}{R}\right) \mathrm{e}^{-\frac{R(t-2)}{L}} - \left(\frac{E}{R}\right) \mathrm{e}^{-\frac{R(t-1)}{L}}$$

$$= \left[10\left(\frac{E}{L}\right) \mathrm{e}^{\frac{2R}{L}} + \left(\frac{E}{R}\right) \mathrm{e}^{\frac{R}{L}} - \left(\frac{E}{R}\right)\right] \mathrm{e}^{-\frac{R(t-1)}{L}}$$

$$= \frac{E}{R}\left[10\left(\frac{R}{L}\right) \mathrm{e}^{\frac{2R}{L}} + \mathrm{e}^{\frac{R}{L}} - 1\right] \mathrm{e}^{-\frac{R(t-1)}{L}}$$

From this we see that just before time $t = 3$ the current is approaching

$$\frac{E}{R}\,[\mathrm{e}^{\frac{R}{L}} - 1]\mathrm{e}^{-\frac{2R}{L}}$$

whereas just afterwards the current is decreasing from

$$\frac{E}{R}\left[10\left(\frac{R}{L}\right) \mathrm{e}^{\frac{2R}{L}} + \mathrm{e}^{\frac{R}{L}} - 1\right] \mathrm{e}^{-\frac{2R}{L}}$$

Consequently the current surge is

$$\frac{E}{R}\left[10\left(\frac{R}{L}\right)\right] = 10\left(\frac{E}{L}\right)$$

Using the formula $P = i^2R$ for the power in the resistor we see that the power surge is given by

$$100R\left(\frac{E}{L}\right)^2$$

■

The bending of beams

We can consider a light uniform beam, fixed at one end, as being represented by a line. The deflection y of the beam, at a distance x from the fixed end, is given by

$$EI\frac{\mathrm{d}^2y}{\mathrm{d}x^2} = M$$

where E is Young's modulus, I is the moment of inertia of a cross-section of the beam about the neutral axis and M is the bending moment.

Moreover if w is the loading per unit length on the beam then, given that E and I are constant and that there is no transverse loading,

$$EI\frac{\mathrm{d}^4y}{\mathrm{d}x^4} = w$$

These formulae can be applied to beam problems and it is then possible to solve them, using Laplace transforms, to obtain the deflection. This will form part of the Further exercises at the end of this chapter.

After the summary you will be able to put your skills to the test. Several of the further exercises give an indication of further developments which can be undertaken.

We shall conclude our discussion of Laplace transforms by listing, without proof, some other properties. As usual we have $F(s) = \mathcal{L}[f(t)]$ and we shall pre-suppose that the appropriate limits and integrals exist.

1 The Laplace transform of an integral

$$\mathcal{L}\left[\int_0^t f(u)\ \mathrm{d}u\right] = \frac{1}{s}\ F(s)$$

2 The initial value theorem

$$\lim_{t\to 0+} f(t) = \lim_{s\to\infty} sF(s)$$

3 The final value theorem

$$\lim_{t\to\infty} f(t) = \lim_{s\to 0} sF(s)$$

4 The Laplace transform of a periodic function
Suppose f is a periodic function with period T

$$\mathcal{L}[f(t)] = \frac{1}{1-\mathrm{e}^{-sT}}\int_0^T \mathrm{e}^{-st}f(t)\ \mathrm{d}t$$

5 Division and multiplication by t

$$\mathcal{L}\left[\frac{f(t)}{t}\right] = \int_s^\infty F(u)\ \mathrm{d}u$$

$$\mathcal{L}[t\ f(t)] = -F'(s)$$

The initial and final value theorems can be useful if we are not particularly interested in the general response of a system but more specifically we wish to know what happened initially or what will happen eventually. These theorems enable us to determine the information without having to obtain the inverse Laplace transform first.

SUMMARY

- The Laplace transform of $f(t)$ is defined by

$$F(s) = \mathcal{L}[f(t)] = \int_0^{\infty} \mathrm{e}^{-st} f(t)\ \mathrm{d}t$$

- We investigated the basic properties of the Heaviside unit function $H(t)$ and saw how to apply the shifting theorems

$$\begin{aligned} F(s+a) &= \mathcal{L}[\mathrm{e}^{-at}\ f(t)] \\ e^{-as}\ F(s) &= \mathcal{L}[f(t-a)\ H(t-a)] \end{aligned}$$

- We investigated the Dirac delta function $\delta(t)$ and discovered its three main properties

$$\begin{aligned} \textstyle\int_{-\infty}^{\infty} \delta(t)\ \mathrm{d}t &= 1 \\ \mathcal{L}[\delta(t-a)] &= e^{-as} \\ \delta(t) &= 0 \quad t \neq 0 \end{aligned}$$

- We showed how to solve ordinary differential equations using Laplace transforms.

EXERCISES

1 Obtain the Laplace transforms of each of the following

a $3t+1$
b $2\sin t + \cos t$
c $(t+2)^3$
d $\sinh 3t - \cosh 3t$
e $\mathrm{e}^{-3t}\cos 4t$

2 Show that the Laplace transform of $(t+2)^2\mathrm{e}^t$ is

$$\frac{4s^2 - 4s + 2}{(s-1)^3}$$

3 Obtain the inverse Laplace transforms of each of the following:

a $1/(s^2 - 4s + 5)$
b $s/(s^2 - 4s + 5)$
c $1/(s+3)^2$
d $\mathrm{e}^{-s}/(s+3)^2$

4 Solve the following differential equations where $x = x(t)$, $\dot{x} = \mathrm{d}x/\mathrm{d}t$, $\ddot{x} = \mathrm{d}^2x/\mathrm{d}t^2$ and $\dot{x}(0) = \lim_{t\to 0+} \dot{x}(t)$.

a $\ddot{x} + 3\dot{x} - x = 0$ where $x(0) = 1, \dot{x}(0) = 0$
b $\ddot{x} - 2\dot{x} - 3x = 1$ where $x(0) = 1, \dot{x}(0) = 0$
c $\ddot{x} + 2\dot{x} = H(t-1) - H(t-2)$ where $x(0) = 1, \dot{x}(0) = 1$
d $\ddot{x} + 2x = \delta(t)$ where $x(0) = 1, \dot{x}(0) = 0$

ASSIGNMENT

1 Obtain the Laplace transforms of each of the following:

a $t\mathrm{e}^{-at}$
b $t^n\mathrm{e}^{-at}$
c $3t^2 - \mathrm{e}^{2t}$
d $(2\mathrm{e}^{3t} - 1)^2$

2 Obtain the inverse Laplace tranforms of each of the following:

a $1/(s+5)^2$
b $s/[(s-1)^2+1]$
c $1/[(s-1)(s-2)]$
d $(s-5)/(s^2-10s+29)$

3 Obtain the inverse Laplace transform of each of the following:

a $(2s+3)/(s^2+2s+5)$
b $s^2/[(s-2)(s+1)(s-3)]$
c $16/(s^2+6s+13)^2$
d $16\mathrm{e}^{-s}/(s^2+6s+13)^2$

4 Solve the following differential equations where $x = x(t)$, $\dot{x} = \mathrm{d}x/\mathrm{d}t$, $\ddot{x} = \mathrm{d}^2x/\mathrm{d}t^2$ and $\dot{x}(0) = \lim_{t\to 0+} \dot{x}(t)$.

a $\ddot{x} - 3\dot{x} + 2x = 0$ where $x(0) = 1, \dot{x}(0) = 2$
b $\ddot{x} - \dot{x} - 6x = 1$ where $x(0) = 1, \dot{x}(0) = 0$
c $\ddot{x} + 2\dot{x} - 3x = \mathrm{e}^{-t}$ where $x(0) = 0, \dot{x}(0) = 1$

FURTHER EXERCISES

1 Determine the inverse Laplace transforms of each of the following:

a $(3s-1)/(s^2+2s+10)$
b $(s^2+1)/[(s-2)(s+1)(s+3)]$
c $s/(s^2+6s+13)^2$
d $16\mathrm{e}^{-s}/(s^2-6s+13)^2$

2 The function f has period 2π and is defined for one period by

$$f(t) = \begin{cases} \sin t & 0 \leqslant t \leqslant \pi \\ 0 & \pi < t < 2\pi \end{cases}$$

Show that

$$F(s) = \mathcal{L}[f(t)] = \frac{1}{(1-\mathrm{e}^{-\pi s})(s^2+1)}$$

3 Initially, at time $t = 0$, a constant electromotive force E was applied to a quiescent circuit consisting of an inductance L henries and a resistance R ohms in series. Show that subsequently the current is given by

$$i = \frac{E}{R}\,(1 - \mathrm{e}^{-Rt/L})$$

4 A square wave is defined by the equations

$$f(t) = \begin{cases} E & 0 < t < T \\ -E & T < t < 2T \\ f(t - 2T) & t > 2T \end{cases}$$

Show that if $F(s) = \mathcal{L}[f(t)]$ then

$$F(s) = \frac{E}{s}\left[1 + 2\sum_{r=1}^{\infty}(-1)^r\,\mathrm{e}^{-rsT}\right]$$

5 The charge $q(t)$ on a capacitor, where $i(t)$ is the current at time t, is given by

$$q(t) = \int_0^t i(u)\,\mathrm{d}u$$

Show that, if $I = \mathcal{L}[i(t)]$,

$$Q(s) = \mathcal{L}[q(t)] = \frac{I}{s}$$

During the time interval from $t = a$ to $t = b$ a constant electromotive force E is applied to a quiescent (R, C) series circuit. Show that the current at time t is given by

$$i = \frac{E}{R}\,\{\mathrm{e}^{-(t-a)/RC}\;H(t-a) - \mathrm{e}^{-(t-b)/RC}\;H(t-b)\}$$

6 An electromotive force E is applied instantaneously to an (L, C) series circuit. If there is no initial charge on the capacitor, show that subsequently the current i is given by

$$i = \frac{E}{L}\cos\left(\frac{t}{\sqrt{LC}}\right)$$

7 An electrical device consists of an (L, R) series component and an (R, C) series component arranged in parallel. The inductor is tuned so that the effect of this device is that of a pure resistor.

By considering an applied electromotive force E, with currents i_1 and i_2 in each component, and taking the Laplace transform of a pair of simultaneous equations, or otherwise, show that $L = CR^2$.

8 A sequence of impulsive electromotive forces is applied at intervals of 1 second, beginning at time $t = 1$, to an initially quiescent (L, R, C) series circuit. The strength of the rth impulse is E/r. By using the final value theorem, or otherwise, determine the behaviour of the current in the circuit as $t \to \infty$.

9 A circuit results in the following simultaneous differential equations:

$$\begin{aligned} L\,\frac{\mathrm{d}i_1}{\mathrm{d}t} + R\,i_1 + R(i_1 - i_2) &= E \\ L\,\frac{\mathrm{d}i_2}{\mathrm{d}t} + R\,i_2 + R(i_2 - i_1) &= 0 \end{aligned}$$

where L, R and E are constant and initially (when $t = 0$) the currents i_1 and i_2 are both zero. Obtain I_1 and I_2, the Laplace transforms of i_1 and i_2, and thereby determine the steady state values of these currents after a sufficiently long time.

10 The current i in a circuit at time t can be obtained from the equation

$$\frac{\mathrm{d}^2 i}{\mathrm{d}t^2} + 2\frac{\mathrm{d}i}{\mathrm{d}t} + ki = E\cos t$$

where k is a positive constant. The circuit is initially quiescent. Obtain the Laplace transform of i and thereby show that as $t \to \infty$

$$i \to \frac{E}{\sqrt{k^2 - 2k + 5}}\cos(t - \theta)$$

where $\tan\theta = 2/(k-1)$.

11 When a light wooden beam of length l is clamped horizontally at each end the deflection y at a point distance x from one end is given by

$$EI\frac{\mathrm{d}^4 y}{\mathrm{d}x^4} = L(x)$$

where E and I are constants and $L(x)$ is the load on the beam at the point x. The boundary conditions can be expressed by

$$y = 0 \qquad \frac{\mathrm{d}y}{\mathrm{d}x} = 0$$

at $x = 0$ and $x = l$.

Show that if $L(x)$ consists of a concentrated load W at a distance a then

$$y = \frac{W}{6EI}\,(x-a)^2 H(x-a) + A\,\frac{x^2}{2} + B\,\frac{x^3}{6}$$

Show further that

$$B = -\frac{W}{EI}\,\frac{(l-a)^2(l+2a)}{l^3} \qquad \text{and} \qquad A = \frac{W}{EI}\,\frac{(l-a)^2}{l^3}$$

12 A light horizontal beam of length l is clamped at each end and carries a load w per unit length from $x = a$ to $x = b$ where x is the distance measured from one end. Show that the deflection y of the beam at x is given by

$$y = A\,\frac{x^2}{2} + B\,\frac{x^3}{6} + \frac{w}{24EI}\,[(x-a)^4 H(x-a) - (x-b)^4 H(x-b)]$$

Show further that

$$\begin{aligned} A &= \frac{w}{12EI}\,\frac{(l-a)^3(l+3a) - (l-b)^3(l+3b)}{l^2} \\ B &= -\frac{w}{2EI}\,\frac{(l-a)^2(l^2-a^2) - (l-b)^2(l^2-b^2)}{l^3} \end{aligned}$$

23 Descriptive statistics

This chapter represents a complete change of mood. We leave the crystal world of mathematics, where there is order and clarity, to visit the opaque world of statistics, where there is randomness and uncertainty.

After working through this chapter you should be able to
- ☐ Use the basic terminology of statistics;
- ☐ Distinguish between population statistics and sample statistics;
- ☐ Present data in a pictorial form to highlight its features;
- ☐ Calculate the basic measures of location – mean, mode and median;
- ☐ Calculate the basic measures of spread – range, mean absolute deviation and variance.

At the end of the chapter we look at a practical problem in production testing.

23.1 TERMINOLOGY

The subject of statistics arises in everyday conversation, in newspapers and on television. Whenever statistics are mentioned they are usually accompanied by the word 'data'. Strictly speaking 'data' is a plural word, the singular being 'datum', but nowadays it is common practice to use it as if it is singular.

Data is the information with which we start, the outcome of the activity or experiment. For example, data could consist of heights, weights, colours, temperature, lifetimes or examination scores. Very often data is in a numerical form; you will be able to think of many other examples. Data, when it has been obtained and not modified in any way, is usually called **raw data**.

The moment we calculate something from the data we have produced a **statistic**.

Once more, just to get the terminology clear:

1 The information which has been collected is called data;
2 Anything which we calculate from the data is known as a statistic.

The measurement in which we are interested – height, temperature or whatever – is often called the **variate**. The set of all the values which the variate takes is known as the **population**. For example, the population could consist of

1 The tensile strengths of hawsers produced by a certain process;
2 The times taken for gauges on fuel tanks to register correctly after they are first switched on;
3 The numbers of faulty bricks in each production batch from a brickworks.

In industrial applications it is not usually feasible to collect data from the whole population. For instance, to obtain the data an item may have to be tested to destruction. No manufacturer would allow his entire output to be destroyed! Even if destruction is not involved it may be too expensive to collect the data corresponding to the population.

23.2 RANDOM SAMPLES

Of course, to say anything with certainty we should need the data from the entire population. However, the theory of statistics enables us to say something with a specified probability by analysing **samples** selected at random from the population.

The procedure by which a **random sample** is selected is fraught with danger, and so a few words are required to clarify things. To choose a random sample from the population we have to select a sample of, say, 100 items at random. How do we select at random? We must ensure that

1 The entire population is available to us;
2 The selection process is in no way likely to bias the results.

For example, suppose a newspaper wishes to predict the result of an election the next day. The editor may ask his reporters to dial numbers at random and ask the people who reply how they intend to vote.

This would not be a random sample of the electorate. First, he will have restricted the population to those people who are telephone subscribers. Secondly, by telephoning at a set time he has restricted that population to those who may be at home then: self-employed, mothers with children, retired people, unemployed etc. The editor will therefore have made a number of fundamental statistical errors. However, this will not deter him from publishing his results and maybe getting a correct prediction!

So then in engineering and science we shall be concerned with **sample statistics**. The data which we have will usually be a random sample taken from the population, and we shall wish to calculate statistics from this data which will enable us to make statements about the underlying population. Our interest is not in the sample but in the population.

23.3 POPULATION STATISTICS

There are fundamental differences between the theory of population statistics and the theory of sample statistics, and to reinforce this difference we shall consider an example. Remember that the purpose for which we are examining the data is the thing of overriding importance. We must ask ourselves at each stage: 'What are we trying to find out, and why?'

In the simplest of all situations we have all the data available and we wish merely to obtain information concerning it. For example, suppose a class of students sits an examination. Then there are several statistics which may be of interest: the class average, the range of marks or the top mark. As far as the examiner is concerned his interest in the marks may begin and end with the class of students. The population is in this instance the marks obtained by these students in the examination, and the statistics obtained are population statistics.

Suppose now we consider a public examination, such as 'A' level, where candidates sit the examination at a number of examination centres. In this situation if an examiner were given random samples of scripts it may be possible, using the data obtained, to estimate various statistics for the population. The resulting statistics are sample statistics. The population here, of course, consists of the complete set of marks from all the candidates.

23.4 DATA

Numerical data falls broadly into two categories: discrete data and continuous data.

Discrete data is data which can only take *isolated* values. Usually, but not always, discrete data consists of natural numbers. Examples include

1 The number of cars parked on consecutive days at a certain time in a car park;

2 The number of defective microchips produced by a machine process each week.

Continuous data is data which could take any value in some specified *interval* or set of intervals. Examples are:

1 The weight of ball bearings produced by a machine under normal working conditions;

2 The heights of a group of students in a class.

In the second of these examples of continuous data, if the shortest student has height 1.43 m and the tallest has height 2.04 m then there is no reason in theory why a student in the class could not have height 1.76 m. Of course practical considerations limit the accuracy to which we can measure any height, and so in practice there is often little distinction between discrete and continuous data. However, there are some important theoretical distinctions to be made and so we should decide at the outset whether the data is discrete or continuous.

We shall not consider situations in which the data is a mixture of discrete and continuous because to do so would involve very advanced mathematical ideas.

23.5 PICTORIAL REPRESENTATIONS

Faced with a collection of data, the statistician usually wishes to display the information in a clear easily understood and unbiased way. A table of data is often very difficult to assess and so pictorial methods have been devised. By carefully selecting the pictorial representation it is often possible to present data in a way which highlights certain characteristics and suppresses others.

To begin with we confine our attention to discrete data; it is a simple matter to modify things for continuous data. A single item of data is called a **data point**; the **frequency** of a data point is the number of times it occurs in the data.

23.6 PIE CHARTS

A pie chart represents the data as a 'pie'. To each distinct data point is assigned a slice of pie with an area proportional to the point's frequency. Although a pie chart is a satisfactory representation if there are only a few distinct data points, it loses much of its visual impact when there are many.

□ A selection of motorists were asked the question: 'Should traffic lights, wherever possible, be replaced by roundabouts?' The results were 63% 'yes', 21% 'no' and 16% 'don't know'. These are displayed in the pie chart to good visual effect (Fig. 23.1). ■

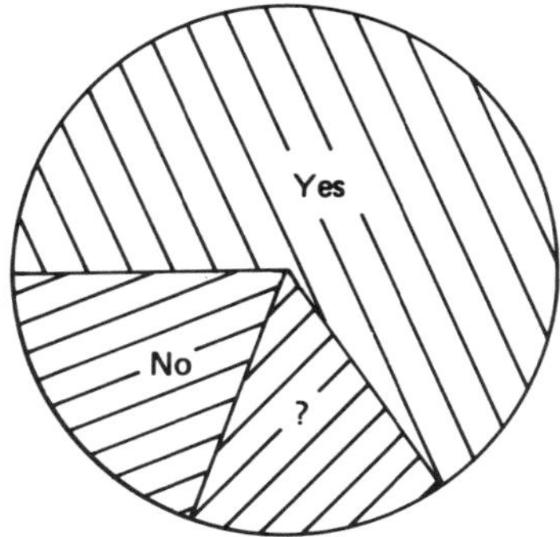

Fig. 23.1 A good pie chart.

□ Twelve television programmes were each assigned a number, and members of the public were asked to select the one they liked most. Those who could not decide were excluded from the sample. The results are shown in the pie chart (Fig. 23.2), but this does not give a very good visual effect because there are too many slices. ■

23.7 BAR CHARTS

A better way of presenting discrete data visually, when there are many distinct data points, is to construct a bar chart.

To do this we use rectangular cartesian axes and assign an x value to each distinct data point. Vertical bars are constructed joining each point (x, f) to the x-axis, where f is the frequency of the data point corresponding to x.

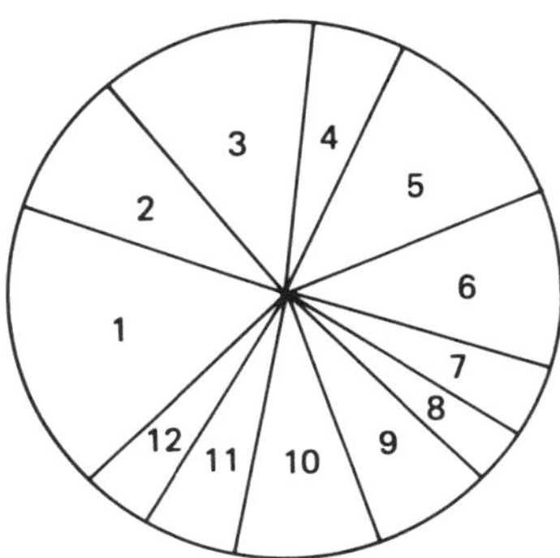

Fig. 23.2 A poor pie chart.

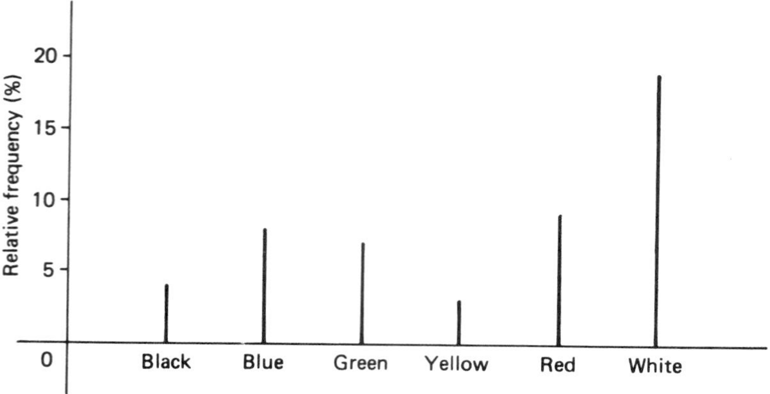

Fig. 23.3 A bar chart.

☐ Fifty potential customers were asked to say which colour car they preferred from those in current production. The results were as follows:

Colour	black	white	red	blue	green	yellow
Frequency	4	19	9	8	7	3

In this case there are many ways we can order the colours. For instance we could arrange them according to popularity, in the order of the spectrum, or alphabetically. The choice of order will depend very much on what we want to show.

With a suitable choice of scales we can produce a bar chart for each of these orders, for example Fig. 23.3. ■

23.8 HISTOGRAMS

If we have continuous data then a bar chart may no longer be appropriate, as most of the data points will be distinct. We therefore need to group the data.

To do this we begin by examining the range of the data – the difference between the highest value which appears and the lowest. Depending on the quantity which we have, we partition the data into a number of **class intervals**, not necessarily of equal length. In practice the number chosen should be not less than four and not more than twenty. The square root of the number of data points gives a good guide to the maximum number of class intervals we should use.

□ Suppose continuous data consists of 100 points, the smallest of which is 2.13 m and the largest 9.87 m. The range of the data is then $9.87 - 2.13 = 7.74$. So, if the data is fairly evenly distributed, it seems sensible to choose eight intervals: 2.00–2.99, 3.00–3.99, . . . , 7.00–7.99. Each of these is called a class interval.

The midpoint of each interval is called its **class mark**. So the class marks in this case are 2.495, 3.495, . . . , 7.495. The end points of each interval are called **class limits**. Notice that the interval 3.00–3.99 will contain any data point x such that $2.995 \leqslant x < 3.995$, so the length of the class interval is 1. These critical numbers 2.995, 3.995, . . . , 6.995 are usually called **class boundaries**. In this way we group the data into class intervals and calculate the frequency – the number of data points in each one.

If we wish to calculate statistics from this grouped data we must regard all the data points in the class interval as concentrated at the class mark. Although it could be misleading to do so, it would be possible to construct a bar chart corresponding to this data.

To construct a histogram we use the rectangular cartesian coordinate system and draw rectangles with intervals between the class boundaries as their bases and each having an area proportional to the frequency. If the class intervals are all of equal length then the heights of these rectangles will also be proportional to the frequencies (Fig. 23.4). ■

A smooth curve drawn through the midpoints of the tops of the rectangles is known as a **frequency curve**. If the midpoints are joined instead by straight lines we obtain a **frequency polygon**. One advantage of having equal class intervals is that the area under a frequency polygon is then the area of the histogram.

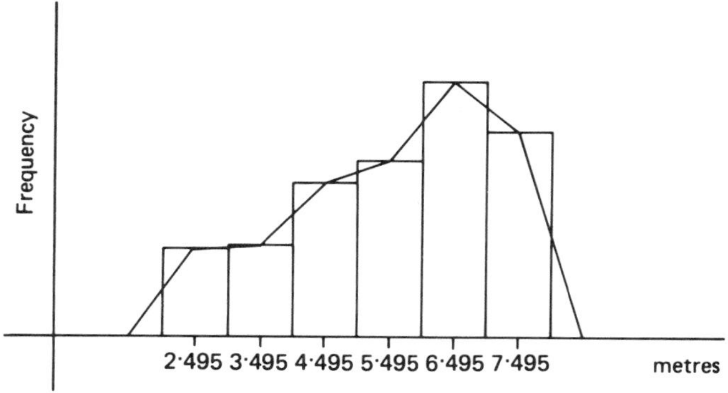

Fig. 23.4 A histogram and frequency polygon.

Table 23.1 Number of cars at 10.00 a.m. on 120 days

51,	53,	55,	56,	57,	58,	61,	62,	62,	65,	67,	69,	70,	70,	72,
73,	73,	74,	74,	75,	75,	76,	76,	76,	77,	77,	77,	78,	78,	78,
79,	79,	79,	80,	80,	80,	80,	81,	81,	81,	82,	82,	83,	83,	83,
83,	83,	83,	84,	84,	85,	85,	85,	85,	85,	85,	86,	86,	87,	87,
87,	88,	88,	89,	89,	89,	90,	90,	90,	90,	90,	90,	91,	91,	91,
92,	92,	92,	93,	93,	93,	93,	93,	94,	95,	95,	95,	95,	95,	96,
96,	97,	97,	98,	98,	98,	99,	99,	100,	101,	104,	105,	106,	107,	107,
107,	108,	108,	109,	110,	110,	111,	111,	112,	114,	116,	117,	119,	119,	120

Grouped data for bar chart production

Interval	*Class mark*	*Frequency*
51–60	55.5	6
61–70	65.5	8
71–80	75.5	23
81–90	85.5	35
91–100	95.5	27
101–110	105.5	12
111–120	115.5	9

23.9 GROUPED DATA

Sometimes in the case of discrete data there are too many observations for a useful bar chart to be constructed without first **grouping** the data.

□ The number of cars parked in a factory car park at 10.00 a.m. was counted on 120 consecutive working days. The results are shown in Table 23.1.

Some graphical indication of how much the car park is used is required. In order to obtain a bar chart we can consider the intervals 51–60, 61–70, . . . , 111–120. The class marks are 55.5, 65.5, . . . , 115.5. The groups are shown in Table 23.1, and the resulting bar chart in Fig. 23.5.

Unfortunately with this choice of interval each of the class marks cannot possibly be a data point. For instance, the car park never has 65.5 cars in it. Although from one point of view this is a disadvantage, it has one advantage: it emphasizes that the data has been grouped.

23.10 CUMULATIVE FREQUENCY DIAGRAMS

If the data is numerical, it is possible to arrange it in ascending order of magnitude. We can then calculate the **cumulative frequency** at each data point. The cumulative frequency is the total number of data points which

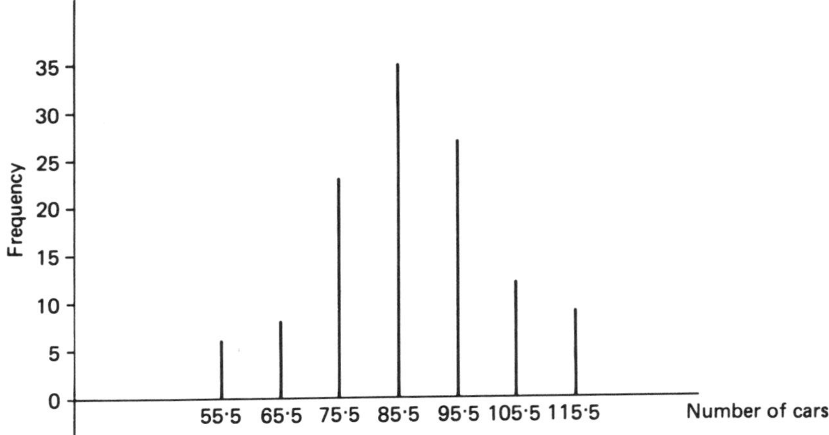

Fig. 23.5 Bar chart.

are less than, or equal to, the chosen point. If we divide the cumulative frequency by the total number of data points we obtain the **relative cumulative frequency**. The relative cumulative frequency will therefore increase from 0 to 1 as we go through the data. An example will show how this works.

□ Consider again the data of the cars in the car park (Table 23.1). The data increases from 50 to 120, and so we have the cumulative and relative frequencies shown in Table 23.2. From this table it is an easy matter to construct a relative cumulative frequency diagram (Fig. 23.6).

Cumulative frequencies are particularly important if we are setting a standard or a **quota**. In the previous example the company may decide to build on the car park and may wish to leave enough space so that 85% utilization of the present car park will be preserved. Table 23.2 could suggest that 105 car-parking spaces will have to be provided. ■

We have seen how data may be presented in a pictorial way, but we have not yet calculated any statistics. This we now do. Let's just go through some of the terminology again to make sure we have it clear in our minds:

1 The variate is the name given to the quantity in which we are interested;
2 Data consists of the results which are available to us;
3 An individual value of the data is called a data point;
4 Numerical data consists of data in numerical form;
5 The frequency of a data point is the number of times the data point appears in the data;
6 Anything which is calculated from the data is called a statistic.

Right! Now that we have that straight we can move ahead.

Table 23.2 Cumulative frequency for cars in car park

Numbers of cars	Cumulative frequency	Relative frequency
50	0	0.000
55	3	0.025
60	6	0.050
65	10	0.083
70	14	0.117
75	21	0.175
80	37	0.308
85	56	0.467
90	72	0.600
95	89	0.742
100	99	0.825
105	102	0.850
110	111	0.925
115	115	0.958
120	120	1.000

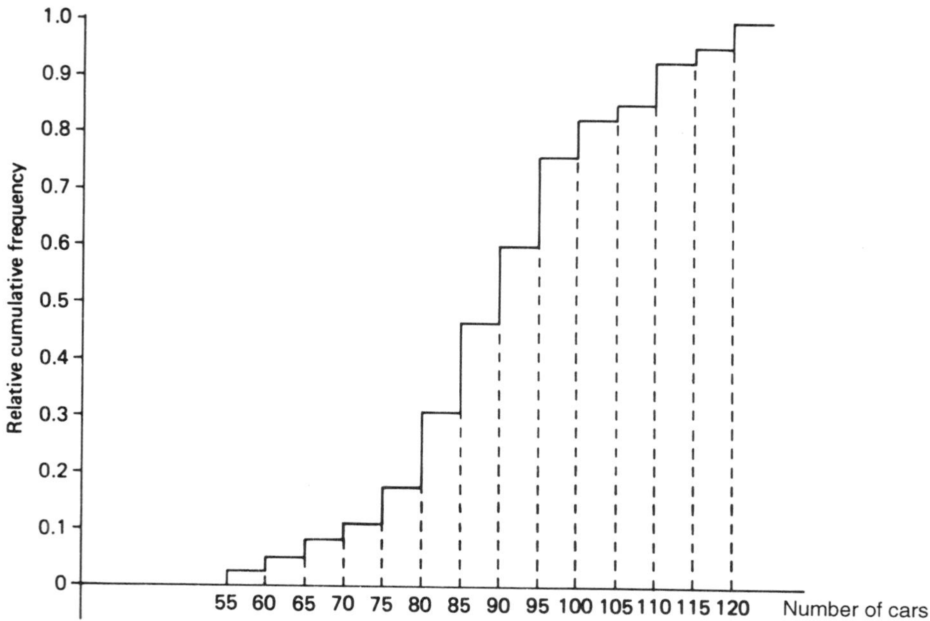

Fig. 23.6 A cumulative frequency diagram.

23.11 MEASURES OF LOCATION AND MEASURES OF SPREAD

Most useful statistics can be classified into either measures of location or measures of spread. We shall now explain what these terms mean.

A measure of **location**, also known as a measure of **central tendency**, attempts to indicate roughly the position around which the data is clustered. The usefulness of this statistic in any instance must be judged by the extent to which it typifies the data. There are three principal measures of location in common use: these are the mean, the mode and the median. Each has the same units as the data points themselves.

A measure of **spread**, also known as a measure of **dispersion**, gives an indication of how widely the data is distributed. Broadly speaking, if the measure of spread is large then the data is widely dispersed, whereas if it is small then it is closely bunched together. There are three principal measures of spread: the range, the mean absolute deviation and the standard deviation.

We shall consider each of these types of statistic in some detail.

23.12 THE MEAN

Suppose we have n distinct data points $x_1, x_2, \ldots, x_n$ and that these appear with frequencies $f_1, f_2, \ldots, f_n$ respectively in the data. Then the (arithmetic) **mean** $\bar{x}$ is obtained by totalling the data and dividing by the total number N of data points:

$$\begin{aligned}\bar{x} &= \frac{1}{N}(f_1x_1 + f_2x_2 + \ldots + f_nx_n)\\ &= \frac{1}{N}\sum_{r=1}^{n} f_rx_r\end{aligned}$$

where

$$N = \sum_{r=1}^{n} f_r$$

Although the mean is not the only measure of location, it is by far the most widely used.

□ Ten students took an examination, and the results were

72, 81, 43, 39, 47, 21, 35, 51, 63, 52

We can easily calculate the mean mark $\bar{x}$:

$$\bar{x} = \frac{1}{10}(72 + 81 + 43 + \ldots + 52) = 50.4$$ ■

Two disadvantages of the mean are immediately apparent:

1 The mean may not be a possible value of the data. This is certainly the case in this example where examination scripts are assigned integer values.

2 If the data is non-numerical, for example colours, then the mean does not exist.

Another disadvantage of relying on the mean as the only statistic is provided by the following cautionary tale.

□ An entertainer is engaged to provide recreation for a mixed party of people and, discovering that the mean age is 14, arranges a disco. However, the party consists of a playgroup and some adult helpers and the ages are

$$5, 3, 4, 4, 3, 5, 2, 38, 25, 51$$

This example illustrates how a measure of location as the sole statistic can sometimes give a misleading impression. ■

23.13 THE MODE

Suppose we have n distinct data points $x_1, x_2, x_3, \ldots, x_n$ and that these occur with frequencies $f_1, f_2, f_3, \ldots, f_n$ respectively.

The point x_r corresponding to the largest frequency f_r is called the **mode**. If there is only one mode then the distribution of data is called **unimodal**, whereas if there are two modes it is called **bimodal**. The terminology may be extended as appropriate. This statistic is particularly valuable if one of the data points occurs with a frequency much greater than any of the others.

□ A total of 100 welders were asked to try four different types of eye shield to say which one they preferred. The results were as follows:

Type	1	2	3	4
Number	19	14	13	54

The mode here is type 4. ■

Although the mode has its uses as a measure of location, it can give a misleading impression. For example if there are several data points each with almost the same frequency, an undue emphasis could be placed on one of them.

23.14 THE MEDIAN

If we have numerical data, it is possible to arrange it in ascending order of magnitude. The data point which appears in the middle is then known as the **median**.

One immediate problem arises: what are we to do if there is an even number of data points? Let's consider some of the options:

1 We could take the mean of the two central points. This compromise, although attractive in some ways, destroys one of the advantages of the median: that it is a data point itself.

2 We could allow two medians, as we do with the mode. This non-uniqueness is the principal disadvantage because, unlike the situation with the mode, the only information it gives is that there is an even number of data points.

3 We could choose a data point at random and discard it. In this way an odd number of data points is obtained together with a unique statistic. The disadvantage is that the original data could produce two different values of this statistic.

4 We could adopt the view that the median is an inappropriate statistic if there is an even number of data points. Which option do you think we choose?

In fact we choose option 1 for the following reason. If there is a significant difference between the two central points then the median is an inappropriate statistic to use as a measure of location. Nevertheless, since the median should represent the middle of the distribution the mean of the two may be taken. If there is no significant difference then it doesn't matter which one is selected. Unless it is important that the median be a typical data point we can take the mean of the two.

Remember when we are calculating statistics we are not simply performing a numerical exercise. We are attempting to represent significant features of the data.

If we draw a frequency curve (Fig. 23.7) we can see that for a unimodal symmetrical distribution the mean, mode and median all coincide, whereas for a skewed distribution they are often quite distinct.

□ Obtain measures of location for the following sample data:

a Numbers of rivets which fail under test conditions:

$$2, 5, 5, 12, 14, 16$$

b Number of errors received in a set of test codewords:

$$2, 2, 2, 2, 3, 3, 3, 3, 3$$

c Voltage measurements:

$$1.1, 1.2, 1.3, 1.3, 1.4, 1.7, 1.8, 1.8$$

We calculate each of the three principal measures of location:

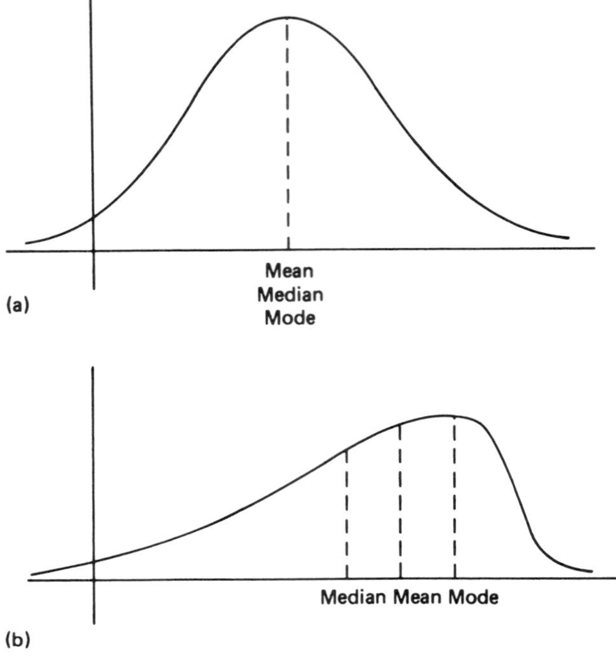

Fig. 23.7 (a) A symmetrical distribution (b) A skewed distribution.

a mean = (2 + 5 + 5 + 12 + 14 + 16)/6 = 9 rivets
mode = 5 rivets
median = $\frac{1}{2}(5 + 12)$ = 8.5 rivets
b mean = $(4 \times 2 + 5 \times 3)/9 = 23/9 \simeq 2.556$ errors
mode = 3 errors
median = 3 errors
c mean = $(1/8)\ (\Sigma) = 11.6/8 = 1.45$ volts
modes = 1.3 and 1.8 volts
median = $\frac{1}{2}(1.3 + 1.4) = 1.35$ volts

Note that very few data points are involved, so if the data had been the entire population it is doubtful if the statistics would have been worth calculating at all. ■

Now let's consider some measures of spread.

23.15 THE RANGE

One of the simplest measures of spread is known as the **range**. This is simply the difference between the largest data point and the smallest. Although in many circumstances this is a perfectly adequate measure of spread, it has one serious drawback. It is unduly affected by freak values of the data.

For example, suppose a manufacturing process resulted in components which usually had a lifetime between 36 and 50 hours, satisfying the retailer's specification. A single faulty component (lifetime 0 hours) would change the range from 14 hours to 50 hours. This could be a very misleading statistic because it might lead to the belief that the process was generally unsatisfactory.

23.16 THE MEAN ABSOLUTE DEVIATION

At first sight it seems a good idea to calculate the average deviation from the mean.

Suppose we have n distinct data points $x_1, x_2, x_3, \ldots, x_n$ and these appear with frequencies $f_1, f_2, f_3, \ldots, f_n$ respectively. The number of data points is N and the mean is $\bar{x}$:

$$N = \sum_{r=1}^{n} f_r$$

$$\bar{x} = \frac{1}{N} \sum_{r=1}^{n} f_r x_r$$

We should then obtain, for the average deviation,

$$\begin{aligned} &\frac{1}{N}[f_1(x_1 - \bar{x}) + f_2(x_2 - \bar{x}) + \ldots + f_n(x_n - \bar{x})] \\ &= \frac{1}{N}(f_1 x_1 + f_2 x_2 + \ldots + f_n x_n) - \frac{\bar{x}}{N}(f_1 + f_2 + \ldots + f_n) \\ &= \bar{x} - \frac{\bar{x}}{N} N = 0 \end{aligned}$$

So the mean deviation is zero whatever the data! Back to the drawing board!

One way round this problem is to consider the absolute value of the deviations and take the mean of these. This is then known as the **mean absolute deviation** (MAD):

$$\text{MAD} = \frac{1}{N}(f_1 |x_1 - \bar{x}| + f_2 |x_2 - \bar{x}| + \ldots + f_n |x_n - \bar{x}|)$$

Although it is easy enough to calculate this statistic, the problems with the modulus signs inhibit theoretical work and so this measure of spread does not play a large part in statistical theory.

23.17 THE STANDARD DEVIATION

Without doubt the most important measure of spread is the standard deviation.

Suppose we have n distinct data points $x_1, x_2, x_3, \ldots, x_n$ and these appear with frequencies $f_1, f_2, f_3, \ldots, f_n$ respectively. As before, the total number N of data points is therefore given by

$$N = \sum_{r=1}^{n} f_r$$

We can calculate the mean $\bar{x}$ of the data, and we wish to obtain a measure of how widely the data is dispersed about the mean. The **standard deviation** s of the data is defined by

$$s = \sqrt{\left[\frac{1}{N-1}\sum_{r=1}^{n} f_r(x_r - \bar{x})^2\right]}$$

This is (almost) the root mean square: the only change is that we are dividing by $N - 1$ instead of N. Since this strange definition can cause some confusion, we shall explain why it is the way it is.

You will remember that we stressed that there was a fundamental difference between population statistics and sample statistics. The statistics we calculate from samples chosen at random from some population are intended to estimate as closely as possible the corresponding statistics for the population. To reinforce this we use different symbols for the statistics corresponding to the population from those corresponding to the sample.

The **population mean** is denoted by μ and the **population standard deviation** is denoted by σ. If we are given a random sample, the mean $\bar{x}$ of the sample is an unbiased estimate for μ. If the data consists of the entire population we have

$$\sigma = \sqrt{\left[\frac{1}{N}\sum_{r=1}^{n} f_r(x_r - \mu)^2\right]}$$

Indeed in general, if the data consists of a random sample from the population and if the population mean μ is known, then the formula

$$s = \sqrt{\left[\frac{1}{N}\sum_{r=1}^{n} f_r(x_r - \mu)^2\right]}$$

would give an **unbiased estimate** for σ.

However, it is relatively rare that μ is known, and so we have to estimate μ by calculating the mean $\bar{x}$ of the sample. When we do this, it can be shown that an unbiased estimate for the population standard deviation is given by

$$s = \sqrt{\left[\frac{1}{N-1}\sum_{r=1}^{n} f_r(x_r - \bar{x})^2\right]}$$

Observe that, unless N is small, the difference between the results obtained by dividing by N and those obtained by dividing by $N - 1$ are negligible.

Most calculators now enable calculations to be made routinely for both the standard deviation of a population and the standard deviation of a sample.

The square of the standard deviation is called the **variance** and is somewhat easier to work with algebraically than the standard deviation. However, the standard deviation has the advantage that it has the same dimension as the data. If we divide the standard deviation by the mean we obtain a dimensionless quantity known as the **coefficient of variation**.

23.18 Workshop

▷ **Exercise** The number of working days lost by each of twenty employees in 1
a small firm during the past twelve months was

$$5, 6, 8, 12, 4, 5, 15, 7, 12, 11,$$
$$6, 0, 2, 4, 5, 5, 8, 10, 12, 11$$

Represent the data using a pie chart and a bar chart.

Have a go at this. It may be necessary to group the data in order to sharpen its impact.

We begin by constructing a frequency table so that we can assess the 2
situation:

Days lost	*Frequency*
0	1
2	1
4	2
5	4
6	2
7	1
8	2
10	1
11	2
12	3
15	1

There are not many data points, and if this had been continuous data we should have had to draw a histogram with not more than four or five class intervals. It therefore seems sensible to group the data into four (or five) intervals here:

Class interval	*Frequency*
0–3	2
4–7	9
8–11	5
12–15	4

Using these we obtain diagrams which reflect the main features of the information (Fig. 23.8).

If all was well, move ahead to step 4. If not, here is a little more practice.

▷**Exercise** Repeat the previous exercise using six class intervals 0–2, 3–5, . . . , 15–17.

3

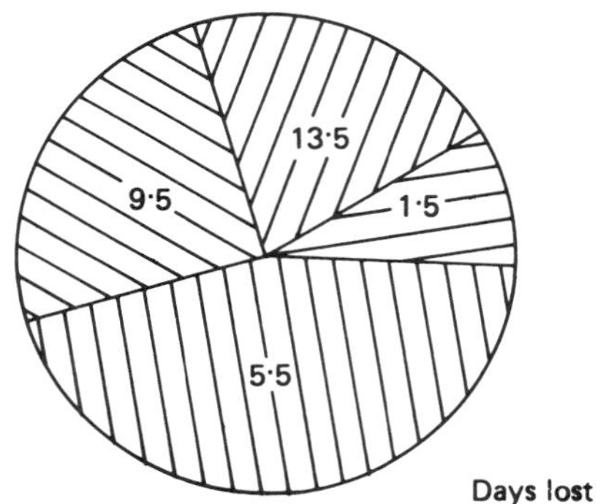

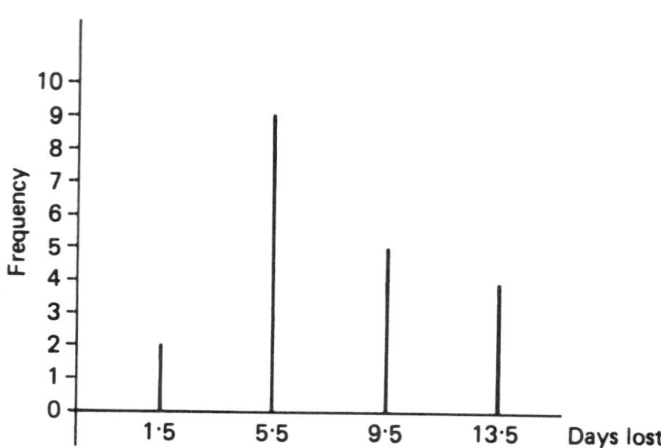

Fig. 23.8 (a) Pie chart (b) Bar chart.

We obtain the following frequency table:

Class interval	*Frequency*
0–2	2
3–5	6
6–8	5
9–11	3
12–14	3
15–17	1

Diagrams can then be produced as before.

Now for a histogram.

4

▷ **Exercise** The following gives the time taken in hours for 30 samples of soil to dry out at room temperature:

5.43, 4.98, 5.24, 5.59, 4.89, 5.01, 4.97, 4.99, 5.11, 5.23,
5.52, 5.61, 5.31, 5.67, 5.51, 5.23, 5.47, 5.55, 4.87, 4.91,
4.84, 5.34, 5.16, 4.86, 5.12, 5.45, 5.48, 5.15, 5.23, 5.42

Group the data into class intervals 4.8–4.9, 5.0–5.1, . . . , 5.6–5.7. Then draw a histogram, construct a frequency polygon, and produce a cumulative frequency diagram.

It's very simple to do all this. Keep an eye on the class boundaries.

5

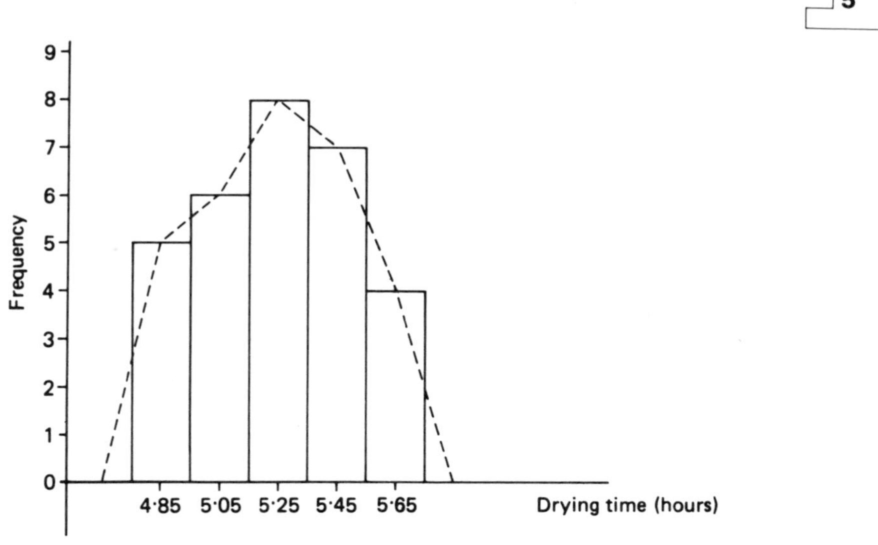

Fig. 23.9 Histogram.

We put the data points into the appropriate class intervals in preparation for drawing the histogram:

Class interval	*Class mark*	*Frequency*
4.8–4.9	4.85	5
5.0–5.1	5.05	6
5.2–5.3	5.25	8
5.4–5.5	5.45	7
5.6–5.7	5.65	4

It's best to make a tally to avoid overlooking data points.

The histogram and frequency polygon are then shown in Fig. 23.9. Notice particularly how by extending the class intervals on either side and giving them frequencies of zero we obtain an area under the frequency polygon which is equal to the area of the histogram. Indeed, had we drawn a histogram using relative frequencies the area enclosed would have been unity. We shall see later that this would imply that the frequency polygon would then be the graph of a probability density function.

We need to note that when we come to the cumulative frequencies we must use the class boundaries 4.95, 5.15, 5.35, 5.55, 5.75. The cumulative frequencies are as follows:

Class boundary	*Cumulative frequency*
4.95	5
5.15	11
5.35	19
5.55	26
5.75	30

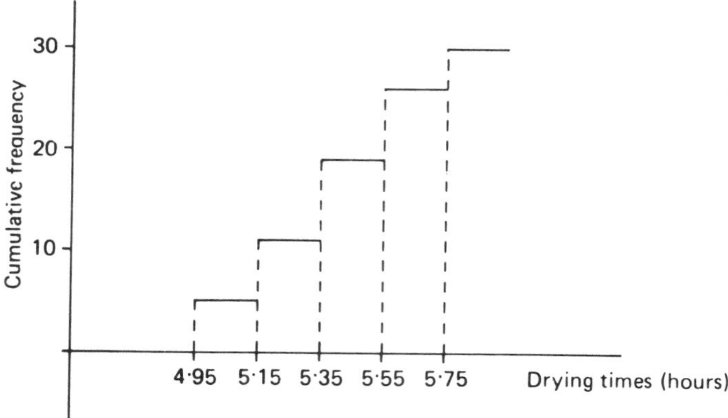

Fig. 23.10 Cumulative frequency diagram.

The cumulative frequency diagram is shown in Fig. 23.10.

Make sure that you had

1 no gaps between your rectangles in your histogram
2 each class mark in the centre of the class interval
3 class boundaries as the boundary markers.

If everything was correct then proceed at full speed to the next section. If not, here is more practice.

▷**Exercise** Draw another histogram using the same data and the following class intervals: 4.80–5.04, 5.05–5.14, 5.15–5.24, 5.25–5.34, 5.35–5.44, 5.45–5.96. Draw a frequency polygon. Is the area under the frequency polygon the same as that of the histogram?

6

Class interval	*Class mark*	*Frequency*
4.80–5.04	4.920	9
5.05–5.14	5.095	2
5.15–5.24	5.195	6
5.25–5.34	5.295	2
5.35–5.44	5.395	2
5.45–5.96	5.705	9

The histogram required is shown in Fig. 23.11, together with the frequency polygon. In this case the area under the frequency polygon is *not* the same as that of the histogram.

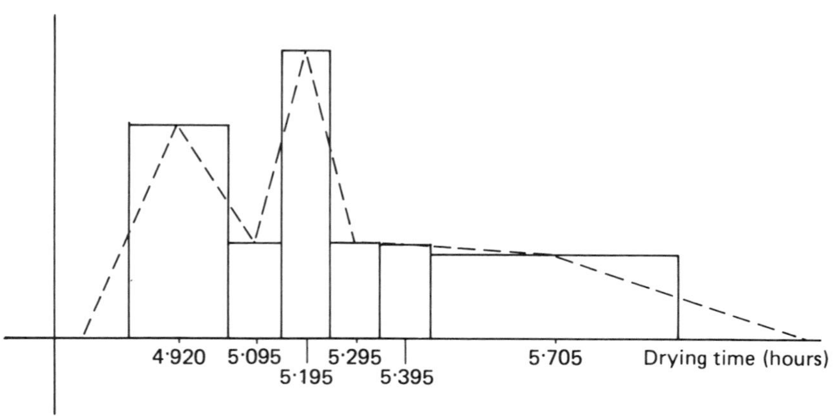

Fig. 23.11 Histogram and frequency polygon.

23.19 Practical

LOAD BEARING

Ten 4-metre girders were taken from a production line and each one was tested for central load bearing when freely supported at each end. The results in kilonewtons were as follows:

4.562, 4.673, 4.985, 4.657, 4.642,
4.784, 4.782, 4.832, 4.637, 4.596

Calculate the mean, mode, median, range, mean absolute deviation and standard deviation for this data.

It's just a question of pressing the buttons really! Work them out and take the next step to see if they are all right.

There are ten observations and the total is 47.150. So the mean is 4.715 kilonewtons.

It is necessary to put the data into class intervals to obtain a meaningful mode. The obvious intervals to choose are 4.55–4.64, 4.65–4.74, 4.75–4.84, 4.85–4.94, 4.95–5.04. Taking the class marks as the representatives, we then have:

Class mark	*Number*
4.595	4
4.695	2
4.795	3
4.895	0
4.995	1

In this way we obtain a mode of 4.595. However, we should indicate the fact that this has been obtained from continuous data by giving fewer decimal places than in the data: so the mode is 4.60. We remark further that this is a poor statistic to use here in view of the almost bimodal nature of the sample.

When we arrange the data in ascending order we obtain

4.562, 4.596, 4.637, 4.642, 4.657,
4.673, 4.782, 4.784, 4.832, 4.985

There are an even number of data points, so we average the middle two: the median is $(4.657 + 4.673)/2 = 4.665$ kilonewtons.

To find the range we must subtract the smallest value which appears in the data from the largest. The range is 4.985 − 4.562 = 0.423 kilonewtons.

To find the mean absolute deviation we begin by subtracting the mean, 4.715, from each of the data points to give the deviations from the mean and the absolute deviations. These are shown in columns 2 and 3 of Table 23.3. The total absolute deviation is 1.046, and so dividing by the total number we obtain MAD = 0.1046 kilonewtons.

Finally we require standard deviation. The squared deviations from the mean are shown in column 4 of Table 23.3. The sum of these is 0.149 950 and the total number is 10. It remains to divide by 9 and take the square root: the standard deviation s is then 0.129 08 kilonewtons.

Did you manage all those?

In the old days we used to insist that students present the numerical work clearly in a tabulated form as shown here. Although there is a lot to be said in favour of this practice, now that calculators and computers are generally available it is usually quicker to tap in the numbers. It's a good idea to do the calculation twice, though, just to check you haven't pressed the wrong buttons!

Table 23.3

Data 1	Deviations from mean 2	Absolute deviations 3	Squared deviations 4
4.562	−0.153	0.153	0.023 409
4.596	−0.119	0.119	0.014 161
4.637	−0.078	0.078	0.006 084
4.642	−0.073	0.073	0.005 329
4.657	−0.058	0.058	0.003 364
4.673	−0.042	0.042	0.001 764
4.782	0.067	0.067	0.004 489
4.784	0.069	0.069	0.004 761
4.832	0.117	0.117	0.013 689
4.985	0.270	0.270	0.072 900
47.149		1.046	0.149 950

SUMMARY

We have seen how to display data pictorially using

☐ pie charts
☐ bar charts
☐ histograms.

We have examined the principal examples of statistics:

☐ Measures of location
  a mean
  b mode
  c median.
☐ Measures of spread
  a range
  b mean absolute deviation
  c standard deviation.

EXERCISES

1 Decide whether the following data is discrete or continuous:
  a Defective batteries in batches
  b Quantity of impurities in water supply
  c Faulty tyres in spot testing
  d Over-stressed components after wind tunnel exposure
  e Anti-cyclones each day in the Northern hemisphere
  f Percentage of pollutants in engine exhaust
  g Percentage of faulty components in production

2 The amount of time devoted to a new piece of research was initially one sixth of the time available. Two other pieces of development were under way, production and testing, and these took equal times. It was decided to increase the amount of time devoted to the new work to 25% and to devote twice as much of the remaining time to production as to testing. Represent this change by means of pie charts.

3 Calculate for each of the following lists of data (i) the mean, (ii) the mode, (iii) the median, (iv) the range:
  a Resistances (ohms)

3, 2, 1.5, 1.5, 2, 1.5, 3, 2, 2, 1.5, 2, 2

  b Percentage of water in soil samples

30.0, 29.6, 21.5, 22.0, 23.5, 18.1, 19.5, 23.0, 24.2, 21.9, 18.7, 14.1, 17.9, 18.3, 19.7

  c Deflections of a beam (metres)

0.113, 0.121, 0.119, 0.110, 0.118, 0.123, 0.121, 0.171, 0.153, 0.161, 0.169, 0.173

In each case give a suitable visual display of the data and comment on the suitability of the statistics you have calculated in reflecting its true nature.

4 Data consists of the digits 0, 1, 2, 3, . . . , 9 used in the decimal representation of the first 100 natural numbers. Obtain the mean, mode, median and range.

5 Data consists of the number of days in each month over a four year period. Obtain the mean, mode, median and range.

ASSIGNMENT

A machine collects measured quantities of soil and deposits them in boxes for analysis in the laboratory. Fifty boxes were taken at random and inspected. It was found that the following numbers of stones were in the boxes:

45, 47, 43, 46, 42, 47, 44, 48, 41, 47,
46, 44, 45, 43, 43, 42, 45, 44, 43, 42,
43, 49, 42, 40, 44, 48, 44, 46, 42, 46,
42, 44, 45, 44, 44, 42, 42, 41, 44, 43,
45, 43, 44, 41, 43, 48, 47, 40, 46, 46

1 State whether the data is discrete or continuous.

2 Without grouping the data, display it as a bar chart.

3 Grouping the data as 40–41, 42–43, 44–45, 46–47, 48–49, present it as a pie chart. Then present it as a histogram, and draw a corresponding frequency polygon.

4 Calculate the mean, mode and median from the raw data.

5 Calculate the range and variance.

FURTHER EXERCISES

1 Two consignments of carbon brushes were examined for defects. Each consignment consisted of 100 boxes each containing 50 brushes. Five boxes from each assignment were selected at random and the number of defective brushes in each one was counted. The results were

a 8, 3, 5, 2, 3

b 20, 5, 2, 3, 5

Calculate for each sample (i) the mean (ii) the mode (iii) the median (iv) the range (v) the standard deviation.

2 Samples of lubricant were chosen and the specific gravity (SG) was measured. The frequencies of samples in SG classes were as follows:

SG	1.11–1.12	1.12–1.13	1.13–1.14	1.14–1.15	1.15–1.16	1.16–1.17	1.17–1.18
Frequency	1	3	8	16	20	11	5

Draw a histogram showing percentage frequency against specific gravity interval.

3 A construction site uses five different grades of sand. On a typical day the number of bags of each type drawn from the store is as follows:

Grade	1	2	3	4	5
Number	12	28	10	16	8

Represent this information using (a) a pie chart (b) a bar chart.

4 The lengths of a sample of 30 steel drive belts produced by a machine were measured in metres to the nearest millimetre. The results were as follows:

2.975, 3.245, 3.254, 3.156, 2.997, 2.995, 3.005, 3.057, 3.046, 3.142,
3.116, 3.052, 3.017, 3.084, 3.119, 3.143, 3.063, 3.158, 3.196, 3.203,
3.225, 3.183, 3.193, 3.174, 3.148, 3.053, 3.202, 3.153, 3.037, 3.048

a Display the data on a histogram by grouping the data into class intervals 2.95–3.00, 3.05–3.10, . . . , 3.25–3.30.
b Write down the class boundaries.
c Using the same class intervals, draw a relative frequency diagram.
d Calculate the mean and standard deviation of (i) the raw data (ii) the grouped data.

5 Suppose data consists of n distinct data points $x_1, x_2, \ldots, x_n$ with frequencies $f_1, f_2, \ldots, f_n$ respectively. Suppose also that N is the total number of data points. Show that, if s is the standard deviation,

$$(N-1)s^2 = \sum_{r=1}^{n} f_r x_r^2 - N\bar{x}^2$$

(In the days before electronic calculators this formula could be used to ease the arithmetical burden.)

6 The following data shows the number of unsatisfactory motherboards produced by a company in a given month:

Number defective	0	1	2	3	4	5	6
Frequency	28	25	24	12	9	1	1

Display this data on a suitable diagram and calculate its mean and variance.

Probability 24

To take our story any further requires some probability theory, and that is the subject of this chapter.

After completing this chapter you should be able to

☐ Use the terminology of statistics correctly – experiment, sample space, event, random variable etc.;
☐ Use the rules of probability correctly;
☐ Obtain the mean and variance of a probability distribution;
☐ Use the binomial, Poisson and normal distributions;
☐ Approximate the binomial and Poisson distributions by the normal distribution;
☐ Use normal probability paper to estimate the mean and variance of a distribution.

At the end of this chapter we look at a practical problem of statistics in engineering.

24.1 CONCEPTS

In order to use sample statistics effectively we need to employ some of the theory of probability. It is necessary first to fix some of the terminology:

1 The word **experiment** is used to denote any activity which has an outcome.
2 The set of all possible outcomes of an experiment is called the **sample space** S.
3 Each of the possible outcomes is called a **sample point**.
4 An **event** E is any collection of sample points.

☐ The outcomes of the activity of throwing two dice can be represented as ordered pairs of numbers. So the sample space may be represented as follows:

$$\begin{array}{llllll}(1,1), & (1,2), & (1,3), & (1,4), & (1,5), & (1,6)\\ (2,1), & (2,2), & (2,3), & (2,4), & (2,5), & (2,6)\\ (3,1), & (3,2), & (3,3), & (3,4), & (3,5), & (3,6)\\ (4,1), & (4,2), & (4,3), & (4,4), & (4,5), & (4,6)\\ (5,1), & (5,2), & (5,3), & (5,4), & (5,5), & (5,6)\\ (6,1), & (6,2), & (6,3), & (6,4), & (6,5), & (6,6)\end{array}$$

There are 36 sample points in the sample space.

If E is the event 'the total sum is greater than 7', then

$$\begin{aligned}E &= \{(x,y) \mid x + y > 7\}\\ &= \{(2,6), (3,5), (3,6), (4,4), (4,5), (4,6), (5,3),\\ &\quad (5,4), (5,5), (5,6), (6,2), (6,3), (6,4), (6,5), (6,6)\}\end{aligned}$$

So E consists of 15 sample points. ■

If there are no sample points in an event E then $E = \varnothing$, the empty set. If every sample point is in the event E then $E = S$, the sample space.

We shall define **probability** in such a way that every event will have a probability in the interval $[0, 1]$. If an event is certain to occur we say it has probability 1, whereas if an event cannot occur we say it has probability 0. So:

1 The event S has probability 1 because one of the points in the sample space must be the result of the experiment.

2 The event $\varnothing$ has probability 0 since by definition the experiment must have an outcome.

It often helps to picture things by using a Venn diagram.

In this it is usual to represent the sample space S by means of a large rectangle. The sample points are then shown inside the rectangle, and an event is represented by means of a loop; the interior of the loop represents the points in the event (Fig. 24.1).

If x is a sample point in S, the sample space, we denote by $P(x)$ the probability that x will occur. If E is an event we then define $P(E)$ by

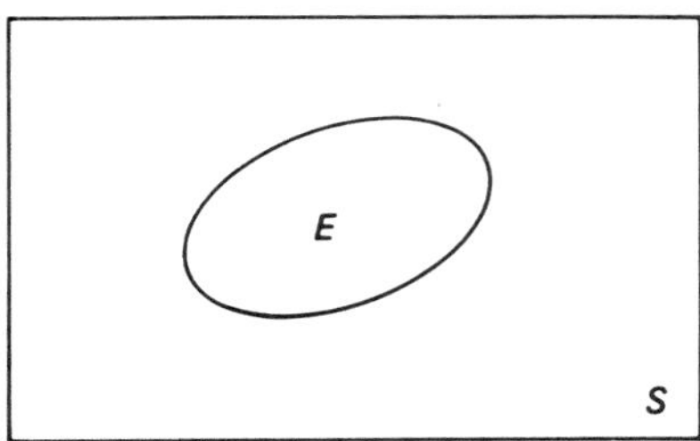

Fig. 24.1 A Venn diagram.

$$P(E) = \sum_{x \in E} P(x)$$

The probability of an event E is the sum of the probabilities of the sample points in E.

In the Venn diagram the probability of an event E is the proportion of the sample space covered by E. We see at once that $P(S) = 1$ and $P(\varnothing) = 0$, which is consistent with the definition. Now we have already seen that

$$P(S) = \sum_{x \in S} P(x) = 1$$

and so we may write

$$P(E) = \frac{\sum_{x \in E} P(x)}{\sum_{x \in S} P(x)}$$

Although Venn diagrams are very useful there is one snag: they may lead us to draw conclusions which are false. For example, not every sample space is bounded.

24.2 THE RULES OF PROBABILITY

We can use Venn diagrams, together with the interpretation we have put on probability, to deduce the basic rules of probability. In the sections that follow we shall suppose that E and F are any two events in the sample space S (see Fig. 24.2).

To begin with we must employ the basic terminology of set theory

1 Union: $E \cup F = \{x \mid x \in E \text{ or } x \in F, \text{ or both}\}$

2 Intersection: $E \cap F = \{x \mid x \in E \text{ and } x \in F\}$

3 Complement: $E' = \{x \mid x \in S \text{ but } x \notin E\}$

There are rules of set theory which can be deduced formally from these definitions. However, for our purposes they can be inferred easily from Venn diagrams. Here are the rules:

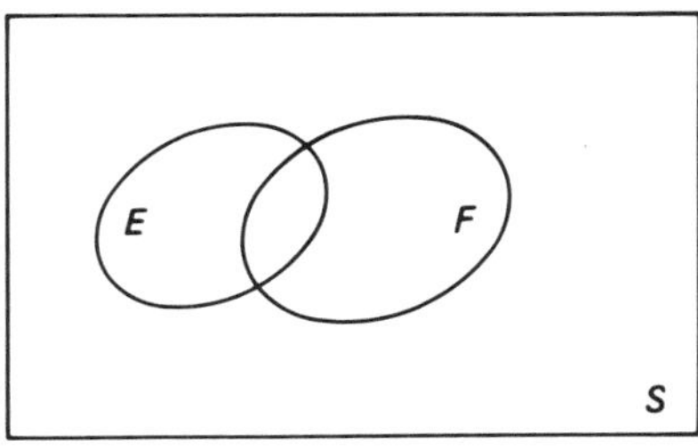

Fig. 24.2 Two events E and F.

$$E \cup F = F \cup E$$
$$E \cap F = F \cap E$$
$$(E \cup F)' = E' \cap F'$$
$$(E \cap F)' = E' \cup F'$$
$$E \cap (F \cup G) = (E \cap F) \cup (E \cap G)$$
$$E \cup (F \cap G) = (E \cup F) \cap (E \cup G)$$

You might like to draw a few Venn diagrams to convince yourself of the truth of these.

Now for the first rule of probability.

24.3 THE SUM RULE

The probability that either the event E or the event F (or both) will occur is the sum of the probability that E will occur with the probability that F will occur less the probability that both E and F will occur:

$$P(E \cup F) = P(E) + P(F) - P(E \cap F)$$

To see this we merely need to note that the area enclosed by both E and F is the area enclosed by E, together with the area enclosed by F, but less the area of the overlap $E \cap F$, which we would otherwise have counted twice.

□ The probability that a drilling machine will break down is 0.35. The probability that the lights will fail is 0.28. It is known that the probability that one or the other (or possibly both) will occur is 0.42. Obtain the probability that both the machine will break down and the lights will fail.

Let E be 'the drilling machine will break down' and F be 'the lights will fail'. Then $P(E) = 0.35$, $P(F) = 0.28$ and $P(E \cup F) = 0.42$. Now

$$P(E \cup F) = P(E) + P(F) - P(E \cap F)$$

Therefore

$$\begin{aligned} P(E \cap F) &= P(E) + P(F) - P(E \cup F) \\ &= 0.35 + 0.28 - 0.42 = 0.21 \end{aligned}$$ ■

24.4 MUTUALLY EXCLUSIVE EVENTS

Sometimes two events E and F in a sample space S have no sample points in common, so that $E \cap F = \emptyset$. In such circumstances the events E and F are said to be **mutually exclusive** events. For mutually exclusive events the addition law of probability becomes simplified:

$$P(E \cup F) = P(E) + P(F)$$

24.5 CONDITIONAL PROBABILITY

We write $P(E \mid F)$ for the probability that the event E will occur, given that the event F does occur.

If we think about this for a few seconds, we see that the precondition that the event F does occur effectively reduces the sample space that we are considering to the points in the event F. We require the proportion of those which are in the event E. Consequently

$$P(E \mid F) = \frac{\sum_{x \in E \cap F} P(x)}{\sum_{x \in F} P(x)} = \frac{P(E \cap F)}{P(F)}$$

□ A car mechanic knows that the probability of a vehicle having a flat battery is 0.24. He also knows that if the vehicle has a flat battery then the probability that the starter motor needs replacing is 0.47. A vehicle is brought in for his attention. What is the probability that it both has a flat battery and needs a new starter motor?

Suppose E is 'the starter motor needs replacing' and F is 'the vehicle has a flat battery'. We know $P(E \mid F) = 0.47$ and $P(F) = 0.24$ and require $P(E \cap F)$. Using

$$P(E \mid F) = \frac{P(E \cap F)}{P(F)}$$

we deduce that $P(E \cap F) = 0.47 \times 0.24 = 0.1128$. ■

24.6 THE PRODUCT RULE

From the equation

$$P(E \mid F) = \frac{P(E \cap F)}{P(F)}$$

we obtain, on multiplying through by $P(F)$,

$$P(E \cap F) = P(E \mid F)P(F)$$

By symmetry therefore we also have

$$P(E \cap F) = P(E)P(F \mid E)$$

The probability that both the event E and the event F will occur is the product of the probability that E will occur with the probability that F will occur, given that E does occur.

24.7 INDEPENDENT EVENTS

Two events E and F are said to be **independent** if $P(E) = P(E \mid F)$, because the event F has no effect whatever on E as far as probability is concerned. Whenever two events E and F are independent the product rule becomes simplified to

$$P(E \cap F) = P(E)P(F)$$

At first sight the condition that two events E and F are independent looks asymmetrical. However, it is a simple algebraic matter to deduce that this is equivalent to

$$P(F) = P(F \mid E)$$

Can you see why? Give it a whirl and then see if you are right.

Suppose that E and F are independent, so that $P(E) = P(E \mid F)$. We have $P(E \cap F) = P(E)P(F)$ but $P(E \cap F) = P(E)P(F \mid E)$. Equating these two expressions and dividing by $P(E)$ gives

$$P(F) = P(F \mid E)$$

The argument fails if $P(E) = 0$, but this would imply $E = \varnothing$. So $P(F) = P(F \mid E)$ since there is no condition to satisfy.

24.8 COMPLEMENTATION RULE

The events E and E' satisfy $E \cup E' = S$ and $E \cap E' = \varnothing$. Therefore $P(E) + P(E') = P(S) = 1$, from which

$$P(E') = 1 - P(E)$$

□ The probability that a telephone switchboard is jammed is 0.25. The probability that a customer will attempt to telephone is 0.15. These events are known to be independent. However, if a customer telephones but fails to get connected the probability that an order will be lost is 0.75. Calculate

a The probability that a customer will telephone while the switchboard is jammed, resulting in a lost order;

b The probability that the order will not be lost even though the telephone switchboard is jammed and the customer tried to telephone.

You might like to try this on your own.

We begin by identifying the events. Let E be 'the telephone switchboard is jammed', F be 'a customer will attempt to telephone' and G be 'an order will be lost'.

We know $P(E) = 0.25$, $P(F) = 0.15$. Now $E \cap F$ is the event 'a customer will attempt to telephone and the switchboard is jammed'. Since E and F are independent,

$$P(E \cap F) = P(E)P(F) = 0.25 \times 0.15 = 0.0375$$

For **a** we require $P[(E \cap F) \cap G]$. We know that $P[G \mid (E \cap F)] = 0.75$, so

$$\frac{P[G \cap (E \cap F)]}{P(E \cap F)} = 0.75$$

Now

$$\begin{aligned} P[(E \cap F) \cap G] &= P[G \cap (E \cap F)] \\ &= P(E \cap F) \times 0.75 \\ &= 0.0375 \times 0.75 = 0.028\,125 \end{aligned}$$

For **b**, G' is the event 'the order will not be lost'. We require $P[G' \mid (E \cap F)]$. We have

$$\begin{aligned} P[G' \mid (E \cap F)] &= 1 - P[G \mid (E \cap F)] \\ &= 1 - 0.75 = 0.25 \end{aligned}$$

■

Although we have seen how to use the rules of probability, we have yet to define fully what is meant by probability.

24.9 A RANDOM VARIABLE

A random variable X is a numerically valued function defined on the sample space S; so $X: S \to \mathbb{R}$. For example, if we consider the experiment of tossing a coin then we could define X by

$$\begin{aligned} \text{heads} &\to 1 \\ \text{tails} &\to 0 \end{aligned}$$

However, there is no restriction on the way we define X. We could, if we wished to be perverse, define X instead by

$$\begin{aligned} \text{heads} &\to \pi \\ \text{tails} &\to \text{e} \end{aligned}$$

The important thing is that X assigns numerical values to the outcome of an experiment. In this way outcomes which we regard as equivalent to one another can be assigned the same numerical value, whereas those which are regarded as distinct can be assigned different values.

In a slight misuse of the function notation, we write $X = r$ if the random variable X has value r at the sample point. We also write $P(X = r)$ for the probability that X has the value r at the sample point.

There are three basic approaches to probability; these are described in the next three sections.

24.10 THE ANALYTICAL METHOD

We begin by looking at an example.

□ Suppose that a box contains 100 microcomputer discs and that 15 of them are defective in some way. If one of the discs is selected at random from the box, what is the probability that it is defective?

Of course we know that it will be either good or defective, so the question we are really asking is: what proportion of the discs is defective? The answer to this is clear; there are 15 defectives and 100 discs altogether, and so the proportion of defectives is $15/100 = 3/20 = 0.15$.

This then is what we define as the probability p of selecting a defective:

$$p = \frac{\text{number of defectives}}{\text{total number}}$$

We observe that if every disc in the box is defective we shall obtain $p = 1$, whereas if none of the discs is defective we shall obtain $p = 0$; this is consistent with our earlier definition.

Looked at in this way, we see we can define a random variable X on the sample space S consisting of each of the discs:

$$\begin{aligned} \text{bad disc} &\to 0 \\ \text{good disc} &\to 1 \end{aligned}$$

Then $P(X = 0) = 0.15$ and $P(X = 1) = 0.85$, so that $P(X = 0) + P(X = 1) = 1$; a disc is either defective or satisfactory. ■

Now let's analyse what we have done. Suppose the discs were numbered $1, 2, 3, \ldots, 100$. Then we can represent the discs as $D_1, D_2, D_3, \ldots, D_{100}$ of which we know there are 15 defective.

Now if we consider these labelled discs, there are 100 sample points in S because each of the labelled discs is a possible outcome of the experiment. Moreover, the selection procedure is random so each disc has the same probability of being chosen.

If E is the event that 'the disc is defective' then there are 15 sample points in this event because there are 15 defective discs. So we obtain

$$P\text{ (disc defective)} = \frac{|E|}{|S|} = \frac{15}{100}$$

where $|A|$ denotes the number of elements in the finite set A.

In general, if each point in a finite sample space is equally likely the probability of an event E is

$$P(E) = \frac{\text{number of points in } E}{\text{number of points in } S} = \frac{|E|}{|S|}$$

There are two problems which arise with this approach:

1 We may not know the number of sample points in the event E.

2 The number of sample points in S may not be finite.

24.11 THE RELATIVE FREQUENCY METHOD

We may not know how many defective discs are in the box. For example, a machine could be making and packaging the discs and it may not be known how many are defective.

One way of proceeding is to select each disc in turn and test it. We then obtain

$$P(n) = \frac{\text{number of defective discs}}{\text{number of discs tested}}$$

where n is the number of discs tested. This would give an estimate of the probability p, and we could argue that as $n \to N$ (the total number) we should have $p(n) \to p$, the probability of a defective. Indeed, we could extend this idea to an infinite population and then obtain $p(n) \to p$ as $n \to \infty$.

□ A coin is thrown to test the probability that it will show heads. Given the following results, taken in order, show how an estimate of the probability varies:

H H T H T H H H T H T T T H H T

where H denotes heads and T denotes tails.

We can construct a table of the relative frequency $p(n)$, that is the number of heads which have shown in n throws:

n	$p(n)$	n	$p(n)$
1	1/1	9	6/9
2	2/2	10	7/10
3	2/3	11	7/11
4	3/4	12	7/12
5	3/5	13	7/13
6	4/6	14	8/14
7	5/7	15	9/15
8	6/8	16	9/16

What are we to make of this? We can argue that if the coin is fair then, if we throw it $2m$ times, we should expect for large m that there would be m heads. However, we don't know anything about the coin in question here.

We could perhaps turn the argument round and use this as a method of testing whether or not the coin is fair. However, this leaves a number of open questions, such as 'How many times do we need to throw the coin to establish the probability?' ■

The major problem with the relative frequency method is that $p(n)$ changes as n changes, and consequently a fluke situation could give misleading results.

24.12 THE MATHEMATICAL METHOD

We must remember that a sample space S consists of all possible outcomes of an experiment and that a random variable assigns a number to each sample point. Therefore if we assign probabilities to the set of values $X(S)$ of the random variable, we automatically assign probabilities to the sample points:

$$S \to X(S) \to \mathbb{R}$$

It is convenient on some occasions to think of probabilities as assigned to the sample space, and on others to think of them as assigned to the values of the random variable.

Suppose S is a sample space and X is a random variable. A **probability density function** (PDF) is a real-valued function with domain $X(S)$ such that

1 If $r \in X(S)$ then $P(r) \geqslant 0$;
2 $\Sigma P(r) = 1$.

where the sum is taken over every element of $X(S)$. So a probability density function is a function which assigns weights to the values of the random variable in such a way that they are all non-negative and total to 1.

Whenever we have a probability density function we say that P defines a **probability distribution**.

□ Obtain h if P defines a probability distribution on $\{1,2,3\}$ as follows:

r	1	2	3
$P(X = r)$	1/4	h	h^2

Obtain also

a the probability that X is greater than 1
b the probability that X is not equal to 2.

We have $\Sigma P(X = r) = 1$ for all values of the random variable X, and so

$$\begin{aligned} \tfrac{1}{4} + h + h^2 &= 1 \\ (h + \tfrac{1}{2})^2 &= 1 \\ h + \tfrac{1}{2} &= \pm 1 \\ h &= -\tfrac{1}{2} \pm 1 \end{aligned}$$

However, we can reject the negative sign because all probabilities must be positive. We conclude therefore that $h = 1/2$.

a $P(X > 1) = P(X = 2) + P(X = 3) = 1/2 + (1/2)^2 = 1/2 + 1/4 = 3/4$

b $P(X \neq 2) = 1 - P(X = 2) = 1 - 1/2 = 1/2$ ■

In the case of a discrete random variable X which can take values $x_1, x_2, \ldots, x_n, \ldots$ with probabilities, $p_1, p_2, \ldots, p_n, \ldots$ respectively we obtain

$$p_r \geqslant 0 \quad \text{for all } r \in \mathbb{N}$$
$$\Sigma p_r = 1$$

where the sum is taken over all $r \in \mathbb{N}$.

There are many discrete probability distributions. However there are two, the binomial distribution and the Poisson distribution, which have many applications. We shall discuss them briefly.

We need one extra piece of terminology first. A single occurrence of an experiment is called a **trial**.

24.13 THE BINOMIAL DISTRIBUTION

Before describing the binomial distribution we shall state the circumstances in which it can be used. It is most important to be sure that these conditions hold before attempting to apply the binomial distribution.

The binomial distribution may be applied whenever an experiment occurs with the following characteristics:

1 There are only two possible outcomes of each trial. For reference purposes we shall call these 'success' and 'failure'.

2 The probability p of success in a single trial is constant. Note that this implies that the probability q of failure is constant too, because $p + q = 1$.

3 The outcomes of successive trials are independent of one another.

We can think of many examples where these conditions hold, such as tossing a coin with outcome heads or tails, or rolling dice to obtain a six. A third example is selecting, one by one at random, electrical components from a box and then testing and replacing them. In this case, if we do not replace a component then the probability of choosing a defective next time will change. However, for a large quantity of components in the box the binomial conditions will be satisfied approximately.

In general, suppose there are n trials and that we define the random variable X as the number of successes. We shall examine the possibilities.

If there is just one trial, we then have only two possibilities: F (failure) or S (success). We know that $P(\text{F}) = q$ and $P(\text{S}) = p$ where $p + q = 1$, so

$$P(X = 0) = q \qquad P(X = 1) = p$$

If there are two trials then the possibilities are FF, SF, FS, SS. So

$$P(X = 0) = q^2$$
$$P(X = 1) = pq + qp = 2pq$$
$$P(X = 2) = p^2$$

Now you list the possible outcomes for three trials and thereby calculate $P(X = 0)$, $P(X = 1)$, $P(X = 2)$ and $P(X = 3)$.

Here are the possible outcomes:

FFF, FFS, FSF, SFF, FSS, SFS, SSF, SSS

From these,

$$P(X = 0) = q^3$$
$$P(X = 1) = 3pq^2$$
$$P(X = 2) = 3p^2q$$
$$P(X = 3) = p^3$$

In the general situation where there are n trials we have

$$P(X = r) = \binom{n}{r} p^r q^{n-r}$$

You will observe that this is the general term in the expansion of $(p + q)^n$ using the binomial theorem (see Chapter 1). Indeed this observation confirms straight away that we have a probability distribution. Each term is positive and the sum of them all is $(p + q)^n$, which is 1 since $p + q = 1$.

☐ A company has eight faulty machines. It is stated by the servicing engineer that if a machine is serviced there is a 75% probability that it will last a further three years. The company has all eight machines serviced. If the servicing engineer is correct, estimate

a The probability that none of the machines will last a further three years;

b The probability that at least six of the machines will last a further three years;

c The probability that at least one of the machines will last a further three years.

The probability that if a machine is serviced it will last a further three years is $p = 0.75$, and we may suppose that the lifetimes of the machines are independent of one another.

The conditions for a binomial distribution are satisfied with $n = 8$. So if the random variable X is defined as the number of machines which will last a further three years, we have

$$P(X = r) = \binom{8}{r} (0.75)^r (0.25)^{8-r}$$

a We must calculate

$$\begin{aligned} P(X = 0) &= (0.75)^0(0.25)^8 \\ &= (0.25)^8 = 0.000\,015\,26 \end{aligned}$$

This is negligible.

b We must obtain

$$\begin{aligned} P(X \geqslant 6) &= P(X = 6) + P(X = 7) + P(X = 8) \\ &= \binom{8}{6} (0.75)^6 (0.25)^2 + \binom{8}{7} (0.75)^7 (0.25) + (0.75)^8 \\ &= 0.311\,462\,4 + 0.266\,967\,8 + 0.100\,112\,9 = 0.678\,543\,1 \\ &= 0.6785 \text{ to four decimal places} \end{aligned}$$

c We require

$$P(X > 0) = 1 - P(X = 0) = 1 - 0.000\,015\,26 = 0.999\,984\,74$$

So it is almost certain that at least one of the machines will last a further three years. ■

24.14 THE MEAN OF A PROBABILITY DISTRIBUTION

In much the same way as we defined the mean of a population, we define the mean of a discrete probability distribution by

$$\mu = \Sigma P(X = r)r$$

where the sum is taken over all possible values of the random variable X. Here we can think of $P(X = r)$ as the relative frequency with which the random variable X attains the value r in a long sequence of trials.

The mean μ of the probability distribution is also known as the **expectation** of random variable.

□ A businessman knows that if he sends a letter to a householder there is a 0.5% probability that he will receive an order for new windows which will give him a profit of £600. If he doesn't receive an order the cost to him in postage and administration is 30p. What is his expected gain?

Writing S for success and F for failure we have $S \rightarrow 600$, $F \rightarrow -0.3$. So defining the random variable X as his expected win, we have $P(X = 600) = 0.005$ and $P(X = -0.3) = 0.995$. Therefore

$$\mu = 600 \times 0.005 + (-0.3) \times 0.995 = 2.7015$$

So if he sends out a lot of letters, on average he will expect to gain £2.70 for each one. ■

□ Obtain the mean μ corresponding to the binomial distribution.

We have

$$P(X = r) = \binom{n}{r} p^r q^{n-r}$$

for $r \in \{0, 1, 2, \ldots, n\}$. Then

$$\begin{aligned}
\mu &= \sum_{r=0}^{n} \binom{n}{r} p^r q^{n-r} r \\
&= \sum_{r=1}^{n} \frac{n(n-1)\ldots(n-r+1)}{1 \times 2 \times 3 \times \ldots \times r} p^r q^{n-r} r \\
&= np \sum_{r=1}^{n} \frac{(n-1)\ldots(n-r+1)}{1 \times 2 \times 3 \times \ldots \times r} p^{r-1} q^{n-r} r
\end{aligned}$$

If we put $s = r - 1$ we obtain

$$\begin{aligned}
\mu &= np \sum_{s=0}^{n-1} \frac{(n-1)\ldots([n-1]-s+1)}{1 \times 2 \times 3 \times \ldots \times (s+1)} p^s q^{n-s-1}(s+1) \\
&= np \sum_{s=0}^{n-1} \frac{(n-1)\ldots([n-1]-s+1)}{1 \times 2 \times 3 \times \ldots \times s} p^s q^{n-s-1}
\end{aligned}$$

Now s is a dummy variable, and so we can call it what we like. Therefore we shall revert to using r; the old r is dead and gone! If you object to this practice, give it some thought and you will realize you are simply being sentimental about good old r. Then

$$\begin{aligned}
\mu &= np \sum_{r=0}^{n-1} \frac{(n-1)\ldots([n-1]-r+1)}{1 \times 2 \times 3 \times \ldots \times r} p^r q^{n-r-1} \\
&= np \sum_{r=0}^{n-1} \binom{n-1}{r} p^r q^{n-1-r} \\
&= np(p+q)^{n-1} = np
\end{aligned}$$

We knew this anyway, didn't we? We knew that if a single trial has constant probability p of success, then if we perform n trials the mean will be np.

■

24.15 THE VARIANCE OF A PROBABILITY DISTRIBUTION

The variance of a probability distribution is defined as the expected value of $(X - \mu)^2$. As with the mean, this is consistent with the definition of a population variance.

Recall that

$$\sigma^2 = \frac{\Sigma f_r (x_r - \mu)^2}{\Sigma f_r}$$

where there are n distinct data points, the sums are taken for $r \in \{1, 2,$

$\ldots, n\}$ and $\Sigma f_r = N$, the total number of data points. If we divide through each term in the numerator by Σf_r we obtain instead of each frequency f_r a relative frequency p_r, so that

$$\sigma^2 = \Sigma p_r(x_r - \mu)^2$$

So for the variance we have

$$\begin{aligned} V(X) = \sigma^2 &= E(X - \mu)^2 = \Sigma P(X = r)(r - \mu)^2 \\ &= \Sigma P(X = r)(r^2 - 2\mu r + \mu^2) \\ &= \Sigma P(X = r)r^2 - 2\mu\, \Sigma P(X = r)r + \mu^2\, \Sigma\, P(X = r) \end{aligned}$$

where, of course, the sums are taken over all possible values of the random variable.

Now we have a probability distribution, and consequently

$$\begin{aligned} \Sigma P(X = r) &= 1 \\ \Sigma P(X = r)r &= \mu \end{aligned}$$

So that substituting these into the expression for V we obtain

$$\begin{aligned} V(X) = \sigma^2 &= \Sigma P(X = r)r^2 - 2\mu\mu + \mu^2 \\ &= \Sigma P(X = r)r^2 - \mu^2 \end{aligned}$$

Therefore we have shown that

$$V(X) = \sigma^2 = E(X - \mu)^2 = E(X^2) - [E(X)]^2$$

This formula can be useful when calculating the variance of a probability distribution. It can be shown using elementary algebra that for the binomial distribution the variance is npq. The standard deviation is therefore $\sqrt{(npq)}$.

In summary, for the binomial distribution, mean $= np$ and variance $= npq$, where n is the number of experiments and p is the probability of success in a single trial.

24.16 THE POISSON DISTRIBUTION

Suppose we have a situation in which incidents occur randomly. Then we have a Poisson distribution if the following conditions are satisfied:

1 On average there are λ incidents in a unit time interval, where λ is a constant.

2 In a small time interval δT, the probability of two or more incidents occurring is zero.

3 If two time intervals have no points in common, the number of incidents occurring in each one is independent of the other.

If the random variable X is defined as the number of incidents which occur in a time interval of length T, then we obtain the following probability distribution:

$$P(X = r) = e^{-\mu} \frac{\mu^r}{r!} \quad \text{for } r \in \mathbb{N}_0$$

where $\mu = \lambda T$. (You will recall that $\mathbb{N}_0 = \{0, 1, 2, 3, \ldots\}$ is the set of non-negative integers.) To see that this is a probability density function, we note first that all its values are positive and secondly that their sum is

$$\sum_{r=0}^{\infty} e^{-\mu} \frac{\mu^r}{r!} = e^{-\mu} e^{\mu} = 1$$

□ A company has three telephone lines and receives on average six calls every five minutes. Assuming a Poisson distribution, what is the probability that more than three calls will be received during a given two-minute period?

We have six calls every five minutes on average, and so $\lambda = 1.2$ per minute. The time interval T in which we are interested is of length 2, and so $\mu = \lambda T = 2.4$. If X is the number of calls which are being received in the two-minute interval, we have

$$P(X = r) = e^{-1.2} \frac{(1.2)^r}{r!} \quad \text{for } r \in \mathbb{N}_0$$

So we require

$$\begin{aligned} P(X > 3) &= 1 - P(X \leqslant 3) \\ &= 1 - \{P(X = 0) + P(X = 1) + P(X = 2) + P(X = 3)\} \\ &= 1 - e^{-1.2}\left[1 + 1.2 + \frac{(1.2)^2}{2} + \frac{(1.2)^3}{6}\right] \\ &= 1 - 0.966 = 0.034 \end{aligned}$$

■

It is a simple algebraic exercise to show that the expectation of the Poisson random variable is μ. We do not therefore have a conflict of notation, as would otherwise be the case. Why not try and deduce this for yourself?

We have

$$P(X = r) = e^{-\mu} \frac{\mu^r}{r!} \quad \text{for } r \in \mathbb{N}_0$$

So

$$\begin{aligned} E(X) &= \sum_{r=0}^{\infty} P(X = r)\, r = \sum_{r=0}^{\infty} e^{-\mu} \frac{\mu^r}{r!} r \\ &= \sum_{r=1}^{\infty} e^{-\mu} \frac{\mu^r}{(r-1)!} \end{aligned}$$

Putting $s = r - 1$,

$$E(X) = \mu \sum_{s=0}^{\infty} e^{-\mu} \frac{\mu^s}{s!}$$
$$= \mu$$

Note that the sum of all the probabilities of the random variable is 1. A similar but more involved algebraic exercise can be used to show that the variance is also μ.

In summary, for the Poisson distribution, mean $= \mu$, variance $= \mu$ and $\mu = \lambda T$, where λ is the number of incidents per unit time interval and T is the length of the time interval.

24.17 APPROXIMATION FOR THE BINOMIAL DISTRIBUTION

If n is large there is a problem in calculating the coefficients of the binomial expansion. However, if p is also small then for the binomial distribution

$$\sigma^2 = npq = np(1 - p) \simeq np = \mu$$

so that the mean and variance are approximately equal.

It is not difficult to show that if $n \to \infty$ and $p \to 0$ in such a way that np remains constant, then the Poisson distribution is a good approximation for the binomial distribution. Clearly we shall require $np^2 \simeq 0$, and this will hold provided μ^2 is much smaller than n. If $n > 20$ and $\mu = np < 5$ then the approximation will be good enough for most purposes.

□ It is known that 5% of all bricks manufactured at a brick works are substandard. A customer buys 30 which are selected randomly. What is the probability that at least four are substandard?

Here $n = 30$ and $p = 0.05$, so that $np = 0.6$ and the Poisson distribution is certainly appropriate. Let the random variable X denote the number of defective bricks in the sample. Then

$$P(X = r) = e^{-0.6} \frac{(0.6)^r}{r!}$$

We are dealing with an approximation, and so we must ignore the fact that these probabilities are defined for r greater than 30. Then

$$P(X \geqslant 4) = 1 - P(X \leqslant 3)$$
$$= 1 - e^{-0.6}\left[1 + 0.6 + \frac{(0.6)^2}{2!} + \frac{(0.6)^3}{3!}\right] = 0.003\,36 \quad ■$$

24.18 CONTINUOUS DISTRIBUTIONS

Given a discrete probability distribution, we can represent it pictorially by means of a histogram with unit area. The probability of each value of the

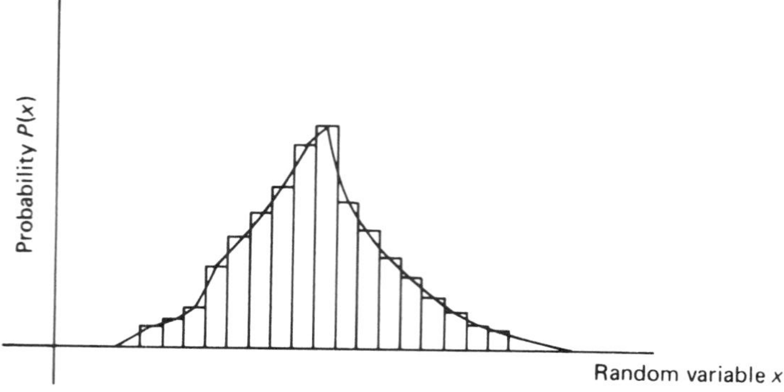

Fig. 24.3 Histogram and frequency polygon.

random variable is then shown as the corresponding area of the histogram. If the random variable has a large number of values within each closed interval, the frequency polygon will approach that of a smooth curve (Fig. 24.3). We can use this to picture probability density functions corresponding to continuous random variables.

In order to extend the idea of a probability density function to *continuous* random variables, we shall need to employ calculus. Suppose $f: \mathbb{R} \to \mathbb{R}$ is a probability density function corresponding to a continuous random variable X. Then f satisfies the condition

$$f(x) \geqslant 0 \quad \text{for all } x \in \mathbb{R}$$

Given any $a, b \in \mathbb{R}$, $a < b$, we require $P(a < X < b)$. Suppose we partition the interval $[a, b]$ into an equal number of subintervals each of length δx (Fig. 24.4). Then, selecting an arbitrary point x,

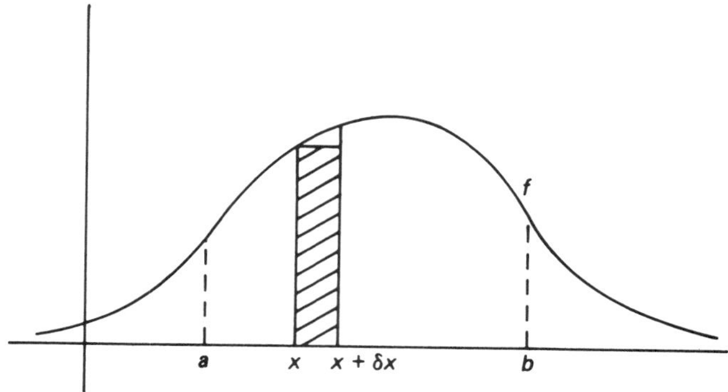

Fig. 24.4 Graph of a probability density function.

$$P(x < X < x + \delta x) \simeq f(x)\,\delta x$$

using a discrete approximation. So

$$P(a < X < b) \simeq \sum_{x=a}^{x=b} f(x)\,\delta x$$

Moreover, the approximation becomes good as $\delta x \to 0$, so that

$$P(a < X < b) = \int_a^b f(x)\,dx$$

We know also that, taken over all possible values of the random variable X, the total probability must be 1. Consequently

$$P(-\infty < X < \infty) = \int_{-\infty}^{\infty} f(x)\,dx = 1$$

It's best to think of the probability that X takes on a value between $X = a$ and $X = b$ as the *area* under the probability curve $y = f(x)$ between $x = a$ and $x = b$. Notice in particular that the area under a point is zero, and so for a continuous distribution $P(X = a) = 0$ for every $a \in \mathbb{R}$. Consequently

$$P(a < X < b) = P(a \leqslant X \leqslant b)$$

You should note that this is certainly not so for discrete distributions. Later we shall be using a continuous distribution as an approximation to a discrete distribution, and we shall have to take account of this difference then.

Remember the two conditions which need to be satisfied if $f:\mathbb{R} \to \mathbb{R}$ is to be a probability density function:

1 $f(x) \geqslant 0$ for all $x \in \mathbb{R}$

2 $\displaystyle\int_{-\infty}^{\infty} f(x)\,dx = 1$

□ The function $f:[0, 1] \to \mathbb{R}$ defined by $f(x) = 3kx^2$ is known to be a probability density function. Obtain the value of k. Obtain also the probability that if an observation were chosen at random it would be (a) less than 1/2 (b) between 1/4 and 1/2.

We note that because the domain of f is an interval $[0, 1]$, we have a continuous random variable. We require

$$f(x) \geqslant 0 \quad \text{for all } x \in [0, 1]$$

and this implies that $3kx^2 \geqslant 0$ and so $k \geqslant 0$. Next we have

$$\int f(x)\,dx = 1$$

where the integral must be taken over the domain of f. In this case this is the interval $[0, 1]$. So

$$1 = \int_0^1 3kx^2 \, dx = [kx^3]_0^1 = k$$

Therefore $k = 1$.

Consequently

a $P\left(X < \frac{1}{2}\right) = \int_0^{1/2} 3x^2 \, dx = [x^3]_0^{1/2} = \frac{1}{8}$

b $P\left(\frac{1}{4} < X < \frac{1}{2}\right) = \int_{1/4}^{1/2} 3x^2 \, dx = [x^3]_{1/4}^{1/2} = \frac{1}{8} - \frac{1}{64} = \frac{7}{64}$ ■

24.19 MEAN AND VARIANCE

The formula for the mean of a continuous distribution involves an integral instead of a sum. The form of this can be deduced, using the calculus, from the formula for a discrete distribution. We obtain

$$\begin{aligned} \mu &= E(X) \\ &= \int_{-\infty}^{\infty} f(x)\, x \, dx \end{aligned}$$

We defined the variance of a discrete distribution in terms of expectation, and this formula will hold good for continuous distributions too:

$$\begin{aligned} V(X) = \sigma^2 &= E(X - \mu)^2 \\ &= \int_{-\infty}^{\infty} (x - \mu)^2 f(x) \, dx \\ &= \int_{-\infty}^{\infty} x^2 f(x) \, dx - 2\mu \int_{-\infty}^{\infty} xf(x) \, dx + \mu^2 \int_{-\infty}^{\infty} f(x) \, dx \\ &= E(X^2) - 2\mu\mu + \mu^2 \\ &= E(X^2) - \mu^2 \\ &= E(X^2) - [E(X)]^2 \end{aligned}$$

□ Obtain the mean and variance of the probability density function $f: [0, 1] \to \mathbb{R}$ defined by $f(x) = 3x^2$.

We have already shown that we have a PDF, and so we need only calculate what is required:

$$\begin{aligned} \mu = E(X) &= \int_0^1 3x^2 x \, dx \\ &= 3 \int_0^1 x^3 \, dx \\ &= \frac{3}{4} [x^4]_0^1 = \frac{3}{4} \end{aligned}$$

Also

$$E(X^2) = \int_0^1 3x^2x^2 \, dx$$
$$= 3\int_0^1 x^4 \, dx$$
$$= \frac{3}{5}[x^5]_0^1 = \frac{3}{5}$$

So

$$V(X) = E(X^2) - \mu^2$$
$$= \frac{3}{5} - \frac{9}{16} = \frac{3}{80}$$ ■

Although there are many continuous distributions, one in particular – the normal distribution – is of great importance and application. We shall discuss this distribution now.

24.20 THE NORMAL DISTRIBUTION

The probability density function for the normal distribution is an ugly-looking beast. Luckily we shall not need to handle it at all because the indefinite integral cannot be obtained explicitly in terms of elementary functions. Therefore tables have had to be constructed so that the probabilities can be calculated. Part of our task will be acquiring the skill necessary to use the tables. For the sake of completeness, and for your general edification, here is the probability density function itself:

$$f(x) = \frac{1}{\sigma\sqrt{(2\pi)}} \exp\left[-\frac{(x-\mu)^2}{2\sigma^2}\right] \qquad x \in \mathbb{R}$$

where $\sigma > 0$ is the standard deviation and μ is the mean.

Although the graph is symmetrical about $x = \mu$ and its height is $1/\sigma\sqrt{(2\pi)}$, its shape depends on σ. For instance, if we take $\mu = 0$ then for $\sigma = 3$ we have a low-humped curve, whereas for $\sigma = 1$ we obtain the more familiar bell-shaped curve (Fig. 24.5).

The **standard normal distribution** has $\mu = 0$ and $\sigma = 1$, and it is areas under the standard normal curve which are tabulated. If the continuous random variable X is normally distributed with mean μ and variance σ^2, and if

$$Z = \frac{X - \mu}{\sigma}$$

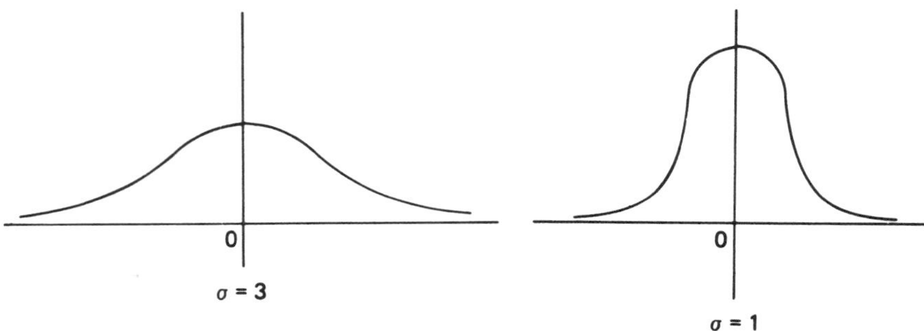

Fig. 24.5 Normal distributions with mean 0.

then Z is normally distributed with mean 0 and variance 1.

There are several ways in which the area under the standard normal curve can be tabulated. However, we confine our attention to just one of them. The curve is symmetrical and so the area under the upper half of the curve will be sufficient. In Table 24.1 we give the area under the upper tail; this is the shaded area shown in Fig. 24.6. When $Z = 0$ the area under the upper tail is 0.5, because the total area is 1 and the curve is symmetrical. Also as Z tends to ∞ the area under the upper tail tends to 0.

Suppose now $a \leqslant b$ and that we require $P(a < Z < b)$. We observe from Fig. 24.6 that

$$P(a < Z < b) = P(Z > a) - P(Z > b)$$

So if $a \geqslant 0$ we can use the table straight away to obtain the required probability. If $a < 0$ then we must use the symmetry of the standard normal curve.

□ A random variable X is normally distributed with mean 0 and variance 1. Obtain the probability that a sample chosen at random will be **a** greater

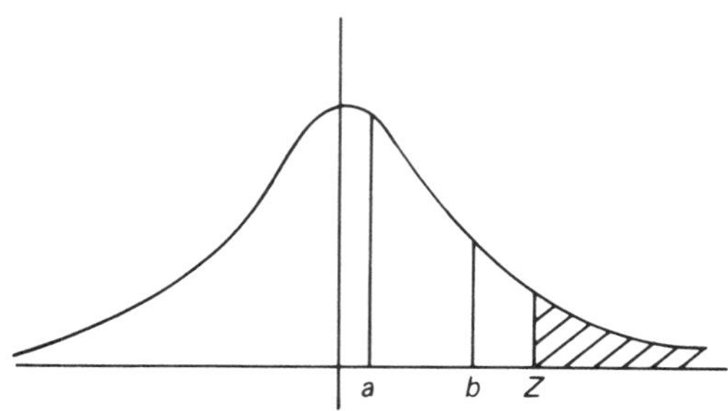

Fig. 24.6 Standard normal distribution.

than 2.12 **b** between 0.55 and 2.15 **c** greater than −1.34 but less than 2.43.

In each case we reduce the probabilities to those corresponding to areas under the upper tail:

a We require $P(X > 2.12) = 0.0170$, directly from the table. Notice how we use the left-hand column to obtain the row corresponding to 2.1 and then move across the columns to find the area corresponding to 2.12.

b To obtain $P(0.55 < X < 2.15)$, we use

$$\begin{aligned} P(0.55 < X < 2.15) &= P(X > 0.55) - P(X > 2.15) \\ &= 0.2912 - 0.0158 = 0.2754 \end{aligned}$$

c For $P(-1.34 < X < 2.43)$ we must use the symmetry and the fact that the total area under the curve is 1:

$$\begin{aligned} P(-1.34 < X < 2.43) &= 1 - P(X > 1.34) - P(X > 2.43) \\ &= 1 - 0.0901 - 0.0075 = 0.9024 \end{aligned}$$

■

The normal distribution applies in many situations, and this can be shown whenever large quantities of data are collected. It is comparatively rare for a non-normal distribution to arise in practice. Here is another example to show how we use standard normal tables when we are solving a problem where the mean and variance are not 0 and 1 respectively.

□ A manufacturing process produces dry cell batteries which have a mean shelf life of 2.25 years and a standard deviation of 3.5 months. Assuming the distribution is normal:

a Obtain the probability that an item selected at random will have a shelf life of at least 2.5 years.

b If three items are selected at random, obtain the probability that at least two will have a shelf life of more than 2.5 years.

We have $\mu = 2.25$ and $\sigma = 0.2917$, so that using the standard normal distribution we put

$$Z = \frac{X - \mu}{\sigma} = \frac{X - 2.25}{0.2917}$$

a When $X = 2.5$ we have $Z = 0.857$, so that

$$P(X \geqslant 2.5) = P(Z \geqslant 0.857) = 0.1957$$

Notice that we have interpolated the value corresponding to 0.857 from those given in Table 24.1:

$$0.85 \to 0.1977 \qquad 0.86 \to 0.1949$$

so the difference is 0.0028. Multiplying by 0.7 gives approximately 0.0020 as the corresponding difference between 0.85 and 0.857.

b If we select three items at random we have a binomial distribution where

Table 24.1 Standard normal distribution: area under upper tail

Z	0.00	0.01	0.02	0.03	0.04	0.05	0.06	0.07	0.08	0.09	
0.0	0.5000	0.4960	0.4920	0.4880	0.4840	0.4801	0.4761	0.4721	0.4681	0.4641	0.0
0.1	0.4602	0.4562	0.4522	0.4483	0.4443	0.4404	0.4364	0.4325	0.4286	0.4247	0.1
0.2	0.4207	0.4168	0.4129	0.4090	0.4052	0.4013	0.3974	0.3936	0.3897	0.3859	0.2
0.3	0.3821	0.3783	0.3745	0.3707	0.3369	0.3632	0.3594	0.3557	0.3520	0.3483	0.3
0.4	0.3446	0.3409	0.3372	0.3336	0.3300	0.3264	0.3228	0.3192	0.3156	0.3121	0.4
0.5	0.3085	0.3050	0.3015	0.2981	0.2946	0.2912	0.2877	0.2843	0.2810	0.2776	0.5
0.6	0.2743	0.2709	0.2676	0.2643	0.2611	0.2578	0.2546	0.2514	0.2483	0.2451	0.6
0.7	0.2420	0.2389	0.2358	0.2327	0.2296	0.2266	0.2236	0.2206	0.2177	0.2148	0.7
0.8	0.2119	0.2090	0.2061	0.2033	0.2005	0.1977	0.1949	0.1922	0.1894	0.1867	0.8
0.9	0.1841	0.1814	0.1788	0.1762	0.1736	0.1711	0.1685	0.1660	0.1635	0.1611	0.9
1.0	0.1587	0.1562	0.1539	0.1515	0.1492	0.1469	0.1446	0.1423	0.1401	0.1379	1.0
1.1	0.1357	0.1335	0.1314	0.1292	0.1271	0.1251	0.1230	0.1210	0.1190	0.1170	1.1
1.2	0.1151	0.1131	0.1112	0.1093	0.1075	0.1056	0.1038	0.1020	0.1003	0.0985	1.2
1.3	0.0968	0.0951	0.0934	0.0918	0.0901	0.0885	0.0869	0.0853	0.0838	0.0823	1.3
1.4	0.0808	0.0793	0.0778	0.0764	0.0749	0.0735	0.0721	0.0708	0.0694	0.0681	1.4
1.5	0.0668	0.0655	0.0643	0.0630	0.0618	0.0606	0.0594	0.0582	0.0571	0.0559	1.5
1.6	0.0548	0.0537	0.0526	0.0516	0.0505	0.0495	0.0485	0.0475	0.0465	0.0455	1.6
1.7	0.0446	0.0436	0.0427	0.0418	0.0409	0.0401	0.0392	0.0384	0.0375	0.0367	1.7
1.8	0.0359	0.0351	0.0344	0.0336	0.0329	0.0322	0.0314	0.0307	0.0301	0.0294	1.8
1.9	0.0287	0.0281	0.0274	0.0268	0.0262	0.0256	0.0250	0.0244	0.0239	0.0233	1.9

	0.00	0.01	0.02	0.03	0.04	0.05	0.06	0.07	0.08	0.09	
2.0	0.0228	0.0222	0.0217	0.0212	0.0207	0.0202	0.0197	0.0192	0.0188	0.0183	2.0
2.1	0.0179	0.0174	0.0170	0.0166	0.0162	0.0158	0.0154	0.0150	0.0146	0.0143	2.1
2.2	0.0139	0.0136	0.0132	0.0129	0.0125	0.0122	0.0119	0.0116	0.0113	0.0110	2.2
2.3	0.0107	0.0104	0.0102	0.0099	0.0096	0.0094	0.0091	0.0089	0.0087	0.0084	2.3
2.4	0.0082	0.0080	0.0078	0.0075	0.0073	0.0071	0.0069	0.0068	0.0066	0.0064	2.4
2.5	0.0062	0.0060	0.0059	0.0057	0.0055	0.0054	0.0052	0.0051	0.0049	0.0048	2.5
2.6	0.0047	0.0045	0.0044	0.0043	0.0041	0.0040	0.0039	0.0038	0.0037	0.0036	2.6
2.7	0.0035	0.0034	0.0033	0.0032	0.0031	0.0030	0.0029	0.0028	0.0027	0.0026	2.7
2.8	0.0026	0.0025	0.0024	0.0023	0.0023	0.0022	0.0021	0.0021	0.0020	0.0019	2.8
2.9	0.0019	0.0018	0.0018	0.0017	0.0016	0.0016	0.0015	0.0015	0.0014	0.0014	2.9
3.0	0.0014	0.0013	0.0013	0.0012	0.0012	0.0011	0.0011	0.0010	0.0010	0.0010	3.0
3.1	0.0010	0.0009	0.0009	0.0009	0.0008	0.0008	0.0008	0.0008	0.0007	0.0007	3.1
3.2	0.0007	0.0007	0.0006	0.0006	0.0006	0.0006	0.0006	0.0005	0.0005	0.0005	3.2
3.3	0.0005	0.0005	0.0005	0.0004	0.0004	0.0004	0.0004	0.0004	0.0004	0.0004	3.3
3.4	0.0003	0.0003	0.0003	0.0003	0.0003	0.0003	0.0003	0.0003	0.0003	0.0002	3.4
3.5	0.00023	0.00022	0.00022	0.00021	0.00020	0.00019	0.00019	0.00018	0.00017	0.00017	3.5
3.6	0.00016	0.00015	0.00015	0.00014	0.00014	0.00013	0.00013	0.00012	0.00012	0.00011	3.6
3.7	0.00011	0.00010	0.00010	0.00010	0.00009	0.00009	0.00008	0.00008	0.00008	0.00008	3.7
3.8	0.00007	0.00007	0.00007	0.00006	0.00006	0.00006	0.00006	0.00005	0.00005	0.00005	3.8
3.9	0.00005	0.00005	0.00004	0.00004	0.00004	0.00004	0.00004	0.00004	0.00003	0.00003	3.9

$p = 0.1957$. Using Y as the random variable, suppose Y is the number selected with a shelf life of at least 2.5 years. Then

$$P(Y = r) = \binom{3}{r}(0.1957)^r(0.8043)^{3-r}$$

We require

$$\begin{aligned} P(Y \geq 2) &= P(Y = 2) + P(Y = 3) \\ &= 3 \times (0.1957)^2 \times (0.8043) + (0.1957)^3 \\ &= 0.0924 + 0.0075 = 0.0999 \end{aligned}$$

■

24.21 DISCRETE APPROXIMATIONS

We have seen that when p is small and n is large the binomial distribution can be approximated by the Poisson distribution. Of course if p is close to 1 then the approximation can still be used since then $q = 1 - p$ is small.

Problems arise when n is large and neither p nor q is small. In such circumstances the normal distribution becomes a good approximation to the binomial distribution where

$$\begin{aligned} \mu \text{ (normal)} &= \mu \text{ (binomial)} &= np \\ \sigma^2 \text{ (normal)} &= \sigma^2 \text{ (binomial)} &= npq \end{aligned}$$

In a similar way, when $\mu > 10$, the Poisson distribution can be approximated by the normal distribution. Once again it is the mean and variance of the distributions which enable the approximation to be effected:

$$\begin{aligned} \mu \text{ (normal)} &= \mu \text{ (Poisson)} &= \mu \\ \sigma^2 \text{ (normal)} &= \sigma^2 \text{ (Poisson)} &= \mu \end{aligned}$$

24.22 CONTINUITY CORRECTION

We noticed earlier that one of the differences between discrete and continuous distributions is that, if X is a random variable, the probability that X takes a particular value is always zero for a continuous distribution. For example, if X is a Poisson random variable we have

$$P(X < 9) = P(X \leqslant 8)$$

whereas if X is a normal random variable we have

$$P(X < 8) = P(X \leqslant 8)$$

To compensate for this difference when we use these approximations, we apply a continuity correction and find $P(X < 8.5)$ instead of $P(X \leqslant 8)$ or $P(X < 9)$.

□ Of the fire extinguishers produced by a factory, 25% are known to be faulty. If 30 extinguishers are selected at random, what is the probability that at least 17 will be satisfactory?

We have a binomial distribution with $n = 30$ and $p = 0.25$, so that $q = 0.75$. Let the random variable X be the number of faulty fire extinguishers selected. Then we require $P(X < 14)$. There are too many terms to handle, and so we approximate using the normal distribution. X is approximately normally distributed with mean $\mu = 30 \times 0.25 = 7.5$ and $\sigma = \sqrt{(7.5 \times 0.75)} = 2.372$.

We require $P(X < 13.5)$, using the continuity correction, and when $X = 13.5$ we have

$$Z = \frac{X - \mu}{\sigma} = \frac{13.5 - 7.5}{2.372} = 2.53$$

$$\begin{aligned} P(X < 13.5) &= P(Z < 2.53) = 1 - P(Z > 2.53) \\ &= 1 - 0.0057 = 0.9943 \end{aligned}$$

■

24.23 NORMAL PROBABILITY PAPER

One way of checking whether or not data is normally distributed is to use special graph paper known as normal probability paper. If we draw the cumulative frequency curve for the normal distribution we shall obtain an S-shaped curve (Fig. 24.7). You can do this yourself if you like because the values which you require are simply the areas under the probability curve, which we already know from our table of the standard normal distribution (Table 24.1).

Normal probability paper distorts the y-axis in such a way that the

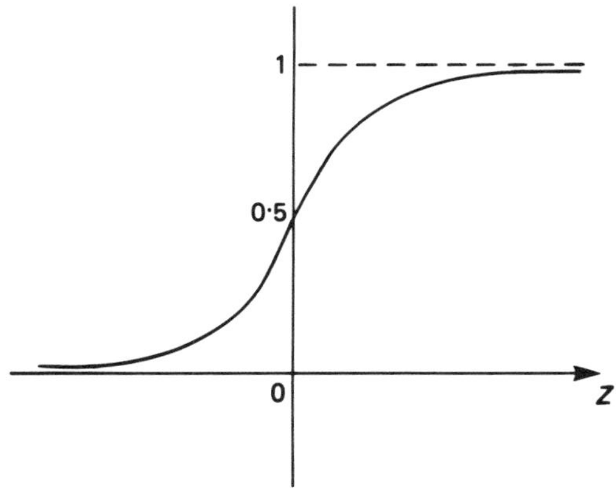

Fig. 24.7 Cumulative frequency curve for the normal distribution.

cumulative frequency curve becomes a straight line. Therefore by calculating the relative cumulative frequencies we can see if our data is approximately normally distributed. We shall illustrate this using some data which we displayed in Chapter 23.

□ The number of cars parked in a factory car park at 10.00 a.m. was counted on 120 consecutive working days. The results were given in Table 23.1.

We constructed Table 23.2 to give the relative cumulative frequencies. If we multiply these by 100 they become percentages. From this data we can draw Fig. 24.8.

If the data is normally distributed, the mean can be obtained from this graph. It is the median of the distribution – the value of the random variable X corresponding to a relative cumulative frequency on the axis of symmetry. This is marked as 50% on the graph paper. Here $\mu = 87$.

The table of the standard normal distribution (Table 24.1) shows that when $Z = 1$ the area under the upper tail is 0.1587, and so the cumulative frequency is about 0.84. We can therefore estimate the standard deviation by obtaining the differences between the two values of the random variable X with relative cumulative frequencies of 50% and 84% respectively:

$$\sigma = 102 - 87 = 15$$ ■

24.24 Workshop

1

Exercise Two brothers work for a large multinational company. It is announced that 25% of the workforce are to be made redundant. A message is received to say that one of the brothers (it is not known which one) will not be losing his job. Assuming that each employee has an equal probability of being made redundant and that these events are independent of one another, what is the probability that both brothers will remain in employment?

Watch out! In this problem you have to be extra careful. When you have decided on your answer, move on to step 2.

2

At first sight you might think this is ridiculously easy and argue as follows: each employee has 0.75 probability of remaining in employment; one brother is already secure; so for both to be safe the probability must be 0.75.

However, this argument is faulty; the error is quite subtle. The point is that we do not know which brother rang home. If the problem had named the brothers as Jim and Tom and had said that Jim rang home then indeed the probability of both holding their jobs would have been 0.75.

If you made that error try again.

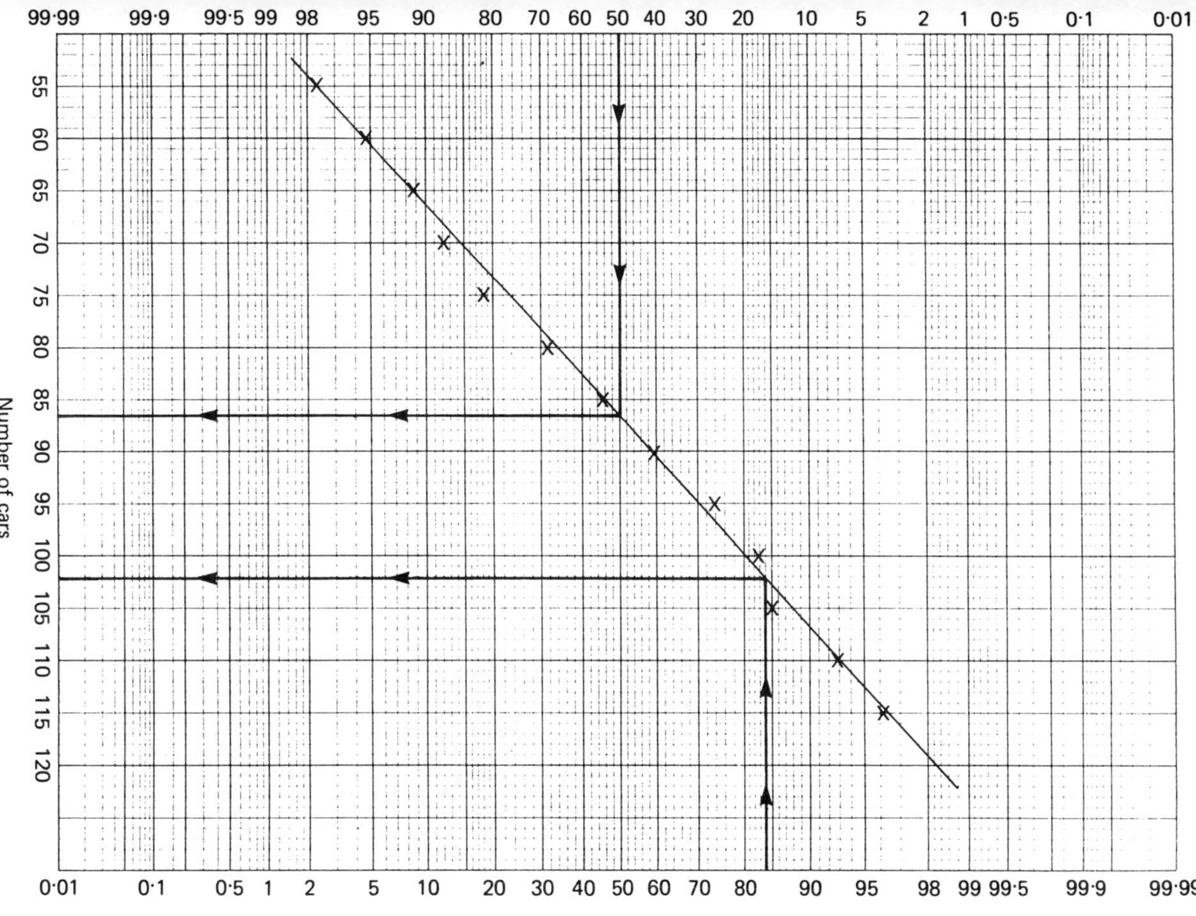

Fig. 24.8 Normal probability paper with plots for the cars.

3

Let's call the brothers Jim and Tom. Let E be 'Jim retains his job' and F be 'Tom retains his job'. We want $P(E \cap F \mid E \cup F)$. Now

$$P(E \cup F) = P(E) + P(F) - P(E \cap F)$$

and since E and F are independent events,

$$P(E \cap F) = P(E)\,P(F)$$

However, $P(E) = P(F) = 0.75$, so that

$$P(E \cup F) = 0.75 + 0.75 - 0.75 \times 0.75$$

Now

$$\begin{aligned} P(E \cap F \mid E \cup F) &= \frac{P[(E \cap F) \cap (E \cup F)]}{P(E \cup F)} \\ &= \frac{P(E \cap F)}{P(E \cup F)} = \frac{(0.75)^2}{0.75 + 0.75 - (0.75)^2} = 0.6 \end{aligned}$$

Here is another problem.

▷**Exercise** In a builders' yard there are two boxes: one contains rods and the other contains clamps. The rods should fit into the clamps. However, 10% of the rods are slightly bent and unusable, and 25% of the clamps are twisted and also unusable. A workman rushes into the yard and selects a rod and a clamp at random from the boxes. Obtain the probability that **a** they are both usable **b** at least one is usable.

Try this and then move on.

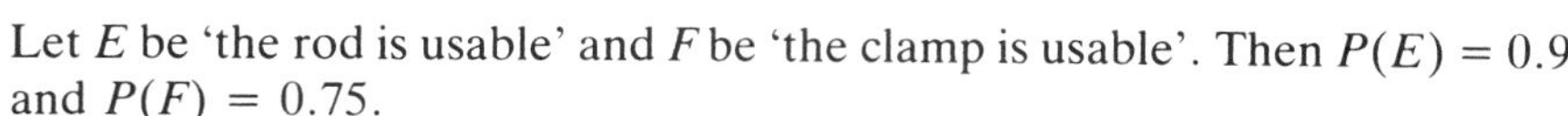

4

Let E be 'the rod is usable' and F be 'the clamp is usable'. Then $P(E) = 0.9$ and $P(F) = 0.75$.

a We require $P(E \cap F)$, and since by the nature of the problem E and F are independent we have

$$P(E \cap F) = P(E)\,P(F) = 0.9 \times 0.75 = 0.675$$

b We require $P(E \cup F)$, and so we use

$$\begin{aligned} P(E \cup F) &= P(E) + P(F) - P(E \cap F) \\ &= 0.9 + 0.75 - 0.675 = 0.975 \end{aligned}$$

If you managed that then move ahead to step 6. If you didn't make it, then try this.

▷**Exercise** In the previous exercise, obtain the probability that neither rod nor clamp is usable.

We require $P(E' \cap F')$. As before, E' and F' are independent events, and so 5

$$P(E' \cap F') = P(E')\,P(F') = 0.1 \times 0.25 = 0.025$$

Now we move on to probability density functions.

6

▷ **Exercise** The following gives the probability distribution for the random variable X. Obtain h and k if the mean is known to be 3.4.

X	1	3	5	7
$P(X)$	$h + k$	h	$2h - k$	k

When you have done this, step forward.

For a probability distribution we have that total probability must be 1. Therefore 7

$$P(1) + P(3) + P(5) + P(7) = 1$$

So

$$\begin{aligned}(h + k) + h + (2h - k) + k &= 1\\ 4h + k &= 1\end{aligned}$$

We are also told that the mean is 3.4. Now $\mu = E(X)$, so

$$\begin{aligned}1P(1) + 3P(3) + 5P(5) + 7P(7) &= 3.4\\ (h + k) + 3h + 5(2h - k) + 7k &= 3.4\\ 14h + 3k &= 3.4\end{aligned}$$

From this pair of simultaneous equations we have $h = 0.2$ and $k = 0.2$. Finally, we need to check that all the probabilities are positive; they are.

If you could not do that exercise, don't worry. The next one is similar for continuous distributions.

▷ **Exercise** The function $f: [0, 2] \to \mathbb{R}$ defined by

$$f(x) = k(x + a) \qquad 0 \leqslant x \leqslant 2$$

is known to be a probability density function. The mean of the distribution is 7/6. Obtain the constants a and k.

When you have made an attempt, move on to the next step.

We must obtain the definite integral of f over the interval $[0, 2]$ and equate it to 1 if we are to have a PDF: 8

$$\int_0^2 k(x + a)\,\mathrm{d}x = \frac{k}{2}\,[(x + a)^2]_0^2$$

Equating to 1 we obtain

$$\begin{aligned} 2 &= k[(2 + a)^2 - a^2] \\ &= k(4a + 4) \\ 1 &= 2k(a + 1) \end{aligned}$$

The expectation determines the mean, and so we have

$$\mu = \frac{7}{6} = \int_0^2 xf(x)\,\mathrm{d}x = \int_0^2 k(x^2 + ax)\,\mathrm{d}x$$

$$\frac{7}{6} = k\left[\frac{x^3}{3} + \frac{ax^2}{2}\right]_0^2$$

from which

$$7 = k(16 + 12a)$$

We have the two equations

$$\begin{aligned} 1 &= 2k(a + 1) \\ 7 &= 4k(3a + 4) \end{aligned}$$

Eliminating k we have

$$4(3a + 4) = 14(a + 1)$$

It follows that $a = 1$. Substituting back into either of the equations for k gives $k = 1/4$.

▷**Exercise** There is a fixed probability that every time a record-making machine operates it will produce a record which is warped. The records are packaged in boxes of five, and 1000 boxes were chosen at random and tested. The numbers warped (0, 1, 2, 3, 4 or 5) were obtained, and the results are as follows:

Number faulty	0	1	2	3	4	5
Number of boxes	41	143	284	343	169	20

Fit a binomial distribution to this data and calculate the corresponding theoretical frequencies.

When you have done this problem, take another step.

---

9 Let the random variable X be the number of warped discs in a package of five. We know that the total number of observations N is 1000, and we can determine $\bar{x}$ from the data which estimates μ, the mean of the binomial distribution. So

$$\frac{1}{1000}(41 \times 0 + 143 \times 1 + 284 \times 2 + 343 \times 3 + 169 \times 4 + 20 \times 5) = 2.516$$

Now $\mu = np$ and $n = 5$, so we can obtain an estimate for $p = 0.5032$, the probability that the machine will produce a warped disc each time it operates. Therefore

$$P(X = r) = \binom{5}{r}(0.5032)^r(0.4968)^{5-r}$$

for $r \in \{0, 1, 2, 3, 4, 5\}$.

If we calculate these and multiply them by 1000 we shall obtain the theoretical frequencies:

X	0	1	2	3	4	5
$P(X)$	0.030	0.153	0.311	0.315	0.159	0.032

You will notice that we have had to 'fiddle' the arithmetic so that the probabilities add up to 1 as they must. This is because we have chosen to display the probabilities to only three decimal places. Strictly speaking we should leave the probabilities exact, working out the arithmetic to as many places as necessary, and adjust the theoretical frequencies to bring them into line with reality.

If something went wrong with this, press ahead nevertheless – provided you are sure that you understand it.

▷**Exercise** A company finds that on average there is a claim for damages which it must pay seven times in every ten years. It has expensive insurance to cover this situation. The premium has just been increased, and the firm is considering letting the insurance lapse for 12 months as it can afford to meet a single claim. Assuming a Poisson distribution, what is the probability that there will be at least two claims during the year?

See how you get on with this.

Assuming a Poisson distribution, we use a time interval of one year to obtain $\lambda = 0.7$, $T = 1$ and so $\mu = \lambda T = 0.7$. We have 10

$$P(X = r) = e^{-0.7}\frac{(0.7)^r}{r!}$$

for $r \in \mathbb{N}_0$.

Writing $P(r)$ for $P(X = r)$ we require

$$\begin{aligned} P(X \geqslant 2) &= 1 - P(0) - P(1) \\ &= 1 - e^{-0.7}(1 + 0.7) = 0.1558 \end{aligned}$$

▷**Exercise** A company manufactures carpet tacks which it sells in boxes. The number of tacks in each box is a random variable which is normally distributed with mean 35.5 and standard deviation 2.35. The company prints on each box a figure indicating the average minimum contents, and wants this figure to be such that 95% of the boxes have at least this number of tacks in them. What figure should be printed on the boxes?

Try this one before you take the next step.

---

11 Suppose the random variable X is the number of tacks in each box. Then X is normally distributed with mean 35.5 and standard deviation 2.35. We use the standard transformation

$$Z = \frac{X - 35.5}{2.35}$$

so that then Z is normally distributed with mean 0 and standard deviation 1.

We are looking for a number A such that $P(Z > A) = 0.95$. To find this number we must look at the areas under the normal curve. We observe that

$$P(Z > 1.645) = 0.05$$

and so by symmetry

$$P(Z > -1.645) = 0.95$$

Now if

$$Z = \frac{X - 35.5}{2.35} > -1.645$$

we have

$$X - 35.5 > -1.645 \times 2.35$$
$$X > 35.5 - 1.645 \times 2.35 = 31.63$$

Consequently if the company prints the figure of 31 on the packet at least 95% of the packets will have at least the contents stated.

Did you manage that? If you did you can do a hop, skip and a jump to step 13. For those who made an error, here is a supplementary question.

▷ **Exercise** The company goes ahead and prints the figure 31 on the box. What percentage contains fewer than 31 matches?

Tally ho!

---

12 We are looking for $P(X > 31)$, and we already have the transformation we require to the standard normal distribution:

$$Z = \frac{X - 35.5}{2.35}$$

When $X = 31$ this gives

$$Z = \frac{31 - 35.5}{2.35} = -1.915$$

$$\begin{aligned} P(Z > -1.915) &= P(Z < 1.915) \\ &= 1 - P(Z > 1.915) = 1 - 0.0278 = 0.9722 \end{aligned}$$

So that less than 3% contain fewer than 31 matches.

13

▷ **Exercise** A company finds that occasionally an export order has been lost through poor communications. It keeps a record of events of this kind and discovers that on average 15 orders each year are lost in this way. Assuming a Poisson distribution, what is the probability that at least 20 orders will be lost during the current year due to poor communications?

You will need to use the normal approximation here unless you have itchy fingers. Don't forget the continuity correction.

14

Suppose the random variable X is the number of orders lost due to poor communications. Then with $\lambda = 15$ and $T = 1$ we have $\mu = 15$ lost per year.

Since $\mu > 10$ we can use the normal approximation and assert that X is approximately normally distributed with mean 15 and variance 15. We require $P(X \geqslant 20) = P(X > 19)$, and so using the continuity correction we shall determine $P(X > 19.5)$. Transforming to the standard normal distribution:

$$Z = \frac{X - 15}{3.873}$$

So when $X = 19.5$ we have $Z = 1.162$, and consequently

$$P(X > 19.5) = P(Z > 1.162) = 0.1226$$

We conclude that there is about a 12¼% probability that at least 20 orders will be lost in the current year.

If you managed that then you can leap ahead to step 16. Those who are left can try this.

▷ **Exercise** A company repairs second-hand television sets which it then guarantees. The probability that any one set will have to be returned to it within twelve months for repair is 0.35. A retailer orders 40 sets which he then sells. Determine the probability that at least half of these sets will have to be returned for repair within twelve months.

You may assume we have a binomial distribution. Try hard before looking at the solution.

15

We have $p = 0.35$ and $n = 40$, so we use the normal approximation: $\mu = np = 14$ and $\sigma^2 = npq = 9.1$. The random variable X is the number of faulty sets in the sample, and we require $p(X \geqslant 20) = P(X > 19)$. So we determine using the continuity correction $P(X > 19.5)$. Transforming to standard normal,

$$Z = \frac{X - 14}{3.017}$$

we see that when $X = 19.5$, $Z = 1.823$. So

$$P(X > 19.5) = P(Z > 1.823) = 0.0342$$

So the probability that at least half will have to be returned is rather less than 3.5%.

There it is then.

16

Now for some practice at drawing. You will need a sheet of normal probability paper. Make sure you know what it looks like. It is not unknown for students in the panic of examinations to use a sheet of logarithmic graph paper by mistake!

▷ **Exercise** The following data represents the lifetimes in hours of 60 dry cell batteries which were tested to destruction:

28.2, 23.4, 21.1, 26.3, 22.7, 25.2, 25.3, 22.7, 24.3, 25.3,
26.3, 24.5, 24.3, 24.8, 26.7, 27.4, 25.6, 22.6, 21.3, 22.6,
20.6, 25.1, 27.4, 25.9, 27.9, 22.1, 25.0, 24.3, 23.6, 24.8,
23.5, 21.8, 27.7, 24.3, 26.4, 20.6, 24.8, 25.1, 23.5, 24.6,
23.7, 25.4, 25.3, 24.7, 23.5, 26.5, 24.6, 24.9, 24.3, 25.4,
25.1, 22.5, 23.6, 25.1, 24.7, 24.9, 25.1, 23.2, 24.7, 23.1

Using class boundaries of

20, 21, 22, 23, 24, 25, 26, 27, 28, 29

show, using normal probability paper, that the data is approximately normally distributed. From your graph estimate the mean and standard deviation.

17

We must begin by calculating the relative cumulative frequencies:

Class boundary	*Cumulative frequency*	*Relative frequency*
21	2	0.033
22	5	0.083
23	11	0.183
24	20	0.333
25	36	0.600
26	50	0.833
27	55	0.917
28	59	0.983
29	60	1.000

Using this we are able to plot the necessary points on a sheet of normal probability paper (Fig. 24.9). The line of central symmetry marked 50% gives the estimate for the mean as 24.5. The difference between the 84% percentile and the 50% percentile gives an estimate for the standard deviation as $26.2 - 24.5 = 1.7$.

It is interesting to compare these with the values which we obtain from the raw data using a calculator; $\bar{x} = 24.498$ and $s = 1.7198$.

24.25 Practical

FAULTY SCAFFOLDING

A box contains a large number of clamps, 40% of which are defective. It is impossible for a construction worker to distinguish visually between the good ones and the bad ones. He selects ten clamps at random and constructs a piece of scaffolding using two clamps for each section. The scaffolding will be dangerous if

1 Either both the clamps on any one section are defective;
2 Or four or more clamps are defective.

Otherwise the scaffolding will be safe. Obtain the probability that the scaffolding will be dangerous.

We shall solve this problem stage by stage so that you may join in the solution at whichever stage you can. First try to analyse the problem to see what is needed.

We shall define events E as 'both the clamps on a section are defective' and F as 'at least four clamps are defective'. We require $P(E \cup F)$. Now

$$\begin{aligned} P(E \cup F) &= 1 - P[(E \cup F)'] \\ &= 1 - P[E' \cap F'] \end{aligned}$$

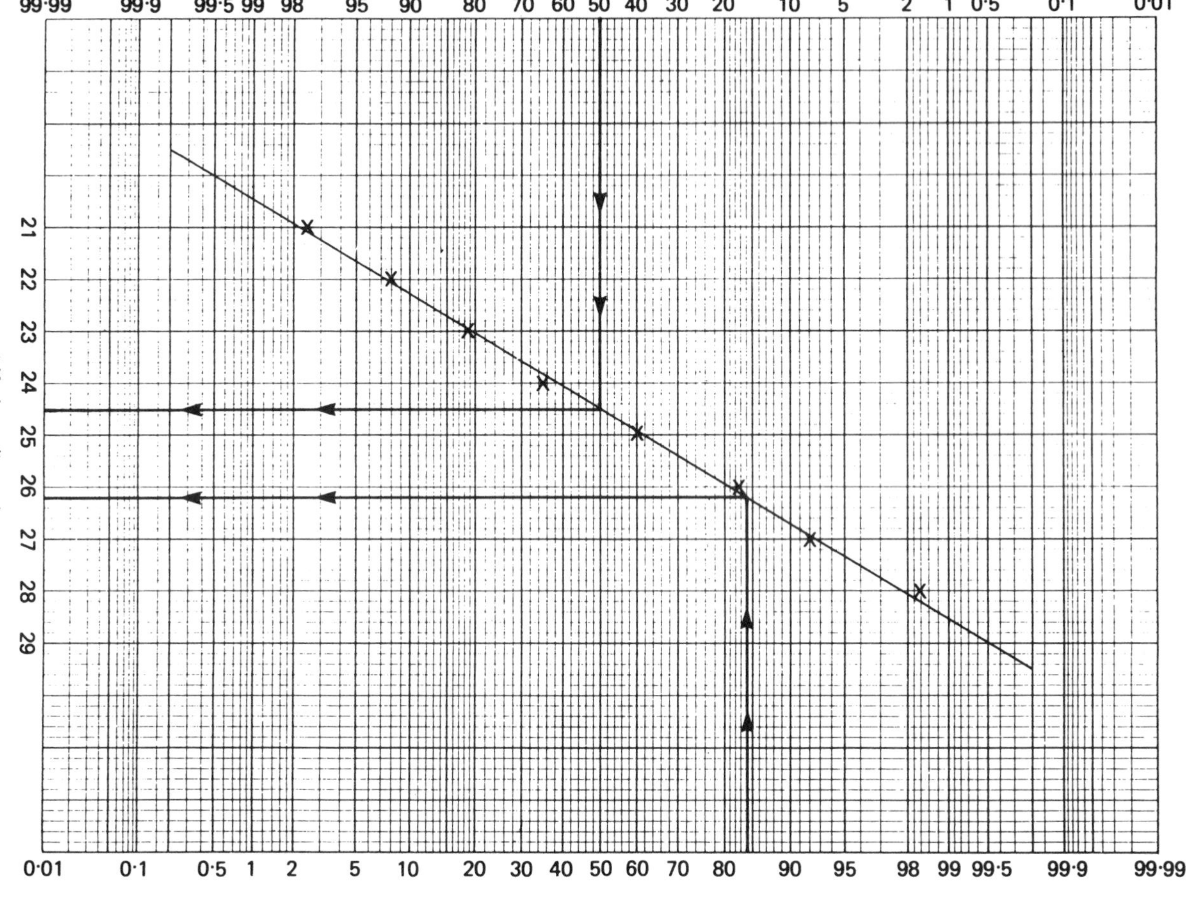

Fig. 24.9 Normal probability paper with plots for lifetimes.

where E' is 'at least one clamp on each section is good' and F' is 'at most three clamps are defective'.

Now let X be 'the number of defective clamps'. Can you carry on?

We require

$$P(E' \cap F') = P(E' \cap [X = 0]) + P(E' \cap [X = 1]) + P(E' \cap [X = 2]) + P(E' \cap [X = 3])$$

To continue with the solution you will need to remember that if A and B are any two events,

$$P(A \cap B) = P(A)P(B|A)$$

Given that there is at most one defective clamp chosen, at least one clamp on every section will be good. Therefore only the binomial probabilities come into the first two terms:

$$\begin{aligned} P(E' \cap [X = 0]) &= P(X = 0)P(E'|X = 0) \\ &= (0.6)^{10} \times 1 = (0.6)^{10} = 0.006\,047 \\ P(E' \cap [X = 1]) &= P(X = 1)P(E'|X = 1) \\ &= [10 \times (0.4) \times (0.6)^9] \times 1 = 0.040\,311 \end{aligned}$$

Things get a little more complicated when two or more clamps are defective.

We consider first $P(E'|X = 2)$. There are five sections and two defective clamps. There are ten ways of placing the first clamp, which leaves nine ways of placing the second. Therefore the probability that a given section contains two defective clamps is

$$\frac{2}{10} \times \frac{1}{9}$$

Now there are five sections, and so the total probability that any one of the sections has two defective clamps (given that there are exactly two defective clamps in the pile) is

$$5 \times \frac{2}{10} \times \frac{1}{9} = \frac{1}{9}$$

Therefore

$$P(E'|X = 2) = 1 - \frac{1}{9} = \frac{8}{9}$$

Try now to complete the calculation of $P(E' \cap [X = 2])$.

We have

$$P(E' \cap [X = 2]) = P(X = 2)P(E' \mid X = 2)$$

$$= \binom{10}{2}(0.4)^2(0.6)^8\left(\frac{8}{9}\right) = 0.107\,495$$

In order to calculate $P(E' \mid X = 3)$ you need to use an argument similar to that just used. See if you can do it yourself.

We have

$$P(E' \mid X = 3) = 1 - 5 \times \frac{3}{10} \times \frac{2}{9} = \frac{2}{3}$$

$$P(E' \cap [X = 3]) = P(X = 3)P(E' \mid X = 3)$$

$$= \binom{10}{3}(0.4)^3(0.6)^7\left(\frac{2}{3}\right) = 0.143\,327$$

You can certainly finish it off now.

If we add up the probabilities which we have calculated we obtain

$$P(E' \cap F') = 0.297\,18$$

Therefore the required probability is

$$P(E \cup F) = 0.702\,82$$

This is the last of our chapters and if you have completed your studies of the material contained in this book you may soon face your sessional examination. You should be able to do this confidently and calmly certain in the knowledge that you have at your fingertips a variety of experience and technique. In the second and subsequent years of your course you will be able to build on this and develop into an engineer who is able to use mathematics to his or her advantage and does not need to fight shy of mathematical methods.

SUMMARY

There has been quite a lot to learn in this chapter, and so we will just summarize the main points.

PROBABILITY RULES

E and F are events in the sample space S:

$$P(E \cup F) = P(E) + P(F) - P(E \cap F)$$
$$P(E \cap F) = P(E)P(F|E)$$

If $E \cap F = \varnothing$ then E and F are mutually exclusive events. For mutually exclusive events,

$$P(E \cup F) = P(E) + P(F)$$

If $P(E|F) = P(E)$ then E and F are independent events. For independent events,

$$P(E \cap F) = P(E)P(F)$$

BINOMIAL DISTRIBUTION

$$P(X = r) = \binom{n}{r} p^r q^{n-r}$$

where there are n trials and p is the constant probability of success in each one; $p + q = 1$.

POISSON DISTRIBUTION

$$P(X = r) = \mathrm{e}^{-\mu} \frac{\mu^r}{r!}$$

where λ is the average number of incidents in unit time, T is the length of the time interval and $\mu = \lambda T$.

NORMAL DISTRIBUTION

If the random variable X is normally distributed with mean μ and variance σ^2 and if

$$Z = \frac{X - \mu}{\sigma}$$

then Z is normally distributed with mean 0 and variance 1.

continued overleaf

continued from previous page

APPROXIMATIONS

If $n > 20$ and $p < 0.3$,

$$\text{binomial} \rightarrow \text{Poisson } (\mu = np)$$

If $n > 20$ and $0.3 \leqslant p \leqslant 0.7$,

$$\text{binomial} \rightarrow \text{normal } (\mu = np, \sigma^2 = npq)$$

If $\mu > 10$,

$$\text{Poisson} \rightarrow \text{normal } (\sigma^2 = \mu)$$

EXERCISES

1 The probability that an electrical component will function properly after n hours of operation is $10/(10 + n)$. Obtain the probability that the component will still function properly after 200 hours given that it was functioning properly after 150 hours.

2 A lifting apparatus has five cables which can be put under strain. In a single lift at least one cable is under strain. Suppose that the probability that n cables are put under strain is a/n^2, where a is a constant. Determine (a) the value of a (b) the probability that there are at least three cables under strain (c) the probability that there are four cables under strain, given that there are three cables under strain.

3 A machine produces square sheets of chipboard with mean side length 4 metres and standard deviation 0.05 m. Obtain the mean area of each sheet.

4 A probability density function is $f(x) = \log_k x (1 \leqslant x \leqslant 2)$. Obtain (a) the value of k (b) the mean of the distribution.

5 The probability that a glass fibre will shatter during an experiment is believed to follow a Poisson distribution. In an apparatus containing 100 glass fibres it was found that on average seven shattered. Obtain the probability that during a single demonstration of the experiment (a) two glass fibres will shatter (b) at least one glass fibre will shatter. How would your answers differ if the distribution was thought to be binomial?

6 The probability that the nth stage of a production process will be completed on time is $p(n) = 1/(n + 1)$. A certain production process has five stages. Obtain the probability that (a) it will be completed on time (b) it will be completed on time given that the first three stages were completed on time (c) none of the stages will be completed on time. Obtain the corresponding results if $p(n) = n/(n + 1)$.

ASSIGNMENT

1 The number of collisions requiring garage services on site on a busy stretch of motorway in any one week is known to be a Poisson random variable with mean 9.43. If there are more than two collisions in a day an auxiliary truck needs to be hired. Determine the probability that an extra truck will need to be hired.

2 A factory assembly is responsible for putting together three independent parts of a microcomputer: the case, the keyboard and the circuit board. The probabilities that these components are substandard are 0.1, 0.15 and 0.07 respectively. Determine the probability that, if a computer is examined, two or more of these components will be substandard.

3 The probability that a factory medical officer will be away is 0.4. The probability that medical assistance will be required is 0.2. The probability that both the medical officer will be away and medical assistance will be required is 0.1.

a If medical assistance is required, what is the probability that the medical officer is away?

b If the medical officer is away, what is the probability that medical assistance will be required?

4 There is a probability 0.005 that a welding machine will produce a faulty joint when it is operated. The machine welds 1000 rivets. Determine the probability that at least three of these are faulty.

5 The function defined by $p(x) = kx$ when $x \in [0, 2]$ is known to be a probability density function. Determine (a) k (b) the mean of the distribution (c) the variance of the distribution.

FURTHER EXERCISES

1 Illustrate by means of diagrams the following laws of probability:

$$P(E \cup F) = P(E) + P(F) - P(E \cap F)$$
$$P(\sim E) = 1 - P(E)$$
$$P[\sim(E \cup F)] = P[(\sim E) \cap (\sim F)]$$

where E and F are two events.

2 a Calculate the mean and variance of the first n natural numbers.

b A fair die is thrown twice and the scores shown are r and s. Represent the sample space of this experiment and show the subspaces corresponding to the following events:

$$E = \{(r, s) : |r - s| = 1\}$$
$$F = \{(r, s) : r + s > 6\}$$
$$G = \{(r, s) : rs \leqslant 6\}$$

Calculate $P(E \cap F)$, $P(F \mid E)$, $P(E \cup F \cup G)$, and $P[(E \cup F) \cap G]$.

3 The probability that a cement mixer will break down during a shift is 0.04. If there are six cement mixers working on site at the start of the shift, obtain the probability that by the end of the shift

a none will have broken down

b at least two will have broken down.

4 Consignments of bricks are subjected to the following inspection procedure. N bricks are selected at random from the consignment and tested. If the number of defective bricks is less than four the consignment is accepted. Otherwise it is rejected. Determine the smallest value of N if the probability of acceptance of a large consignment in which 50% are defective is to be no more than 0.1.

5 The average number of lorries per hour delivering cement to a building site during an 8-hour shift is 0.5. The workforce can handle up to three loads per shift. Further loads must be redirected to another site. Obtain, assuming a Poisson distribution,

a the probability that a lorry arriving during the shift will be redirected to another site

b the tonnage of cement which the workers on the site expect to handle during the shift if each lorry carries 6 tonnes of cement.

6 The numbers of people telephoning a certain number for advice on 40 consecutive working days was recorded as follows:

3	0	0	1	0	2	1	0	1	1
0	3	4	1	2	0	2	1	3	1
1	0	1	2	0	2	1	0	1	2
3	1	1	0	2	1	0	3	1	2

Calculate the median and mode for this data and represent it by a bar chart. Fit a Poisson distribution to this data and compare the theoretical frequencies with those actually obtained.

7 The weights of ball bearings are normally distributed with a mean of 0.845 newtons and a standard deviation of 0.025 newtons. Determine the percentage of ball bearings with weights

a between 0.800 N and 0.900 N

b greater than 0.810 N.

8 The probability that a loom will break down and require attention during a shift is 0.04. If ten looms are in working order at the start of the shift, determine the probability that during the shift

a none will break down

b not more than two will break down.

9 The number of employees in a firm required to appear before magistrates on driving summonses is a Poisson random variable with mean 4.5 per month. Determine the probability that in any one week at least one employee will receive a summons.

10 The number of telephone calls received by a receptionist between

9.00 a.m. and 9.30 a.m. follows a Poisson distribution with mean 12. Determine the probability

a that on any given day at that time there will be fewer than 8 calls

b that the total number of calls on three consecutive days at that time will be less than 30.

11 Over a ten-week period the number of weeks $f(r)$ in which r employees forgot to clock off was recorded. Fit a Poisson distribution to the data and compare the observed frequencies with the actual ones. The data is as follows:

r	0	1	2	3	4	5
$f(r)$	4	4	1	0	1	0

12 A laboratory has a large number of electrical heaters each one of which heats a tank. The probability that a number n of these heaters will malfunction in any day is given by

$$\frac{\mu^n}{n!} e^{-\mu}$$

where $\mu = 2$. The technicians can replace three of these heaters in any one day but if more than this number fail to function it is necessary to move the tank concerned to another laboratory. Determine the percentage of the heaters which malfunction which result in tanks being moved to the other laboratory.

13 A large batch of video display units produced in a workshop is examined by taking samples of five of them. It is found that the number of samples containing 0, 1, 2, 3, 4, 5 defective units are found to be

Number of defectives	0	1	2	3	4	5
Frequency	58	32	7	2	1	0

Determine the mean and variance of this data. Assuming that these defectives conform to a Poisson distribution of mean 0.56 calculate the theoretical frequencies and compare them with the true ones.

14 The average number of faxes which are received by a large engineering company between 9.00a.m. and 9.10a.m. is found to be 4. Which probability distribution is likely to provide a suitable model for the distribution of these messages in this period?

On the basis of this model determine the probability of each of the following:

a Not more than one fax is received in this period.

b At least five faxes are received in this period.

15 An electrician has a box containing 5 different resistances. If any two resistors are taken from the box it is found that the ratio of the larger resistance to that of the smaller is at least 10. Determine the number of different total resistances which the electrician could make by connecting one or more of these resistors in series.

16 A machine makes rubber belts with a mean breaking strength of 50 newtons and a variance of 4 N^2. If the distribution is normal, determine an appropriate tolerance level if not more than 0.1% are to fail to meet specification.

Hints and solutions

CHAPTER 1

EXERCISES

1 Products: (1) (3 decimal places) **a** 7.1536×10^1
b 2×10^{-3}
c 0
d 1.540081×10^2
(2) (4 significant figures) **a** 7.154×10^1
b 1.827×10^{-3}
c 2.731×10^{-8}
d1.540×10^2
Sums: (1) (3 decimal places) **a** 1.7708×10^1
b 1.6254×10^1
c 1×10^{-3}
d 1.55×10^2
(2) (4 significant figures) **a** 1.771×10^1
b 1.625×10^1
c 5.176×10^{-4}
d 1.550×10^2

2 (1) (5 significant figures) **a** 217.38
b 0.00028430
c 11.100
d 432.49
e 1.0000
f 1.0000
g 1.0001
(2) (5 decimal places) **a** 217.38500
b 0.00028
c 11.10000
d 432.49500
e 1.00005
f 1.00005
g 1.00005

3 **a** $a^2 + ab - 6b^2$
b $u^2w + v^2u + w^2v - u^2v - v^2w - w^2u$
c $9xyz + 2xz^2 + 2yx^2 + 2zy^2 + 4x^2z + 4y^2x + 4z^2y$
d $a^4 - b^4$
4 **a** $a(a+b)(a+2b)$ **b** $(x-2y)(x+2y)(x+y)$
c $(u-v)(u^2 - 6uv + v^2)$ **d** $x^2y^2(x-y)(x+y)$
5 **a** $x = y/(y-1)$ **b** $x = (y+1)/(y-1)$ **c** $x = y^3/(1-y^2)$ **d** $x = y/(y-1)$ or $x = y$
6 **a** $x = 3, y = 2$ **b** $u = 3, v = 5$ **c** $p = 7, q = 5$ **d** $h = 2, k = 1$

8 a Integers $\{1, -2\}$ **b** no real roots **c** real roots $-1 \pm \sqrt{3}$
d natural numbers $\{1, 2\}$ **e** integers $\{-1, -2\}$ **f** rational numbers $\{1/2, 1\}$

9 a $1 - 10x + 40x^2$ **b** $3^7 + 7 \times 3^6 x + 21 \times 3^5 x^2$
c $3^8 - 16 \times 3^7 x + 112 \times 3^6 x^2$ **d** $2 - 5x/4 - 25x^2/64$
e $3^{2/3} + 10 \times 3^{-4/3} - 25 \times 3^{-10/3} x^2$ **f** $1/9 + 2x/27 + x^2/27$

ASSIGNMENT

1 a $x \in \{-2, 3\}$; integers **b** $y \in \{1/3, 2\}$; rational numbers **c** $u \in \{-1, 1\}$; integers
d $v \in \{\pm\sqrt{3}, \pm\sqrt{5}\}$; real numbers **e** $x = 1/2$ repeated; rational

2 $1 + x$

3 $r = r_1 r_2/(r_1 + r_2)$

4 a $u = 1/3, v = 1/2$ **b** $u = 1/4, v = 1/2$ **c** $u = 1/2, v = 1$ **d** $u = 1/3, v = 1/2$ or $u = 1/2, v = 1$

5 a identity **b** $x = (-1 \pm \sqrt{5})/2$ **c** identity

6 $k = \pm 1/3$

7 70

FURTHER EXERCISES

1 a $\{1,\ 3/2\}$, rational numbers **b** complex numbers **c** $\pm\sqrt{3}$, real numbers
d $\{1,\ 2\}$, natural numbers

2 a $(a - 2b)(a + 2b)(a - b)(a + b)$ **b** $(2u - v)(2u + v)(u - 2v)(u + 2v)$ **c** $2^4 uv^2 w$
d $x(x - 1)(x - 2)(x - 3)(x - 4)$

4 a $1 + 10x + 45x^2 + 120x^3$
b $1 - x + x^2 - x^3$
c $1 + x/2 + 3x^2/8 + 5x^3/16$
d $1 + 5x + 20x^2 + 220x^3/3$
e $8 - 21x + 147x^2/16 + 343x^3/128$

5 $b = a(h - 3d)/(3d - 2h)$

6 $S = (R^2 - r^2)\pi$

7 $E_1[1 \pm \sqrt{\{1 - 4(E_2 i_1)/(E_1 i_2)\}}]/(2i_1)$

8 $x = (2ka)/(v^2 a + 2k)$

9 $A[1 \pm \sqrt{\{1 - (4X^2T^2)/A^2\}}]/(2x^2 T)$

10 Hint: Show $ab = (3h - 8d)h^2/(6h - 8d)$ and by obtaining $a^2 + b^2$
that $(a/b) + (b/a) = 8d/(3h - 8d)$
Hence deduce $a/b = 4d[1 \pm \sqrt{\{1 - (3h - 8d)^2/(4d)^2\}}]/(3h - 8d)$

12 3.5 %

13 1.5% decrease

14 5% decrease

CHAPTER 2

EXERCISES

1 a 8 **b** $2^{9/2} \times 3^{11/4}$ **c** $(1 + x)^2(1 + x^2)/(1 - x)^2$ **d** $(a - b)^6(a + b)^4$

2 a $x = \ln 5$ **b** $x = 0$ or $x = 2$ **c** $x = 2$ or $x = 3$ **d** $x = \ln 5/\ln 3$ or $x = \ln 3/\ln 5$

3 a $x = \ln 2$, equation **b** $x = 1 \pm \sqrt{2}$, equation **c** identity

4 a $x = (a+1)/(a-1)$ **b** $x = 4a$ **c** $x = a - 2$ **d** $x = -a$

5 a $x < -2$ or $x > 2$ **b** $-3/2 < x < 3$ **c** $x > 1$ or $0 < x < 1/2$

6 a $-1/[2(x-1)] + 1/[2(x-3)]$ **b** $-3/(x+3) + 4/(x+4)$
c $1/(6x) + 3/[2(x+2)] - 5/[3(x+3)]$
d $-1/[18(x-5)] + 4/[3(x+1)^2] + 1/[18(x+1)]$

7 Hints: **a** $2n + 2m = 2(n+m)$, **b** $(2n+1) + (2m+1) = 2(n+m+1)$
c $(2n+1) + 2m = 2(n+m) + 1$

ASSIGNMENT

2 $(1-2x)^{-2}$

3 $x = 1$ or $x = 2$

7 a $-2 < x < 0$ or $x > 2$ **b** $-2 \leqslant x \leqslant 3$ **c** $-1 < x < 1$

8 a $1/x + 1/(x-1) - 1/(x+1)$ **b** $3 - 1/x^2 + 2/(x+1)$
c $4 + 1/(x^2+1) + 1/[2(x-2)] + 1/[2(x+2)]$

9 Hint: $p/q + r/s = (ps + qr)/qs$

10 Hint: If $p/q + x = r/s$ then $x = r/s - p/q$

FURTHER EXERCISES

2 $x = \ln 2$

3 a identity **b** $x = (1+\sqrt{5})/2$ **c** identity **d** $x = 0$ or $x = 2$ **e** identity

4 $-2 < x < -1$ or $1 < x < 2$

5 a $3/(x-2) - 2/(x-1) + 1/(x-1)^2 + 4/(x-1)^3$
b $1/[4(x-1)] - 1/[4(x+1)] + 1/[4(x-1)^2] + 1/[4(x+1)^2] + 1/[2(x-1)^3] - 1/[2(x+1)^3]$

6 a $\mathbb{R} \setminus \{1, -1\}$ **b** $|x| > 1$ **c** Not defined anywhere

7 Hint: $(p/q) \times (r/s) = (pr/qs)$

8 Hint: If $(p/q)x = r/s$ then $x = (qr)/(ps)$

11 a $\mathbb{R} \setminus \{0\}$; $\{0, 1\}$
c $i(t) = tH(t) + 2(1-t)H(t-1) - (2-t)H(t-2)$

12 $n = 6$

14 $(x+\sqrt{2})/[2\sqrt{2}(x^2 + x\sqrt{2} + 1)] - (x - \sqrt{2})/[2\sqrt{2}(x^2 - x\sqrt{2} + 1)]$

CHAPTER 3

EXERCISES

2 a $\{\pi/6, 5\pi/6, 3\pi/2\}$
b $\{3\pi/16, 7\pi/16, 11\pi/16, 15\pi/16, 19\pi/16, 23\pi/16, 27\pi/16, 31\pi/16\}$
c $\{0, \pi/2\}$
d $\{\pi/4, 3\pi/4, 5\pi/4, 7\pi/4\}$
e $\{0, 2\pi/3, 4\pi/3\}$
f $\{0, \pi\}$

3 a $3\cos(\theta - \alpha)$ where $\alpha = \cos^{-1}(1/3)$
b $4\cos(\theta - \alpha)$ where $\alpha = \pi - \sin^{-1}(3/4)$
c $5\cos(\theta - \alpha)$ where $\alpha = -\cos^{-1}(4/5)$
4 a Circle, centre (0, 0), radius 5
b rectangular hyperbola, centre (1, 5)
c pair of straight lines; $y = x + 1$, $y = -x + 1$
d rectangular hyperbola, centre (1, 2)
5 a Circle, centre (1, −1), radius 1
b rectangular hyperbola, centre (2, 1)
c ellipse, centre (0, 0), lengths of axes $2\sqrt{2}$ and 4 (minor axis on x-axis)
d rectangular hyperbola, centre (−1, 1)
6 a $y - 3x = 5$
b $y + 5x = 1$
c $6y + x = 32$
d $2y + x = 7$
e $3y - 5x = 15$
f $2y = 3x - 6$
g $y = 3x - 5$
h $2x + y = 4$
i $x^2 + y^2 - 2x - 4y - 11 = 0$
j $x^2 + y^2 - 4x + 6y - 12 = 0$
7 a Slope = −1/4, x intercept = 12, y intercept = 3
b slope = −2/3, x intercept = −3, y intercept = −2
c slope = −2/5, x intercept = 11/2, y intercept = 11/5
d slope = −4/3, x intercept = 7/2, y intercept = 14/3
8 a Centre = (−2, −3), radius = 2
b centre = (−3, −4), radius = 2
c centre = (1, −2), radius = 3
d centre = (3, 1), radius = 5

ASSIGNMENT

2 $\{n\pi/3 + (-1)^n\pi/18 : n \in \mathbb{Z}\}$
3 $\{\pi/4, 3\pi/4, \pi, 5\pi/4, 7\pi/4\}$
4 Hint: put in terms of 2θ. Answers:
$\{n\pi \pm \pi/2 : n \in \mathbb{Z}\} \cup \{(n\pi/2) + (-1)^n(\pi/12) : n \in \mathbb{Z}\}$
5 a Circle centre 0, radius $4/\sqrt{5}$ **b** ellipse $(x + 15)^2 + 5y^2 = 225$
c circle centre (1/2, 0), radius $(1/2)\sqrt{65}$
d pair of straight lines $y = \pm 4$
e three straight lines $y = 3x$, $x + y = 1$, $y = x + 2$
6 a $\sqrt{(2)}\cos(\theta - \pi/4)$ **b** $2\cos(\theta - \pi/6)$
7 Hyperbola $4y^2 - 3x^2 = 5$

FURTHER EXERCISES

2 a $\theta = n\pi$ or $\theta = n\pi + \pi/4$ **b** $\theta = n\pi \pm \pi/3$ **c** no roots
d no roots **e** $n\pi - \pi/4 - (-1)^n \pi/4$

8 $y = 1; (-15/8, 1)$

9 18.75 metres

10 12 metres

11 No: 55/16 metres

12 $2axl/(x^2 + a^2)$

15 $\tan\beta = r(r+t)[\tan(\theta+\alpha) - \tan\theta]/\{(r+t)^2 + r^2\tan\theta\tan(\theta+\alpha)\}$

16 $d = h\sin\alpha/[\sin(\theta-\alpha)\sin\theta]$

17 $l = (3/4)\sqrt{h^2 + r^2}$

CHAPTER 4

EXERCISES

1 a $6x + 5$
b $3x^2 - 4x$
c $x^{-1/2}/2 - x^{-3/2}/2$
d $6(x+2)^5$
e $3\cos(3x+4)$
f $6\tan 3x \sec^2 3x$
g $4x/(2x^2+1)$
h $2x^3\cos x^2 + 2x\sin x^2$

2 a $(t^2+1)(t+2)\{2t/(t^2+1) + 1/(t+2) - 2t/(t^2+2) - 1/(t+1)\}/$
$(t^2+2)(t+1)$
b $(t+1)^3(t+2)^3\{3/(t+1) + 3/(t+2) - 2/(t+3)\}/(t+3)^2$
c $\cos 4t$
d $-2e^t/(e^t-1)^2$

3 a $3x^2\alpha$
b $2\alpha x\cos x^2$
c $\alpha\cot x$
d $\pm\alpha e^{x/2}/2$

4 a 1
b 1/2
c 2/3
d 3/2

ASSIGNMENT

1 a $-1/\sqrt{2}$ **b** hint: $\sin 3x = \sin x\,(4\cos^2 x - 1)$; answer $1/\sqrt{3}$
c $-\infty$ **d** 2/3

2 a Hint: use half-angle formula; answer 0
b hint: $\cos^3 x - 1 = (\cos x - 1)(\cos^2 x + \cos x + 1)$; answer $-2/3$

4 a $2\ln x\ x^{\ln x-1}$ **b** $(2\sec^2 2x + 3\tan 2x)\,e^{3x}$
6 $dy/dx = (1 - \sin t)/(1 + \cos t)$

FURTHER EXERCISES

1 a $16x - 10$ **b** $3x^2 - 12x + 11$ **c** $x(x^2 - 1)^{-1/2}$
d $-12x(x^2 - 3)^{-2}$ **e** $bmnx^{m-1}(a + bx^m)^{n-1}$ **f** $a\sec^2(ax + b)$
g $-x\cot x^2 \sqrt{\operatorname{cosec} x^2}$
2 a $9x^8, 72x^7, 504x^6, 3024x^5$
b $(x + 1)^{-1/2}/2, -(x + 1)^{-3/2}/4, 3(x + 1)^{5/2}/8, -15(x + 1)^{-7/2}/16$
c $-2\sin x\cos x = -\sin 2x, -2\cos 2x, 4\sin 2x, -8\cos 2x$
d $x^2e^x + 2xe^x, x^2e^x + 4xe^x + 2e^x, x^2e^x + 6xe^x + 6e^x,$
$x^2e^x + 8xe^x + 12e^x$
4 $-6\cot t - 2\cot^3 t$
6 Hint: $(dy/dx)(dx/dy) = 1$, differentiate this
9 a -1 **b** -2
11 $kA^{3/2}/4\sqrt{\pi}$
14 $w = -EI\,e^{-x}(x - 6)(x - 2)$
15 $6\varrho/h$
18 a $xu(x^2 + h^2)^{-1/2}$ **b** decreasing at $hu/(x^2 + h^2)$
21 a $x[\tan(\theta + h) - \tan\theta]$
22 a $x^2(2x\cos 2x + 3\sin 2x)$ **b** $-(1 + t\cos t)$
c $2x/[e^y(y + 1) - 2y] = 2xy/[x^2(y + 1) + y^2(y - 1)]$

CHAPTER 5

EXERCISES

2 a $x = \ln 5$
b $x = 0$ or $x = 1$
c $x = \ln 3$ or $x = \ln 4$
d $x = \pm\ln(2/3)$
3 a $-3\operatorname{sech} 3t\tanh 3t$
b $2t^2\cosh 2t + 2t\sinh 2t$
c $\operatorname{sech} 2t\cosh t(1 - 2\tanh t\tanh 2t)$
d $-\operatorname{sech} t\operatorname{cosech} 2t(2\coth 2t + \tanh t)$
e $-t\tanh t^2\sqrt{\operatorname{sech} t^2}$
f $-(\operatorname{sech}^2\sqrt{t}\tanh\sqrt{t})/\sqrt{t}$
4 a $2/\sqrt{(t^2 + 1)}$
b $-2/t(t^2 + 2)$
c $(\sinh t)/\sqrt{(t^2 - 1)} + \cosh t\cosh^{-1} t$
d $(t^2 - 1)^{-1/2}/\cosh^{-1} t$
e $1/t\sqrt{[(\ln t)^2 - 1]}$
f $-\{(\sinh^{-1} t)^2\sqrt{(t^2 + 1)}\}^{-1}$

ASSIGNMENT

1 **a** $x = (1/2)\ln 3$ **b** $x = 0$ or $x = \pm(1/3)\ln 2$
3 $x = \frac{1}{2}\ln 3,\ x = \frac{1}{3}\ln 2$
4 $x = \ln 2$ or $x = 0$

FURTHER EXERCISES

1 **a** 1 **b** 1/2
3 **a** $3\sin x\,(1 - 9\cos^2 x)^{-1/2}$ **b** $2(x^2 + 1)^{-1}$ **c** $\cos x\,(1 + \sin^2 x)^{-1}$
7 **a** $x = \ln(2 \pm \sqrt{3})$ **b** $x = \ln(\sqrt{2} - 1)$ **c** $x = 0$
8 Hint: $r = \alpha[(t-1)^2 + 1]$ **a** min. $r = \alpha$ **b** $[1, 2]$ **c** 1, 1, 8
9 Hint: $h = \sqrt{(2)}\cos(2t - \pi/4) + 2$ **a** $2 \pm \sqrt{2}$ **b** π seconds
c $5\pi/8$
10 $I \in [\pi + \cos^{-1}(1/4), 2\pi]$
11 $n = 341$, ‘β’ = 231, ‘γ’ = 110
12 $x = 2$ or $x = 4/3$
13 $x = \ln(3 \pm 2\sqrt{2})$

CHAPTER 6

EXERCISES

1 **a** $y = x + 1$
b $y + 1 = 0,\ y - 1 = 0$
c $y = -\pi x/2$
d $y = 0$
2 **a** $y = x$
b $5y - 8x + 11 = 0,\ 5y + 3x + 4 = 0$
c $9y - 3x = 10$
d $y - 1 = \pi x$
3 **a** $-17^{3/2}/32$
b 1
c $5\sqrt{5}/6,\ -\sqrt{2}/3$
4 **a** 1
b $-1/\sqrt{2}$
c infinite
d $2\sqrt{2}$

ASSIGNMENT

1 $\varrho = 1/2$, centre $(0, 3/2)$
3 Tangent $y = 1$; normal $x = 0$

4 $\varrho = 1/2$
6 Tangent $y(t^2 + 2t^3 + 5) + 3t^2x = t^4 + 10t$; normal
$9t^3(y - t) - 3t(2t^3 + t^2 + 5)x = (t^3 + t^2 - 5)(2t^3 + t^2 + 5)$
7 Tangent $y(t - 1) - x(t + 1) = t^2$; normal $y(t + 1) + x(t - 1) = t(t^2 + 2)$

FURTHER EXERCISES

3 $(71/6, -1/6)$
6 a Rectangular hyperbola centre $(2, 1)$
b pair of straight lines $x = 4$, $y = 3$
c ellipse centre $(-2, -3)$, major axis 4, minor axis $2\sqrt{(2)}$
d hyperbola centre $(1, 3)$ **e** circle centre $(4, 3)$ radius 4
7 $X = -2\sin^3\theta$, $Y = -2\cos^3\theta$; $X^{2/3} + Y^{2/3} = 2^{2/3}$
10 $(\cos^6 p + \sin^6 p)^{3/2}/(\cos^6 p + \sin^6 p + 1)$
14 $4a$
16 864 m/s^2
17 a $x^2\cosh x/(1 + \sinh x) + 2x\ln(1 + \sinh x)$ except for $x \leqslant \ln(-1 + \sqrt{2})$
b $-1/x^2 - 4/x^3 - 9/x^4$ except for $x = 0$ **c** $2(1 - 2x)/(1 - x + x^2)^2$
18 $\mathrm{d}y/\mathrm{d}x = 1/(1 + x^{1/2})^2$, $\mathrm{d}^2y/\mathrm{d}x^2 = 2/[(1 - x)(1 + x^{1/2})^2]$,
$\mathrm{d}^3y/\mathrm{d}x^3 = 6/[(1 - x)^2(1 + x^{1/2})^2]$
19 $\mathrm{d}y/\mathrm{d}x = (x + 1)^{-2}$; $\mathrm{d}z/\mathrm{d}x = (2x + 1)^{-2}$

CHAPTER 7

EXERCISES

1 a $f_x = 3x^2 + 2xy$, $f_y = x^2$
b $f_x = -\sin xy \sin(x + y) + y\cos(x + y)\cos xy$
$f_y = -\sin xy \sin(x + y) + x\cos(x + y)\cos xy$
c $f_x = 3(x + 2y)^2$, $f_y = 6(x + 2y)^2$
d $f_x = (-y\sin xy + \cos xy)\exp(x + y)$
$f_y = (-x\sin xy + \cos xy)\exp(x + y)$
e $f_x = \sinh(x + y)/2\sqrt{\cosh(x + y)} = f_y$
f $f_x = (1/y)\sinh(x/y)$, $f_y = -(x/y^2)\sinh(x/y)$
2 a $f_{xy} = -xy/(x^2 + y^2)^{3/2} + \cos xy - xy\sin xy$
b $f_{xy} = 2x - (x + 2y)^{-3/2}/2$
c $f_{xy} = -12x^2\sin(3x + 4y) + 8x\cos(3x + 4y)$
d $f_{xy} = (6x^3/y^7)\sin(x^2/y^3) - (6x/y^4)\cos(x^2/y^3)$
3 a $z_x = (z_u - z_v)/2uv(v - u)$, $z_y = (uz_u - vz_v)/(u - v)$
b $z_x = -z_u/3 + 2z_v/3$, $z_y = 2z_u/3 - z_v/3$
c $z_x = [(u - 1)z_u - vz_v]/[u - v - 1]$, $z_y = [uz_u - (1 + v)z_v]/[u - v - 1]$
d $z_x = (uz_u + vz_v)/(u^2 + v^2)^{1/2}$, $z_y = (v^3z_u - uv^2z_v)/(u^2 + v^2)$

ASSIGNMENT

1 a $f_x = \cos x \cos y + y^2$; $f_y = -\sin x \sin y + 2xy$
b $f_x = e^x \cos y + e^x \sin y$; $f_y = -e^x \sin y + e^x \cos y$
c $\partial z/\partial x = x/(x^2 + y^2)$; $\partial z/\partial y = y/(x^2 + y^2)$
d $\partial z/\partial x = 2(x - 2y)^4(3x + 4y)$; $\partial z/\partial y = -8(x - 2y)^4(x + 3y)$
e $\partial z/\partial u = 4u(u^2 - v^2) \cos (u^2 - v^2)^2$;
$\partial z/\partial v = -4v(u^2 - v^2) \cos (u^2 - v^2)^2$
3 $e^{-u} \cos v\,(\partial z/\partial u + \partial z/\partial v) + e^{-u} \sin v\,(\partial z/\partial u - \partial z/\partial v)$
4 Approximately 12%

FURTHER EXERCISES

8 Decrease of 3δ%
12 1%
13 $(3/4)[2(a/b) + 3(b/a) + 3] \geqslant 3\sqrt{3/2} + 9/4 > 5\%$

CHAPTER 8

EXERCISES

1 a $1 + x + x^2/2$
b $1 - x^2/2 + 5x^4/24$
c $x - 3x^2/2 + 11x^3/6$
2 a $x = 0$, max; $x = \pm\sqrt{5}$, min
b $x = (-1 + \sqrt{5})/2$, min; $x = (-1 - \sqrt{5})/2$, max
c $x = 0$, min; $x = 1$, max; $x = -1$, max
d $x = 0$, point of inflexion
3 a 0
b -1
c 1 (take logarithms)
d 3
4 a 1/2
b e (take logarithms)
c 1
d 1

ASSIGNMENT

1 a $(1/2) \sec x$ **b** $(\sin x)^x [x \cot x + \ln (\sin x)]$ **c** $6 \tan 3x \sec^2 3x$
d Hint: put $x = \sin t$, then $dy/dx = 2 [\sin^{-1} x + x(1 - x^2)^{-1/2}]$
2 $dy/dx = [2y \cos 2x - y \ln y - y^2 \exp (xy)]/[xy \exp (xy) + x]$
4 $1 + x - \frac{1}{3}x^3$

5 a Stationary points at ± 1, points of inflexion at $0, \pm 1$
b stationary points at 0, 1, 2; 0 (min.), 1 (max.), 2 (min.)
6 Diameter = twice height
9 a n **b** 1/2 **c** $\pi/2$ **d** 1/2
10 a 0 **b** Hint: put $u = 1/x$; $\ln a$

FURTHER EXERCISES

6 $3[(7/3)^{2/3} - 1]^{3/2}$ metres; approximately 2 metres.
7 Length = 2 × breadth
8 a $\sqrt{2}/4$ **b** $\sqrt{2}/2$
10 $14 - 8\sqrt{3}$ ohms
11 $i(t) = t - t^2/2 + t^3/3 + \dots$
15 Hint: use l'Hospital's rule
16 1/2
17 $x = 1$ (repeated), $x = -1/2$; $-27/(8e)$; $y + 1 = x$; $y + 1 = -x$
18 p, 0
19 $[\lambda \ln \sqrt{1 + \theta}]^{-1}$
20 Procedure invalid! $t > 2/3$ which is not sufficiently small for the approximations to hold

CHAPTER 9

EXERCISES

1 a 1
b 1
c 2/3
d 2 (take logarithms)
2 a $n/(2n + 1)$
b $1 - 1/(n + 1)^2$
c $1 - 1/\sqrt{(n + 1)}$
d $-\operatorname{cosech} 1 (\coth n - \coth 1)$

ASSIGNMENT

1 Divergent: divergence test
2 Convergent: comparison test
3 Convergent: ratio test
4 Divergent: divergence test
5 Convergent: comparison test
6 Divergent: ratio test
7 $x \leqslant 0$ convergent, $x > 0$ divergent: ratio and comparison tests
8 Divergent: comparison test

9 Convergent: ratio test
10 Convergent: s_n explicitly
11 $R = 2$
12 $R = \infty$
13 $R = \sqrt{3}$
14 $R = 1$
15 $R = \infty$
16 $R = 1/2$
17 $R = 0$
18 $R = 1/\mathrm{e}$
19 $R = 1$
20 $R = 1$

FURTHER EXERCISES

1 a Divergent **b** divergent **c** divergent **d** divergent
2 a Absolutely convergent **b** absolutely convergent
c conditionally convergent
4 a Hint: $n + nx > 1 + nx$; divergent **b** divergent
c divergent (if $|x| > 1$ then $|x|^n \to \infty$, if $|x| < 1$ then $|x|^{-n} \to \infty$)
6 a Convergent **b** convergent
c convergent when $x = -1$, divergent when $x = 1$
d convergent when $x = -1$, divergent when $x = 1$
10 a $V - v + v/2^n$ **b** $V = v$ **c** $v/V = 0.25024$; just over 25%
11 Yes; $5v/3$; $100[1 - 5^{1/20}/10^{1/5}] = 31.62\%$; below specification; $v(10^{1/5}/5^{1/20}) \approx 1.4624v$
12 $v(1 - v/V)^n$; $V[1 - (1 - v/V)^n]$; yes

CHAPTER 10

EXERCISES

1 a $54 + 29\mathrm{i}$
b $(13 - 9\mathrm{i})/25$
c $(7 - \mathrm{i})/10$
d $\mathrm{e}^2 \cos 4 + \mathrm{i}[\mathrm{e}^2 \sin 4 + 1]$
2 a $2\left[\cos\left(\frac{-\pi}{4}\right) + \mathrm{i}\sin\left(\frac{-\pi}{4}\right)\right]$
b $2[\cos \pi + \mathrm{i} \sin \pi]$
c $2[\cos \pi/2 + \mathrm{i} \sin \pi/2]$
d $1[\cos 1 + \mathrm{i} \sin 1]$
3 a $-2 \pm \mathrm{i}$
b $0, \mathrm{i}$
c $\pm 1 + \mathrm{i}\sqrt{2}, \pm 1 - \mathrm{i}\sqrt{2}$

d $i(-1 \pm \sqrt{2})$
4 a Circle, centre (0, 3), radius 5
b ellipse, foci $(0, \pm 1)$, semi-axes 2, $\sqrt{3}$
c circle, centre (0, 0), radius 1

ASSIGNMENT

1 $(195 - 104i)/221$
2 $(1/2)(\cos \pi/2 + i \sin \pi/2)$
3 $z = i(w + 1)/(w - 1)$ where $w = \exp i(\pi/6 + 2k\pi/3)$, $k \in \{-1, 0, 1\}$
4 $\cos 11\theta - i \sin 11\theta$
5 $\cos 4\theta = \cos^4\theta - 6\cos^2\theta\sin^2\theta + \sin^4\theta$,
$\sin 4\theta = 4\cos^3\theta\sin\theta - 4\cos\theta\sin^3\theta$
6 $(\tan x \operatorname{sech}^2 y + i \tanh y \sec^2 x)/(1 + \tan^2 x \tanh^2 y)$
7 $z = n\pi + \ln[(-1)^n + \sqrt{2}]$
8 $z = x + iy$ lies on the circle $x^2 + y^2 = 1$

FURTHER EXERCISES

1 a Semicircle $x^2 + y^2 = 16$ $(y \geqslant 0)$
b $x^2/9 + y^2/16 = 1$ (ellipse); $z = 4i$
2 a $-1 + i\sqrt{3}$ **c** $-2 + 5i, -1, 4 + i$
3 a $2\exp(\pi i/4), 2\exp(3\pi i/4), 2\exp(-\pi i/4), 2\exp(-3\pi i/4)$
b $(\pm 7 \pm i\sqrt{31})/2$
4 a 2^{12} **b** $z = \pm 1, \pm i; 0, \pm i$
5 a $(\pm 5 \pm i\sqrt{3})/2$ **b** Hint: square and use $|z|^2 = z\bar{z}$

CHAPTER 11

EXERCISES

1 a $\begin{bmatrix} 2 & 5 \\ -3 & 7 \end{bmatrix}$ **b** $\begin{bmatrix} -1 & -16 \\ 16 & -34 \end{bmatrix}$ **c** $\begin{bmatrix} 2 & 5 \\ -3 & 7 \end{bmatrix}$

d $\begin{bmatrix} -3 & 6 \\ -10 & 20 \end{bmatrix}$ **e** $\begin{bmatrix} 2 & -4 \\ -4 & 8 \end{bmatrix}$ **f** $\begin{bmatrix} 5 & -5 \\ -5 & 5 \end{bmatrix}$

2 $(U + V + W)^2 = U^2 + V^2 + W^2 + UV + UW + VU + VW + WU + WV$

3 a $\begin{bmatrix} 0 & 1/2 \\ -1/2 & 1/2 \end{bmatrix}$ **b** $\begin{bmatrix} 0 & -2 \\ 1 & -2 \end{bmatrix}$ **c** $\begin{bmatrix} 0 & -4 \\ 4 & 1 \end{bmatrix}$

d $\begin{bmatrix} -1.4 & -2.2 \\ 0.9 & -3.3 \end{bmatrix}$ **e** $\begin{bmatrix} 2 & 1 \\ -1 & 4 \end{bmatrix}$ **f** $\begin{bmatrix} 4 & -6 \\ 3 & -2 \end{bmatrix}$

4 $\operatorname{diag}\{\pm 1, 0\}$

ASSIGNMENT

1 $x = \sin w, y = -\cos w$ or $x = -\sin w, y = \cos w$
3 a $I\,(2 \times 2)\; O\,(3 \times 3)$ **b** $A^n = \mathrm{diag}\,\{a^n, b^n, c^n, d^n\}$
4 a $(A + B)^2 = A^2 + AB + BA + B^2$
b $(A + 2B)^2 = A^2 + 2AB + 2BA + 4B^2$
5 $a = 2, b = -1, c = 8$
6 $a = 0, b = 0, c = 0, d = -3, e = 2, f = -6$
7 $x = 3, y = 3, z = -1$ or $x = \mp 4\sqrt{2/3}, y = 2/3, z = \pm 2\sqrt{2}$

FURTHER EXERCISES

2 Hint: use $A(BC) = (AB)C$ to deduce $A = O$
3 a $x = 2, y = 1, z = -1$ **b** $x = 4, y = 2, z = -2$
5 a $\begin{bmatrix} 1 + Z_1/Z_2 & Z_1 \\ 1/Z_2 & 1 \end{bmatrix}$
b $\begin{bmatrix} 1 + Z_1/Z_2 & Z_1 + Z_3 + Z_1Z_3/Z_2 \\ 1/Z_2 & Z_3/Z_2 + 1 \end{bmatrix}$
c $\begin{bmatrix} 1 + Z_2/Z_3 & Z_2 \\ 1/Z_1 + 1/Z_3 + Z_2/Z_1Z_3 & Z_2/Z_1 + 1 \end{bmatrix}$
6 $n = 3$
8 c i $x = 39, y = 35, z = 21$ **ii** $u = 18, v = 47, w = 20$
9 $E_1 = (1/3)\,(e_1 + e_2 + e_3), E_2 = (1/3)(e_1 + \alpha e_2 + \alpha^2 e_3)$,
$E_3 = (1/3)(e_1 + \alpha^2 e_2 + \alpha e_3)$

CHAPTER 12

EXERCISES

1 a $x = -3$ or $x = 5$
b $x = -2$ or $x = 7$
c $x = 3/2$
d $x = 0$
2 a 5 **b** -24 **c** 9 **d** $-9x - 18$
3 a 5775 **b** 18,000
4 a $x = 0, y = 1/a\sqrt{(1 + a^2)}$
b $x = \exp u, y = \exp(-v), z = \exp w$

ASSIGNMENT

1 $x = \pm 4$
2 8
3 $x = 1/2$

4 $x = 5$

5 $M = \begin{bmatrix} 2 & -2 & -3 \\ -12 & 7 & 14 \\ -9 & 6 & 11 \end{bmatrix}$

6 $C = \begin{bmatrix} -1 & -2 & 2 \\ 2 & 7 & -6 \\ -2 & -9 & 7 \end{bmatrix}$

FURTHER EXERCISES

1 An identity; true for all w
2 $x = \pm 1$
3 $k = 1/a + 1/b + 1/c + 1/d$
4 $x = 0$ or $x = \pm 6$
6 $x = -1, x = -2, x = -2$
7 -12
8 a Hint:

$$D = \frac{1}{xyz}\begin{vmatrix} x & 1 & y+z \\ y & 1 & z+x \\ z & 1 & x+y \end{vmatrix}$$

b Hint: $C_1 - C_2$
9 $u = \cos w$, $v = \tan w$
10 Hint: $\Delta = 0$
11 $\lambda = -3,\ 0,\ 3$
12

$$i = \frac{(R_1R_4 - R_2R_3)E}{(R_1R_2R_3 + R_2R_3R_4 + R_3R_4R_1 + R_4R_1R_2) + R(R_1R_2 + R_2R_3 + R_3R_4 + R_4R_1)}$$

CHAPTER 13

EXERCISES

1 a $\begin{bmatrix} -5 & 2 & 4 \\ 2 & -1 & -1 \\ 0 & 1 & -2 \end{bmatrix}$ **b** $\begin{bmatrix} 3 & 1 & -7 \\ -2 & 1 & 1 \\ -1 & -1 & 4 \end{bmatrix}$

c $\begin{bmatrix} 15 & 1 & -12 \\ -8 & 1 & 6 \\ -6 & -1 & 5 \end{bmatrix}$ **d** $\begin{bmatrix} 3 & 12 & -11 \\ -2 & -1 & 2 \\ -1 & -7 & 6 \end{bmatrix}$

2 a $x = 2$ **b** $x = 1$ **c** $x = 3$ or $x = -3 \pm \sqrt{13}$
d $x = 1/3$ or $x = 5$
4 a $x = 35, y = -6, z = -9$
b $x = -6, y = 13, z = 1$
c $x = 4, y = 17, z = -35$
d $x = 3, y = 5, z = -4$

ASSIGNMENT

1 $\begin{bmatrix} -3 & -2 & 13 \\ 0 & 1 & -3 \\ 1 & 0 & -2 \end{bmatrix}$

3 diag $\{27, 54, -81\}$
4 Hint: pre-multiply and post-multiply AB by the expression suggested

FURTHER EXERCISES

1 $(\text{adj } A)^{-1} = (1/|A|)A$
2 $|\text{adj } A| = |A|^{n-1}$ where n is the order
3 Grade, number: 1, 10; 2, 15; 3, 20
4 Hint: if a matrix is non-singular it has an inverse
5 Hint: verify that $A^{\mathrm{T}}(A^{-1})^{\mathrm{T}} = (A^{-1})^{\mathrm{T}}A^{\mathrm{T}} = I$
6 Example: $x + y + 1 = 0, 2x + 2y + 2 = 0, 2x + 2y + 3 = 0$
7 $k = 1, k = 2, k = 3$
9 a $M^{-1} = \begin{bmatrix} \cosh a & -Z \sinh a \\ -(1/Z) \sinh a & \cosh a \end{bmatrix}$
b Use induction
10 $x = 3, y = -2, z = 1$
11 $\lambda = 3, 5; \begin{pmatrix} 3 \\ 1 \end{pmatrix}, \begin{pmatrix} 1 \\ 1 \end{pmatrix}; P^{-1}AP = \begin{pmatrix} 3 & 0 \\ 0 & 5 \end{pmatrix}$

CHAPTER 14

EXERCISES

1 a $\mathbf{a} + \mathbf{b} = 3\mathbf{i} - 2\mathbf{j} + 4\mathbf{k}, \mathbf{a} \cdot \mathbf{b} = 2, \mathbf{a} \times \mathbf{b} = 10\mathbf{i} + \mathbf{j} - 7\mathbf{k}$
b $\mathbf{a} + \mathbf{b} = 3\mathbf{i} + 6\mathbf{j} - 2\mathbf{k}, \mathbf{a} \cdot \mathbf{b} = -5, \mathbf{a} \times \mathbf{b} = 22\mathbf{i} - 13\mathbf{j} - 6\mathbf{k}$
c $\mathbf{a} + \mathbf{b} = 5\mathbf{j} - 5\mathbf{k}, \mathbf{a} \cdot \mathbf{b} = 11, \mathbf{a} \times \mathbf{b} = 5\mathbf{i} - 5\mathbf{j} - 5\mathbf{k}$
d $\mathbf{a} + \mathbf{b} = -2\mathbf{i} + 5\mathbf{j} + \mathbf{k}, \mathbf{a} \cdot \mathbf{b} = -1, \mathbf{a} \times \mathbf{b} = -9\mathbf{i} - 5\mathbf{j} + 7\mathbf{k}$
2 a $\pm(1/3\sqrt{6})\{5\mathbf{i} - 5\mathbf{j} - 2\mathbf{k}\}$
b $\pm(1/3\sqrt{5})\{4\mathbf{i} - 5\mathbf{j} + 2\mathbf{k}\}$

c $\pm(1/\sqrt{2})\{\mathbf{i} + \mathbf{j}\}$
d $\pm(1/\sqrt{2})\{\mathbf{i} - \mathbf{k}\}$

3 a 2 **b** 4 **c** −14 **d** 0

4 a $\mathbf{a} \times (\mathbf{b} \times \mathbf{c}) = \mathbf{j} - \mathbf{i}$, $(\mathbf{a} \times \mathbf{b}) \times \mathbf{c} = \mathbf{k} - \mathbf{i}$
b $\mathbf{a} \times (\mathbf{b} \times \mathbf{c}) = 5\mathbf{j} - 5\mathbf{k}$, $(\mathbf{a} \times \mathbf{b}) \times \mathbf{c} = -5\mathbf{i} + 5\mathbf{j}$
c $\mathbf{a} \times (\mathbf{b} \times \mathbf{c}) = 9\mathbf{i} + 22\mathbf{j} - 17\mathbf{k}$, $(\mathbf{a} \times \mathbf{b}) \times \mathbf{c} = -26\mathbf{i} + 43\mathbf{j} - 3\mathbf{k}$
d $\mathbf{a} \times (b \times c) = -\mathbf{i} - \mathbf{j} + 2\mathbf{k}$, $(\mathbf{a} \times \mathbf{b}) \times \mathbf{c} = \mathbf{i} - 2\mathbf{j} + \mathbf{k}$

5 a $\dot{\mathbf{r}} = -\sin t\,\mathbf{i} - \cos t\,\mathbf{j} + \mathbf{k}$
b $\dot{\mathbf{r}} = -3\sin 3t\,\mathbf{i} + 5\cos 5t\,\mathbf{j} + 2t\,\mathbf{k}$
c $\dot{\mathbf{r}} = 4t(1 + t^2)\mathbf{i} + 9t^2(1 + t^3)^2\mathbf{j} + 16t^3(1 + t^4)^3\mathbf{k}$
d $\dot{\mathbf{r}} = -t^2(3 + 5t^2)\mathbf{i} - 2t(1 - 3t^4)\mathbf{j} + t^2(3 + 5t^2)\mathbf{k}$

ASSIGNMENT

1 (a) 1 (b) 1/6 (c) $3\mathbf{i} - 5\mathbf{j} + \mathbf{k}$ (d) $\sqrt{35}$
2 $\pm(1/5)(3\mathbf{j} + 4\mathbf{k})$
5 $\mathbf{x} = (1/|\mathbf{a}|^2)[\mathbf{a} - \mathbf{a} \wedge \mathbf{b}]$
6 $t = -4$

FURTHER EXERCISES

2 (a) $\pi/4$ (b) $\pm(2\mathbf{i} - 10\mathbf{j} - 11\mathbf{k})/15$
3 $\sqrt{2}(\mathbf{i} - 2\mathbf{j} + \mathbf{k})$
4 (a) Hint: put $\mathbf{e} = \mathbf{c} \wedge \mathbf{d}$
(b) Hint: $(\mathbf{x} \wedge \mathbf{y}) \cdot (\mathbf{z} \wedge \mathbf{w}) = (\mathbf{x} \cdot \mathbf{z})(\mathbf{y} \cdot \mathbf{w}) - (\mathbf{y} \cdot \mathbf{z})(\mathbf{x} \cdot \mathbf{w})$
5 $\mathbf{r} \cdot \mathbf{r} = 2\cos^2\theta$; $\mathbf{r} \wedge \mathbf{r} = 0$
(a) $\cos\theta = 0 \Rightarrow \theta = \pi/2 \Rightarrow \mathbf{r} = \mathbf{0}$
(b) $\cos^2\theta = 1 \Rightarrow \mathbf{r} = \mathbf{i} \pm \mathbf{k}$ (no calculus needed)
9 $\sqrt{5}$ and 2

14 a $\dot{\mathbf{r}} = (\cos t - t\sin t)\mathbf{i} + (\sin t + t\cos t)\mathbf{j} + \mathbf{k}$ **b** $t = 3$
c $|\ddot{\mathbf{r}}| = \sqrt{13}$, $2(\mathbf{k} - \mathbf{i})$

15 a $\mathbf{r} = 2\mathbf{i} + 3\mathbf{j} - \mathbf{k}$,
$\mathbf{s} = 3\mathbf{i} - \mathbf{j} + 3\mathbf{k}$
b $\mathbf{r} \times \mathbf{s} = 8\mathbf{i} \div 9\mathbf{j} - 11\mathbf{k}$,
$\mathbf{u} = \pm(1/\sqrt{26})\{-4\mathbf{i} + 3\mathbf{j} + \mathbf{k}\}$
c Volume is 6 units

CHAPTER 15

EXERCISES

1 a $x + x^2 + 3x^4/4 + C$

b $e^x - 2e^{-x} + C$
c $x + \sinh^{-1} x + C$
d $\tan x + \sec x + C$
2 a $(1 + 2x)^8/16 + C$
b $(1/3) \sin 3x + C$
c $-(1/3) \cos x^3 + C$
d $(1/2) \tan^{-1} (x/2) + C$
3 a $(1/7) \ln |x + 2| - (1/7) \ln |3x - 1| + C$
b $(1/3) \ln |x| - (1/6) \ln (x^2 + 3) + C$
c $-(1/4) \ln |x| + (1/8) \ln |x^2 - 4| + C$
d $(1/3) \tan^{-1} x - (1/6) \tan^{-1} (x/2) + C$
4 a $(2x/3) \sin 3x + (2/9) \cos 3x + C$
b $(x/3) \exp 3x - (1/9) \exp 3x + C$
c $\{(x^2 - 1)/2\} \exp x^2 + C$
d $(x^3/3) \ln (x^2 + 1) - 2x^3/9 + 2x/3 - (2/3) \tan^{-1} x + C$
5 a $x^2/4 + \ln x^2 + C$
b $-\cot x + \operatorname{cosec} x + C$
c $(1/3) \exp x^3 + C$
d $(1/2) \tan x^2 + C$

ASSIGNMENT

1 $x + \ln |x| + C$
2 $x - \ln |x + 1| + C$
3 $\tan x - x + C$
4 $x + \ln (x - 1)^2 + C$
5 $\ln |x - 1| + C$
6 $-(1/2) \exp (-x^2) + C$
7 $\ln (x + 1)^2 + C$
8 $\ln (e^x + 1) + C$
9 $(1/3) \sin^3 x + C$
10 $-\tan^{-1} (\cos x) + C$
11 $\tan^{-1} (e^x) + C$
12 $x \sin^{-1} x + \sqrt{(1 - x^2)} + C$
13 $(1/3)(x^2 + 4)^{3/2} + C$
14 $-2\sqrt{(\cos^2 x + 9)} + C$
15 $(x^2/2) \ln x - x^2/4 + C$
16 $[(x^2 + 1)/2] \tan^{-1} x - x/2 + C$
17 $5 \ln |x - 3| - 4 \ln |x - 2| + C$
18 $(1/2) \tan^2 x + \ln (\cos x) + C$
19 $(1/4) \ln |(x - 1)/(x + 1)| - (1/2) \tan^{-1} x + C$
20 $(1/4) \ln |x^4 - 1| + C$
21 $(1/2)(x^4 + 1)^{1/2} + C$

FURTHER EXERCISES

1 **a** $(1/2)\sin x^2 + C$ **b** $2\sqrt{(x+3)} + C$ **c** $2\ln(1+e^t) - t + C$
d $(1/4)\ln(1+x^4) + C$
2 **a** $\ln(\ln u) + C$ **b** $-(\cos\theta - \sin\theta) + (1/3)(\cos^3\theta - \sin^3\theta) + C$
c $t\sec^{-1}t - \cosh^{-1}t + C$ **d** $(1/2)(\ln x)^2 + C$
3 **a** $u\tan^{-1}u - (1/2)\ln(1+u^2) + C$ **b** $\sin x/(\cos x - 1) + C$
c $(1/4)\ln[(x+1)/(x-1)] - 1/2(x-1) + C$
d $\ln\{(1-\sin\theta)(1-\cos\theta)\} + 2/(1-\sin\theta-\cos\theta) + C$
4 $i = (E/R)[1 - \exp(-Rt/L)]$
6 $wx^2(x-L)^2/24EI$
7 $30/(1-2^{-1/3})$ minutes; 2 hours 25 minutes (approx.)
8 $(\cos 2\theta + \sin 2\theta)/(\cos\theta + \sin\theta) + C$
9 $u = \ln\{\tan[\pi/4 + (\alpha/2)(1-e^t)]\}$

CHAPTER 16

EXERCISES

1 **a** $e^x + x - 2\ln(e^x+1) + C$
b $2\sin x + C$
c $x\ln x - x + x^2/2 + C$
d $\sinh^{-1}x + C$
2 **a** $\ln\sqrt{\sec(1+x^2)} + C$
b $-\cos\sqrt{(1+x^2)} + C$
c $\ln(1+\sin^2 x) + C$
d $-(1/14)\cos 7x + (1/2)\cos x + C$
3 **a** $\sin x - \ln(1+\sin x) + C$
b $(3/2)\sin^{-1}x + (x/2)\sqrt{(1-x^2)} + C$
c $(x/4)\{\sqrt{(x^2+1)} - \sqrt{(x^2-1)}\}$
$+ (1/4)\ln|[x+\sqrt{(x^2+1)}]\cdot[x+\sqrt{(x^2-1)}]| + C$
d $(x/2)\{\sqrt{(1+x^2)} + \sqrt{(1-x^2)}\} + \ln\sqrt{[x+\sqrt{(1+x^2)}]}$
$+ (1/2)\sin^{-1}x + C$
4 **a** $(1/25)\{3\ln[(1+t^2)/(2t^2+3t-2)] + 8\tan^{-1}t\} + C$
where $t = \tan(x/2)$
b $(1/13)\ln|[5+\tan(x/2)]/[1-5\tan(x/2)]| + C$
5 **a** $(1/7)\cosh^6 x\sinh x + (6/35)\cosh^4 x\sinh x + (8/35)\cosh^2 x\sinh x$
$+ (16/35)\sinh x + C$
b $x^5\sin x + 5x^4\cos x - 20x^3\sin x - 60x^2\cos x + 120x\sin x$
$+ 120\cos x + C$

ASSIGNMENT

1 $\ln(1-\sin x) - \cos x/(1-\sin x) + C$
2 $\exp\sin x + C$

3 $(1/3)\ln(x-1) - (1/6)\ln(x^2+x+1) + (1/\sqrt{3})\tan^{-1}[(2x+1)/\sqrt{3}] + C$
4 $-\ln(\sec x + \tan x) + C$
5 $\exp \sin^2 x + C$
6 $(1/2)\ln(x^2-1) + C$
7 $-\cos x + (2/3)\cos^3 x - (1/5)\cos^5 x + C$
8 $3x/8 + (1/4)\sin 2x + (1/32)\sin 4x + C$
9 $-(2/3)\cos^3 x + \cos x + C$
10 $\tan x + (2/3)\tan^3 x + (1/5)\tan^5 x + C$
11 Hint: obtain the integral of $\sec^3 x$ first. Answer:
$(1/4)\sec^3 x \tan x + (3/8)\sec x \tan x + (3/8)\ln(\sec x + \tan x) + C$
12 $(1/5)\sec^5 x + C$
13 $(1/12)(x + 1/x)^{12} + C$
14 $(1/8)(1 + \sin x)^8 + C$

FURTHER EXERCISES

1 (a) Hint: put $\operatorname{cosec}^2 x = 1 + \cot^2 x$; answer
$-(1/2)\operatorname{cosec} x \cot x + (1/2)\ln|\operatorname{cosec} x - \omega tx| + C$
(b) hint: put $\tanh^2 x = 1 - \operatorname{sech}^2 x$; answer $\ln(\cosh x) + (1/2)\operatorname{sech}^2 x + C$
(c) hint: standard t substitution; answer
$(1/5)\ln\{[1 + 3\tan(\theta/2)]/[3 - \tan(\theta/2)]\} + C$
2 (a) Hint: put $u = 1 - 1/x$; answer $(2/3)(1 - 1/x)^{3/2} + C$
(b) hint: put $u = \cos\theta$; answer $\ln[(1+\cos\theta)^2/\cos\theta] + C$
(c) hint: express $(1+\sin\theta)/(1+\cos\theta)$ in terms of $\theta/2$; answer
$\exp\theta \tan\theta/2 + C$
3 (a) $\ln(1 + \sin x \cos x) + C$ (b) hint: put $u = \tan\theta$; answer
$(12/169)\ln(5\cos\theta + 12\sin\theta) + (5\theta/169) + C$
(c) $x/2(x^2 + 2x + 2) + (1/2)\tan^{-1}(x+1) + C$
4 (a) $(1/\sqrt{2})\ln\{[(\sqrt{2}+1) + \tan x/2]/[(\sqrt{2}-1) - \tan x/2]\} + C$
(b) Hint: put $u = \cos x$; answer $(1/12)\ln[(1-\cos x)^2(2+\cos x)] -$
$(1/12)\ln(1+\cos x)^2(2-\cos x) + C$
(c) $(1/16)(\sin 8x + 4\sin 2x) + C$
5 Hint: $\sec^n x = \sec^{n-2} x \sec^2 x$
6 Hint: $\tan^n x = \tan^{n-2} x \tan^2 x$
7 Hint: integrate by parts
8 Hint: show that $d^2s/d\psi^2 = 1$

9 a $-(t\cos 2t)/2 + (\sin 2t)/4 + C$
b $\ln(t^2 + 3t + 1) + C$
c $(t/3)(2t+1)^{3/2} - (1/15)(2t+1)^{5/2} + C$

10 a $(2/3)\tan^{-1}(1/3)$
b $(1/2)\ln 3 - (1/4)\ln 5$
c $\pi/8 - (5/6)\tan^{-1}(1/3)$

11 a 16/35 **b** 8/105

CHAPTER 17

EXERCISES

1 **a** $\pi/4$ **b** $5/6$ **c** $\ln\sqrt{2}$ **d** 2
2 **a** $3/10$ **b** $1/12$ **c** $1/6$ **d** $(3-\mathrm{e})/2$
3 **a** $16\pi/15$ **b** $\pi\mathrm{e}^2/4 - 7\pi/4$ **c** $\pi[9\ln(2+\sqrt{3}) - 6\sqrt{3}]$
d $[72(\ln 2)^2 - 264\ln 2 + 149]\pi/27$
4 **a** $1/2$ **b** $(\mathrm{e}^2-1)/(2\mathrm{e})$ **c** $2\ln 2 - 1$ **d** $(2/\pi)\ln 2$
5 **a** $(2+\sinh 2)^{1/2}/2$ **b** $(4/\pi - 1)^{1/2}$ **c** $1/\sqrt{2}$ **d** $(1019/120)^{1/2}$

ASSIGNMENT

1 $\ln(\sqrt{2}+1)$
2 $3c/4$
3 $3\pi/8$
4 π
5 $\pi[3 - (5/4)\ln 2 - (1/16)(\ln 2)^2]$
6 On the axis of symmetry, $h/4$ from the centre of the base
7 (a) $ma^2/4$ (b) $ma^2/4$
8 $13Ma^2/20$

FURTHER EXERCISES

2 $a^2/15$; $(\bar{x},\bar{y}) = (3a/8, 15a/28)$
3 $15/4 + \ln 2$
4 $25/32$; $l = \frac{1}{8} + \frac{1}{2}\ln 3$
6 $3Ma^2/10$
7 (a) Does not exist: $x^2(\ln x^2 - 1) \to \infty$ as $x \to \infty$
(b) 1: $xe^{-x} + e^{-x} \to 0$ as $x \to \infty$, $xe^{-x} + e^{-x} \to 1$ as $x \to 0$
(c) $\pi/2$
9 Hint: use Pappus's theorem
12 $-4/\pi^2 + 24/\pi^4$
13 $(3/2) - 2\ln 2$; $25\pi/3 - 12\pi\ln 2$

CHAPTER 18

EXERCISES

1 **a** $x = 1.849$ **b** $x = 0.6486$
2 $c = \{ab(b^2\mathrm{e}^b - a^2\mathrm{e}^a) + b - a\}/\{b^3\mathrm{e}^b - a^3\mathrm{e}^a\}$, $a = 0$, $b = 1$
3 $x_{n+1} = \{x_n x_{n-1}(x_n + x_{n-1}) + 5\}/\{x_n{}^2 + x_n x_{n-1} + x_{n-1}{}^2 + 1\}$
$x_0 = 1$, $x_1 = 2$
4 **a** $x = 0.273\,89$ **b** $x = 3.146\,19$

5 a 1.468 14 **b** 0.737 43
6 a 2.030 10 **b** 1.005 71

ASSIGNMENT

1 At most 17 steps (in fact 11 will do); 0.619
2 $x_{n+1} = (x_n - 1)/(1 - 3\exp[-x_n])$; 0.619 (4 steps)
3 $x_{n+1} = (x_{n-1}\exp x_n - x_n \exp x_{n-1})/[\exp x_n - \exp x_{n-1} - 3(x_n - x_{n-1})]$:
$x_2 = 0.780\,202\,717$, $x_3 = 0.496\,678\,604$, $x_4 = 0.635\,952\,246$
4 $(1+h)f(x+h) - (2-h^2)f(x) + (1-h)f(x-h) \simeq h^2 \sec x$
5 0.859 533 8 (a) 0.859 166 6 (b) 0.859 140 9

FURTHER EXERCISES

2 0.69
3 2.187
4 0.443
5 0.5671
7 (a) $T = 0.784\,75$ (b) $S = 0.785\,40$
8 Hint: consider volume of rotation using $x^2 + y^2 = r^2$ between ordinates a and b
9 $u(t+h) = (1-h)u(t)$; $u(0.2r) \simeq (0.8)^r$, so
a $u(1) \simeq 0.327\,68$ **b** 10.93% underestimate
10 0.4603
11 1.95
12 0.589 (radians!)
13 0.416, $23^0 50'$, 23.835^0

CHAPTER 19

EXERCISES

1 a $y = Axe^x$
b $y + \ln(y/x) + 1/x = C$
c $(y-1)e^{x+y} + x + 1 = Ce^x$
2 a $xy = (x-1)e^x + C$
b $y = x(x-1)e^x + Cx$
c $y = (1/2)\sin x + C \operatorname{cosec} x$
3 a $\ln\sqrt{(2x^2 + y^2)} = (1/\sqrt{2})\tan^{-1}(y/x\sqrt{2}) + C$
b $(x+y)^3(x-4y)^2 = A$
c $\ln(xy) + y/x = A$
4 a $y = (\exp -x^2)/(C - x)$
b $x/y = C - x^3/3$
c $y^2 + 2xy - 2x^2 - 6y = C$

ASSIGNMENT

1 $e^y = \ln(x^2 + 1) + C$
2 $\ln x = \tan t + C$
3 $\ln(y^2 + 1) = \frac{1}{2}\sin 2u + C$
4 $y^2 - 1 = Axe^x$
5 $y = \cot\theta + C\operatorname{cosec}\theta$
6 $y\ln x = (x^2 + 1)^{1/2} + C$
7 $x = e^t + Ce^{-t}$
8 $xy = x^2 + C$
9 $Ax = \exp(y/x)$
10 $s^2 = t^2(\ln t + C)$
11 $x^2\ln x = y^2 - xy + Cx^2$
12 $\sec(y/x) + \tan(y/x) = Ax$

FURTHER EXERCISES

1 $\ln x = y^3/3x^3 - y/x + C$
2 $\ln x = -x/y + C$
3 $x\sin t = C\cos t + 1$
4 $x^2 + y^2 = A\sin^2 x$
5 $u = 5/2 + A(v - 3)^2$
6 $x^2 - 12xy - y^2 = C$
7 $\ln[(q - 3)^2 + (p - 2)^2] = 6\tan^{-1}[(p - 2)/(q - 3)] + C$
8 $x = k\sin(y/x)$
9 $y = x\ln x/(1 + \ln x)$
10 $r = \sec\theta$
12 Hint: obtain a differential equation for q first; answer $EC[\sin\omega t + \omega RC\{\exp(-t/RC) - \cos\omega t\}]/[1 + (RC\omega)^2]$

CHAPTER 20

EXERCISES

1 a $x = Ae^{t/2} + Be^{3t}$
b $x = e^t\{A\cos 3t + B\sin 3t\}$
c $y = (A + Bx)e^{4x/3}$
2 a $x = Ae^{-t/2} + Be^{5t} - t/5 + 9/25$
b $y = Ae^{2x/3} + Be^{2x} + xe^{2x}/4$
c $u = Ae^{2x/3} + Be^{x/3} + e^x/2$
d $y = e^{2x/5}[A\cos(x/5) + B\sin(x/5)] - [\cos x + \sin x]/8$
3 a $x = (2/39)[1 - e^{3t}]\cos t + (1/13)[1 + e^{3t}]\sin t$
b $x = (3/196)[e^{t/3} - e^{5t}] + (t/14)e^{5t}$
c $x = (t^3/150)e^{3t/5}$

ASSIGNMENT

1 $y = (A + Bx)e^{x/2} + e^x$
2 $y = e^{-2t}(A \cos 2t + B \sin 2t) + (1/20) \cos 2t + (1/10) \sin 2t$
3 $x = Ae^t + Be^{-3t} + (1/5)e^{2t}$
4 $u = (A + Bv)e^{4v} + v^2/16 + v/16 + 3/128$
5 $s = e^{-3t}(A \cos t + B \sin t) + (1/13) \cos t + (2/39) \sin t$
6 $u = Ae^{2t} + Be^{5t} + (1/10) + (1/3)te^{5t}$
7 $y = (A + Bx)e^{-3x} + (1/2)x^2e^{-3x} + (1/16)e^x$
8 $y = Ae^{5u} + Be^{-2u} - (1/24)e^{2u} - (1/14)ue^{-2u}$
9 $y = e^{-x}(A \cos 3x + B \sin 3x) + (x/6)e^{-x} \sin 3x$
10 $y = Ae^w + Be^{-2w} - (3w/10) \cos w + (w/10) \sin w + (1/25) \cos w + (11/50) \sin w$

FURTHER EXERCISES

1 Hint: complex roots of auxiliary equation ⇒ oscillations
2 Hint: obtain $x = x(t)$, differentiate and eliminate t to obtain the second constant; answer $x = \sqrt{[(u/\lambda)^2 + a^2]} \sin \lambda t$
7 Hint: express as a differential equation involving x and θ only and solve

CHAPTER 21

EXERCISES

1 a $\pi/8 + (2/\pi) \sum_{n=1}^{\infty}(1/n^2)[(-1)^n - \cos(n\pi/2)] \cos nx$
b $3\pi/8 + (2/\pi) \sum_{n=1}^{\infty}(1/n^2)[\cos(n\pi/2) - \cos n\pi] \cos nx$

2 a $\sum_{n=1}^{\infty}(1/n^2)[2 \sin(n\pi/2) + n\pi \cos(n\pi/2)] \sin nx$

b $(4/\pi) \sum_{n=1}^{\infty}[\sin(n\pi/2)/n^2] \sin nx$

3 $2/\pi - (2/\pi) \sum_{n=1}^{\infty}[(1 + \cos n\pi)/(n^2 - 1)] \cos nx$

4 $5\pi/18 + (2/\pi) \sum_{n=1}^{\infty}[(\pi/3n) \sin(n\pi/3) + (1/n^2) \cos(2n\pi/3) + (2\pi/[3n]) \sin(2n\pi/3) - (1/n^2) \cos(n\pi/3) - (\pi/[3n]) \sin(n\pi/3)] \cos nx$
$S(0) = \pi/3,\ S(\pi/3) = \pi/3,\ S(2\pi/3) = \pi/3,\ S(\pi) = 0,\ S(2\pi) = \pi/3$

ASSIGNMENT

1 a $l/2 - (2l/\pi) \sum_{n=1}^{\infty}(1/n) \sin(n\pi/2) \cos(n\pi x/l)$

b $l/3 + (2l/\pi) \sum_{n=1}^{\infty}(1/n)[\sin(2n\pi/3) - \sin(n\pi/3)] \cos(n\pi x/l)$

2 a $l/2 + (2l/\pi) \sum_{n=1}^{\infty}(1/n)[1 - \cos n\pi] \sin(n\pi x/l)$

b $2l/3 + (l/\pi) \sum_{n=1}^{\infty}(\cos(n\pi x/l)/n)[\sin(2n\pi/3) - \sin(4n\pi/3)] + (l/\pi) \sum_{n=1}^{\infty}(\sin(n\pi x/l)/n)[\cos(4n\pi/3) - \cos(2n\pi/3)]$

3 $(2/\pi) \sum_{n=1}^{\infty}[n(\cos n\pi + 1)/(n^2 - 1)] \sin(n\pi x/l)$

4 Cosine series converges to $t - \pi$ when $\pi < t < 2\pi$, sine series converges to $\pi - t$ when $\pi < t < 2\pi$

FURTHER EXERCISES

1 a $\pi/8 + (1/2\pi)\sum_{n=1}^{\infty}[(1 - \cos n\pi)/n^2]\cos 2nx - (1/2)\sum_{n=1}^{\infty}(1/n)\sin 2nx$

b $3\pi/8 - (1/2\pi)\sum_{n=1}^{\infty}[(1 - \cos n\pi)/n^2]\cos 2nx + (1/2)\sum_{n=1}^{\infty}(1/n)\sin 2nx$

c $\pi/8 - (1/2\pi)\sum_{n=1}^{\infty}[(1 - \cos n\pi)/n^2]\cos 2nx + (1/2)\sum_{n=1}^{\infty}[(\cos n\pi)/n]\sin 2nx$

d $\pi/4 - (1/\pi)\sum_{n=1}^{\infty}[(1 - \cos n\pi)/n^2]\cos 2nx$

4 a $3\pi/4 + (4/\pi)\sum_{n=1}^{\infty}[(\cos(n\pi/2) - 1)/n^2]\cos nx$

$S(\pi) = \pi,\ S(-\pi) = \pi$

b $3\pi/8 + (1/2\pi)\sum_{n=1}^{\infty}[(\cos n\pi - 1)/n^2]\cos 2nx$

$S(0) = \pi/2,\ S(\pi) = \pi/2,\ S(-\pi) = \pi/2$

6 $1/2$

7 $1/\pi + (1/2)\sin(w\pi t) - (1/\pi)\sum_{n=2}^{\infty}[(1 + \cos n\pi)/(n^2 - 1)]\cos n\pi wt$

$S(1/w) = 0,\ S(2/w) = 0,\ S(k/w) = 0$

8 a $1 + (4/\pi)\sum_{n=1}^{\infty}[\sin(n\pi/2)/n]\cos(n\pi x/2)$

b $(4/\pi)\sum_{n=1}^{\infty}[(1 - \cos(n\pi/2))/n]\sin(n\pi x/2)$

CHAPTER 22

EXERCISES

1 a $3/s^2 + 1/s$ **b** $(s + 2)/(s^2 + 1)$ **c** $6/s^4 + 12/s^3 + 12/s^2 + 8/s$
d $-(s - 3)/(s^2 - 9)$ **e** $(s + 3)/[(s + 3)^2 + 16]$

3 a $e^{2t}\sin t$ **b** $e^{2t}(\cos t + 2\sin t)$
c te^{-3t} **d** $H(t - 1)(t - 1)e^{-3(t-1)}$

4 a $x = e^{-3t/2}[\cosh(t\sqrt{13}/2) + 3/\sqrt{13}\sinh(t\sqrt{13}/2)]$
b $x = e^{-t} + e^{3t}/3 - 1/3$
c $x = 3/2 - e^{-2t}/2 + H(t - 1)(1 - e^{-2(t-1)})/2 + H(t - 2)(1 - e^{-2(t-2)})/2$
d $x = \cos(t\sqrt{2}) + \sqrt{2}\sin(t\sqrt{2})$

ASSIGNMENT

1 a $1/(s + a)^2$ **b** $n!/(s + a)^{n+1}$
c $6/s^2 - 1/(s - 2)$ **d** $4/(s - 6) - 4/(s - 3) + 1/s$

2 a te^{-5t} **b** $e^t(\cos t + \sin t)$
c $e^{-t} + e^{2t}$ **d** $e^{5t}\cos 2t$

3 a $e^{-t}(4\cos 2t + \sin 2t)/2$
b $-4e^{2t}/3 + e^{-t}/12 + 9e^{3t}/4$
c $e^{-3t}(\sin 2t - 2t\cos 2t)$
d $H(t - 1)e^{-3(t-1)}[2\cos 2(t - 1) - 2t\cos 2(t - 1) + \sin 2(t - 1)]$

4 a $x = e^{2t}$ **b** $x = -1/6 + 7e^{-2t}/10 + 7e^{3t}/15$
c $x = 3e^t/8 - e^{-t}/4 - e^{-3t}/8$

FURTHER EXERCISES

1 a $e^{-t}[3\cos 3t - 4\sin 3t]$
b $e^{2t}/3 - e^{-t}/3 + e^{-3t}$
c $e^{-3t}[6t\cos 2t + 4t\sin 2t - 3\sin 2t]/16$
d $H(t-1)e^{3(t-1)}[2\cos 2(t-1) - 2t\cos 2(t-1) + \sin 2(t-1)]$

8 $I = [ECs/(LCs^2 + RCs + 1)]\sum_{r=1}^{\infty} e^{-rs}/r \quad i \to 0$

9 $i_1 \to EL/(3R^2), \; i_2 \to 0$

10 $I = Es/[(s^2+1)\{(s+1)^2 + (k-1)\}]$

CHAPTER 23

EXERCISES

1 a Discrete **b** continuous **c** discrete **d** discrete **e** discrete
f continuous **g** discrete

3 a Mean = 2 ohms, mode = 2 ohms, median = 2 ohms, range = 1.5 ohms
b mean = 21.47%, equimodal but for grouped data approximately 19%, median = 19.7%, range = 15.9%
c mean = 0.1377 m, mode (ungrouped) = 0.121 m but grouped data is bimodal, median = 0.122 m, range = 0.063 m

4 Mean = 4.69, mode = 1, median = 5, range = 9

5 Mean = 30.4375, mode = 31, median = 31, range = 3

ASSIGNMENT

1 Discrete **4** 44.1, 44, 44 **5** 9, 4.8265

FURTHER EXERCISES

1 a(i) 4.2 (ii) 8 (iii) 3 (iv) 6 (v) 2.39
b(i) 7 (ii) 20 (iii) 5 (iv) 18 (v) 7.38

4 b 2.925, 3.025, 3.125, 3.225, 3.325 **d**(i) $\bar{x} = 3.115$, $s = 0.0813$ (raw data) (ii) $\bar{x} = 3.118$, $s = 0.0898$ (grouped data)

5 Hint: expand $\Sigma f_r(x_r - \bar{x})^2$

6 Mean 1.56, variance 1.806

CHAPTER 24

EXERCISES

1 16/21

2 **a** $a = 3600/5269$
b 769/5269
c 9/16
3 16.0025 m^2 $E(x^2) = V(x) + [E(x)]^2$
4 **a** 4/e **b** $1 + 1/[4 \ln 4 - 4]$
5 **a** 2.28×10^{-3} **b** 0.0676 (answers respectively 2.26×10^{-3} and 0.0676)
6 **a** 1/6! **b** 1/30 **c** 1/6 (**a** = 1/6, **b** = 2/3, **c** = 1/6!)

ASSIGNMENT

1 0.154 (note $\mu = 1.347$)
2 0.0304
3 **a** 0.5 **b** 0.25
4 0.875
5 (a) $k = 1/2$ (b) $\mu = 4/3$ (c) $\sigma^2 = 2/9$

FURTHER EXERCISES

2 **a** $\mu = (n+1)/2$, $\sigma^2 = (n^2 - 1)/12$ (population statistics)
b $P(E \cap F) = 1/6$, $P(F \mid E) = 3/5$, $P(E \cup F \cup G) = 11/12$, $P[(E \cup F) \cap G] = 1/6$
3 **a** 0.7828 **b** 0.0216
4 Hint: we require $(0.1)2^N > 1 + N + N(N-1)/2 + N(N-1)(N-2)/6$, and so $N = 12$
5 **a** 0.001 75 **b** 2.96 tonnes
6 Median = mode = 1; actual frequencies (10, 15, 9, 5, 1); theoretical frequencies (11, 14, 9, 4, 1)
7 **a** 95% **b** 92%
8 **a** 0.6648 **b** 0.9938
9 0.6753
10 **a** 0.0968 **b** 0.14
11 $\mu = 1$; {4, 4, 2, 1, 0, 0}
12 Hint: $P(X \geqslant 4)/P(X > 0)$, 16.52%
13 0.56, 0.626 (57, 32, 9, 2, 0, 0)
14 Poisson **a** 0.0916 **b** 0.3712
15 $2^5 - 1 = 31$
16 $43.46 < X < 56.54$ newtons

Subject index

Symbols index

Entries are in chronological order